W0107024

68

Fortschritte der Chemie organischer Naturstoffe

Progress in the Chemistry of Organic Natural Products

Founded by
L. Zechmeister

Edited by
W. Herz, G. W. Kirby,
R. E. Moore, W. Steglich,
and Ch. Tamm

Author:
G. W. Gribble

SpringerWienNewYork

Prof. W. HERZ, Department of Chemistry,
The Florida State University, Tallahassee, Florida, U.S.A.

Prof. G. W. KIRBY, Chemistry Department,
The University, Glasgow, Scotland

Prof. R. E. MOORE, Department of Chemistry,
University of Hawaii at Manoa, Honolulu, Hawaii, U.S.A.

Prof. Dr. W. STEGLICH, Institut für Organische Chemie der Universität
München, München, Federal Republic of Germany

Prof. Dr. CH. TAMM, Institut für Organische Chemie der Universität Basel,
Basel, Switzerland

Softcover reprint of the hardcover 1st edition 1996

Library of Congress Catalog Card Number AC 39-1015

Typesetting: Macmillan India Ltd., Bangalore-25
Printing: Novographic, Ing. W. Schmid, A-1238 Wien
Cover design: Ecke Bonk
Printed on acid-free and chlorine-free bleached paper

ISSN 0071-7886
ISBN-13:978-3-7091-7424-1 e-ISBN-13:978-3-7091-6887-5
DOI: 10.1007/978-3-7091-6887-5

Contents

List of Contributors

GRIBBLE, Prof. Dr. G. W., Department of Chemistry, 6128 Burke Laboratory, Dartmouth College, Hanover, New Hampshire 03755-3564, USA.

Naturally Occurring Organohalogen Compounds – A Comprehensive Survey*

G. W. Gribble, Department of Chemistry, Dartmouth College, Hanover, New Hampshire 03755, USA

Contents

* Dedicated to the memory of my father, Waldron Boger Gribble, 1907–1980.

Author's Note

Only naturally occurring halogenated compounds are numbered. However, in a few cases, it was necessary to number a nonhalogenated compound to avoid a gap in the numbering sequence. These unavoidable exceptions are approximately balanced by a few late-addition halogenated compounds that have to share a number. Thus, the total number of compounds in this review is essentially equal to the number of known naturally occurring organohalogen compounds. This numbering scheme allows one readily to determine that there are, for example, 189 natural halogenated indoles and 208 halogenated marine sesquiterpenes.

1. Introduction

Natural chlorine-, bromine-, iodine-, and fluorine-containing organic chemical compounds are abundant on our planet. Occurring on land, in the oceans, and in the atmosphere, they pervade every form of life. They are produced by marine and terrestrial plants, bacteria, fungi, insects, marine animals, and even mammals. Natural combustion processes, such as volcanoes and other geothermal events, and forest and brush fires contribute large quantities of halogenated compounds to the environment. In 40 years, the number of known natural organohalogens has multiplied 200 times, from a dozen in 1954 to nearly 2400 today (Table 1) (*1–12*).

In addition to the reviews cited in Table 1, a number of other excellent reviews of portions of this field are available (*13–40*), especially of marine products. More specific reviews will be cited as appropriate. However, no single review to date covers natural organohalogen compounds of all types, from all sources, over the entire period of human discovery. An attempt was made to cover the literature through mid-1994, but the sheer enormity of this task almost guarantees omissions. Since the author felt that it was more important to document all known organohalogens, the discussion of most individual compounds is severely limited. Compound characterization and synthesis are omitted. Biosynthetic pathways and proposals are discussed only to a limited extent. Man-made organohalogens are presented only when they also have a natural origin. Organohalogen isolation artifacts and "forced" metabolites are generally not included.

The organization follows an earlier, limited review (*11*), and is by structural type. Classification is sometimes arbitrary, especially for those substances that do not fall into a well-defined biogenetic or structural group. For example, 2-bromoindole-containing cyclic peptides are "Cyclic Peptides" rather than "Indoles". References are generally in chronological sequence within each section.

Table 1. *Discovery of Naturally Occurring Organohalogen Compounds*

Year	Reviewer(s)	Summary	Reference
1954	Bracken	12 organochlorine natural products	(*1*)
1961	Petty	29 organochlorines	(*2*)
1968	Fowden	30 organochlorines and a few containing bromine, iodine, and fluorine	(*3*)
1971	Turner	70 chlorinated fungal metabolites	(*4*)
1973	Siuda and DeBernardis	200 organohalogens from all sources (150 organochlorines, 50 organobromines)	(*5*)
1976	Minale	100 organobromines (all but one from marine sources)	(*6*)
1978	Thomson	150 organohalogens from marine sources	(*7*)
1981	Fenical	400 organohalogens from marine sources	(*8*)
1983	Turner and Aldridge	80 new chlorinated fungal metabolites	(*9*)
1986	Engvild	130 organochlorines from higher plants	(*10*)
1992	Gribble	611 new organohalogens from all sources isolated between 1980–1991	(*11*)
1993	Naumann	500 organochlorines from all sources are discussed	(*12*)
1994	Gribble	2,450 organohalogens from all sources	(this work)

Due to the intense scrutiny under which organohalogens – particularly those containing chlorine – have come in recent years, it seemed urgent to review comprehensively the entire field of naturally occurring organohalogen compounds. It is hoped that this review will provide important background material for regulatory and government agencies worldwide to facilitate informed decisions regarding the use of halogenated chemicals in our society.

2. Origins

The halide salts of the four common halogens are abundant to various degrees on earth (Table 2). As discussed in Biohalogenation (Sect. 4), enzymatic oxidation of chloride, bromide, and iodide is the initial biochemical event leading to the production of natural organohalogens. Whereas halide in the oceans (Cl^-, Br^-, I^-) serves as the origin of the vast number of marine organohalogens to be discussed, it is believed that chloride ion as sea salt spray can also be oxidized to free chlorine by ozone and thence to chlorine atoms and HCl (*44, 45*). The concentration of sea level gaseous chlorine (Cl_2) is 1 ppb (*45*), which may explain the low concentration of ozone near the surface of the ocean. The major source of natural HCl and HF in the atmosphere is volcanoes (*43, 46–56*), and several recent eruptions have been studied with regard to their gas emissions. For example, the Mt. Pinatubo eruption in 1991 released 4.5×10^6 tons of total chlorine (HCl and Cl^-) into the atmosphere (*48*), and the 1976 Mt. Augustine eruption in Alaska emitted 0.6×10^6 tons of chlorine, of which up to 0.2×10^6 tons was HCl (*49–51*). Likewise, Hekla in 1970 (*52*), El Chichón in 1982 (*53, 54*), Guatemala volcanoes in 1978 (*55*), and the ten-year Kilauea eruptions all produced large quantities of HCl, HF and perhaps HBr (*56*). For example, the El

Table 2. *Distribution of Halides in the Environment*

Halide	Oceans (*41, 2245*) mg/l	Sedimentary Rocks (*39, 2245*) mg/kg	Fungi (*42*) mg/kg	Wood Pulp (*43*) mg/kg	Plants (*70, 2245*) mg/kg
Cl^-	19,000	10–320		70–2100	200–10,000
Br^-	65	1.6–3	100		
I^-	0.05	0.3			
F^-	1.4	270–740			

Chichón eruptions produced 9% of the global HCl (0.04×10^6 tons); over a large part of the globe there was a 40% increase in the amount of gaseous HCl (*54*). Hydrogen chloride has also been detected outside of our solar system (*57, 58*).

In addition to the presence of I^- and Br^- in the oceans, free iodine was identified in marine organisms long before organoiodides were characterized. The algae *Asparagopsis armata, Falkenbergia doubleti*, and *Bonnemaisonia asparagoides* contain iodine (*59*). Fifty species of Cyanophyceae, Chlorophyceae, Phaeophyceae, and Rhodophyceae contain iodide (or organic iodides) (*60*), and, of 46 species of sponges examined, all contained iodine and all but one contained bromine (*61*) (probably organic halides). The total halogen content (dry weight) of the common algae *Laurencia pacifica* and *Plocamium pacificum* is 0.29% and 0.64%, respectively. These algae are found in all of the oceans (tropical, temperate, and polar), and *Plocamium* occurs from the intertidal regions to depths of greater than 35 meters (*62*).

Although the concentration of fluoride in seawater is only 1.3 mg/l (*41*), the sponge *Halichondria moorei* contains potassium fluorosilicate (K_2SiF_6) to the extent of 10% (dry weight) (*64*). Whereas this animal is able to sequester and concentrate fluoride, another sponge species from the same taxonomic order and in a nearby location contains no fluoride.

It is clear that ample halide and/or halogen is available for incorporation by marine and terrestrial organisms into specific organohalogens by enzymatic processes and for conversion into organohalogens by geothermal processes. To what extent these natural sources of halogen affect the ozone layer is still a matter of debate (*45, 47, 49, 65, 66*), but unfortunately, this area is outside the scope of the present review.

3. Occurrence

3.1. Simple Alkanes

Large numbers of simple chloro-, bromo-, and iodoalkanes, and their mixed derivatives, are found in nature, being produced by both plants and animals and formed during natural combustion processes. Many of these same compounds are also anthropogenic and are used industrially (*67, 68*). Table 3 lists the known natural simple halogenated alkanes.

Table 3. *Simple Halogenated Alkanes and Alkenes that Occur Naturally*

Compound	Source (Ref)
CH_3Cl (**1**)	see text
CH_2Cl_2 (**2**)	see text
$CHCl_3$ (**3**)	see text
CCl_4 (**4**)	see text
CH_3Br (**5**)	giant kelp (*79, 119*), volcanoes (*94–96*), biomass combustion (*151*), Arctic atmosphere (*116, 117*), other atmospheres (Point Barrow, Mauna Loa, Samoa, New Zealand) (*118*) (Japan) (*120*), Antarctic ice algae (*121*), oceans (*126*)
CH_3I (**6**)	giant kelp (*79, 119, 123*), volcanoes (*94–95*), marine algae (*77, 102, 122, 123, 2210*), oceans (*125, 126, 147*), Antarctic atmosphere and seawater (*124*)
CH_2Br_2 (**7**)	marine algae (*122, 123, 132, 133, 137, 2210*), Arctic atmosphere (*116, 117, 135*), Atlantic ocean (*122, 134, 147*), Japan atmosphere (*120*), Arctic seawater (*135*), phytoplankton (*136*), Antarctic air and seawater (*138*)
$CHBr_3$ (**8**)	marine algae (*77, 102, 122, 123, 132, 133, 137, 141–143, 2210*), Arctic atmosphere (*116, 135*), Arctic seawater (*112, 135, 140, 141*), Atlantic air and seawater (*122, 134*), atmosphere over Pt. Barrow, Mauna Loa, Samoa, and New Zealand (*118*), atmosphere over Japan (*120*), open ocean (*144, 147*), Arctic and Antarctic microalgae (*121, 136, 150*), Antarctic air and seawater (*138*)
CBr_4 (**9**)	marine algae (*102, 132, 133, 145*)
CH_2ClBr (**10**)	marine algae (*121, 122*), Arctic air and water (*117, 135*), Atlantic air and water (*134*), Japan air (*120*)
CH_2ClI (**11**)	marine algae (*102, 137, 2210*), phytoplankton (*136*), Antarctic air and water (*124*), Atlantic air and water (*134, 146*), open ocean (*144*), Arctic water and air (*135*)
CH_2BrI (**12**)	marine algae (*132, 133*)
CH_2I_2 (**13**)	marine algae (*77, 122, 132, 133, 137, 2210*), Atlantic air and water (*122, 134*), open ocean (*137, 144*), Arctic air and water (*135*)
CHI_3 (**14**)	marine algae (*132, 133, 145*)
$CHCl_2Br$ (**15**)	marine algae (*77, 102, 121, 122, 134, 2210*), oceans (*77, 122, 134, 138, 147*)
$CHClBr_2$ (**16**)	marine algae (*77, 102, 121, 122, 132, 133, 134, 137, 145, 2210*), oceans (*77, 122, 134, 137, 138, 147*), phytoplankton (*136*)
$CHBr_2I$ (**17**)	marine algae (*102, 132, 133, 145*)
$CHBrI_2$ (**18**)	marine algae (*132, 133, 145*)
$CHClBrI$ (**19**)	marine algae (*132, 133, 145*)
CH_3CH_2Br (**20**)	marine algae (*122, 134*), mine air (*103*)
CH_3CH_2I (**21**)	marine algae (*77, 122, 134, 137*), oceans (*77, 137*)
$BrCH_2CH_2I$ (**22**)	marine algae (*132, 133*)
$CH_3CH_2CH_2Br$ (**23**)	marine algae (*77*), oceans (*77*)
$CH_3CH_2CH_2I$ (**24**)	marine algae (*77, 137*), oceans (*77, 135, 137, 144*)

Table 3 (*continued*)

Compound	Source (Ref)
$(CH_3)_2CHI$ (**25**)	marine algae (*77, 137*), oceans (*77, 135, 137, 144*)
$CH_3CH_2CH_2CH_2I$ (**26**)	marine algae (*77, 137, 2210*), oceans (*77, 137, 144*)
$CH_3CH_2CH_2CH_2CH_2Br$ (**27**)	marine algae (*77*), oceans (*77*)
$CH_3CH_2CH_2CH_2CH_2I$ (**28**)	marine algae (*77*), oceans (*77*)
$CH_3CH_2CH(CH_3)I$ (**29**)	oceans (*144*)
$Cl_2C{=}CHCl$ (**30**)	volcanoes (*93*), marine algae (*1653, 2210*)
$Cl_2C{=}CCl_2$ (**31**)	volcanoes (*86, 1183*), marine algae (*1653*)
CH_3CCl_3 (**32**)	oceans (*112*)
$CHFCl_2$ (**33**)	volcanoes (*86, 93, 1183*)
CHF_2Cl (**34**)	volcanoes (*93*)
$CFCl_3$ (**35**)	volcanoes (*86, 93, 1183*), drill wells (*106*), mine air (*103*)
CF_2Cl_2 (**36**)	volcanoes (*86, 1183*), drill wells (*106*), mine air (*103*)
$F_2C{=}CF_2$ (**37**)	volcanoes (*93*)
$FClC{=}CF_2$ (**38**)	volcanoes (*93*)
CCl_2FCClF_2 (**39**)	volcanoes (*93*)
$CF_3CF{=}CF_2$ (**40**)	volcanoes (*93*)
$(CH_3)_2SiF_2$ (**41**)	volcanoes (*86, 1183*)
CF_4 (**42**)	unknown (*148, 149*)
$Br_2C{=}CHCHCl_2$ (**43**)	marine algae (*132, 133, 145*)
$Br_2C{=}CHCHClBr$ (**44**)	marine algae (*132, 133, 145*)
$Br_2C{=}CHCHBr_2$ (**45**)	marine algae (*132, 133, 145*)
$BrIC{=}CHCHBr_2$ (**46**)	marine algae (*132, 133, 145*)

3.1.1. Chloromethane

Chloromethane (**1**) is probably the most abundant natural organohalogen and accounts for 25% of the chlorine in the atmosphere (*63*). It is estimated that the natural production of global CH_3Cl is 5 million tons per year (*69, 70*), whereas the level of anthropogenic CH_3Cl is only 26,000 tons per year (*70*). A number of biogenic and combustion sources contribute to the former total (*71*). Chloromethane is produced by several species of wood-rotting fungi which occur worldwide, including *Fomes* (*F. conchatus, F. occidentalis, F. pomaceus, F. ribis, F. rimosus, F. vinosus*) (*72*), *Phellinus* (*P. pomaceus, P. ribis, P. occidentalis, P. robiniae, P. ignarius, P. lundelli, P. pachyphloes, P. pini, P. populicola, P. trivialis*) (*70, 73–75*), and *Inonotus* (*I. andersonii, I. hispidus*) (*74*) in 75–90% yield based on chloride. It is also found in cultivated mushrooms (*Agaricus bisporus*) (*76*), potato tubers (*Solanum tuberosum*) (*100*), phytoplankton (*77*), marine algae (*Endocladia muricata*) (*78*), giant kelp (*Macrocystis pyrifera*) (*79*), a bryozoan (*Biflustra per-*

fragilis) from Tasmania (*80*), the ice plant (*Mesembryanthemum crystallium*) (*78*), the pencil cedar (*81*), the northern white cedar (*81*), and the evergreen cypress (*81*).

Of 63 species of wood-rotting *Phellinus*, *Hymenochaete*, *Fomitoporia*, *Inonotus*, *Omnia*, and *Phaeolus* fungi, 34 were found to produce CH_3Cl (*74*) (Hymenochaetaceae). However, of 27 species of the Ganodermataceae and Polyporaceae only one, *Fomitopsis cytisina*, was capable of CH_3Cl biosynthesis (*74*). These results suggest that CH_3Cl production by wood-rotting fungi is massive, especially since *Phellinus* and *Inonotus*, for example, are widely distributed in both tropical and temperate regions (*70, 74, 75*). Biosynthetic studies reveal that *Phellinus pomaceus* utilizes methionine (probably *S*-adenosylmethionine) as the methyl donor to synthesize CH_3Cl (*73, 82*), and that CH_3Cl serves as a methyl donor for the biosynthesis of esters and anisoles (*e.g.*, veratryl alcohol) in this fungus (*83*) and others (*Phanerochaete chrysosporium*, *Phlebia radiata*, *Coriolus versicolor*) (*84, 85*). Since veratryl alcohol plays a key role in the degradation of wood lignin, its biosynthesis involving CH_3Cl is of vital importance to these wood-rotting fungi (*84, 85*).

Chloride ion is normally present in plants, wood, and soil, as well as minerals (*70, 86, 103*), in levels ranging from 200–10,000 ppm (*70*), and the combustion of plant material leads to the formation of organochlorines such as CH_3Cl. Forest fires (*71, 87, 88*), brush fires (*71, 89–91*), tobacco smoke (*71, 92*), and volcanoes (*43, 93–97*) produce CH_3Cl. Chloromethane in volcanic gases can have two origins: hot lava flowing over chloride-rich vegetation as in Kilauea lava flows (*96*), and the combustion of vegetation, peat, or minerals during explosive eruptions such as those of Mt. St. Helens (*97*) and Santiaguito (*93*). More recently, CH_3Cl has been detected in mines and mineral processing plants and is present in several ores and minerals (*103*).

Since natural fires have occurred on earth since land plants evolved 350–400 million years ago (*88*), it is clear that CH_3Cl has been a natural component of our atmosphere and stratosphere for eons. A recent study of Australian brush fires reveals that the earlier estimates of the global release of CH_3Cl from vegetation combustion may be underestimated (*91*). This particular source of CH_3Cl is viewed skeptically by some (*98, 99*).

3.1.2. Dichloromethane

Dichloromethane (CH_2Cl_2) (**2**), which is a widely used industrial solvent and is classified as a Priority Pollutant (*67, 68*), is found in the atmosphere in low concentrations (*101*) and has been identified in volcanic gases (*86*), barley (*86*), a bryozoan (*Biflustra perfragilis*) (*80*), several species of brown (*Ascophyllum nodosum, Fucus vesiculosis*), green (*Enteromorpha linza, Ulva lacta*) and a red alga (*Gigartina stellata*) (*77*). It is also present in mine air, minerals, and the gas emissions of mining industry plants (*103*).

3.1.3. Chloroform

Not only is chloroform ($CHCl_3$) (**3**) a widely-used industrial chemical (*38, 67, 68*), it is also dispersed in the environment (*36, 101, 104*) and appears to have several natural sources. Chloroform has been detected in deciduous moss (*81*), the northern white cedar (*81*), volcanoes (*86, 1183*), drill wells (*86, 106*), mine gas and minerals (*103*), lemon (*86*), orange (*86*), barley (*86*), mushroom (*Cantharellus cibarius*) (*107*), marine algae (*Asparagopsis taxiformis, A. armata*) (*102*), (*Meristiella gelidium*) (*2210*), terrestrial plants (*Farsetia aegyptia, F. ramosissima*) (*108*), sapodilla fruit (*Achras sapota*) (*109*), shrimp paste (*Sergia lucens*) (*110*), Malabar Nightshade (*Basella rubra*) (*111*). As will be seen later, the haloform reaction *in vivo* probably accounts for a huge percentage of $CHCl_3$ in the oceans, as well as in ground and surface water from the natural decomposition of humic substances.

3.1.4. Carbon Tetrachloride

Carbon tetrachloride (CCl_4) (**4**) which is more toxic than the other chloromethanes is also an important industrial chemical (*67, 68*) and is found in the oceans (*36*) and the atmosphere (*101, 114*). Natural sources of CCl_4 include volcanoes (*86, 1183*), drill wells (*86, 106*), mine gas and minerals (*103*), marine algae (*102*), terrestrial plants (*108*), arctic ocean (*112*), including the bottom (*113*), and perhaps the atmospheric chlorination of CH_4 (*105, 114*).

3.1.5. Other Simple Haloalkanes

Table 3 summarizes the other simple haloalkanes that are known to have one or more natural sources, and Table 4 lists the marine algal sources of simple haloalkanes. Methyl bromide (**5**) has a large number of sources, both marine and combustion, and is a widely used soil fumigant

(*115*). Methyl iodide (**6**) also has a massive marine component. For example, the concentration of CH_3I in seawater near *Laminaria* kelp beds is 1000 times higher than in the open ocean (*89*) and CH_3I is produced by brown, red, and green algae alike (*123*). This compound is by far the most dominant volatile organoiodide in the environment (*124*) although it is believed unlikely to play a role in destroying tropospheric ozone (*125*). Methyl iodide is a product of both kelp and the bacteria associated with it (*119*). It has been suggested that CH_3I plays a role in cycling iodide between land and sea (*105, 126*), and CH_3Br may function similarly (*126–129*). The bromine and iodine concentrations in the Arctic atmosphere peak just after the Arctic dawn, corresponding to the oceanic bloom of algae (*130, 131*). Indeed, the Arctic troposphere has the highest bromine concentration in the world (*130*).

It is interesting that bromide and iodide, but not fluoride, are also converted into CH_3Br and CH_3I by *Fomes* (*72*) and *Phellinus* fungi (*75*).

Bromoform (**8**) is enormously abundant in the oceans, produced and secreted into the ocean by algae and plankton. For example, the major (80%) of the nearly 100 compounds present in *Asparagopsis taxiformis*, "limu kohu," the favorite edible seaweed of Hawaiians, is $CHBr_3$ (*132, 133*). Another source of $CHBr_3$ may be the bromination of free organic matter in the oceans by hypobromite (*139*) since $CHBr_3$ is present at all depths in the Arctic Ocean (*140, 141*). Arctic microalgae are a significant source of $CHBr_3$ and the surface destruction of ozone in the Arctic is proposed to be due to $CHBr_3$ and other brominated compounds (*150*).

Compounds that have not been shown to have a clear biogenic source, such as $BrCH_2CH_2Br$ and $CClBrF_2$, are not included in Table 3 even though they are detected in the oceans and the air over the oceans (*116, 117, 134*). Although tetrafluoromethane (**42**) is believed to have a large natural component (650,000 tons), its source is unknown (*148, 149*). Due to its extreme stability ($>$ 10,000 years), it could have originated long ago from a weak geological source (*148*).

In a most astounding result, it has been found that trichloroethylene (**30**) and perchloroethylene (**31**) – both compounds long thought to be exclusively anthropogenic pollutants – are produced by several species of macroalgae and one microalgae (*Porphyridium purpureum*) (*1653*). Of the 27 species of macroalgae (temperate, subtropical and tropical) that produce these compounds, the ones producing the highest amounts were *Falkenbergia hillebrandii*, *Asparagopsis taxiformis*, and *Gracilaria cornea* (*1653*). The emissions of these two compounds from the oceans into the atmosphere may be of the same magnitude as anthropogenic levels.

The Freons (**33–36**) and related CFCs (**37–40**) which are produced by volcanoes (*86, 93, 1183*) are of particular interest since it had been widely believed that these were only anthropogenic. ISIDOROV (*86, 1183*) has estimated that 75% of the world's 2000 active volcanoes have a mineral composition and geologic configuration suitable for producing halogenated compounds. In some cases, the concentrations of the CFCs emerging from solfataric vents from the Kamchatka volcanoes are 400 times that of background.

Compounds **35** and ClF_2CCF_2Cl have been reported in the aroma of the Malabar nightshade (*Basella rubra*) (*111*), but it seems likely that they are artifacts in the headspace.

3.2. Simple Functionalized Acyclic Organohalogens

Asparagopsis taxiformis and *A. armata* were two of the first seaweeds to be examined for their organohalogen content. In monumental efforts, MOORE (*132, 133, 145, 152*) and FENICAL (*102, 153*) identified 100 organohalogens from these two red algae (Tables 3–6). The former alga, "limu kohu" (supreme seaweed), is prized by Hawaiians for its flavor and aroma. The fact that this mind-boggling array of chemicals includes many haloacetones, which are known to be powerful lachrymators and alkylating agents, is cause for concern. Indeed, FENICAL has described the "discomfort" of the researchers performing chromatographic analyses with these polyhaloketones and also reported the fact that they display strong antimicrobial activity (*153*).

Although the chemical makeup of *A. taxiformis* from Hawaii and California differs significantly, several metabolites are common to both specimens (Tables 5, 6). The red alga *Falkenbergia rufolanosa*, a tetrasporophyte of *A. armata*, produces several new halo carbonyls (**130–134**) in addition to known compounds (Table 7). The esters are apparently not artifacts (*154*).

A detailed examination of the red alga *Bonnemaisonia hamifera* has resulted in the isolation of an array of C-7 bromo ketones, alcohols, acetates, and other compounds (*155, 156, 158*) (Table 8). The iodo metabolites **137** and **142** may be artifacts formed by an S_N2 reaction with iodide during the isolation process (*156*). The persistent sweet aroma of this alga is due to the various polybrominated 2-heptanones (*155*). The major metabolite isolated from the red alga *Trailliella intricata* is **135** (*158*).

In contrast to *B. hamifera*, *B. asparagoides* produces an array of polyhalogenated 1-octen-3-ones (**156–162**) (*158*) (Table 9) which have antibacterial activity. The biogenesis of these compounds is proposed to

Table 4. *Distribution of Halomethanes in Marine Algae*

Alga	Haloalkanes	Refs.
Brown algae		
Macrocystis pyrifera	CH_3Cl, CH_3Br, CH_3I, CH_2Br_2, $CHBr_3$	*79, 119, 123*
Laminaria sp.	CH_3I	*89*
Eisenia arborea	CH_3I, $CHBr_3$, CH_2Br_2	*119, 123*
Egregia menziesii	CH_3I, $CHBr_3$, CH_2Br_2	*119, 123*
Laminaria farlowii	CH_3I, $CHBr_3$, CH_2Br_2	*119, 123*
Pterygophora californica	CH_3I	*119*
Dictyota binghamie	CH_3I	*123*
Cystoseira osmundacea	CH_3I, $CHBr_3$, CH_2Br_2	*123*
Fucus vesiculosis	CH_2Br_2, $CHBr_3$, $CHBr_2Cl$, CH_2Cl_2, $CHBrCl_2$, CH_2I_2, CH_3I, CH_2ICl	*77, 137*
Ascophyllum nodosum	$CHBr_3$, $CHBr_2Cl$, CH_2Cl_2, $CHBrCl_2$, CH_2I_2, CH_3I	*77*
Fucales sargassum	CH_3I, CH_2I_2, $CHBr_3$, CH_2Br_2, $CHBr_2Cl$, $CHBrCl_2$, CH_2BrCl	*122, 134*
Red algae		
Bonnemaisonia hamifera	$CHBr_3$	*142*
Asparagopsis taxiformis	CH_2Br_2, $CHBr_3$, CBr_4, $CHBr_2I$, CH_2BrI, $CHBrI_2$, CHI_3, CH_2I_2, $CHBrClI$, $CHBr_2Cl$, CH_3I, $CHCl_3$, CCl_4, CH_2ClI	*102, 132, 133*
Asparagopsis armata	$CHCl_3$, CCl_4, $CHBr_3$, CBr_4, $CHBr_2I$, CH_3I, CH_2ClI, $CHBrCl_2$, $CHBr_2Cl$	*102*
Gigartina stellata	$CHBr_3$, $CHBr_2Cl$, CH_2Cl_2, $CHBrCl_2$, CH_2I_2, CH_3I	*77*
Corallina officinalis	CH_3I, $CHBr_3$, CH_2Br_2	*123*
Pterocladia capillacea	CH_3I, $CHBr_3$, CH_2Br_2	*123*
Rhodymenia californica	CH_3I, $CHBr_3$, CH_2Br_2	*123*
Meristiella gelidium	CH_3I, $CHBr_3$, CH_2Br_2, $CHBr_2Cl$, $CHCl_3$, CH_2I_2, CH_2ClI	*2210*
Green algae		
Penicillus capitatus	$CHBr_3$	*143*
Enteromorpha linza	$CHBr_3$, $CHBr_2Cl$, CH_2Cl_2, $CHBrCl_2$, CH_2I_2, CH_3I	*77*
Ulva lacta	$CHBr_3$, $CHBr_2Cl$, CH_2Cl_2, $CHBrCl_2$, CH_2I_2, CH_3I	*77*
Ulva sp.	CH_3I, $CHBr_3$, CH_2Br_2	*123*
Enteromorpha intestinalis	CH_3I, $CHBr_3$, CH_2Br_2	*123*
Phytoplankton		
Nitzschia sp.	$CHBr_3$, $CHBr_2Cl$, CH_2Br_2, CH_2ClI	*136*
Porosira glacialis	$CHBr_3$, $CHBr_2Cl$, CH_2Br_2, CH_2ClI	*136*
Nitzschia stellata	CH_3Br, CH_2Br_2, $CHBr_3$, CH_2BrCl, $CHBrCl_2$, $CHBr_2Cl$	*121*
Porosira pseudodenticulata	CH_3Br, CH_2Br_2, $CHBr_3$, CH_2BrCl, $CHBrCl_2$, $CHBr_2Cl$	*121*

Table 5. *Simple Functionalized Organohalogens Isolated from the Hawaiian Alga Asparagopsis taxiformis* (*132, 133, 145, 152*)

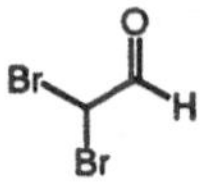

47

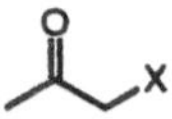

48 X = Br
49 X = I

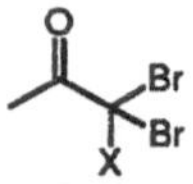

50 X = H
51 X = Br
52 X = Cl

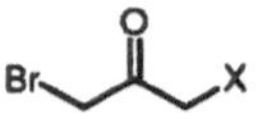

53 X = Br
54 X = I

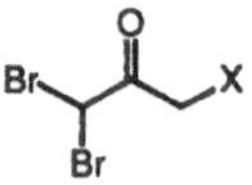

55 X = Br
56 X = I

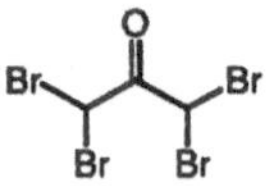

57

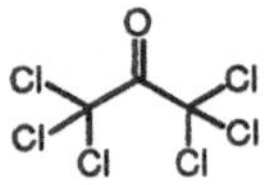

58

59

60

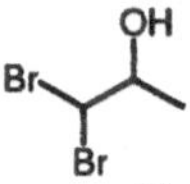

61

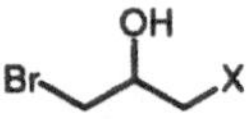

62 X = Br
63 X = I

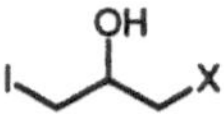

64 X = Cl
65 X = I

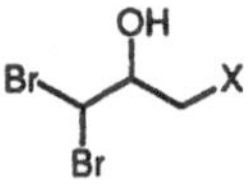

66 X = Br
67 X = Cl
68 X = I

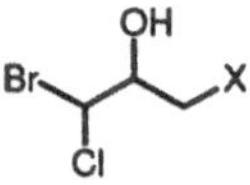

69 X = Cl
70 X = Br
71 X = I

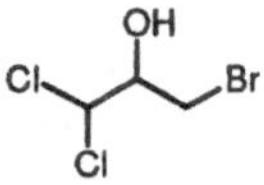

72

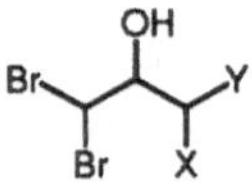

73 X = Y = Br
74 X = Y = Cl
75 X = Y = I
76 X = Cl, Y = Br
77 X = Br, Y = I

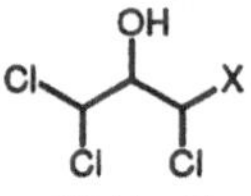

78 X = Cl
79 X = Br

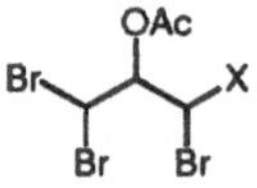

80 X = H
81 X = Br

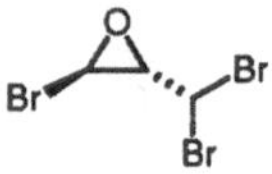

82

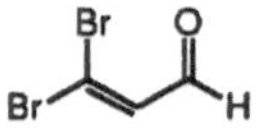

83

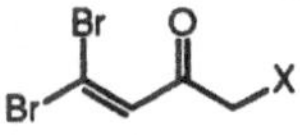

84 X = H
85 X = Br
86 X = I

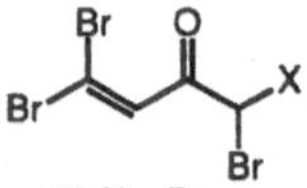

87 X = Br
88 X = Cl

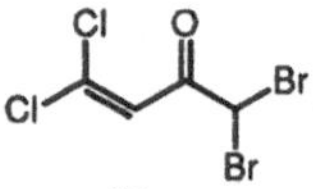

89

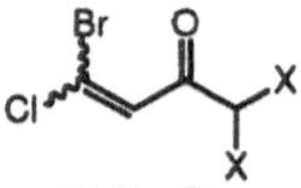

90 X = Br
91 X = Cl

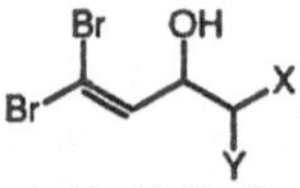

92 X = H, Y = Br
93 X = H, Y = Cl
94 X = Y = Br
95 X = Cl, Y = Br
96 X = Y = Cl

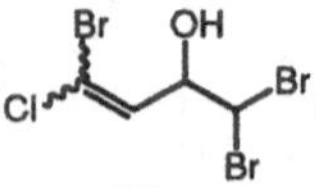

97

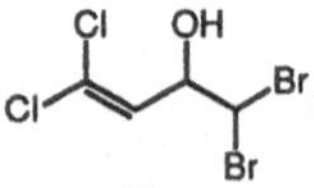

98

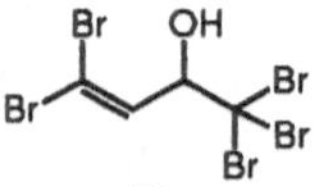

99

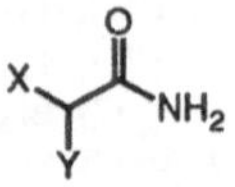

100 X = Y = Br
101 X = Y = I
102 X = Cl, Y = Br
103 X = Br, Y = I
104 X = Cl, Y = I

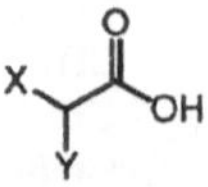

105 X = Y = Br
106 X = Y = I
107 X = Br, Y = I

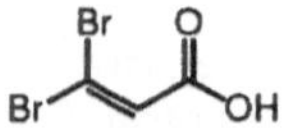

108

Table 6. *Simple Functionalized Organohalogens Isolated from the Gulf of California Asparagopsis taxiformis and Mediterranean A. armata* (*102, 153*)[a]

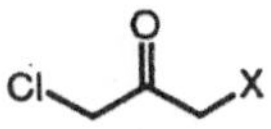

CI X
109 X = Cl
110 X = Br
111 X = I

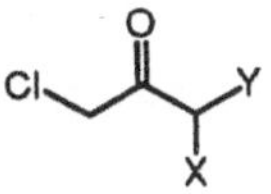

CI Y X
112 X = Y = Cl
113 X = Y = Br
114 X = Cl, Y = Br

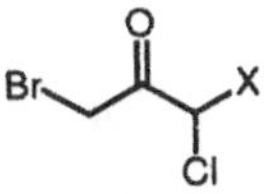

Br X Cl
115 X = Cl
116 X = Br

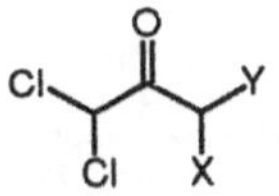

Cl Y Cl X
117 X = Y = Cl
118 X = Y = Br
119 X = Cl, Y = Br

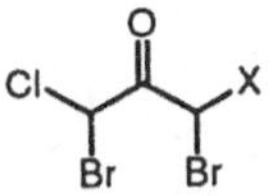

Cl X Br Br
120 X = Br
121 X = Cl

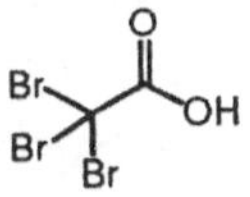

Br Br OH Br
122[b]

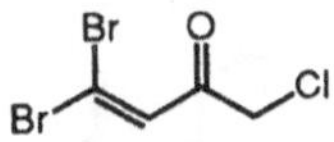

Br Br Cl
123

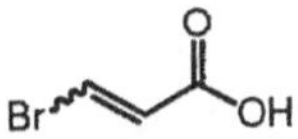

Br OH
124[b]

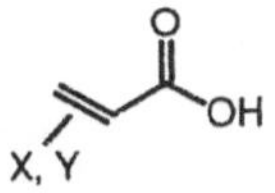

OH X, Y
125[b] X = H, Y = I
126[b] X = Y = I
127[b] X = Cl, Y = Br
128[b] X = Br, Y = I

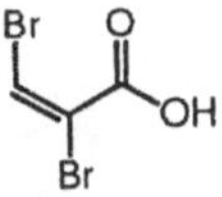

Br OH Br
129[b]

[a] Also isolated were **53, 53–57, 85, 87, 88, 105, 107, 108**
[b] Isolated as the ethyl ester; the acid was shown to be present

Table 7. *Organohalogens Isolated from the Red Alga Falkenbergia rufolanosa* (*154*)[a]

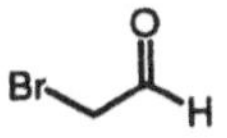

Br H
130

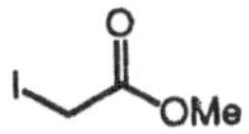

I OMe
131

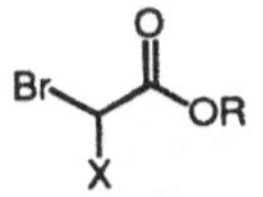

Br OR X
132 X = Br, R = Me
133 X = Br, R = Et
134 X = I, R = Et

[a] Also isolated were **8, 17, 47, 48, 50, 53, 55, 57, 112, 115, 116, 118–121**

Table 8. *Organohalogens Isolated from the Red Alga Bonnemaisonia hamifera (155, 156, 158)*

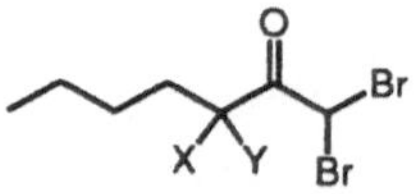

Br Br X Y Br Br X Br Br Br X

135 X = Y = Br (major)
136 X = H, Y = Br
137 X = H, Y = I
138 X = Y = H

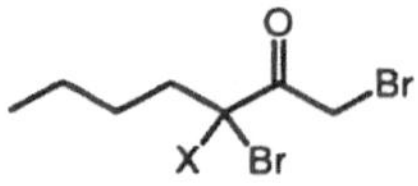

139 X = Br
140 X = H

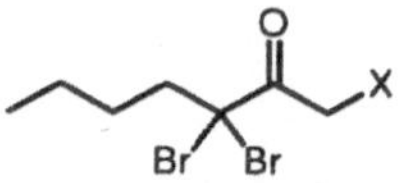

141 X = H
142 X = I

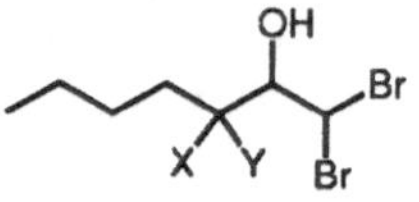

143 X = Y = Br (major)
144 X = H, Y = Br
145 X = Y = H

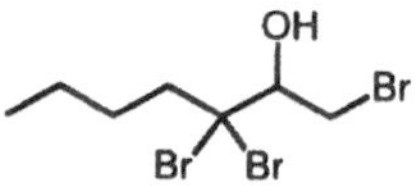

146

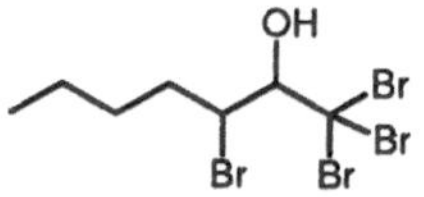

147

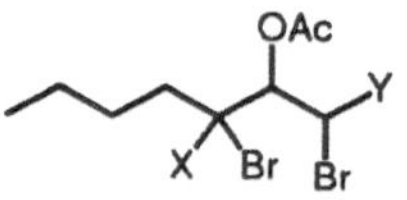

148 X = Y = Br
149 X = Br, Y = H
150 X = H, Y = Br

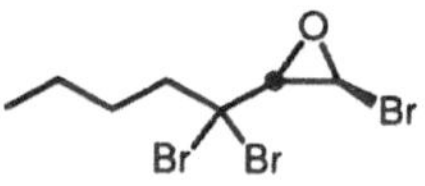

151 (major)

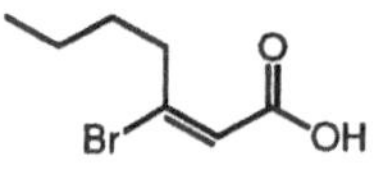

152[a]

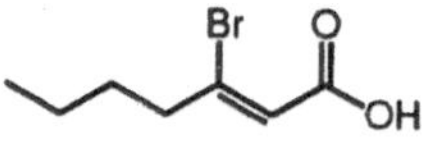

153[a]

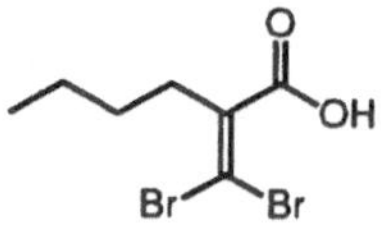

154[a]

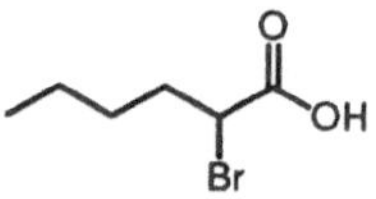

155[a]

[a] Isolated as the ethyl ester

Table 9. *Organohalogens Isolated from the Red Alga Bonnemaisonia asparagoides (157, 158)*

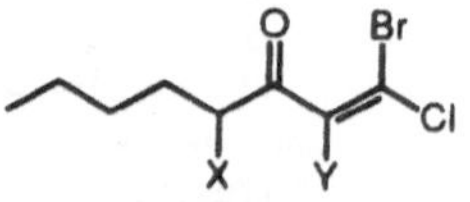

156 X = H, Y = Cl
157 X = Y = Cl
158 X = Cl, Y = Br
159 X = Y = Br

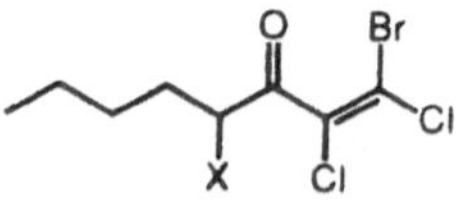

160 X = H
161 X = Cl

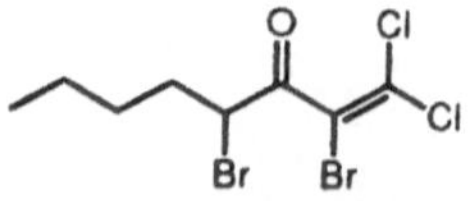

162

Table 10. *Organohalogens Isolated from the Red Alga Bonnemaisonia nootkana (158, 159)*[a]

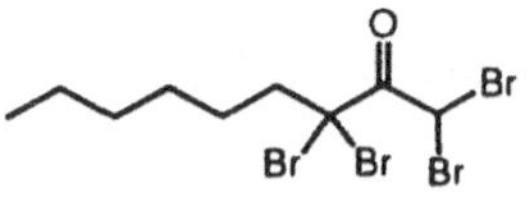

163

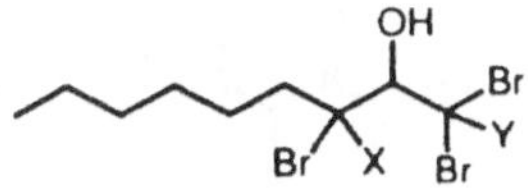

164 X = Br, Y = H
165 X = H, Y = Br

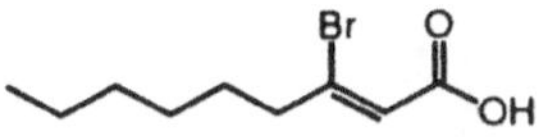

166[b]

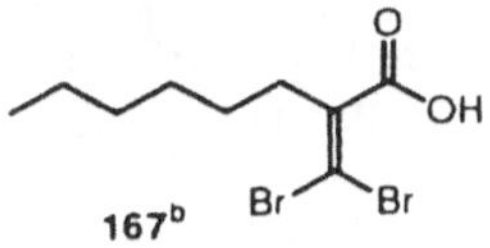

167[b]

[a] Also isolated were **143**, **149**, **151**, **153**[b], **154**[b]
[b] Isolated as the ethyl ester

Table 11. *Organohalogens Isolated from the Red Alga Delisea fimbriata (160)*[a] *and D. pulchra (161)*[b,c]

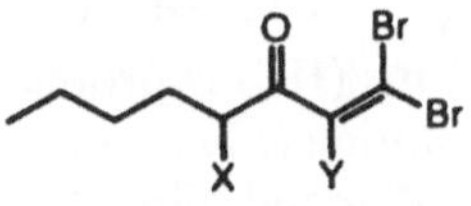

168[a,b] X = H, Y = Br
169[a] X = Y = Br
170[a] X = Cl, Y = Br
171[a] X = H, Y = I
172[a] X = Cl, Y = I

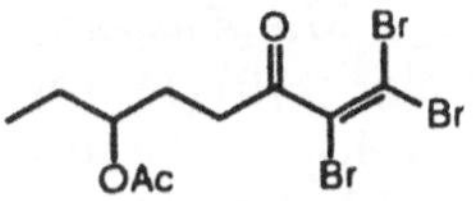

173, 174 (racemic)[b]

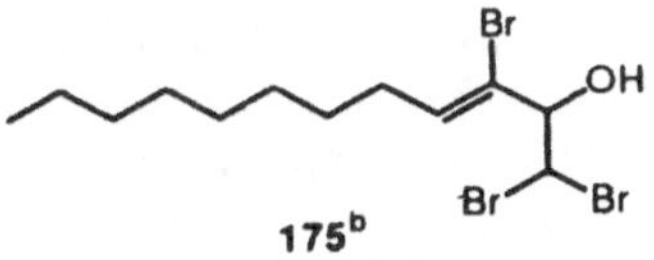

175[b]

[a] *D. fimbriata (160)*
[b] *D. pulchra (161)*
[c] Furanone metabolites are described separately

involve Favorskii rearrangements (*157*). The red alga *B. nootkana* was found to contain several new C-9 polybrominated ketones, alcohols, and carboxylic acids (**163–167**) as well as several known compounds from *B. hamifera* (Table 10) (*158, 159*). In addition to containing several furanones which will be discussed in Sect. 3.11, the red algae *Delisea fimbriata* and *D. pulchra* produce a set of 1-octen-3-ones (**168–174**) (Table 11) (*160, 161*) different from those of *Bonnemaisonia asparagoides*

Table 12. *Bromo Ketones Isolated from the Red Alga Ptilonia australasica* (*162*)[a]

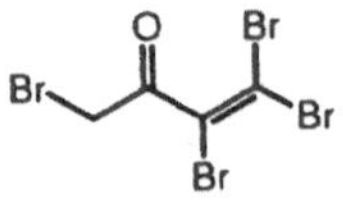

176

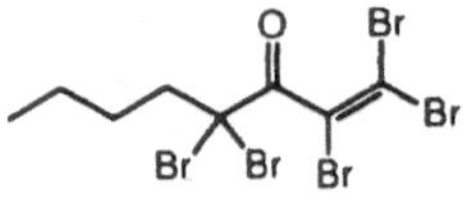

177

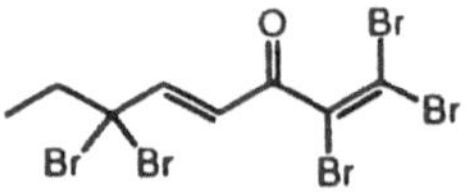

178

[a] Also isolated were **168, 169**

(Table 9). The red alga *Ptilonia australasica* produces yet another complement of new polybromo ketones (**176–178**) (Table 12) (*162*), including two 1-octen-3-ones, in addition to two brominated pyrones (see a later section).

The red alga *Marginisporum aberrans* has been found to contain dichloroacetamide (**179**) (*163*), while fluoroacetone (**180**) is produced by homogenates of *Acacia georginae* exposed to fluoride (*164*), so the latter compound could be considered to be a "forced" metabolite. Antarctic krill contains 1,3-dichloro-2-propanol (**181**) (*165*), and several chlorinated alcohols have been isolated from the fungus *Caldariomyces fumago* (*166*), including **78** and **182–184**. The red alga *Ptilonia magellanica* produces the dibromoheptanediol derivative **185** (*167*). An extraordinary find is the isolation of bromoester **186** from the cerebrospinal fluid of normal humans (*168*). This spectacular development – which is discussed further in Sect. 6 – is the first instance of a human halogenated metabolite to be identified, other than thyroxine and related iodinated hormones. Compound **186** whose structure was confirmed by synthesis is biologically active and is postulated to have a physiological function in humans. The blue-green algae *Schizothrix calcicola* and *Oscillatoria nigroviridis* (extracted together) produce the interesting vinyl chloride diol **187** (*169*). This compound, which probably has the (*R*, *R*) configuration, is active against *Mycobacterium smegmatis*. The two polyenones, neocarzilins A (**188**) and B (**189**), were isolated from *Streptomyces carzinostaticus* (*170*), and **188** whose absolute configuration was established by synthesis from L-leucine (*171*) is highly cytotoxic against K562 leukemia cells (IC_{50} = 0.06 μg/ml) (*170*).

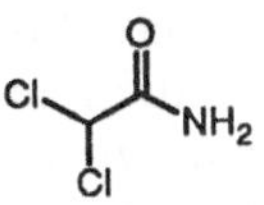

179 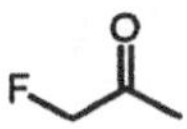 **180**

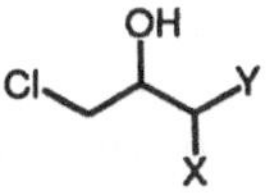

181 X = H, Y = Cl
182 X = Y = H
183 X = Y = Cl

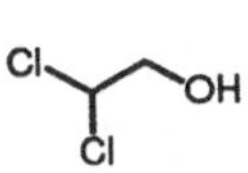

184

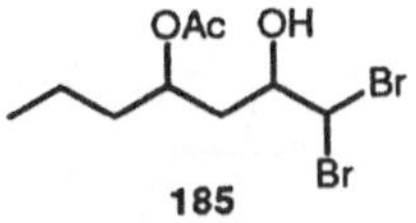

185 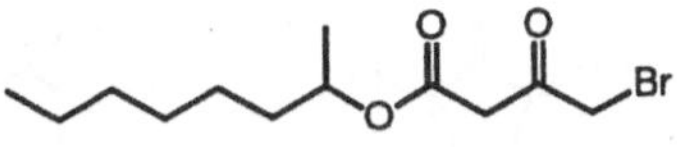 **186**

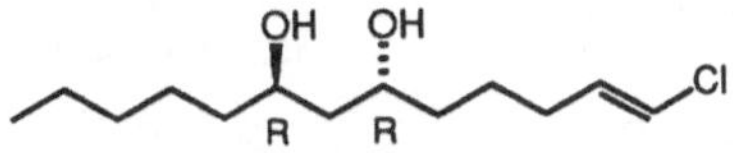

187

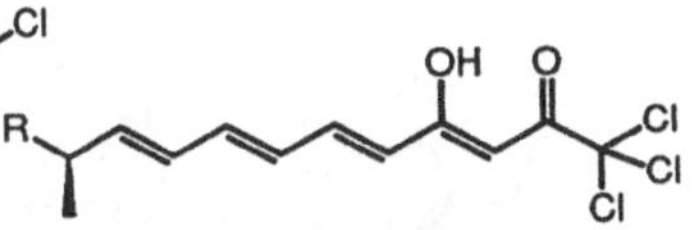

188 R = Et (neocarzilin A)
189 R = Me (neocarzilin B)

3.3. Simple Functionalized Cyclic Organohalogens

3.3.1. Cyclopentanes

In this section are listed several halometabolites that do not easily fit into a recognizable category or biogenetic pathway. Most of these are fungal metabolites.

The Western Australian red alga *Vidalia spiralis* produces the novel dihydrofulvene dibromide **190** (*172*). One of the first naturally occurring organochlorines to be discovered was the *Caldariomyces fumago* fungal metabolite caldariomycin (**191**) in 1940 (*173*). The original structure was confirmed by synthesis (*174*) and, later, the absolute configuration was established (*175*). The dibromo analogue is produced when this fungus is grown in the presence of bromide (*176*). More recently, the calmodulin inhibitors KS-504a (**192**), KS-504b (**193**), KS-504d (**194**), and KS-504e (**195**) have been isolated from the fungus *Mollisia ventosa* (*177, 178*). These novel cyclopentanes contain up to 69% chlorine by weight! The structures of metabolites **192, 193**, and **195** were established by X-ray analysis (*178*). The cyclopentenone derivatives **196–200** are produced by several fungi. Cryptosporiopsin (**196**) is produced by *Sporormia affinis* (*179*), along with **197** and **198**, and **196** is also found in cultures of *Periconia macrospinosa* (*180*), *Cryptosporiopsis* sp. (*181–183*), and

Phialophora asteris f. sp. *helianthi* (*184*). The latter study has also tentatively identified a stereoisomer of **196** and a dehydration product. Cryptosporiopsin inhibits the sunflower pathogen *Sclerotinia sclerotiorum* (*184*). *Cryptosporiopsis* sp. also produces **199** (*183*) and *P. macrospinosa* has yielded cyclopentendiol **200** (*180*).

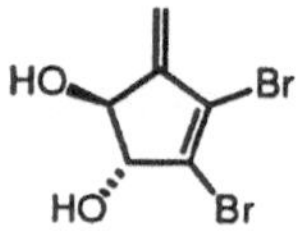

190

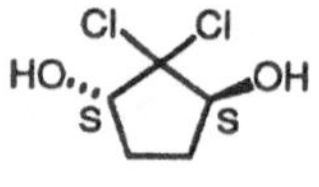

191 (caldariomycin)

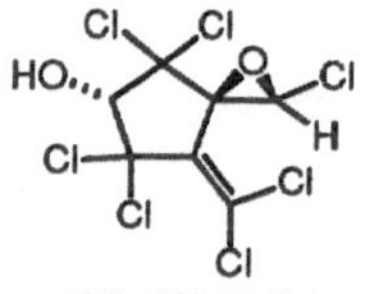

192 (KS-504a)

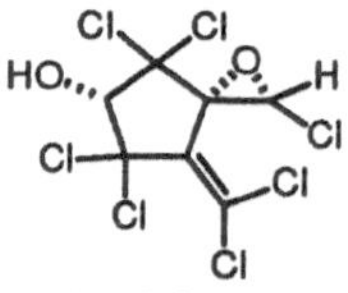

193 (KS-504b)

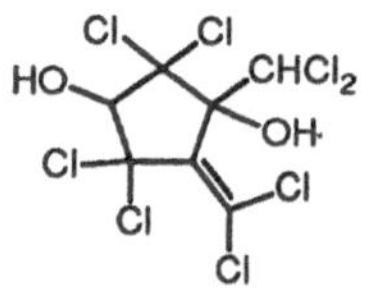

194 (KS-504d)

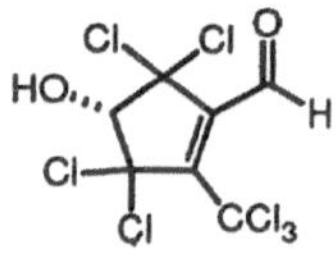

195 (KS-504e)

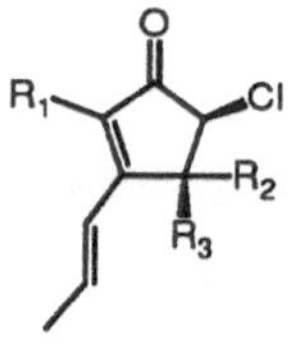

196 R_1 = Cl, R_2 = OH, R_3 = CO_2Me (cryptosporiopsin)

197 R_1 = H, R_2 = OH, R_3 = CO_2Me

198 R_1 = H, R_2 = CO_2Me, R_3 = OH

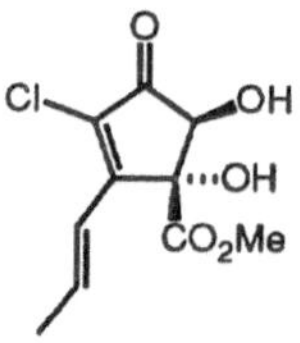

199

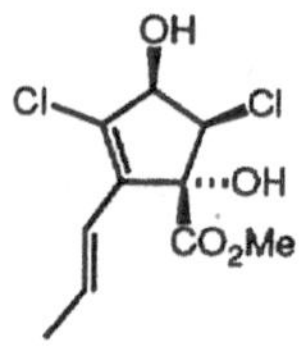

200

3.3.2. Cyclitols and Benzoquinones

Several organohalogens which, while containing a cyclohexane ring, are not terpene-derived are presented here. Some originate from the shikimate pathway while others are polyketide products. Included in this section are benzoquinone-related metabolites and cyclitols.

The cyclitol metabolite pipoxide chlorohydrin has been isolated from *Piper hookeri* (*185*), *P. nigrum* (*186*), and *P. attenuatum* (*276*) and assigned structure **201** (*186*). The culture broth of *Phyllosticta* sp. produces two chlorohydrins (**202, 203**) in addition to epoxydon (phyllosinol) and chlorogentisyl alcohol (**204**) (*187, 188*). The biosynthesis of **204** involves epoxydon → **203** → **202** → **204** (*189*). The chlorotoluquinone **205** has been isolated from *Aspergillus terreus* (*190*), and *A. fumigatus* produces fumigatin chlorohydrin (**206**) (*4*). The ascomycete *Lachnum papyraceum* produces lachnumon (**207**) and lachnumol A (**208**), both of which have nematicidal, cytotoxic, and antimicrobial activities (*191*, *192*). Also isolated were mycorrhizin A and chloromycorrhizin A (*vide infra*).

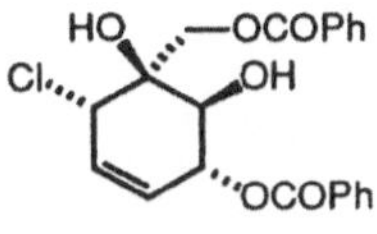

201

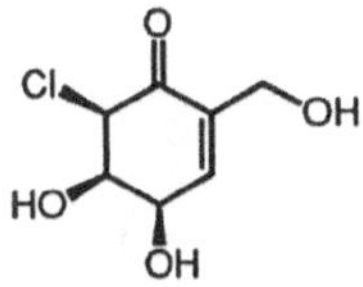

202

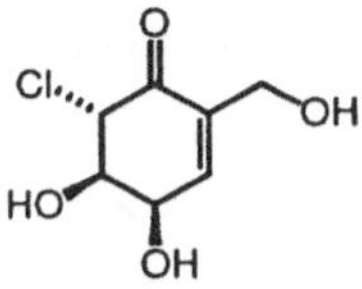

203

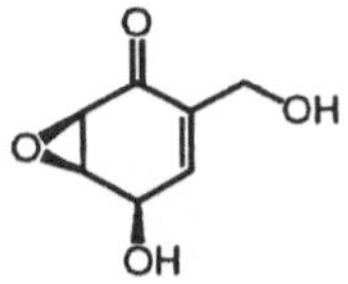

epoxydon

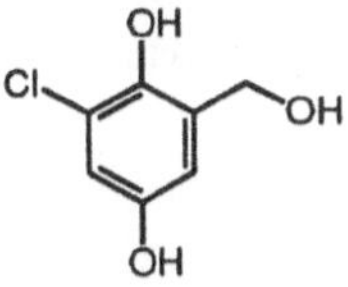

204 (chlorogentisyl alcohol)

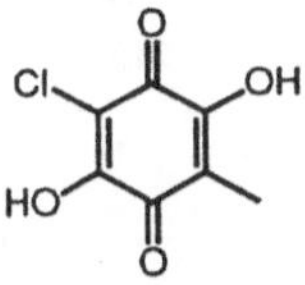

205

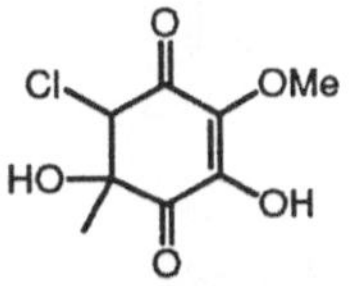

206 (fumigatin chlorohydrin)

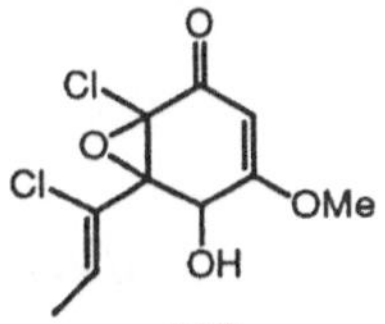

207 (lachnumon)

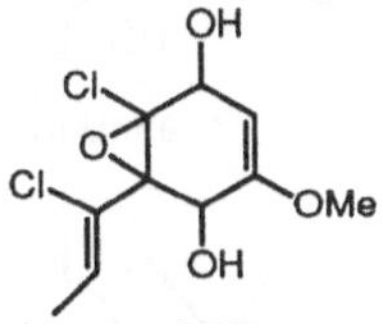

208 (lachnumon A)

Marine acorn worms are a rich source of brominated metabolites including benzoquinones, phenols, and related compounds. Thus, *Ptychodera flava laysanica* produces benzoquinones **209**–**212** (*193*), while another *Ptychodera* species, collected from deep caves on Maui, contains the bromocyclohexenes **213**–**217** (*194*). Epoxide **214** is highly cytotoxic (IC_{50} = 10 ng/ml against P388 cancer cells *in vitro*). Epoxides **213** and **214** have also been isolated from a Florida *Ptychodera bahamensis* (reclassified as *Balanoglossus* sp.) (*195*).

Several cyclopropane-containing fungal metabolites have been isolated and characterized. Mikrolin (**218**) is produced by *Gilmaniella humicola* (*196*, *197*) and mycorrhizin A (**219**) and chloromycorrhizin A (**220**) are found in the fungus *Monotropa hypopitys* (*198*, *199*), while **219** is also present in *G. humicola* (*200*) and both **219** and **220** are produced by *Lachnum papyraceum* (*191*, *192*). These structures have been established

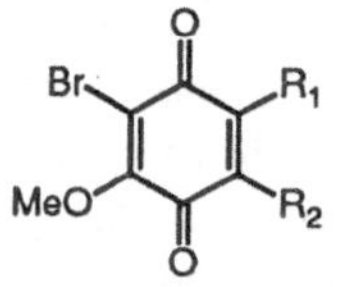

209 R_1 = Br, R_2 = OMe
210 R_1 = R_2 = OMe
211 R_1 = R_2 = Br

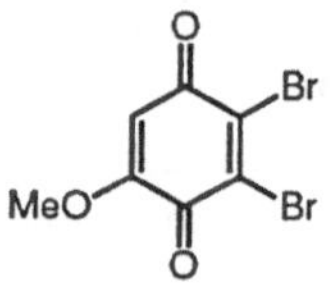

212

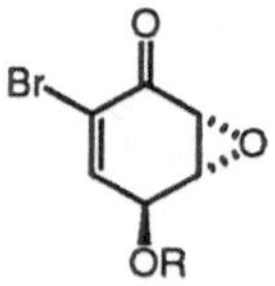

213 R = H
214 R = Ac

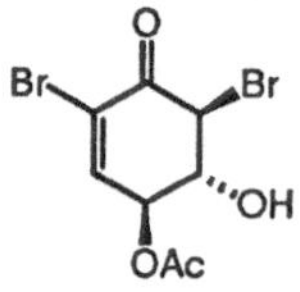

215

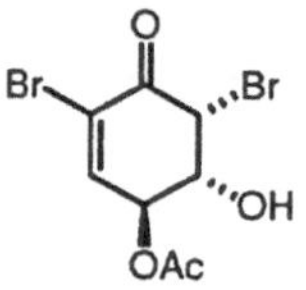

216

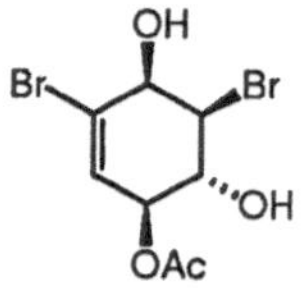

217

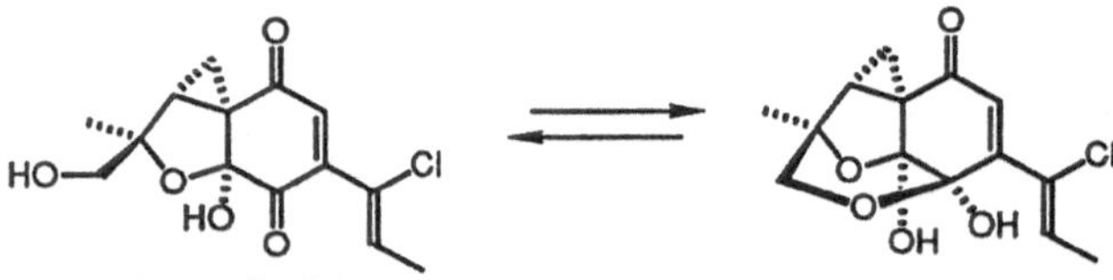

218 (mikrolin)

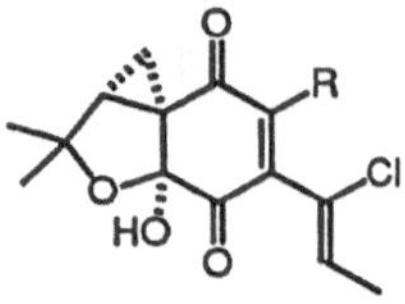

219 R = H (mycorrhizin A)
220 R = Cl (chloromycorrhizin A)

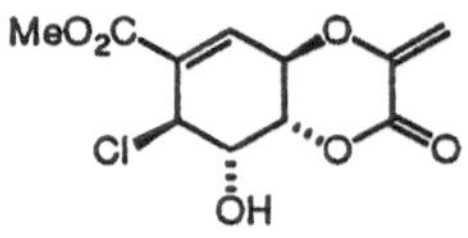

221

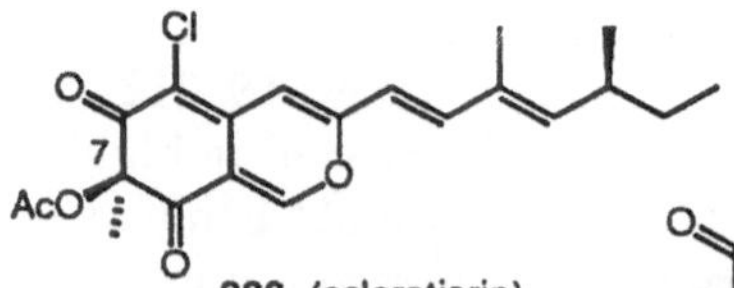

222 (sclerotiorin)
223 (7-epimer)

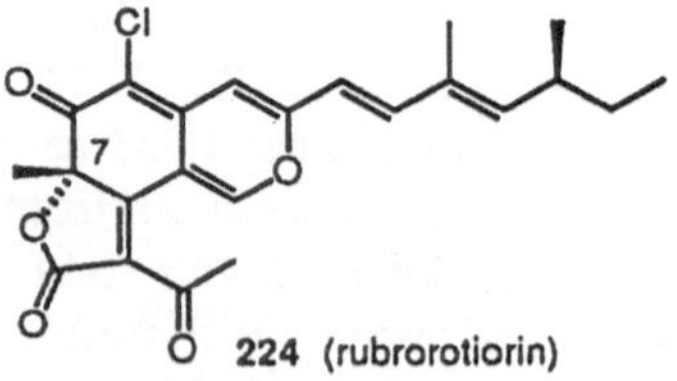

224 (rubrorotiorin)

by X-ray crystallography (*197, 199, 200*), and **219** and **220** are highly active against *Fomes annosus* (*198*). The novel chloroshikimate metabolite **221** has been isolated from two unidentified fungi (*201*). The fungal metabolite sclerotiorin (**222**) was first isolated independently by two groups from *Penicillium sclerotiorum* (*202, 203*) and later from *P. multicolor* (*203*). The 7-epimer **223** is also a known metabolite (*P. hirayamae*) (*204–206*). The structure of **222** was finally established as shown (*207, 208*). The related rubrorotiorin (**224**) is also found in *P. hirayamae* (*209, 210*), although the full structure was not established until later (*208*).

The fungus *Chaetomium globosum* var. *flavo-viridae* produces the novel azaphilones chaetovirdins A–D (**225–228**, respectively). The absolute configuration of **225** has been established (*211*). The related metab-

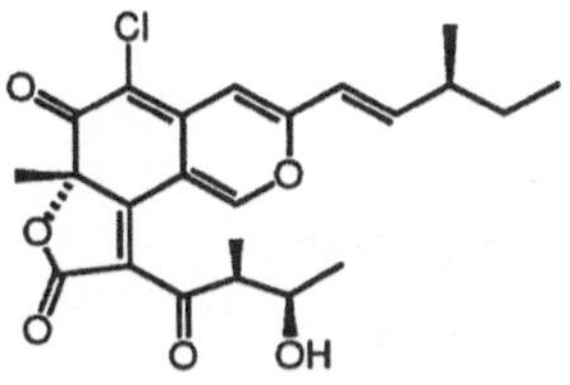

225 (chaetovirdin A)

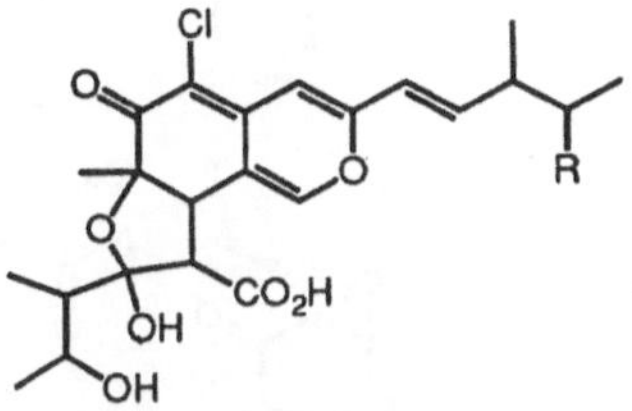

226 R = H (chaetovirdin B)
228 R = OH (chaetovirdin D)

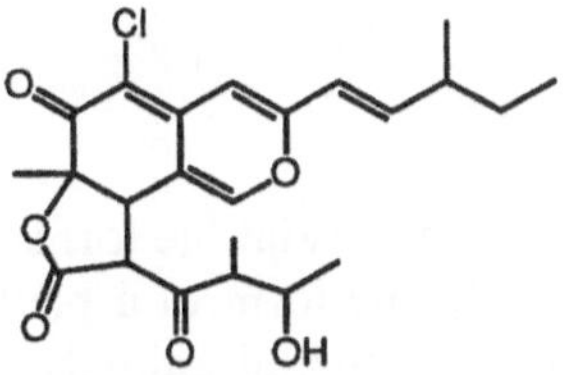

227 (chaetovirdin C)

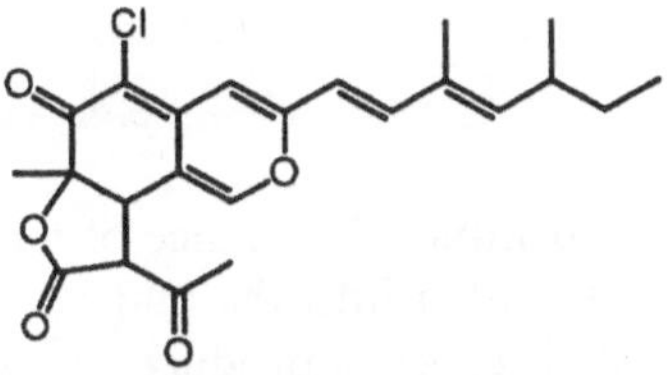

229 (isochromophilone I)

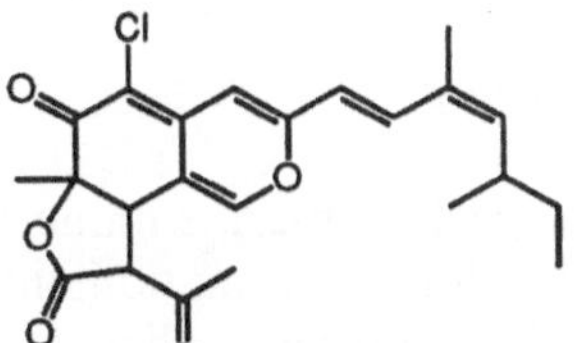

230 (isochromophilone I)

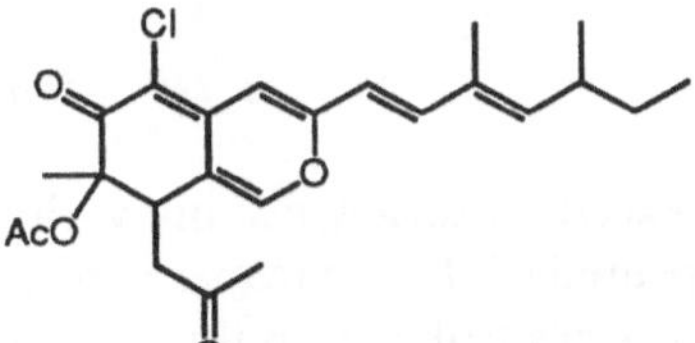

231 (isochromophilone II)

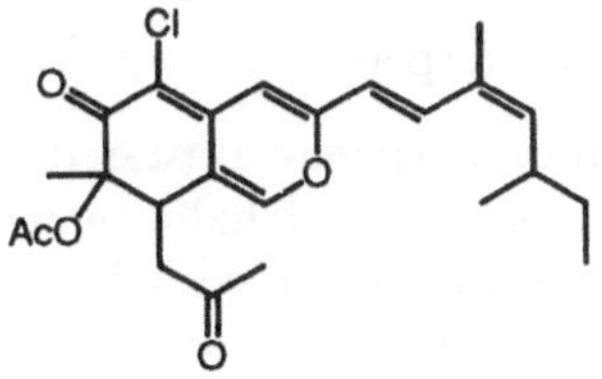

232 (isochromophilone II)

olites isochromophilones I (**229**, **230**) and II (**231**, **232**), which are mixtures as shown, have been isolated from a *Penicillium* sp. and are inhibitors of gp-120-CD4 binding of HIV cells (*212*). Isomer **229** is a dihydrorubrorotiorin.

The two novel metabolites stylocheilamide (**233**) and **234** are produced by the herbivorous sea hare *Stylocheilus longicauda* (*213*).

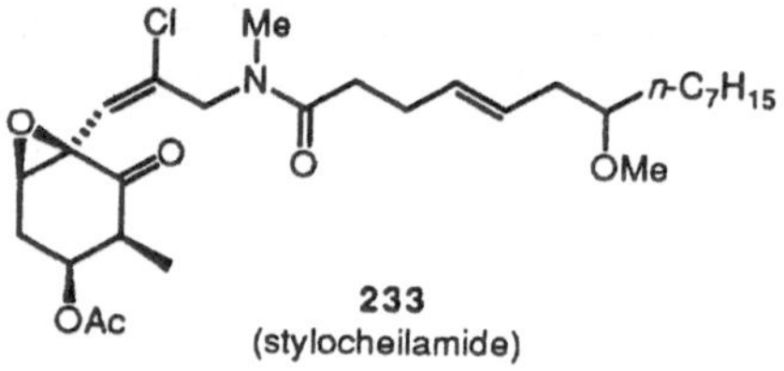

Cl Me N O O O n-C_7H_{15} OMe OAc

233
(stylocheilamide)

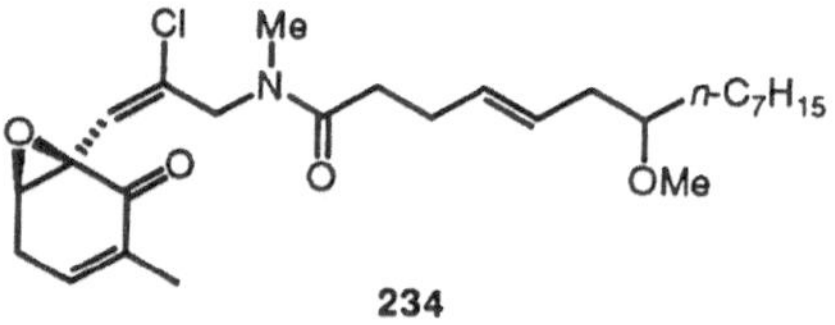

Cl Me N O O O n-C_7H_{15} OMe

234

3.4. Terpenes

The enormous abundance of terpenes in terrestrial life forms on our planet has spilled into the marine domain in the form of a plethora of halogenated terpenes produced by oceanic plants and animals.

3.4.1. Monoterpenes

An excellent review dealing with all aspects of marine monoterpenes has appeared (*214*); the reader is referred to it for information on marine monoterpenes isolated up through 1982. In those cases where structural revisions have been necessary, only the corrected structure is depicted.

3.4.1.1. Acyclic Monoterpenes

The first reports of halogenated monoterpenes appeared in 1973 when FAULKNER reported the isolation of the polyhalogenated 1-octen-3-ol **235** (*215*) and 1,5-octadiene **236** (*216*) from the sea hare *Aplysia californica*. The latter compound was described as the first naturally occurring vinyl bromide monoterpene (*216*). FAULKNER recognized that since **235** was also present in the red alga *Plocamium coccineum* (*217*), which is the diet of *A. californica*, the animal probably acquires these monoterpenes from the alga. The structure of **235** was established in part by conversion to the corresponding epoxide with base (*215, 218*), and that of **236** by X-ray crystallography (*216*). Furthermore, since this sea hare enjoys the absence of predators, it is likely that these halogenated monoterpenes serve in chemical defense (*217*). The following year, CREWS isolated cartilagineal (**237**) from the red alga *Plocamium cartilagineum* (*219*). This monoterpene was subsequently found to be toxic to mosquito larvae.

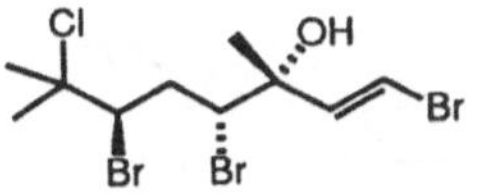

235

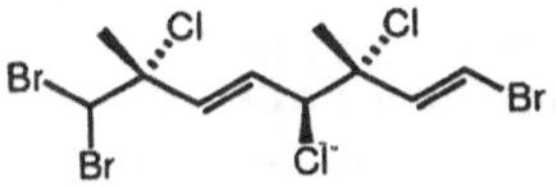

236

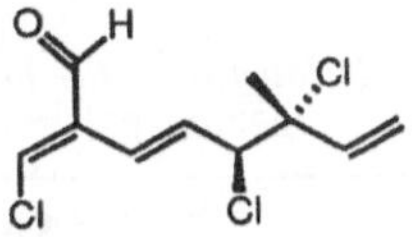

237
(cartilagineal)

There soon followed the isolation of a multitude of marine halogenated acyclic monoterpenes, mainly from red algae and/or from sea hares (*25–31, 214*). The red alga *Desmia* (*Chondrococcus*) *hornemanni* produces an array of halogenated acyclic monoterpenes (**238–247**) derived from myrcene (Table 13) (*220*), but *D.* (*Chondrococcus*) *japonicus* contains the two α-myrcenes **248** and **249** (*221*). It was found necessary to use acetone in the extraction process since methanol produced methyl ether solvolysis products (*221*).

In addition to **236**, several polyhalogenated octatrienes (**250–260**) are produced by *Plocamium cartilagineum* (*222*) (Table 14). The sea hare *Aplysia limacina*, which feeds on *P. cartilagineum*, contains the new octatriene **261** in addition to **251** in its digestive gland (*223*). The structure of **261** was revised by CREWS (*214, 224*) who also isolated octatriene

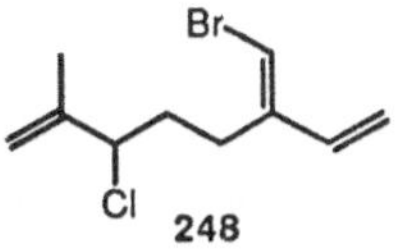

248

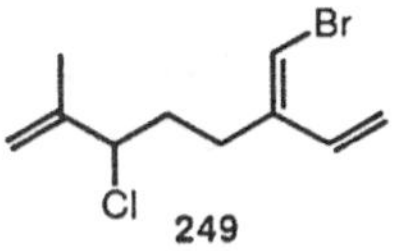

249

Table 13. *Halogenated Myrcenes Isolated from the Red Alga Desmia (Chondrococcus) hornemanni (220)*

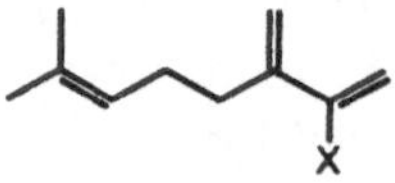

238 X = Cl
239 X = Br

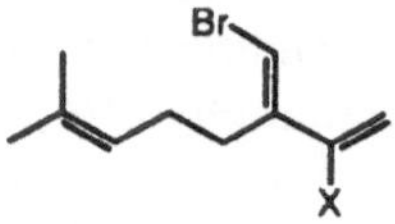

240 X = H
241 X = Cl

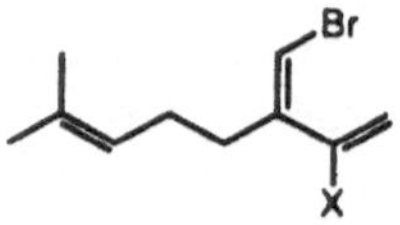

242 X = H
243 X = Cl

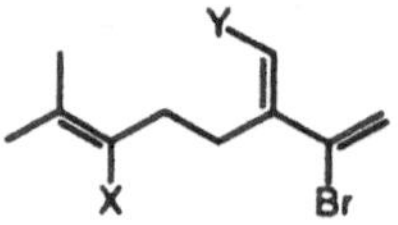

244 X = Cl, Y = Br
245 X = Br, Y = Cl

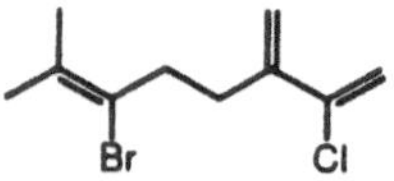

246[a]

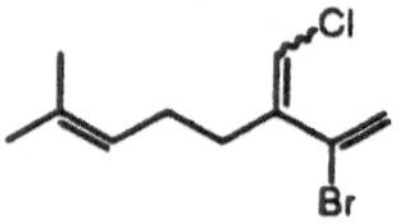

247[a]

[a] Tentative

Table 14. *Halogenated Octatrienes Isolated from the Red Alga Plocamium cartilagineum (222–225)*[a]

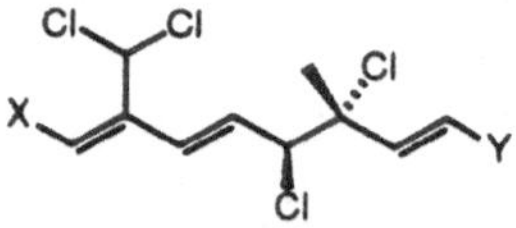

250 X = Y = H
251 X = Cl, Y = H
252 X = Cl, Y = Br

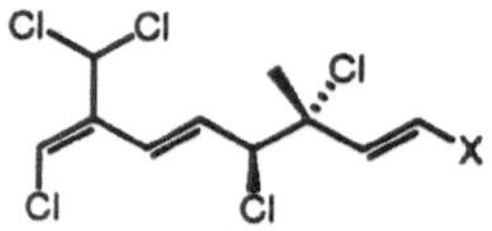

253 X = H
254 X = Br

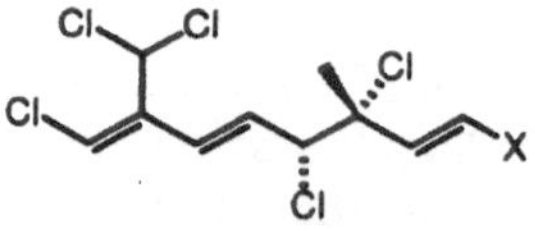

255 X = H
256 X = Br

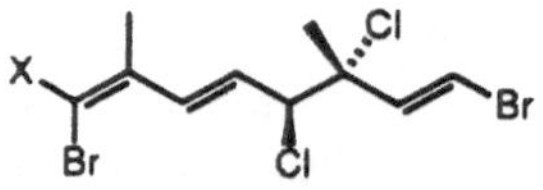

257 X = H
258 X = Br

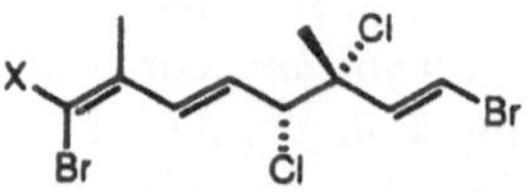

259 X = H
260 X = Br

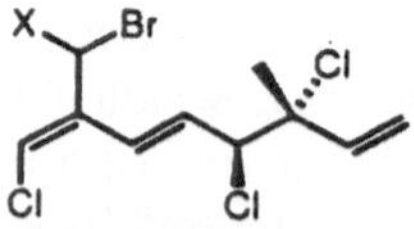

261 X = Br
262 X = Cl

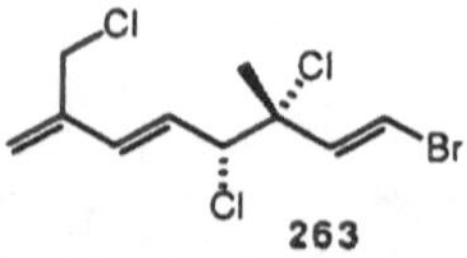

263

[a] Also isolated was **236** (*222*)

262 from this alga (*224*). More recently, **263** was isolated from *P. cartilagineum*; this new monoterpene was found to have antifungal, antibacterial, and molluscicidal activity (*225*).

A *Plocamium* sp. from the Antarctic, identified as *P. cartilagineum* in an earlier study from the same location (*226*), was found to contain monoterpenes **264–267** (*227*) which are strongly mutagenic in the Ames assay against TA100 and TA1535 (*239*). The structure of another monoterpene, also isolated from the digestive gland of the sea hare *Aplysia californica* (*228*), was revised to **268** (*214, 227*). The earlier study also reported the isolation of acetate **269** (*228*) from *A. californica*, which feeds on *Plocamium cartilagineum*. The content of halogenated monoterpenes in New Zealand *P. cartilagineum* varies from sample to sample; the novel C_8 metabolites **270** and **271** were found in one sample, whereas **272–280** were isolated from the second sample (*229*).

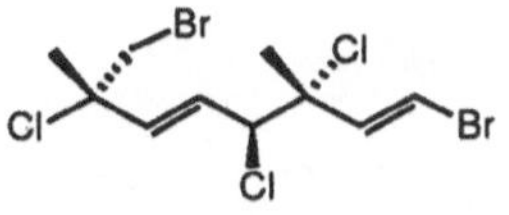

264 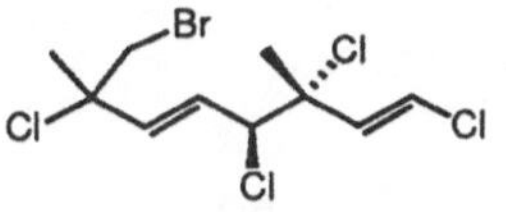 **265**

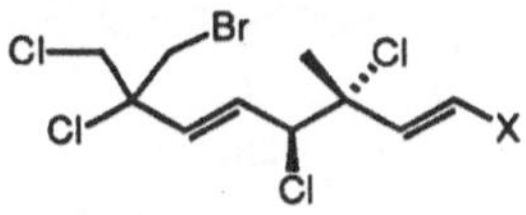

266 X = Cl
267 X = Br (oregonene A)

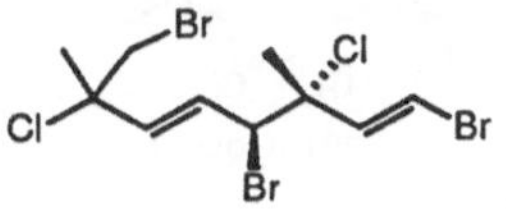

268

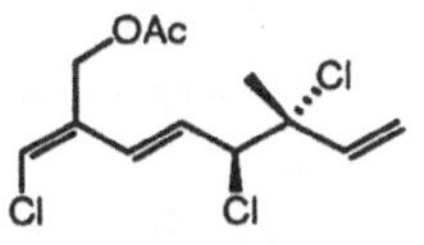

269

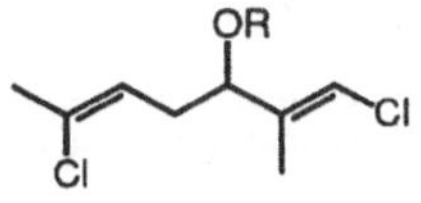

270 R = H (major)
271 R = Ac

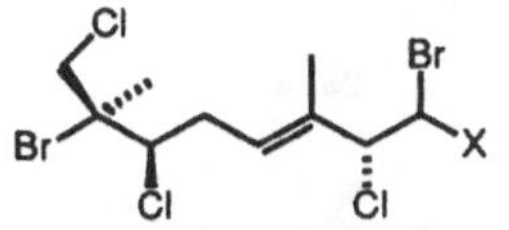

272 X = Br
273 X = H

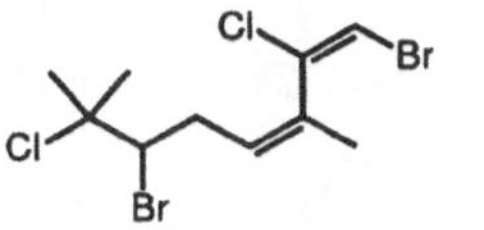

274

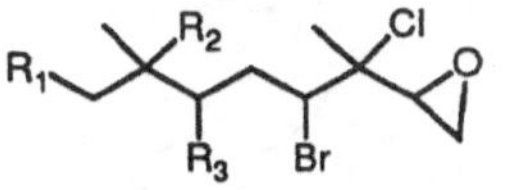

275 R_1 = H, R_2 = Br, R_3 = Cl
276 R_1 = H, R_2 = Cl, R_3 = Br
277 R_1 = Cl, R_2 = αBr, R_3 = βCl

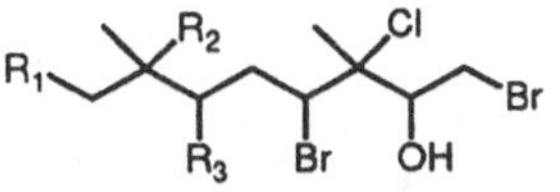

278 R_1 = H, R_2 = Br, R_3 = Cl
279 R_1 = H, R_2 = Cl, R_3 = Br
280 R_1 = Cl, R_2 = αBr, R_3 = βCl

Plocamium costatum contains costatol (**281**) (*230*), and *P. oregonum* produces **267** ("oregonene A"), which was later isolated from *P. cartilagineum* (*225*), as well as **257, 259, 264**, and **282** (*231*), while *P. violaceum* contains the octadienes preplocamenes A–C (**283–285**) (*214, 232*). The New Zealand *P. cruciferum* produces the bisnormonoterpene **270** as the major metabolite, along with **272**, which is the first example of a 3-ene monoterpene to be reported (*233, 234*). Plocamenone A (**286**), a unique halogenated monoterpene and a potent mutagen, was isolated from a *Plocamium* sp. (*235*). This enone undergoes an interesting methoxide induced Favorsky rearrangement-elimination to afford **286a**. *Plocamium hamatum* produces the novel metabolites **287** and **288** (*236*). Collections of this alga from different sites yielded different compounds. The red alga *P. angustum* produces the new monoterpene **289** (*237*) and the previously known **290** (*237, 238*).

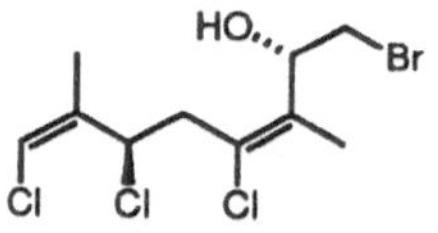

281 (costatol)

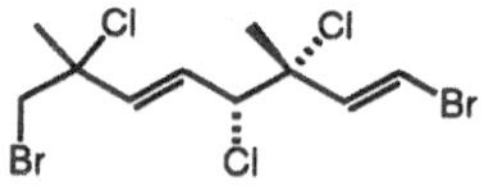

282

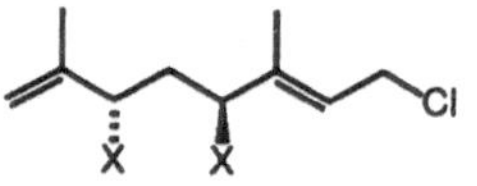

283 X = Br (preplocamene A)
284 X = Cl (preplocamene B)

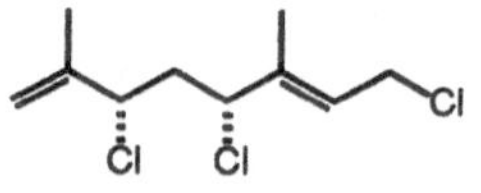

285 (preplocamene C)

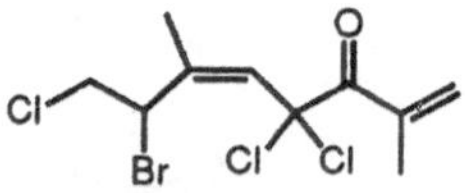

286 (plocamenone A)

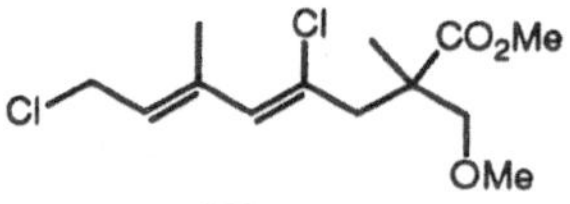

286a

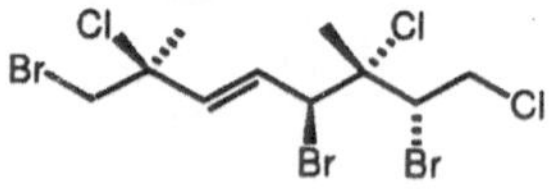

287

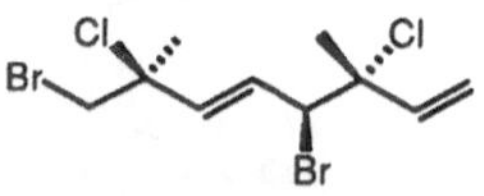

288

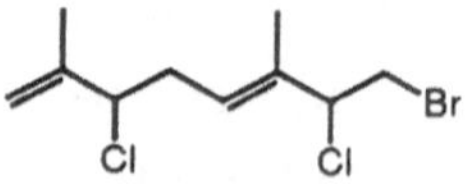

289

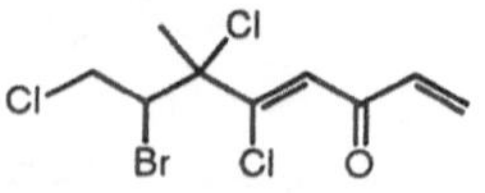

290

The red alga *Chondrococcus hornemanni* was one of the earliest marine plants to be intensively studied. The Hawaiian version of this alga produces several new halogenated myrcene derivatives **291–297** and the known myrene **246** (*240*). It is proposed that these dihalomyrcenes are biogenic precursors of the halogenated myrcenes **238–247** (Table 13) (*240*). For example, *in vivo* didehalogenation of **297** would afford **240** and dehydrochlorination of **295** would give **241** and **243**. A subsequent study reported the isolation of the new halogenated myrcene **298** and the known **238, 240–242** (*241*). The Sri Lankan *C. hornemanni* contains dihalomyrcene **299**, which may be the biogenic precursor of **240** and **241** due to dehydrochlorination (*242*).

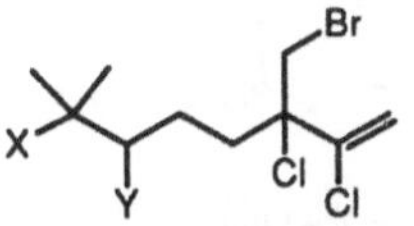

291 X = Br, Y = Cl
292 X = Cl, Y = Br

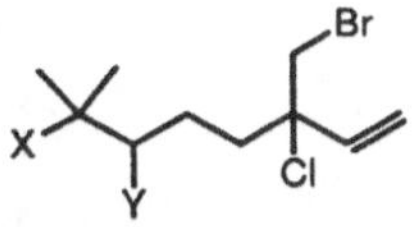

293 X = Br, Y = Cl
294 X = Cl, Y = Br

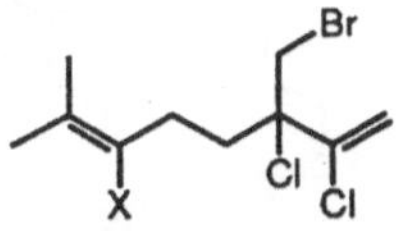

295 X = H
296 X = Cl

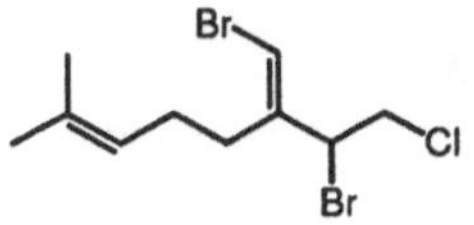

297

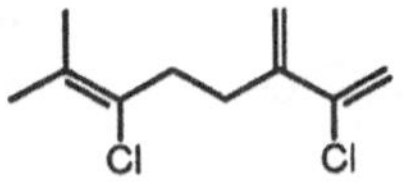

298

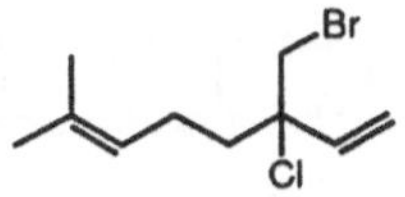

299

Several collections of *C. hornemanni* from the Great Barrier Reef and North Queensland were found to contain myrcene derivatives **300–303** (*243*), and **304–310** (*244*), respectively. The chemical content of these red algae samples varies depending on location and time of the collection (*244*). Although allylic bromides are extremely susceptible to alcoholic solvolysis (*221*), the authors of this study believe that **305–310** are not artifacts (*244*). Four new halometabolites, **311–314**, were isolated from the red alga *Portieria hornemannii* along with several known monoterpenes (*245*). As noted earlier, sea hares and other opistobranchs can acquire some or all of their halogenated metabolites from an algal diet. For example, the sea hare *Aplysia kurodai* contains the related acyclic monoterpenes kurodainol (**315**) (*246*) and aplysiaterpenoid B (**316**) (*247, 248*).

Four marine hydroids (*Aglaophenia pluma, Halocordyle disticha, Eudendrium glomeratum, Sertularella crassicaulis*) contain the halogenated octatrienes **251, 253, 255**, and **261** (*249*), previously isolated from *Plocamium cartilagineum* (Table 14).

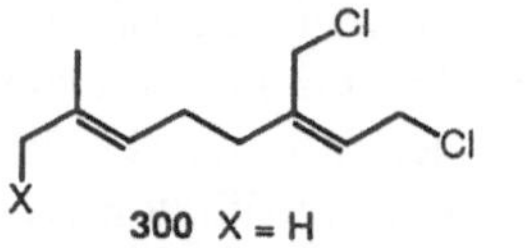

300 X = H
301 X = Cl

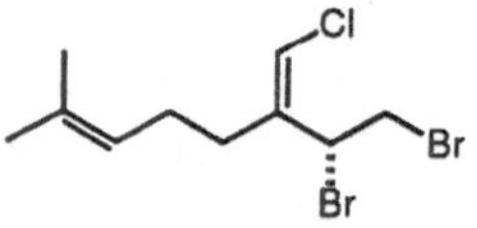

302

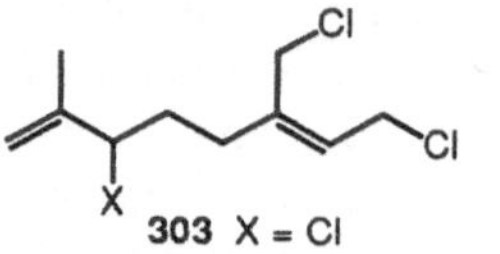

303 X = Cl
304 X = Br
305 X = OMe
306 X = OH

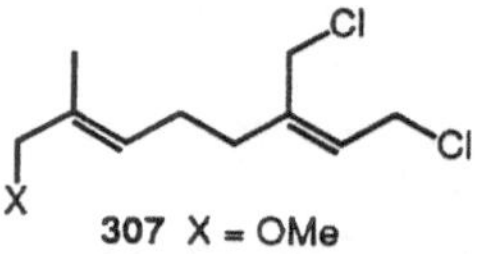

307 X = OMe
308 X = OH

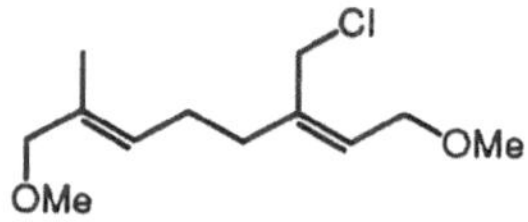

309

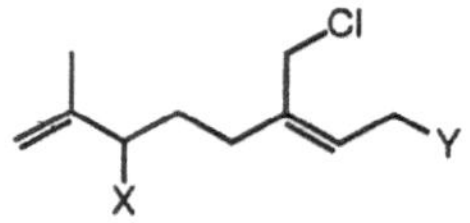

310 X = Y = OMe
311 X = OH, Y = OMe
312 X = OMe, Y = OH

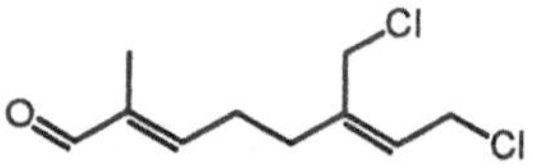

313

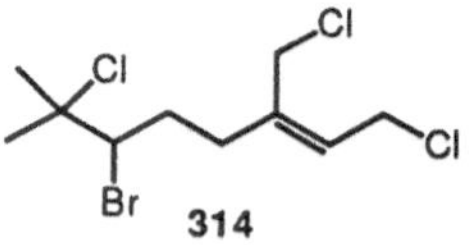

314

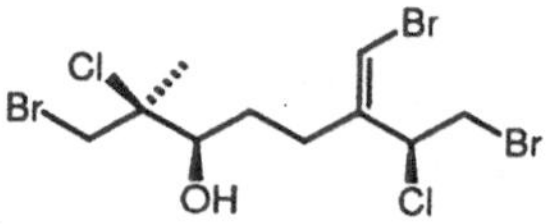

315
(kurodainol)

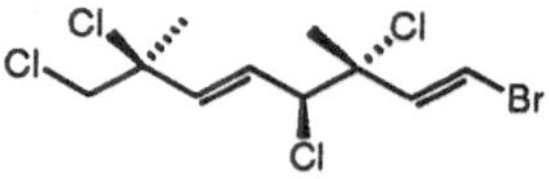

316
(aplysiaterpenoid B)

3.4.1.2. Alicyclic Monoterpenes

The biogenetic precursors of the halogenated alicyclic marine terpenes are thought to be the acyclic monoterpenes (*214*). Consequently, most of the red algae discussed in the previous section also contain monocyclic monoterpenes. Although these alicyclic metabolites are obviously less conformationally flexible than their acyclic counterparts and, therefore theoretically more amenable to structural elucidation, the presence of multiple halogens in these compounds provides for challenging structural problems nevertheless!

The red algae *Plocamium* sp. have proved to be a rich source of halogenated monocyclic monoterpenes. However, the different chemical types of, for example, *Plocamium violaceum* and *P. cartilagineum* from the west coast of California (*214, 250, 251*) have made this area of research very tedious. Moreover, several structural revisions have been required over the past 15 years and absolute configurations are only known in a few cases. The structures shown here are those considered to be correct based on the latest information and are depicted so as to indicate the proposed biogenetic relationship **A** → **B** (*214*). Monoterpenes of type **C** are discussed separately.

The first marine cyclic halogenated monoterpene to be isolated was violacene by FAULKNER from *Plocamium violaceum* (*252*), shown to have structure **317** (as revised by CREWS from the original **317a**) (*253, 254*). Table 15 illustrates the monocyclic halogenated monoterpenes **317**–**323** from this red alga. Several of these compounds, such as the plocamenes

Table 15. *Halogenated Monocyclic Monoterpenes Isolated from the Red Alga Plocamium violaceum*

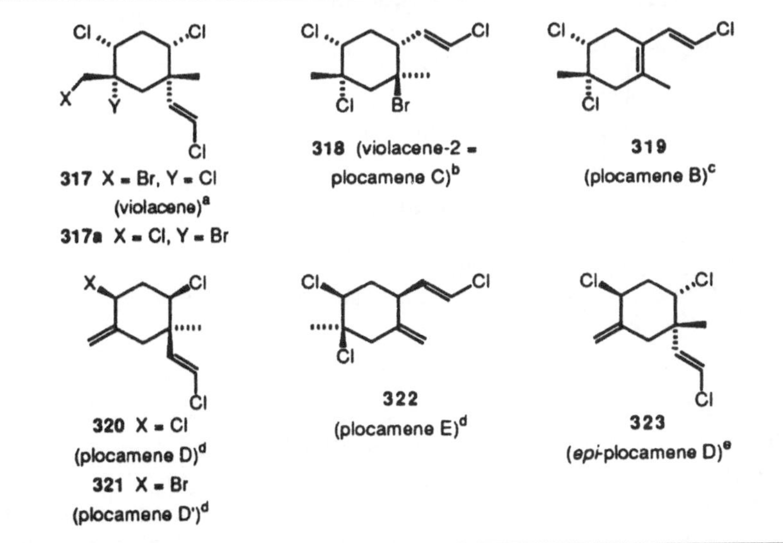

[a] Ref. *252–254* [b] Ref. *255* [c] Ref. *256* [d] Ref. *254* [e] Ref. *224*

Table 16. *Halogenated Monocyclic Monoterpenes Isolated from the Red Alga Plocamium cartilagineum*[a]

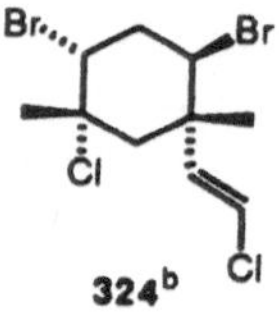

324[b] 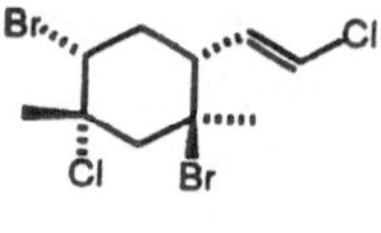 **325**[b,c,d,g] 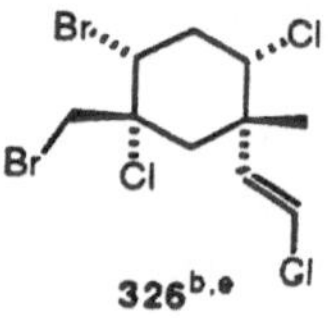 **326**[b,e]

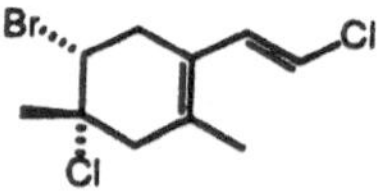

327[b] 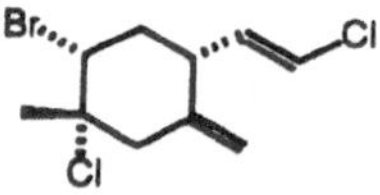 **328**[b] 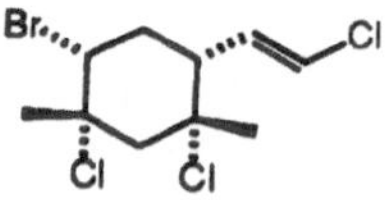 **329**[c,f]

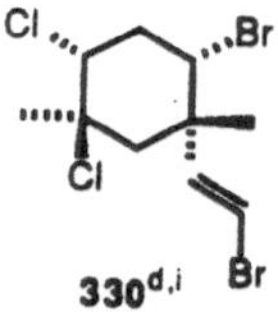

330[d,i] 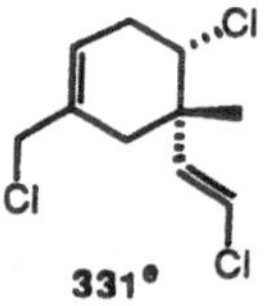 **331**[e] 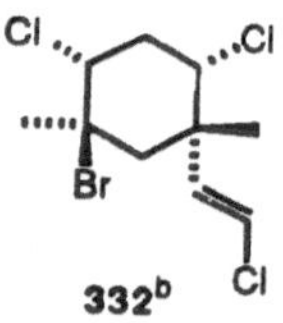 **332**[b]

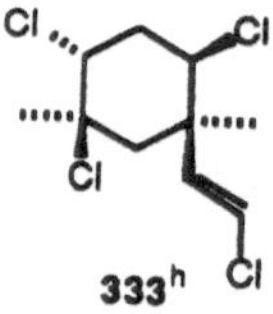

333[h] 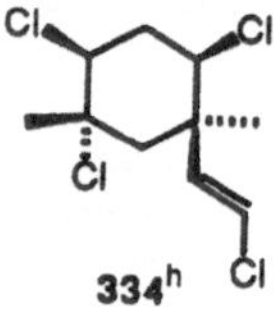 **334**[h] 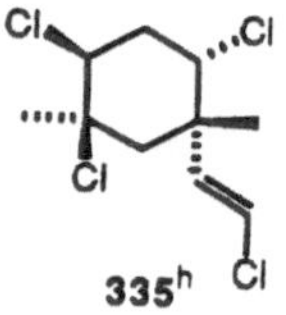 **335**[h]

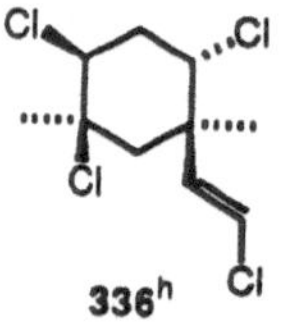

336[h] 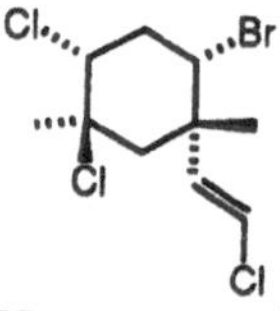 **337 (mertensene)**[h]

[a] Also isolated were **321** and **323** (*226*)
[b] Ref. *257* [c] Ref. *258* [d] Ref. *259* [e] Ref. *226* [f] Ref. *224* [g] Ref. *225*
[h] Ref. *260, 261* [i] Ref. *266*

320–322, have potent activity against mosquito larvae at levels down to 0.03 ppm (*254*). For example, plocamene B was three times more effective than the commercial insecticide lindane against mosquito larvae (*272*).

Even more numerous have been the halogenated monocycles **324–337** isolated from the ubiquitous *Plocamium cartilagineum* as summarized in Table 16. In addition to **333–336**, the Chilean *P. cartilagineum* contains violacene (**317**), plocamene D (**320**), *epi*-plocamene D (**323**) and mertensene (**337**) (*260, 261*). These metabolites are also present in the red alga *Shottera nicaensis* (*260*) and a study of their insecticidal and acaricidal activity has been reported (*261*). Of the group, violacene (**317**) is the most potent insecticide.

Mertensene (**337**) was first isolated from *Plocamium mertensii* (*258*) but the original structure has been revised several times (*214, 224, 236*). It has also been found in *P. hamatum* along with **338** and **339** (*236*). The Western Australian *P. mertensii* produces **340**, which is the enantiomer of **328** (*262*), and *P. telfairiae* produces telfairine (**341**), a very potent insecticide, along with the known **338** (*263, 279*). The Senegalese *P. coccineum* contains **342** (*264*); this red alga from Northwest Spain produces coccinene (**343**) (*265, 266*), an epimer of mertensene (**337**), along with the known metabolites. The red algae *Microcladia borealis*, *M. californica*, and *M. coulteri* contain violacene (**317**), plocamene B (**319**), C (**318**), and D (**320**) (*268*). A biosynthesis study supported the reductive pentose pathway for the origin of these monocyclic terpenes (*267*).

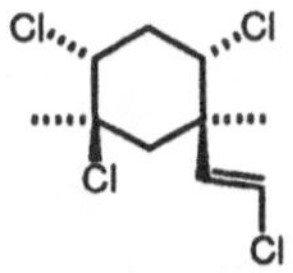

Cl Cl Cl Cl
338 (aplysiaterpenoid A; gelidene)

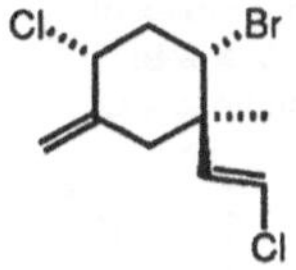

Cl Br Cl
339

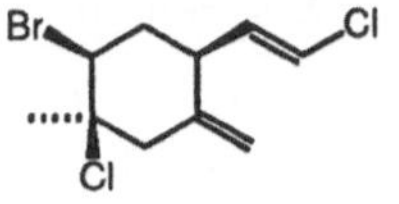

Br Cl Cl
340

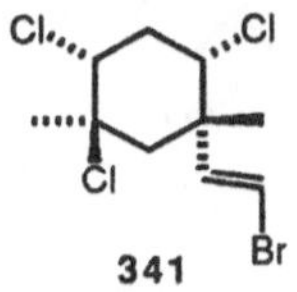

Cl Cl Cl Br
341 (telfairine)

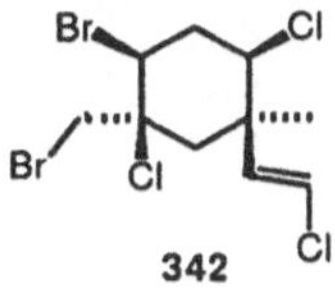

Br Cl Br Cl Cl
342

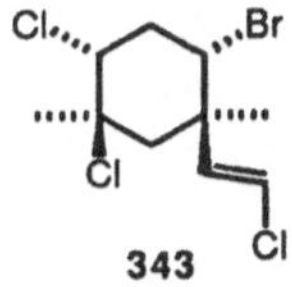

Cl Br Cl Cl
343 (coccinene)

Sea hares, presumably mainly through their diet, also contain halogenated monocyclic terpenes. *Aplysia kurodai* contains aplysiaterpenoid A (**338**) (*248*) which was also isolated from *Plocamium hamatum* (*236*). Aplysiaterpenoid A is extremely active against German cockroaches, mosquito larvae and L1210 cancer cells *in vitro* (*279*). This metabolite was also isolated from the red alga *P. leptophyllum* and found to be an extremely potent antifeedant for several marine herbivores. A dose of 40 μg stopped the feeding of all test animals (*269*). The red alga *Gelidium sesquipedale* contains gelidene (*270*), a compound apparently having the same structure as aplysiaterpenoid A (**338**). *Aplysia kurodai* also contains mertensene (**337**) (*247*).

A large number of marine halogenated monoterpenes of general structure C (*vide supra*) is also known. The first examples of this class to be isolated were chondrocole A (**344**), B (**345**), and C (**346**) and **347** from the red alga *Chondrococcus hornemanni* (*240, 241*) (structure revisions in *214, 224, 271*). This alga also contains chondrocolactone (**348**) (*271*), which can be synthesized from **344** with chromic oxide. *C. hornemanni* from Tahiti contains ochtodene (**349**) and the new ochtodane **350** and its (impure) epimer (**351**) (*272*). This collection also resulted in isolation of cartilagineal (**237**), violacene (**317**), plocame B (**319**), D (**320**), and E (**322**). The new ochtodanes **352** and **353** were found in *C. hornemanni* from the Great Barrier Reef (*243*).

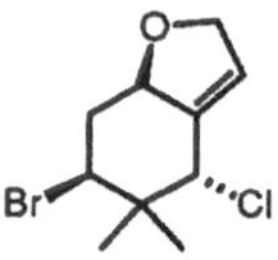

344 (chondrocole A)

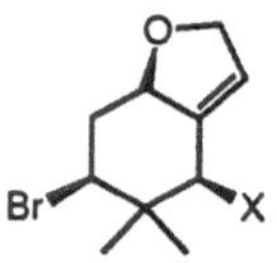

345 X = Cl (chondrocole B)
346 X = Br (chondrocole C)

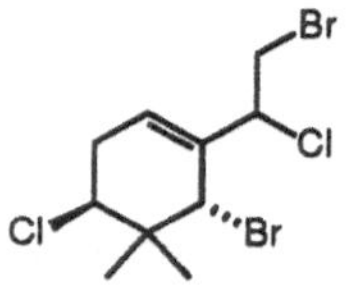

347

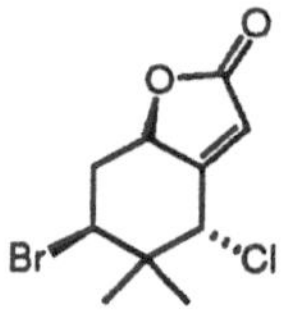

348 (chondrocolactone)

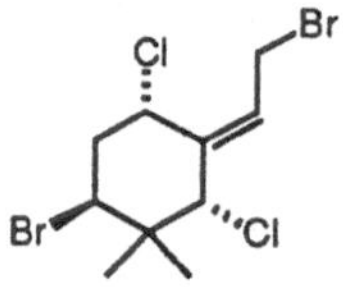

349 (ochtodene)

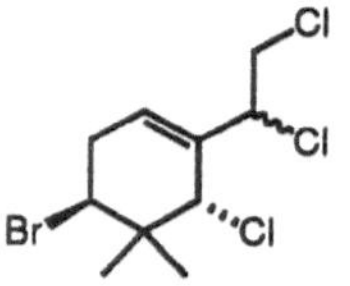

350, 351

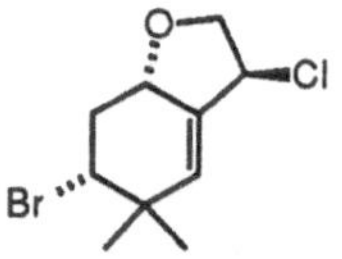

352

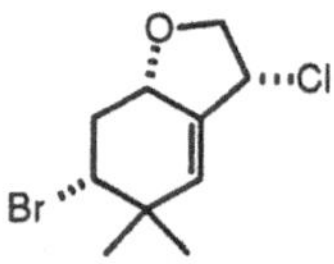

353

Ochtodene (**349**) was first isolated from the Caribbean red alga *Ochtodes secundiramea* along with ochtodiol (**354**) and chondrocole A (**344**) (*273*). Ochtodene shows strong antibacterial activity. Two other collections of this alga yielded **355** (*224*) and **356** (*274*). The richest source of this ring system has been the red *Ochtodes crockeri* from Galapagos, which has furnished chondrocoles A (**344**) and C (**346**) and the new compounds **357**–**364** (*275*).

354 355 356

357 358 α-OH 359 β-OH 360

361 362 α-OH 363 β-OH 364

There is another small group of cyclic monoterpenes with a pyran ring instead of a cyclohexane ring. For example, *Plocamium costatum* contains costatone (**365**) (*230, 277*) and costatolide (**366**) (*277*). The sea hare *Aplysia kurodai* contains aplysiapyranoids A–D (**367–370**, respectively) (*278*).

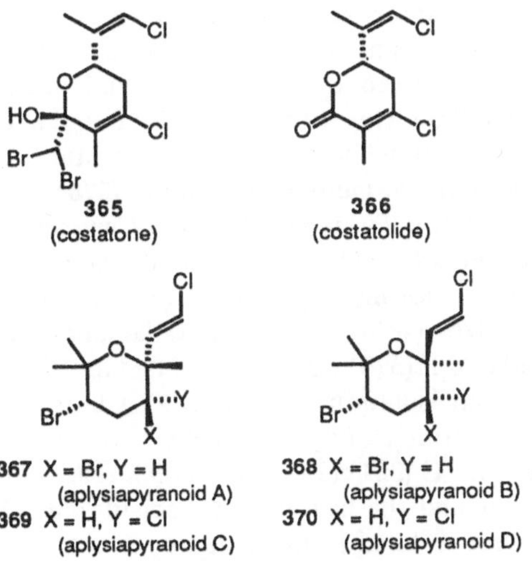

365 (costatone)

366 (costatolide)

367 X = Br, Y = H (aplysiapyranoid A)

368 X = Br, Y = H (aplysiapyranoid B)

369 X = H, Y = Cl (aplysiapyranoid C)

370 X = H, Y = Cl (aplysiapyranoid D)

The aromatic phenolic monoterpenes of mixed biogenesis are discussed under "Complex Phenols" (Sect. 3.22), and the monoterpenoid iridoids are presented later (Sect. 3.7).

3.4.2. Sesquiterpenes

Unlike the halogenated monoterpenes which are only found in the oceans, the halogenated sesquiterpenes have a significant terrestrial source – the plant kingdom. For the most part, these particular sesquiterpenoids are based on the guaianolide sesquiterpene skeleton. Since many, if not most of these chlorosesquiterpenes are apparently biogenetically derived from the corresponding epoxide, the obvious possibility arises that these metabolites are isolation artifacts. However, several of these studies have shown either that the epoxide can survive the isolation procedure unchanged or that the chlorine metabolite is present by direct examination (*e.g.*, thin layer chromatography) of plant material. In most cases, the absolute configuration is not known.

3.4.2.1. Terrestrial Sesquiterpene Lactones

KUPCHAN isolated the first examples of chlorinated sesquiterpenoids in 1968 from the composite *Eupatorium rotundifolium* (tribe Eupatorieae) which yielded eupachlorin (**371**), eupachlorin acetate (**372**) and eupachloroxin (**373**) (*280, 281*). These three compounds showed significant antitumor activity *in vitro* and *in vivo* (*280, 281*). All other chlorinated sesquiterpene lactones reported subsequently have also come from members of the family Compositae. Centaurepensin (**374**) was isolated from *Centaurea repens* (Russian knapweed, tribe Cardueae, subtribe Centaureinae), a noxious weed toxic to horses and other livestock (*282*, structure revised in *283*). This substance is also found in *C. solstitialis*, the yellow star thistle (*284*), while *C. repens* also contains chlororepdiolide (**375**) (*285*). The related lactone acroptilin (**376**) is found both in *C. hyrcanica* and *Acroptilon repens* DC. (= *Centaurea repens* L.) (*286*, structure revised in *287*). These compounds were also isolated from *C. hyssopifolium* as chlorohyssopifolins A (= centaurepensin, **374**) and C (= acroptilin, **376**) (*288–290*). The new lactones chlorohyssopifolin B (**377**), D (**378a**) and E (**378**) were also isolated in these studies. Lactone **378a** is probably an isolation artifact (*289*). Chlorohyssopifolins A–E were also isolated from *C. linifolia* together with the new linichlorins A (**379**), B (**380**) and C (**381**) (*291*). *Acroptilon repens* also contains **374** and **376**; the latter and the new lactones solstitiolide (**383**) and episolstitiolide (**384**) were isolated from *Centaurea solstitialis*.

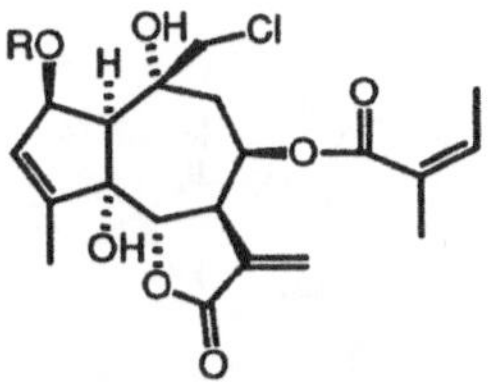

371 R = H (eupachlorin)
372 R = Ac

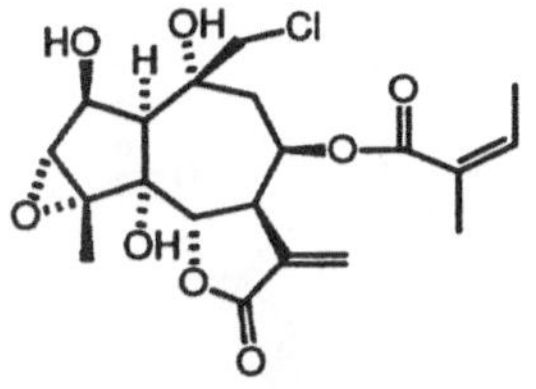

373
(eupachloroxin)

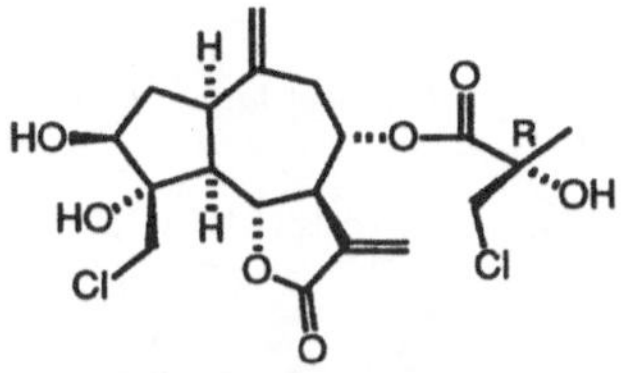

374 (centaurepensin = chlorohyssopifolin A)

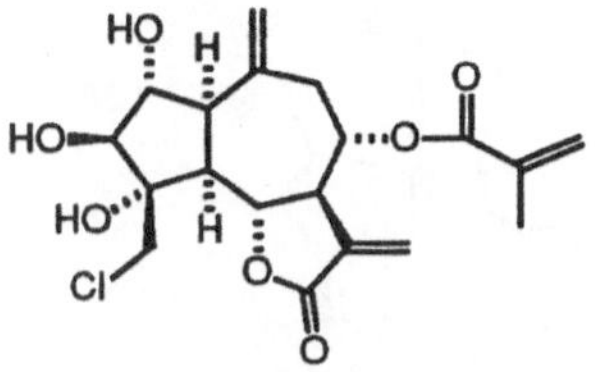

375 (chlororepdiolide = cebellin E)

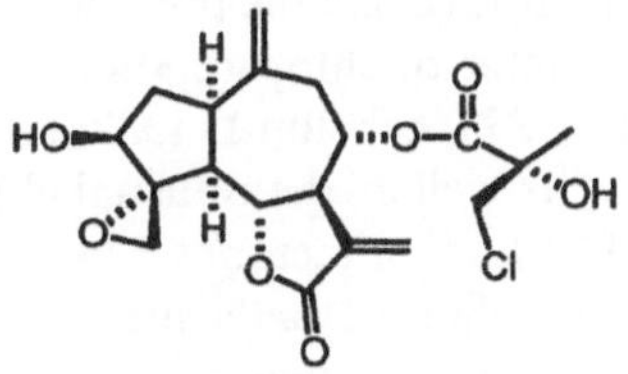

376 (acroptilin = chloro-hyssopifolin C)

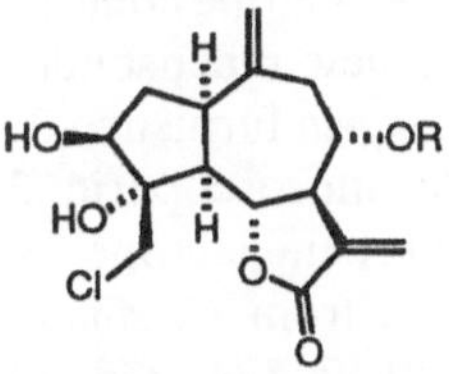

377 R = H (chlorohyssopifolin B)
378 R = $COC(CH_3)OHCH_2OH$ (chlorohyssopifolin E)
378a R = $COC(CH_3)OEtCH_2OH$ (chlorohyssopifolin D)

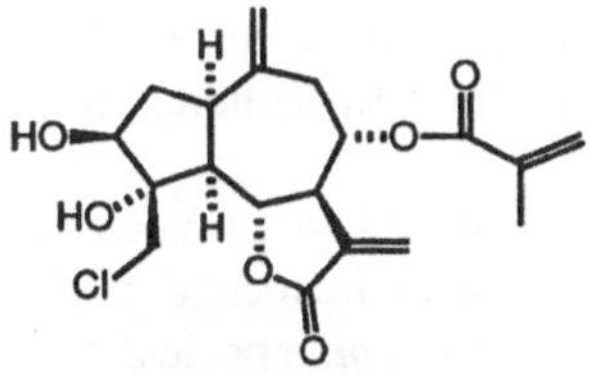

379 (linichlorin A = 19-desoxychlorojanerin)

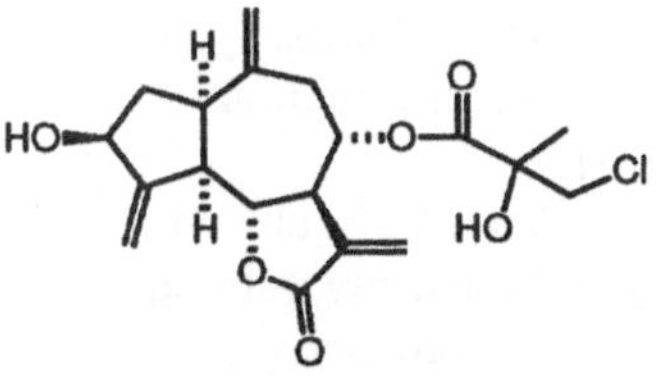

380 (linichlorin B)

Various other *Centaurea* species have been examined for their sesquiterpene lactone content. Thus *C. janeri* contained the new chlorojanerin (**385**) while *C. nigra* produced centaurepensin (**374**) (*296*). *C. aegyptica* produced not only **374**, **377** and **385**, but two compounds which, while reported as new, were linichlorin A (**379**), solstitiolide (**383**) and episol-

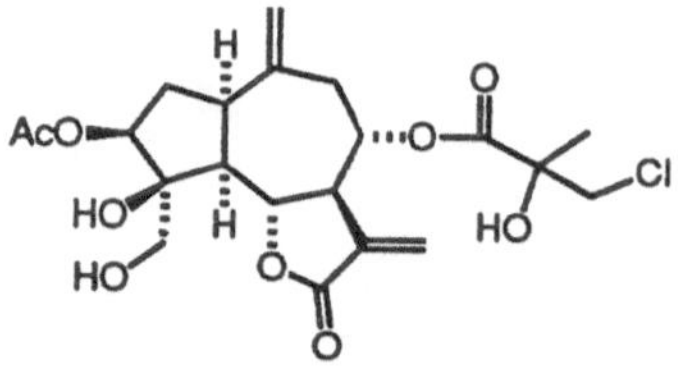

381 (linichlorin C)

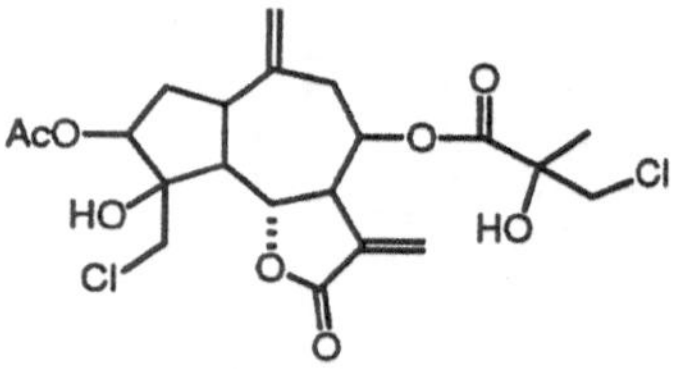

382 (salegin)

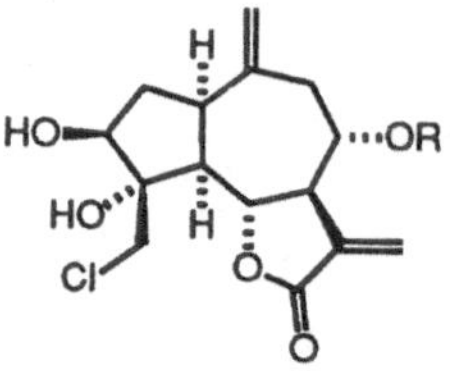

383 R = CO– (solstitiolide)

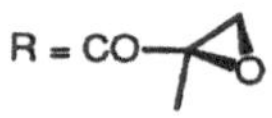

384 R = CO– (*epi*-solstitiolide)

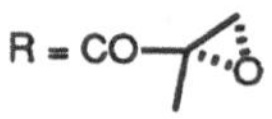

stitiolide (**384**) (*297*). Subsequently **378** was also found in this species. Study of an Argentine collection of *C. repens* revealed the presence of **379**, **385** and the new repensolide (**386**), an epimer of chlororepdiolide (**375**) (*299*). *C. scoparia* furnished **379**, **385** (*300*), **377**, cebellin D (**387a**) and its new acetate chloroscoparin (**387b**) (*2279*) as well as the unusual chlorinated sesquiterpene lactone **388** (*2280*). The known lactones **374** and **375** were isolated from *C. sinaica* (*301*, *2281*) while *C. bella* furnished **379** (*302*) as well as **374**, **375**, **376**, **385**, **387** and the new **389** (cebellin J) which is a C-2′-epimer of centaurepensin (**374**), *C. adjarica* providing additionally **383** or **384**, but no **379** or **389** (*305*, *2282*, *2283*). *C. hermannii* contained **379** and **385** (303) whereas *C. imperialis* furnished the then new cebellin D (**387a**) as well as **374** (*304*). *C. incana* contained **376** and **383** or **384** (*2284*) and *C. kotschyi* gave **380** (*2285*). *C. phaeopappoides* and *C. thracica* each contained **385** while *C. marshalliana* furnished **376**, **385** and **387a** (*305*).

Three *Chartolepis*, eleven *Psephellus* and three *Leuzia* species belonging to genera closely related to *Centaurea* also furnished chlorinated sesquiterpene lactones (*2283*). All three *Chartolepis* species, *C. biebersteinii*, *glastifolia* and *pterocaula*, contained **374**, **376** and **385**; in addition *C. glastifolia* produced the 2′-*S*-epimer of **374** cebellin J (**389**) as well as linichlorin C (**381**) and its 2′-*S*-epimer (*2283*, *2286*). Lactones **374**, **376** and **380** were found in *Psephellus carthalinicus*, *colchicus*, *daghestanicus*, *dealbatus*, *karabaghensis*, *nogmovii*, *somcheticus* and *zangezuri*, whereas *P. declinatus*, *hypoleucus* and *leucophyllus* furnished only **380** (*2283*, *2287*). Lastly, three *Leuzia* taxa, *L. carthamoides*, *L. rhapontica* ssp. *helenifolia* and *L. rhaponticoides* contained **385**, with *L. carthamoides* additionally furnishing **388** (*2283*, *2288*).

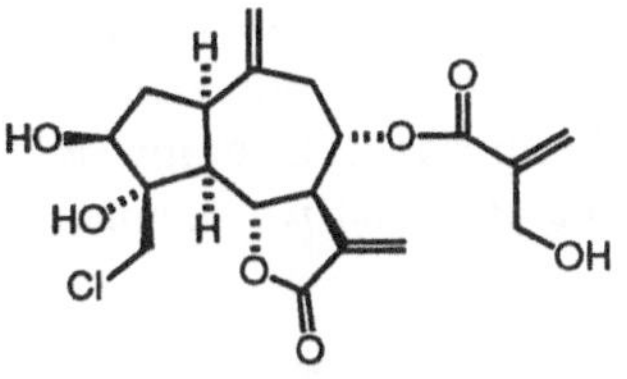

385
(chlorojanerine)

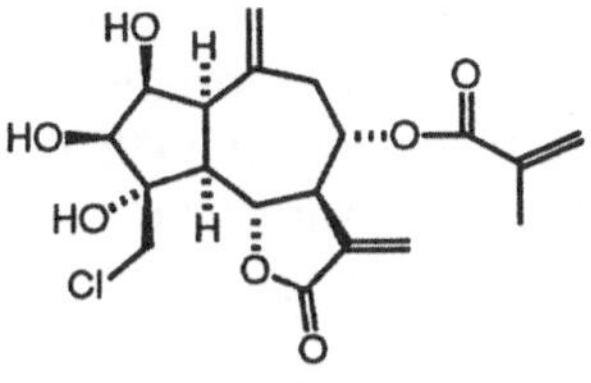

386
(repensolide)

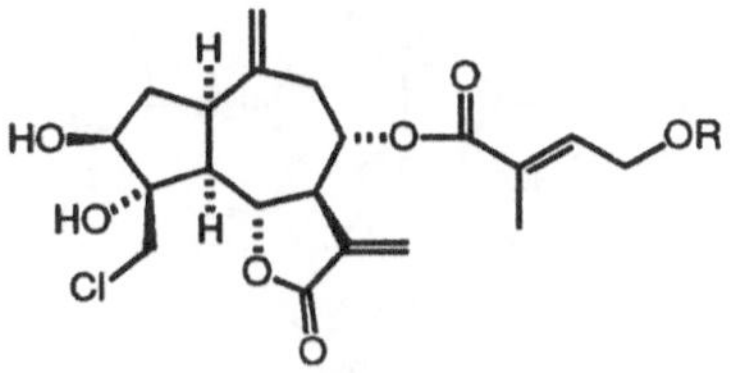

387a R = H (cebellin D)
387b R = Ac (chloroscoparin)

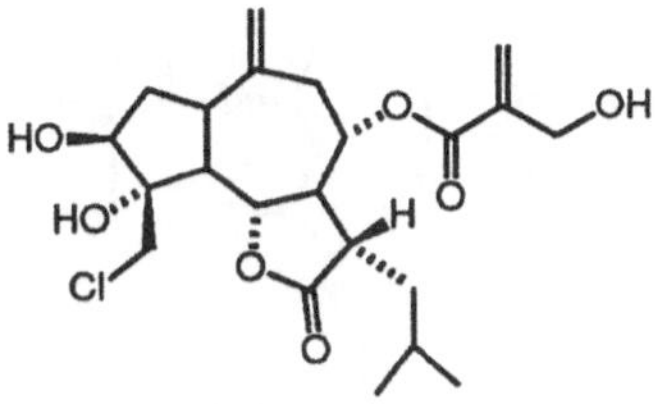

388

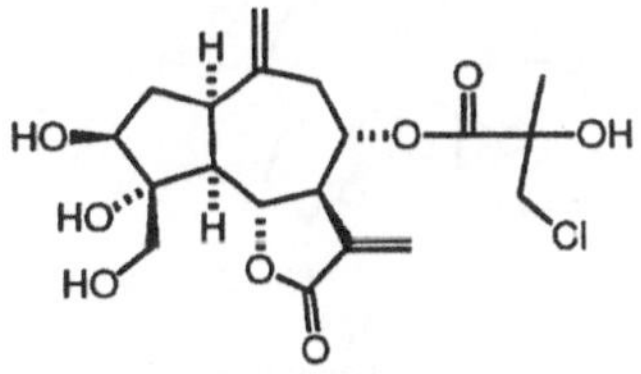

389 (cebellin J)

Three *Saussurea* species (tribe Cardueae, subtribe Carduinae) are reported as containing chlorinated sesquiterpene lactones. *S. candicans* yielded centaurepensin (**374**), **379** and **385** (*306*), *S. elegans* gave elegin (presumably linichlorin A, **379**) (*469*) and salegin (**382**) of uncertain stereochemistry at C-3, C-4 and C-8 (*294*), and *S. lipshitzii* from Mongolia produces **385** (*307*). A study of two *Cousinia* species in the same subtribe revealed the presence of **379** in both *C. canescens* and *C. piptocephala* and lactone **385** in the latter, as well as the new chlorosesquiterpene lactone **392** in both species (*309*). *Gutenbergia marginata* (tribe Vernonicae) produces angelate **390** and the novel dihydrochloroangelate **391** (*308*).

Biebsanin (**393**) was isolated from *Achillea biebersteinii* (tribe Anthemideae) (*310*). The Brazilian species *Lasiolaena santonii* (tribe Eupatorieae) contains the three guaianolides **394**–**396** (*311*), while *L. morii* contains **395** and the new **397** (*312*) (revised in *313*). The related **398** was found in *Trichogonia gardneri* (*313*) and the related lactones **399** and **400** were identified in *Mikania vitifolia* (*314*), both in the same tribe. The first example of a chloroguaianolide reported from the genus *Artemisia* (tribe Anthemideae) was chloroklotzchin (**401**) from *A. klotzchiana* (*315*). Two compounds, **402** and **403** that would appear to be epimeric deoxybiebsanins have been isolated from *Chrysanthemum parthenium*, also in Anthemideae (*316*).

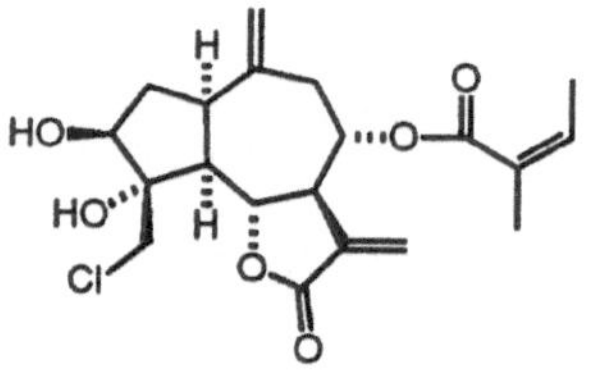

390

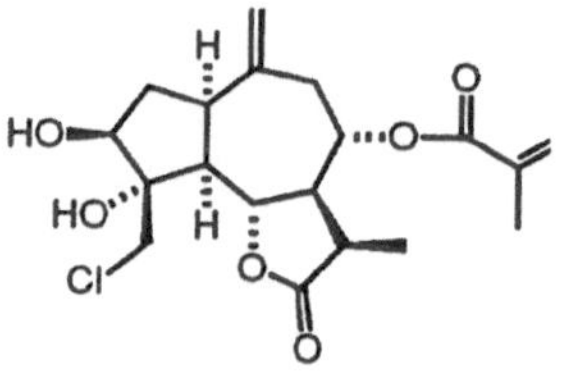

391

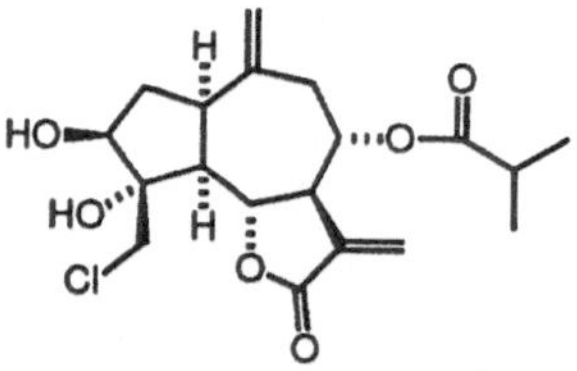

392

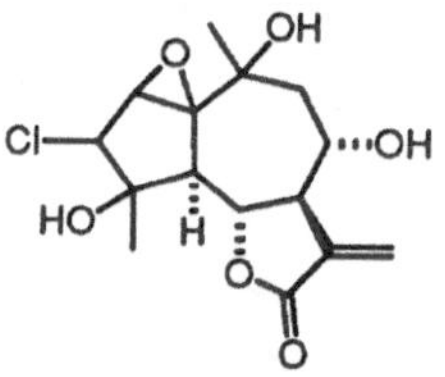

393 (biebsanin)

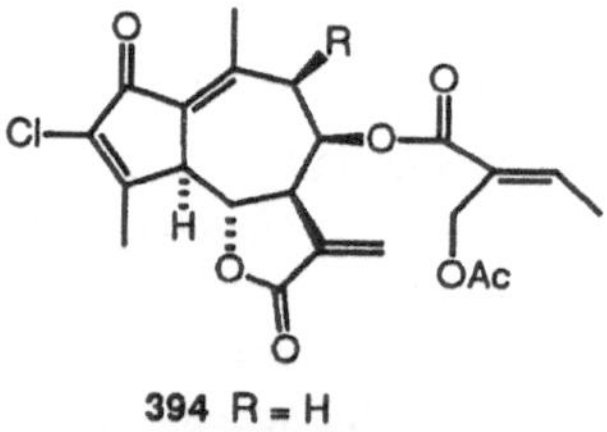

394 R = H
395 R = OH
396 R = OAc

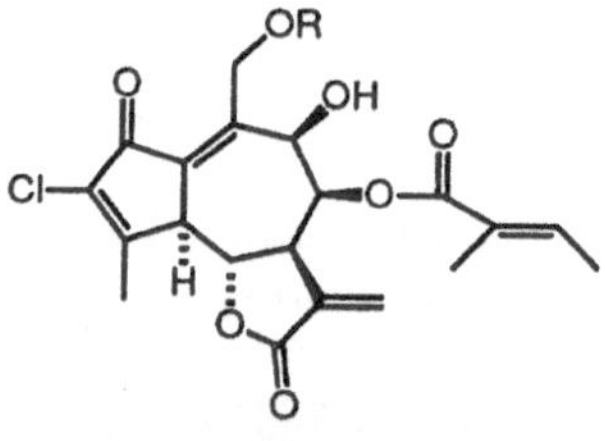

397 R = H
398 R = Ac

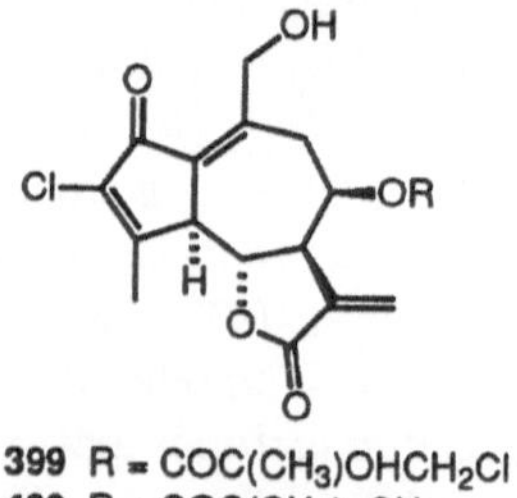

399 R = COC(CH_3)OHCH_2Cl
400 R = COC(CH_3)=CH_2

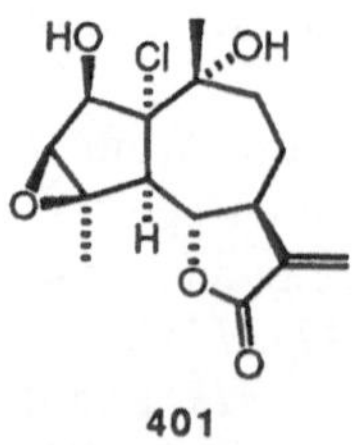

401 (chloroklotzchin)

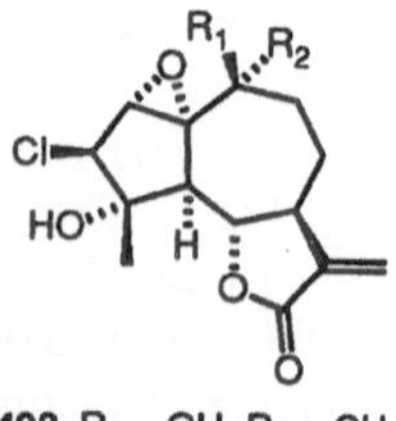

402 R_1 = OH, R_2 = CH_3
403 R_1 = CH_3, R_2 = OH

The Egyptian medicinal plant *Ambrosia maritima* (Heliantheae, subtribe Ambrosiinae) contains chloroambrosin (**404**) (*317*), and the Argentinian species *Stevia sanguinea* (tribe Eupatorieae) produces the chloro analogue **405** of a guaianolide from *Helianthus microcephalus* (Heliantheae, subtribe Helianthinae) (*318*). The closely related 3-chlorodehydroleucodin (**406**) is found in *Kaunia lasiophthalma* (Eupatorieae) (*319*). The unusual C-13 chlorine-containing 13-chlorosolstitialin (**407**) and its acetate **408** have been isolated from *Cynara humilis* (Cardueae, subtribe Carduinae) (*320*). The herb *Venidium fastuosum* (≡ *Arctotis fastuosa*, tribe Arctodideae) is reported to contain chlorovenidol (**409**) and chlorovenidin (**410**) (*321*). The C-5 stereochemistry of these two lactones differs from that assigned (*321*) because the reference compound arctolide also has an α-orientated C-5 hydroxyl.

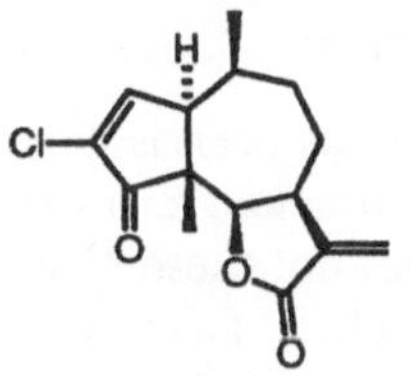

404 (chloroambrosin)

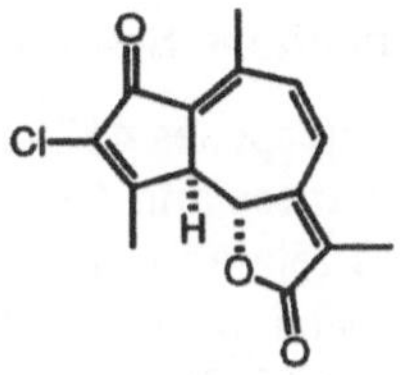

405

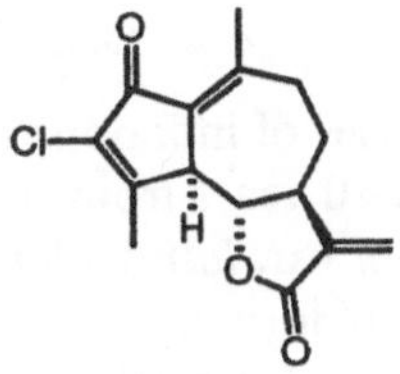

406

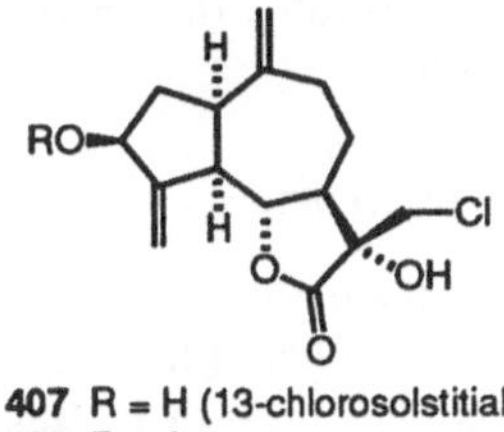

407 R = H (13-chlorosolstitialin)
408 R = Ac

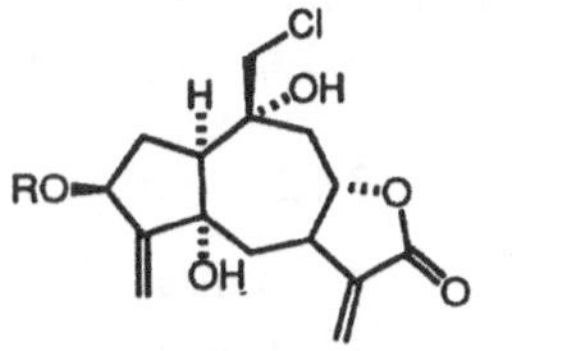

409 R = H (chlorovenidol)
410 R = $COCH_2CH_3$ (chlorovenidin)

The chlorofuranoheliangolides **411** and **412** have been isolated from *Calea morii*, while **411** and **413** are found in *C. pilosa* (*322*), and **414** is found in *C. villosa* (*323*), all in tribe Heleniantheae, subtribe Neurolinae. Lactone **415** was isolated from *Liatris acidota* (tribe Eupatorieae), but the authors suggest that it conceivably could be an artifact (*324*). The enormous abundance of chlorinated sesquiterpene lactones would suggest otherwise, and the authors have no evidence either way. The Ecuadorian species *Critoniopsis huaircajana* (tribe Vernonieae) produces poskeanolide (**416**), a novel seco-germacranolide (*325*).

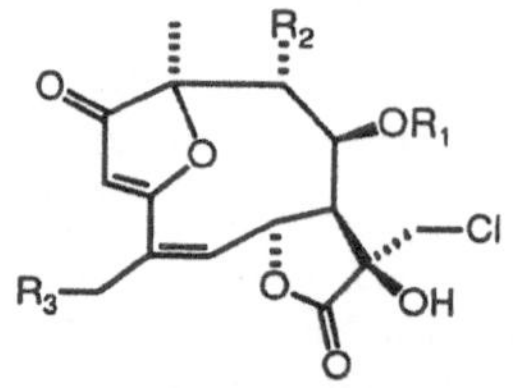

411 R_1 = $COC(CH_3)$=$CHCH_3$ (Z)
R_2 = R_3 = H
412 R_1 = $COC(CH_3)$=$CHCH_3$ (E)
R_2 = R_3 = H
413 R_1 = $COC(CH_3)$=$CHCH_3$ (Z)
R_2 = OH, R_3 = H
414 R_1 = $COC(CH_3)$=$CHCH_3$ (Z)
R_2 = H, R_3 = OH

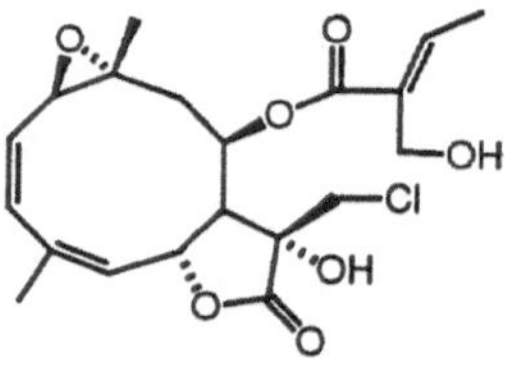

415

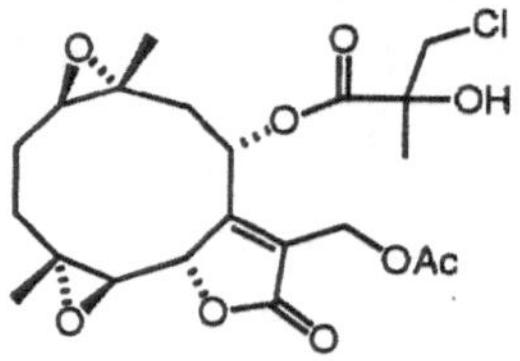

416 (poskeanolide)

3.4.2.2. Indanone Sesquiterpenes

A series of indanone sesquiterpenes **417–424a** has been isolated over the past 20 years from the Japanese and Chinese bracken fern *Pteridium aquilinum* var. *latiusculum* and related ferns. Several of these "pterosins" contain chlorine, as summarized in Table 17. These ferns have been responsible for cattle poisoning (*333*), and the pterosins have cytotoxic effects on sea urchin embryos and a ciliate (*334*). *Pteris podophylla* is a Costa Rican fern (*336*). A review of these metabolites makes reference to the unpublished histiopterosin B (**424a**) (*466*).

3.4.2.3. Other Terrestrial Sesquiterpenes

The two eudesmanes **425** and **426** have been described as being present in *Pluchea odorata* (Compositae, tribe Plucheae) from El Salvador (*326*). Since the corresponding epoxide is also present in this plant and because boling $CHCl_3$ (30 hours) was used in the extraction process, there is a strong possibility that **425** and **426** are artifacts. Control experiments were not reported. The Pakistani species *P. arguta* contains the epimeric **427** and **428** (*340*). The drimane-type sesquiterpene **429** was isolated from liverwort *Makinoa crispata* (*341*), and jaeschkenol (*430*) was found in the Himalayan plant *Ferula jaeschkeana* (Umbelliferae) (*342*).

3.4.2.4. Marine Sesquiterpenes

Not only is there a myriad of known halogenated marine sesquiterpenes, but the diversity of ring systems present in these metabolites is

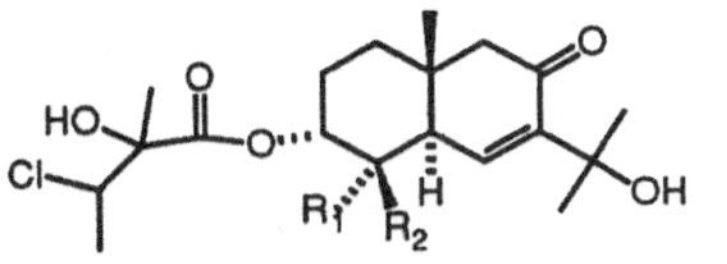

425 R_1 = OH, R_2 = CH_3
426 R_1 = OAc, R_2 = CH_3
427 R_1 = CH_3, R_2 = OH
428 R_1 = CH_3, R_2 = OAc

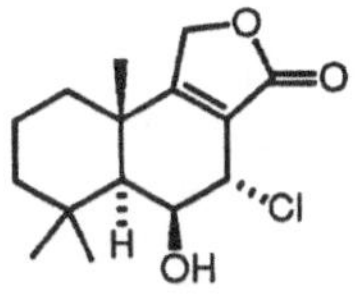

429

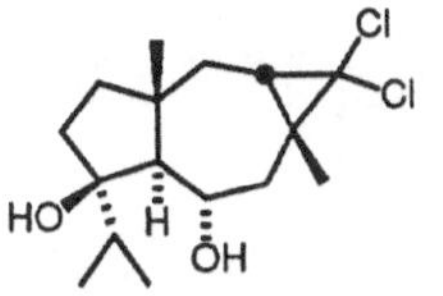

430 (jaeschkenol)

truly astonishing. The extensive nature of this subtopic requires that it be divided up into several sections. For excellent reviews, see (*343*, *344*).

3.4.2.4.1. Monocyclic and Other Simple Sesquiterpenes. As will be seen, *Laurencia* is a treasure trove of halogenated metabolites (*344*) and a few monocyclic sesquiterpenes have been isolated from species of this red alga. In several cases, the absolute configuration was determined.

Preintricatol (**431**), which is proposed to be the biogenetic precursor of halogenated chamigrenes (*vide infra*), has been found in *Laurencia intricata* (*345*), and also in *L.* sp. cf. *L. gracilis* from New Zealand (*346*). The related bisabolane puertitols A (**432** and B (**433**) (*L. obtusa*) (*347*) and **434** and **435** (*L. caespitosa*) (*348*) have been isolated from *Laurencia* from the Canary Islands.

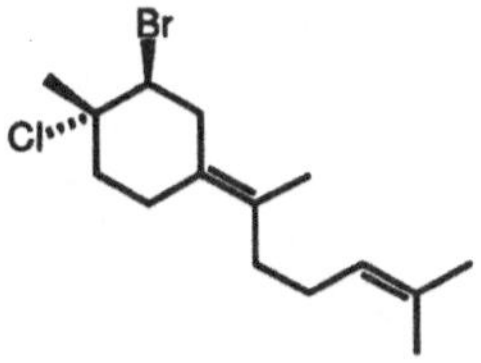

431 (preintricatol)

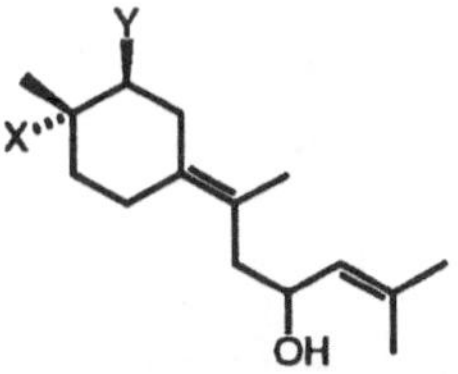

432 X = Cl, Y = Br
(puertitol A)
433 X = Br, Y = Cl
(puertitol B)

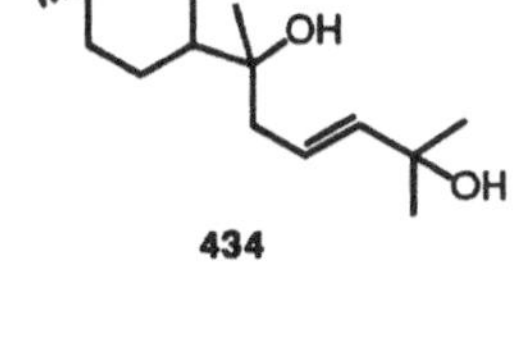

434

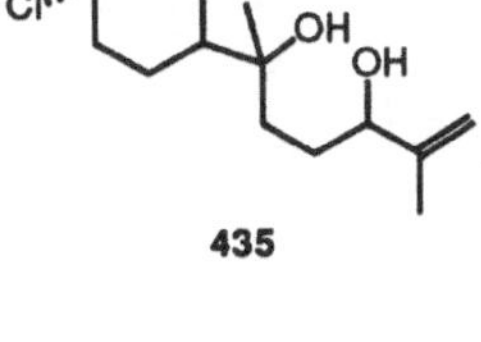

435

Table 17. *Chlorinated Indanone Sesquiterpenes*

Compound	Name	Plant	Ref(s).
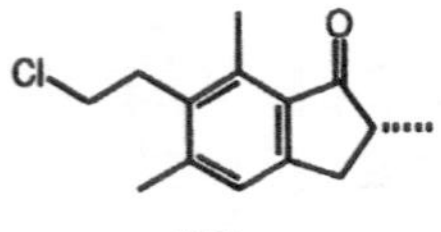 417	pterosin F	*Pteridium aquilinum* var. *latiusculum*	*327, 329, 332, 333, 334*
		Pteridium aquilinum subsp. *wightianum*	*337*
		Microlepia strigosa	*338*
		Microlepia substrigosa	*338*
		Pteris tremula	*339*
		Pteris dactylina	*339*
		Pteris multifida	*339*
		Pteris cretica	*339*
		Pteris angustipinna	*339*
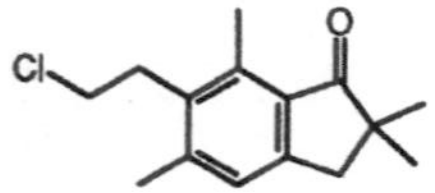 418	pterosin H (= hypolepin A)	*Dennstaedtia scaba*	*330*
		Hypolepis punctata	*328*
		Pteridium aquilinum	*334*
		Pteridium aquilinum subsp. *wightianum*	*337*
		Microlepia speluncae	*338*
		Microlepia trapeziformis	*338*
		Microlepia obtusiloba	*338*
		Microlepia substrigosa	*338*
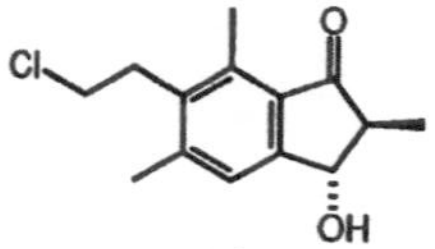 419	pterosin J	*Pteridium aquilinum* var. *latiusculum*	*329, 332*
		Pteris tremula	*339*

Structure	Compound	Source	Ref.
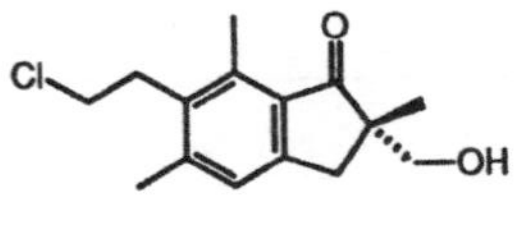 **420**	pterosin K	*Pteridium aquilinum* var. *latiusculum* *Dennstaedtia scaba* *Hypolepis punctata*	*329, 332* *330* *331*
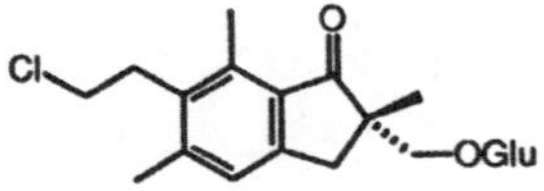 **421**	pteroside K	*Pteridium aquilinum* var. *latiusculum*	*333*
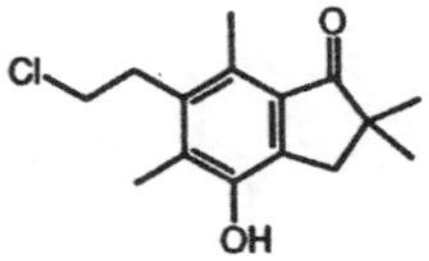 **422**	pterosin R	*Cibotium barometz*	*335*

Table 17 (*continued*)

Compound	Name	Plant	Ref(s).
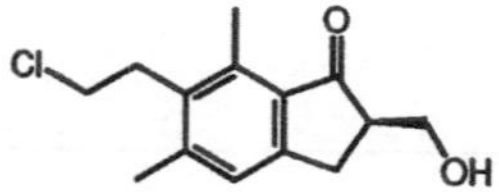 Cl OH O **423**		*Pteris podophylla*	*336*
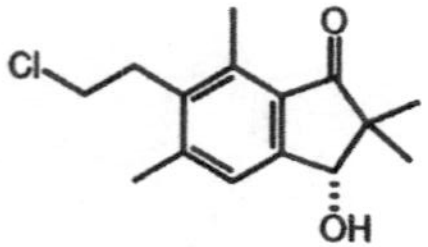 Cl O OH **424**		*Pteridium aquilinum* subsp. *wightianum*	*337*
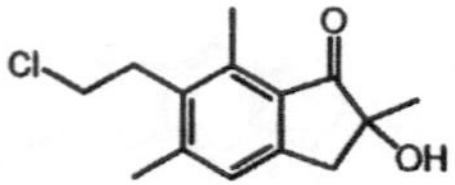 Cl O OH **424a**	histiopterosin B	*Histopteris incisa*	*466*

The isomeric α- (**436**) and β-snyderols (**437**) were present in *L. obtusa* (Spain) and (*L. snyderae*) (California), respectively (*349*). *L. obtusa* from the Red Sea also contains β-snyderol (**437**) along with acetate **438** as a minor product (*350*). The red alga *L.* cf. *palisada* also contains a minor product, palisol (**439**) (*351*), while *L. caespitosa* has yielded **440** and **441**, the latter of which possesses the unprecedented γ-snyderol system (*352*). The first two examples of the rare carbonimidic functionality, **442** and **443**, were isolated from the sponge *Pseudaxinyssa pitys* (*353*). It is proposed that these metabolites arise by enzymatic chlorination of the corresponding isonitriles (*353*), which are also well known from marine sources (*vide infra*). For a review of isonitrile natural products see (*2211*).

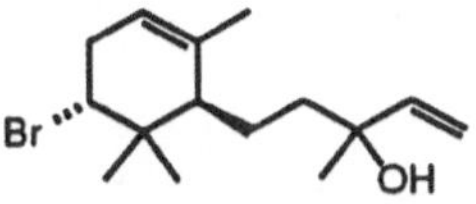

436 (α-snyderol)

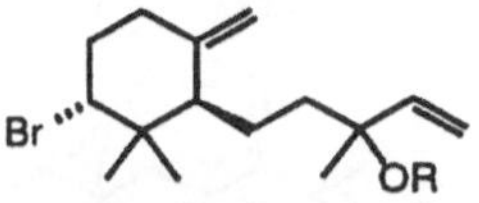

437 R = H (β-snyderol)
438 R = Ac

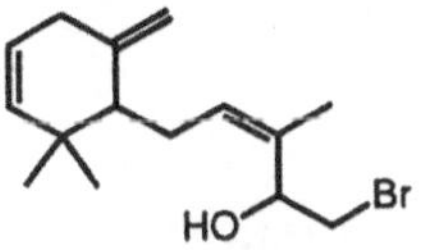

439 (palisol)

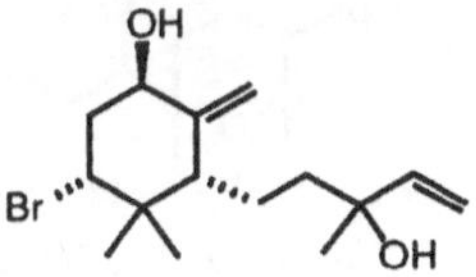

440

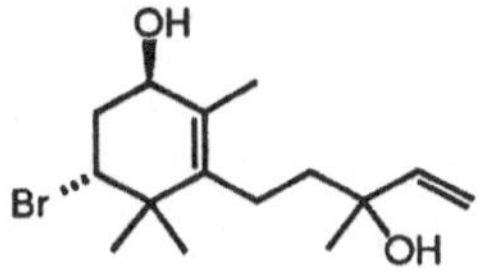

441

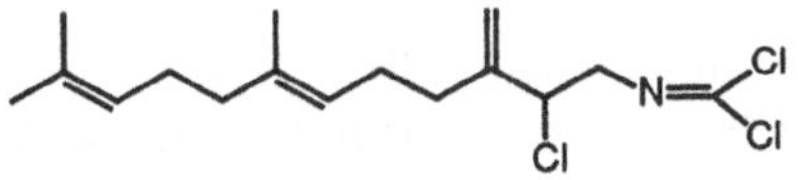

442

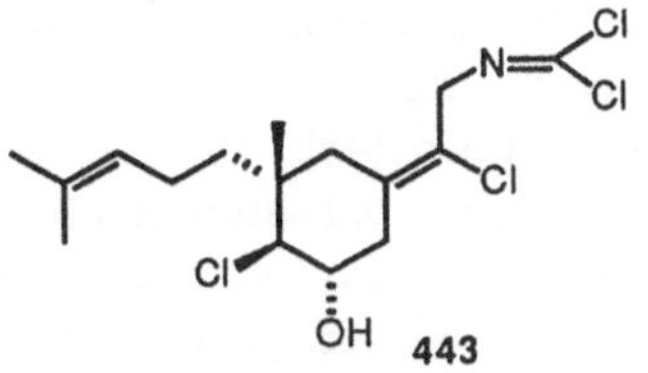

443

Several cyclic halogenated ether metabolites related to the monocyclic sesquiterpenes discussed above are known. It would appear that these compounds arise biogenetically by bromonium ion-triggered cyclization of the corresponding alcohol (*2212*). For example, the monocyclic pyran derivative obtusenol (**444**) (*Laurencia obtusa*) (*354*) might arise from the sesquiterpene alcohol nerolidol as shown in Scheme I (*343*, *344*). These bromonium ion cyclizations, involving either hydroxyl groups or olefins, are well known in the laboratory (for an early example, see *355*), and those that occur *in vivo* with the assistance of bromoperoxidase will be discussed later.

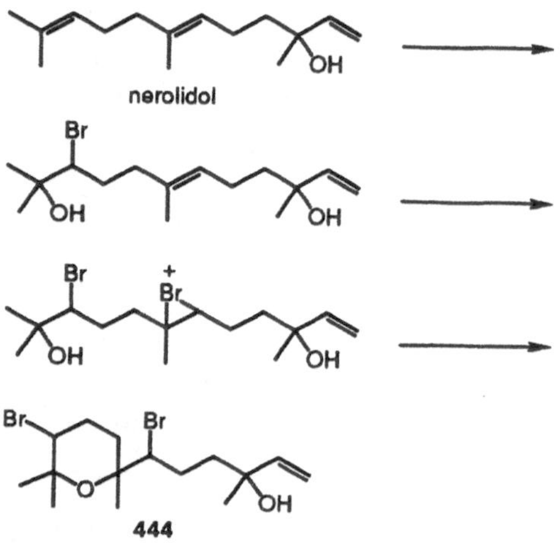

Scheme I. Possible biogenesis of obtusenol (**444**)

The major metabolite of *Laurencia obtusa* from the English Channel is bicyclic ether **445** which is clearly related to (and perhaps formed from) the snyderols (**436, 437**) (*356*). The red alga *L.* cf. *palisada* contains a group of bicyclic and tricyclic ethers 5-acetoxypalisadin B (**446**), palisadin B (**447**), **448**, and palisadin A (**449**), the major constituent (*351*). PAUL and FENICAL also isolated aplysistatin (**450**) in this study, but suggested that it was probably formed from palisadin A (**449**) by air oxidation, since this transformation could be demonstrated in the laboratory (*351*). Aplysistatin (**450**) had been isolated previously from the Australian sea hare *Aplysia angasi* by PETTIT (*357*). This and the 6β-hydroxy derivative **451** were found in *Laurencia filiformis* (*358*), while *L. flexilis* from the Philippines contains the new palisadins **452–456** as well as the known **446–450** (*359*, *360*).

The first halogenated norsesquiterpene, napalilactone (**457**), was isolated from the soft coral *Lemnalia africana* from Pohnpei (*361*). A biogenesis originating from the sesquiterpene aristolene was proposed (*361*).

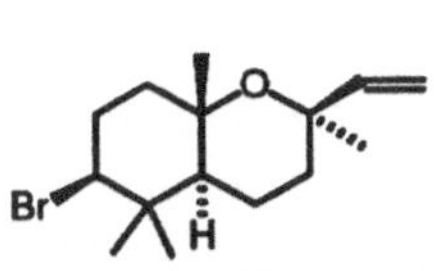

445

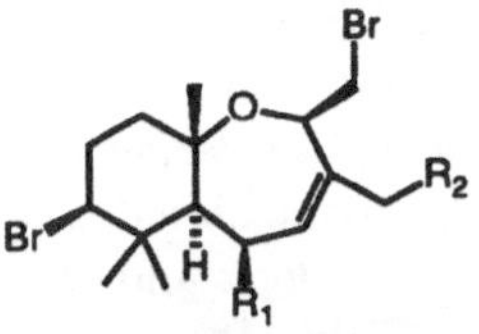

446 R_1 = OAc, R_2 = H
447 R_1 = R_2 = H (palisadin B)
448 R_1 = H, R_2 = OH
452 R_1 = H, R_2 = Br
453 R_1 = OH, R_2 = H

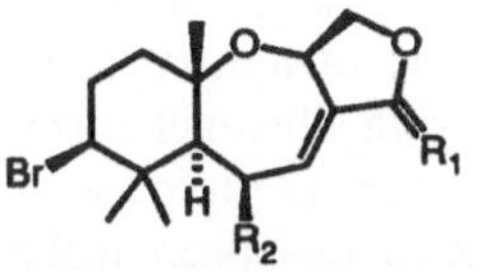

449 R_1 = H_2, R_2 = H (palisadin A)
450 R_1 = O, R_2 = H (aplysistatin)
451 R_1 = O, R_2 = OH
454 R_1 = H_2, R_2 = OAc

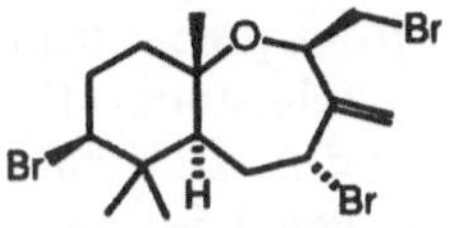

455 (palisadin C)

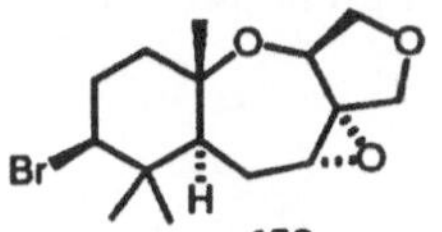

456

The interesting furan metabolites furocaespitane (**458**) and isofurocaespitane (**459**) have been reported from *Laurencia caespitosa* (*362*, *363*) and more recently the related C_{12} metabolites **460** and **461** (isolated as the methyl ester) have been isolated from this red alga (*364*).

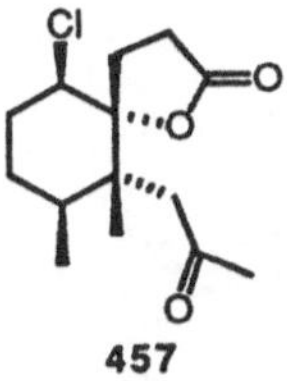

457 (napalilactone)

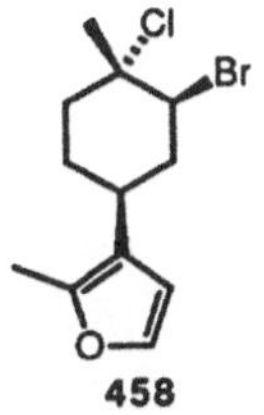

458 (furocaespitane)

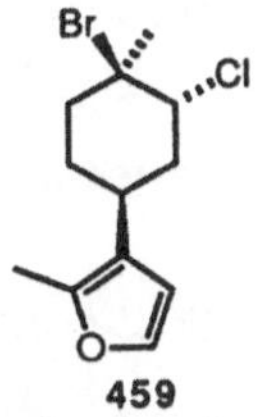

459 (isofurocaespitane)

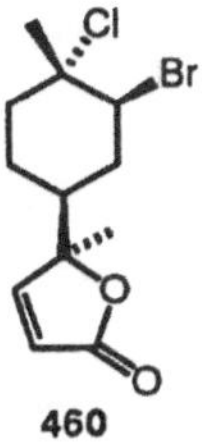

460

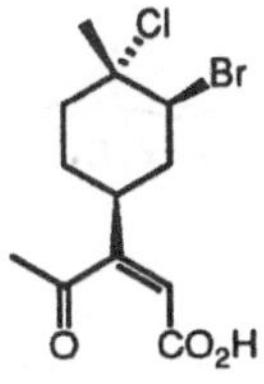

461

3.4.2.4.2. Chamigrene and Related Types. Halogenated spirochamigrenes represent perhaps the most commonly encountered type of sesquiterpene in *Laurencia* red algae (*343, 344*). The first such marine metabolite to be isolated was spirolaurenone (**462**) from *Laurencia glandulifera* (*365, 366*), and several others (**463–468**) soon followed from this red alga (Table 18).

Pacifenol (**469**) was the first natural product to contain both chlorine and bromine when it was isolated from *L. pacifica* in 1971 (*371*), although in this study it was an artifact formed during silica gel chromatography. Pacifenol (**469**) which is found in *L. tasmanica* (*372*) forms readily (silica gel or heat) from prepacifenol (**470**) which has been isolated from *L. filiformis* (*372*). Johnstonol (**471**) which is related to pacifenol is present in *L. johnstonii* and *L. okamurai* (*373*). The former alga also contains trace amounts of pacifenol. Several other halogenated chamigrenes are present in *L. pacifica*, including 10-bromo-α-chamigrene (**467**) (*374*) which has

Table 18. *Halogenated Chamigrenes Isolated from the Red Alga Laurencia glandulifera*

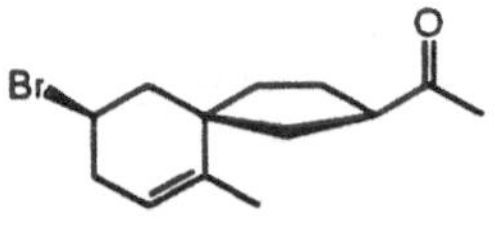

462 (spirolaurenone)[a,e]

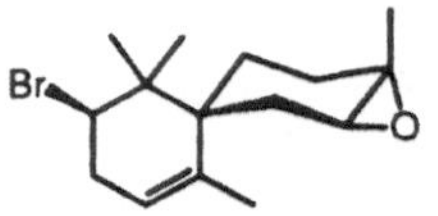

463[b,c,e]

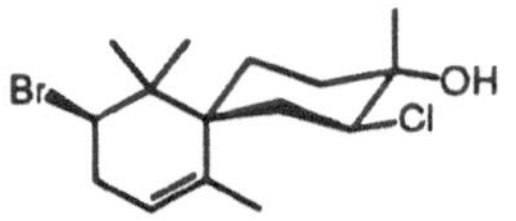

464 (glanduliferol)[c,d,e]

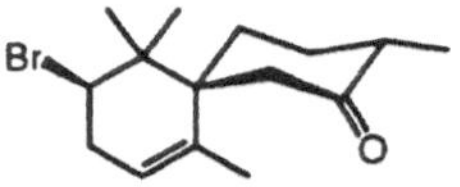

465[b,c,e]

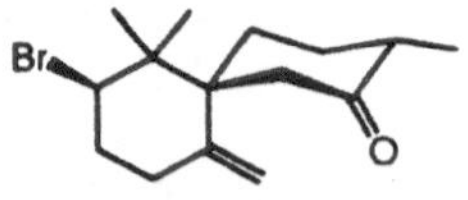

466[b,c]

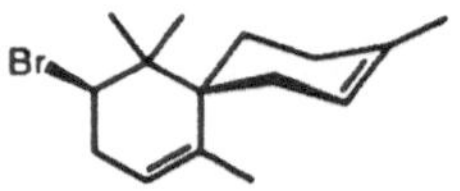

467 (10-bromo-α-chamigrene)[c,e]

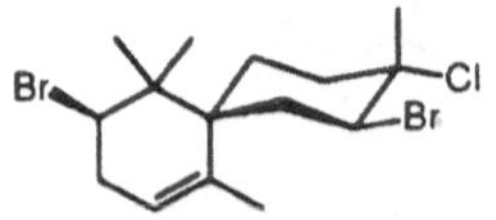

468[c,e,f]

[a] Ref. *365, 366* [b] Ref. *367* [c] Ref. *368* [d] Ref. *369* [e] Ref. *370* [f] Ref. *379*

been synthesized in a biomimetic manner (*375*, *380*). Mexican *L. pacifica* contains **472** (*376*, *377*). Kylinone (**473**), which represents a new ring system "kylinane," is found in *L. pacifica* (*378*). A New Zealand *Laurencia* species cf. *L. gracilis* produces several sesquiterpenes and C_{15} acetogenins (Sect. 3.6) (*346*), of which all of the former are previously known compounds pacifenol (**469**), prepacifenol (**470**), deoxyprepacifenol (**474**),

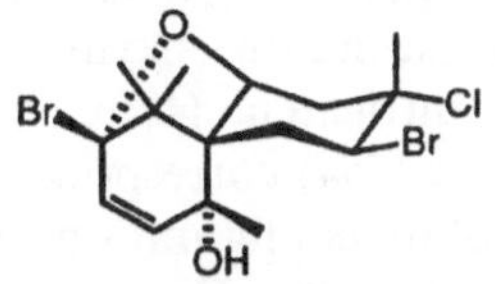

469 (pacifenol)

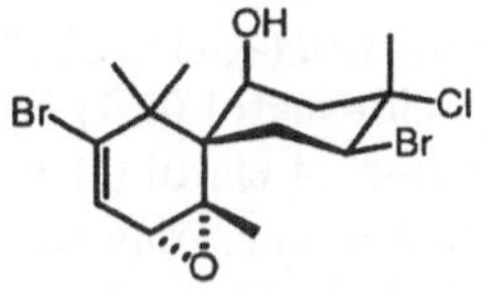

470 (prepacifenol)

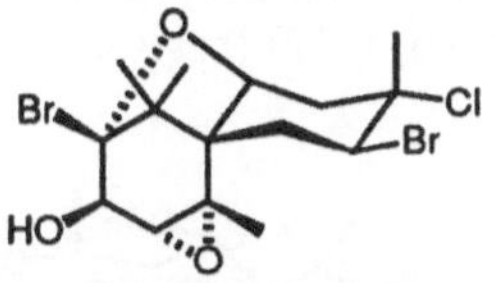

471 (johnstonol)

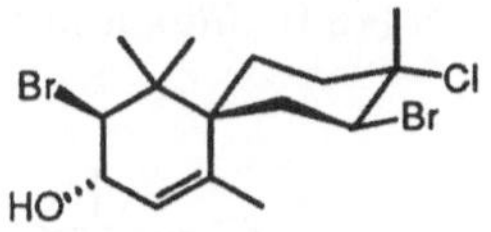

472

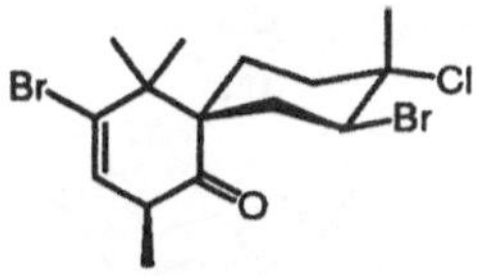

473 (kylinone)

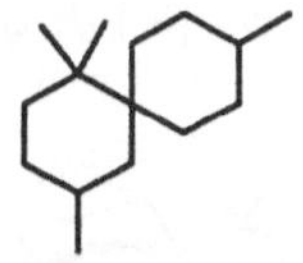

(kylinane)

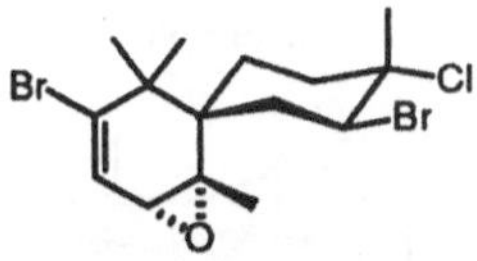

474 (deoxyprepacifenol)

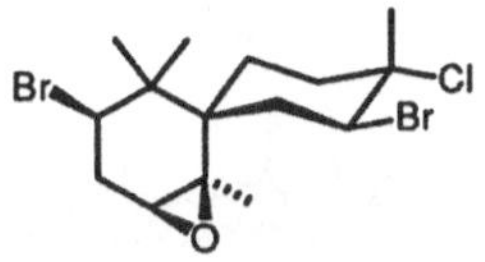

475

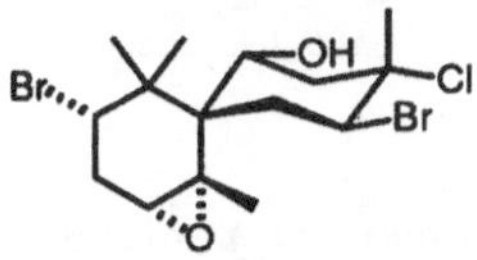

476

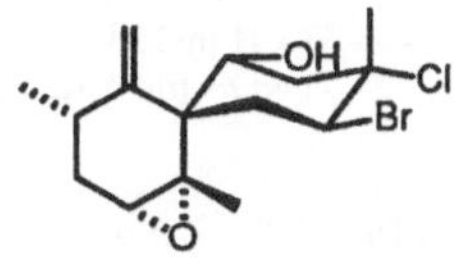

477

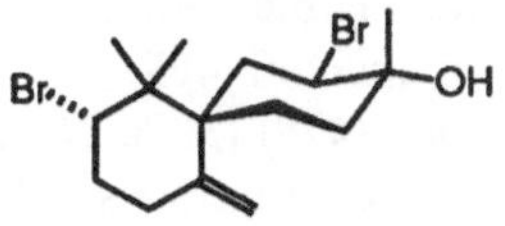

478

467, and **468**. Deoxyprepacifenol (**474**) was first isolated from the sea hare *Aplysia californica* (*228*). The metabolites **468** and the new **475** are found in *Laurencia* sp. from the Gulf of California (*379*), and **476** and **477** from Florida (*381*). The structure of **475** from *L. scoparia* has been established by X-ray crystallography (*437*). The synthetic compound **478** has now been isolated from *L. flexilis* (*438*).

The red alga *Laurencia intricata* contains a set of halogenated chamigrenes: intricatene (**479**), acetoxyintricatol (**480**), and cyclodebromointricatol (**481**) (*345, 383*). It can be seen how enzymatic bromination of preintricatol (**431**) could lead to intricatene (**479**). The absolute configuration of elatol (**482**) from *L. elata* (*384*) corresponds to that of (+)-β-chamigrene, opposite to the absolute configuration of the other halogenated chamigrenes presented thus far. The Hawaiian *L. nidifica* contains nidificene (= intricatene (**479**) ?) (*385*), nidifidiene (**483**) (*385*), nidifidienol (**484**) (*386*), and nidifocene (**485**) (*387*) (revised in *388*).

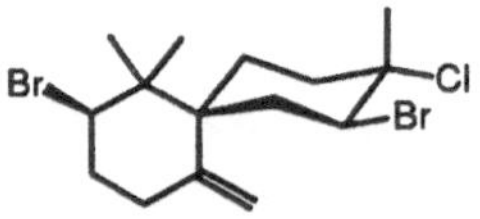

479 (intricatene = nidificene)

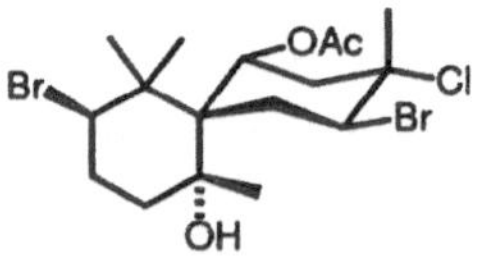

480

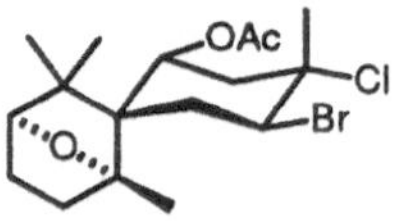

481

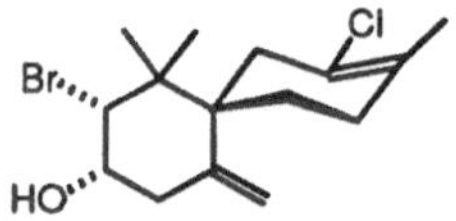

482 (elatol)

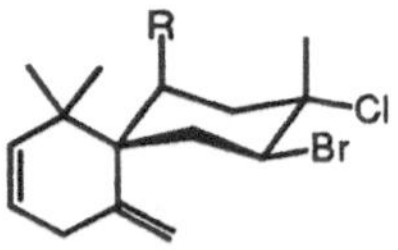

483 R = H (nidifidiene)
484 R = OH (nidifidienol)

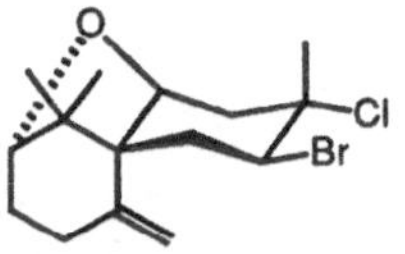

485 (nidifocene)

The Japanese version of *Laurencia nipponica* Yamada contains an extraordinary array of halogenated sesquiterpenes (**486–502**) which have been uncovered mainly by Suzuki *et al.* (Table 19). Several known compounds are also found in this alga, i.e. pacifenol (**469**) (*389*), **468** (*391*), **475** (or epoxide isomer) (*391*) and nidificene (**479**) (*392*). Noteworthy is the isolation of several new chamigrenes, secochamigrenes (**493–496**) and the ring contracted laurencial (**501**). Also, the structure of epoxy-

Table 19. *Halogenated Chamigrenes Isolated from the Red Alga Laurencia nipponica Yamada*

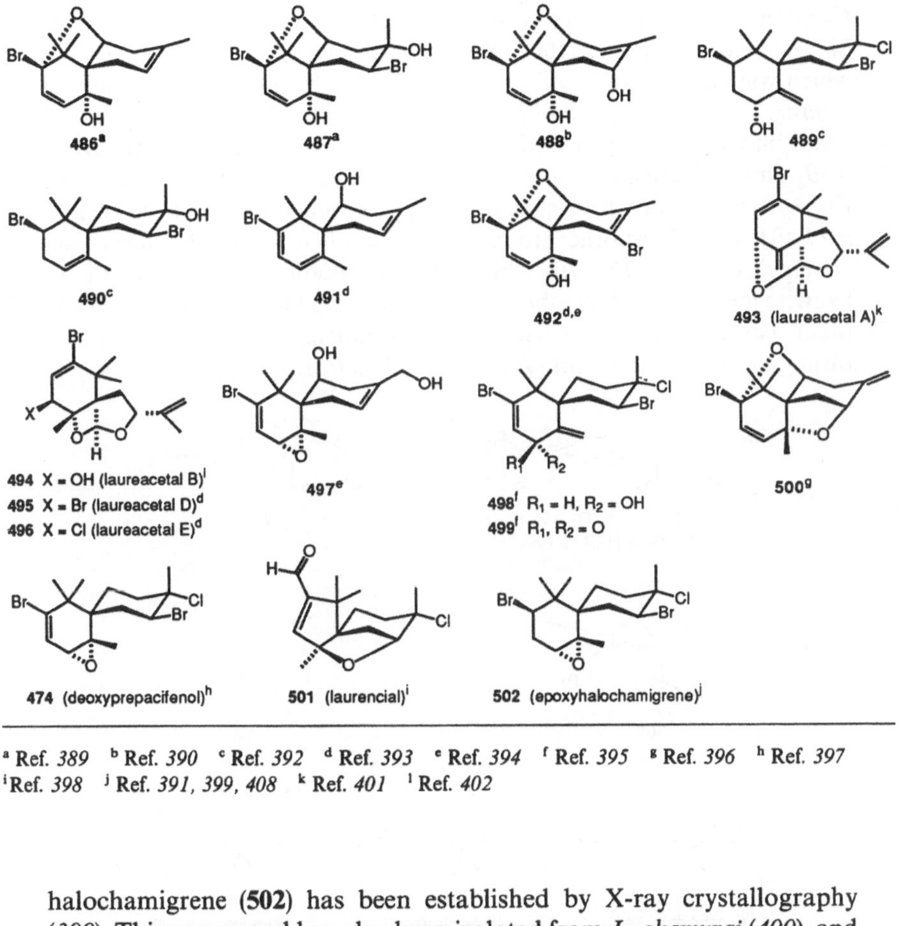

486[a] 487[a] 488[b] 489[c]

490[c] 491[d] 492[d,e] 493 (laureacetal A)[k]

494 X = OH (laureacetal B)[l]
495 X = Br (laureacetal D)[d]
496 X = Cl (laureacetal E)[d]

497[e]

498[f] R_1 = H, R_2 = OH
499[f] R_1, R_2 = O

500[g]

474 (deoxyprepacifenol)[h] 501 (laurencial)[i] 502 (epoxyhalochamigrene)[j]

[a] Ref. *389* [b] Ref. *390* [c] Ref. *392* [d] Ref. *393* [e] Ref. *394* [f] Ref. *395* [g] Ref. *396* [h] Ref. *397* [i] Ref. *398* [j] Ref. *391, 399, 408* [k] Ref. *401* [l] Ref. *402*

halochamigrene (**502**) has been established by X-ray crystallography (*399*). This compound has also been isolated from *L. okamurai* (*400*), and is the epoxide epimer of **475**. Although it has been suggested by SUZUKI that *L. glandulifera* should be revised to *L. nipponica* (*392*), these algae are kept separate in this article.

The red alga *Laurencia majuscula* also contains several new halogenated chamigrenes. As FENICAL and NORRIS have stated, each *Laurencia* species seems to be characterized by one or more halogenated metabolites unique to that species (*400*).

Japanese *L. majuscula* Harvey contains the chamigrenes **503**, **504** (*370*, *403*, *404*) and **505** (*370*). Mediterranean *L. majuscula* has been found to contain pacifenol (**469**) (*405*), which is the first report of this metabolite from this alga and from the Mediterranean Sea. The labile **506** (dehydrochloroprepacifenol) has been isolated from *L. majuscula* (*406*), and the Great Barrier Reef species contains **503**, **505**, and the new metabolites **507**, and isorhodolaureol (**508**), in addition to isoobtusadiene (**511**) (*407*), which was previously found in *L. obtusa* (*vide infra*). An earlier study of *Laurencia* sp. found rhodolaureol (**509**) and rhodolauradiol (**510**) (*439*). The synthetic conversion of isoobtusol (**516**) to **510** has been described (*440*), thus establishing the possible biogenetic link between these two ring systems. Some of the structures reported in (*407*) have been revised in (*409*) (e.g., **507**). Another study of *L. majuscula* from North Queensland waters discovered **512** (*410*), while still other studies of this red alga have likewise found several metabolites previously isolated from *L. obtusa* and other *Laurencia* species such as isoobtusadiene (**511**), isoobtusol, obtusane, obtusol, and elatol (*411*, *412*) (*vide infra*).

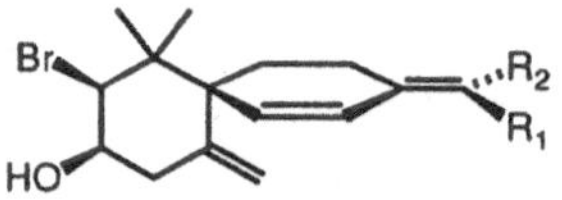

503 R_1 = Br, R_2 = H
505 R_1 = H, R_2 = Br

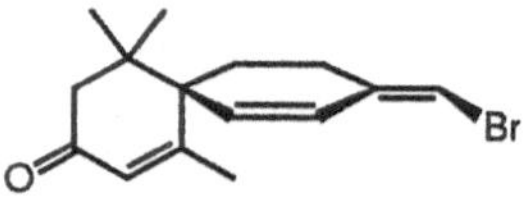

504

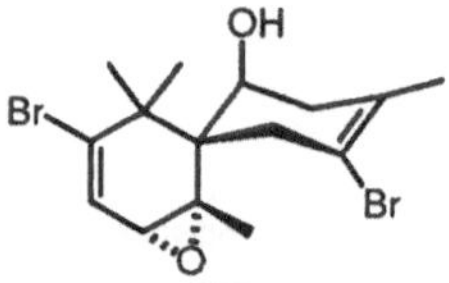

506

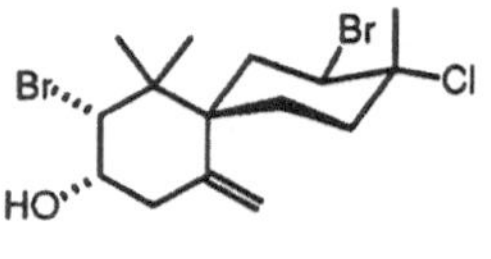

507

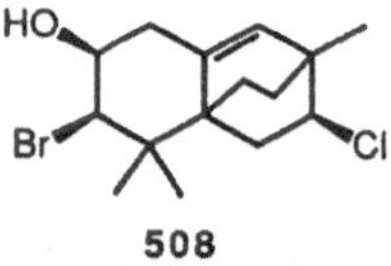

508 (isorhodolaureol)

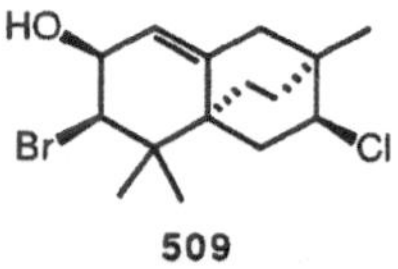

509 (rhodolaureol)

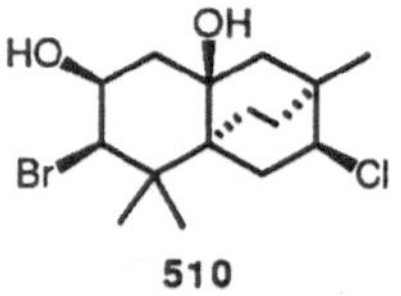

510 (rhodolauradiol)

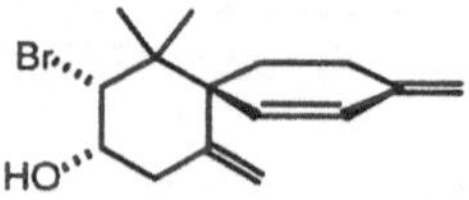

511 (isoobtusadiene)

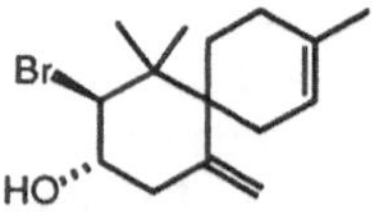

512

Laurencia obtusa has provided a rich and at times frustratingly complex set of sesquiterpene metabolites of the chamigrene class (**513**–**531**) as summarized in Table 20. This situation is complicated by the fact that, in some cases, both enantiomers are known and that the absolute configuration is known for only a few compounds. In most cases, only relative configurations are depicted in Table 20. For a good discussion of this situation, see (*344*).

Jamaican *L. obtusa* gives the unusual *cis*-chlorohydrin **521**, which is formed from **522** upon base treatment (*418*). Another collection of *L. obtusa* has yielded the novel laurencenones A–D (**524**–**527**) in addition to the commonly seen elatol (**482**) and isoobtusol (**516**) (*419*). Likewise, the Canary Islands version of *L. obtusa* affords the novel **528**–**530**, in addition to several known chamigrenes (*420*), while a Red Sea collection provided the interesting chlorohydrin hurgadol (**531**), (*350*), an epimer of glanduliferol (**464**). Several other *Laurencia* species have yielded either new or previously known chamigrenes. Thus, *L. perforata* contains **472** as a minor metabolite (*421*), while *L. glomerata* from South Africa produces the novel epoxide **532** in addition to several known compounds

Table 20. *Halogenated Chamigrenes Isolated from the Red Alga Laurencia obtusa*

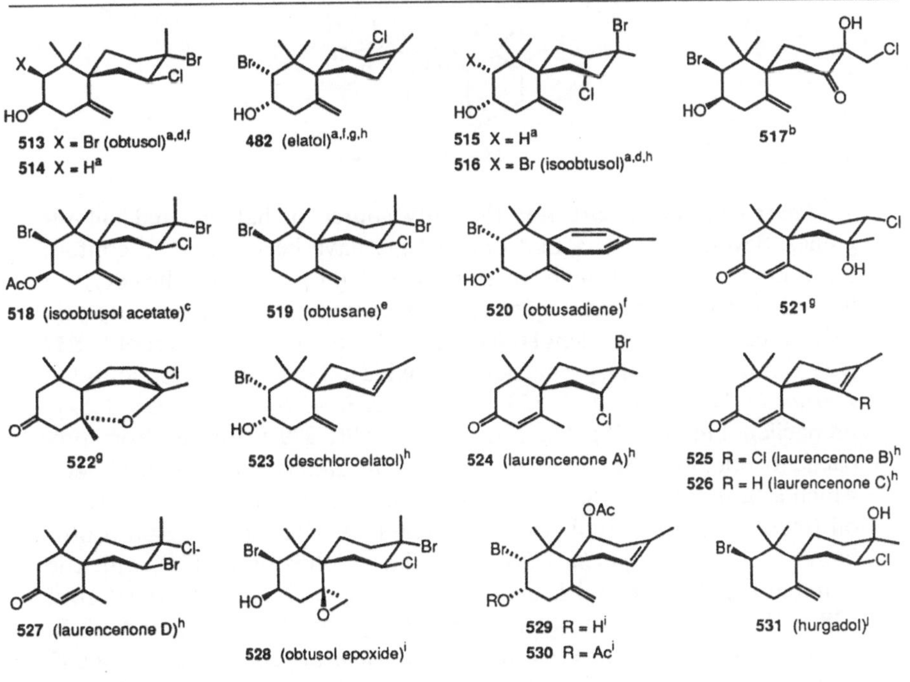

[a] Ref. *413* [b] Ref. *414* [c] Ref. *415* [d] Ref. *416* [e] Ref. *363*
[f] Ref. *417* (isoobtusadiene (**510**) was also isolated). [g] Ref. *418* [h] Ref. *419* [i] Ref. *420* [j] Ref. *350*

(**468**, **472**, **502**) (*422*). *Laurencia pinnatifida* from the Karachi coast has also provided a rich array of halogenated sesquiterpenoids, including **475** (*424*), **502** and the new **533** (*425*), pinnatifidone (**534**) (*426*), pinnatifenol (**535**), the known **477** (*427*), and the bridged cyclic ether pinnatazane (**536**) (*428*) which is structurally similar to "ichthytoxin" (**476**).

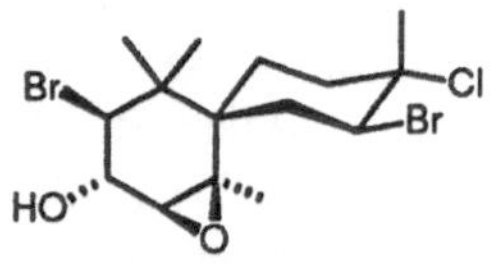

532

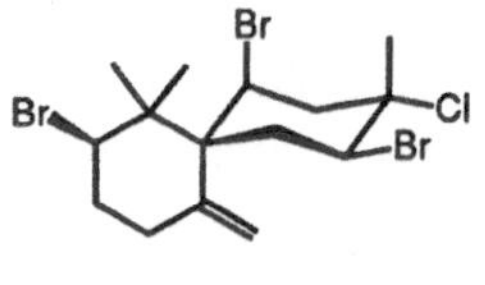

533

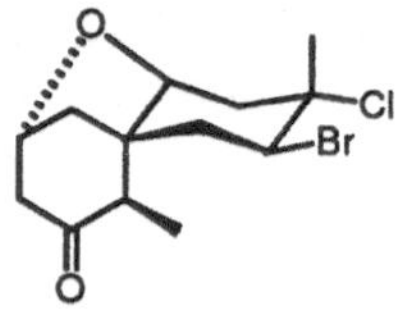

534 (pinnatifidone)

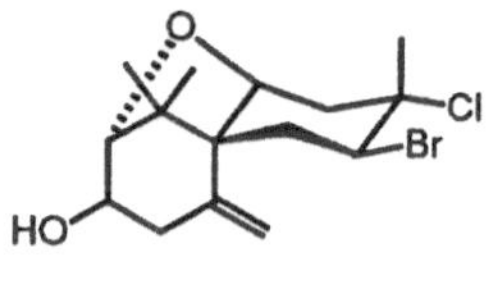

535 (pinnatifenol)

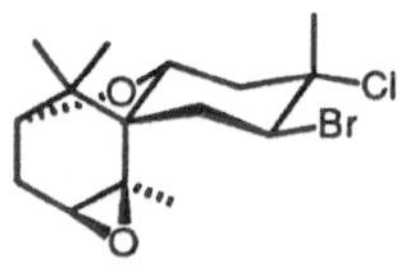

536 (pinnatazane)

Laurencia species are not the only source of halogenated chamigrenes. Several sea hares and one sponge have been found to contain these metabolites although, in some cases, they are probably derived from a diet of *Laurencia* seaweed. For example, the sea hare *Aplysia californica* contains pacifenol (**469**) (*217*, *429*) and the new metabolite **537** (*429*), johnstonol (**471**), prepacifenol epoxide (**538**), which is also found in *Laurencia johnstonii*, and **539** (minor) (*228*, *430*), as well as deoxyprepacifenol (**474**) (*228*). The minor metabolite **539** is formed from prepacifenol epoxide (**538**) upon acid treatment. Likewise, pacifidiene (**540**), which is found in the digestive gland of the sea hare, can also be formed on treatment of pacifenol (**469**) with acid (*429*). Pacifidiene (**540**) is also found in *Laurencia pacifica* (*378*). Similarly, johnstonol (**471**), which is found in *L. johnstonii* and *L. okamurai* (*vide supra*), is readily formed from **538** with acid and may form from **538** in the digestive gland of *A. californica* (*217*, *430*). The sea hare *A. dactylomela* produces **541** and **542** (431), the chamigrenes **543**–**546** (*432*), and the known **487** (*433*). The structures assigned to **543** and **544** were supported by synthesis (*434*). The marine sponge *Spongia zimocca* contains rogiolol acetate (**547**) (*435*), which is the first and only report of a chamigrene from a sponge. This sponge also contains **541** and **542** which, along with alcohol **507** are found in *Laurencia microcladia*, the diet of the sponge (*436*). Interestingly, the sponge appears to be able to acetylate the secondary alcohol **507**, to form **547**, but not the tertiary alcohols **541** and **542**, consistent with human experience in the laboratory!

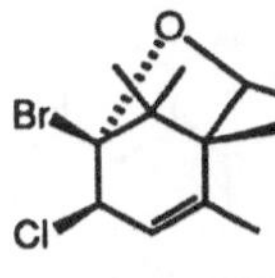

537

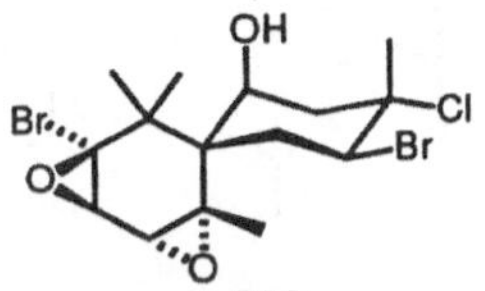

538 (prepacifenol epoxide)

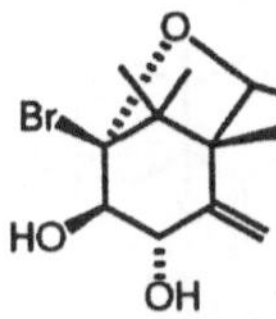

539

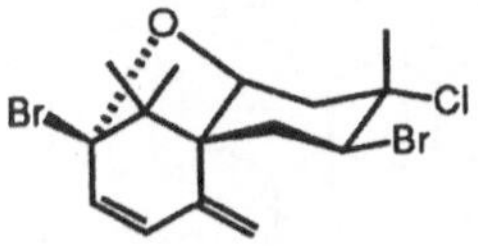

540 (pacifidiene)

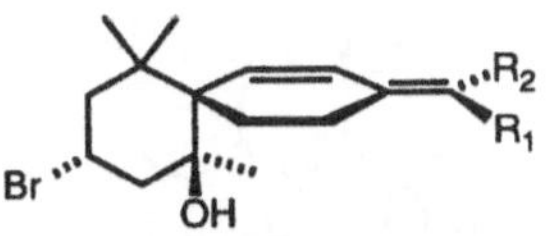

541 R_1 = Br, R_2 = H
542 R_1 = H, R_2 = Br

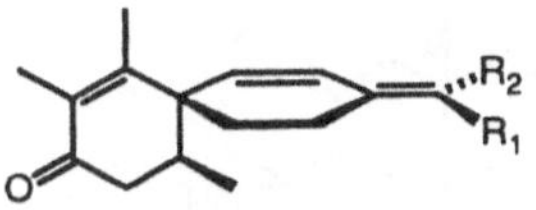

543 R_1 = Br, R_2 = H
544 R_1 = H, R_2 = Br

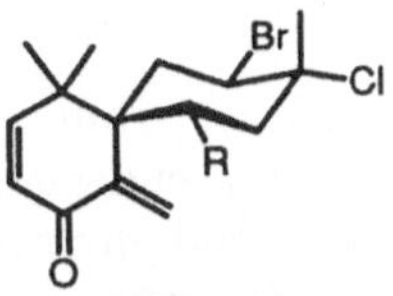

545 R = H
546 R = OAc

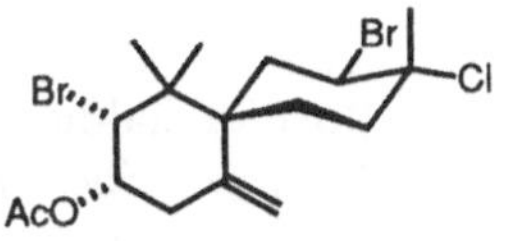

547 (rogiolol acetate)

3.4.2.4.3. Eudesmane and Other Types. Several simple halogenated eudesmanes have been isolated from *Laurencia* species. Thus, the selinane sesquiterpenoid heterocladol (**548**) has been extracted from *Laurencia filiformis* f. *heteroclada* (*441*), and the related **549** was found in *Laurencia* from the Gulf of California (*442, 443*), the absolute configuration of which was established by conversion to (−)-(δ)-selinene (**550**).

The chloro derivative **551** is also known (*444*). The sea hare *Aplysia brasiliana* contains the feeding deterrents brasudol (**552**) and isobrasudol (**553**) (*445*). The structure of **552** was supported by reduction (Li/NH_3) to β-eudesmol (**554**). Western Australian *Laurencia filiformis* contains austradiol acetate (**555**) and diacetate (**556**) (*446*).

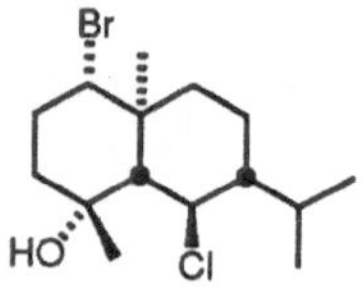

548 (heterocladol) **549** 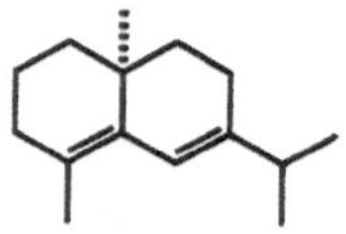 **550** ((–)-(δ)-selinene)

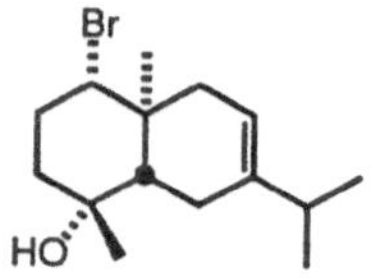

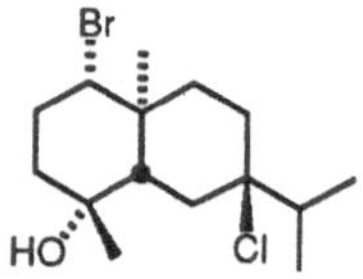

551 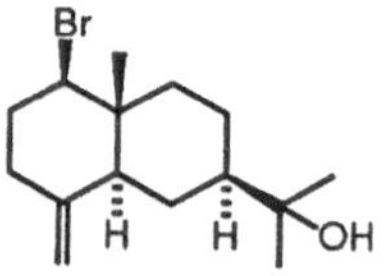 **552** (brasudol) **553** (isobrasudol)

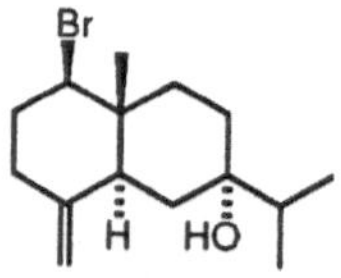

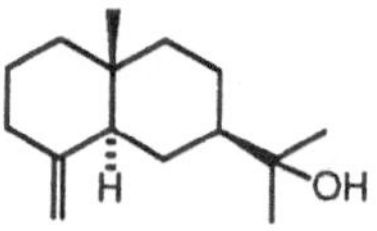

554 (β-eudesmol)

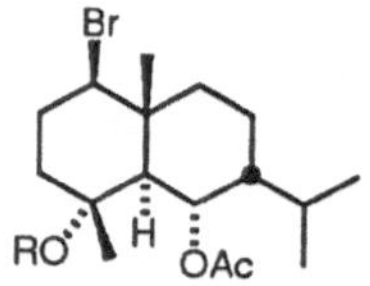

555 R = H (austradiol acetate)
556 R = Ac (austradiol diacetate)

The first report of halogenated sesquiterpenes from a green alga (*Neomeris annulata*) describes neomeranol (**557**), a eudesmane **558** and a cycloeudesmane **559** which have both cytotoxic and phytotoxic activity (*447*). The absolute configuration of neomeranol (**557**) has been determined (*448*). The novel sesquiterpene hydroquinone peyssonol A (**560**) has been isolated from an alga *Peyssonnelia* sp. and the sponge *Hyatella intestinalis* (*449*). This compound is the first bromosesquiterpene hydroquinone to be described. The sponge *Acanthella* sp. from British Columbia contains acanthene A (**561**) as well as a number of known compounds including violacene (**317**) (*450*), and the sponge *Pseudaxinyssa pitys* produces the four carbonimidic dichlorides **562**–**565** (*451, 452*).

Oppositol (**566**) with a novel carbon skeleton has been discovered in *Laurencia subopposita* (*453*) together with the related metabolites **567** and **568** (*454*), while *L. perforata* from the Canary Islands contains

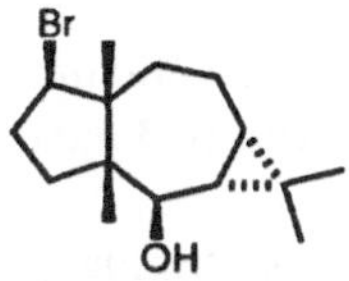

Br OH
557 (neomeranol)

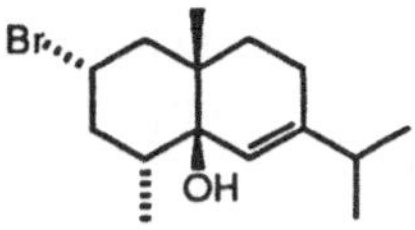

Br OH
558

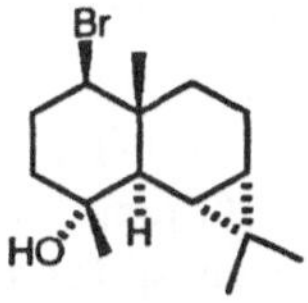

Br HO H
559

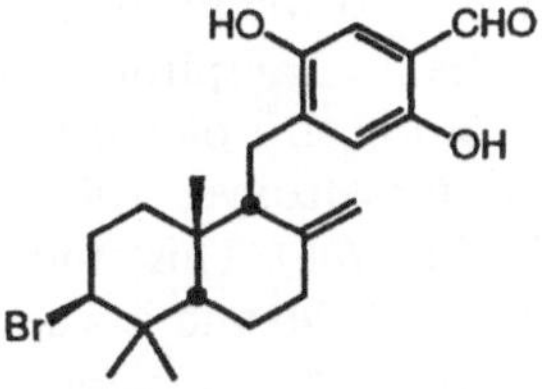

HO CHO OH Br
560 (peyssonol A)

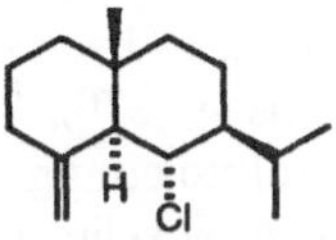

H Cl
561 (acanthene A)

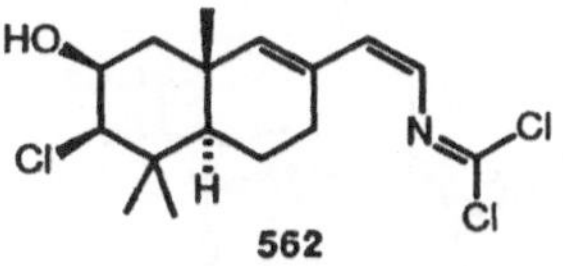

HO Cl H N Cl Cl
562

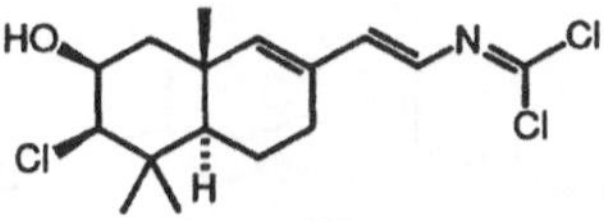

HO Cl H N Cl Cl
563

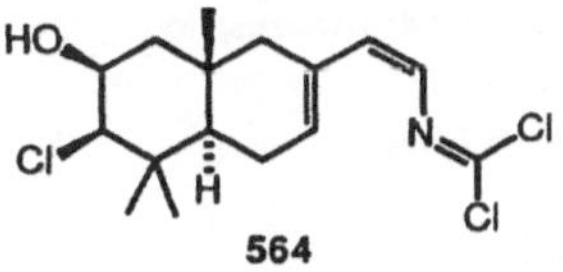

HO Cl H N Cl Cl
564

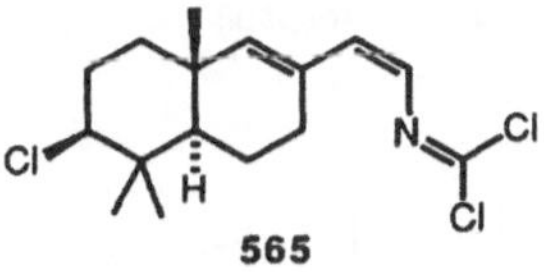

Cl H N Cl Cl
565

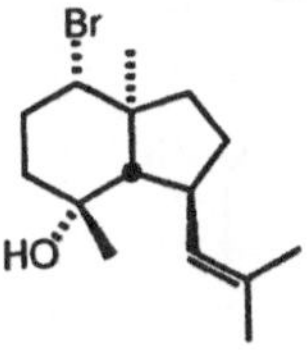

Br HO
566 (oppositol)

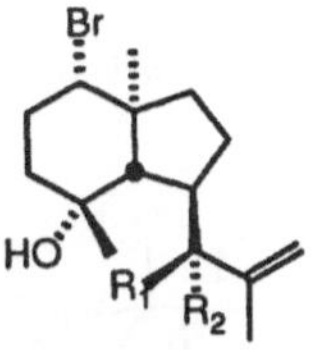

Br HO R_1 R_2
567 R_1 = OH, R_2 = H
568 R_1 = H, R_2 = OH

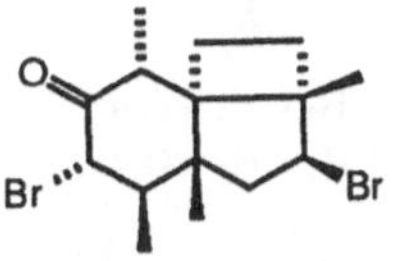

O Br Br
569 (perforatone)

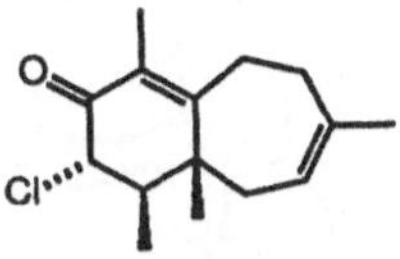

O Cl
570 (perforenone B)

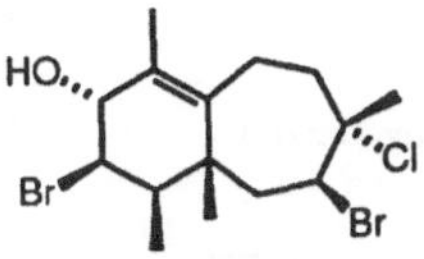

HO Br Cl Br
571 (perforenol)

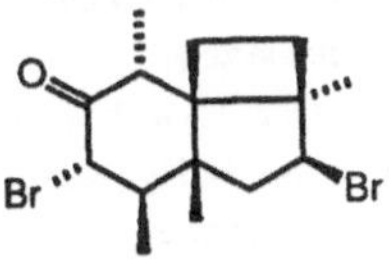

O Br Br
572

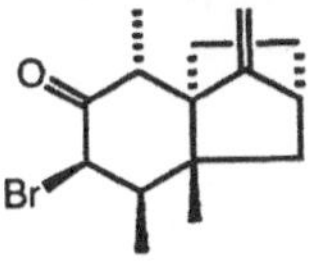

O Br
573

perforatone (**569**) (*455*), perforenone B (**570**) (*455*), and perforenol (**571**) (*456*), the structures of which have been confirmed by synthetic interconversions (*467*, *468*). The related sesquiterpenoids **572** and **573** have been isolated from *L. tenera* (*457*).

Although caespitol from *Laurencia caespitosa* was originally proposed to be a chamigrene (*382*), the structure was soon revised to that of a bisabolane **574** (*423*). Isocaespitol (**575**) was also found in this alga (*458*); interestingly, this metabolite rearranges to caespitol (**574**) on melting (*423*). The deoxy analogue **576** has also been isolated from *L. caespitosa* (*459*) as have been caespitane (**577**), 6-hydroxycaespitol (**578**) and laucapyranoids A (**579**), B (**580**) and C (**581**) (*460*). This study also confirmed the absolute configurations of caespitol (**574**) and isocaespitol (**575**) as shown. The bicyclic ether **582** has been found in *L. obtusa* (*461*).

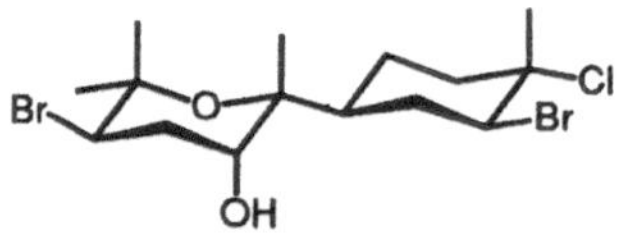

574 (caespitol)

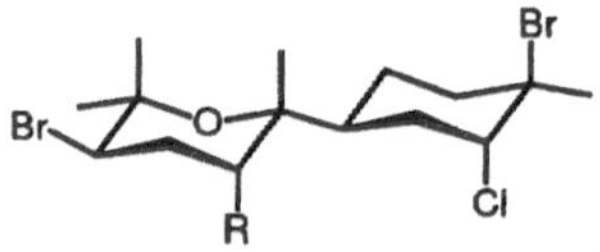

575 R = OH (isocaespitol)
576 R = H
586 R = OAc

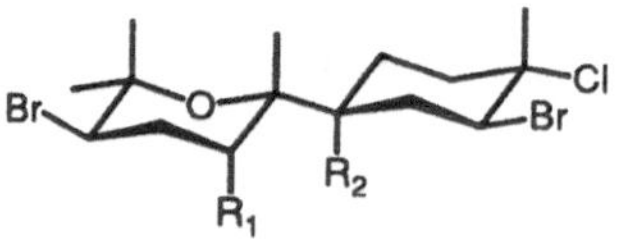

577 $R_1 = R_2 = H$ (caespitane)
578 $R_1 = R_2 = OH$

579 (laucapyranoid A)

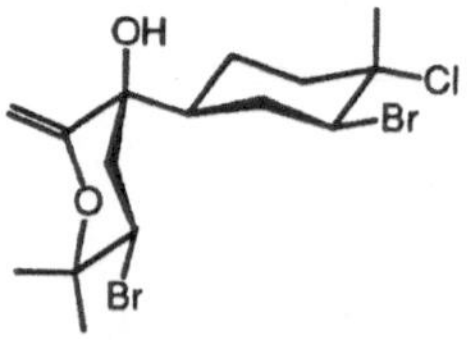

580 (laucapyranoid B)

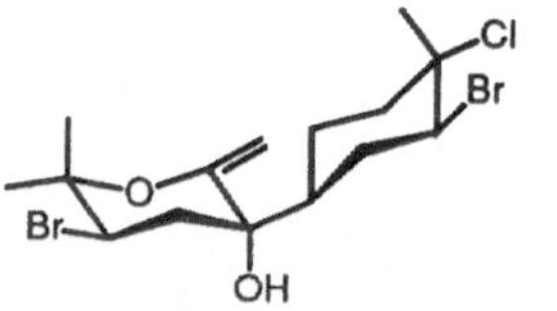

581 (laucapyranoid C)

582

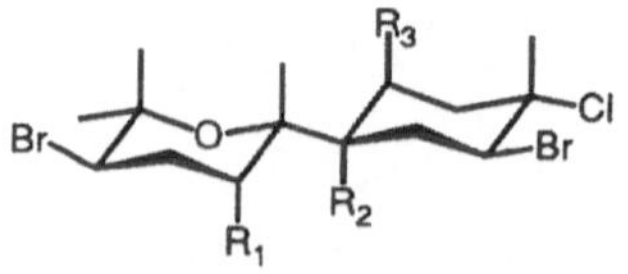

583 $R_1 = R_2 = H$, $R_3 = OH$ (deodactol)
584 $R_1 = OAc$, $R_2 = R_3 = OH$
585 $R_1 = R_3 = H$, $R_2 = OH$ (isodeodactol)

The sea hare *Aplysia dactylomela* has been found to contain several bisabolene sesquiterpenes such as deodactol (**583**) (*462*), **584** (*463*), and isodeodactol (**585**) (*464*) as well as isocaespitol acetate (**586**) (*465*). This latter study also studied the cytostatic and antibacterial activity of these (**574**, **575**, **586**) and other sesquiterpenes as well as some monoterpenes.

3.4.2.4.4. Cuparene, Laurene, and Other Aromatic Types. Although the aromatic marine sesquiterpenes are generally considered to be rearranged chamigrenes (*343*, *344*), they are presented here in a separate section. Terrestrial aromatic (phenolic) sesquiterpene hybrids are discussed in Sect. 3.22.9.

Although the first examples of marine aromatic sesquiterpenes, which are of the cuparene type, were isolated from the sea hare *Aplysia kurodai*, it is *Laurencia* red algae that have furnished most of these metabolites, and, of course, probably also furnished them to the sea hare!

Aplysin (**587**) and aplysinol (**588**) along with debromoaplysin were isolated from *A. kurodai* (*470*). Later, these metabolites along with laurinterol (**589**) were found in *Laurencia okamurai* (*471*). Their structures were confirmed by X-ray crystallography, including the absolute configurations (*472–474*). *Aplysia californica* also contains aplysin and laurinterol in the digestive gland (*217*). Other species of the genus *Laurencia* have yielded aromatic sesquiterpenes. Laurenisol (**590**) and isolaurinterol (**591**) are found in *L. nipponica* (*475*) and *L. intermedia* (*476*), respectively. This latter alga also contains laurinterol (**589**) which cyclizes to aplysin (**587**) upon acid treatment (*476*). This transformation is also shown to occur in the digestive gland of the sea hare *Aplysia californica* (*429*). The parent hydrocarbon laurene has been isolated from *L. glandulifera* (*482*). Hawaiian *L. nidifica* also contains aplysin and laurinterol (*385*), while the first true cuparene types, α-bromocuparene (**592**) and α-isobromocuparene (**593**), were extracted from *L. glandulifera* and **593** from *L. nipponica* (*477*). The absolute stereochemistry of these metabolites was established by their synthesis from the corresponding cuparenols (*477*). *L. flexilis* from the Philippines also contains **592** (*438*). South Australian *L. filiformis* f. *heteroclada* produces the major metabolite allolaurinterol (**594**) which is slowly converted on standing to filiformin (**595**) (more rapidly so in trifluoroacetic acid) (*478*). Filiformin (**595**), filiforminol (**596**), and several other known laurenes are also found in this alga (*411*, *441*, *478*). Allolaurinterol (**594**) (10-bromo-7-hydroxylaurene) is also found in *L. subopposita* (*454*). The similar brominated ethers **597**–**599** have been identified from *L. glandulifera* (*479*, *486*). A study of *L. nana* (*480*) (renamed as *L. caraibica* (*481*)) describes the discovery of the first iodin-

cuparene

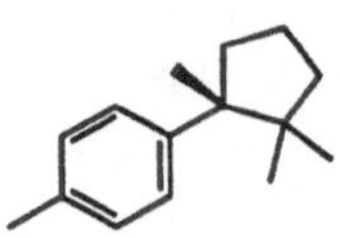

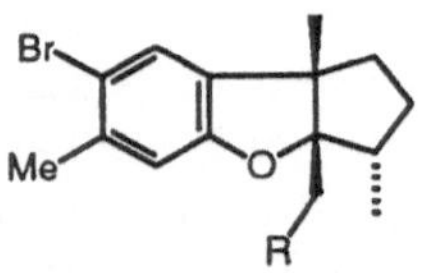

587 R = H (aplysin)
588 R = OH (aplysinol)

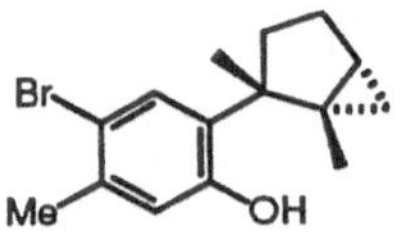

589 (laurinterol)

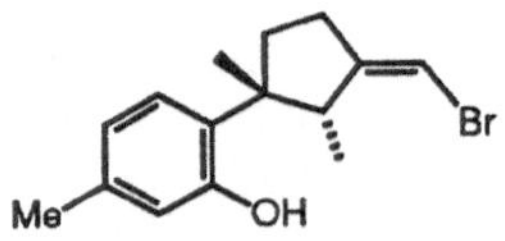

590 (laurenisol)

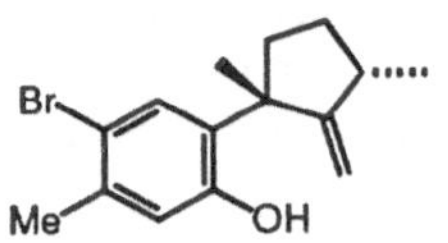

591 (isolaurinterol)

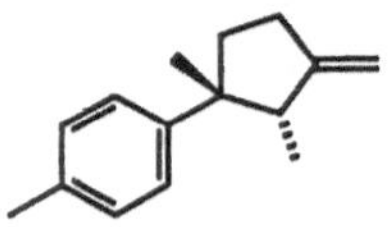

laurene

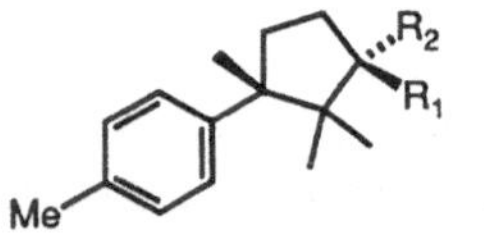

592 R_1 = Br, R_2 = H (α-bromocuparene)
593 R_1 = H, R_2 = Br (α-isobromocuparene)

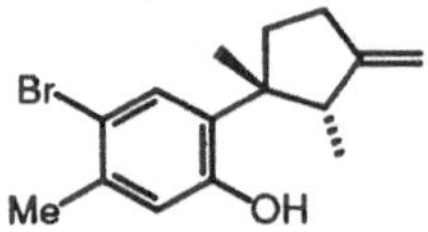

594 (allolaurinterol)

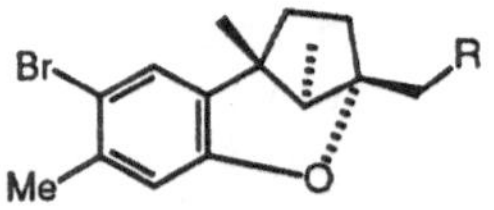

595 R = H (filiformin)
596 R = OH (filiforminol)
597 R = Br
600 R = I

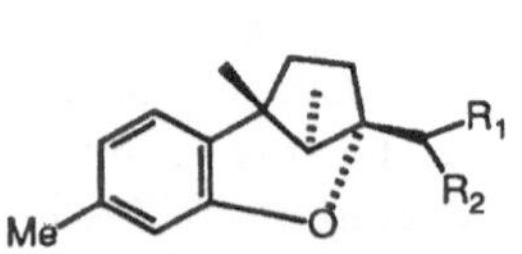

598 R_1 = Br, R_2 = H
599 R_1 = R_2 = Br

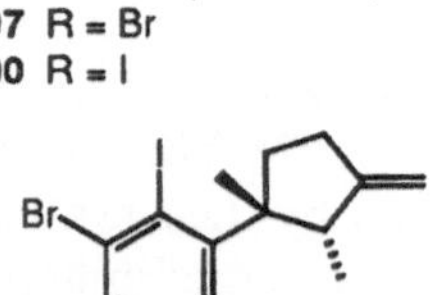

601

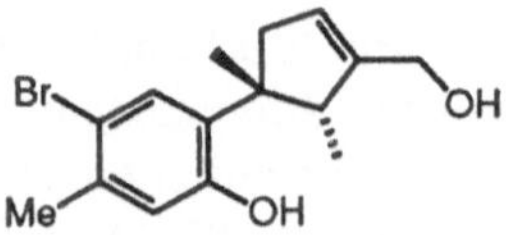

602

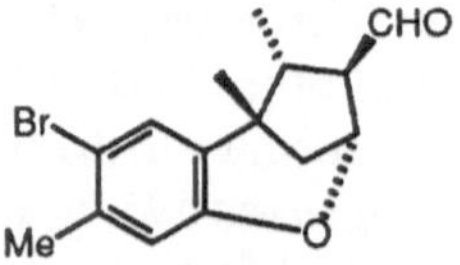

603 (caraibical)

ated sesquiterpenes **600** and **601**, together with another new laurene **602** and the known **592**, **594**, **595**, **597** (*480*). This same alga also produces caraibical (**603**), the formation of which may involve cyclization (and oxidation) of **602** (*481*).

The Japanese *Laurencia okamurai* continues to be a rich source of halogenated terpenes. In addition to many known metabolites several new laurenes have been reported, including neolaurinterol (**604**), isoaplysin (**605**) as well as **606**–**608** (*483, 484*). This latter paper contains an excellent summary of the halogenated sesquiterpene content of nine different *Laurencia* species (*484*). In fact, gas chromatography-mass spectrometry has been used to characterize *Laurencia* species by identifying

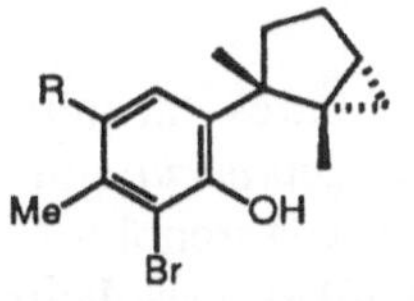

604 R = H (neolaurinterol)
606 R = Br

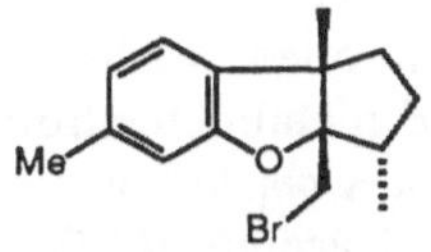

605 (isoaplysin)

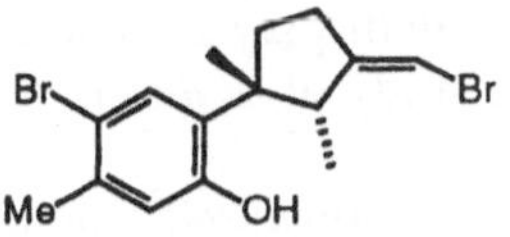

607

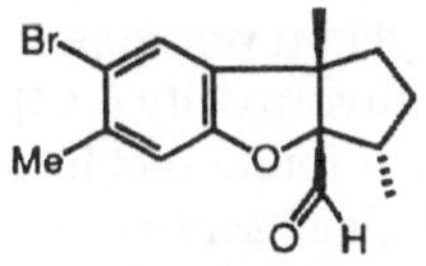

608 (aplysinal)

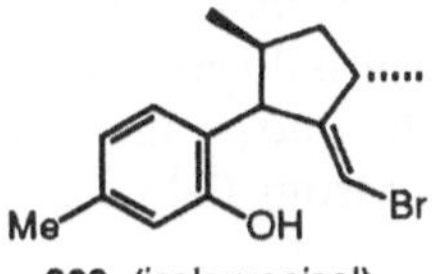

609 (isolaurenisol)

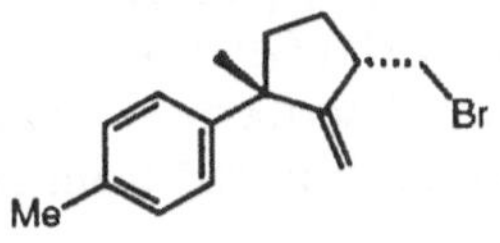

610

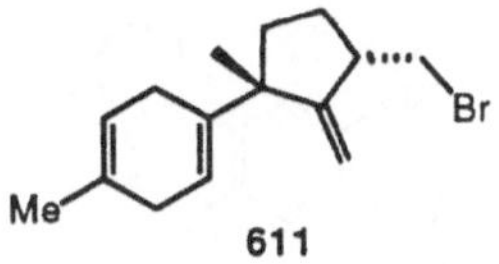

611

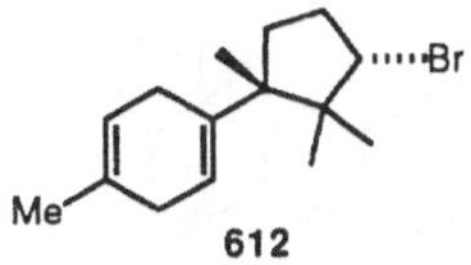

612

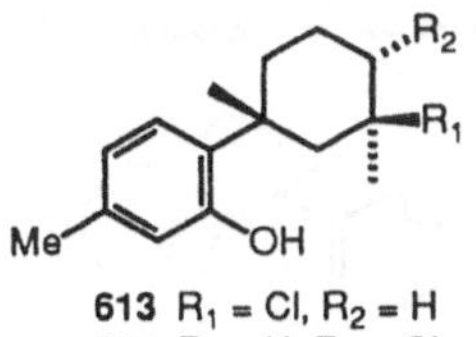

613 R_1 = Cl, R_2 = H
614 R_1 = H, R_2 = Cl

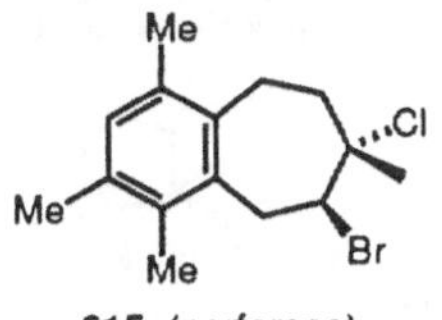

615 (perforene)

the sesquiterpenes (and other metabolites) that are present (*485*). The New Zealand *L. distichophylla* contains isolaurenisol (**609**) which on treatment with trifluoroacetic acid gives isoaplysin (**605**) (*487*). Isolaurenisol (**609**) was also isolated from *L. gracilis* and assigned the *E* configuration from NOESY data (*346*). *L. pinnatifida* from the Atlantic ocean contains the novel deoxy **610** and its apparent precursor **611**, since the latter is slowly oxidized to **610** on standing (*488*), while *L. majuscula* from the area of a deep-water coral reef in Queensland contains the known α-isobromocuparene (**593**) and the novel cyclohexadiene **612** (*489*). This alga also contains the novel chloro metabolites **613** and **614** (*412*). A similarly novel aromatic chloro metabolite, perforene (**615**), is found in *L. perforata* (*490*). This compound is obtained upon acid treatment of chamigrene **472** and might also arise biogenetically from perforenol (**571**) (*421*).

As mentioned earlier, *Aplysia* sea hares acquire numerous halogenated metabolites from their diet of *Laurencia* and other algae. Thus, *A. dactylomela* contains the novel cupalaurenol (**616**) as well as cupalaurenol acetate (**617**), cyclolaurenol (**618**), cyclolaurenol acetate (**619**), and the parent hydrocarbon cyclolaurene, the latter isolated for the first time (*491*). These halogenated laurenes are the first to have bromine and hydroxyl in reversed position. They all display antibacterial, antifungal, and ichthyotoxicity (*491*). Another study of *A. dactylomela* from the Indian Ocean found dichloro **620** (*433*).

The red algae *Marginisporum aberrans*, *Amphiroa zonata*, and *Corallina pilulifera* all contain aplysinal (**608**), isolated here for the first time, in addition to the known aplysin (**587**), aplysinol (**588**), laurinterol (**589**) and isolaurinterol (**591**) (*492*). The first report of halogenated terpenes from genus *Hypnea* described the isolation of the known filiformin (**595**) and filiforminol (**596**) and **602** from *H. pannosa* from Pakistan (*493*).

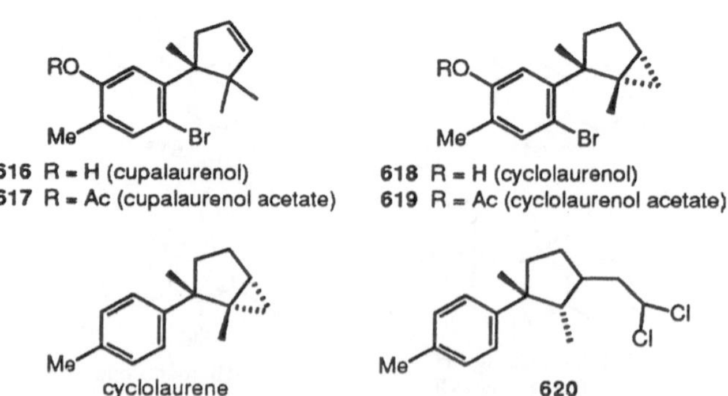

616 R = H (cupalaurenol)
617 R = Ac (cupalaurenol acetate)

618 R = H (cyclolaurenol)
619 R = Ac (cyclolaurenol acetate)

cyclolaurene

620

The green calcareous alga *Cymopolia barbata* produces an array of prenylated bromohydroquinones and related cyclic ethers. Included in this group are cymopol (**621**), **622**, cyclocymopol (**623**), **624**, cymopolone (**625**), isocymopolone (**626**), and the major metabolite cymopochromenol (**627**) (*494*). Later studies reported the isolation of cyclocymopol diastereomers **628** and **629** (*495*), and the hydroxycymopolone **630** (*496*).

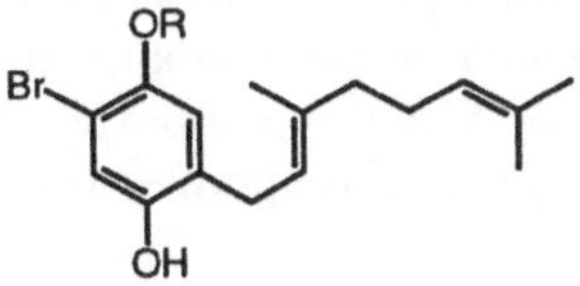

621 R = H (cymopol)
622 R = Me

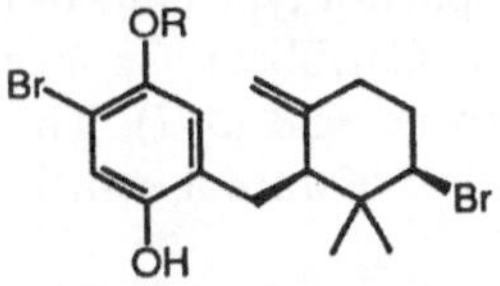

623 R = H (cyclocymopol)
624 R = Me

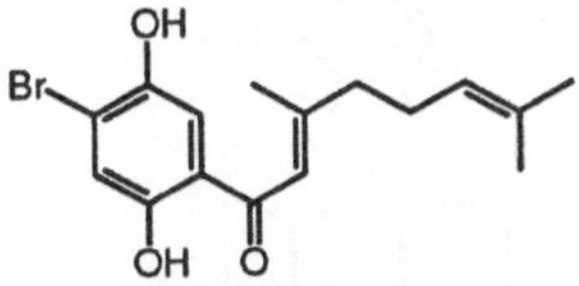

625 (cymopolone)

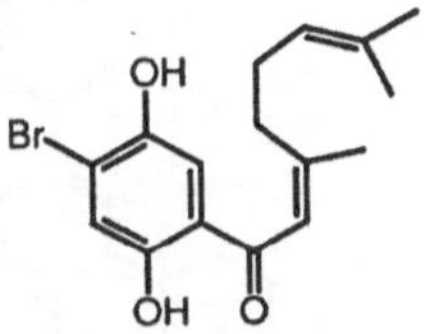

626 (isocymopolone)

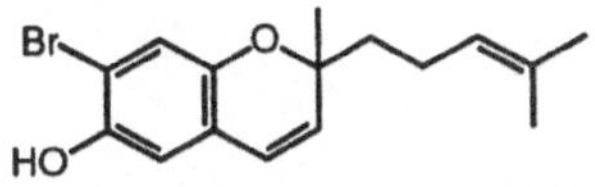

627 (cymopochromenol)

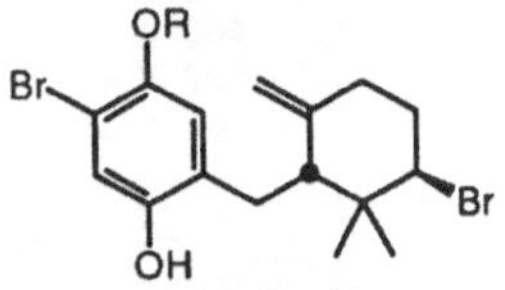

628 R = H
629 R = Me

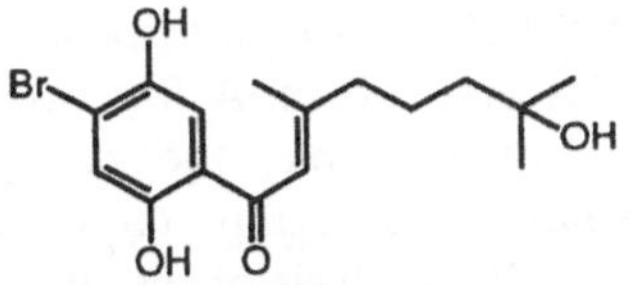

630

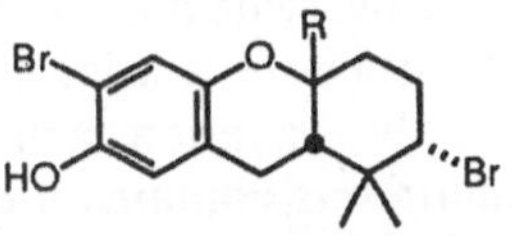

631 R = β-Me (cymobarbatol)
632 R = α-Me (4-isocymobarbatol)

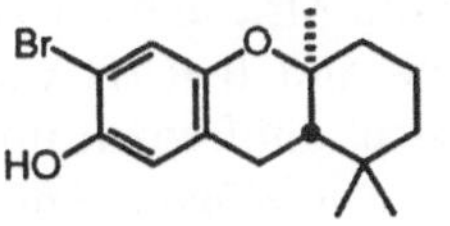

633
(debromoisocymobarbatol)

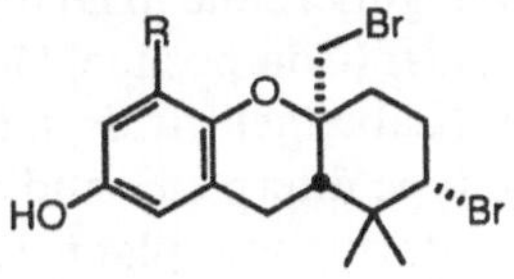

634 R = H
(tribromocacoxanthene)
635 R = Br
(tetrabromocacoxanthene)

The tricyclic cymobarbatol (**631**) and 4-isocymobarbatol (**632**), which are highly antimutagenic, are also found in *C. barbata* (*497*), as is the feeding deterrent debromoisocymobarbatol (**633**), isolated from *C. barbata* from the Florida Keys (*498*). Interestingly, a sponge belonging to the genus *Cacospongia* produces the related metabolites tribromocacoxanthene (**634**) and tetrabromocacoxanthene (**635**) (*499*).

An unidentified yellow encrusting sponge produces the novel drimane-phenol type chloro- (**636**) and bromopuupehenones (**637**) (*500*), while the Caribbean sponge *Smenospongia aurea* contains the new sesquiterpene **638** (*501*). The structure of **638** was supported by the chlorination of aureol, which is also present in this sponge, to give **638** (*501*).

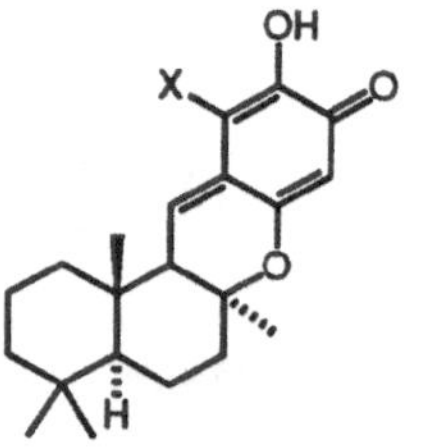

636 X = Cl
637 X = Br

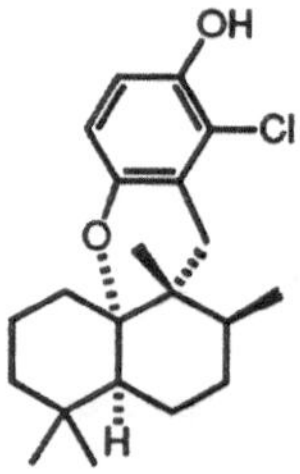

638

3.4.3. Diterpenes

3.4.3.1. Terrestrial Diterpenes

Although the overwhelming number of halogenated diterpenes occurs in marine life, a few terrestrial examples are known. A few others are considered to be isolation artifacts. For example, the report that *Phaseolus multiflorus* contains a chlorinated gibberellin "compound b" (*502*) is probably due to an artifact since GC analysis of the plant seeds shows none of this compound (*503*).

However, gutierolide (**639**) has been isolated from the herb *Gutierrezia dracunculoides* (Compositae) (*504*), and the South African bush *Teucrium africanum* (Labiatae) yields tafricanin A (**640**) and B (**641**) which are clerodane-type diterpenes and which have insect antifeedant properties (*505*). The Portuguese plant *T. polium* subsp. *vincentinum* contains the related chlorohydrin teuvincentin A (**642**). Careful control experiments involving subjecting the corresponding epoxide to the isolation conditions demonstrated that **642** is not an artifact (*506*). Similarly, a fresh

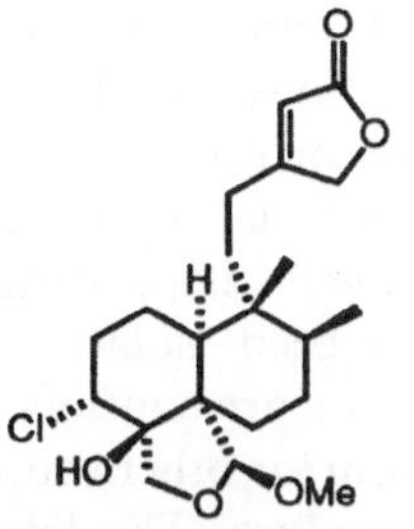

639 (gutierolide)

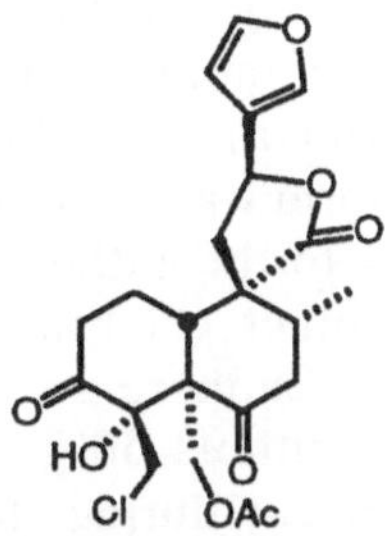

640 (tafricanin A)

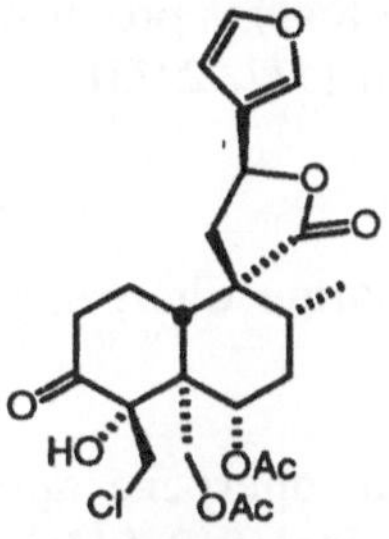

641 (tafricanin B)

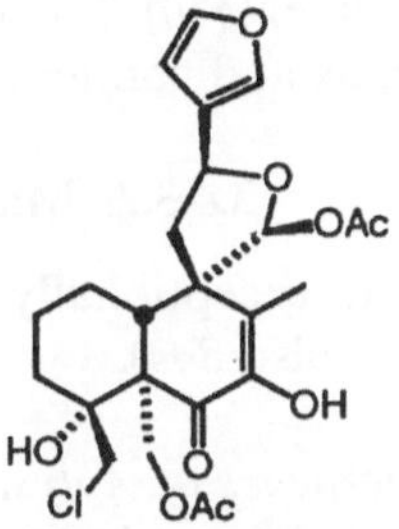

642 (teuvincentin A)

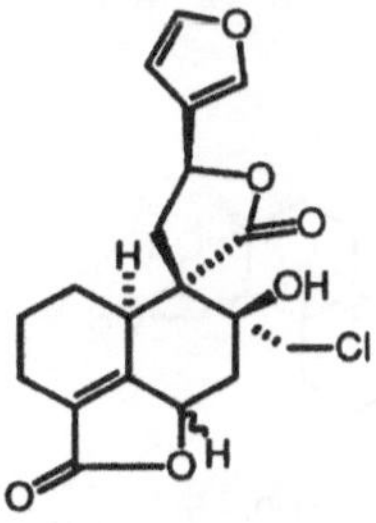

643 (teuperin D)

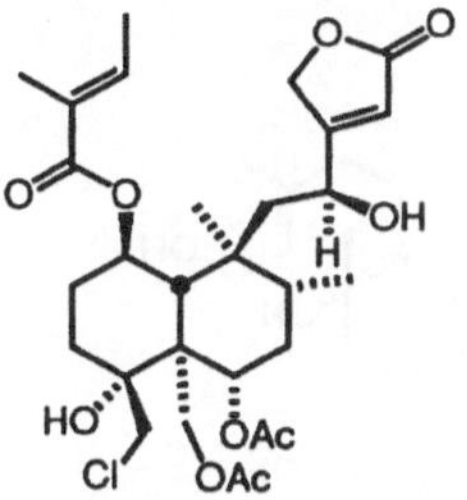

644 (ajugamarin chlorohydrin)

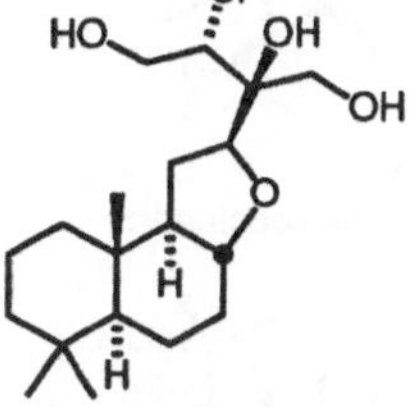

645 (chlorosilphanol A)

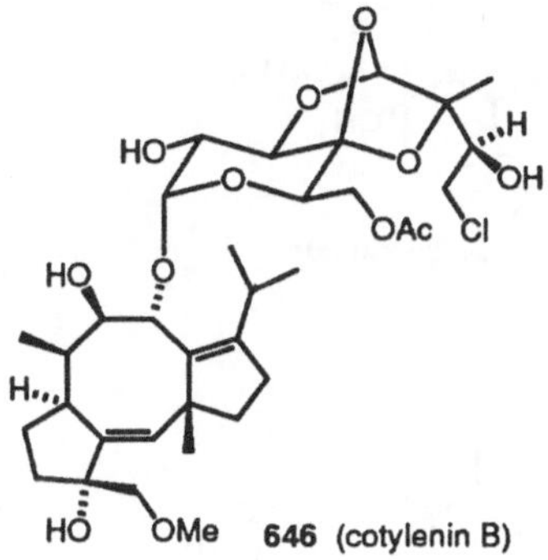

646 (cotylenin B)

acetone extract of *T. pernyi* shows the presence of teuperin D (**643**) by thin layer chromatography (*507*). Leaves of *Ajuga nipponensis* (Labiatae) have yielded ajugamarin chlorohydrin (**644**) and the corresponding epoxide. Since chloroform was used in the isolation protocol, **644** may be an artifact according to the authors (*508*), but no evidence is presented to this effect. Two other chlorohydrins which have been isolated from *Euphorbia fidjiana* (Euphorbiaceae) and are not shown are considered to be artifacts by the authors (*509*), although chloroform (or other source of acid) was not present during the extraction procedure. The labdane diterpene chlorosilphanol A (**645**) has been isolated from *Silphium perfoliatum* (Compositae) (*510*). The novel cotylenin B (**646**) is produced by an identified fungus and contains an unusual sugar (*511*, *2213*).

3.4.3.2. Marine Diterpenes

A review of the exceptionally large field of marine diterpenes has appeared (*526*, see also *344*).

3.4.3.2.1. Diterpenes of Aplysia. The first marine diterpene, aplysin-20 (**647**), was isolated from the sea hare *Aplysia kurodai* (*512*, *513*) which also contains isoaplysin-20 (**648**) (*514*), the structure of which was

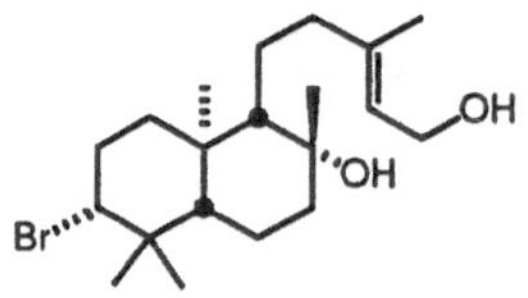

647 (aplysin-20)

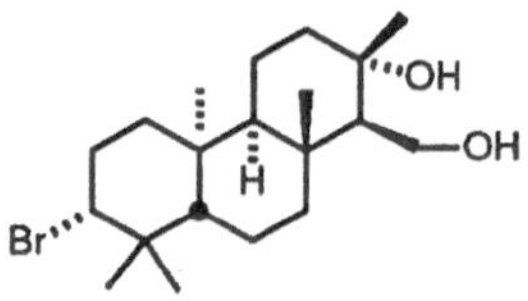

648 (isoaplysin-20)

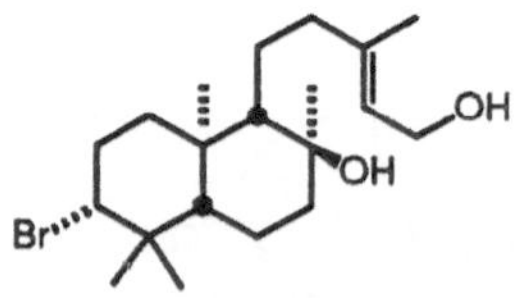

649 (*epi*-aplysin-20)

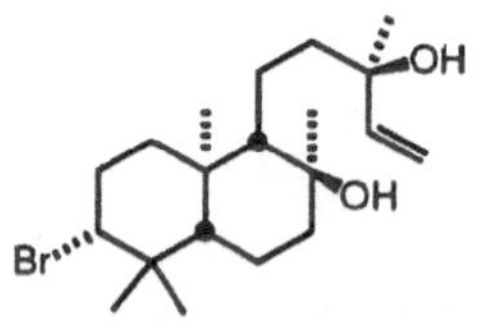

650 (*ent*-isoconcinndiol)

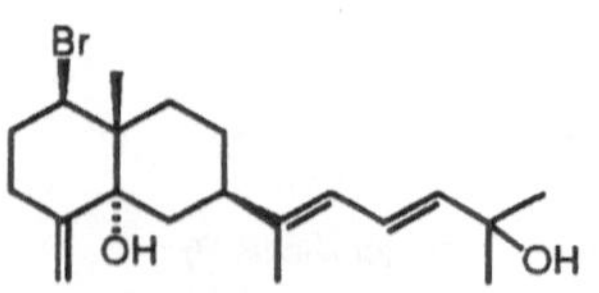

651 (aplysiadiol)

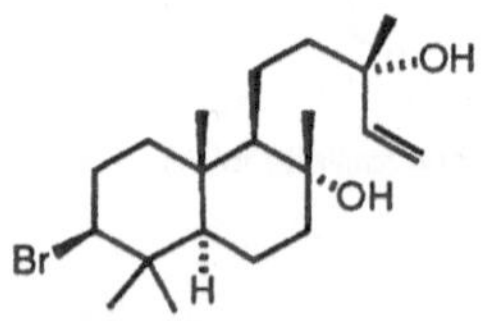

652 (isoconcinndiol)

clinched by synthesis and X-ray crystallography (*515*) following an initial erroneous structure revision (*516*). Isoaplysin-20 has an unusual *anti-syn-anti* ring fusion with a skew-boat B-ring. More recently, *epi*-aplysin-20 (**649**) and *ent*-isoconcinndiol (**650**) were found in this sea hare (*517*) as was aplysiadiol (**651**) (*518*). The corresponding side chain methyl ether was also isolated (not shown) but it was considered to be an isolation artifact since methanol was used in the extraction. Isoconcinndiol (**652**), which was actually first isolated from *Laurencia* sp. (*vide infra*) is also found in *A. dactylomela* (*519*). The originally proposed structure was revised as a result of synthesis (*520*).

Dactylomelol (**653**), a new structural diterpene type, was also isolated from the sea hare *Aplysia dactylomela* (*521*) as was the unusual regular isoprenoid diterpene **654** (*522*). This latter metabolite has marginal antitumor activity. Also isolated from this sea hare were parguerol (**655**), deoxyparguerol (**657**), isoparguerol (**658**), and acetates **656** and **659** (*523*).

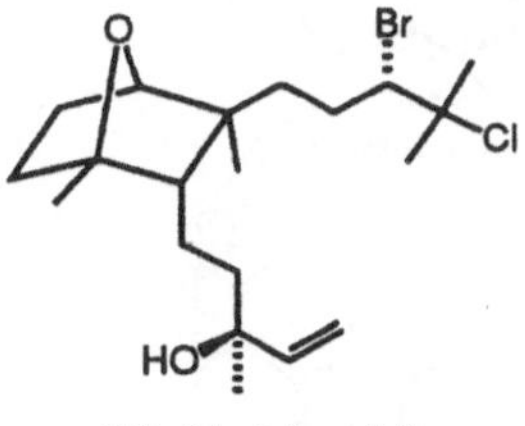

653 (dactylomelol)

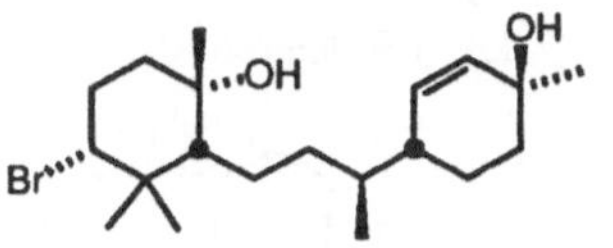

654

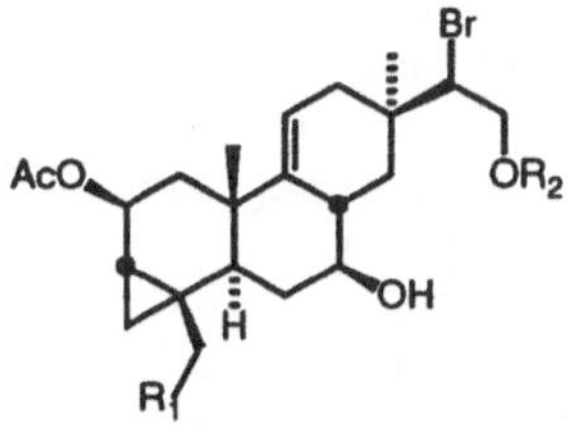

655 R_1 = OH, R_2 = H (parguerol)
656 R_1 = OH, R_2 = Ac
657 R_1 = R_2 = H (deoxyparguerol)

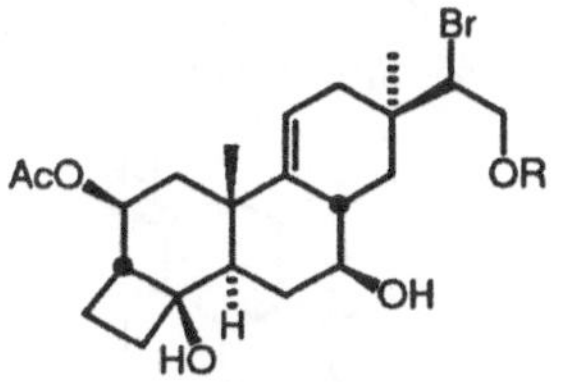

658 R = H (isoparguerol)
659 R = Ac

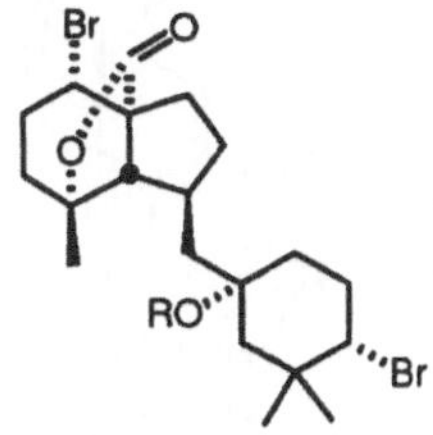

660 R = H (angasiol)
661 R = Ac

This study also identified several known sesquiterpenes in this sea hare (elatol, isoobtusol acetate, and others). The sea hare *A. angasi* contains the novel lactone angasiol (**660**) (*524*), while angasiol acetate (**661**) is found in *A. juliana*, a sea hare living in the Arabian Ocean (*525*).

3.4.3.2.2. Diterpenes of Laurencia. Several diterpene structural types are present in *Laurencia* red algae (*344*, *526*). Concinndiol (**662**) from *L. concinna* was the first such bromoditerpene to be isolated (527). This metabolite is also found in *L. snyderiae* (*528*). Isoconcinndiol (**652**) was first extracted from *L. snyderiae* var. *guadalupensis* (*529*). Pinnatols A–D (**663–666**) are found in *L. pinnata* (*530*), and venustanol (**667**) has been isolated from *L. venusta* (531). The three diterpenes **668–670** have been

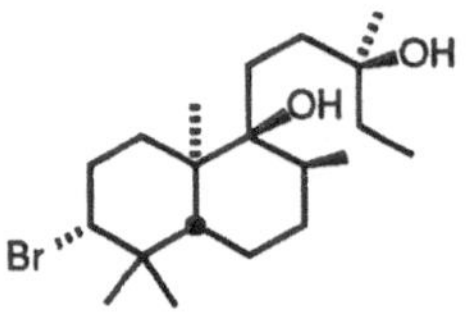

662 (concinndiol)

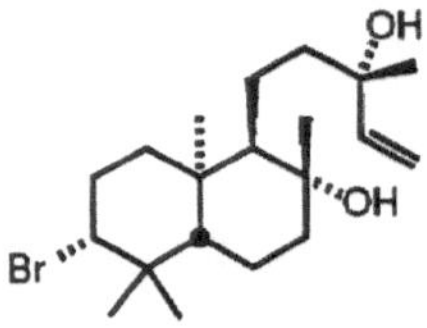

663 (pinnatol A)

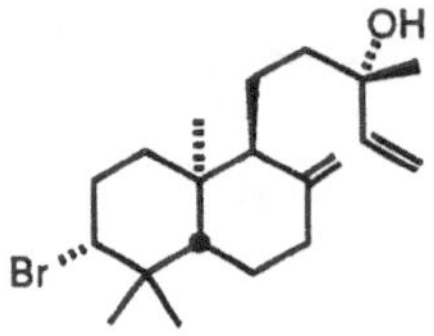

664 (pinnatol B)

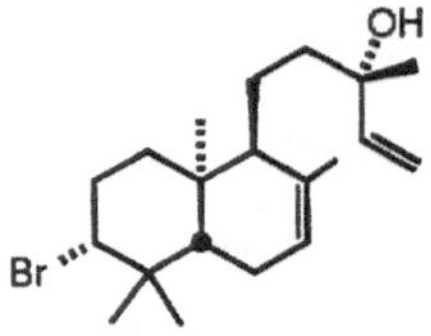

665 (pinnatol C)

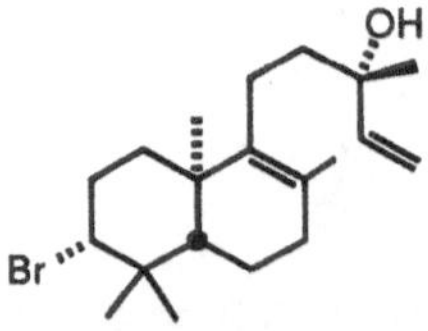

666 (pinnatol D)

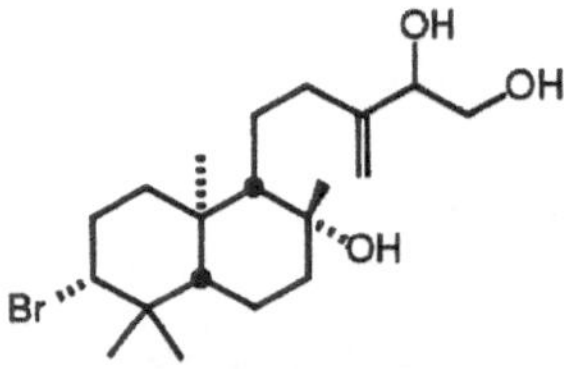

667 (venustanol)

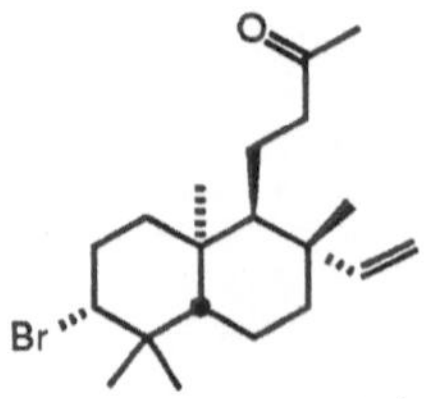

668

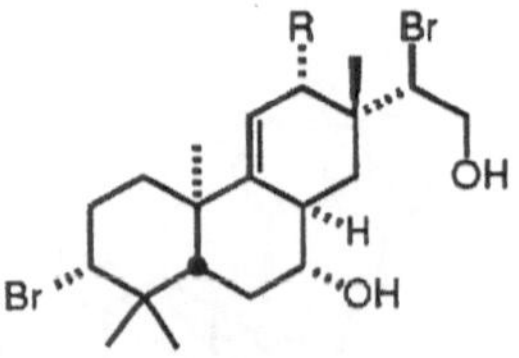

669 R = H
670 R = OH

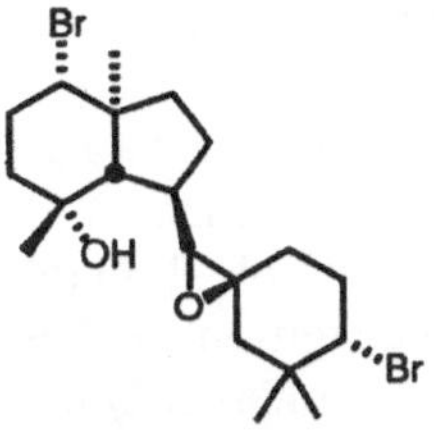

671 (irieol A)

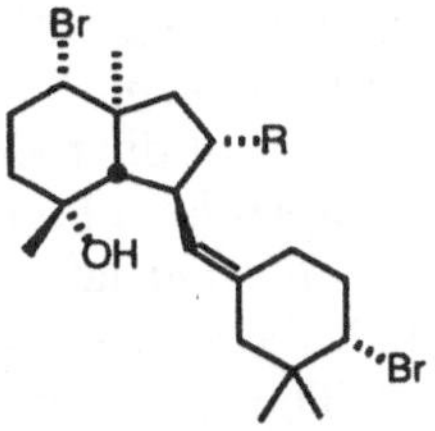

672 R = OH (iriediol)
673 R = H (irieol)

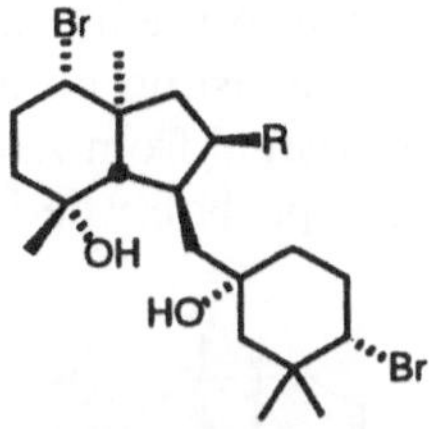

674 R = OH (irieol B)
675 R = H (irieol C)

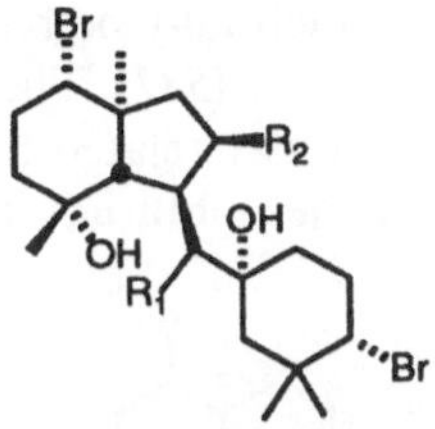

676 R_1 = OH, R_2 = OAc (irieol D)
677 R_1 = OH, R_2 = H (irieol E)
678 R_1 = R_2 = OH (irieol F)
679 R_1 = OAc, R_2 = OH (irieol G)

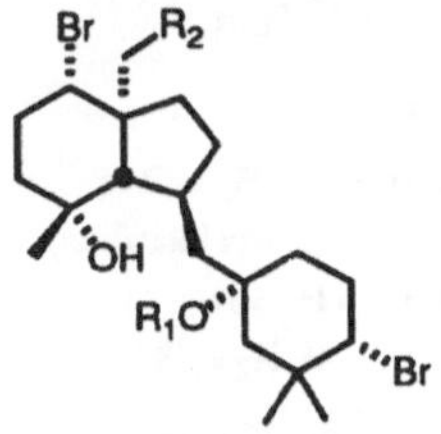

680 R_1 = R_2 = H (pinnaterpene A)
681 R_1 = Ac, R_2 = OH (pinnaterpene B)

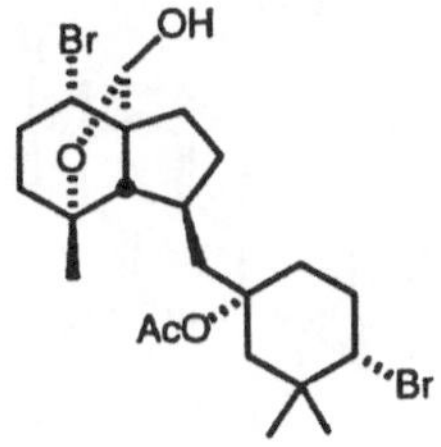

682 (pinnaterpene C)

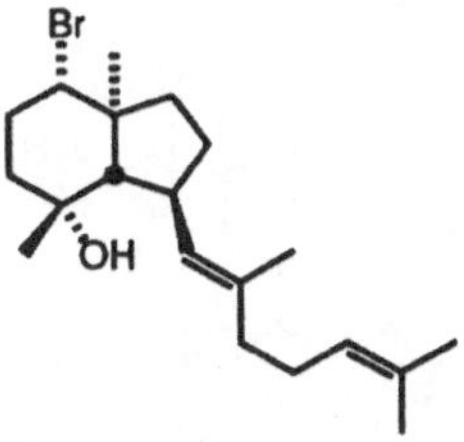

683 (prepinnaterpene)

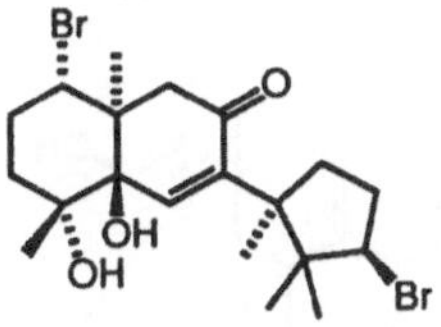

684 (neoirieone)

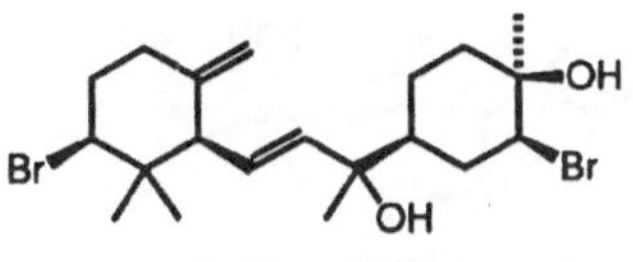

685 (obtusadiol)

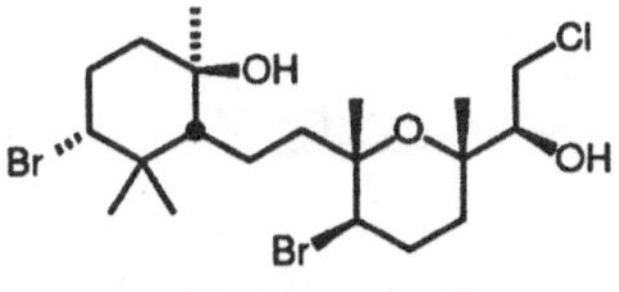

686 (laurencianol)

found in *L. perforata* along with isoaplysin-20 (**648**) (*532*), and **669** is found in *L. microcladia* (*436*).

A new skeletal class of diterpenes was discovered in *Laurencia irieii*, and nine such examples have been reported, including irieol A (**671**) and iriediol (**672**) (*533, 534*), and, later, irieol (**673**), and irieols B–G (**674**–**679**) (*534*). For a detailed review of the structure elucidation of these compounds, see (*526*). The related pinnaterpenes A–C (**680**–**682**) have been described from *L. pinnata* (*535*), along with prepinnaterpene (**683**) (*536*). This latter study also found irieol (**673**) in this red alga. The novel neoirieone (**684**) which may be biogenetically related to the irieanes has been found in *L. irieii* (*537*). The biogenetically interesting obtusadiol (**685**) which is obviously related to **654** has been isolated from *L. obtusa* (*538*). Like the irieanes, obtusadiol (**685**) and **654** are probably derived by

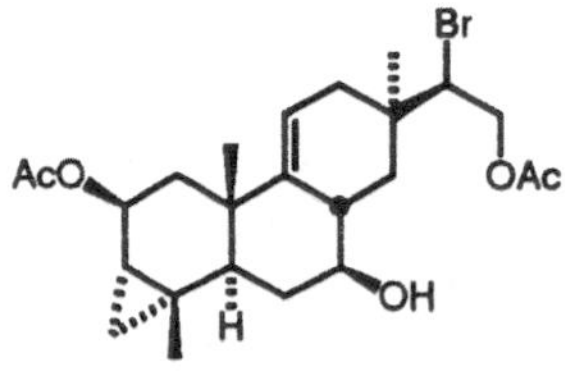

687

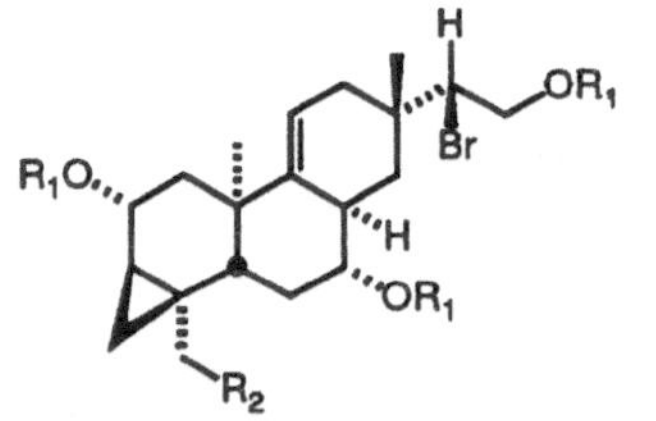

688 R_1 = Ac, R_2 = OAc (parguerol triacetate)
689 R_1 = R_2 = H
690 R_1 = H, R_2 = OH

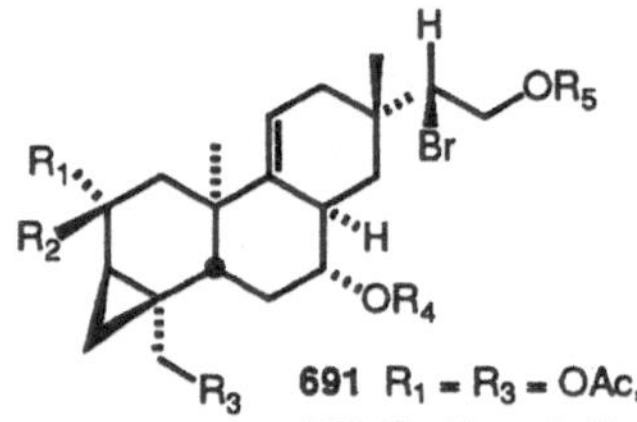

691 R_1 = R_3 = OAc, R_2 = R_4 = H, R_5 = Ac
692 R_1, R_2 = O, R_3 = OH, R_4 = R_5 = H
693 R_1 = R_2 = R_3 = R_4 = R_5 = H

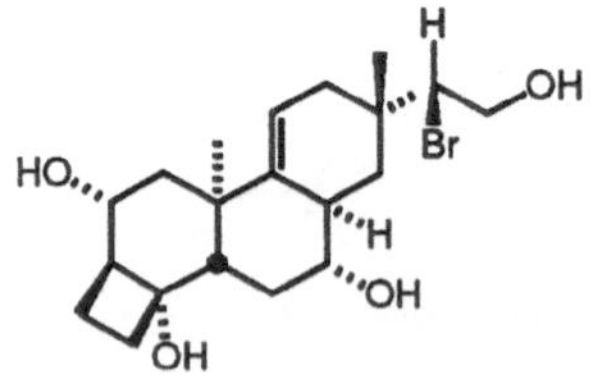

694

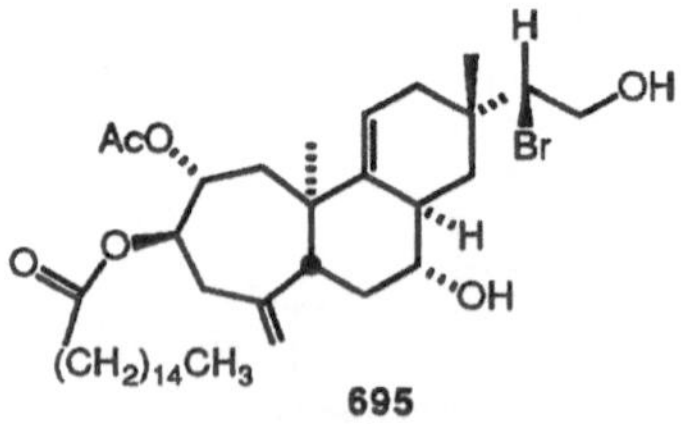

695

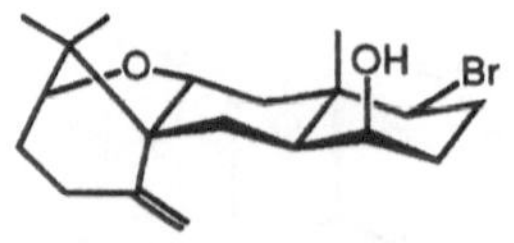

696 (kahukuene A)

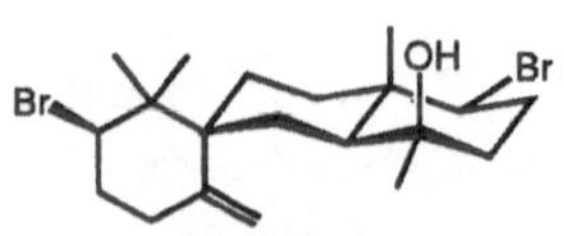

697 (kahukuene B)

a bromonium ion-induced cyclization of geranylgeraniol (*344*). Laurencianol (**686**) has been isolated from *L. obtusa* from Eastern Sicily and was found to have antibacterial activity (*E. coli, Bacillus subtilis*) (*539*).

Laurencia obtusa has also been the source of several diterpenes of the parguerane and isoparguerane types. Thus, **687** (*540*), parguerol triacetate (**688**) and triol **689** (*541*) have been reported and the absolute configuration of the latter two metabolites is described (*541*). Parguerol acetate (**688**) which has cytotoxic activity comprises 10% of the neutral oil. Subsequent extractions of *L. obtusa* yielded pargueranes **690**–**693** and isoparguerane **694** in addition to the known **657** and **659** (*542*) and the novel neoparguerane structure **695** which is cytotoxic (*543*). The isolation of **695**, along with the pargueranes and isopargueranes, completes the carbocation triad of the cyclopropylcarbinyl-cyclobutyl-homoallylic cation system. Hawaiian *L. majuscula* contains the novel prenylated chamigrenes kahukuenes A (**696**) and B (**697**) (*544*).

3.4.3.2.3. Sphaerococcus and Other Red Algae Diterpenes. The Mediterranean red alga *Sphaerococcus coronopifolius* has been a rich source of several novel rearranged brominated diterpenes, the possible biogenesis of which has been discussed (*526*).

Sphaerococcenol A (**698**) was isolated from the Spanish version of *S. coronopifolius* (*545*), while at about the same time bromosphaerol (**699**) was reported as being present in a Sicilian collection of this alga together with **698** (*546*). Bromosphaerodiol (**700**) was also found as a minor component (*547*) as was 12*S*-hydroxybromosphaerol (**701**), which, it was suggested, is a biogenetic intermediate in the conversion of bromosphaerol (**699**) to sphaerococcenol A (**698**) (*548*). Oxidation and dehydrobromination of **701** with PCC affords **698** (*548*), thus linking the two independent studies. The epimeric **702** and **703** are also found in this alga (*555*). Another minor metabolite, dihydrobromosphaerodiol **704**, has been extracted from *S. coronopifolius* (*549*).

More recently, bromosphaerenes A (**705**) and B (**706**) have been found in *S. coronopifolius* (*550*). These compounds could be prepared from bromosphaerol (**699**) by the action of $POCl_3$ (*550*). This versatile alga also produces the bicyclic bromocorodienol (**707**) which upon treatment with NBS undergoes cyclization to bromosphaerol (**699**) in 15% yield, providing evidence for a proposed biogenesis (*551*). Cafieri *et al.*, have also isolated the tetracyclic coronopifoliol (**708**) from *S. coronopifolius* and propose a reasonable biogenesis from bromosphaerol (**699**) involving C=C participation and a Wagner–Meerwein rearrangement (*552*). Similarly, bromotetrasphaerol (**709**) is found in this alga and may arise from **699** in a manner similar to that of **708** (*553*). In fact, the formation of

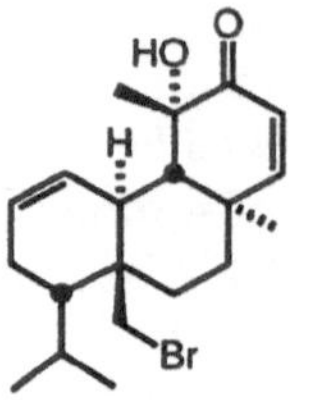

698 (sphaerococcenol A)

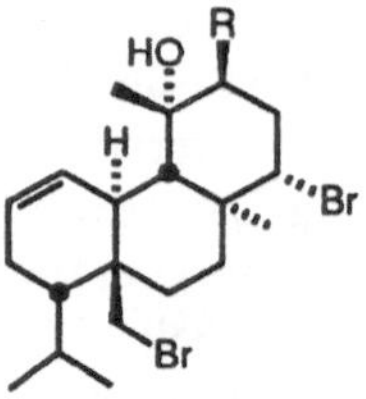

699 R = H (bromosphaerol)
701 R = β–OH
702 R = α–OH

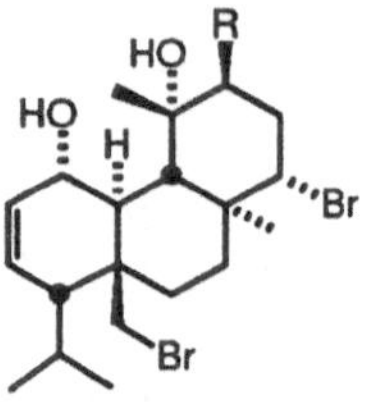

700 R = H (bromosphaerodiol)
703 R = OH

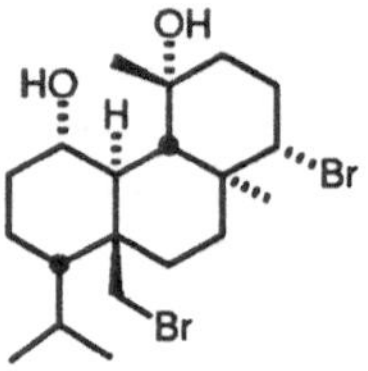

704

709 from **699** may involve the previously unknown generation of a 2-norbornyl cation via "the π-route" from a cyclohexenylmethyl system (*554, 559*), although the generation of the fascinating 2-norbornyl cation via solvolysis of a cyclopentenylethyl substrate is well known (*554, 560*). The related methyl ether **710** as well as **711** have been isolated more recently from *S. coronopifolius* (*556*). A collection of this alga from the North Adriatic Sea has yielded the novel oxetane sphaeroxetane **712** in addition to **698** and **699** (*557*). The isolation of sphaerococcenol A (**698**) has been aided by a contact bioautography assay using *E. coli* and *Bacillus subtilis* (*558*).

The Turkish red alga *Chondria tenuissima* collected in the Sea of Marmara was found to contain *ent*-13-epiconcinndiol (**713**) (*561*).

3.4.3.2.4. Sponge Diterpenes. Only a few halogenated sponge diterpenes are known, relative to the number isolated from marine algae. In one case, a link between an alga and a sponge has been found (*vide infra*).

The Pohnpeian sponge *Chelonaplysilla* sp. (earlier identified as a *Dendrilla* sp.) produces 1-bromo-8-ketoambliol A acetate (**714**) (*562, 563*), while the Mediterranean encrusting sponge *Mycale rotalis* contains rotalin B (**715**) (*564*). An extensive investigation of the South Pacific sponge *Acanthella* spp. from Guam and Fiji has revealed an array of isonitriles and related compounds, the kalihinols, seven of which contain

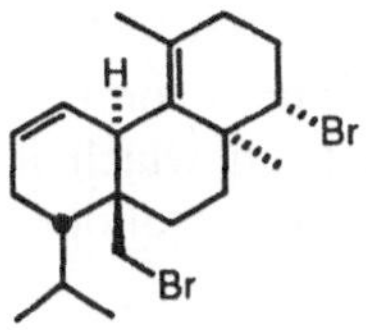

705 (bromosphaerene A)

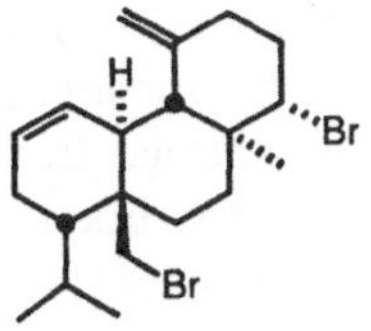

706 (bromosphaerene B)

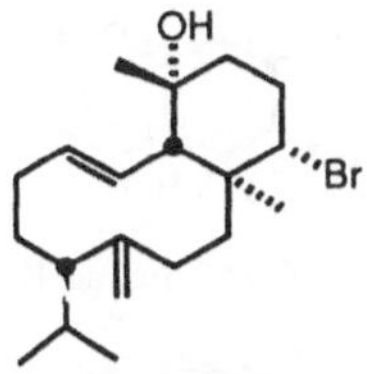

707 (bromocorodienol)

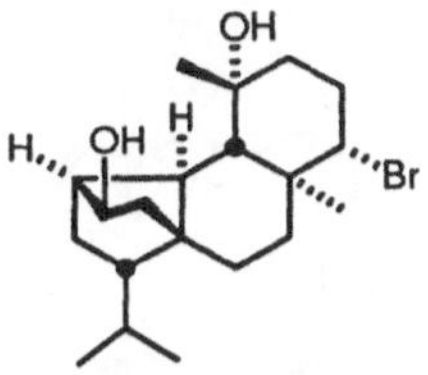

708 (coronopifoliol)

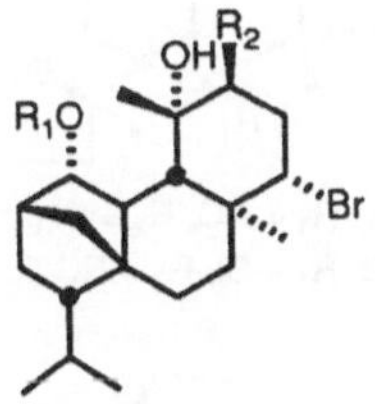

709 $R_1 = R_2 = H$ (bromotetrasphaerol)
710 $R_1 = Me$, $R_2 = OH$

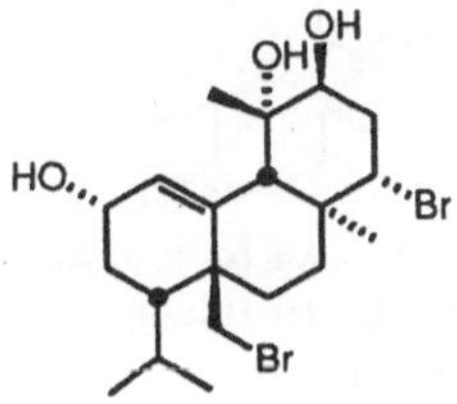

711

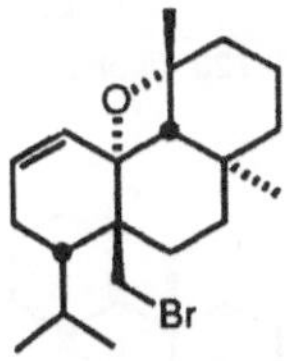

712 (sphaeroxetane)

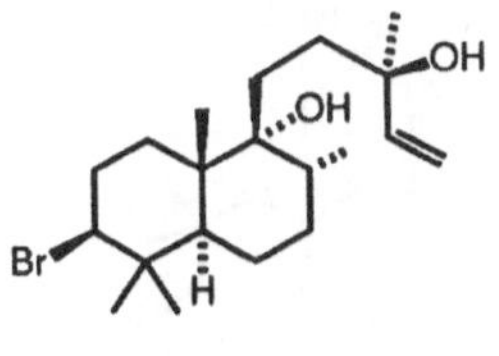

713 (*ent*-13-epiconcinndiol)

chlorine (*565*). The Guam sponge contains kalihinol A (**716**), E (**717**), B (**721**) and D (**722**), while the Fiji variety contains kalihinol X (**718**), Y (**719**), and Z (**720**) (*565*). A collection of *Acanthella klethra* has discovered isokalihinol B (**723**) in addition to kalihinol A (**716**) (*566*). Metabolite **723** is also contained in *A. cavernosa* (*567*), and this same sponge from Thailand produces kalihinol X (**718**) and Y (**719**), in addition to the new kalihinol I (**724**) and J (**725**) (*568*). The latter investigation discovered that kalihinols I, J, and Y have potent antiparasite activity against *Nippostrongylus brasiliensis*. The sponge *Spongia zimocca* contains the acetylated version **726** of diterpene **669** as well as sphaerococcenol A (**698**) and bromosphaerol (**699**). These compounds seem to be transferred to the sponge from the algae on which it feeds, and the secondary alcohol **669**, but not the tertiary alcohols **698** and **699**, is acetylated by the sponge (*436*).

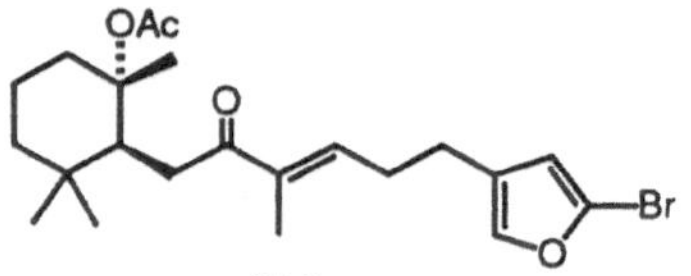

714

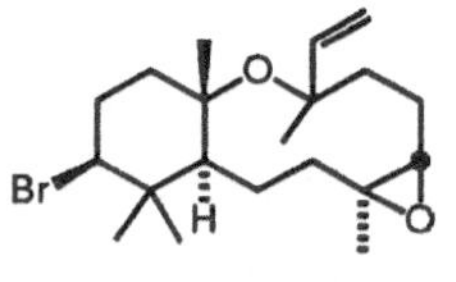

715 (rotalin B)

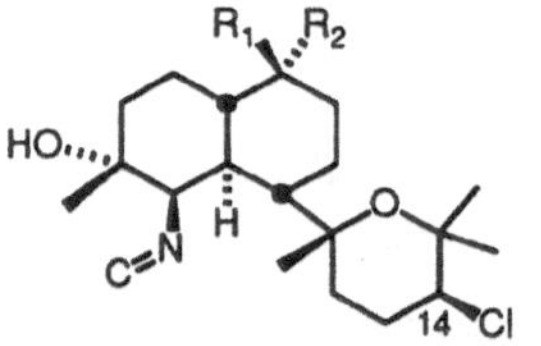

716 R_1 = NC, R_2 = Me (kalihinol A)
717 R_1 = NC, R_2 = Me (14αCl) (kalihinol E)
718 R_1 = Me, R_2 = NCS (kalihinol X)
719 R_1, R_2 = CH_2 (kalihinol Y)
720 R_1 = Me, R_2 = NC (kalihinol Z)

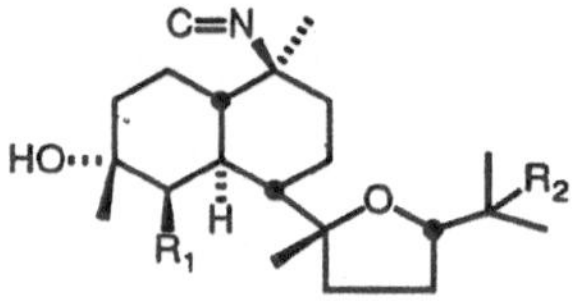

721 R_1 = NC, R_2 = Cl (kalihinol B)
722 R_1 = Cl, R_2 = NC (kalihinol D)

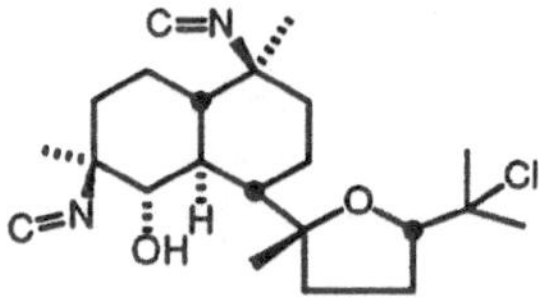

723 (isokalihinol B)

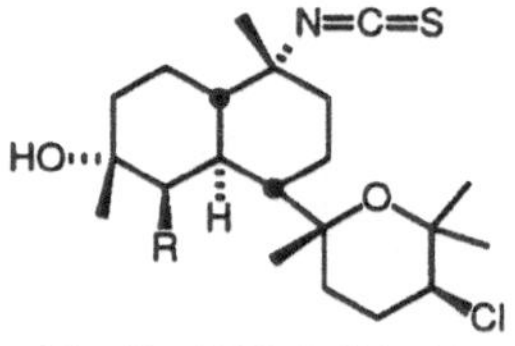

724 R = NCS (kalihinol I)
725 R = NHCHO (kalihinol J)

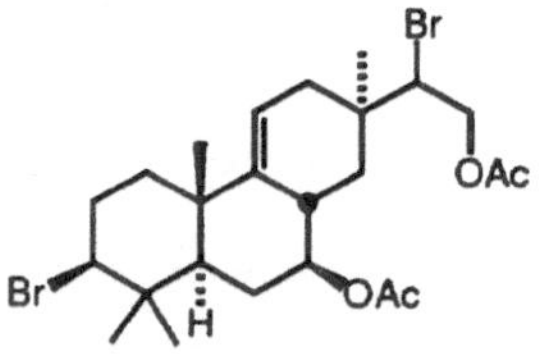

726

3.4.3.2.5. Gorgonian Diterpenes. Marine gorgonians have been the source of a large number of chlorinated diterpenes possessing a unique bicyclo[8.4.0]tetradecane skeleton. For an excellent summary and a discussion of the classification of these marine invertebrates see (*569*). It has been suggested by Fenical (*526*) that a geranylgeraniol precursor can cyclize to a cembrene which can then cyclize further to the bicyclic ring system characteristic of these metabolites.

The Caribbean gorgonian *Briareum asbestinum* produces briarein A (**727**) (*571*), a structure that because of its complexity required eleven years of work to elucidate (*526*). Since then, a number of related metabolites have been found in *Briareum* sp., including brianthein Z (**728**) (*572, 573*), X (**729**) (*448, 573–575*), and Y (**730**) (*573*) from *B. polyanthes*, brianthein V (**731**) from *B. asbestinum*, which also contained Y (**730**) and Z (**728**) (*576*), **732** and **733** from an Australian *Briareum* sp. (*577*), and brianolide (**734**) from an Okinawan *Briareum* sp. (*578*). These metabolites have a wide range of biological activities including cytotoxic (*576*), insecticidal (*573*), antiviral (*576*), and antiinflammatory activity (*578*).

The sea pen *Stylatula* sp. produces stylatulide (**735**) (*579*) and its minor C-17 epimer **736** (*580*), while another sea pen, *Ptilosarcus gurneyi*, produces the toxic ptilosarcone (**737**) (*581*) (structure revised in (*582*)), ptilosarcenone (**738**), ptilosarcol (**739**) and **740–743** (*582*). Interestingly, the nudibranch *Tochuina tetraquetra* has been found to contain ptilosarcenone (**738**) and the new **744**, the former probably acquired by feeding

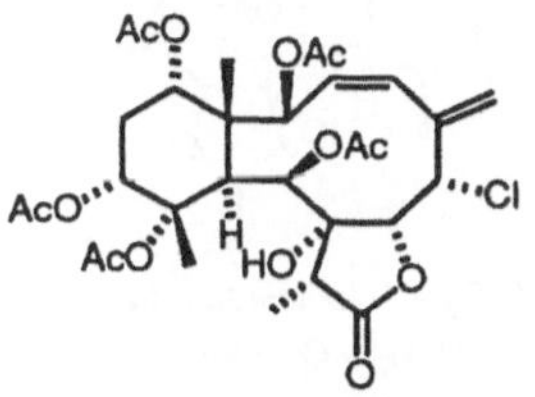

727 (briarein A)

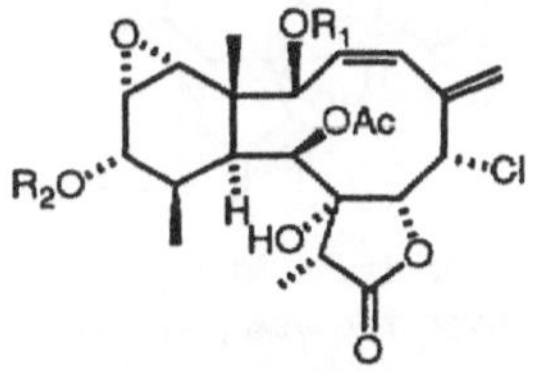

728 $R_1 = R_2 = Ac$ (brianthein Z)
729 $R_1 = H$, $R_2 = Ac$ (brianthein X)
730 $R_1 = COCH_2CH_2CH_3$, $R_2 = Ac$ (brianthein Y)
731 $R_1 = R_2 = COCH_2CH_2CH_3$ (brianthein V)

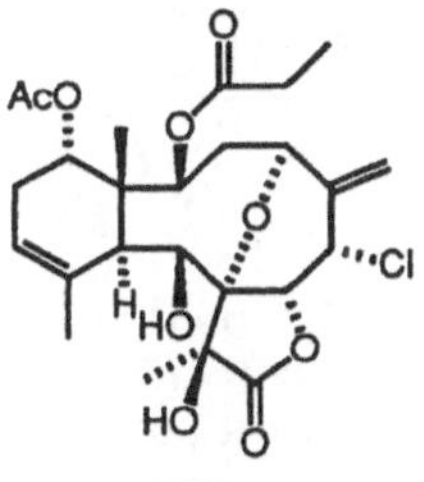

732

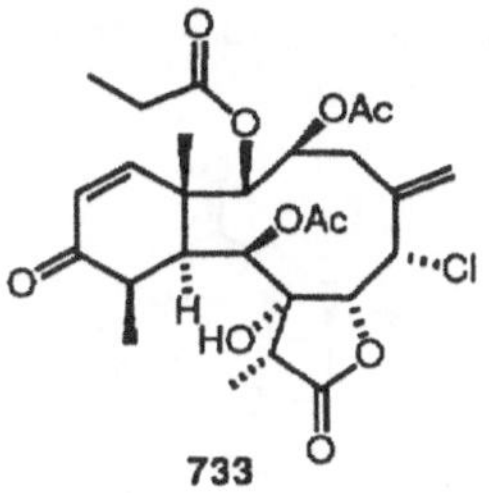

733

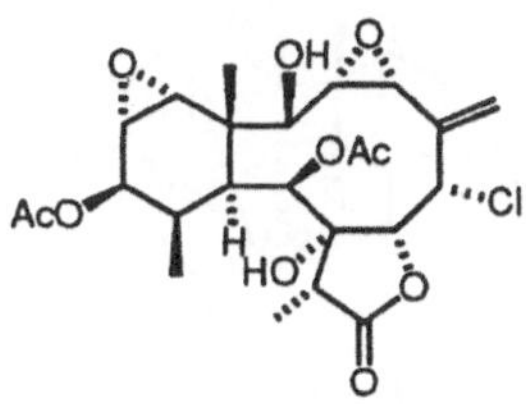

734 (brianolide)

on *P. gurneyi* (*583*). The New Caledonian sea pen *Pteroides laboutei*, a newly discovered species, produces pteroidine (**745**) and **746** (*584*).

The Caribbean octocoral *Erythropodium caribaeorum* has provided an array of chlorinated diterpenes, including the erythrolides A (**747**) and B (**748**), which are interrelated by a naturally occurring "di-π-methane rearrangement," the first of its kind to be found in nature (*585*). Both of these compounds are found in the organism and it is proposed that a photochemical process in the shallow marine environment in which these organisms live converts erythrolide B (**748**) into A (**747**) (*585*). More recently, erythrolides C (**749**), D (**750**), E (**751**), F (**752**), I (**753**), G (**754**), and the nonchlorinated H (**755**) were isolated from this organism (*586*). An examination of *E. caribaeorum* off Tobago has also produced eryth-

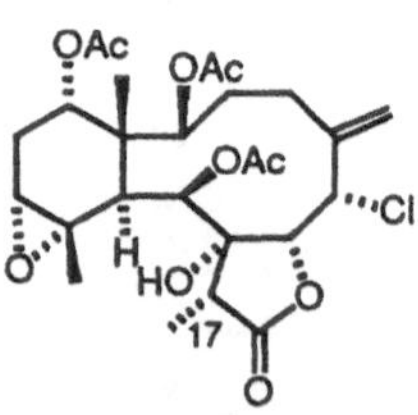

735 (stylatulide)
736 (17-*epi*-stylatulide)

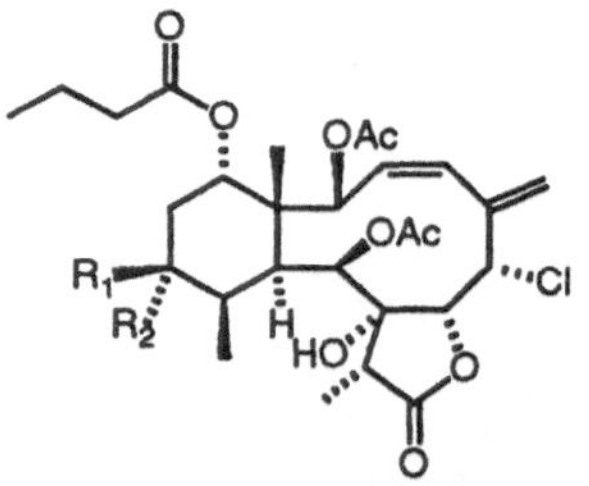

737 $R_1, R_2 = O$ (ptilosarcone)
739 $R_1 = H, R_2 = OH$ (ptilosarcol)

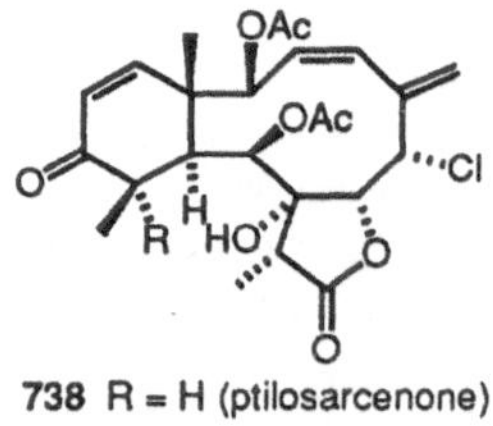

738 R = H (ptilosarcenone)
740 R = OH

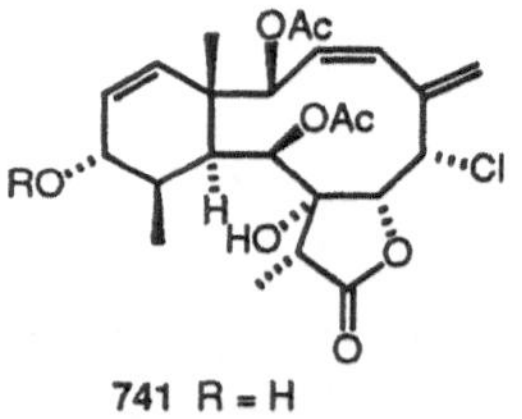

741 R = H
742 R = Ac
743 R = $COCH_2CH_3$

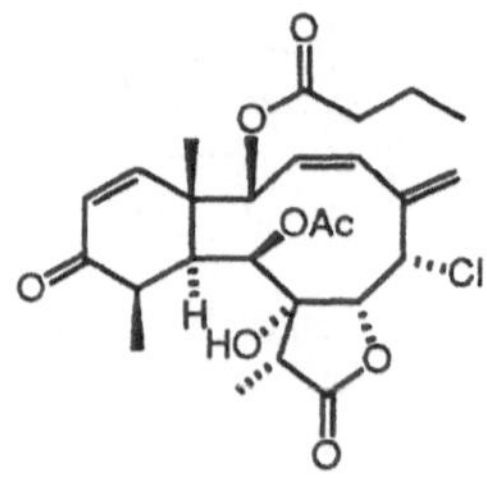

744

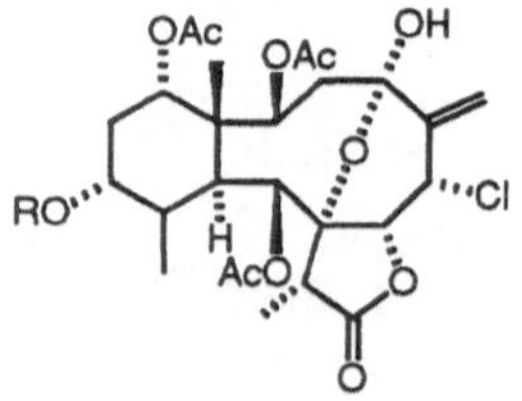

745 R = Ac (pteroidine)
746 R = COPh

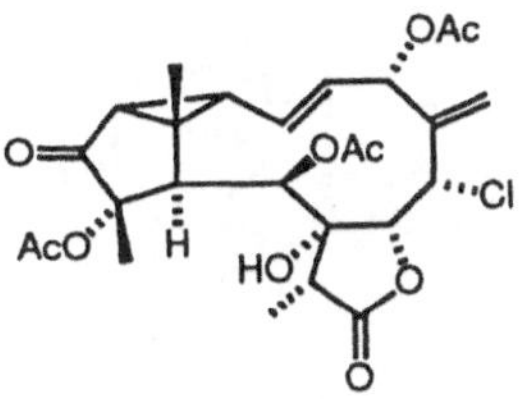

747 (erythrolide A)

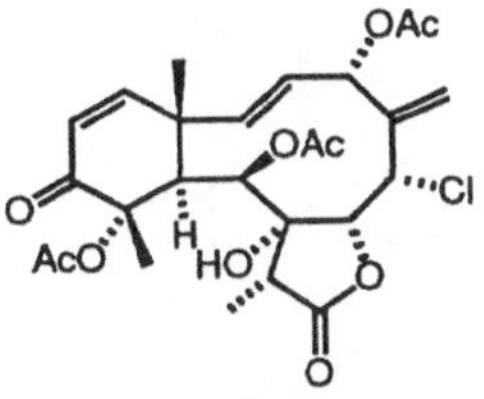

748 (erythrolide B)

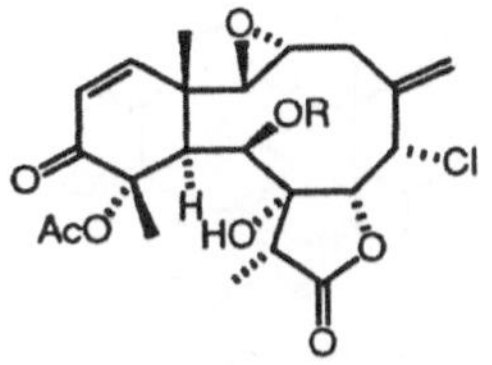

749 R = Ac (erythrolide C)
750 R = $COCH_2OAc$ (erythrolide D)

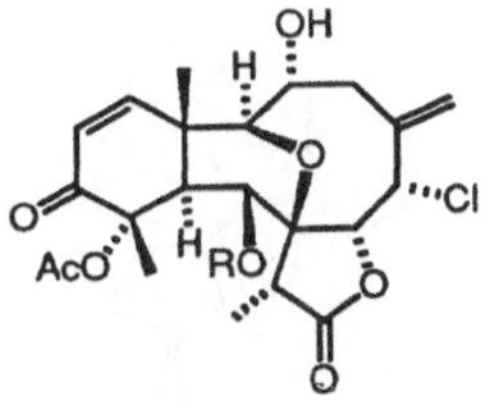

751 R = Ac (erythrolide E)
752 R = $COCH_2OAc$ (erythrolide F)
753 R = $COCH_2OH$ (erythrolide I)

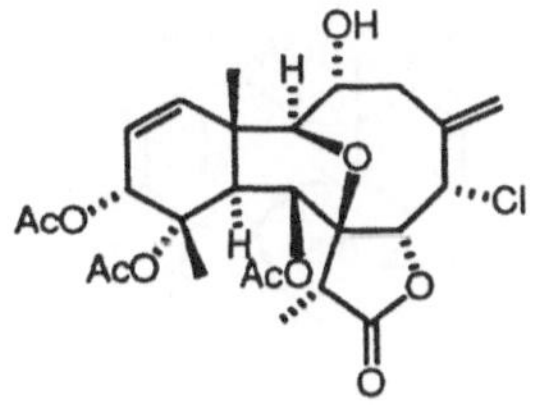

754 (erythrolide G)

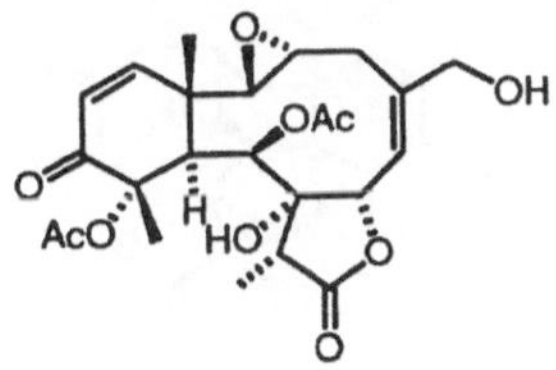

755 (erythrolide H)

rolides A, B, and E as well as several new non-chlorinated metabolites (*587*).

The soft coral *Minabea* sp. from Truk Lagoon in the Eastern Caroline Islands produces the minabeins 1–10, of which 1–5 (**740, 756–759**) contain chlorine (*588*). Minabein-1 appears to be identical with **740**. The octocoral *Solenopodium* sp. produces the solenolides A–E (**760–764**) which have potent antiinflammatory, insecticidal, and antiviral properties (*589*). A gorgonian from the South China Sea, *Plexaureides praelongà*, produces **765** (*590*).

Also from the South China Sea, the gorgonian *Junceella squamata* produces junceellin (**766**) (*591*) and junceellin B (**767**) (592), while *J. fragilis*, the white sea whip, contains the antiinflammatory junceellolides A–C (**768–770**) (*593*). The Red Sea *J. juncea* produces juncins A–F (**771–776**) (*594*). Juncin C (**773**) is not completely characterized but is a diacetate isovalerate derivative of juncin A, and juncin F (**776**) is

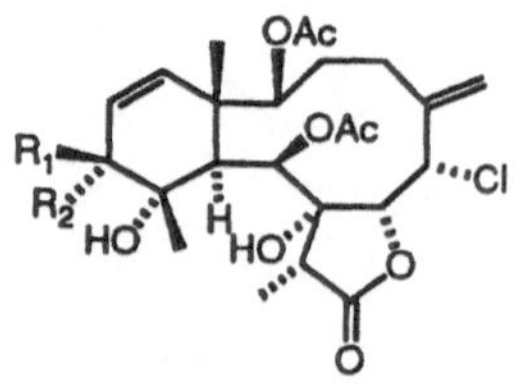

756 R_1, R_2 = O (minabein-2)
757 R_1 = H, R_2 = OAc (minabein-3)

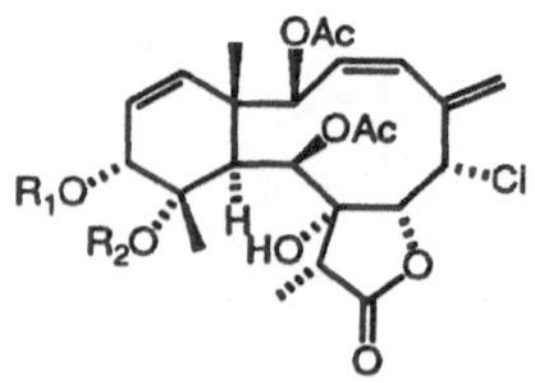

758 R_1 = Ac, R_2 = H (minabein-4)
759 R_1 = R_2 = H (minabein-5)

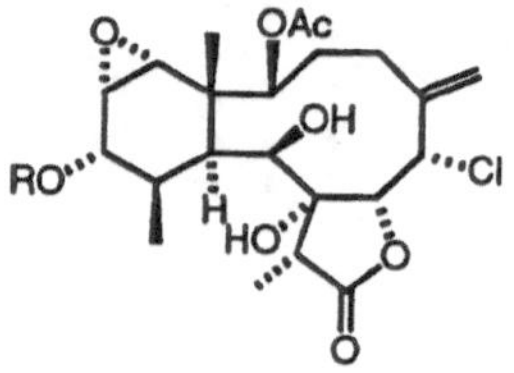

760 R = CO*n*-Pen (solenolide A)
761 R = Ac (solenolide B)

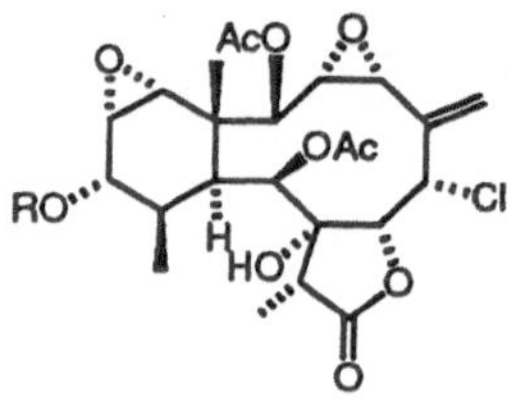

762 R = H (solenolide C)
763 R = Ac (solenolide D)

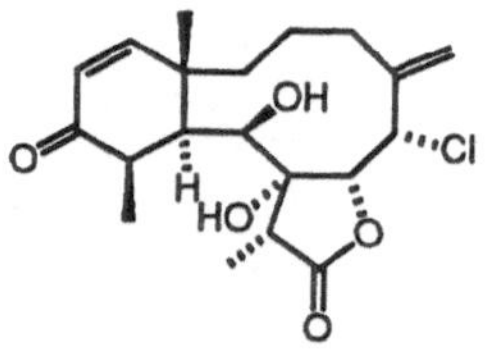

764 (solenolide E)

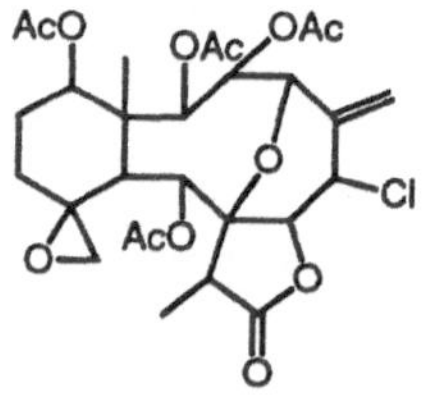

765

a 3,4-dihydro compound containing three acetates and an isobutyrate. Juncins A and B are the *Z* isomers of junceellolides C and B, respectively. The related gemmacolides A–E (**777–781**) have been isolated from *J. gemmacea* (*595*).

The Australian soft coral *Nephthea chabrolii* produces the caryophyllene-type chlorohydrin **782** (*596*), while the New Caledonian soft coral *Xenia membranaceae* contains several novel diterpenes, havannachlorohydrins **783–790** (*597, 598*). The octocoral *Veretillum cynomorium* and its prey, the nudibranch *Armina maculata*, contain verecynarmin D (**791**), a novel briarane diterpene (*599*). The New Caledonian ascidian *Lissoclinum voeltzkowi* produces the cytotoxic labdane dichlorolissoclimide (**792**) (*600*). This compound is active at the ng/ml level against P388 and KB cells, and represents the first chlorinated metabolite to be isolated from Urochordata.

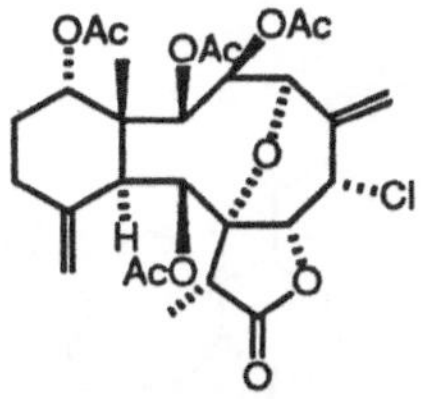

766 (junceellin)

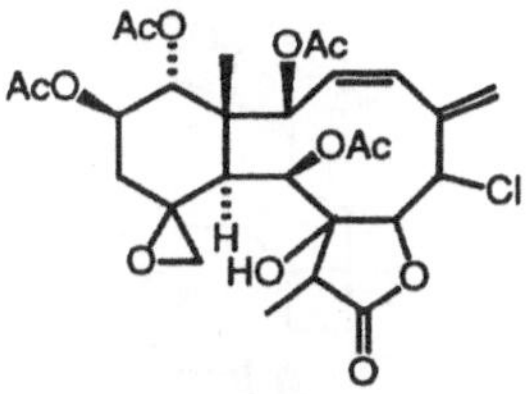

767 (junceellin B)

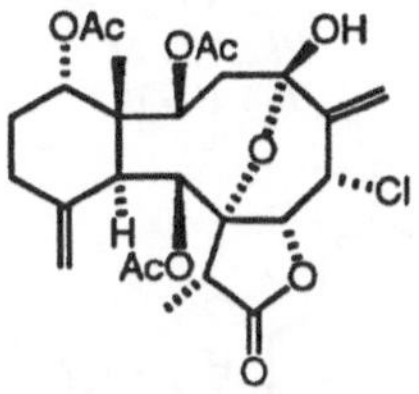

768 (junceellolide A)

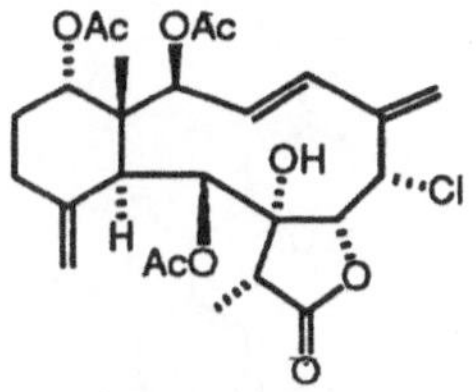

769 (junceellolide B)

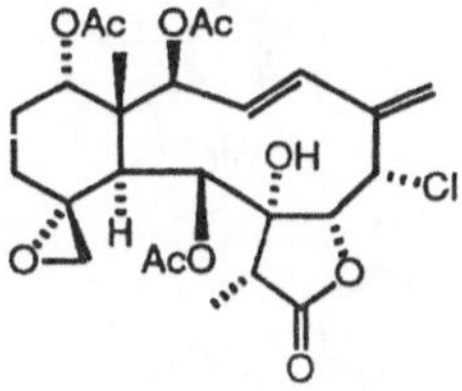

770 (junceellolide C)

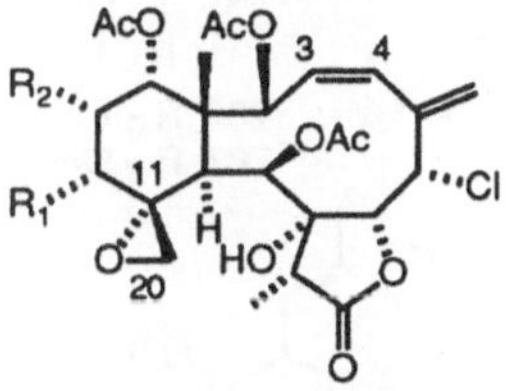

771 $R_1 = R_2 = H$ (juncin A)
772 $R_1 = R_2 = H$ (11,20-deoxy) (juncin B)
774 $R_1 = OAc$, $R_2 = H$ (juncin D)
775 $R_1 = R_2 = OAc$ (juncin E)

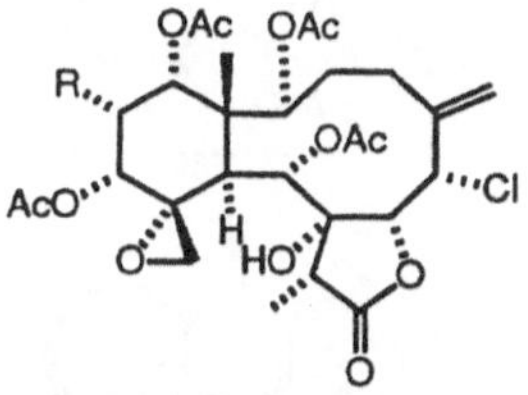

777 R = OAc (gemmacolide A)
778 R = $OCOCH_2CHMe_2$ (gemmacolide B)
779 R = H (gemmacolide C)

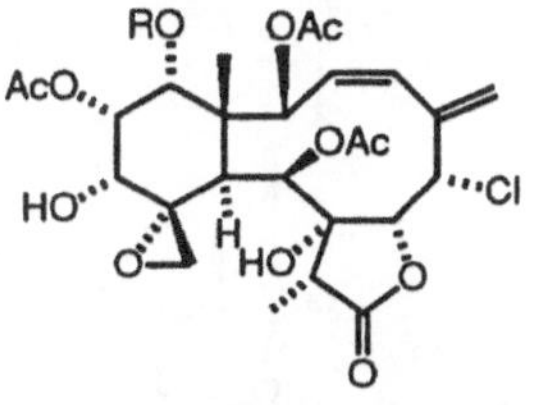

780 R = Ac (gemmacolide D)
781 R = H (gemmacolide E)

3.4.4. *Higher Terpenes*

In comparison with the other terpenes (*vide supra*), there are preciously few halogenated triterpenes and other polyterpenes.

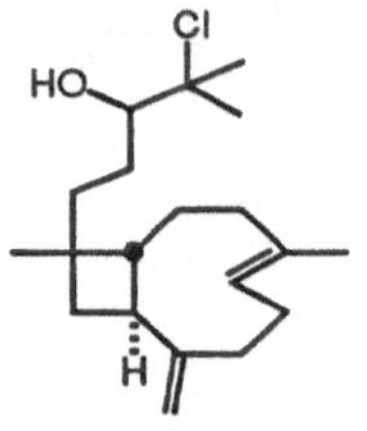

782

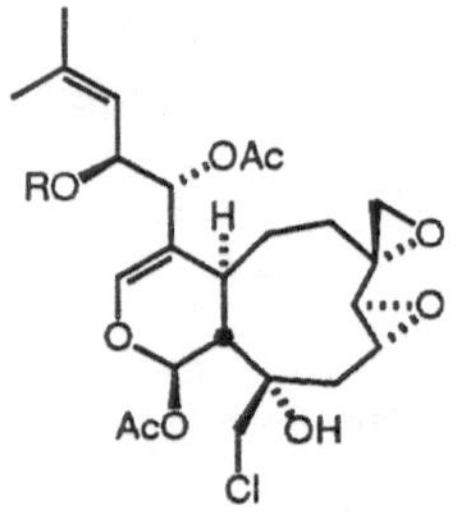

783 R = Ac
784 R = H

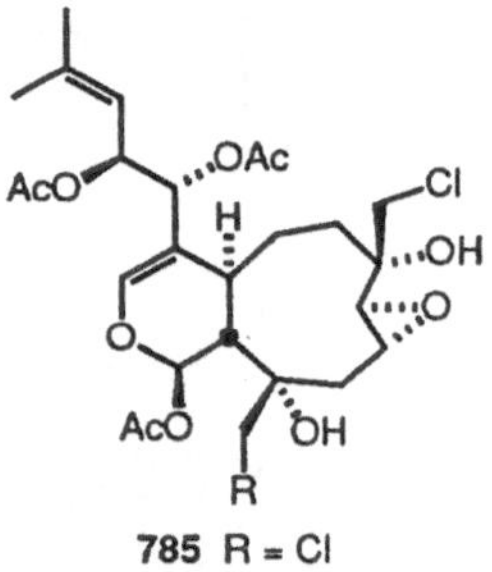

785 R = Cl
787 R = OAc

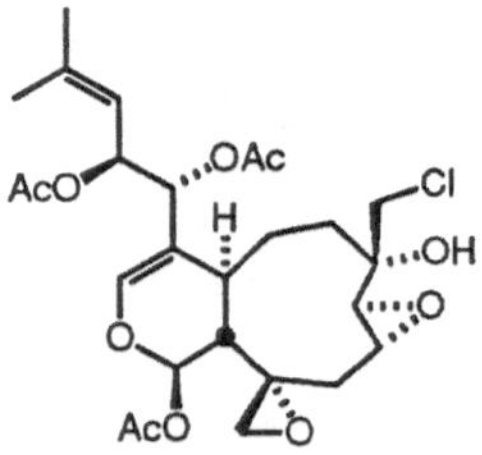

786

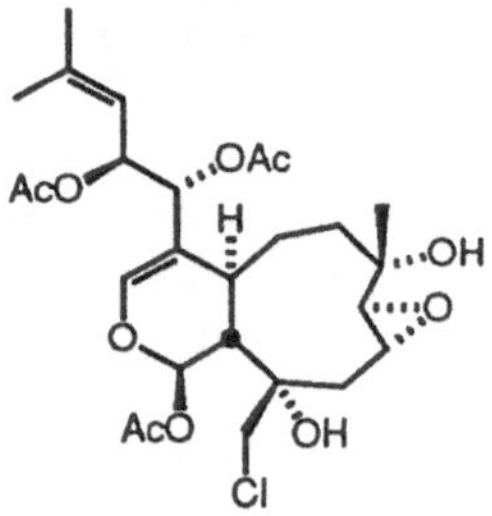

788

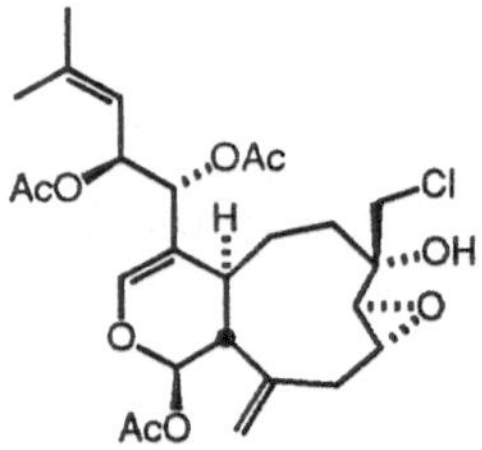

789

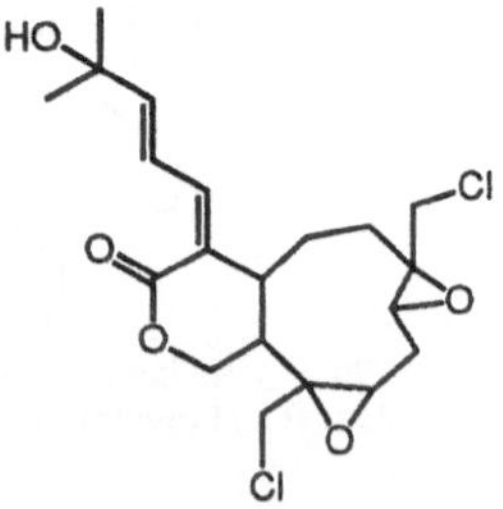

790

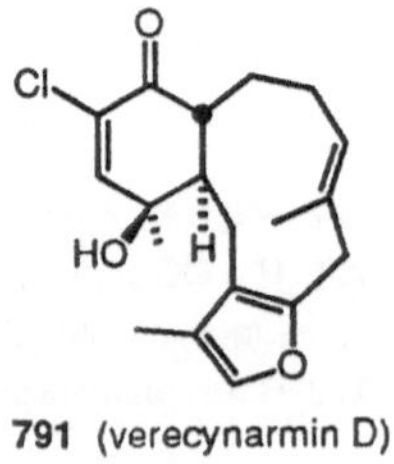

791 (verecynarmin D)

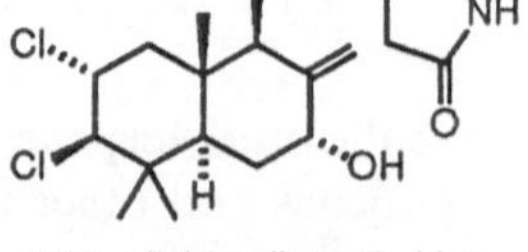

792 (dichlorolissoclimide)

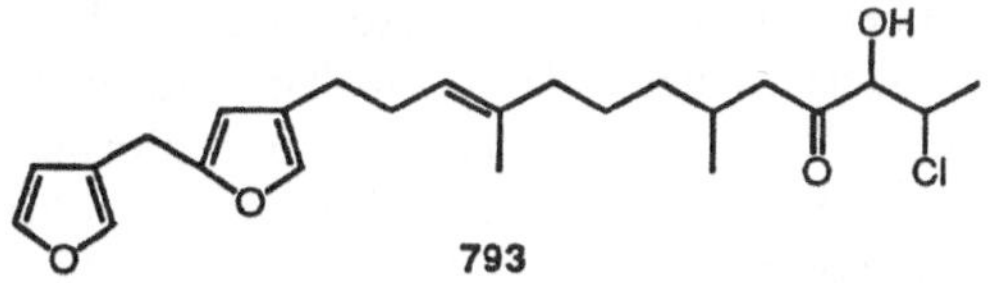

793

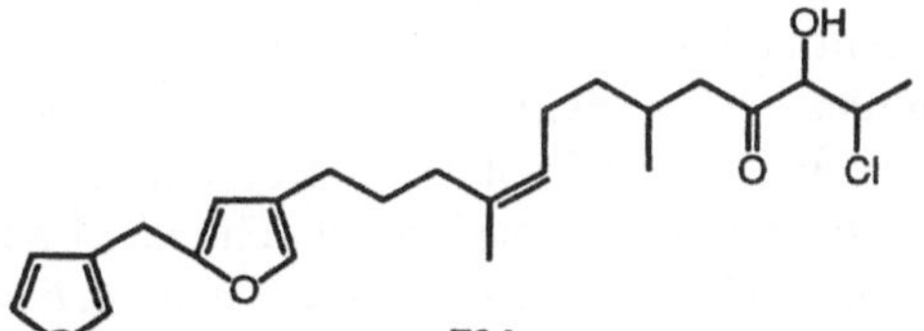

794

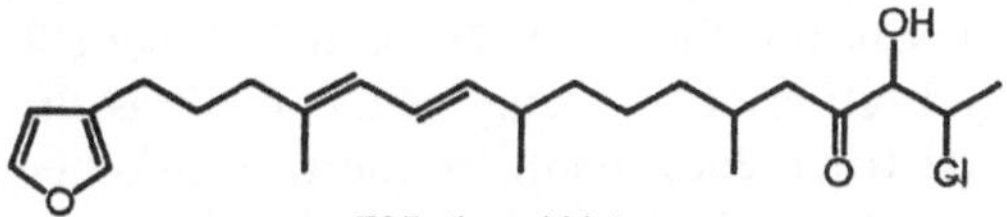

795 (konakhin)

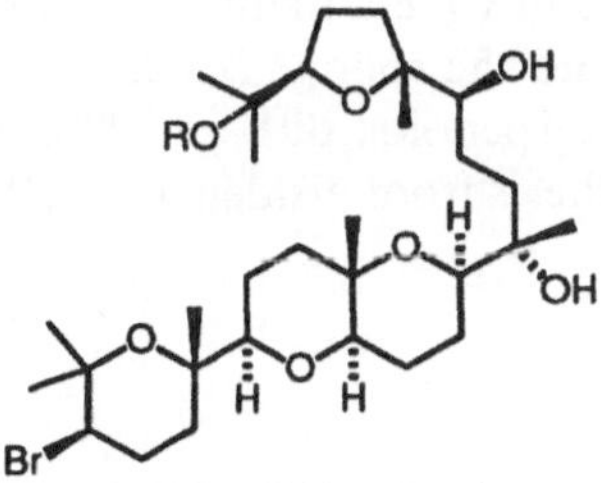

796 R = H (thyrsiferol)
797 R = Ac

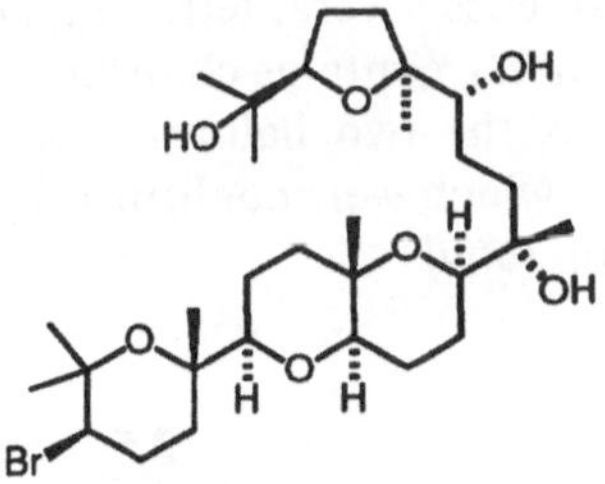

798 (venustatriol)

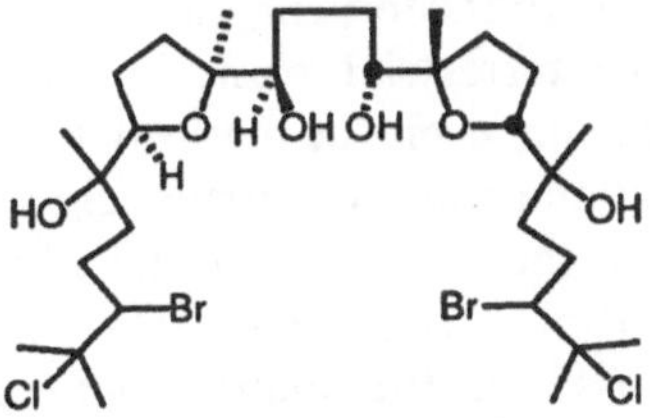

799 (intricatetraol)

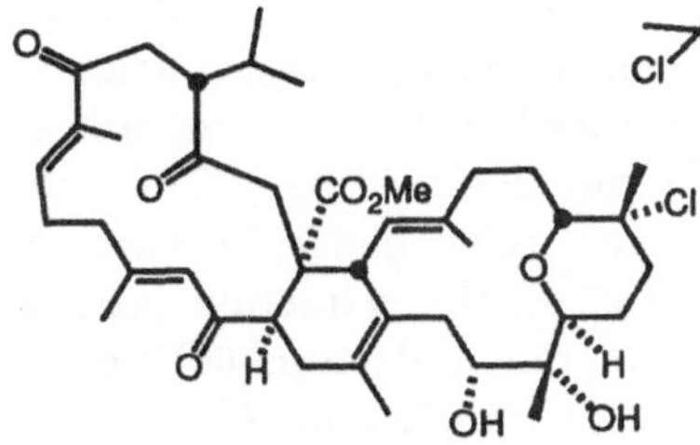

800 (methyl chlorosarcophytoate)

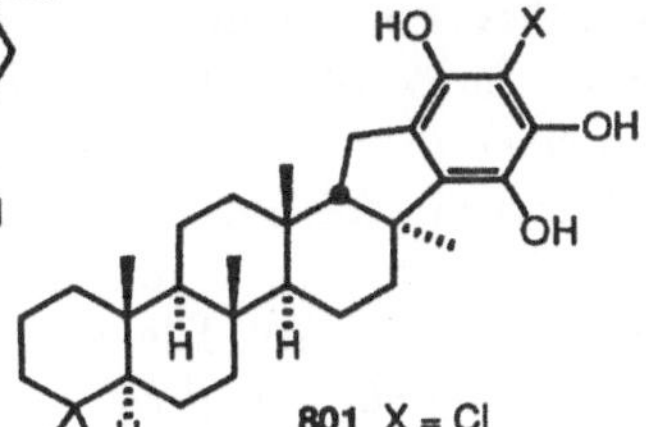

801 X = Cl
802 X = Br

A sponge, *Ircinia oros*, from the North Adriatic Sea, has yielded the two chlorinated norsesterterpenes **793** and **794**. The authors propose that the biogenesis of these compounds proceeds by chlorination of the hydroxyfuranone metabolites, ircinin 1 and 2, which are also present in this sponge (*601*). A closely related compound, konakhin (**795**), was isolated from an unidentified Senegal sponge (*602*).

The red algae *Laurencia* has provided several squalene-derived triterpenes, the first of which is thyrsiferol (**796**) isolated from *Laurencia thyrsifera* (*603*) (structure revised in (*604*, *606*)). The C-23 acetate (**797**) was isolated from *L. obtusa* (*605*) (structure revised in (*604*, *606*)). Both of these metabolites are highly cytotoxic. The antiviral venustatriol (**798**) was found in *L. venusta* together with **796** and **797** (*606*). The authors suggest that the cyclization of squalene pentaepoxide is the biogenetic route to these novel triterpenes (*606*). In contrast, another group has suggested that intricatetraol (**799**), which was isolated from *L. intricata*, is derived from squalene tetraepoxide (*607*). The soft coral *Sarcophyton glaucum* produces several tetraterpenes one of which, methyl chlorosarcophytoate (**800**), contains chlorine (*608*), and the sponge *Disidea pallescens* contains the two halogenated sesterterpenoids **801** and **802**, the structures of which were confirmed by synthesis from disidein (X =H) by halogenation (*609*).

3.5. Steroids

In contrast to the terpenes, most examples of halogenated steroids are found in the terrestrial plant kingdom. All of the known examples contain chlorine, even the few isolated from marine sources.

The first naturally-occurring chlorinated steroids to be identified appear to be the jaborosalactones C (**803**) and E (**804**) from *Jaborosa integrifolia* (Solanaceae) (*610*, *611*), which, like all such compounds, probably arise in the plant from the corresponding epoxide. For example, the withanolide **805** was isolated from *Withania frutescens* (Solanaceae) along with the epoxide withaferin A (*612*). The authors conclude that **805** is not an artifact since no chlorinated solvents, which are a possible source of HCl, were used in the extraction procedure. *Physalis peruviana* (Solanaceae) contains physalolactone (**806**) (*613*), 4-deoxyphysalolactone (**807**) (*614*), and physalolactone C (**808**) (*615*). Physagulin B (**809**) has been found in *P. angulata* together with the corresponding epoxide (*616*). The chlorohydrin of withanolide D (**810**) is found in both *Withania somnifera* and *Acnistus breviflorus* (Solanaceae) along with the previous 4-deoxyphysalolactone (**807**) (*617*). *A. breviflorus*

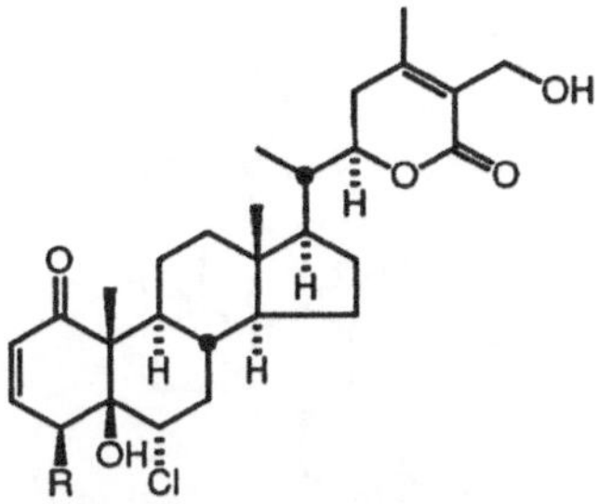

803 R = H (jaborosalactone C)
805 R = OH

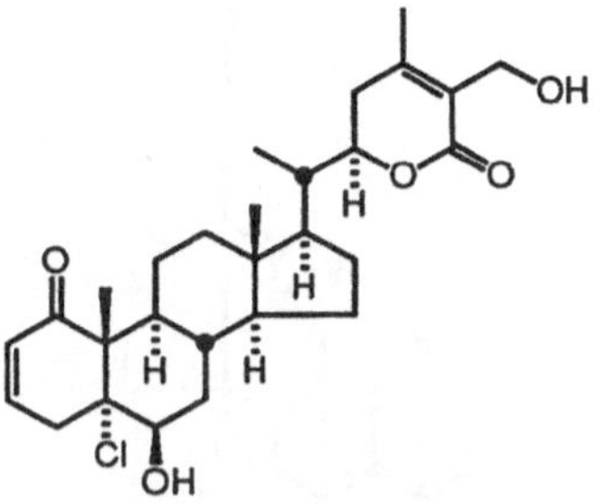

804 (jaborosalactone E)

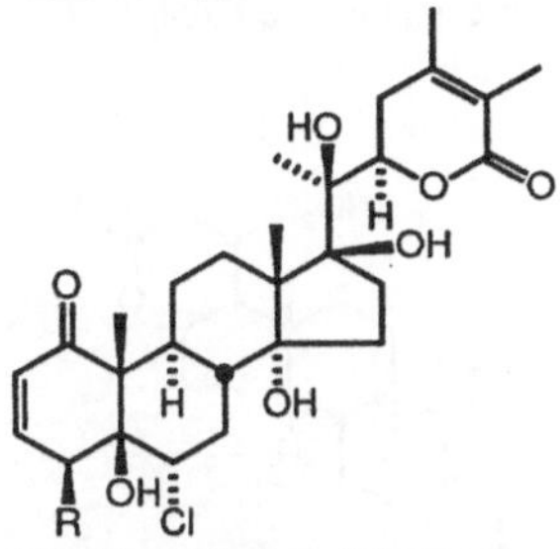

806 R = OH (physalolactone)
807 R = H (4-deoxyphysalolactone)

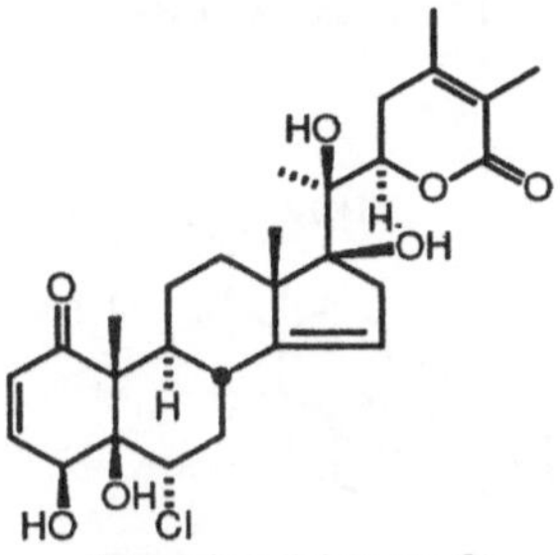

808 (physalolactone C)

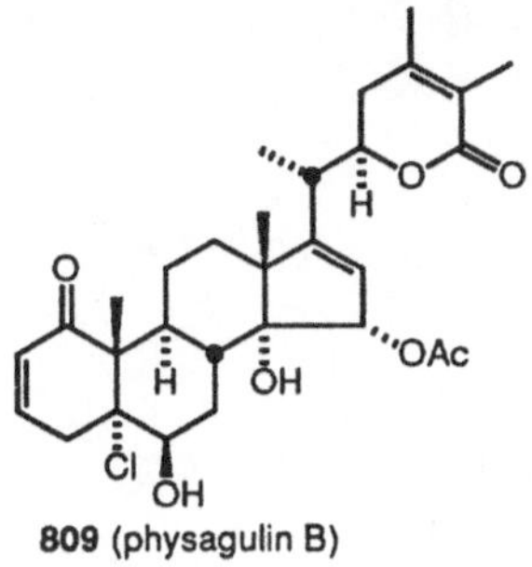

809 (physagulin B)

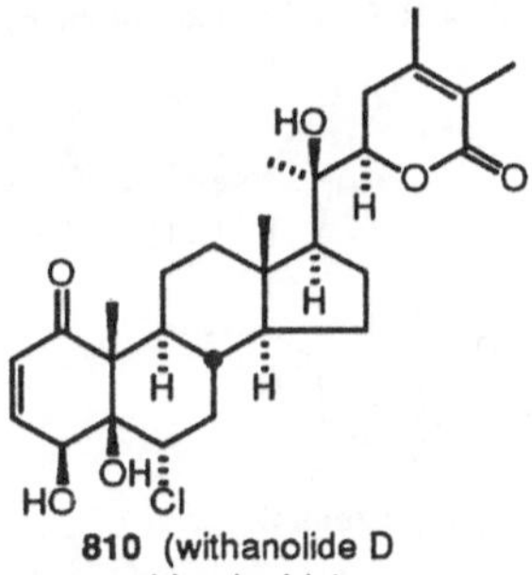

810 (withanolide D chlorohydrin)

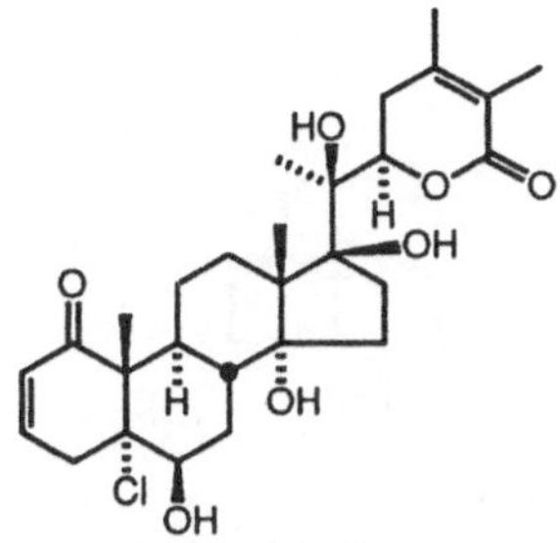

811 (withanolide C)

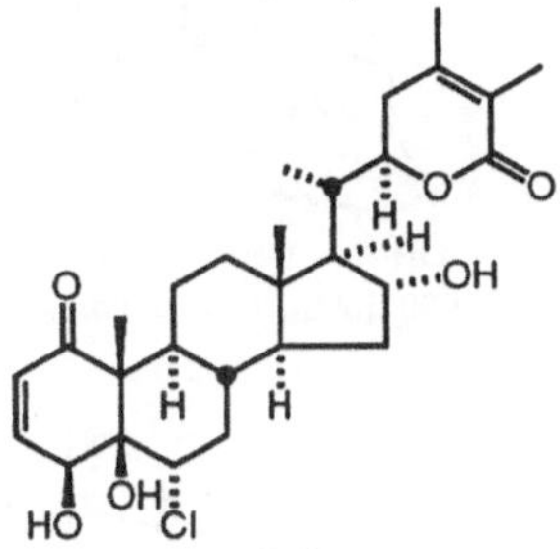

812

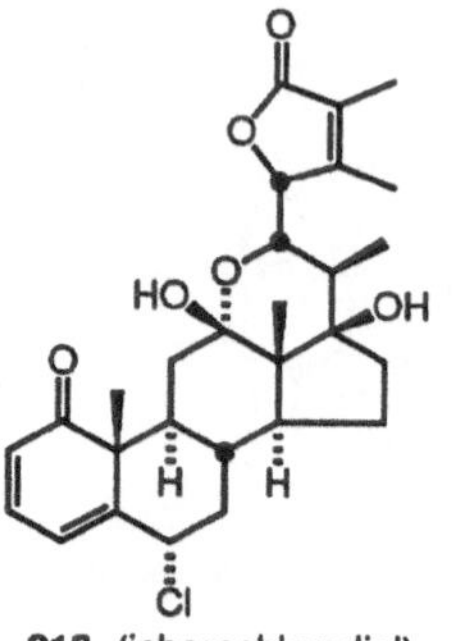

813 (jaborochlorodiol) 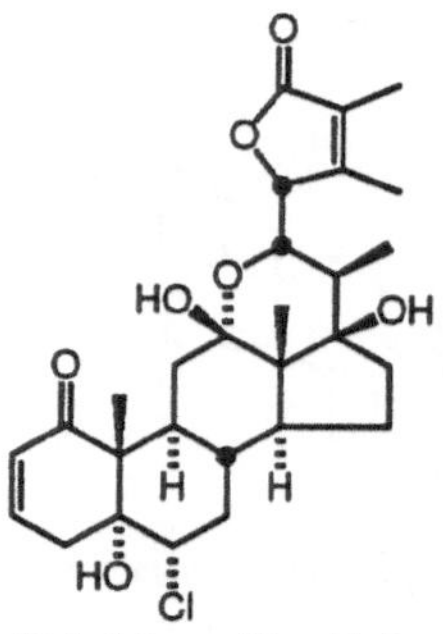 **814** (jaborochlorotriol)

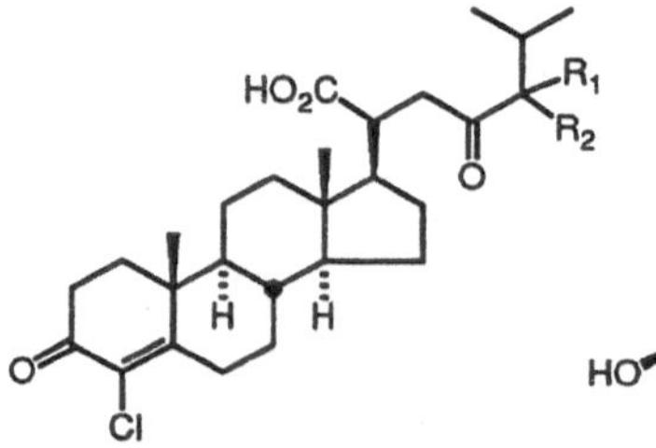

815 $R_1, R_2 = CH_2$ (kiheisterone C)
816 $R_1 = H, R_2 = Et$ (kiheisterone E) 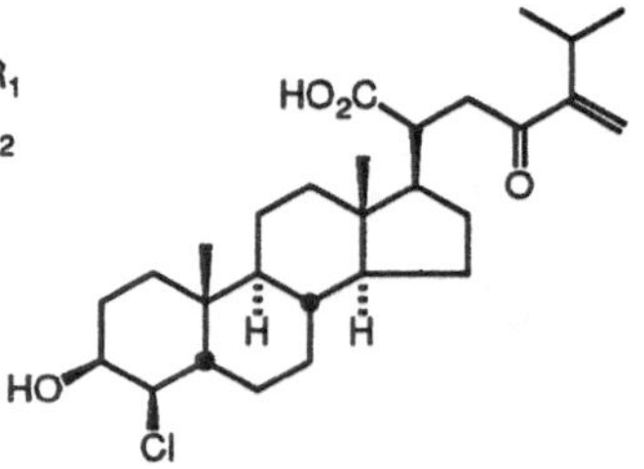
817 (kiheisterone D)

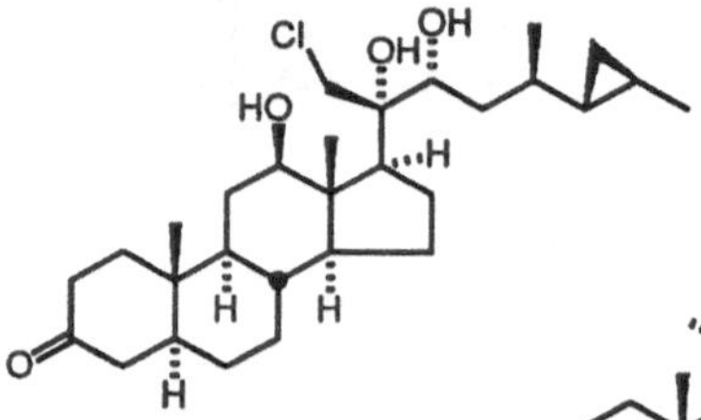

818 (aragusterol C)

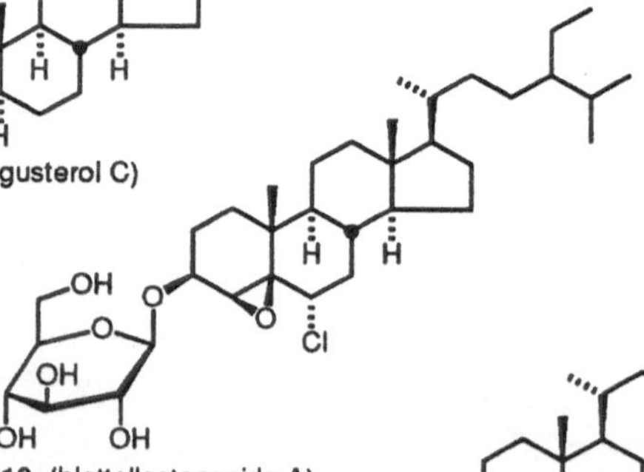

819 (blattellastanoside A)

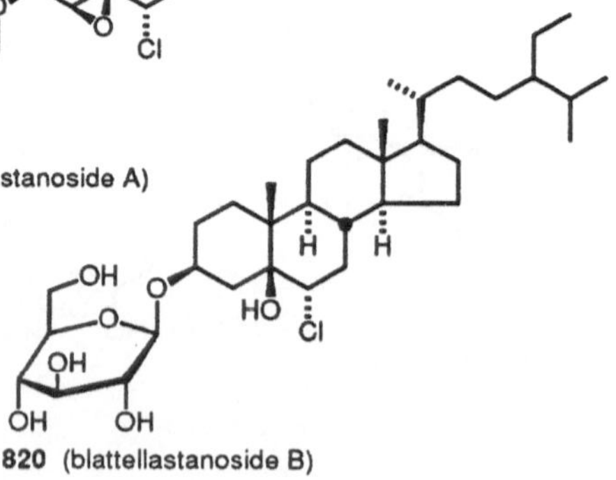

820 (blattellastanoside B)

also contains withaferin A chlorohydrin **805** (*618*, *621*), which has been found to possess significant cytostatic activity (*619*), and also produces jaborosalactone E (**804**) (*620–622*). A later study of *Withania somnifera* has revealed the presence of withanolide C (**811**) along with **807** (*623*), and **812** has been reported as being present in *Dunalia tubulosa* also in Solanaceae (*624*). The novel bridged compounds, jaborochlorodiol (**813**) and jaborochlorotriol (**814**), have been described from *Jaborosa magellanica* (*625*).

Although marine sponge steroids are very well known and abundant (*626*), until recently, no examples of halogenated steroids from any marine source were known. However in 1993 SCHEUER and co-workers reported kiheisterones C (**815**), E (**816**), and D (**817**) from the Maui sponge *Strongylacidon* sp. (*627*). The Okinawan sponge *Xestospongia* sp. produces aragusterol C (**818**), a metabolite that possesses potent antitumor activity (*628*). For example, the KB antiproliferation inhibition is $IC_{50} = 0.041$ μg/ml and the L-1210 *in vivo* survival is $T/C = 257\%$ at 1.6 mg/kg (*628*).

Perhaps even more remarkable than the discovery of chlorinated marine steroids is the finding that the German cockroach, *Blattella germanica*, uses the stigmastane steroids, blattellastanoside A (**819**) and B (**820**), as aggregation pheromones to mark its harboring sites (*629*).

3.6. Marine Nonterpenes – C_{15} Acetogenins

An astoundingly large number of halogenated marine nonterpenoid metabolites – the C_{15} acetogenins – are known, mainly from red algae of the genus *Laurencia* (*344*, *630*). As will be seen, there is good evidence to support the notion (*344*) that these compounds arise from the metabolism of fatty acids.

The first such compound to be identified was the oxocin laurencin (**821**) from *Laurencia glandulifera*, the structure of which, including the absolute stereochemistry, was established as that shown (*631–634*). A biomimetic synthesis of laurencin (**821**) from *trans*-laurediol using bromide, hydrogen peroxide, and lactoperoxidase has been described (*635*, *641*). The initial product in this bromonium ion-induced cyclization is deacetyllaurencin (**822**) which is found in *L. nipponica*, along with (*3E*, *6R*, *7R*)-laurediol and other laurediols (*636*). This latter species of *Laurencia* is a rich source of halogenated nonterpenes and their precursors. Thus, laureatin (**823**) and isolaureatin (**824**) are novel bicyclic oxocane analogs of laurencin (**821**), but of opposite chirality (*637–640*). The Lewis acid-catalyzed rearrangement of laureatin (**823**) to

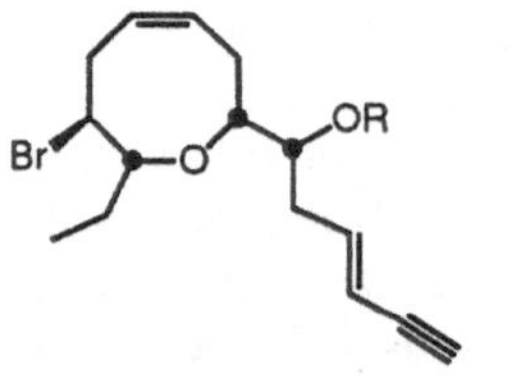

821 R = Ac (laurencin)
822 R = H (deacetyllaurencin)

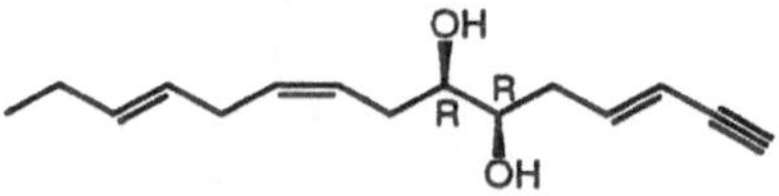

(3*E*,6*R*,7*R*)-laurediol

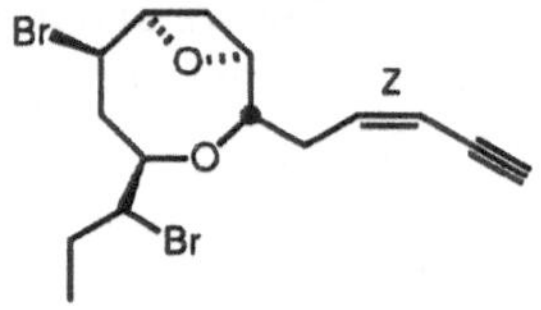

823 (laureatin)
823a ((*E*)-laureatin)

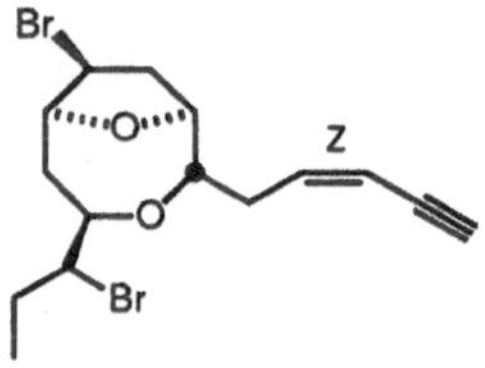

824 (isolaureatin)
824a ((*E*)-isolaureatin)

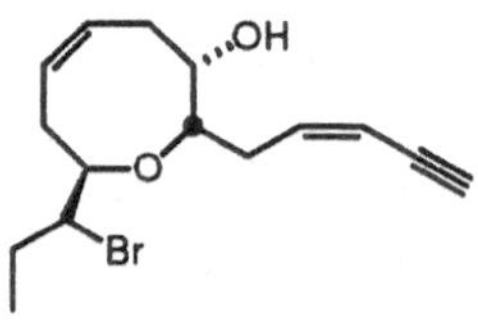

825 (prelaureatin)

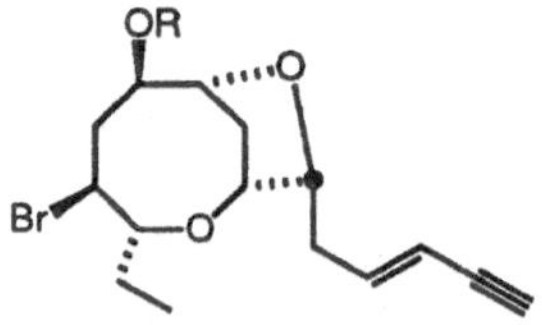

826 R = H (laurefucin)
827 R = Ac (acetyllaurefucin)

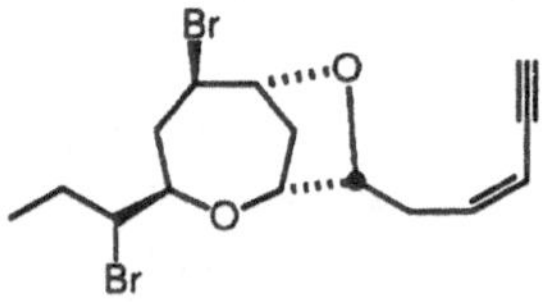

828 (isoprelaurefucin)

isolaureatin (**824**) is known and both compounds are naturally found in the alga (*642*). More recently, prelaureatin (**825**), a suggested biogenetic intermediate, has been isolated from *L. nipponica* (*643*). Indeed, the naturally occurring (*3Z, 6S, 7S*)-laurediol (*636*) has been enzymatically converted (lactoperoxidase, bromide, hydrogen peroxide) into prelaureatin (**825**) and laureatin (**823**) (*644*). The geometric isomers (*E*)-laureatin (**823a**) and (*E*)-isolaureatin (**824a**) have also been found in *L. nipponica* (*636, 667*). Further studies of *L. nipponica* have led to the characterization of laurefucin (**826**) and acetyllaurefucin (**827**) (*645*) (structures revised in (*646*)), and isoprelaurefucin (**828**) (*647, 666*). Laurencin, laureatin, isolaureatin, and laurefucin all inhibit the metabolism of pentobarbitone in mice (*648*), and laureatin and isolaureatin show insecticidal activity against mosquito larvae (*397*).

The first example of a bromoallene from marine algae was laurallene (**829**) from *L. nipponica*, a structure established in part by chemical correlation with laureatin (**823**) (*649*). One can imagine cyclization of prelaureatin (**825**) to laurallene (**829**) via *in vivo* bromination of the

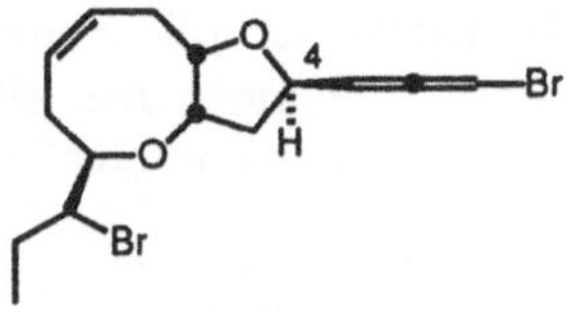

829 (laurallene)
830 (epilaurallene) (at C-4)

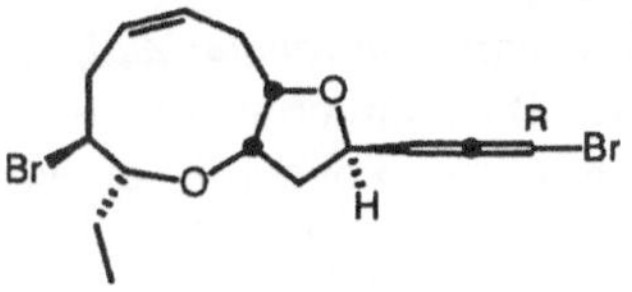

831 (isolaurallene)

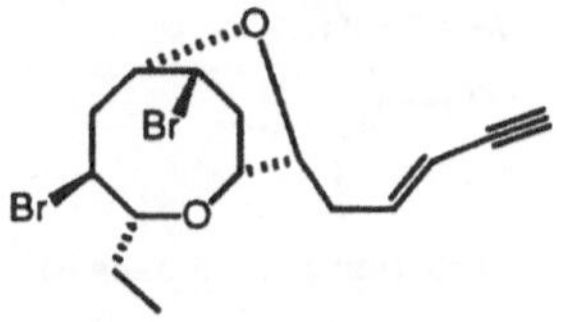

832 (laureoxanyne)

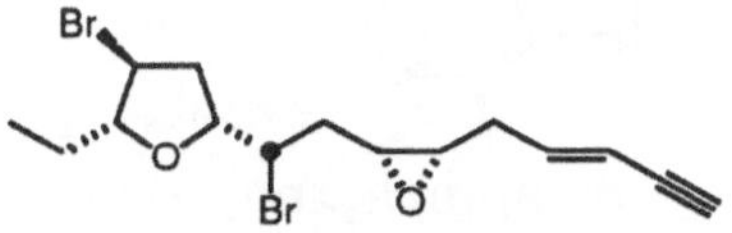

833 (laureepoxide)

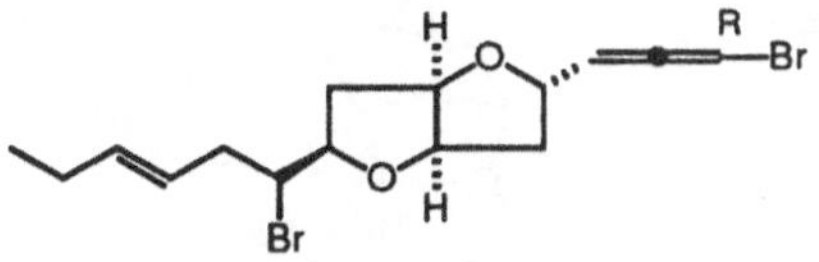

834 (kumausallene)

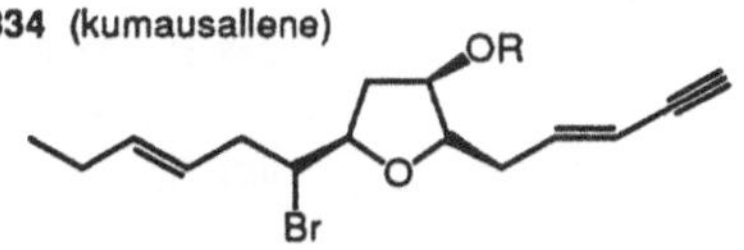

835 R = Ac (*trans*-kumausyne)
837 R = H (*trans*-deacetylkumausyne)

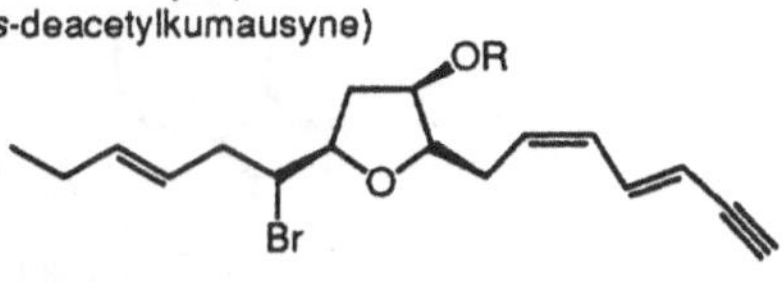

836 R = Ac (*cis*-kumausyne)
838 R = H (*cis*-deacetylkumausyne)

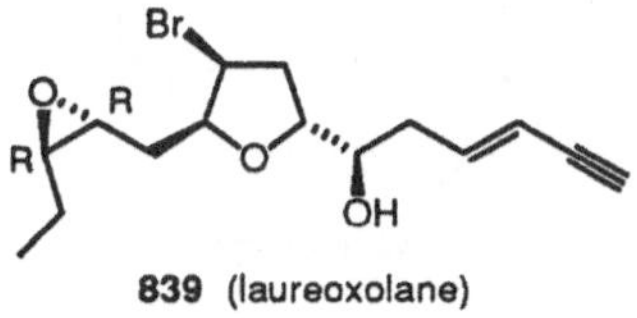

839 (laureoxolane)

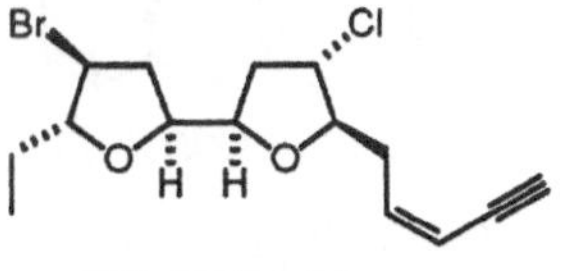

840 (notoryne)

alkyne. Subsequent work with *L. nipponica* has afforded epilaurallene ((*S*)-C4) (**830**) (*650*)), isolaurallene (**831**), a novel oxonane (*651,652*), and laureoxanyne (**832**) (*653*). This Japanese red alga has also provided several bromotetrahydrofurans and related compounds including laureepoxide (**833**) (*654*), kumausallene (**834**) (*655*), *trans*- (**835**) and *cis*-kumausyne (**836**), *trans*- (**837**) and *cis*-deacetylkumausyne (**838**) (*656*), laureoxolane (**839**) (*657*), and the novel *bis*-THF derivative notoryne (**840**) (*658*).

Hawaiian *Laurencia nidifica* has furnished several related compounds, the maneonenes (**841**–**844**) and isomaneonenes (**845, 846**) (*659–661*), the biogenesis of which is certain to be interesting (*344*).

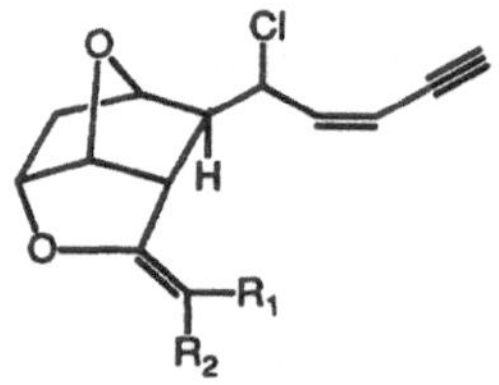

841 R_1 = Br, R_2 = Et (*cis*-maneonene A)
842 R_1 = Et, R_2 = Br (*cis*-maneonene B)

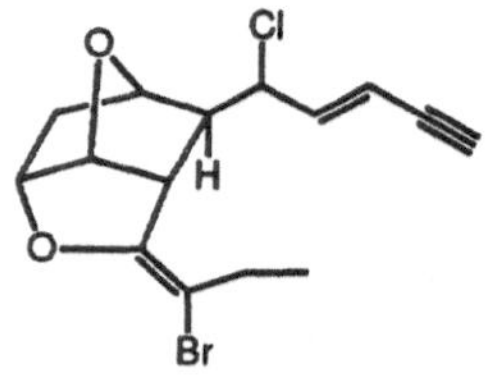

843 (*trans*-maneonene B)

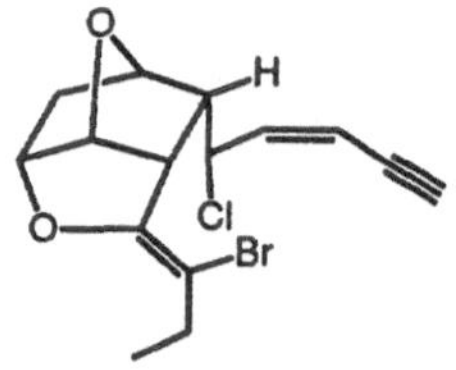

844 (*cis*-maneonene C)

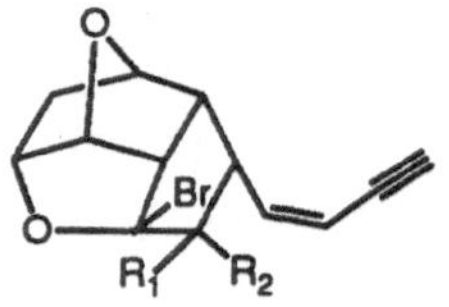

845 R_1 = Et, R_2 = Br (isomaneonene A)
846 R_1 = Br, R_2 = Et (isomaneonene B)

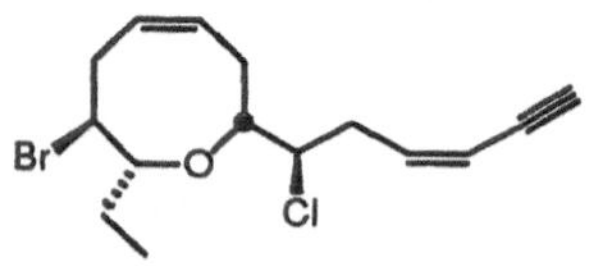

847 (intricenyne)

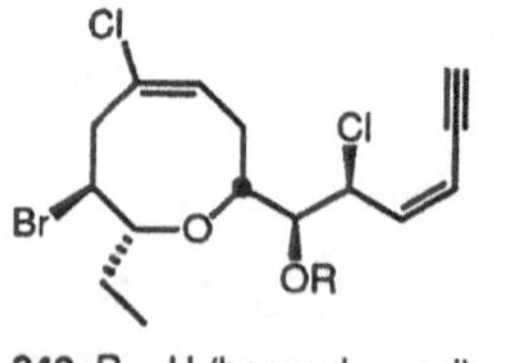

848 R = H (bermudenynol)
849 R = Ac

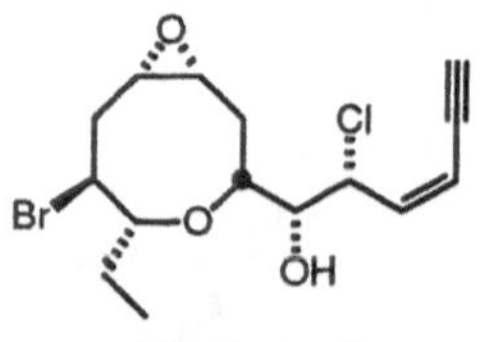

850 (poiteol)

Although its stereochemistry was not established, intricenyne (**847**) was extracted from *L. intricata* and is clearly related to laurencin (**821**) (*662*). This red alga also contains bermudenynol (**848**) and its acetate (**849**) (*663*), and another oxocane chlorohydrin, poiteol (**850**), is found in *L. poitei* (*664*).

Laurencia obtusa has been extensively studied and has furnished a rich set of halogenated nonterpenes. Obtusenyne (**851**) having the novel oxonane structure was simultaneously isolated by two groups from *L. obtusa* collected in different locations (*664, 665*). Whereas *L. obtusa* from Spain yielded obtusin (**852**) (*668*), a collection off Sicily afforded the dichloro laurencienyne (**853**) (*669–671*), and this alga in the Aegean Sea gave the related laurenyne (**854**) (*672*). The novel bicyclic obtusallene I

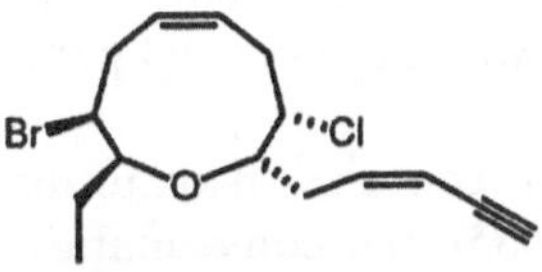

851 (obtusenyne)

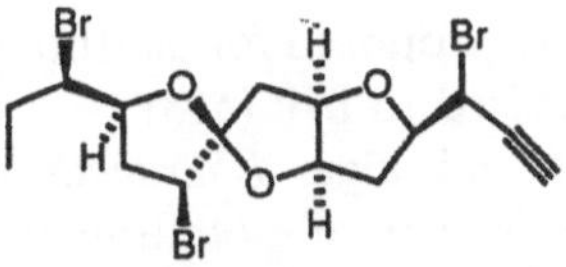

852 (obtusin)

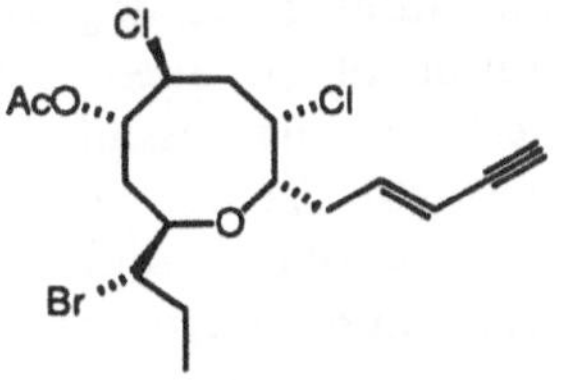

853 (laurencienyne)

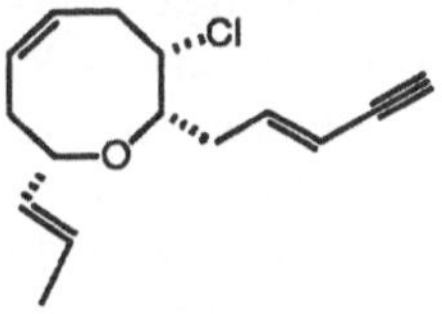

854 (laurenyne)

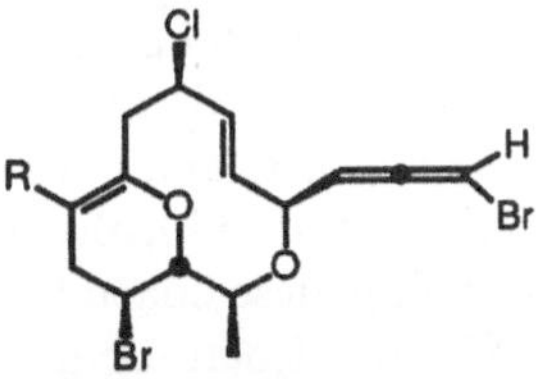

855 R = H (obtusallene I)
856 R = Br

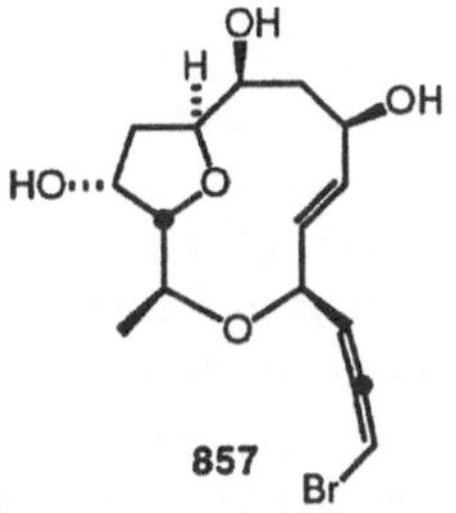

857

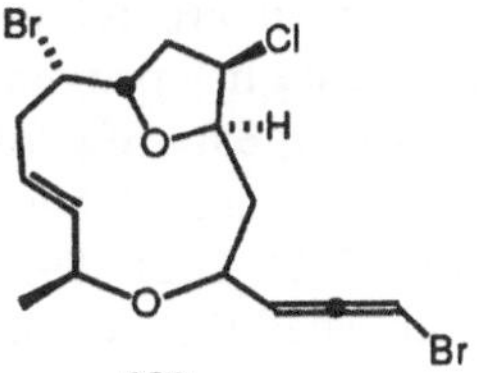

858

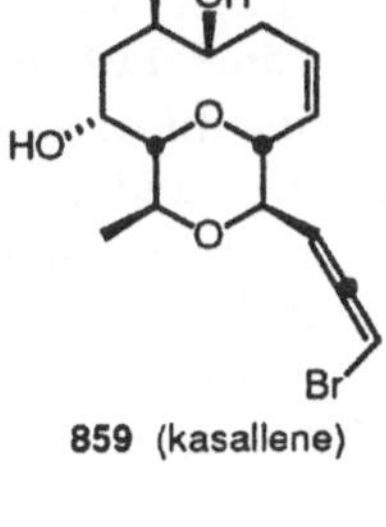

859 (kasallene)

(**855**) was isolated from *L. obtusa* in the Mediterranean Sea (*673, 674*), and this was followed by characterization of several related bicyclic ether bromoallenes also from *L. obtusa*, 10-bromoobtusallene (**856**) (*675*), **857**, **858** (*676*), and kasallene (**859**) (*677*), the latter of which is the first dioxane type.

A collection of *L. obtusa* from the Canary Islands has identified the two new tetrahydrofurans graciosin (**860**) (*678*) (revised in 679), **861** (*678*), and graciosallene (**862**) (*679*). Further study revealed the presence of **863**–**865** in this alga (*680*). Sicilian *L. obtusa* from a different location has yielded neoobtusin (**866**) (*681*). The rhodophytin derivative epoxy-*trans*-isodihydrorhodophytin (**867**) has also been found in *L. obtusa* (*682*). Two unidentified *Laurencia* species from Mexico contain the related epoxyrhodophytin (**868**), *trans*- (**869**) and *cis*-rhodophytin (**870**), *trans*- (**871**) and *cis*-chondrin (**872**), and *trans*-chondriol (**873**) (*683*). The original structure proposed for *cis*-rhodophytin was a cyclic vinyl peroxide (*684*) later revised to **870** (*683*).

The red alga *Laurencia okamurai* contains neolaurallene (**874**), a stereoisomer of isolaurallene (**831**) (*685*); the conformations of both metabolites have been evaluated by force-field calculations as well as by X-ray crystallography (*652, 685*). The kumausallene-analog **875** is also found in this alga (*686*). The configuration of the allene center was assigned using Lowe's Rule (*687*). Interestingly, the possible biogenetic precursors to laurencin (**821**), laurepinnacin (**881**), isolaurepinnacin (**882**), *cis*-maneonene A (**841**) and others, via the appropriate diols, have been found in *L. okamurai*. Thus, *cis*- and *trans*-laurencenyne, and *cis*- and *trans*-neolaurencenyne were isolated in these studies (*688, 689*). From a collection of *L. okamurai*, which is now believed to be *L. intricata*, there were isolated several novel brominated C_{15} metabolites including okamurallene (**876**) (*690*) (revised in *692, 693*), deoxyokamurallene (**877**) and isookamurallene (**878**) (*691*) (revised in *692, 693*), and **879** (*692*). Also found in this alga was **880**, a novel laurallene-type chlorohydrin (*692*).

A version of *Laurencia pinnata* found in Japanese coastal waters has been found to contain laurepinnacin (**881**) and isolaurepinnacin (**882**) (*694*). *L. pinnatifida* produces *cis*- (**883**) and *trans*-pinnatifidienynes (**884**) as well as a series of biogenetically significant chlorohydrins **885**–**891** (*695, 696*). The two oxocanes **892** and **893** which are related to obtusenyne (**851**) were also identified (*696*). *L. venusta* has provided venustin A (**894**) and B (**895**) (*697*) and, in a later study, **896**–**898** (*698*).

A study of *Laurencia thyrsifera* from New Zealand has afforded several chlorinated laurencins (**899**–**903**) (*699, 700*). Diols **899** and **900** were isolated as acetates. This study also established the stereochemistry of intricenyne (**847**) (*700*). Western Australian *L. filiformis* contains *cis*-

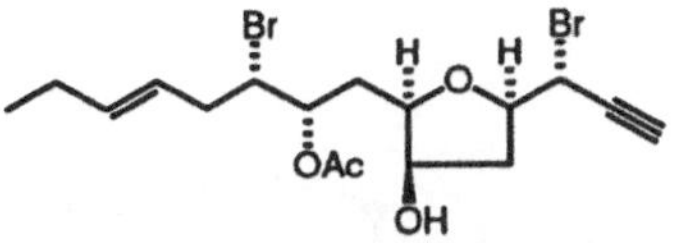

Br H H Br OAc OH
860 (graciosin)

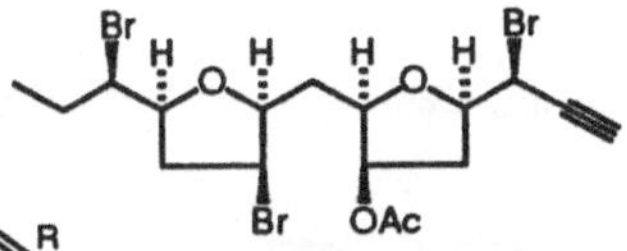

Br H H H H Br Br OAc
861

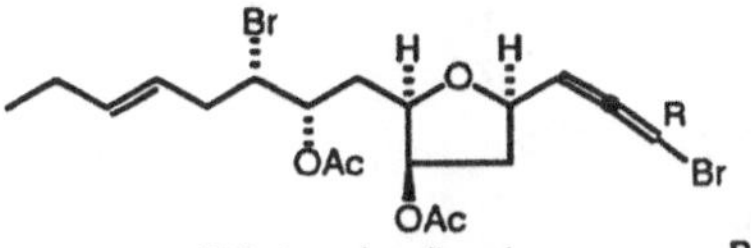

Br H H R OAc OAc Br
862 (graciosallene)

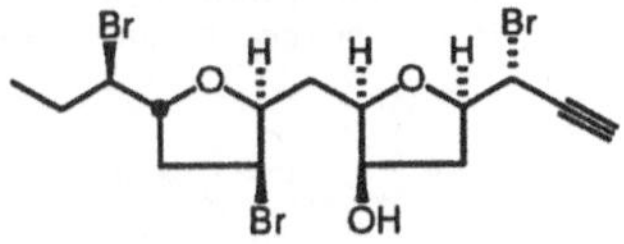

Br H H H H Br Br OH
863

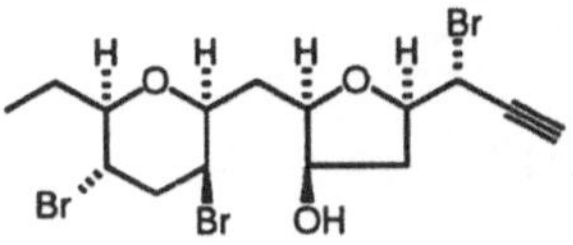

H H H H Br Br Br OH
864

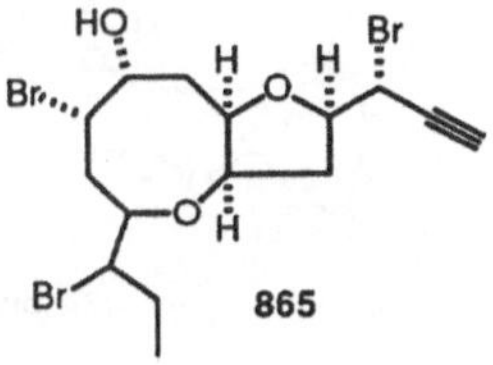

HO H H Br Br O H Br
865

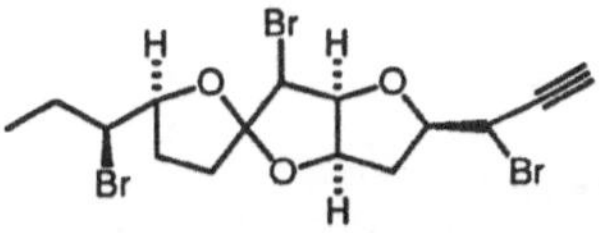

H Br H O O O Br Br H
866 (neoobtusin)

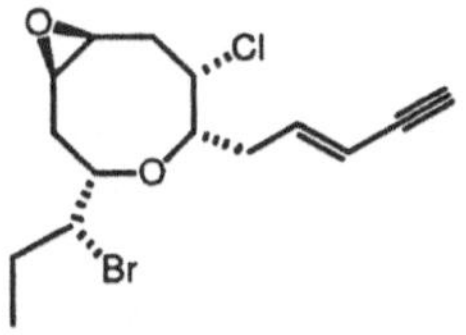

O Cl O Br
867

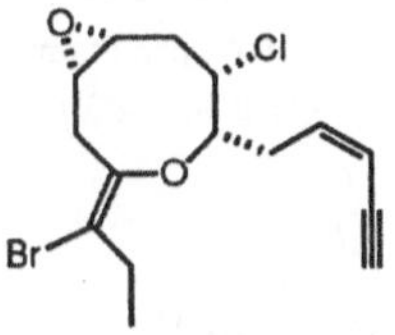

O Cl O Br
868 (epoxyrhodophytin)

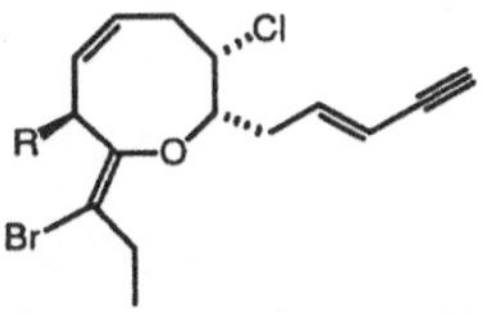

Cl R O Br
869 R = H (*trans*-rhodophytin)
873 R = OH (*trans*-chondriol)

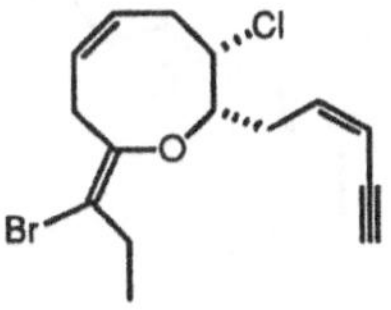

Cl O Br
870 (*cis*-rhodophytin)

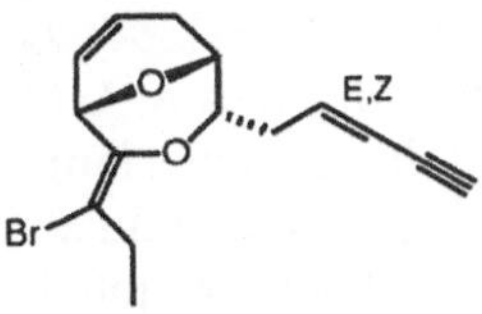

O O E,Z Br
871 *E* (*trans*-chondrin)
872 *Z* (*cis*-chondrin)

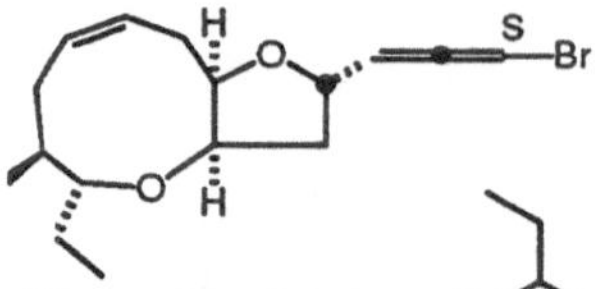

874 (neolaurallene)

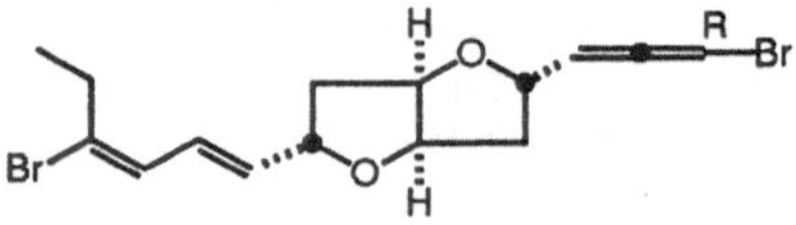

875

R = Et (*cis*-laurencenyne)
R = *n*-Pen (*cis*-neolaurencenyne)

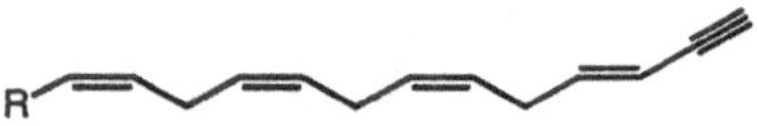

R = Et (*trans*-laurencenyne)
R = *n*-Pen (*trans*-neolaurencenyne)

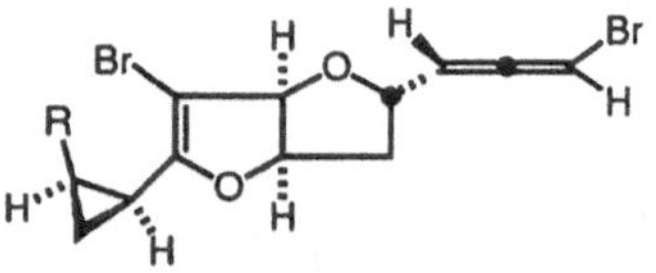

876 R = (epoxide)–Me (okamurallene)

877 R = CHCH$\overset{Z}{=}$CHMe (deoxyokamurallene)

878 R = CH_2Ac (isookamurallene)

879 R = –CH(Cl)CH(OH)Me

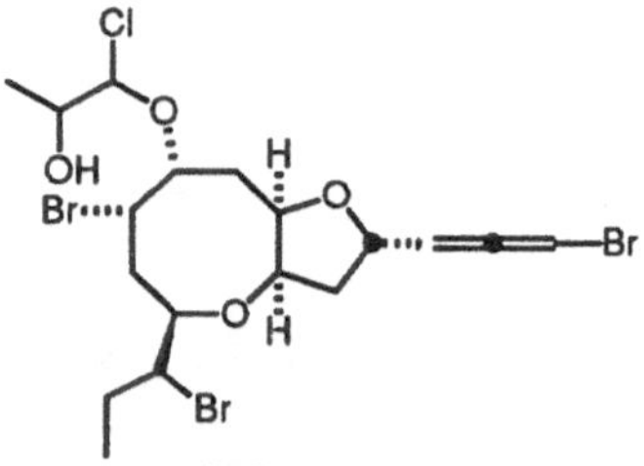

880

dihydrorhodophytin (**904**) and *cis-epi*-dihydrorhodophytin (**905**) (*446*), while *L. snyderae* produces chlorofucin (**906**) (*664*). *Laurencia microcladia* possesses the novel bicyclic microcladallenes A–C (**907–909**) (*701*), while a collection from Italy has yielded the branched lauroxepane rogiolenyne A (**910**) (*702*). A later study of this alga determined the absolute configuration of rogiolenyne A (**910**) and also resulted in the isolation of rogiolenyne B (**911**) and D (**912**) (*436, 703*). The same alga also produces rogioloxepane A (**913**), B (**914**) and C (**915**) (absolute configurations) as

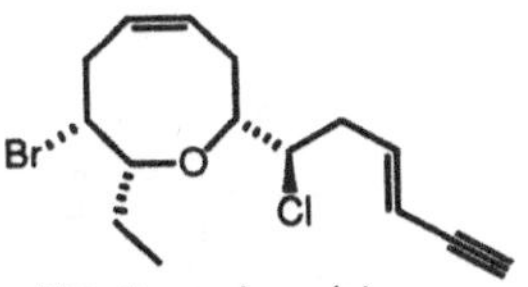

881 (laurepinnacin)

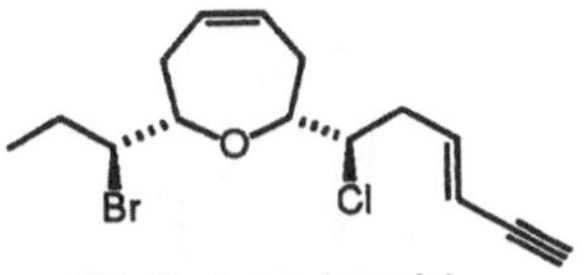

882 (isolaurepinnacin)

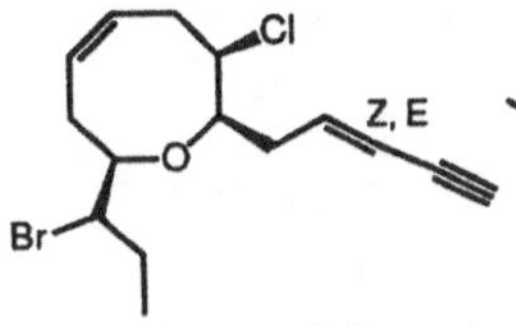

883 *Z* (*cis*-pinnatifidienyne)
884 *E* (*trans*-pinnatifidienyne)

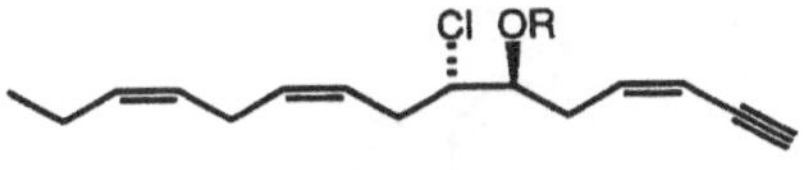

885 R = H
886 R = Ac

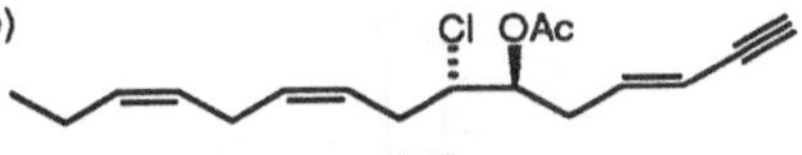

887

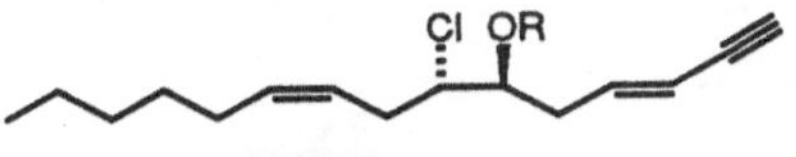

888 R = H
889 R = Ac

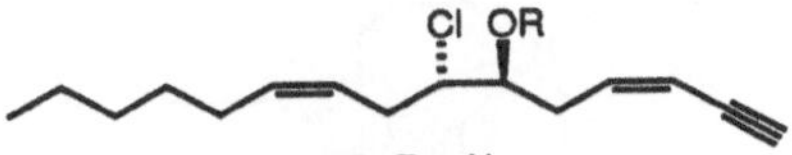

890 R = H
891 R = Ac

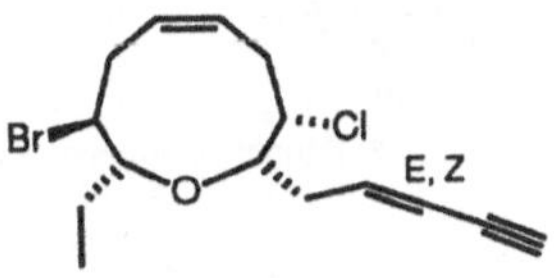

892 *E* isomer
893 *Z* isomer

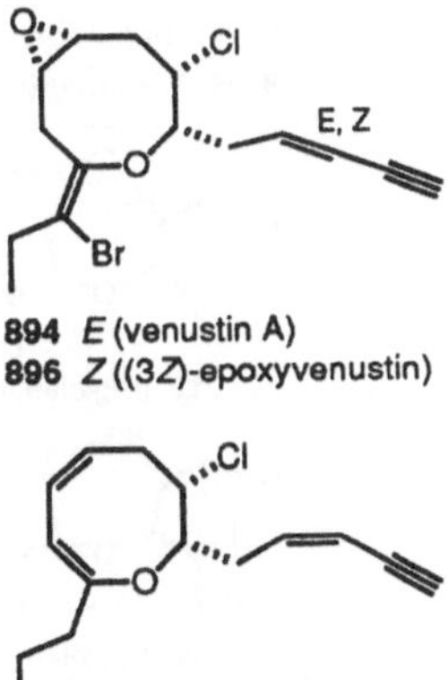

894 *E* (venustin A)
896 *Z* ((3*Z*)-epoxyvenustin)

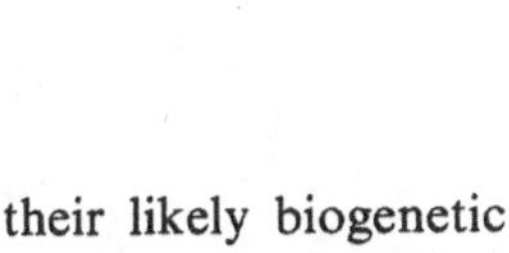

895 *E* (venustin B)
897 *Z* ((3*Z*)-venustin)

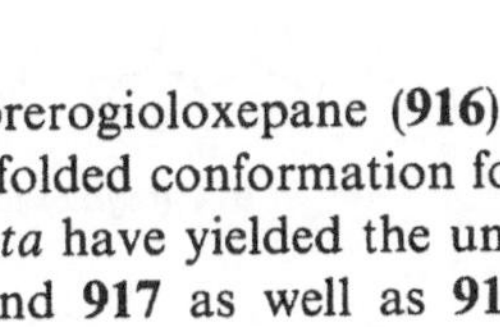

898 ((3*Z*)-venustinene)

well as their likely biogenetic precursor prerogioloxepane (**916**) (*704*). Molecular mechanics calculations reveal a folded conformation for **916**.

Two examinations of *Laurencia implicata* have yielded the unprecedented 1-oxacyclodecane ring in compound **917** as well as **918–922** (*705, 706*) (**920** revised in (*706*)). The known neolaurallene (**874**) was also

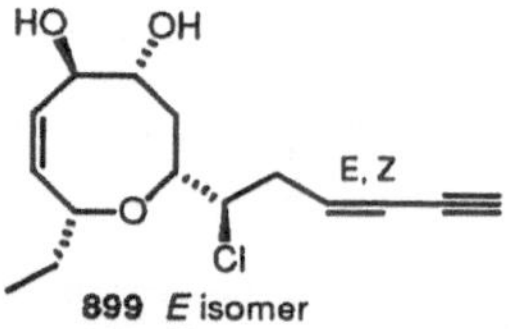

899 *E* isomer
900 *Z* isomer

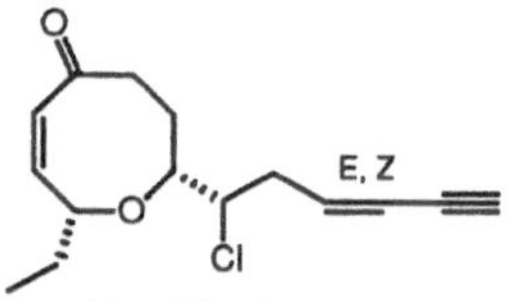

901 *E* isomer
902 *Z* isomer

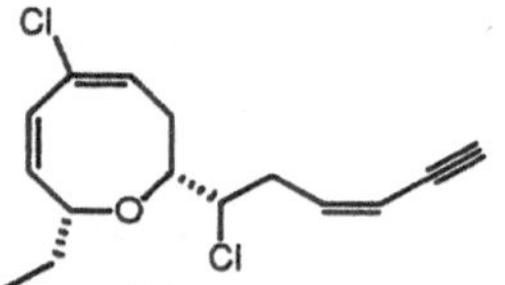

903

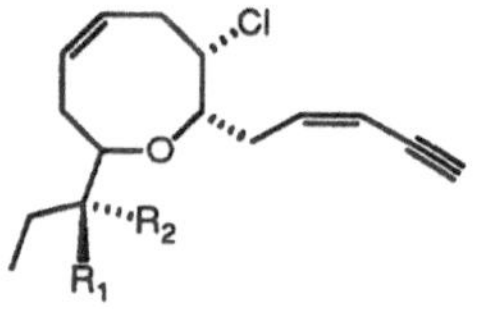

904 R_1 = Br, R_2 = H (*cis*-dihydrorhodophytin)
905 R_1 = H, R_2 = Br (*cis-epi*-dihydrorhodophytin)

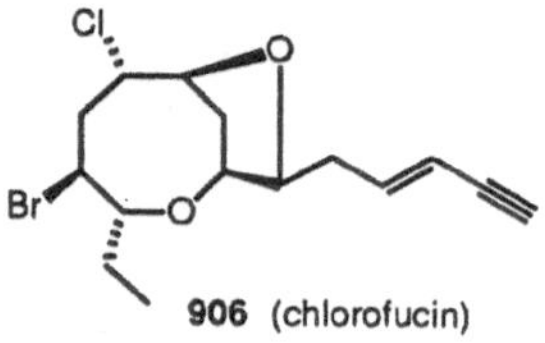

906 (chlorofucin)

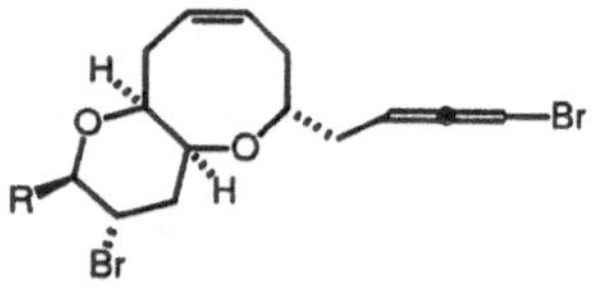

907 R = Et (microcladallene A)
909 R = $CH{=}CH_2$ (microcladallene C)

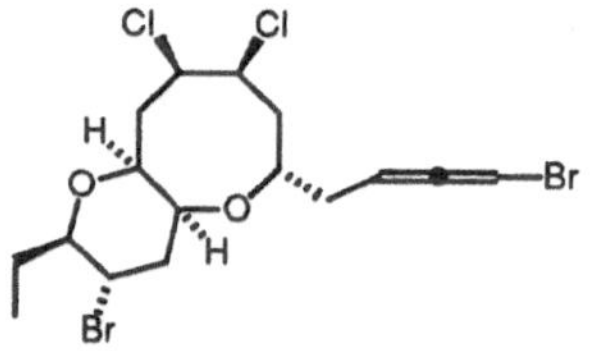

908 (microcladallene B)

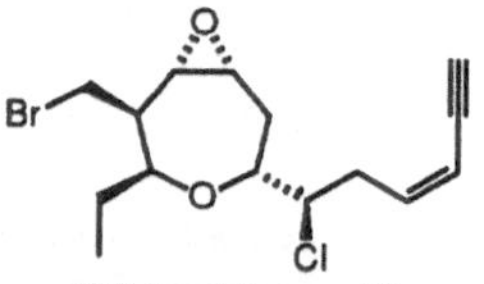

910 (rogiolenyne A)

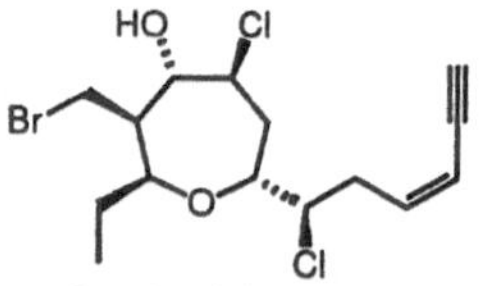

911 (rogiolenyne B)

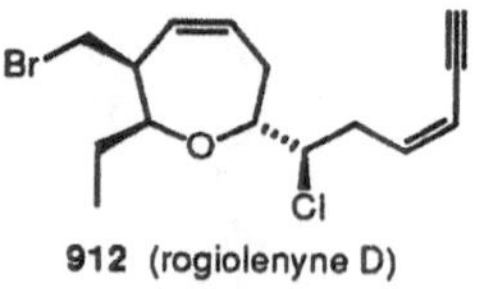

912 (rogiolenyne D)

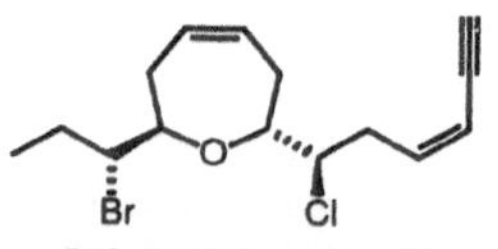

913 (rogioloxepane A)

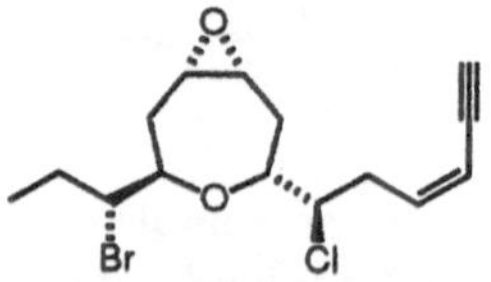

914 (rogioloxepane B)

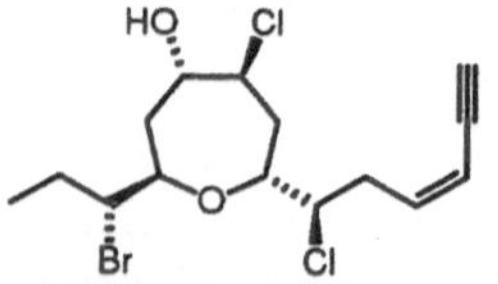

915 (rogioloxepane C)

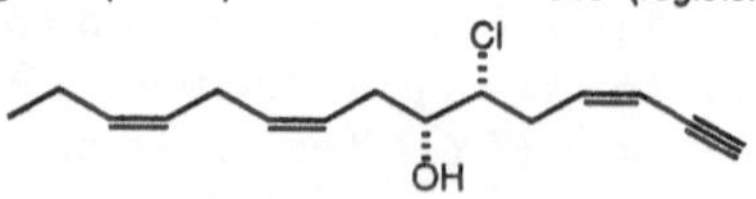

916 (prerogioloxepane)

isolated. Compounds **920** and **921** are closely related to laurallene (**829**). From the Hawaiian *L. majuscula*, the first *bis*-tetrahydrofuran metabolite was isolated, **923** (*707*) which is the dibromo analog of notoryne (**840**) although only the partial relative stereochemistry as shown is known. New Zealand *L. gracilis* contains the four new metabolites **924**–**927** (*346*), which are all related to laurencienyne (**853**) or laurefucin (**826**), but only **924** contains chlorine. Also isolated from this alga were laurefucin (**826**) and deacetyllaurencin (**822**). *L. subopposita* also contains laurefucin

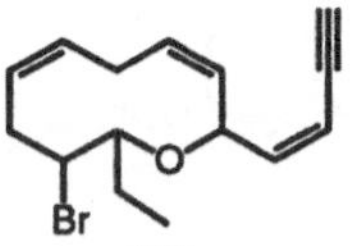

917 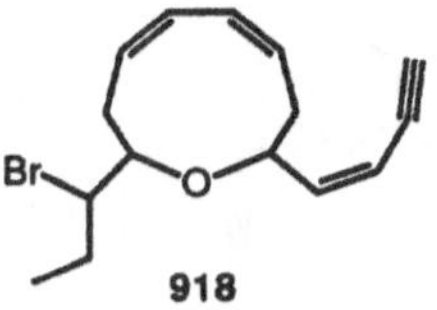 **918**

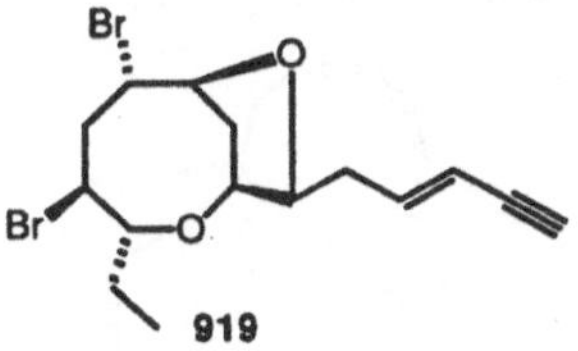

919

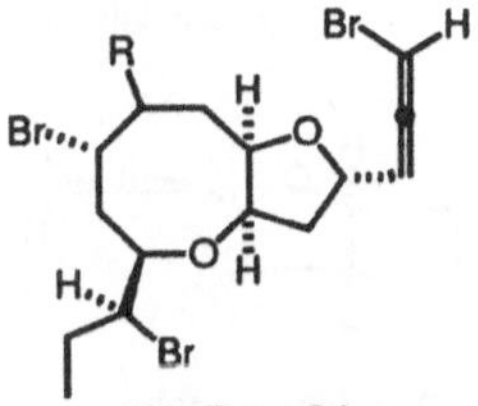

920 R = αOAc
921 R = βCl

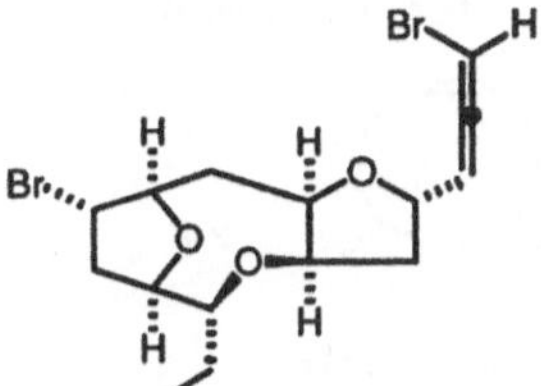

922 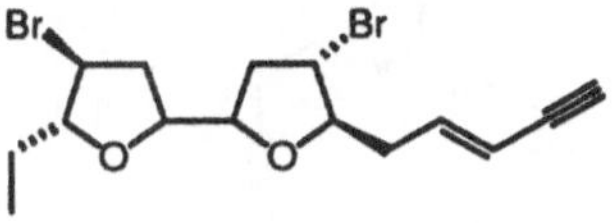 **923**

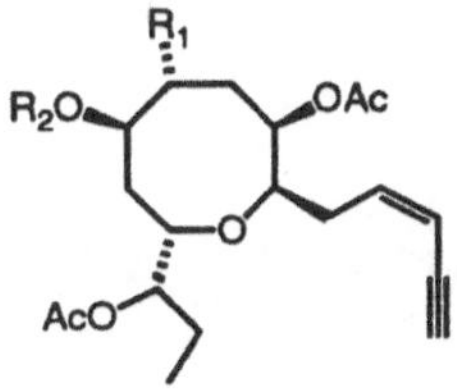

924 R_1 = Cl, R_2 = Ac
925 R_1 = OAc, R_2 = H
926 R_1 = OH, R_2 = Ac

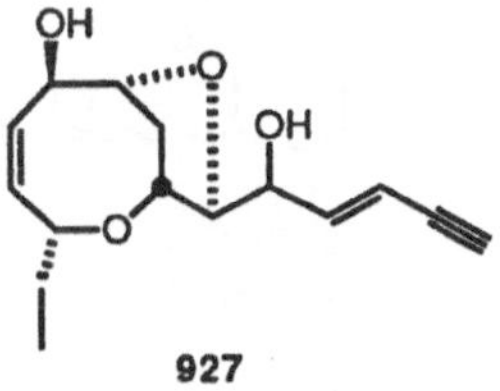

927

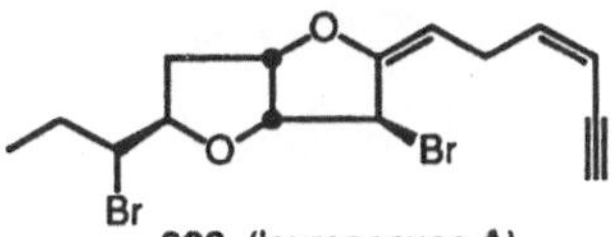

928 (laurenenyne A) 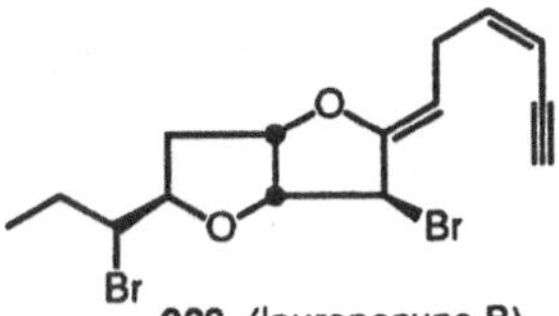 **929** (laurenenyne B)

(**826**) as well as isoprelaurefucin (**828**) (both *E* and *Z*) (*454*). An undescribed *Laurencia* species related to *L. tristicha* has furnished laurenenyne-A (**928**) and -B (**929**) (*708*). These unstable compounds interconvert on standing and slowly decompose to the corresponding furans, even at −18°C.

The red alga *Chondria oppositiclada* which may actually be a *Laurencia* sp. (*400*) produces chondriol (**930**) (*709*) (structure revised in (*710*)), which, in fact, was one of the first C_{15} nonterpenes to be isolated.

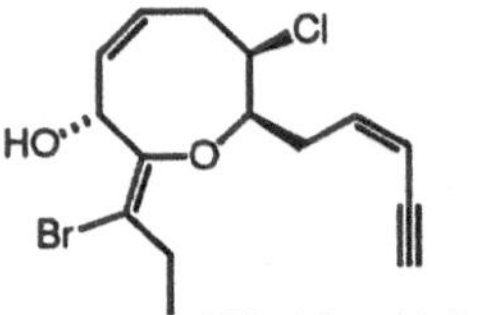

930 (chondriol)

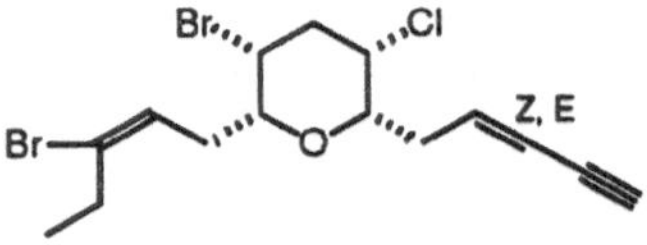

931 *Z* isomer (dactylyne)
932 *E* isomer (isodactylyne)

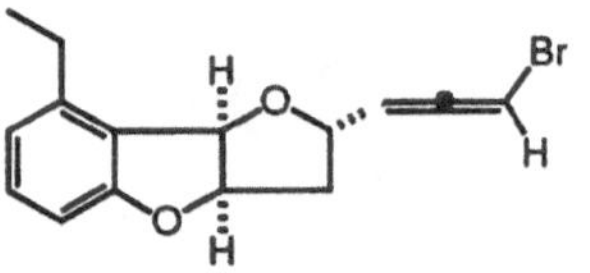

933 (panacene)

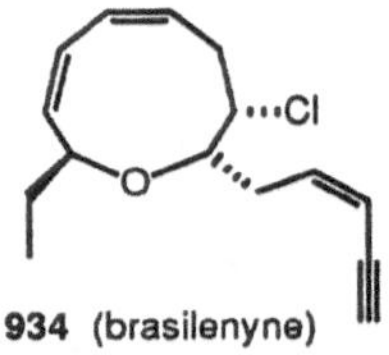

934 (brasilenyne)

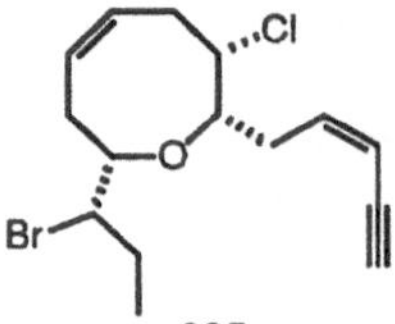

935

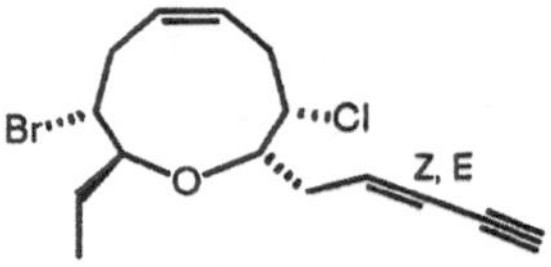

936 *Z* isomer
937 *E* isomer

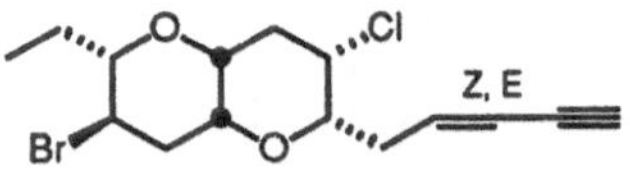

938 *Z* isomer
939 *E* isomer

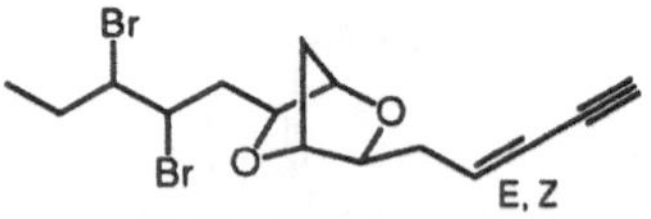

940 (*E*)-ocellenyne
941 (*Z*)-ocellenyne

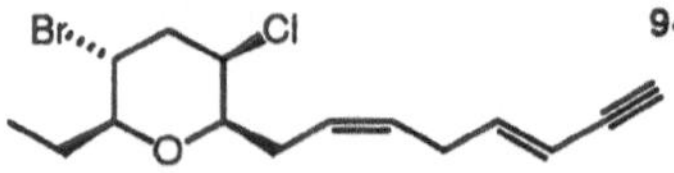

942 (srilankenyne)

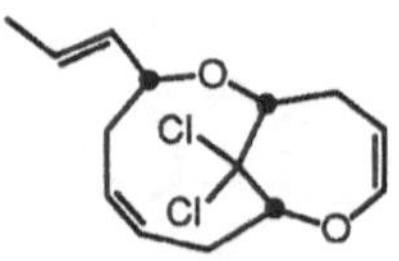

943

Sea hares which feed extensively on *Laurencia* and other algae contain a variety of new or transformed nonterpenes. The Caribbean *Aplysia dactylomela* contains the novel tetrahydropyrans dactylyne (**931**) (*711*) and isodactylyne (**932**) (*712*). Dactylyne inhibits the metabolism of phenobarbital in mice (*713*). Panacene (**933**) from Floridian *A. brasiliana* is a potent antifeedant and this sea hare is rejected by sharks (*714*). Texan *A. brasiliana* contains the alga-derived brasilenyne (**934**), **935** and *cis*-dihydrorhodophytin (**904**) (*715*) which appear to serve as defensive substances (*716*). A more recent study of *A. dactylomela* has yielded the obtusenyne analogs **936** and **937** as well as **938** and **939** (*464*). The Hawaiian sea hare *A. oculifera* contains the novel bicyclic (*E*)- (**940**) and (*Z*)-ocellenyne (**941**) (*717*), and the Sri Lankan animal contains srilankenyne (**942**) (*718*). Italian *A. punctata* has been found to produce the novel bicyclic ether **943** (*570*). Interestingly, the sea hare *Dolabella auricularia* contains two oxocanes, doliculols A and B which are the first nonhalogenated C_{15} acetogenins (*719*) (not shown).

The two tetrahydropyrans **944** and **945** have been extracted from a *Haliclona* sp. sponge (*720*), while the Guam "bubble shell," *Haminoea cymbalum*, releases the potent feeding deterrent kumepaloxane (**946**)

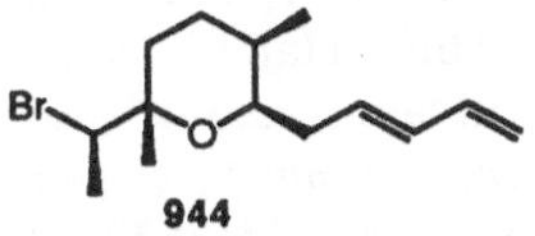

944

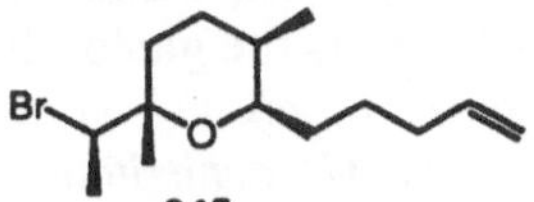

945

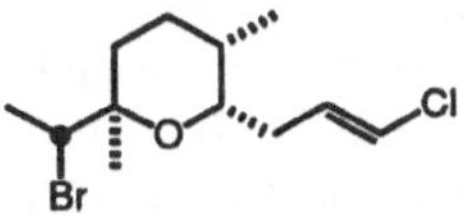

946 (kumepaloxane)

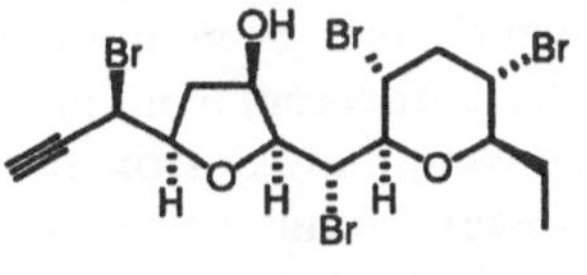

947

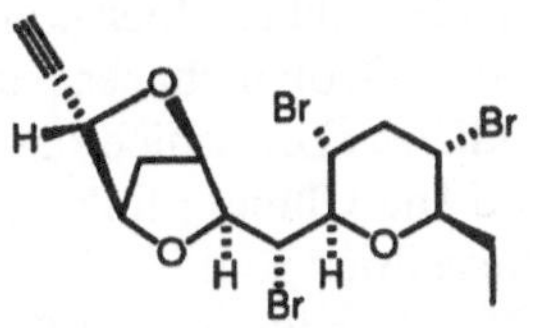

948

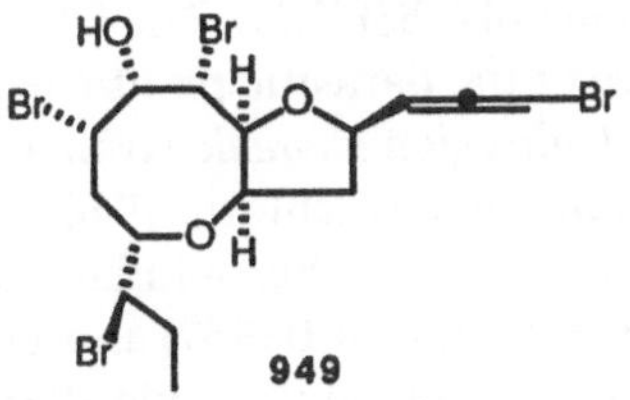

949

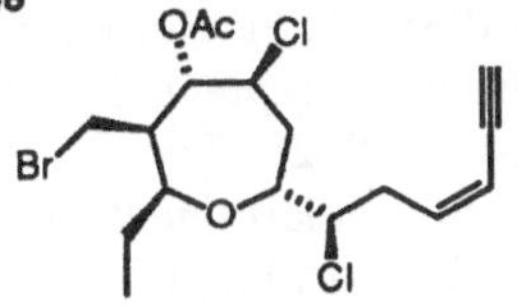

950 (rogiolenyne C)

when this mollusc is disturbed (*721*). The encrusting sponge *Mycale rotalis* produces the structurally interesting **947–949** (*722,723*). Compound **947** is transformed easily into **948** with base, suggesting a possible biogenesis. The sponge *Spongia zimocca* which feeds on the red alga *Laurencia microcladia* contains rogiolenyne C (**950**) in addition to rogiolenyne B (**911**) (*436,702*). The authors suggest that rogiolenyne A (**910**) is transformed into rogiolenyne B (**911**) and then acetylated to rogiolenyne C (**950**) by the sponge (*436,702*).

3.7. Iridoids

A small group of chlorinated plant metabolites comprises chlorinated iridoids which are not terpenoid in the usual sense of the word, but, rather, are mevalonate-derived in origin and isoprenoid in carbon skeleton (*724*). Some iridoids are derived from a set of defensive secretions, such as iridodial, produced by *Iridomyrmex* species, a genus of ants. Many iridoids are plant products such as loganin which plays a key role in the biosynthesis of indole and isoquinoline alkaloids.

The first example of a chloroiridoid was linarioside (**951**) from the Japanese folk medicine *Linaria japonica* (Scrophulariaceae) and is present in the plant as the glucoside (*725,726*). This metabolite is also found in *Cymbalaria muralis* (*727*) of the same plant family and in several other Linaria species (*L. genistifolia, L. dalmatica, L. simplex, L. pelisseriana, L. vulgaris*, and *L. purpurea*) (*728,729*). Thunbergioside (**952**), the structure of which had been misassigned earlier, is found in *Thunbergia fragrans* (Acanthaceae) (*730*) and *Retzia capensis* (Retziaceae) (*731*), the latter also containing 5-deoxythunbergioside (**953**) (*731*). *Asystasia bella* (Acanthaceae) produces asystasioside E (**954**) which is the chlorohydrin of catapol (*732*), and *Mentzelia decapetala* (Loasaceae) has yielded 7-chlorodeutziol (**955**) (*733*). The Chinese drug Cistanchis which consists of a mixture of the parasitic plant *Cistanche salsa* (Orobanchaceae) and the roots of *Haloxylon ammadendron* (Chenopodiaceae) on which the former grows contains cistachlorin (**956**) (*734*), and the Chinese folk medicine *Rehmannia glutinosa* (Scrophulariaceae) whose dried root is still used has afforded rehmaglutin B (**957**) and D (**958**) (*735,736*), jioglutin A (**959**) and B (**960**) (*737*), and glutinoside (**961**) (*738*). Both *Eustoma russellianum*, a Japanese garden flower, and *Gentiana septemfida*, both in Gentianaceae produce the secoiridoid eustoside (**962**) (*739, 740*). Two chlorinated valepotriates, valechlorine (**963**) and valeridine (**964**, structure unknown but closely related to **963**), have been isolated from *Valeriana officinalis* (Valerinaceae) (*741*).

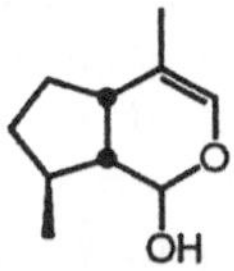

iridodial

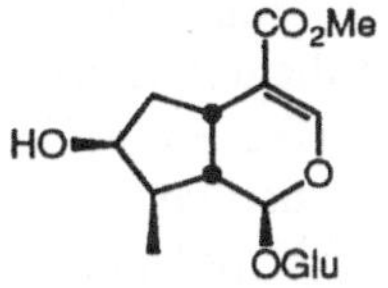

loganin

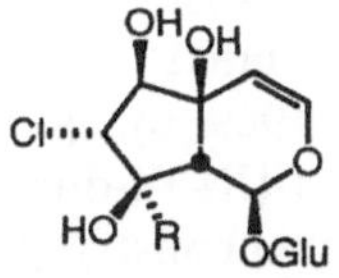

951 R = Me (linarioside)
952 R = H (thunbergioside)

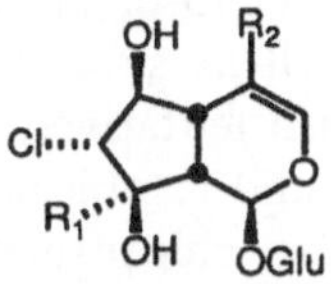

953 $R_1 = R_2 = H$
(5-deoxythunbergioside)
954 $R_1 = CH_2OH$, $R_2 = H$
(asystasioside E)
955 $R_1 = H$, $R_2 = CH_3$
(7-chlorodeutziol)

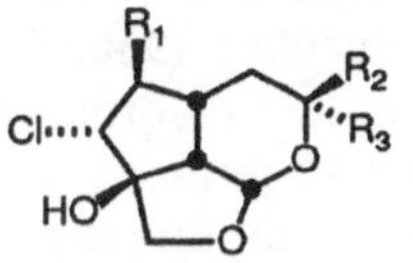

956 $R_1 = R_2 = R_3 = H$
(cistachlorin)
957 $R_1 = R_2 = OH$, $R_3 = H$
(rehmaglutin B)
958 $R_1 = OH$, $R_2 = R_3 = H$
(rehmaglutin D)
959 $R_1 = OH$, $R_2 = OMe$, $R_3 = H$
(jioglutin A)
960 $R_1 = OH$, $R_2 = H$, $R_3 = OMe$
(jioglutin B)

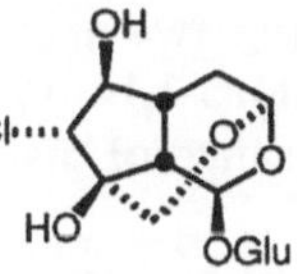

961 (glutinoside)

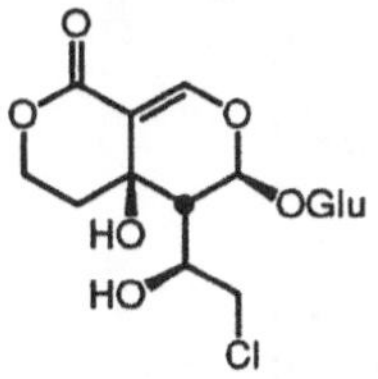

962 (eustoside)

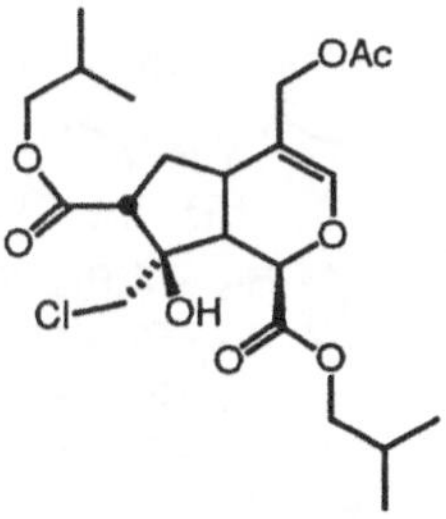

963 (valechlorine)

3.8. Lipids and Fatty Acids

Fatty acids are as essential to life as any other class of natural chemical and are just as ubiquitous. Unsaturated and saturated fatty acids also occur in ester combination with glycerol as fats and oils, with long-chain alcohols as waxes, with sugar derivatives as glycolipids, and with phosphorus-containing compounds as phospholipids and sphingolipids. Not surprisingly, unsaturated fatty acids which contain a reactive carbon-carbon double bond are susceptible to biohalogenation.

A set of chlorinated long-chain alkyl sulfates – "natural detergents" – was isolated from the phytoflagellate *Ochromonas danica* independently in 1969 by two groups (*742, 743*). HAINES *et al.* identified (*R*)-13-chloro-1-(*R*)-14-docosanediol disulfate (**965**) and determined the absolute stereochemistry (*742*). Several polychlorosulfolipids were also found to be present but not characterized. Simultaneously, ELOVSON and VAGELOS described **965** (no stereochemistry) and 11,15-dichloro-1,14-docosanediol disulfate (**966**) from *O. danica* as well as several others containing from 3–6 chlorine atoms (*743*). They also tentatively identified 14-chloro-1,15-tetracosanediol disulfate (**967**) and established that chloride is essential for the biosynthesis of these chlorosulfolipids. Subsequent studies revealed that 2,2,11,13,15,16-hexachloro-1,14-docosanediol disulfate (**968**) is the major chlorosulfolipid in *O. danica* (*744*) and that a number of such lipids are present in this alga of both the docosane-1,14-diol (**965, 966, 968–971**) and tetracosane-1,15-diol types (**967, 972, 973**) (*745–747*). In addition to these compounds, several other chlorosulfolipids have been isolated but not yet positively identified with respect to the position of the chlorines, including two trichloro-1,14-docosanediol disulfates (**974, 975**), two tetrachloro-1,14-docosanediol disulfates (**976, 977**), and a dichloro-1,15-tetracosanediol disulfate (**978**)

965 R_{13} = Cl, R_{2a} = R_{2b} = R_{11} = R_{15} = R_{16} = H
966 R_{11} = R_{15} = Cl, R_{2a} = R_{2b} = R_{13} = R_{16} - H
968 R_{2a} = R_{2b} = R_{11} = R_{13} = R_{15} = R_{16} = Cl
969 R_{2a} = R_{2b} = R_{11} = R_{13} = R_{15} = Cl, R_{16} = H
970 R_{2a} = R_{2b} = R_{11} = R_{13} = R_{16} = Cl, R_{15} = H
971 R_{2a} = R_{2b} = Cl, R_{11} = R_{13} = R_{15} = R_{16} = H

967 R_{14} = Cl, R_{2a} = R_{2b} = R_{12} = R_{16} = R_{17} = H
972 R_{2a} = R_{12} = R_{14} = R_{16} = R_{17} = Cl, R_{2b} = H
973 R_{2a} = R_{2b} = R_{12} = R_{14} = R_{16} = R_{17} = Cl

979 (malhamensilipin A)

(*745, 746*) (not shown). Biosynthetic studies indicate that these lipids are probably first sulfated and then chlorinated (*747, 748*). A complement of the same chlorosulfolipids was identified in *Tribonema aequale* (*749*), and a detailed examination of 30 algae species from a wide range of classes found that all but one of 22 fresh water species contained chlorosulfolipids with from one to six chlorines in the C_{22} series (**965, 966, 968–971, 974–977**), but none of eight marine species showed the presence of these compounds (*750*). In particular, relatively large quantities of these lipids were identified in *O. danica* and *O. malhamensis*. For example, 14% of the total lipids in *O. danica* are chlorinated. Other species containing these chlorosulfolipids are *Ulothrix subtilissima*, *Stigeoclonium tenue*, *Cosmarium botrytis*, *Monodus subterraneus*, *Tribonema aequale*, *Porphyridium aerugineum*, *Cyanidium caldarium*, *Chlorogleopsis fritschii*, *Cyanophora paradoxa*, *Nostoc* sp., *Aphanocapsa* sp., *Porphyridium cruentum*, *Botrydium granulatum*, *Zygnema* sp., and *Elakatothrix viridis* (*750*). The novel protein tyrosine kinase inhibitor malhamensilipin A (**979**), which contains the new 2-chlorovinylsulfate functionality,

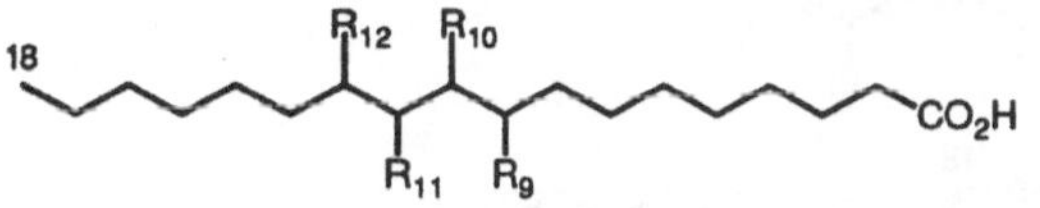

980 $R_9 = R_{10} = Cl$, $R_{11} = R_{12} = H$ (9,10-dichlorostearic acid)
981 $R_9 = Cl$, $R_{10} = OH$, $R_{11} = R_{12} = H$ (9-chloro-10-hydroxystearic acid)
982 $R_9 = OH$, $R_{10} = Cl$, $R_{11} = R_{12} = H$ (10-chloro-9-hydroxystearic acid)
983 $R_9 = R_{10} = H$, $R_{11} = Cl$, $R_{12} = OH$ (11-chloro-12-hydroxystearic acid)
984 $R_9 = R_{10} = H$, $R_{11} = OH$, $R_{12} = Cl$ (12-chloro-11-hydroxystearic acid)

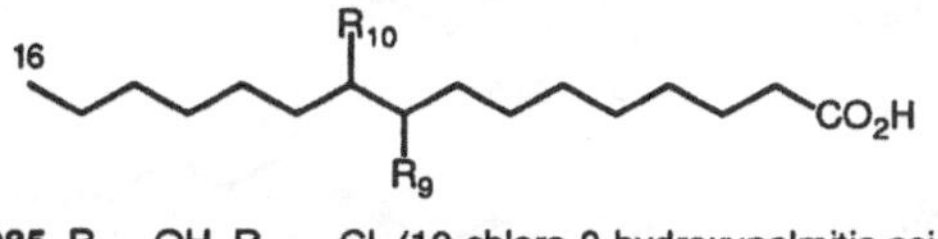

985 $R_9 = OH$, $R_{10} = Cl$ (10-chloro-9-hydroxypalmitic acid)
986 $R_9 = Cl$, $R_{10} = OH$ (9-chloro-10-hydroxypalmitic acid)

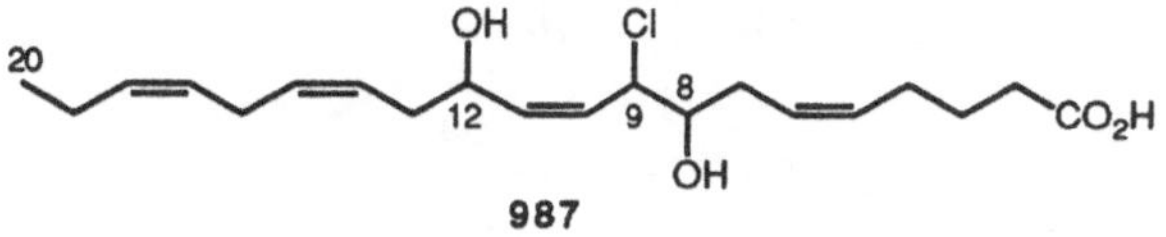

987

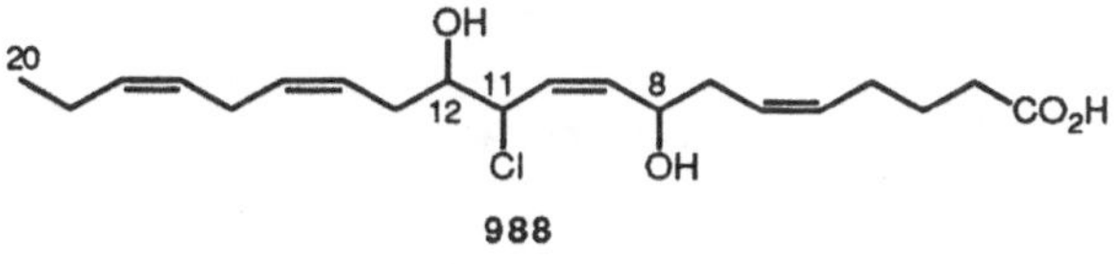

988

has been isolated from the cultured chrysophyte *Poterioochromonas malhamensis* (*751*).

Chlorinated fatty acids themselves have also been detected in natural sources; the first example appears to be 9,10-dichlorostearic acid (**980**) which was identified as a minor lipid in the pathogenic fungus *Verticillium dahliae* and isolated as the methyl ester (*752*). More striking is the observation that six fatty acid chlorohydrins (**981**–**986**) are found in edible jelly fish and in the White Sea Jelly (*Aurita aurita*) (*753*). These compounds comprise 1.4% of the total lipids but only 30% of the organic chlorine present in the lipid fraction. Radioactive chloride was

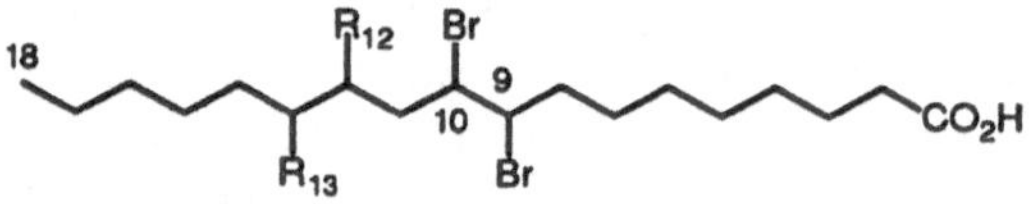

989 $R_{12} = R_{13} = H$ (9,10-dibromostearic acid)
990 $R_{12} = R_{13} = Br$ (9,10,12,13-tetrabromostearic acid)

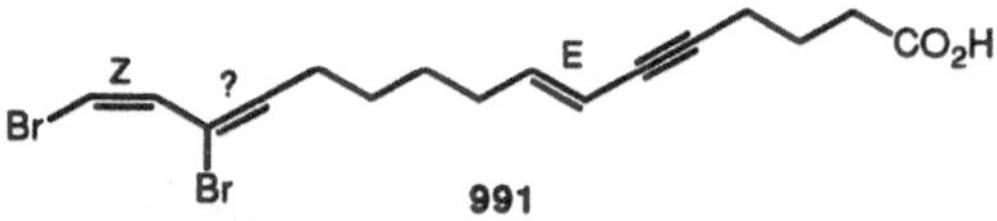

991

992 (18-bromooctadeca-9(*E*),17(*E*)-dien-7,15-diynoic acid)

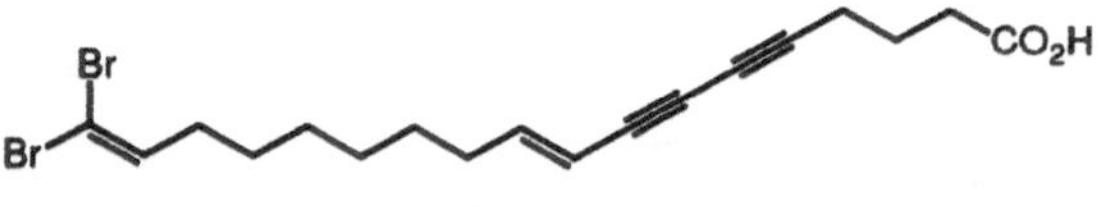

993

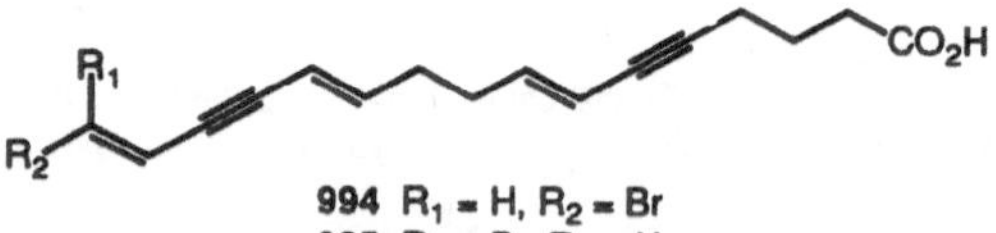

994 $R_1 = H$, $R_2 = Br$
995 $R_1 = Br$, $R_2 = H$

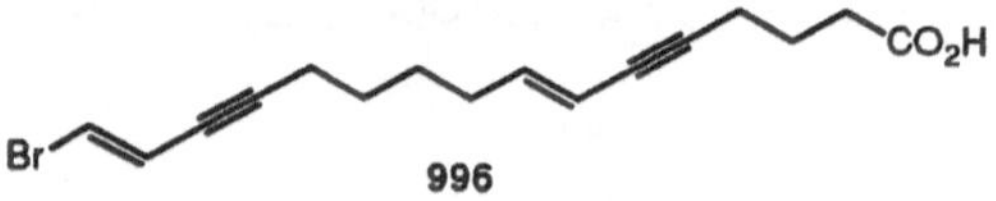

996

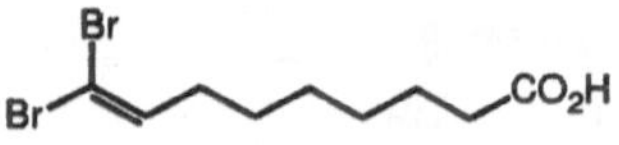

997

not incorporated into these compounds when it was added to the extracts. A survey of North Pacific marine organisms has revealed brominated lipids, mainly fatty acids, in salmon, halibut, sole, crab, oysters, cod, hake, sea lion, walrus, dolphin, and gray whale, to the extent of 40–900 μg/g (*754*). Likewise, chlorinated and brominated fatty acids have been found in fish in Norway although in this case it is believed that the origin of most of the chlorinated fatty acids is effluent from wood pulp bleaching mills (*755*). The compounds tentatively identified in this study were 9,10-dichlorostearic acid (**980**), and di- and tetrabromopalmitic and myristic acids. The question of natural versus anthropogenic origins of chlorinated fatty acids in fish lipids was addressed at the International Conference on Naturally-Produced Organohalogens, but remains unresolved (*756*). The barnacle *Balanus balanoides* contains the novel fatty acids, 9- (**987**) and 11-chloro-8,12-dihydroxy-5,11(9),14,17-eicosatetraenoic acids (**988**), which are "tentatively identified" (*757*).

Two brominated stearic acids **989** and **990** have been isolated from the seed oil of *Eremostachys molucelloides* (*758*). Marine sponges have afforded brominated fatty acids; the first such example is the isolation of dibromohexadecatrienynoic acid **991** from *Xestospongia muta* (*759*). The brominated *bis*-acetylenic acid **992** was found in *X. testudinaria* (*760*), while a study of *Xestospongia* sp. from the Red Sea has uncovered the new fatty acids **993–997** in addition to **992** (*761*). The sponges *Petrosia ficiformis* and *P. hebes* have both yielded the two brominated hexacosadienoic acids **998** and **999** (*762*). Two additional bromo fatty acids,

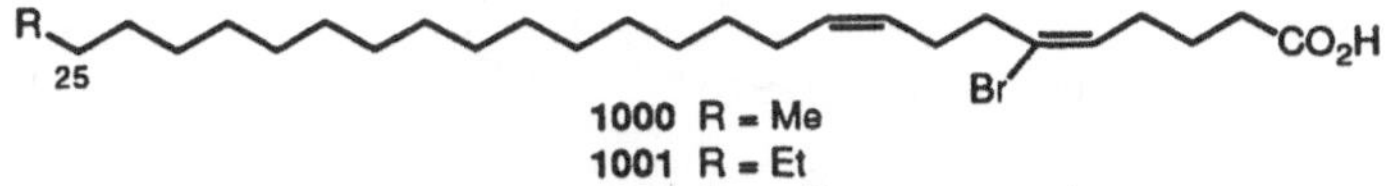

26 Br CO_2H

998

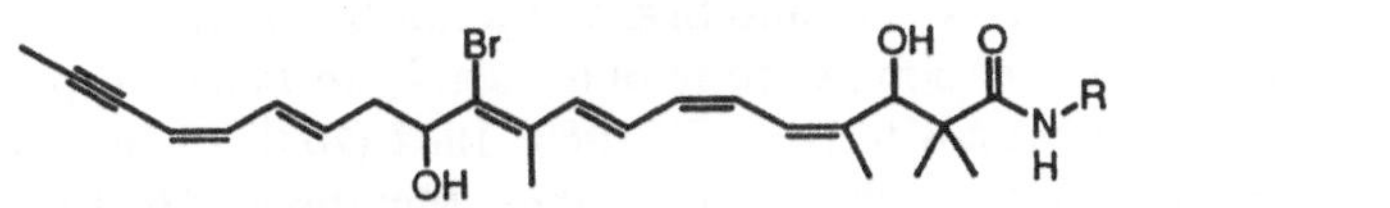

26 Br CO_2H

999

R 25 Br CO_2H

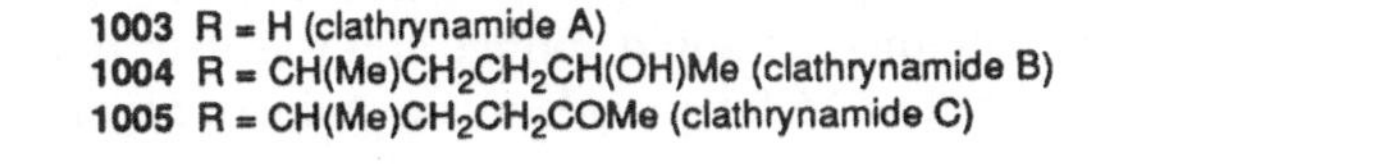

1000 R = Me
1001 R = Et
1002 R = *n*-Pr

Br OH OH O N R H

1003 R = H (clathrynamide A)
1004 R = $CH(Me)CH_2CH_2CH(OH)Me$ (clathrynamide B)
1005 R = $CH(Me)CH_2CH_2COMe$ (clathrynamide C)

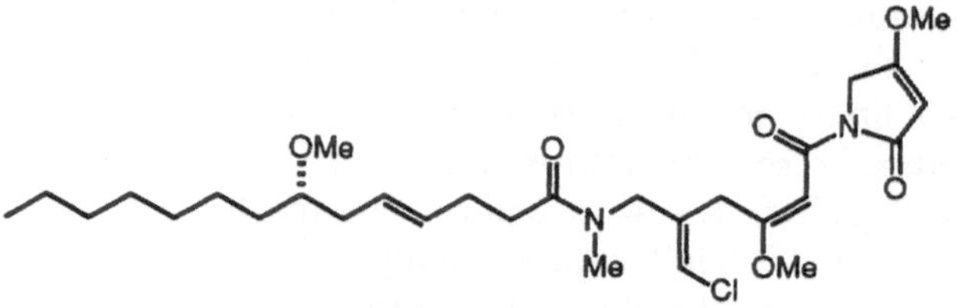

1006 (malyngamide A)

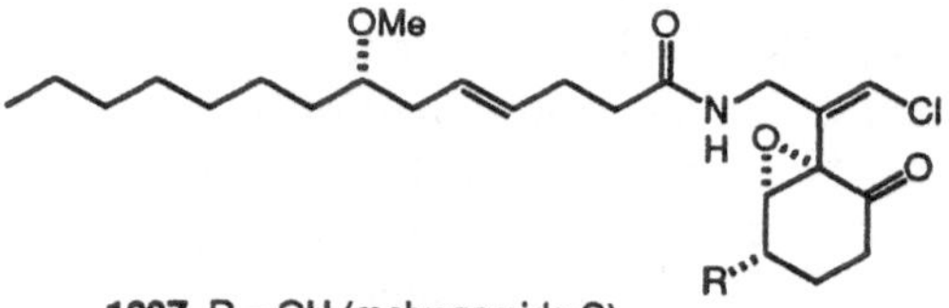

1007 R = OH (malyngamide C)
1008 R = OAc (malyngamide C acetate)
1009 R = H

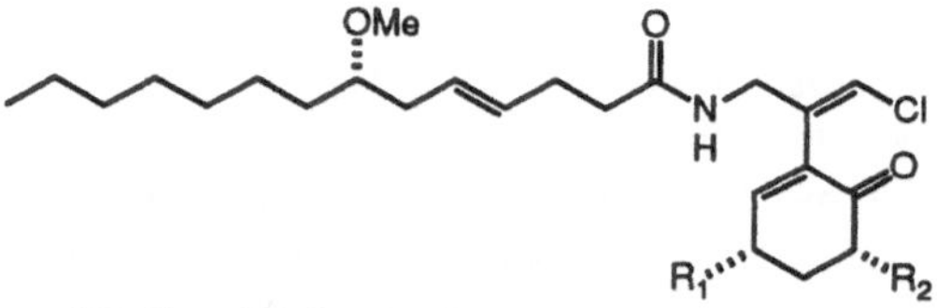

1010 R_1 = OH, R_2 = H (deoxymalyngamide C)
1011 R_1 = R_2 = H (dideoxymalyngamide C)
1012 R_1 = H, R_2 = OH (malyngamide D)
1013 R_1 = H, R_2 = OAc (malyngamide D acetate)

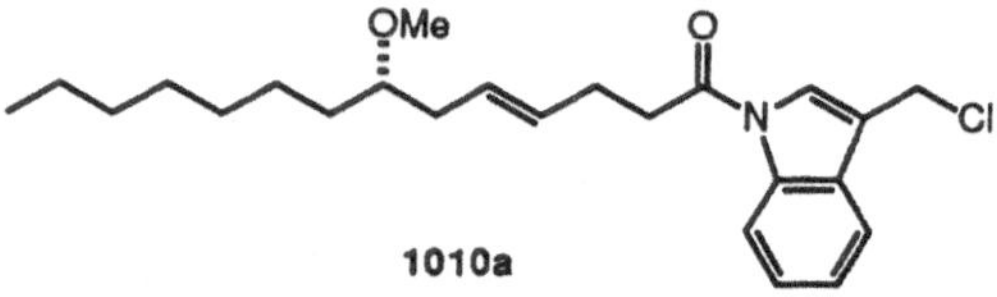

1010a

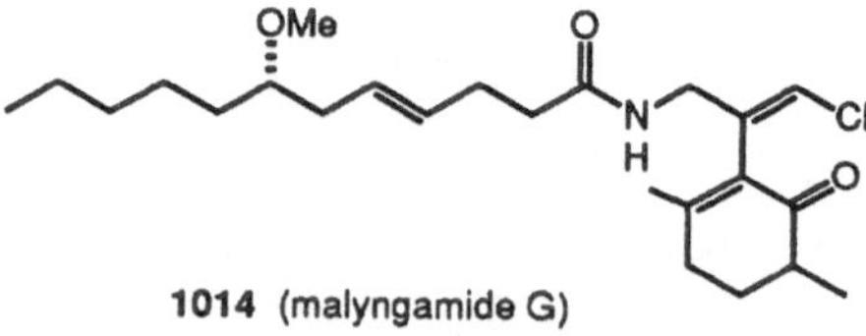

1014 (malyngamide G)

which appear to be the C_{28} homologs, could not be isolated in sufficient quantities for identification. A study of a Caribbean *Petrosia* sp. sponge has characterized bromo fatty acids **1000–1002** (*763*), an investigation that identified 54 fatty acids in this sponge. Metabolite **1000** had been previously found in a Hymeniacidonid sponge and was determined to be the third most abundant of 16 fatty acids in the animal, accounting for

13% by weight of the fatty acids (*764*). This study also provided evidence that bromination is the final step in the biosynthesis of **1000**. The marine sponge *Clathria* sp. produces clathrynamides A (**1003**), B (**1004**), and C (**1005**), metabolites that are potent inhibitors of cell division of starfish eggs (IC_{50} = 6 ng/ml) and are cytotoxic towards K-562 human myeloid cells (IC_{50} = 0.2 μg/ml) (*765*).

The blue-green marine alga *Lyngbya majuscula* has been a rich source of novel fatty acid derived amides. These include malyngamide A (**1006**) (*766*), malyngamide C (**1007**) (*767*), **1008** (*767*), deoxymalyngamide C (**1009**) (*410*) as well as deoxymalyngamides **1010** and **1011** (*767*). A collection of this cyanobacterium from Puerto Rico yielded malyngamide D (**1012**) and acetate **1013** (*768*). The absolute configuration of malyngamide C was established using CD; in this same study it was found that the minor metabolite deoxymalyngamide C (**1010**) undergoes slow conversion ($CHCl_3$, – 20°C) to the indole derivative **1010a** (*767*). The related stylocheilamide (**233**) and **234** were discussed earlier (*213*). A blue-green alga epiphyte of the brown alga *Cystoseira crinita* produces malyngamide G (**1014**) (*769*) and an Australian *Lyngbya* sp. contains the novel brominated lipid **1015** (*770*). Two cytotoxic tetramic acid glycosides, aurantosides A (**1016**) and B (**1017**), have been found in the marine sponge *Theonella* sp. (*771*). These novel dichloropolyenes show good

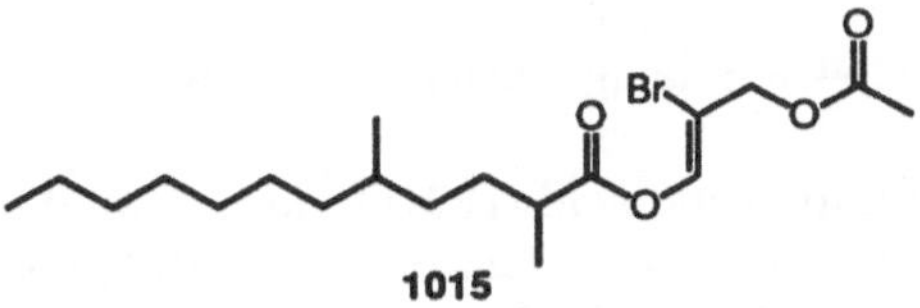

1015

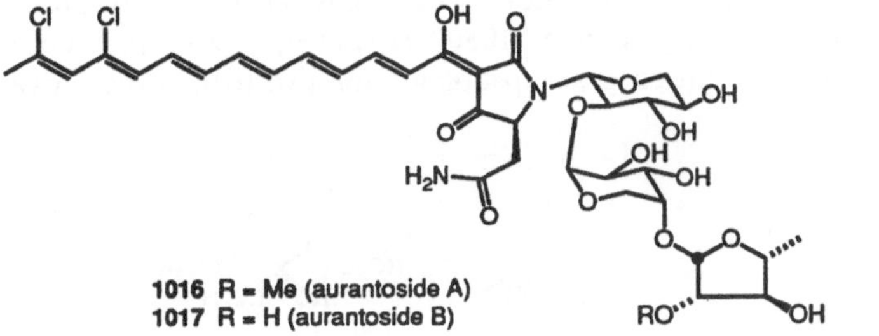

1016 R = Me (aurantoside A)
1017 R = H (aurantoside B)

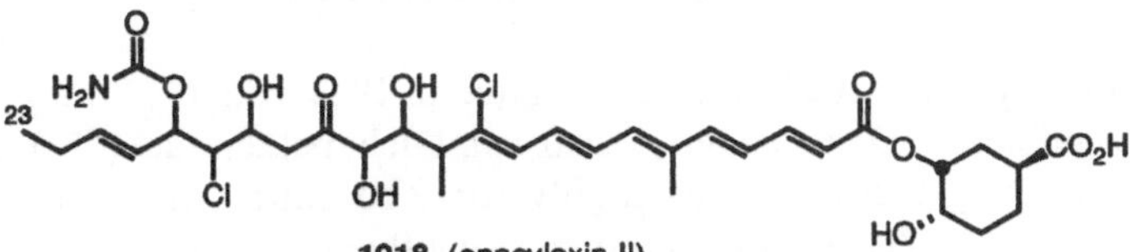

1018 (enacyloxin II)

cytotoxicity in the P-388 and L-1210 assays (IC_{50} = 1.8–3.4 μg/ml). The polyenic antibiotic enacyloxin II (**1018**) is produced by *Gluconobacter* sp. (*772*).

With the exception of the iodine-containing thyroid hormones (*vide infra*) and bromoester **186**, no examples of halogen-containing compounds in mammals are known. However, it has now been discovered that the thyroid gland in dogs contains iodinated lipids (*773*) and two iodolactones (**1019, 1020**) have been identified (*774, 775*). The transformation of arachidonic acid and docosahexaenoic acid by the action of lactoperoxidase, iodide, and hydrogen peroxide into these iodolactones has been demonstrated *in vitro*, and it is suggested that this pathway with thyroid peroxidase occurs *in vivo* (*774, 775*).

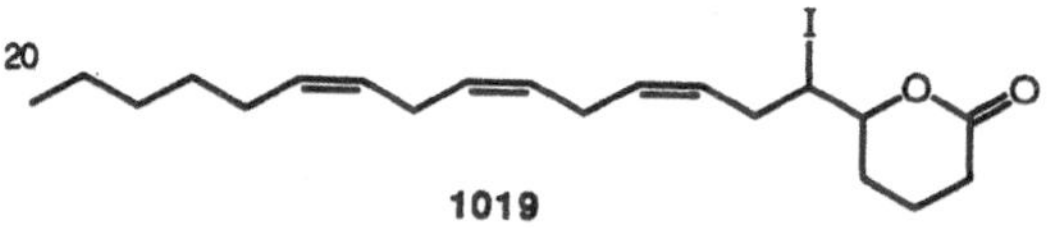

20 I O O

1019

22 I O O

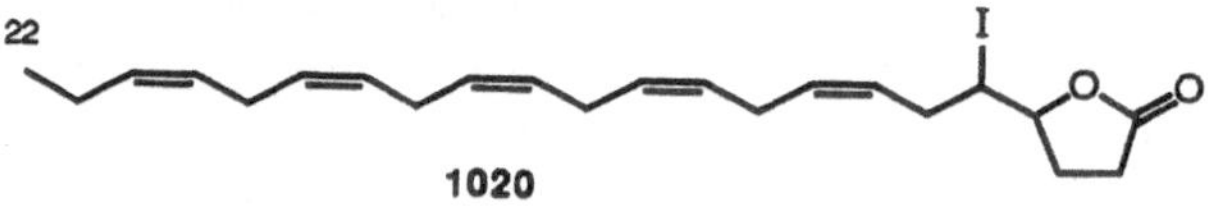

1020

3.9. Fluorine-Containing Carboxylic Acids

Naturally occurring fluoroacetic acid (**1021**) (fluoroacetate) has had a long history of interest and concern because of its powerful toxicity and very interesting mode of action (*776, 777*). Not surprisingly, fluoroacetate has been used as a pesticide, "1080". However, its widespread occurrence in plants has caused grave problems for livestock owners over many years.

FCH_2CO_2H

1021

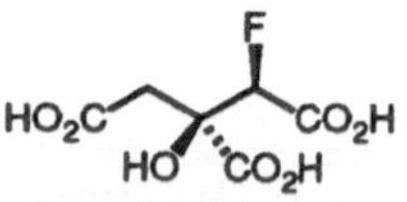

F HO_2C CO_2H HO CO_2H

1022 (fluorocitric acid)

Fluoroacetate (**1021**) was first isolated in 1944 from the South African plant "gifblaar" (*Dichapetalum cymosum*, Dichapetalaceae) (*778*), a plant that was long recognized as highly toxic to cattle and sheep in this country (*776, 777*). Less than an ounce of gifblaar leaves is enough to kill a sheep and it is claimed that one half a leaf is fatal to an ox (*784*). Several other *Dichapetalum* species contain fluoroacetate, viz. *D. stuhlmanii* (*779, 782*), *D. toxicarium* (*780*), *D. heudelotii* (*781*), *D. michelsonii* (*781*), *D. guineense* (*781*), *D. venenatum* (*781*), *D. braunii* (*781, 782*), *D. macrocarpum* (*781*), *D. ruhlandii* (*781*), *D. barteri* (*783*), and *D. edule* (*782*). Whereas plants in general contain only 0.1–10 ppm fluoride, *D. toxicarium* can absorb and store fluoride up to levels of 450 ppm in young leaves from an initial soil concentration of only 1–6 ppm (*780*). The Tanzanian *D. braunii* can reach levels of fluoroacetate of 7200 ppm in young leaves and 8000 ppm in the seeds (*782*). Analytical HPLC techniques have been developed which can detect 0.1 ppm levels of fluoroacetate in plant extracts (*785*), while a ^{19}F NMR method can detect fluoroacetate at levels of 4 ppm (*786*).

The Australian plant *Acacia georginae* (Mimosaceae) which is widespread in Northwest Queensland and the Northern Territory covering 28,000 square miles and which has caused serious sheep and cattle losses also contains fluoroacetate (*787, 788, 2214*). The highly toxic Brazilian plant *Palicourea marcgravii* (Rubiaceae), "rat weed," also contains fluoroacetate (*791*), as does the Australian *Gastrolobium grandiflorum*, which killed 2000 sheep on one ranch in a single episode (*792*). Other fluoroacetate-containing plants are *Oxylobium parviflorum* (Compositae) (*779, 793*), *Spondianthus preussii* (Euphorbiaceae) (*794*) and *Cyamopsis tetragonolobus* (Leguminosae) (*795*). Interestingly, the latter plant, "Guar Gum," is used medicinally in Finland, although most of the fluoroacetate (0.07–1.4 ppm) in the gum is removed in the extraction process (*795*). There does not appear to be a correlation between total fluoride in the soil and the presence of fluoroacetate-containing plants, as none of the afore-mentioned plants grow in high-fluoride soils (*779, 793*).

Biosynthetic studies have shown that fluoride is metabolized by single cell cultures of *Acacia georginae* to fluoroacetate and fluorocitrate (**1022**) (*789, 790*). This investigation also found that tea (*Thea sinensis*) cultures are likewise capable of transforming fluoride into fluorocitrate, and, furthermore, that fluorocitrate is present in commercial tea leaves ($<$ 30 ppm), and oatmeal ($<$ 62 ppm). The latter also contains some fluoroacetate. Single cell cultures of *Glycine max* also convert fluoride into both fluoroacetate and fluorocitrate (*798*). Fluorocitrate is biosynthesized *in vivo* from fluoroacetate and is ultimately responsible for the observed toxic effects, since the former blocks the enzyme aconitase in the citric acid (Krebs) cycle (*776, 777*). It has been proposed, with some supporting evidence, that fluoroacetate may be biosynthesized from fluoride, malonic acid, and fluoroperoxidase in plants (*796*), although the latter enzyme is unknown.

Despite the high toxicity of fluoroacetate to some animals such as the Texan pocket gopher ($LD_{100} < 0.05$ mg/kg), other animals such as the South African clawed toad are remarkably immune to this chemical ($LD_{50} > 500$ mg/kg) (*776*). And whereas the Southern Australian brush-tailed possum (*Trichosurus vulpecula*) is very susceptible to the toxicity of fluoroacetate ($LD_{50} = 0.68$ mg/kg), the Western Australian brush-tailed possum is surprisingly resistant to it ($LD_{50} = 100$ mg/kg) (*797*). It would appear that the latter animal has adapted to the presence of fluoroacetate-containing plants by acquiring a glutathione-defluorination metabolism mechanism (*797*).

Further examination of the seeds of the Sierra Leone shrub ratsbane (*Dichapetalum toxicarium*) revealed the presence of 18-fluorooleic acid (**1023**) (*799–802*), 16-fluoropalmitic acid (**1024**) (*803*), 6-fluorodecanoic acid (**1025**) (*803*), 14-fluoromyristic acid (**1026**) (*803*), and *threo*-18-fluoro-9,10-dihydroxystearic acid (**1027**) (*804*). In accord with the theory of β-oxidation of fatty acids, these even-numbered carbon chain fluorinated fatty acids are equally toxic as fluoroacetate, since they are metabolized to the latter *in vivo* (*776*).

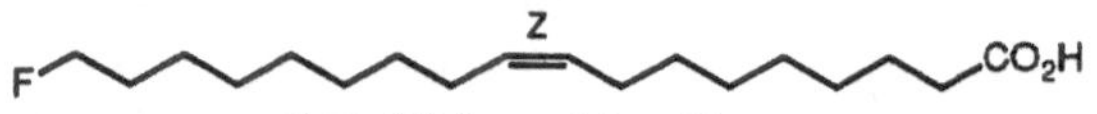

1023 (18-fluorooleic acid)

1024 (16-fluoropalmitic acid)

1025 (6-fluorodecanoic acid)

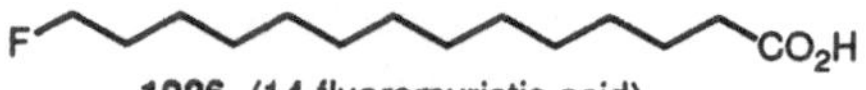

1026 (14-fluoromyristic acid)

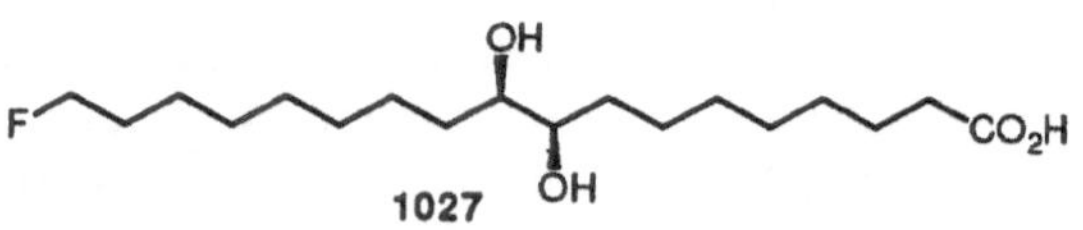

1027

3.10. Prostaglandins

One of the highlights in this field of natural organohalogens occurred ten years ago when chlorinated prostaglandins possessing enormous biological potency were discovered in marine animals.

The striking octocoral *Telesto riisei* (*805*) produces punaglandins 1–4 (**1028–1031**) (*806*) (stereochemistry f C-12 revised in (*807–809*)). Punaglandins 3 and 4 actually exist as both the 7*E*- (**1030**, **1031**) and the 7*Z*-isomers (**1032**, **1033**). The marine stolonifer *Clavularia viridis* produces the similar chlorovulones I–IV (**1034–1037**) (*810*) (absolute configuration in (*811*)), the novel bromovulone I (**1038**) and iodovulone I (**1039**) (*812*), and the epoxy prostanoid **1040** (*813*). All of these compounds have potent antitumor or antiproliferative activity in a range of cell lines. For example, **1040** exhibits $IC_{50} = 0.04$ μg/ml in the human HL-60 leukemia cell line (*813*), and the punaglandins and clavulones have *in vivo* activity

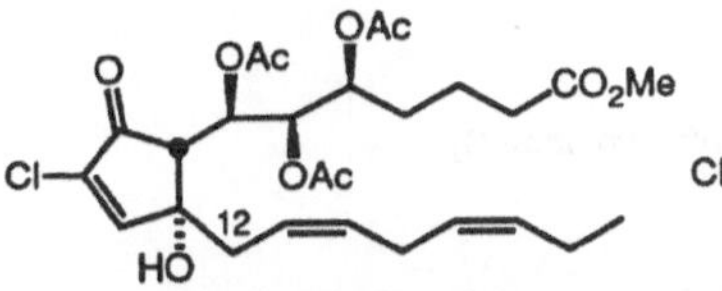

1028 (punaglandin 1)

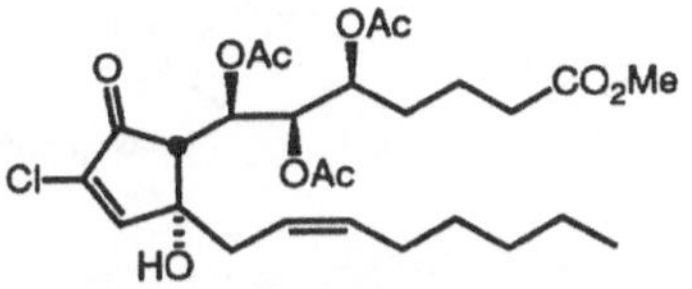

1029 (punaglandin 2)

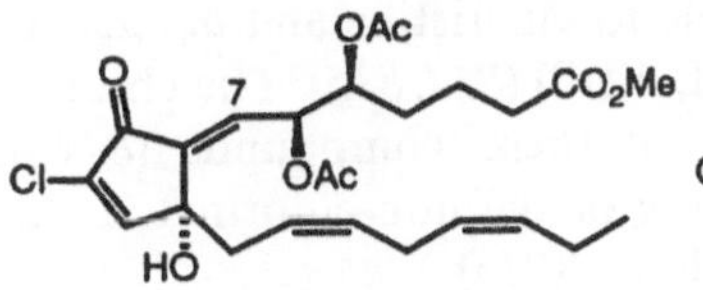

1030 (7*E*-punaglandin 3)

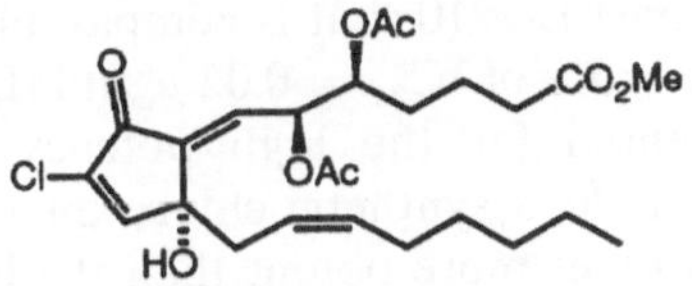

1031 (7*E*-punaglandin 4)

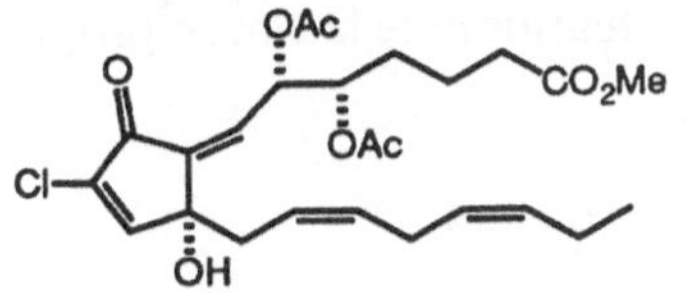

1032 (7*Z*-punaglandin 3)

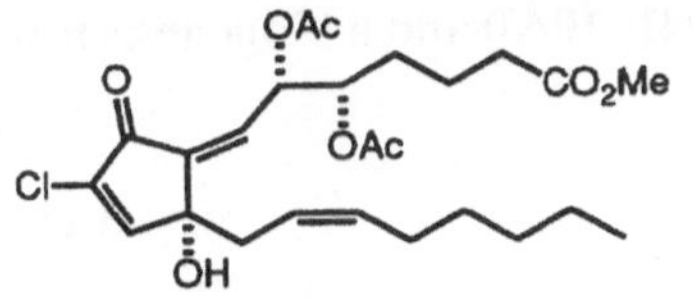

1033 (7*Z*-punaglandin 4)

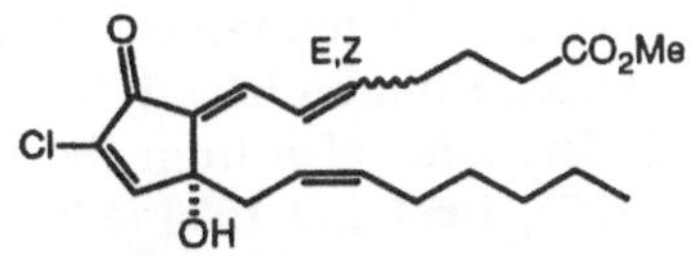

1034 *Z* isomer (chlorovulone I)
1035 *E* isomer (chlorovulone II)

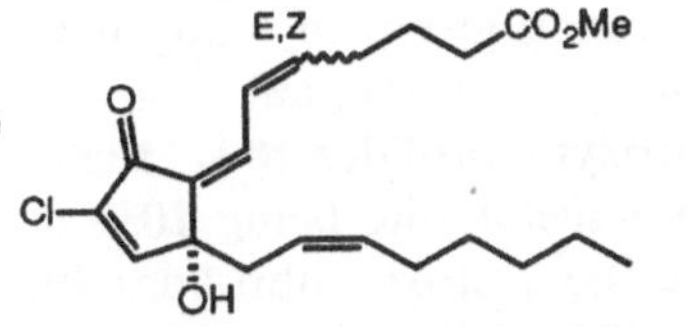

1036 *E* isomer (chlorovulone III)
1037 *Z* isomer (chlorovulone IV)

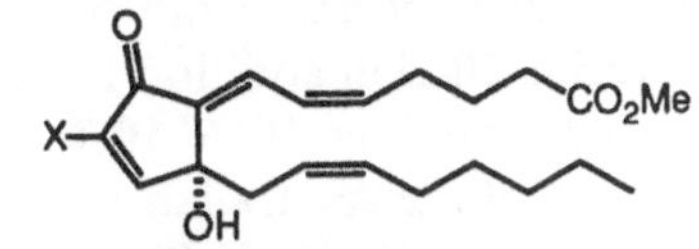

1038 X = Br (bromovulone I)
1039 X = I (iodovulone I)

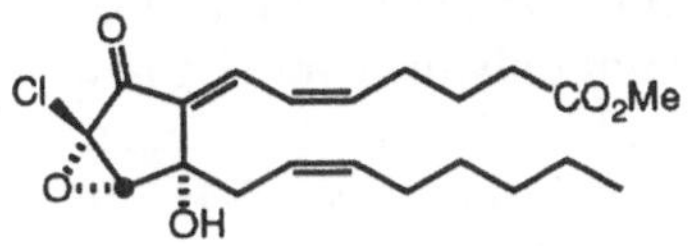

1040

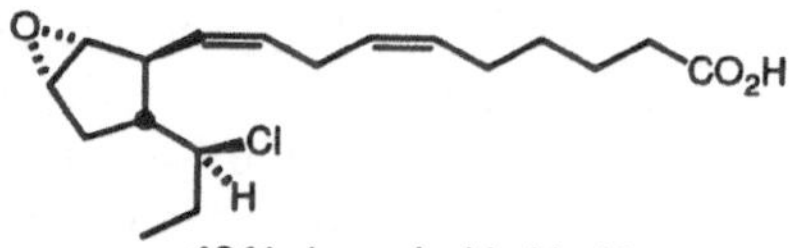

1041 (egregiachloride A)

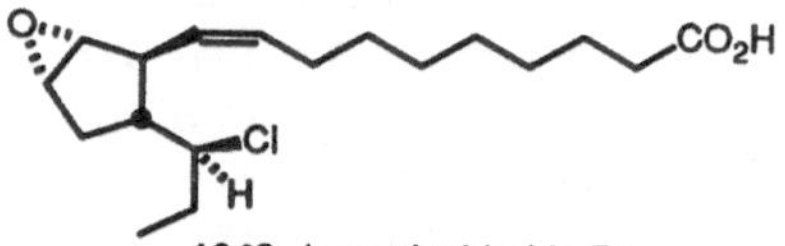

1042 (egregiachloride B)

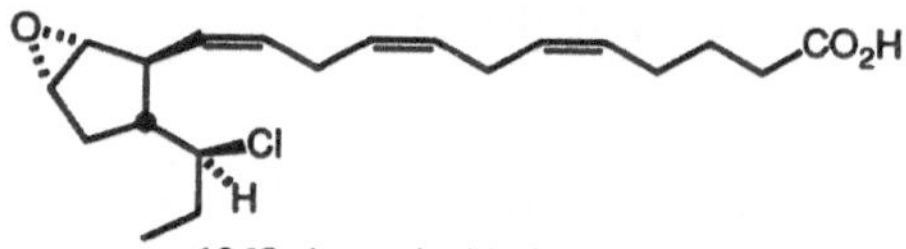

1043 (egregiachloride C)

against L-1210 that is comparable to vincristine and *in vitro* activity on the order of $IC_{50} = 0.02\ \mu g/ml$ (L-1210) (*814,815*). The chlorine atom is essential for the high potency of these compounds (*814,816*). For example, a synthetic chloro analog of the non-chlorinated clavulone is ten times more potent than the latter (*816*).

The brown alga *Egregia menziesii* produces egregiachlorides A–C (**1041**–**1043**) and a biogenesis from stearidonate has been proposed (*817*).

3.11. Furanones

Independently, and from different sources, two groups discovered an interesting group of halogenated lactones in the red alga *Delisea fimbriata.* An Australian collection yielded the fimbrolides **1044**–**1046**, hydroxyfimbrolides **1047**–**1053**, and acetoxyfimbrolides **1054**–**1060**, the major metabolite being **1055** (*818*). The same alga from the Antarctic gave the acetoxyfimbrolides **1054**–**1057**, **1059** and **1060** (*819*). The red alga *Beckerella subcostatum* produces bromo- (**1061**, **1062**) and chlorobeckerelide (**1063**, **1064**) as two epimers each (*820*); the former are also found in *Delisea pulchra* (*161*). A study of *D. elegans* from New Zealand has not only uncovered a new fimbrolide **1065** in addition to the known **1046**, but also a set of dimeric spiro metabolites **1066**–**1070**, including three cyclobutanes (*821*). A recent study of the Australian *D. pulchra* (cf. *fimbriata*) has uncovered seven new furanones (**1071**–**1077**) in addition to most of those already isolated (*vide supra*) (*822*).

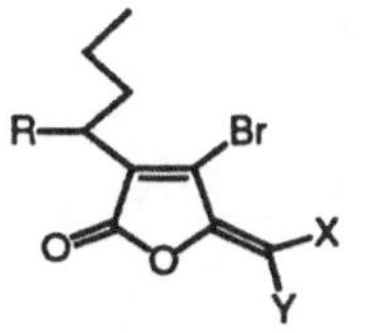

	R	X	Y
1044	H	Br	H
1045	H	H	Br
1046	H	Br	Br
1047	OH	Br	H
1048	OH	H	Br
1049	OH	I	H
1050	OH	H	I
1051	OH	Cl	H
1052	OH	H	Cl
1053	OH	Br	Br
1054	OAc	Br	H
1055	OAc	H	Br
1056	OAc	H	I
1057	OAc	I	H
1058	OAc	Cl	H
1059	OAc	H	Cl
1060	OAc	Br	Br

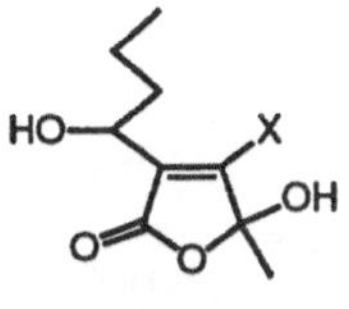

1061, 1062 X = Br (bromobeckerelides)
1063, 1064 X = Cl (chlorobeckerelides)

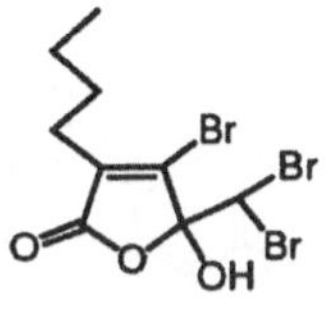

1065

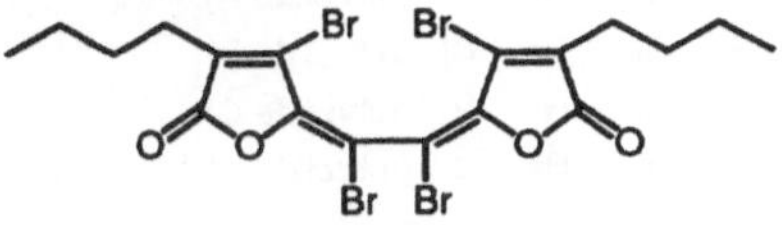

1066

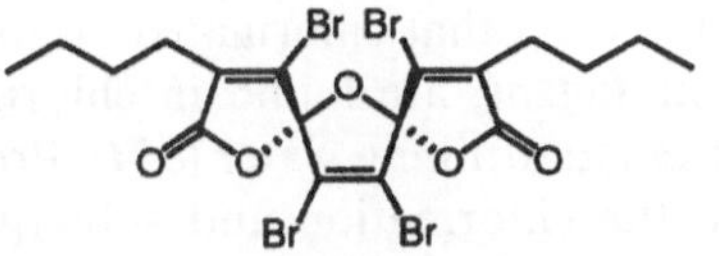

1067

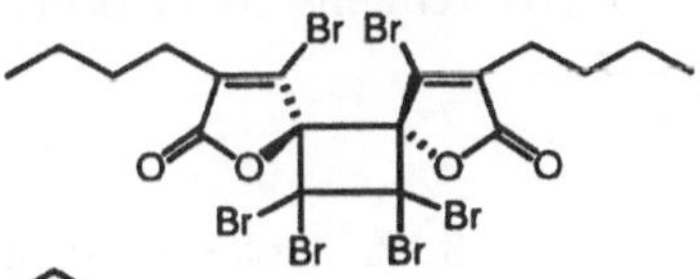

1068

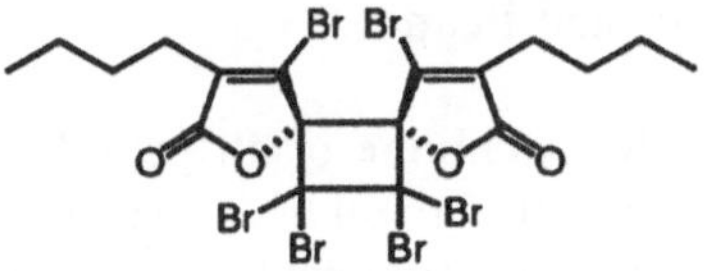

1069

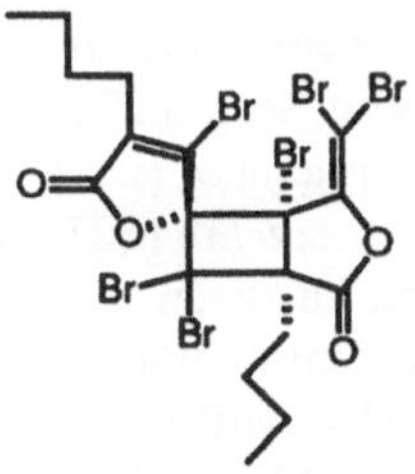

1070

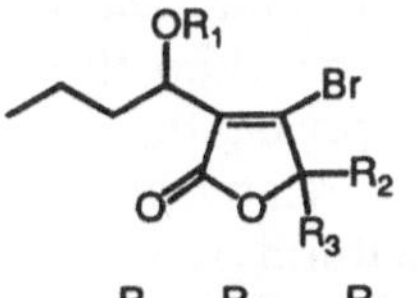

	R_1	R_2	R_3
1071	Ac	OMe	CH_2I
1072	H	OMe	CH_2I
1073	H	CH_2I	OMe
1074	Ac	Me	OMe
1075	Ac	OMe	Me
1076	Ac	$CHBr_2$	OMe
1077	Ac	OMe	$CHBr_2$

The colonial tunicate *Ritterella rubra* produces an array of halogenated diarylfuranones rubrolides A–H, six of them containing halogen (**1078**–**1083**) (*823*). These novel compounds are potent antibacterial agents and show selective phosphatase inhibition.

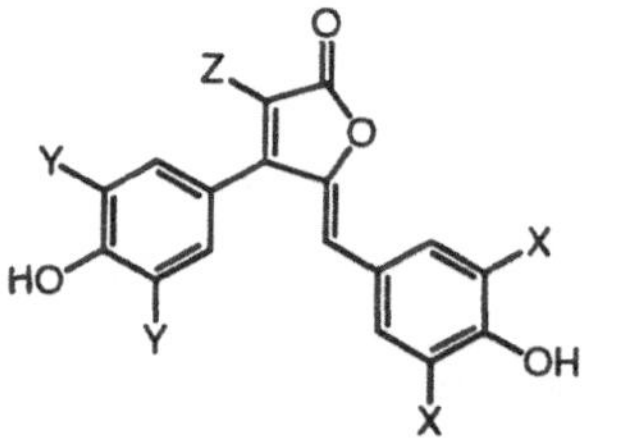

	X	Y	Z	
1078	Br	Br	H	(rubrolide A)
1079	Br	Br	Cl	(rubrolide B)
1080	Br	H	H	(rubrolide C)
1081	H	Br	H	(rubrolide D)

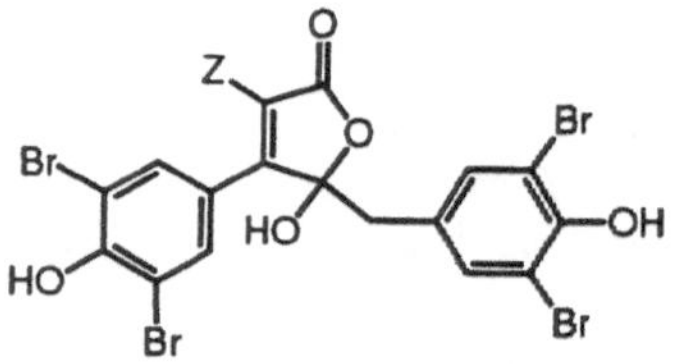

1082 Z = H (rubrolide G)
1083 Z = Cl (rubrolide H)

It is of interest that chlorinated furanones, some of which are potent bacterial mutagens, are found in chlorine-bleached pulp mill effluents and chlorinated drinking water (*824*). Presumably, these compounds are formed in the chlorination and subsequent breakdown of humic and fulvic acids and other phenolics naturally present in wood pulp and drinking water. It remains to be seen if these compounds also have a natural source.

3.12. Amino Acids and Peptides

Given that amino acids provide the backbone of all peptides, proteins, and enzymes, and, therefore, are essential to all life as we know it, it is perhaps surprising that more halogenated amino acids are not known. However, the relatively few such compounds originate from a wide variety of natural sources and the number seems certain to grow. Halogenated tyrosines are discussed separately in Sect. 3.22.3.

Although not an amino acid *per se*, chloramphenicol (**1084**) (chloromycetin) which is produced by *Streptomyces venezuelae* (*825–827*) and the moon snail (*Lunatia heros*) (*2010*) and which has been a commercial antibiotic for many years is on the shikimic acid-amino acid biosynthetic pathway (*828, 829*). Its structure was confirmed by synthesis and shown to have the D-(−)-*threo* configuration (*830, 831*). The amino derivative **1085**, which is thought to be the biosynthetic precursor of chloramphenicol (**1084**), is also produced by *S. venezuelae* (*832*). When bromide ion is present in the culture, brominated analogs of **1084** (bromine in place of chlorine) are produced (*833*) but are not counted here as natural products. The dichloromethyl group is also found in bactobolin (**1086**),

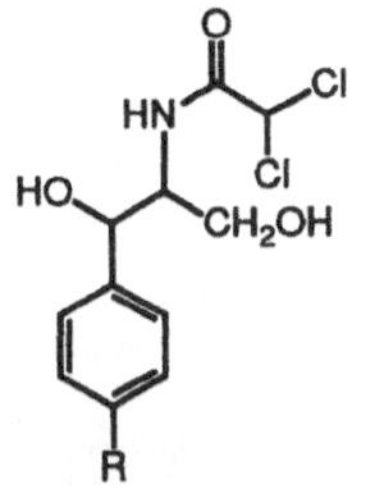

1084 R = NO_2 (chloramphenicol)
1085 R = NH_2

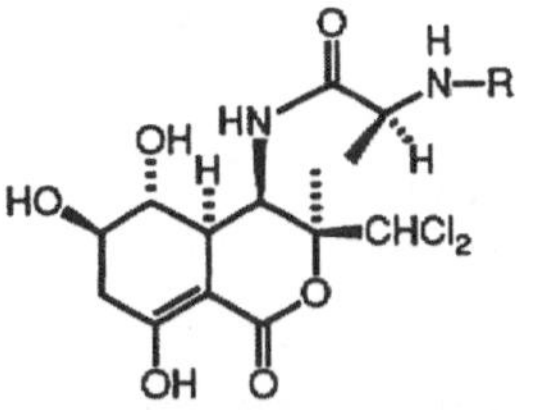

1086 R = H (bactobolin)
1087 R = $CH(CH_3)CO_2H$ (bactobolin B)

an antitumor antibiotic peptide produced by *Pseudomonas yoshitomiensis* (*834,835*) and isolated independently as "BN-183B" from *Pseudomonas* sp. by others (*836*). Subsequent research revealed the L-alanine derivative bactobolin B (**1087**) in this microbe (*837*).

The first simple naturally occurring halogenated amino acid excluding iodinated tyrosine is 2-amino-4,4-dichlorobutyric acid (**1088**) from *Streptomyces armentosus* var. *armentosus*, a metabolite with antibacterial action (*838*). The microbe *S. griseosporeus* produces γ-chloronorvaline (**1089**) which also has antibacterial activity, especially against *Pseudomonas aeruginosa* (*839*). The novel amino acid **1090** is found in *S. viridogenes* (*840*), and the related dipeptide "FR-900148" (**1091**) is produced by both *S. viridogenes* (*840*) and *S. xanthocidicus* (*841,842*) (structure revised in (*843*)). The novel antitumor agent U-43,795 (**1092**) is found in cultures of *S. sviceus* (*844*), and a chloroaminobutyric acid of unknown structure (**1093**) is found in the xylem sap of *Pisum sativum* (Leguminosae) (*845*). The common bacterium *Bacillus subtilis* contains chlorotetaine (**1094**) and bromotetaine (**1095**) (*846,870*). Culturing of *Streptomyces cattleya* in the presence of fluoride leads to the formation of 4-fluorothreonine (*847*) although this is not considered "natural" in the present context. Several chlorinated amino acids are present in mushrooms (*848*). For example, *Amanita solitaria*, which has a high chloride concentration (2000 ppm) compared to terrestrial plants (50–500 ppm), contains 2-amino-5-chloro-4*Z*-hexenoic acid (**1096**) (*849*), and from *A. pseudoporphyria* there has been isolated the antibacterial **1097** (*850, 851*). The mushrooms *A. onusta* and *A. miculifera* both produce **1098** (*852,848*), while *A. gymnopus* contains **1099** (*848*). The toxic "White Mushroom" (*A. abrupta*) has yielded the chlorohydrin **1100** (*853*).

The marine sponge *Dysidea herbacea* has proved to be a rich source of novel amino acid and peptide-related metabolites all of which are discussed in this section. Dysidin (**1101**) (*854*), dysidenin (**1102**) (*855*), and

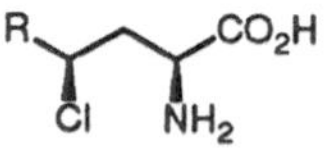

1088 R = Cl
1089 R = CH_3

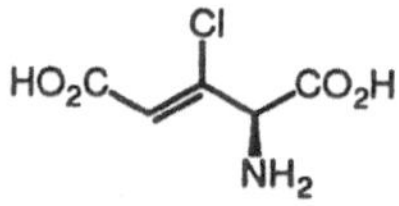

1090

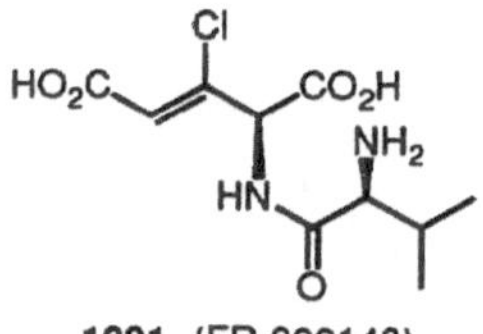

1091 (FR-900148)

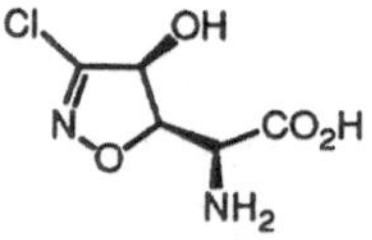

1092 (U-43,795)

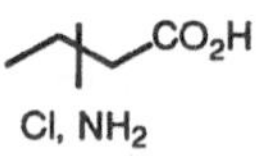

1093

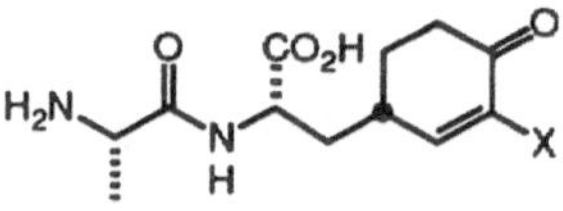

1094 X = Cl (chlorotetaine)
1095 X = Br (bromotetaine)

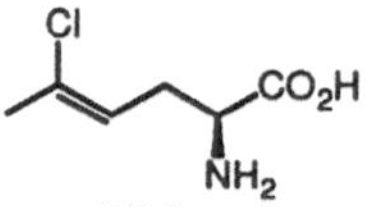

1096

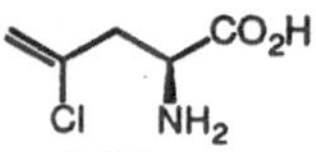

1097

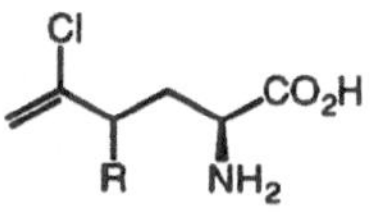

1098 R = H
1099 R = OH

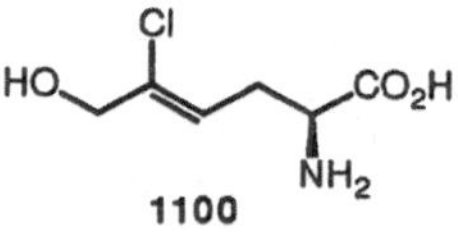

1100

isodysidenin (**1103**) (*856*) were the first metabolites to be isolated from this animal although the structures of **1102** and **1103** needed to be revised more than once (*857–860*). A collection of *D. herbacea* from the Great Barrier Reef identified four new metabolites, **1104–1107** (*861*), and the diketopiperazine **1108** was also found (*862*). A study of this sponge and its cyanobacterial symbiont, *Oscillatoria spongeliae*, reveals that the latter produces diketopiperazine **1109** as well as several other *D. herbacea* metabolites (*863*). The Hainan Island *D. fragilis* produces the new diketopiperazine dysamides A (**1110**), B (**1111**), and C (**1112**), and the known **1108** (*864*). A Red Sea *D. herbacea* contains dysidamide (**1113**) (*865*), the absolute configuration of which has been determined (*866*). This study also discovered dysidamides B (**1114**) and C (**1115**) as minor components (*867*). Another investigation of *D. herbacea* yielded herbaceamide (**1116**) (*868*). This study also confirmed the 5*S*, 13*S* configuration of dysidenin. A study of *D. herbacea* from Pohnpei identified dysideathiazole (**1117**), *N*-methyldysideathiazole (**1118**), and the minor component **1119**, while the same sponge from Palau yielded **1119** as a major component along with **1120** and **1121** (*869*). A different Palau *D. herbacea* gave dysideapyrrolidone (**1122**) (*869*). It remains to be seen if this array of trichloro- and dichloromethyl metabolites from *Dysidea* serves as a source of natural marine chloroform and dichloromethane.

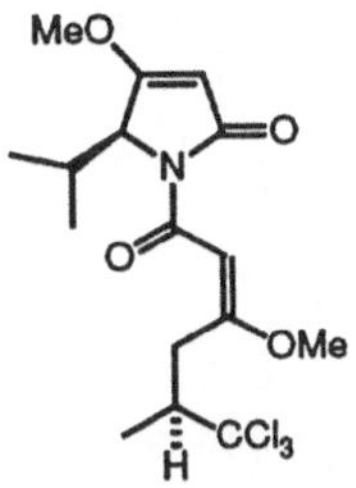

1101 (dysidin)

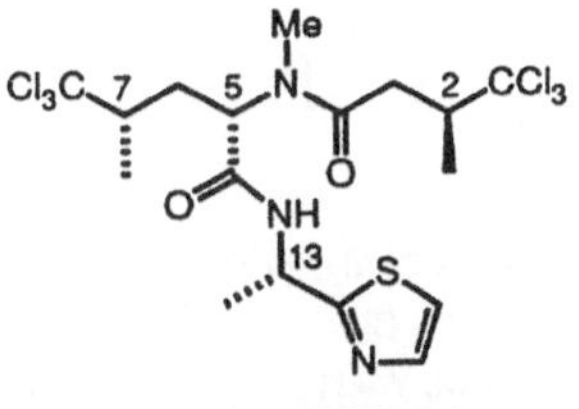

1102 (dysidenin)

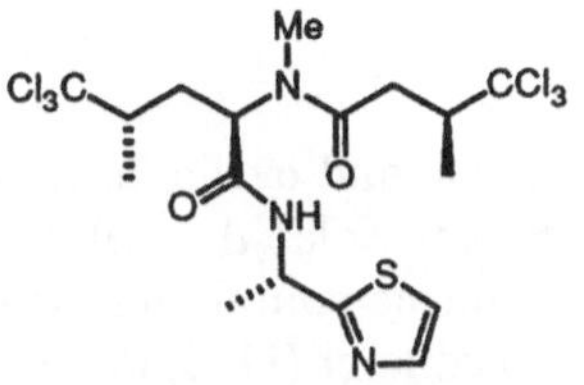

1103 (isodysidenin)

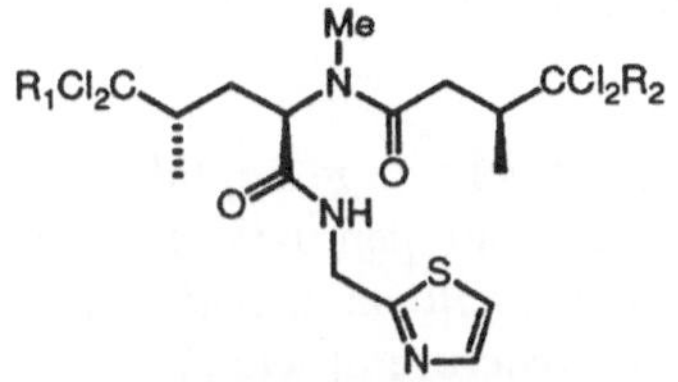

1104 $R_1 = R_2 = Cl$
1105 $R_1 = Cl, R_2 = H$
1106 $R_1 = H, R_2 = Cl$

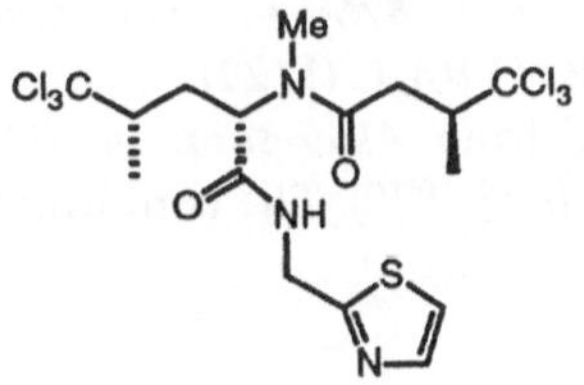

1107

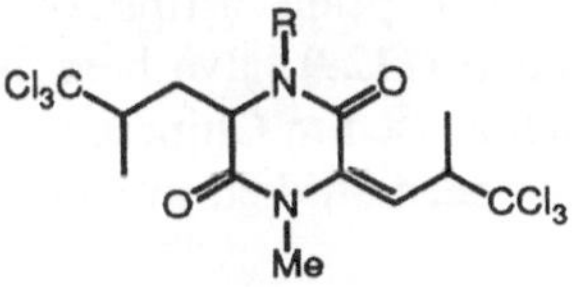

1108 R = Me
1109 R = H

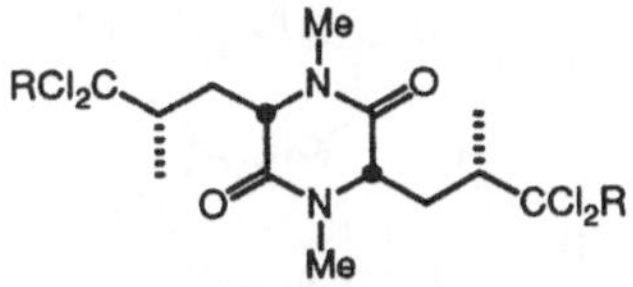

1110 R = Cl (dysamide A)
1111 R = H (dysamide B)

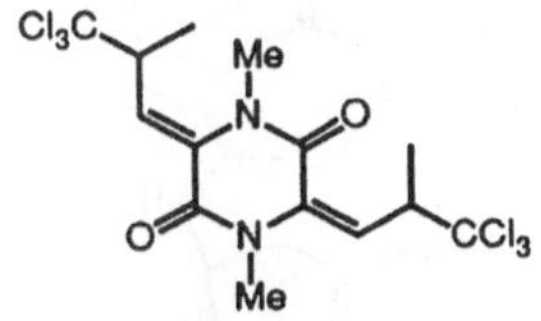

1112 (dysamide C)

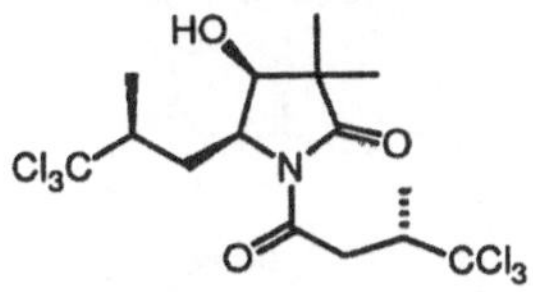

1113 (dysidamide)

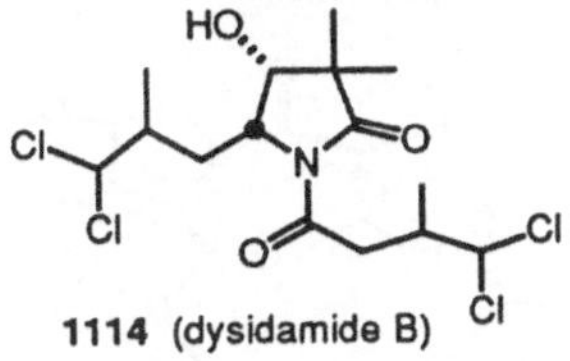

1114 (dysidamide B)

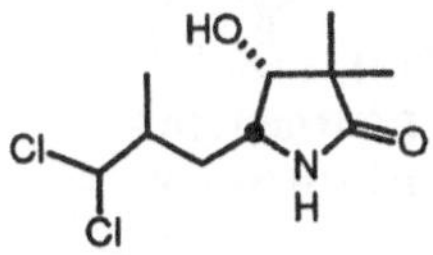

1115 (dysidamide C)

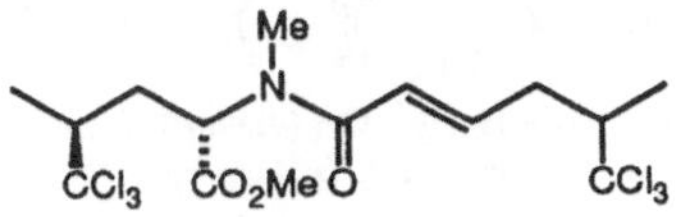

1116 (herbaceamide)

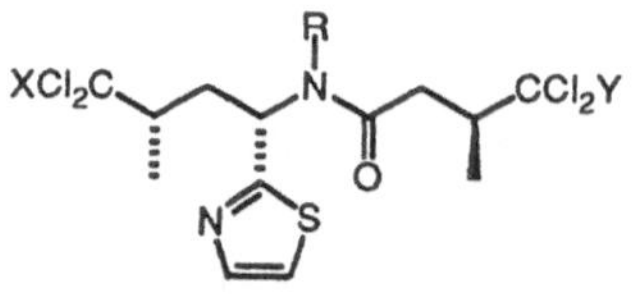

1117 R = H, X = Y = Cl (dysideathiazole)
1118 R = Me, X = Y = Cl
1119 R = Me, X = Cl, Y = H
1120 R = Me, X = Y = H
1121 R = H, X = Cl, Y = H

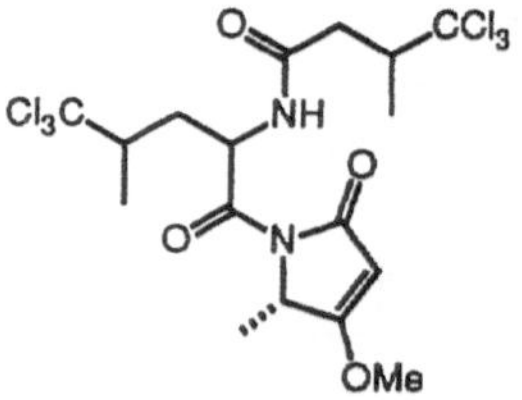

1122 (dysideapyrrolidone)

Several relatively large peptides, both acyclic and cyclic, have been found to contain halogen. Islanditoxin which was isolated in 1955 from a culture of *Penicillium islandicum* is a chlorine-containing peptide (*871–873*) the structure of which was determined later (**1123**) (*874*). This organism has also yielded the related cyclochlorotine (**1124**) which, like islanditoxin, is an infectant of yellowed rice (*875, 876*). The very similar antitumor cyclic peptides astins A (**1125**), B (**1126**), C (**1127**) (= asterin), D (**1128**), and E (**1129**) have been isolated from *Aster tataricus* (Compositae) which is used in Chinese medicine (*877–879*). The conformation of astin B has been studied in some detail (*1816*).

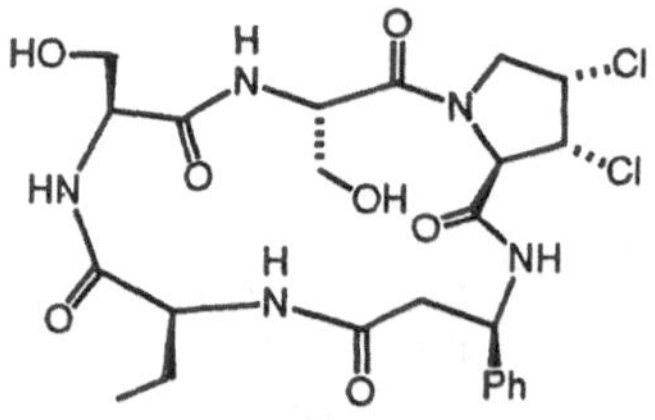

1123 (islanditoxin)

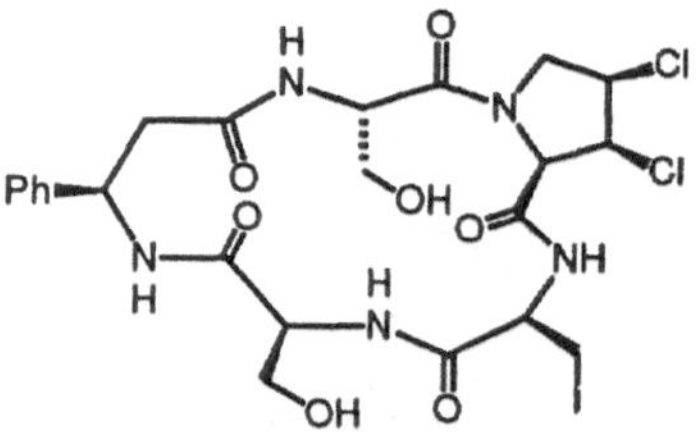

1124 (cyclochlorotine)

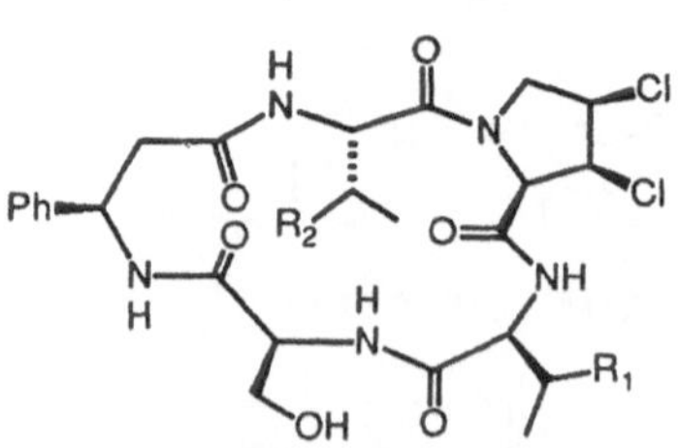

1125 R_1 = H, R_2 = OH (astin A)
1126 R_1 = OH, R_2 = H (astin B)
1127 R_1 = R_2 = H (astin C)

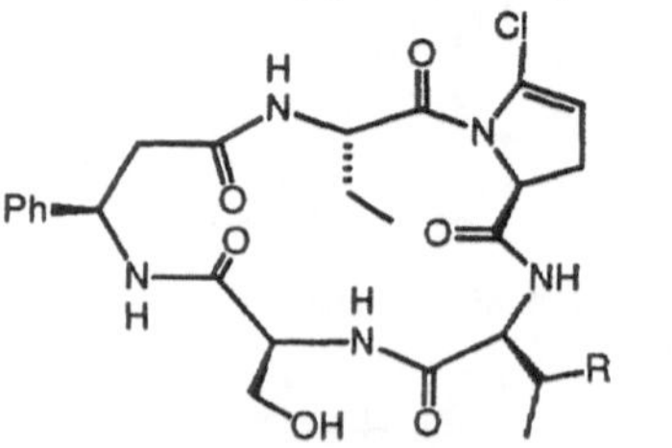

1128 R = H (astin D)
1129 R = OH (astin E)

A *Streptomyces* sp. produces the novel β-lactam **1130** (*880*) while another *Streptomyces* strain produces chlorocardicin (**1131**) (*881–882*). Cultures of *Streptomyces griseoviridus* that are augmented with 4-chloro-L-proline produce chlorinated analogs of viridogrisein (*883*), but these are considered "unnatural" and are not included here. The fungus *Metarhizium anisopliae* produces the chlorohydrin cyclic peptide **1132** (*884*). Interestingly, the corresponding epoxide is not converted to **1132** upon treatment with concentrated hydrochloric acid.

Several haloindole-containing peptides have been isolated in recent years from a variety of sources, but especially marine sponges. For a review of bioactive marine sponge peptides see (*885*). The burrowing sponge *Cliona celata* which can bore into hard coral produces a set of 6-bromotryptophan-derived peptides, from the simple clionamide (**1133**) (*886*) to the more complex celenamides A–C (**1134–1136**) (*887, 888*) which contain the novel amino acid α, β-didehydro-3,4,5-trihydroxyphenylalanine. The antitumor tetrapeptides halocyamine A (**1137**) and B (**1138**) were isolated from the hemocytes of the solitary ascidian *Halocynthia roretzi* (*889*). The dehydrotryptamine unit in both compounds is very unusual. The halocyamines inhibit the growth of fish

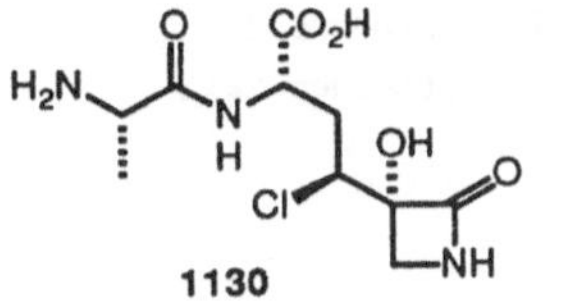

1130

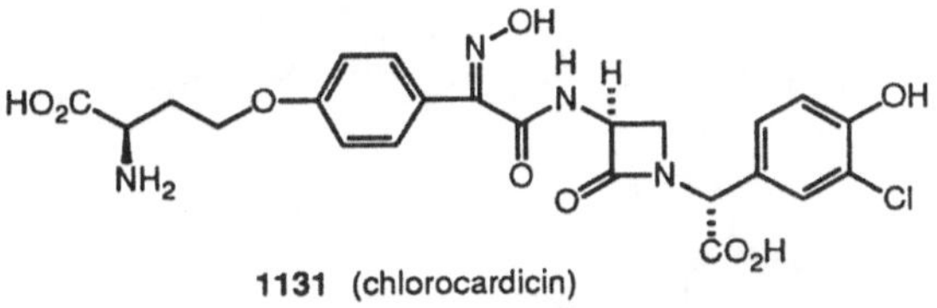

1131 (chlorocardicin)

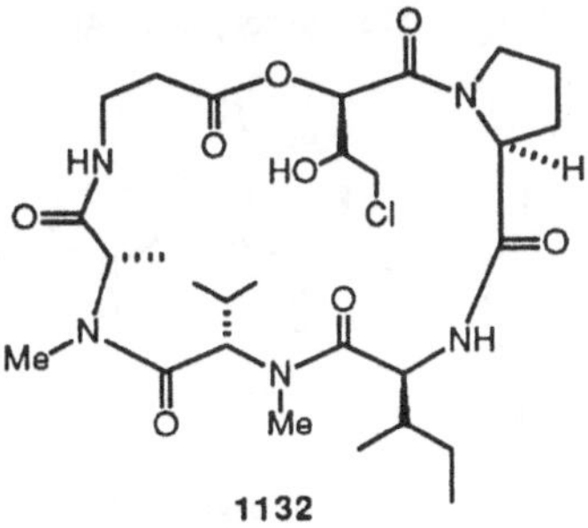

1132

viruses and marine bacteria and thus may have a chemical defense role in the ascidian (*1190*). The 6-bromotryptophan residue is also present in several cyclic peptides such as jaspamide (**1139**) (*890*) (= jasplakinolide (*891*)), which was isolated simultaneously by two groups from the sponge *Jaspis* sp. Jaspamide, which contains the rare amino acids β-tyrosine and

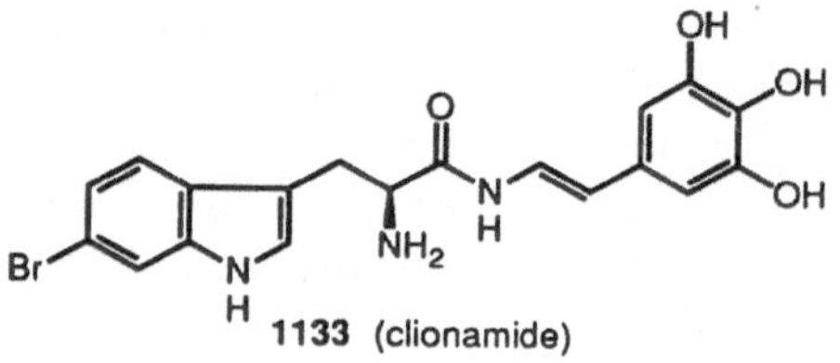

1133 (clionamide)

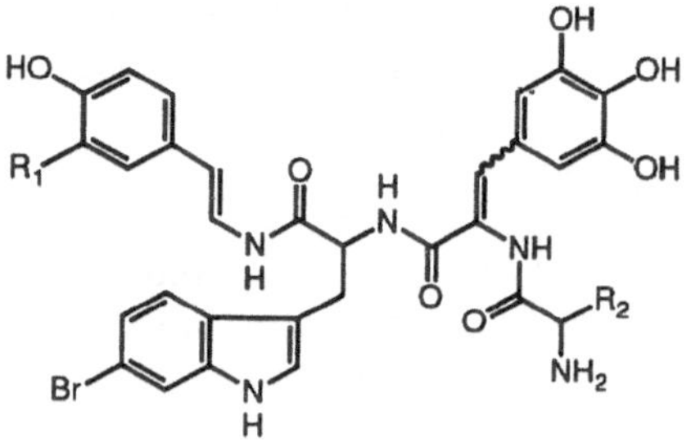

1134 R_1 = OH, R_2 = $CH_2CH(CH_3)_2$ (celenamide A)
1135 R_1 = OH, R_2 = $CH(CH_3)_2$ (celenamide B)
1136 R_1 = H, R_2 = $CH_2CH(CH_3)_2$ (celenamide C)

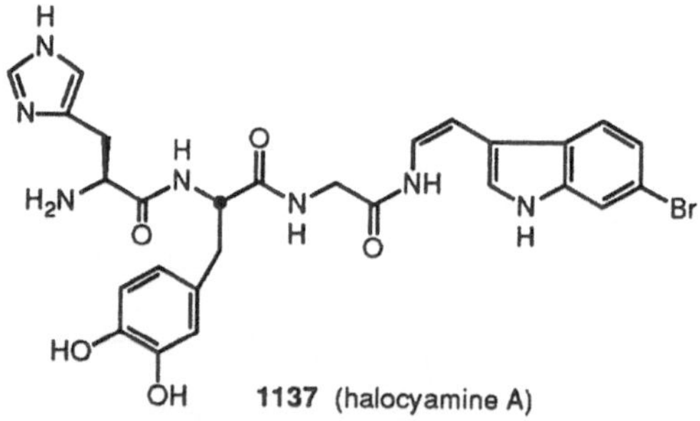

1137 (halocyamine A)

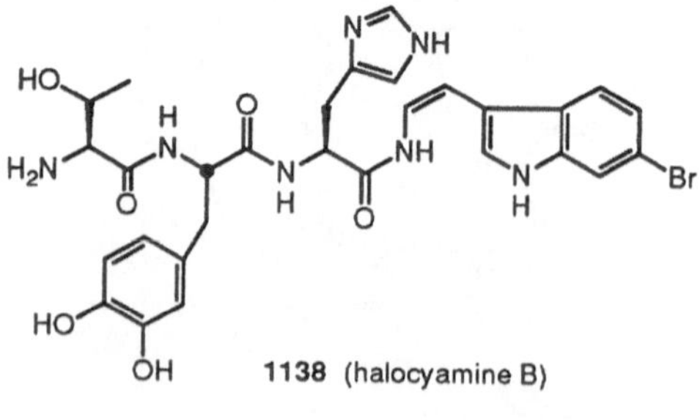

1138 (halocyamine B)

2-bromotryptophan, is also found in *Jaspis johnstoni* (*892*). Both detailed conformational (*893*) and lithium complexation studies (*894*) have been reported for this potent antitumor, antifungal and insecticidal compound. Also reported simultaneously were orbiculamide A (**1140**) (*895*)

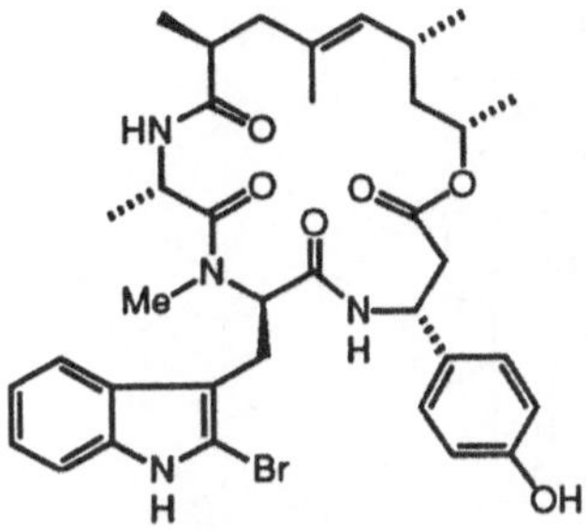

1139 (jaspamide)

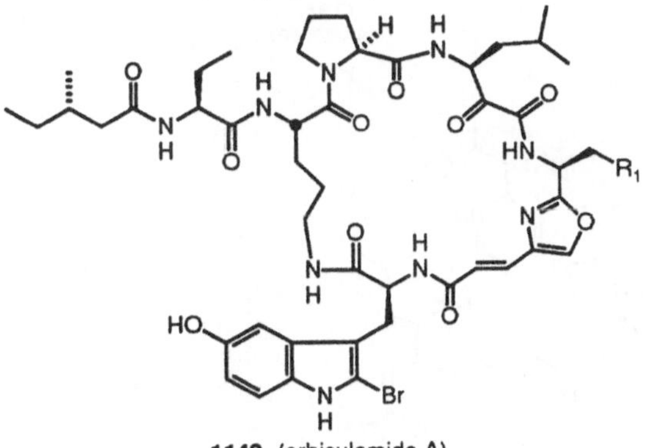

1140 (orbiculamide A)

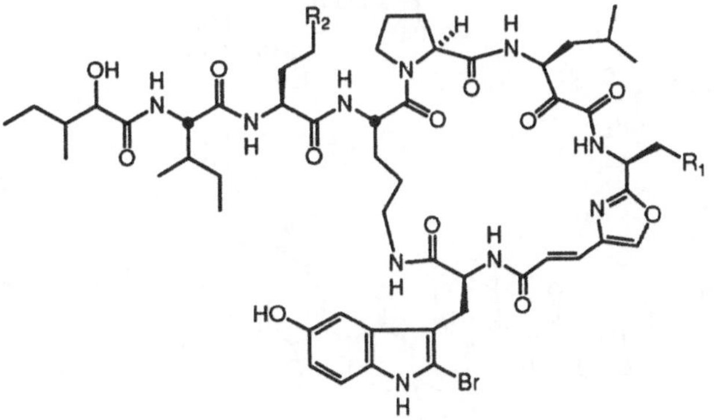

1141 $R_1 = R_2 = CH_3$ (keramamide B)

1142 $R_1 = CH_3$, $R_2 = H$ (keramamide C)

1143 $R_1 = R_2 = H$ (keramamide D)

and the closely related keramamides B–D (**1141–1143**) (*896*), all four of which were found in the sponge *Theonella* and contain the new amino acid 2-bromo-5-hydroxytryptophan. The keramamides inhibit superoxide generation in human neutrophils. The Okinawan sponge *Theonella*

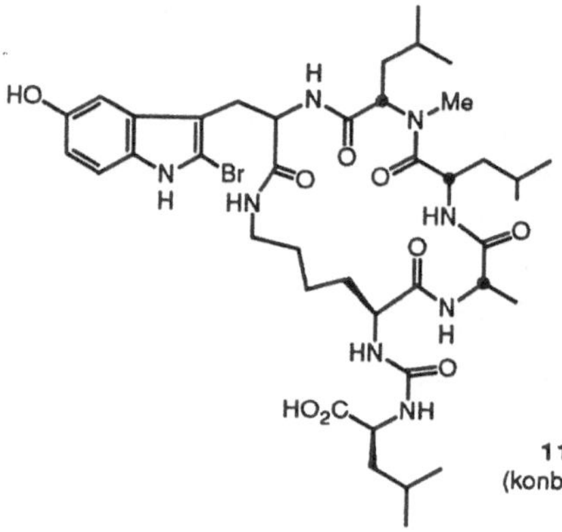

1144
(konbamide)

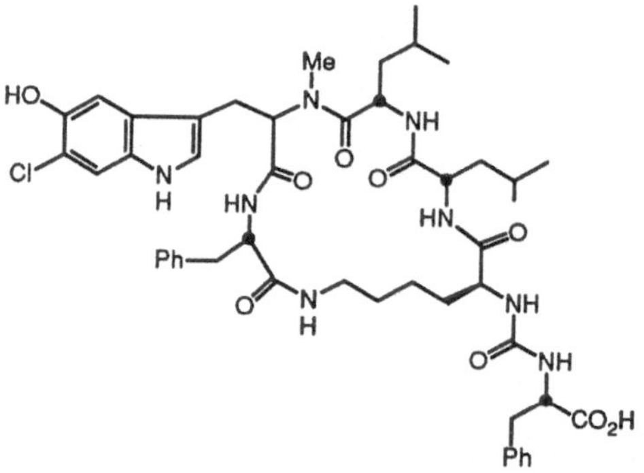

1145
(keramamide A)

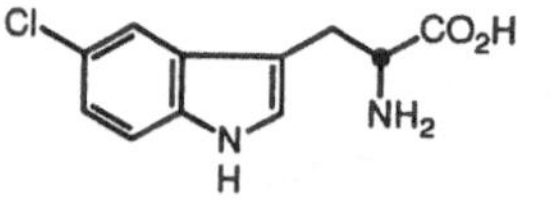

1146

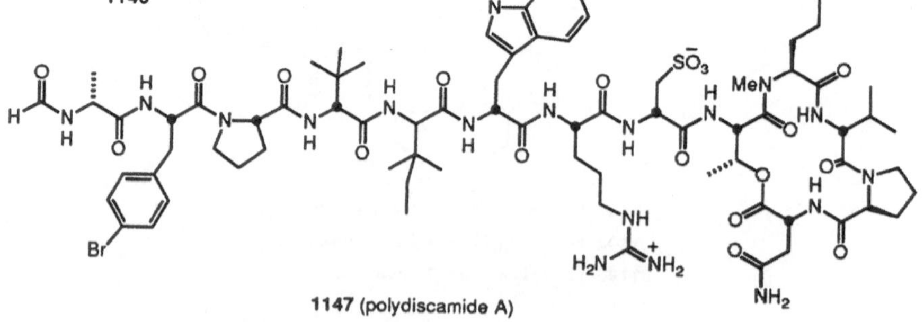

1147 (polydiscamide A)

sp. also contains the similar konbamide (**1144**) (*897*) and keramamide A (**1145**) (*898*), the latter containing the new amino acid 6-chloro-5-hydroxy-*N*-methyltryptophan. Both peptides show calmodulin antagonist activity. The antibiotic longicatenamycin contains the novel amino acid 5-chloro-D-tryptophan (**1146**) as one of its residues, although the complete structure is unknown (*899*). The novel peptide polydiscamide A (**1147**) which contains an indole ring but is brominated in the phenyl ring was isolated from the deep-sea sponge *Discodermia* sp. (*900*). Both the *p*-bromophenylalanine and 3-methylisoleucine are new amino acids. The marine ascidian *Diazona chinensis* secretes the potent cytotoxic diazonamides A (**1148**) and B (**1149**) which have a unique *bis*-oxazole indole structure (*901*).

A *Streptomyces* sp. produces the farnesyl-protein transferase inhibitors pepticinnamins (*902*) such as pepticinnamin E (**1150**) which contains a novel chlorinated tyrosine derivative (*903*). The marine sponge *Geodia* sp. contains the novel iodo- and bromotyrosine-containing cyclic peptides geodiamolides A (**1151**) and B (**1152**) (*904*); these two cyclodepsipeptides as well as C–F (**1153–1156**) were isolated from the marine sponge *Pseudaxinyssa* sp. (*905*). Activity of these compounds against L-1210 *in vitro* is very high ($IC_{50} = 0.002$ μg/ml). The endothelin antagonists cochinmicins II (**1157**), III (**1158**), and IV (**1159**) which contain a novel 2-chloropyrrole moiety are produced by a *Microbispora* sp. (*906, 907*). A marine bacterium associated with the jellyfish *Cassiopeia xamachana* produces salinamide A and B (**1160**) the former of which is the epoxide corresponding to **1160** (*908*). Both metabolites exhibit potent topical anti-inflammatory activity. The potent antitumor depsipeptides cryptophycins have been identified in a *Nostoc* sp. blue-green alga (*909*). The major compound is cryptophycin A (**1161**); the minor metabolites include cryptophycins C (**1162**), E (**1163**), F (**1164**), and G (**1165**), with

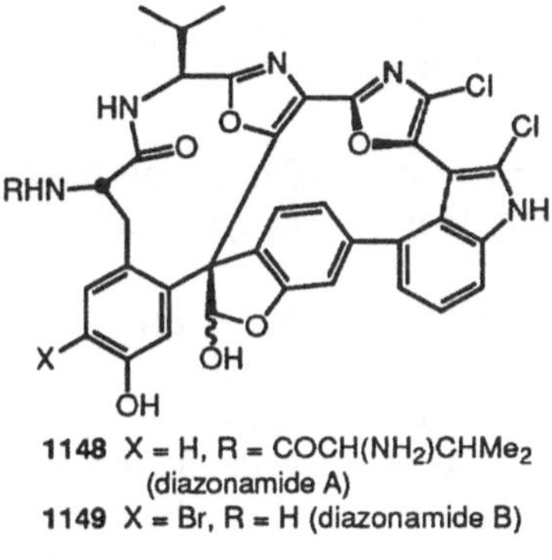

1148 X = H, R = $COCH(NH_2)CHMe_2$
(diazonamide A)
1149 X = Br, R = H (diazonamide B)

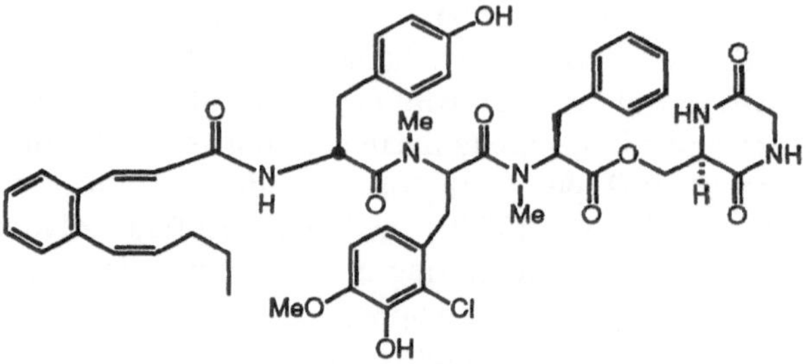

1150 (pepticinnamin E)

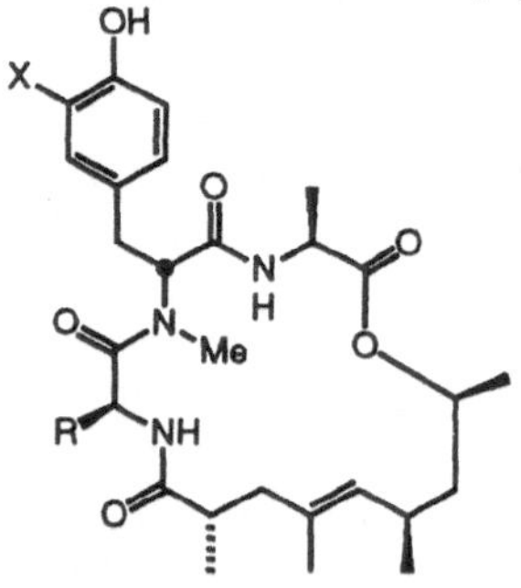

1151 X = I, R = Me (geodiamolide A)
1152 X = Br, R = Me (geodiamolide B)
1153 X = Cl, R = Me (geodiamolide C)
1154 X = I, R = H (geodiamolide D)
1155 X = Br, R = H (geodiamolide E)
1156 X = Cl, R = H (geodiamolide F)

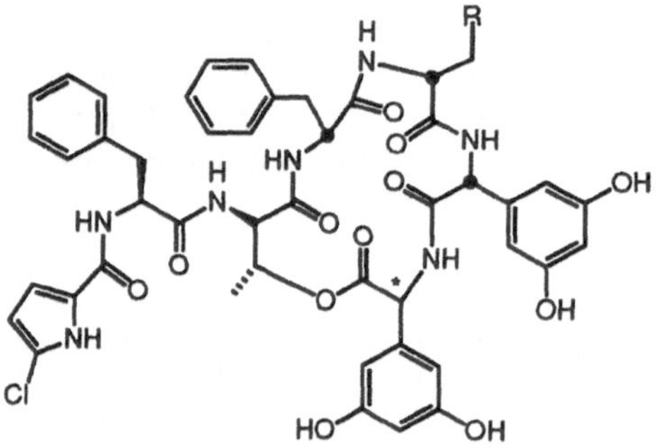

1157 R = H *S (cochinmicin II)
1158 R = H *R (cochinmicin III)
1159 R = OH *S (cochinmicin IV)

E and F isolated as methyl esters (*909*). Cryptophycin A shows excellent *in vivo* antitumor activity in mice and cytotoxicity activities in the range of $IC_{50} = 3–5$ pg/ml.

The Japanese sponge *Theonella* sp. produces the antifungal and cytotoxic cyclic dodecapeptide theonellamide F (**1166**) which incorporates several unusual amino acids, of which the most notable is the histidinoalanine residue (*910*). Likewise, the unprecedented 3-amino-14-chloro-2-hydroxy-4-methylpalmityl amino acid is found in

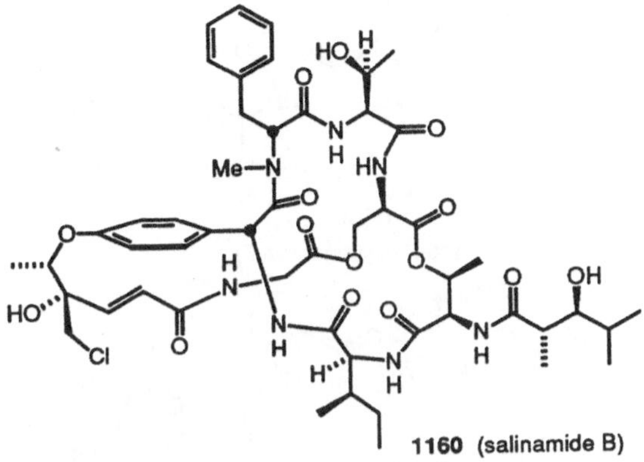

1160 (salinamide B)

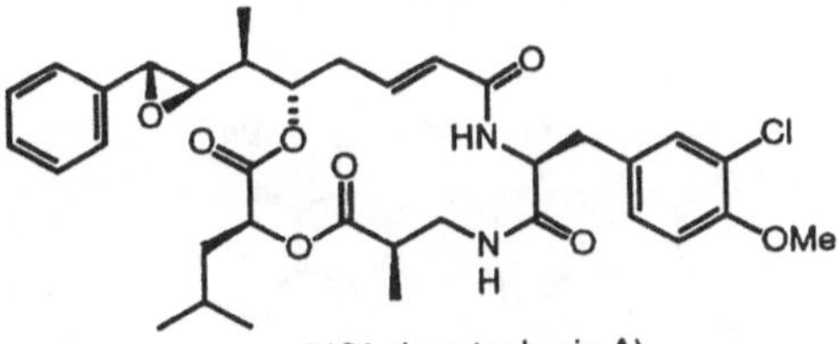

1161 (cryptophycin A)

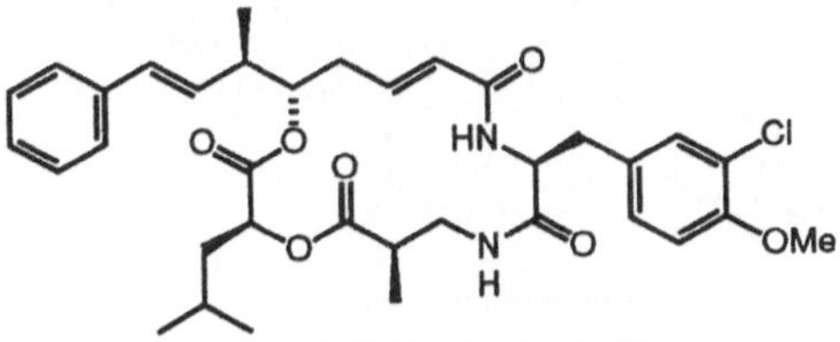

1162 (cryptophycin C)

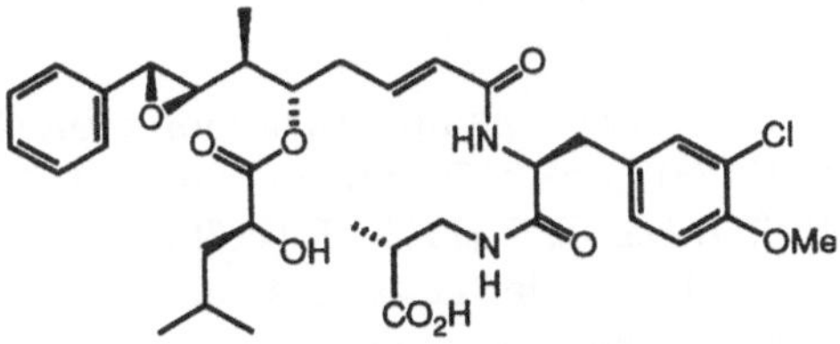

1163 (cryptophycin E)

puwainaphycin C (**1167**) and D (**1168**), which were isolated from cultures of the terrestrial blue-green alga *Anabaena* sp. (*911*). Two closely related antibiotic cyclic peptides, enduracidin A (**1169**) and B (**1170**), have been isolated from *Streptomyces fungicidicus* (*912*). A series of papers has successfully elucidated the structures of several similar cyclic peptides from *Pseudomonas syringae* which is a bacterial pathogen of plants (*e.g.*,

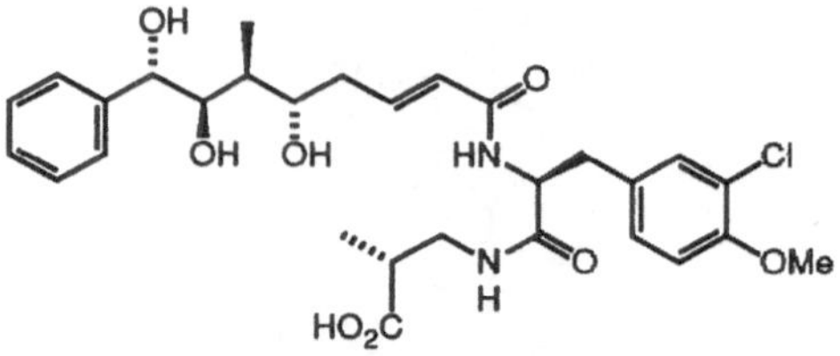

1164 (cryptophycin F)

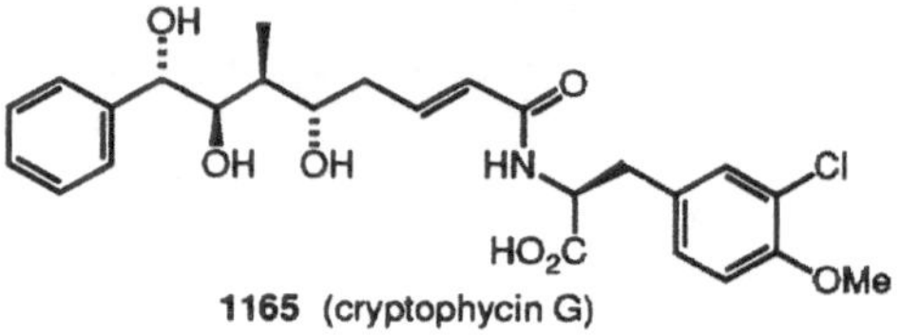

1165 (cryptophycin G)

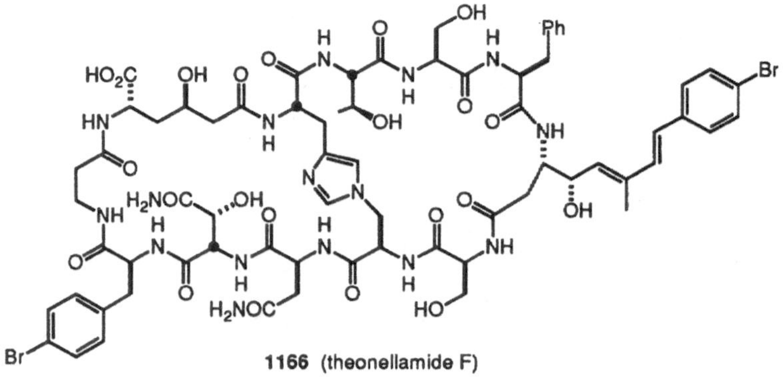

1166 (theonellamide F)

lilac). These include the chlorine-containing syringomycins (**1171**), syringomycin E (**1172**), G (**1173**) (*913–915*), syringotoxin B (**1174**) (*915*), and syringostatins A (**1175**), B (**1176**), C (**1177**), E (**1178**), G (**1179**), H (**1180**) (*914, 916, 917*), all of which contain the *threo*-4-chlorothreonine residue and most of which are fully characterized.

3.13. Alkaloids

Given the multitude of terrestrial plant alkaloids, it is perhaps surprising that only a few halogenated terrestrial alkaloids exist. In fact, nearly all representatives of this class come from the oceans. Indole, pyrrole, carboline, and related common ring systems are covered sepa-

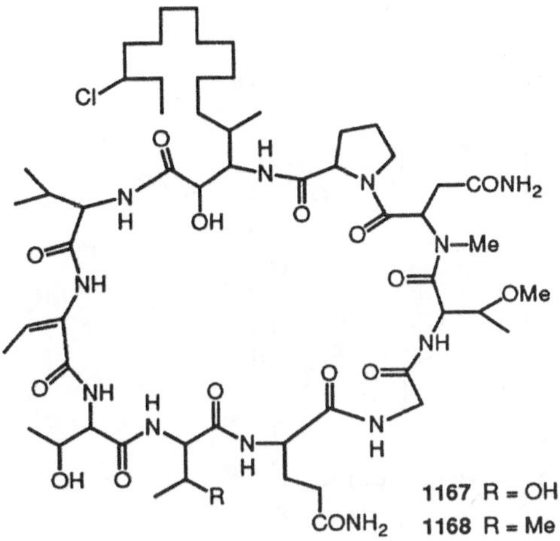

1167 R = OH (puwainaphycin C)
1168 R = Me (puwainaphycin D)

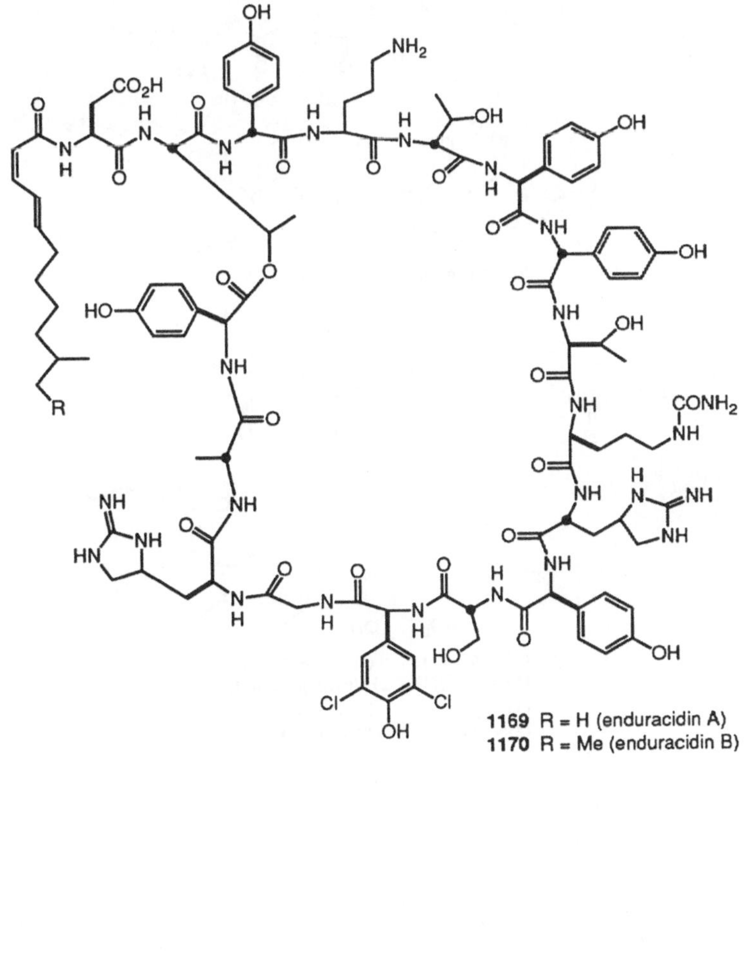

1169 R = H (enduracidin A)
1170 R = Me (enduracidin B)

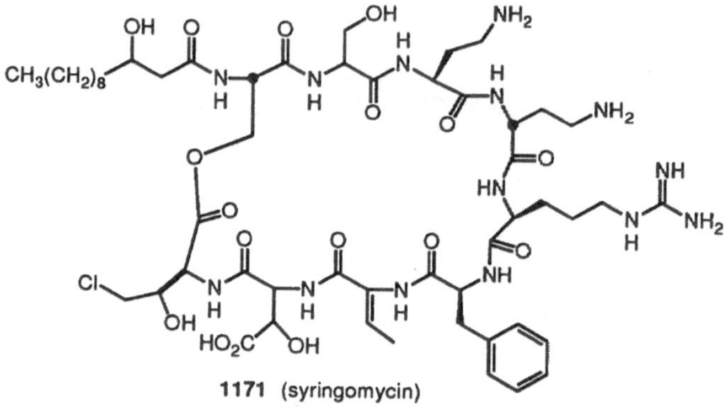

1171 (syringomycin)

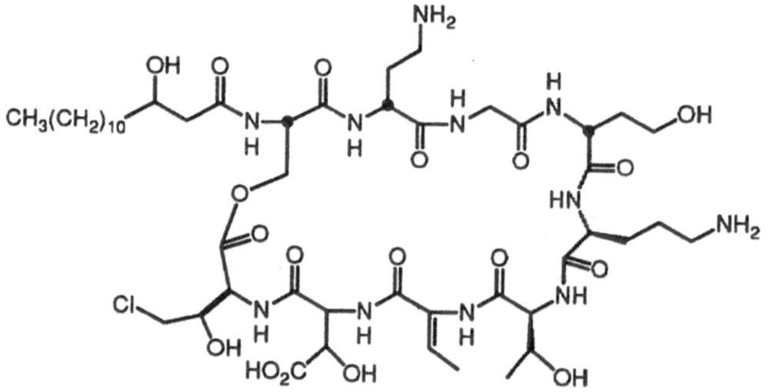

1174 (syringotoxin B)

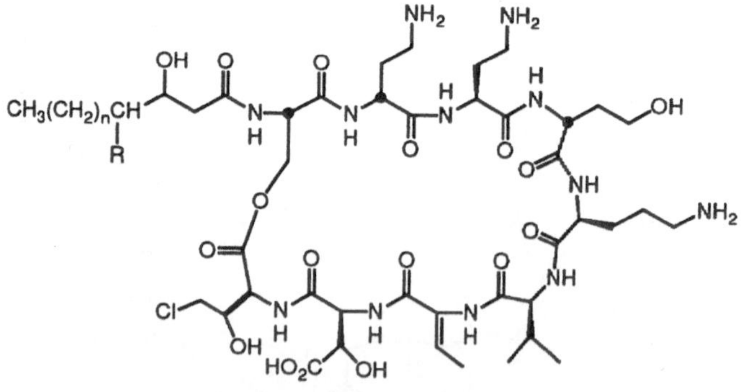

1175 n = 9, R = H (syringostatin A)
1176 n = 9, R = OH (syringostatin B)
1178 n = 7, R = H (syringostatin E)
1180 n = 11, R = OH (syringostatin H)

rately. The first example of a halogen-containing alkaloid from any source appears to be the pyrrolizidine alkaloid jaconine (**1181**) from *Senecio jacobaea* (Compositae) (*918, 940, 941*) although its gross structure was determined later (*919–921*). The toxicity of this plant, "tansy ragwort," and other *Senecio* sp. to cattle and other livestock is well documented and is caused by pyrrolizidine alkaloids (*922, 925*). The related pyrrolizidine alkaloids chlorodeoxysceleratine (**1182**) (from *Senecio*

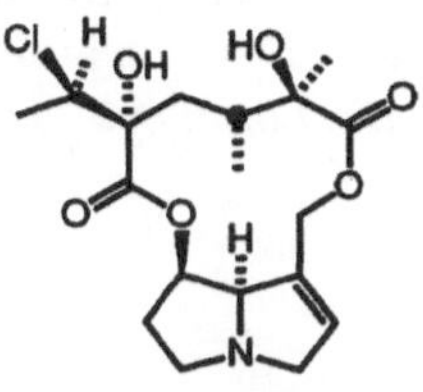

1181 (jaconine)

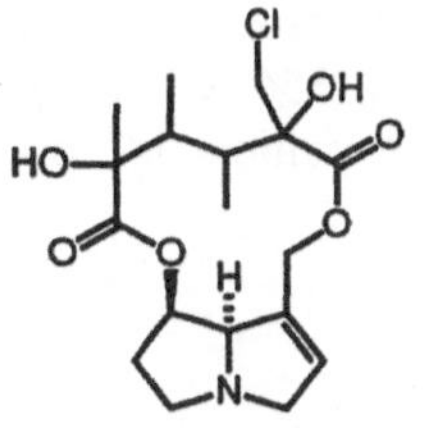

1182 (chlorodeoxysceleratine)

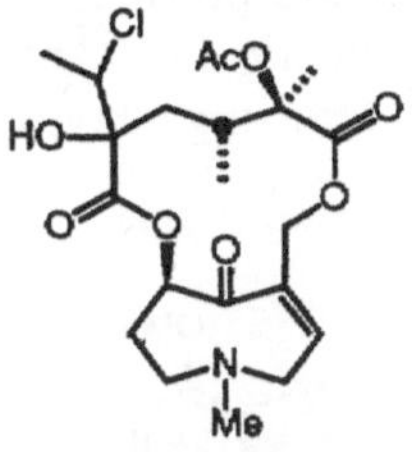

1183 (doronine)

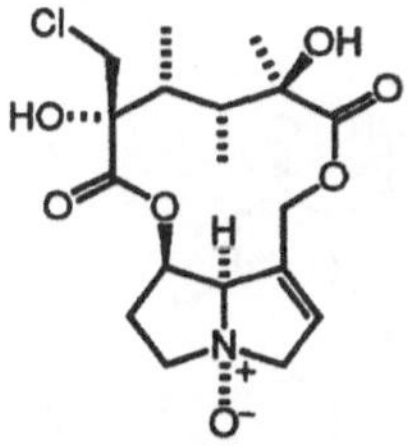

1184 (merenskine *N*-oxide)

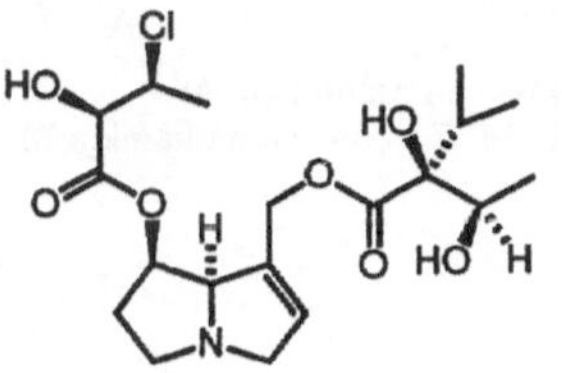

1185

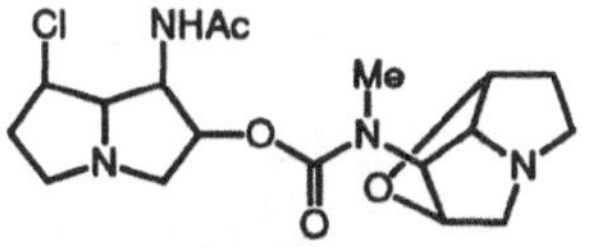

1186 (lolidine)

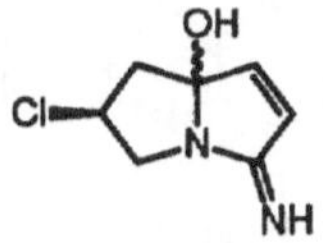

1187 α-OH (clazamycin A)
1188 β-OH (clazamycin B)

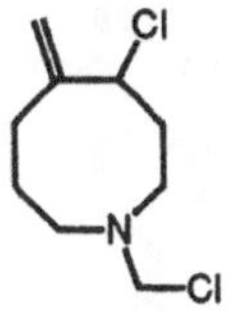

1189 (oxypterine)

1190 R_1 = Me, R_2 = OH (acutumine)
1191 R_1 = H, R_2 = OH (acutumidine)
1192 R_1 = Me, R_2 = H

sceleratus) (*923*) and doronine (**1183**) (from *Doronicum macrophyllum*) (*924*) also contain chlorine. Jaconine (**1181**) has also been detected in *Senecio alpinus* (*926*) and doronine is present in *S. clevelandii* (*927*) and in *S. abrotanifolius* (*928*). In *S. latifolius* is found the new alkaloid merenskine *N*-oxide (**1184**) (*929*), while from both *Cryptantha clevelandii* and *C. leiocarpa* (Boraginaceae) there was isolated the alkaloid **1185** (*930*). The seeds of *Lolium cuneatum* (Gramineae) contain lolidine which has been assigned structure **1186** (*932*). The novel clazamycins A (**1187**) and B (**1188**) are produced by *Streptomyces* sp. (*933, 934*). The latter metabolite is identical to "Antibiotic 354" from *S. puniceus* subsp. *doliceus* (*935*). The interesting oxypterine which is found in *Lotononis oxyptera*

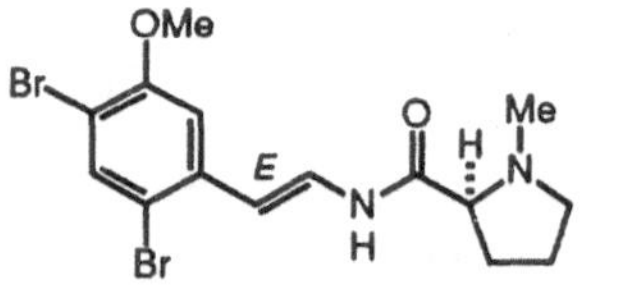

1193 (amathamide A)
1194 *Z*-isomer (amathamide B)

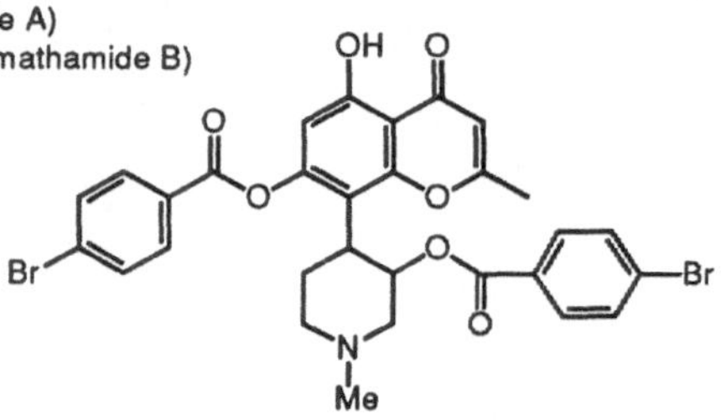

1195 (tubastraine)

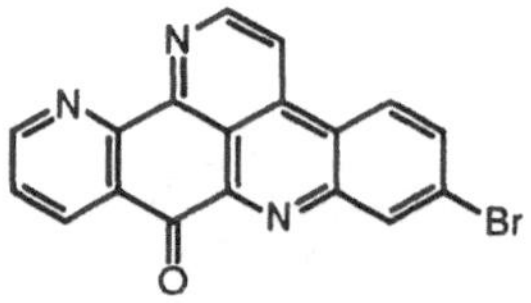

1196 (2-bromoleptoclinidinone)

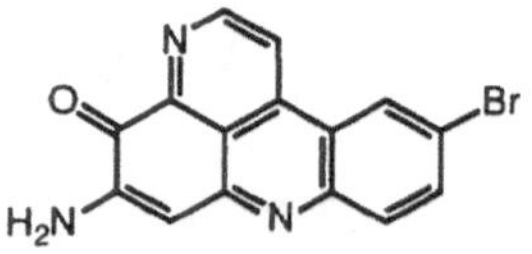

1198 (pantherinine)

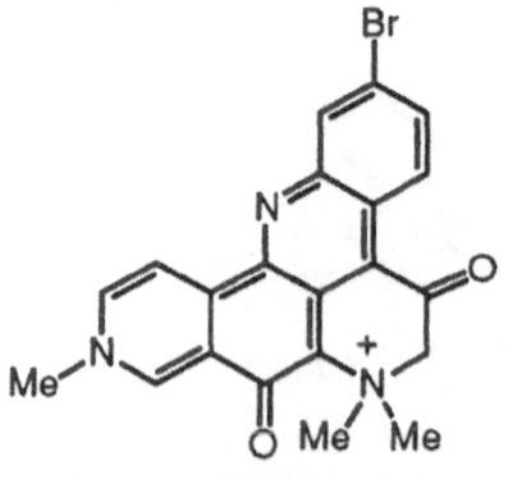

1197 (petrosamine)

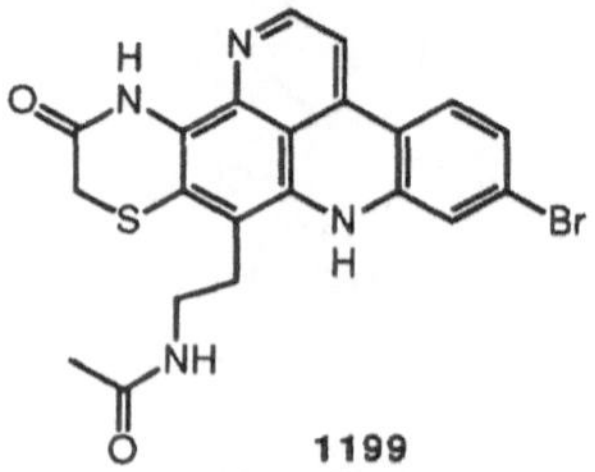

1199 (shermillamine A)

(Fabaceae) is proposed to have the most unlikely α-chloroamine structure **1189** (*936*).

Although acutumine (**1190**) was first isolated in 1929, the structure of it and the related acutumidine (**1191**) from *Sinomenium acutum* and *Menispermum dauricum* (both Menispermaceae) were not elucidated until 1967 (*937–939, 1082*). The related acutuminine (**1192**) was isolated from the latter plant in 1969 (*942*). Radiolabeling evidence has been reported to show that acutumine (**1190**) is not an isolation artifact (*943*). *M. canadense* also contains **1190** and **1191** (*1083*).

The largest group of natural halogenated alkaloids is found in marine organisms (*944, 945*). More importantly, these metabolites display a range of biological activities. The marine bryozoan *Amathia wilsoni* produces amathamide A (**1193**) and B (**1194**) (*946*). The stony coral *Tubastraea micrantha* which is avoided by the destructive Crown-of-Thorns seastar produces tubastraine (**1195**), the first chromone to be isolated from a marine invertebrate (*947*). The first of a series of halogenated pyridoacridine alkaloids to be isolated from marine sources was 2-bromoleptoclinidinone (**1196**), isolated from a Truk Lagoon ascidian *Leptoclinides* sp. (*948*) (structure revised in (*949*)). A ruthenium complex of this alkaloid intercalates and cleaves DNA when irradiated with visible light (*950*). A *Petrosia* sp. sponge from Belize produces the interesting petrosamine (**1197**) which exists as the enol form (not shown) in solution (*951*). The ascidian *Aplidium pantherinum* from Southern Australia produces pantherinine (**1198**) which shows some cytotoxicity against the P388 cell line (*952*). The tunicate *Trididemnum* sp. contains the novel shermilamine A (**1199**) (*953*).

The Okinawan sponge *Theonella* sp. has yielded theoneberine (**1200**), the first tetrahydroprotoberberine alkaloid from marine sources and the first brominated derivative of this alkaloid family to be isolated from any natural source (*954*). The Tasmanian ascidian *Clavelina cylindrica* produces the novel alkaloids cylindricine A (**1201**) and B (**1202**), the latter of which is a novel pyrido[2,1-*j*]quinoline ring system (*955*). Not surprisingly, these metabolites are in equilibrium (3:2, **1201/1202**) *via* an aziridinium ion intermediate in a well-precedented example of neighboring-group participation (*956*). *Melodinus celastroides* (Apocynaceae) contains the novel *bis*-indole alkaloids **1203** and **1204**, which the authors suggest may be artifacts from the dichloromethane used in the extraction process (*957*). The Chinese and Japanese folk medicine plant *Houttuyniae cordata* (Saviuraceae) has furnished the alkaloid 7-chloro-6-demethyl-cepharadione B (**1205**), in an isolation protocol devoid of any source of chlorine (*958*). The soil fungus *Chaetasbolisia erysiophoides* produces the novel angiogenesis inhibitors WF-16775 A_1 (**1206**) and A_2 (**1207**) (*959*).

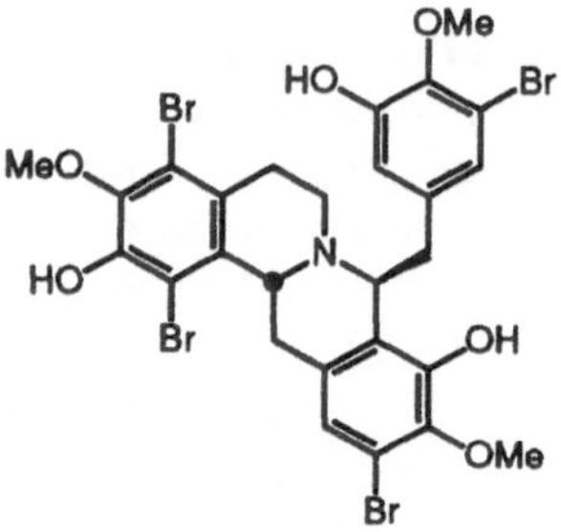

1200 (theoneberine)

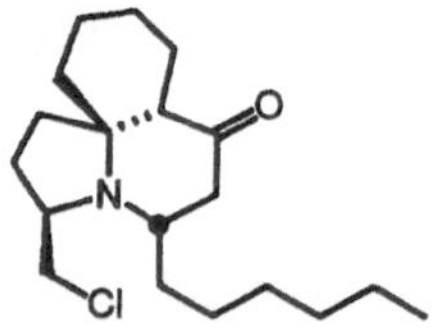

1201 (cylindricine A)

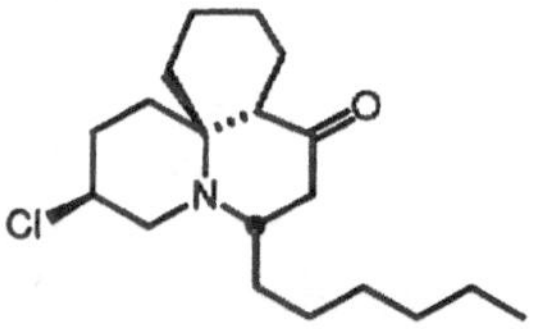

1202 (cylindricine B)

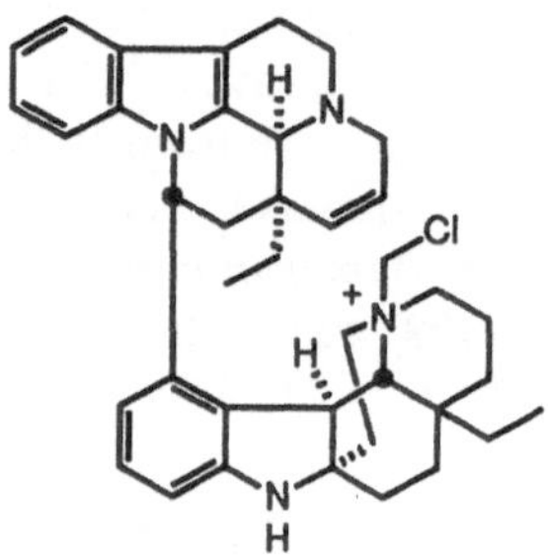

1203

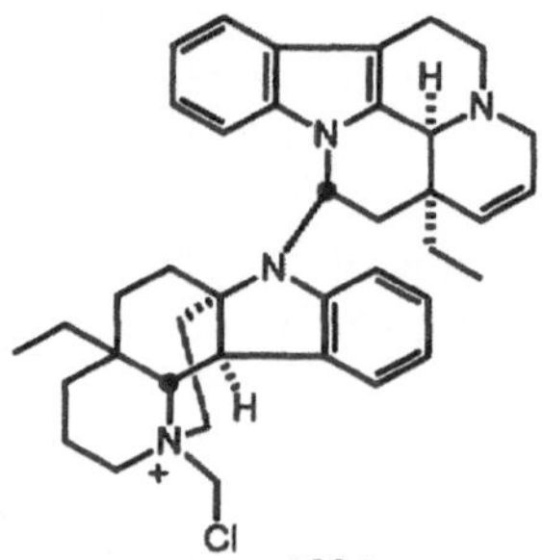

1204

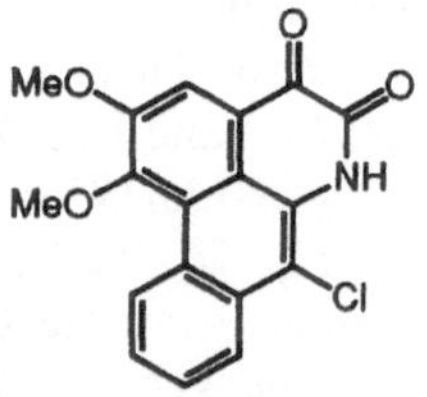

1205

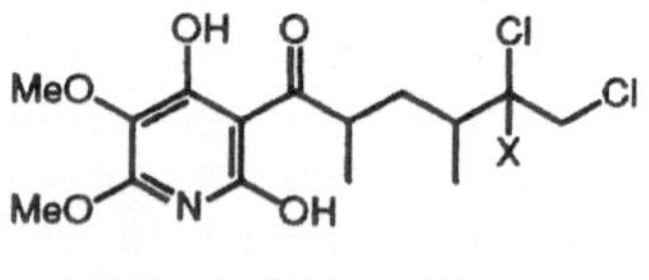

1206 X = H (WF-16775 A_1)
1207 X = Cl (WF-16775 A_2)

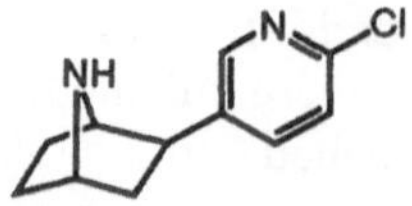

1208 (epibatidine)

Both compounds have potent antiangiogenic activity, which prevents new blood vessel formation. A fitting way to conclude this section is with the recent discovery of epibatidine (**1208**) from the Ecuadoran frog *Epipedobates tricolor* (*960, 961*), a powerful analgesic with a novel mode of action and a compound that is also found at lower levels in *E. anthonyi*, *E. espinosai*, and *E. pictus*. This simple chloropyridino-7-aza-norbornane derivative is 500–1000 times more potent than morphine. Control isolation experiments demonstrate that epibatidine is not an artifact. The absolute configuration has also been established (*962*).

3.14. Heterocycles

3.14.1. Pyrroles

The enormous reactivity of pyrroles in electrophilic substitution – 3×10^{18} times faster than benzene in electrophilic bromination (*963*) –portends an abundance of naturally occurring halogenated pyrroles from both terrestrial and marine sources.

Several halogenated pyrroles are produced by *Pseudomonas* microbes. Pyoluteorin (**1209**) was isolated from *P. aeruginosa* (*964, 965*)

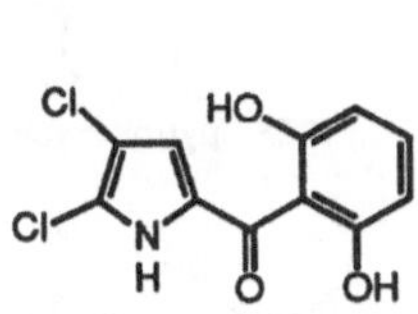

1209 (pyoluteorin) 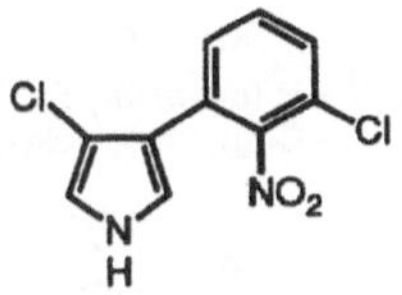 **1210** (pyrrolnitrin)

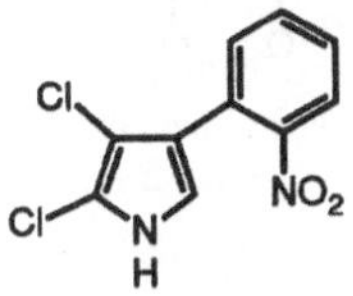

1211 (isopyrrolnitrin) **1212** (oxypyrrolnitrin) **1213** X = H **1214** X = Cl

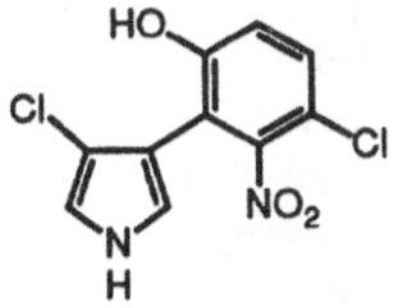

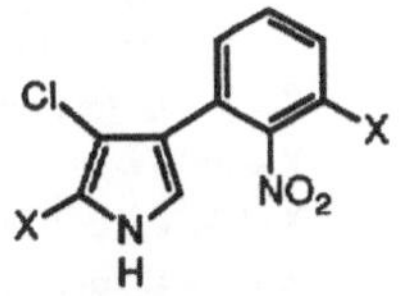

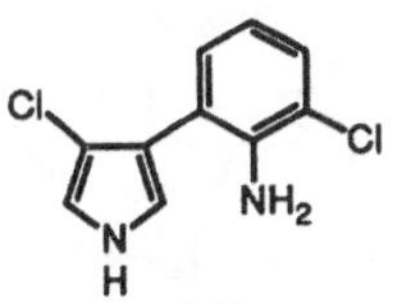

1215 (aminopyrrolnitrin) **1216** X = Cl **1217** X = H

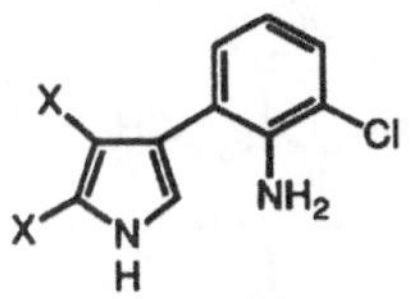

and, later, also from *P. fluorescens* (*966*). Pyrrolnitrin (**1210**), which contains a rare natural nitro group, was isolated from *P. pyrrocinia* (*967, 968*), *P. aureofaciens* (*969, 970*), and *P. cepacia* (*971*). Related metabolites have also been found in these organisms: isopyrrolnitrin (**1211**) (*P. pyrrocinia*) (*972*), oxypyrrolnitrin (**1212**) (*P. pyrrocinia*) (*973*), and **1213** (*P. pyrrolnitrica*) (*974*), **1214** (*P. fluorescens*) (*975*) (*P. cepacia*) (*971*) (*P. aureofaciens*) (*976*), **1215** (*P. cepacia*) (*971*) (*P. aureofaciens*) (*976*), **1216** (*P. cepacia*) (*971*), and **1217** (*P. aureofaciens*) (*970*). Biosynthetic studies support a pathway for the formation of pyrrolnitrin (**1210**) involving 7-chlorotryptophan and aminopyrrolnitrin (**1215**) with oxidation of the

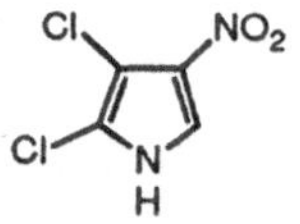

1218 (pyrrolomycin A)

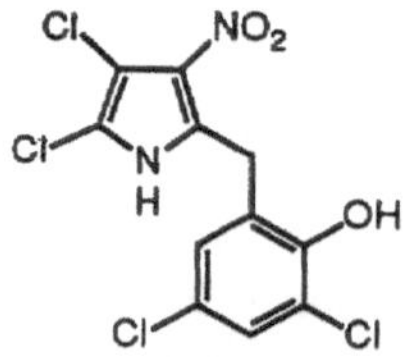

1219 (pyrrolomycin B)

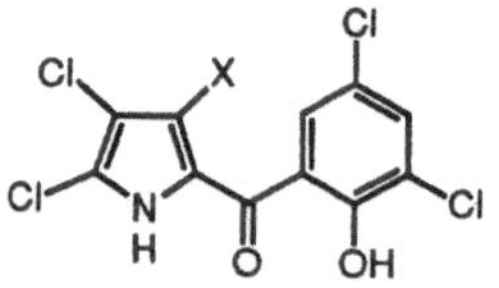

1220 X = H (pyrrolomycin C)
1221 X = Cl (pyrrolomycin D)

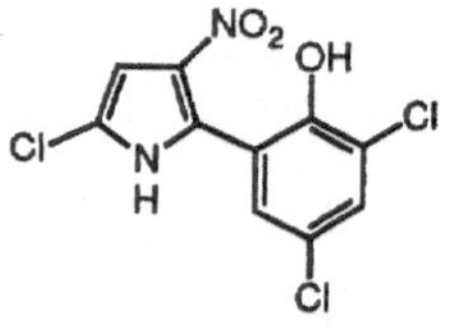

1222 (pyrrolomycin E)

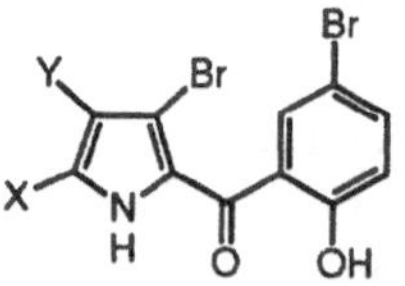

X = Y = Br (pyrrolomycin F_1)
X = Br, Y = Cl (pyrrolomycin F_{2a})
X = Cl, Y = Br (pyrrolomycin F_{2b})
X = Y = Cl (pyrrolomycin F_3)

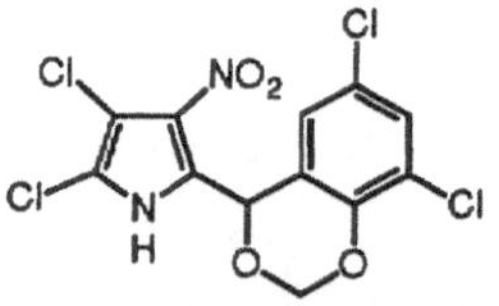

1223 (pyrroxamycin)

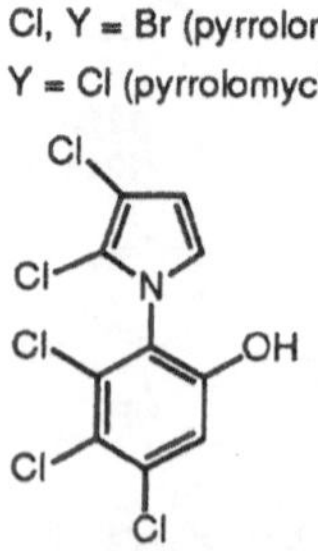

1224 (neopyrrolomycin)

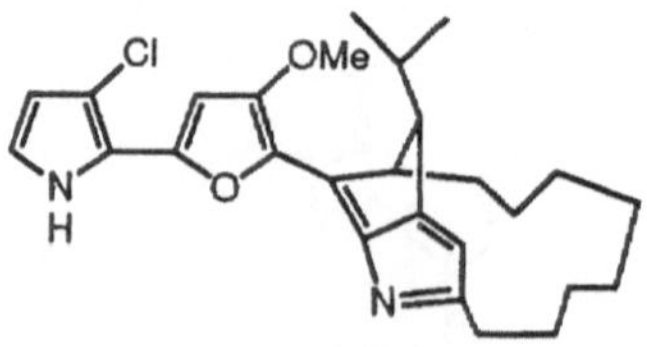

1225 (roseophilin)

amino group to the nitro group by chloroperoxidase in the final step (*970, 977*). Pyrrole **1216** shows activity against apple pathogens (*971*) and aminopyrrolnitrin (**1215**) is a non-steroidal androgen-receptor antagonist (*978*).

An *Actinomyces* sp. produces pyrrolomycin A (**1218**) and B (**1219**) (*979, 980*), the first two of a series of related microbial antibacterial agents to display powerful, perhaps clinical, antibiotic activity. Pyrrolomycin B (**1219**) was also isolated from *Streptomyces* sp. (*981*) and pyrrolomycins C (**1220**), D (**1221**), and E (**1222**) are produced by *Actinosporangium vitaminophilum* (*982, 983*). When bromide is present in the fermentation broth of *A. vitaminophilum*, the brominated pyrrolomycins (F_1, F_{2a}, F_{2b}, F_3) are produced (*984*). These "forced" metabolites, which have stronger antibiotic activity than their chlorinated counterparts (**1218**–**1222**), are not counted here as being natural. Neither iodide nor fluoride is incorporated into similar metabolites by these microorganisms (*984*). Likewise, the addition of ammonium bromide to the culture medium of *Pseudomonas pyrrolnitrica, P. pyrrocinia*, or *P. schuylkilliensis* furnishes bromonitrins A–C, the bromo analogs of pyrrolnitrin (**1210**), **1214**, and an isomer of **1214**, respectively (*1084*). Pyrroxamycin (**1223**) is produced by a *Streptomyces* sp. (*985*) and the optically active neopyrrolomycin (**1224**) has been identified in a culture of another *Streptomyces* sp. (*986*). This antibiotic shows activity against both gram positive and negative bacteria. The novel metabolite roseophilin (**1225**) from *S. griseoviridis* shows significant cytotoxicity against KB cells (IC_{50} = 0.88 μM) and K562 cells (IC_{50} = 0.34 μM) (*987*).

The marine bacterium *Pseudomonas bromoutilis* produces the bromo analog **1226** (*988, 989, 1010*), a metabolite that is also found in the marine bacterium *Chromobacterium* sp. along with the brominated pyrrole **1227** and *bis*-pyrrole **1228** (*990*). Marine bacteria are becoming increasingly important as a new natural products resource (*991*). The marine worm *Polyphysia crassa* produces the antibacterial metabolite 2,3,4-tribromopyrrole (**1229**), a structure which was confirmed by synthesis (*992*). This extremely labile compound undergoes disproportionation at $-18°$ C to **1227** and a dibromopyrrole. The chemical defensive and fish antifeedant compounds tambjamines B (**1230**) and D (**1231**) are two of four metabolites that are secreted by the bryozoan *Sessibugula translucens*. These *bis*-pyrroles are also found in the grazing predatory nudibranchs *Tambje eliora* and *T. abdere* which prey upon *S. translucens* as well as in the carnivorous nudibranch *Roboastra tigris* which in turn preys upon the *Tambje* species (*993, 994*). The *Tambje* animals seem to utilize these compounds both as food attractants and alarm pheromones, depending on the concentration (*994*). The novel iodine-containing pyrrolo-oxazinone lukianol B (**1232**) is found in an unidentified tunicate (*995*), while the interesting polycitone A (**1233**) and polycitrins A (**1234**) and B (**1235**) are produced by a marine ascidian *Polycitor* species (*996*). The fungus *Auxarthron umbrinum* produces rumbrin (**1236**) which prevents membrane lipid peroxidation and calcium overload and thus may be useful in the treatment of myocardial and cerebral ischemia (*997, 998*).

1226

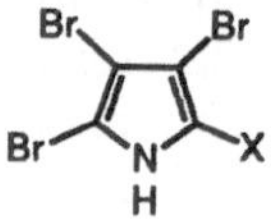

1227 X = Br
1229 X = H

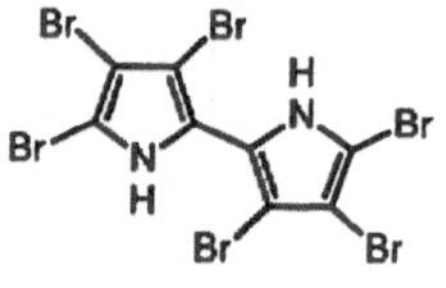

1228

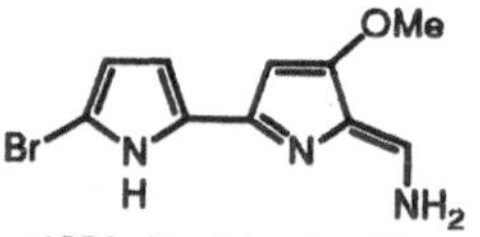

1230 (tambjamine B)

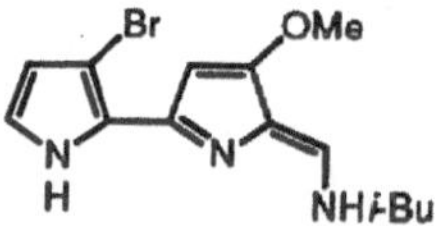

1231 (tambjamine D)

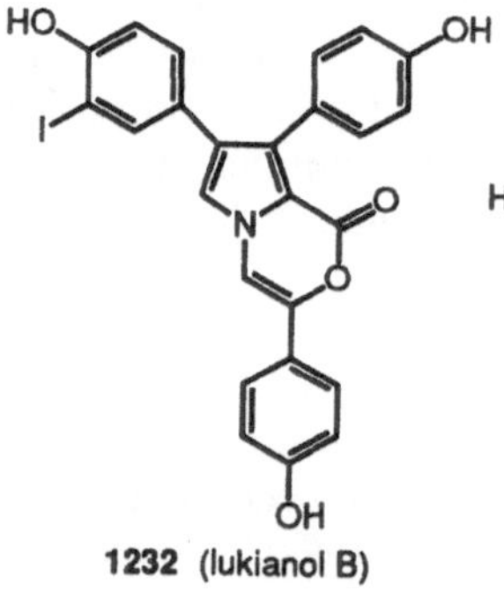

1232 (lukianol B)

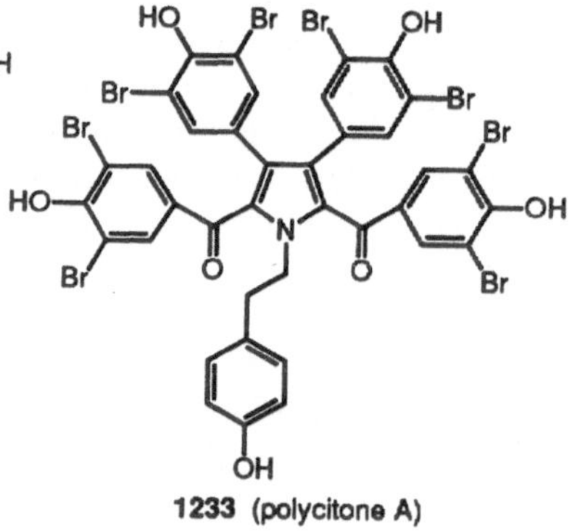

1233 (polycitone A)

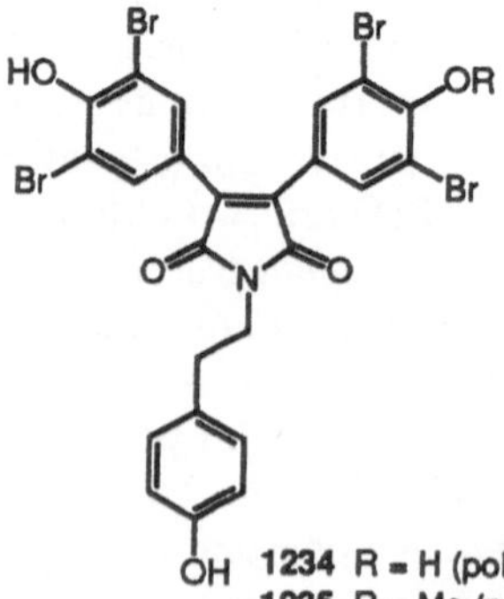

1234 R = H (polycitrin A)
1235 R = Me (polycitrin B)

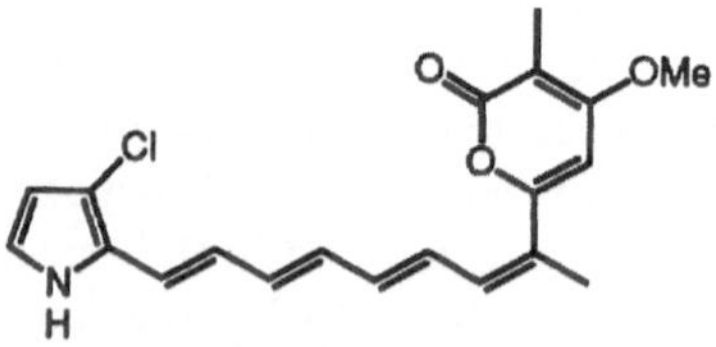

1236 (rumbrin)

Thiazohalostatin (**1237**) which is found in cultures of *Actinomadura* sp. also has a cytoprotective effect and prevents cell death caused by calcium overload (*999, 1000*). The marine ciliate *Pseudokeronopsis rubra* produces the four chemical defense agents keronopsins A_1 (**1238**), A_2 (**1239**), B_1 (**1240**), and B_2 (**1241**) (*1001*). It appears that when this organism is disturbed it converts the two A compounds (**1238, 1239**) into the more toxic B compounds (**1240,1241**) (*1001*). The similarity of rumbrin (**1236**) to the keronopsins is noteworthy.

Marine sponges have provided us with a large assortment of brominated pyrroles. One of the first such metabolites was oroidin (**1242**) isolated from the *Agelas oroides* sponge from the Bay of Naples (*1002*) (structure revised in (*1003*)). This sponge also yielded the simple pyrroles (**1243–1245**) (*1002*), the former of which was also found in *A. flabelliformis* from the Bahamas and a compound that exhibits potent immunosuppressive activity (*1004*). Another *Agelas* sponge contains 2,3-dibromopyrrole (**1246**), as well as **1247–1250** and the known **1243** and **1245** (*1005*). Acid **1249** had been isolated earlier from an identified sponge in the Marshall Islands together with the new midpacamide (**1251**) (*1006*). Both **1251** and 5-debromomidpacamide (**1252**) as well as **1250** were found in *A. mauritiana* (*1007*), and **1253** was identified in another *Agelas* sp. (*1008*). The similar keramadine (**1254**) which is produced by an Okinawan *Agelas* sponge, is a novel antagonist of serotonergic receptors (*1009*). Another study of an Okinawan *Agelas* sp. has revealed the presence of the antileukemic agelasine G (**1255**) (*1011*). Fijian *Agelas mauritiana* contains in addition to the known midpacamide (**1251**) and

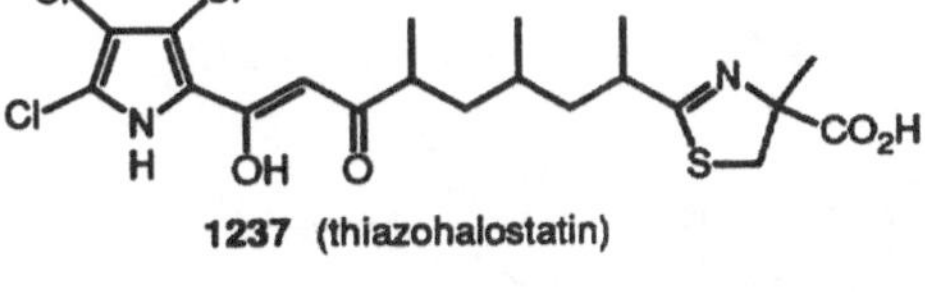

1237 (thiazohalostatin)

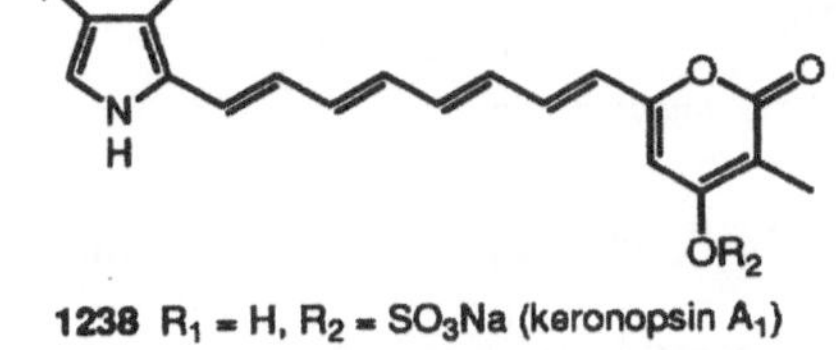

1238 R_1 = H, R_2 = SO_3Na (keronopsin A_1)
1239 R_1 = Br, R_2 = SO_3Na (keronopsin A_2)
1240 R_1 = R_2 = H (keronopsin B_1)
1241 R_1 = Br, R_2 = H (keronopsin B_2)

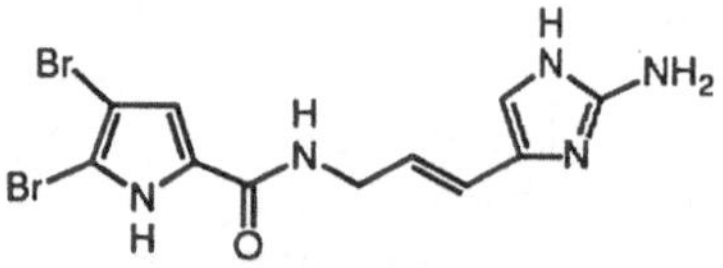

1242 (oroidin)

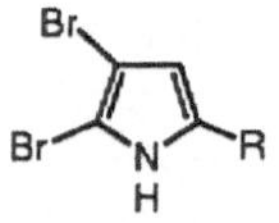

1243 R = CO_2H
1244 R = CN
1245 R = $CONH_2$
1246 R = H
1247 R = CH_2OMe
1248 R = CO_2Me

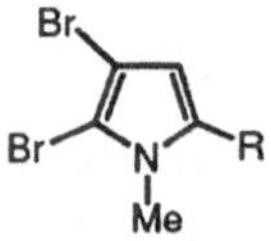

1249 R = CO_2H
1250 R = CO_2Me

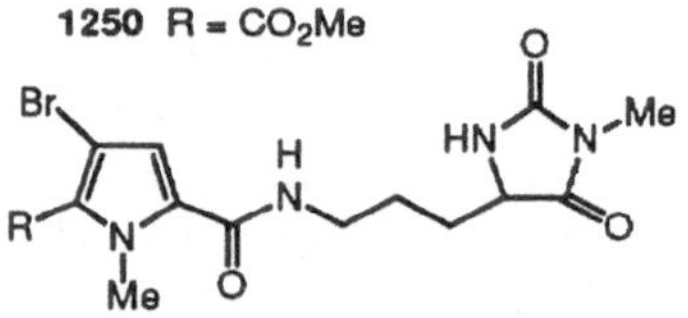

1251 R = Br (midpacamide)
1252 R = H

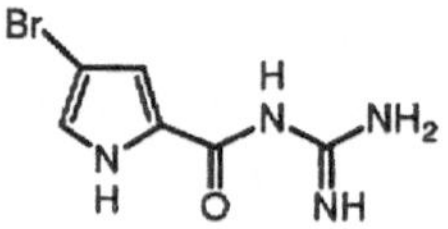

1253

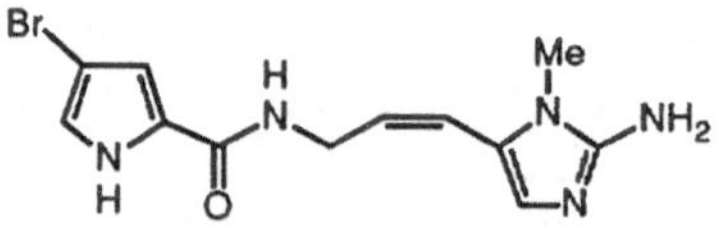

1254 (keramadine)

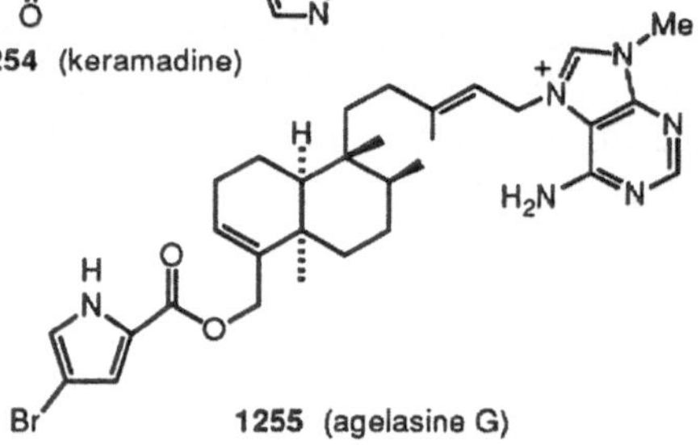

1255 (agelasine G)

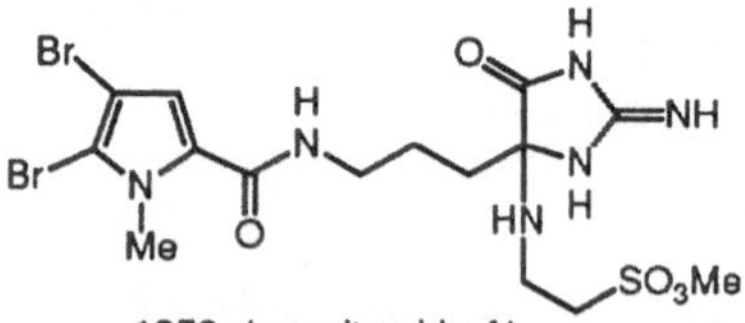

1256 (mauritamide A)

dibromophakellin (**1268**, *vide infra*) the novel mauritamide A (**1256**) (*1012*).

The novel tetracyclic dibromoagelaspongin (**1257**) has been identified in an *Agelas* sponge (*1013*). The similarly highly congested agelastatin A (**1258**) and **1259** are found in *Agelas dendromorpha* from the Coral Sea

(*1014*). The former metabolite is cytotoxic against KB cells (IC_{50} = 0.1–0.5 μg/ml). A biosynthesis of **1258** from an oroidin-like precursor is suggested (*1014*). The first of a series of dimeric metabolites was isolated from *Agelas sceptrum* in the form of the optically active sceptrin (**1260**) (*1015*). This was followed by isolation of the related sceptrins **1261** and **1262** and the related oxysceptrins **1263** and **1264** from *Agelas conifera* (*1016*). The synthetic magic of this Caribbean sponge was further displayed by the isolation of the ageliferins **1265–1267** (*1016, 1017*). The ageliferins are potent actomyosin ATPase activators and are present in

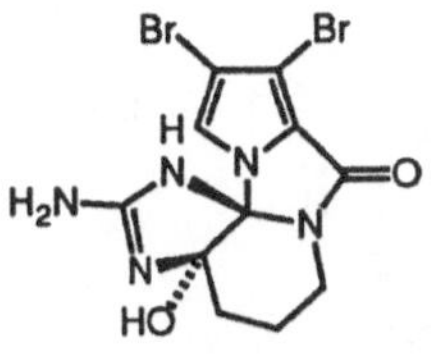

1257

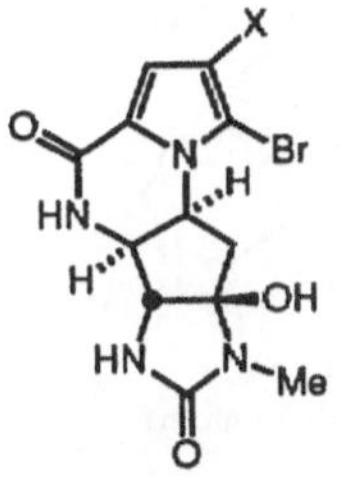

1258 X = H (agelastatin A)
1259 X = Br

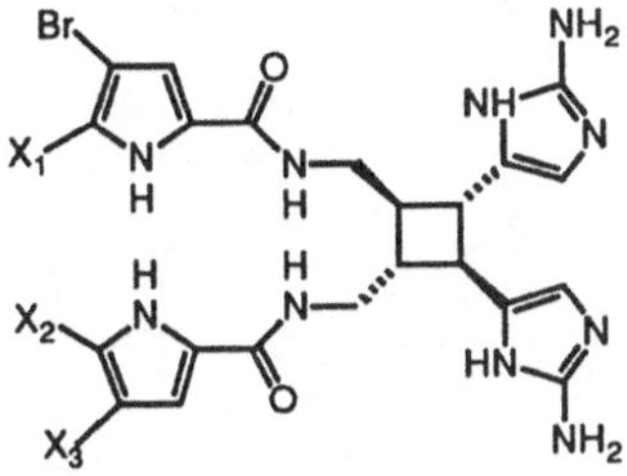

1260 $X_1 = X_2 = H$, $X_3 = Br$ (sceptrin)
1261 $X_1 = X_2 = X_3 = H$
1262 $X_1 = X_2 = X_3 = Br$

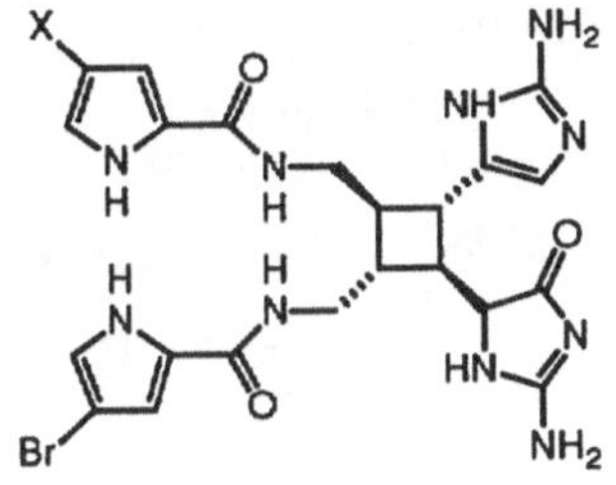

1263 X = H
1264 X = Br

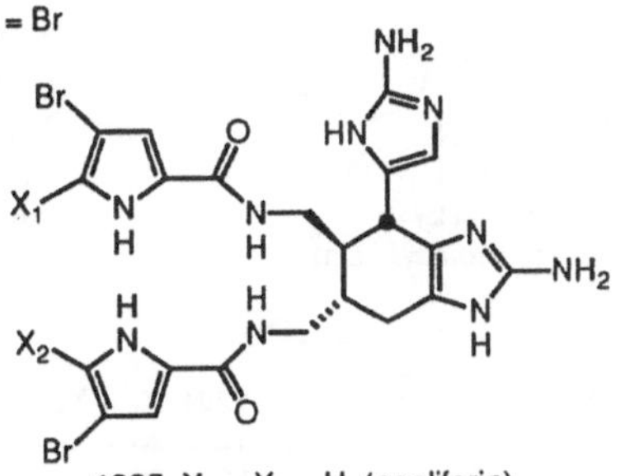

1265 $X_1 = X_2 = H$ (ageliferin)
1266 $X_1 = Br$, $X_2 = H$
1267 $X_1 = X_2 = Br$

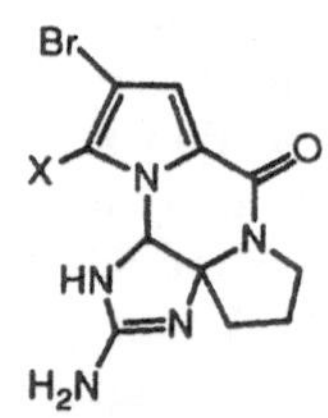

1268 X = Br (dibromophakellin)
1269 X = H (bromophakellin)

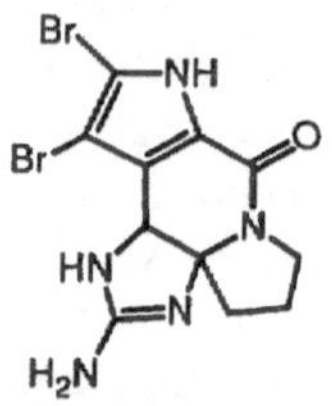

1270 (dibromoisophakellin)

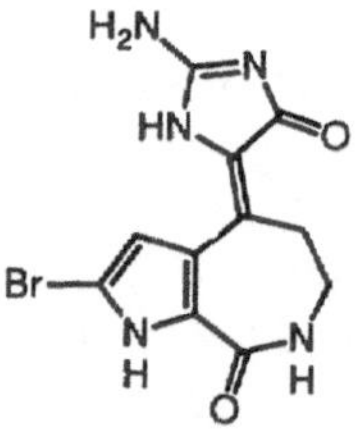

1271 (hymenialdisine)

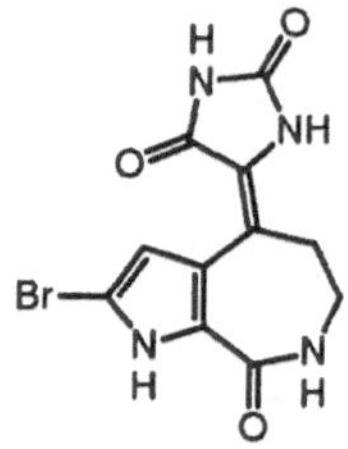

1272 (axinohydantoin)

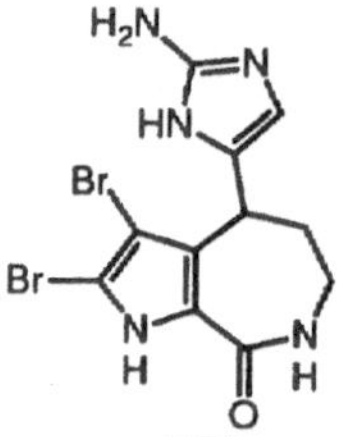

1273 (hymenin)

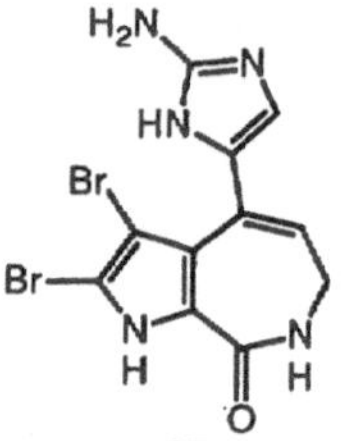

1274 (stevensine)

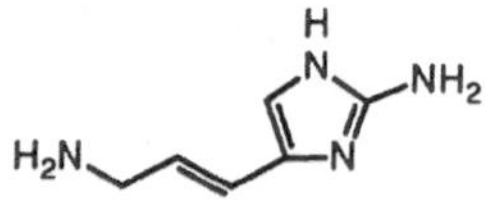

1242a (possible oroidin precursor)

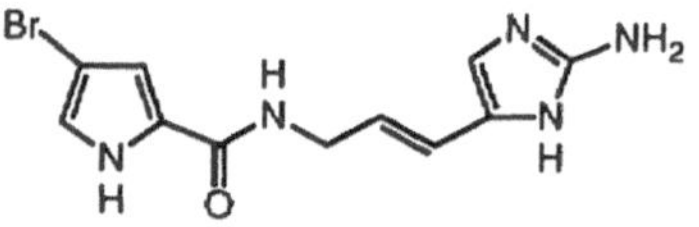

1275 (hymenidin)

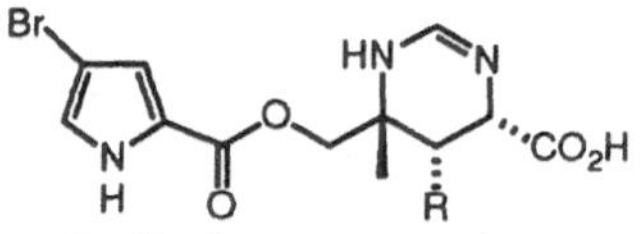

1276 R = H (manzacidin A)
1277 R = OH (manzacidin B)

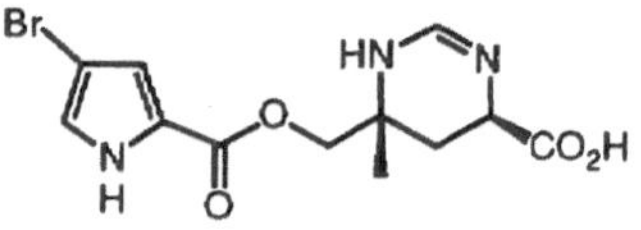

1278 (manzacidin C)

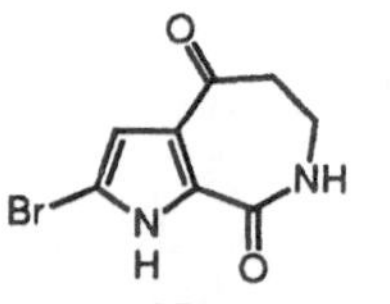

1279 (2-bromoaldisin)

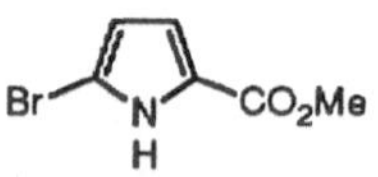

1280

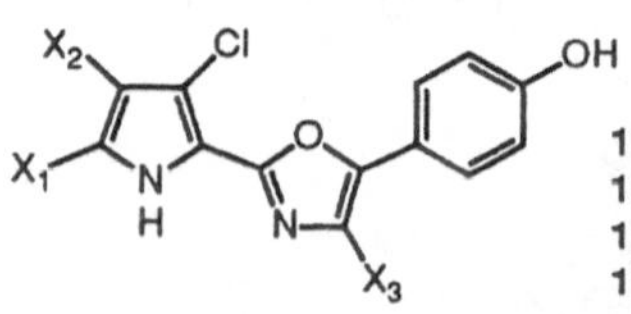

1281 $X_2 = X_3 = Cl$, $X_1 = H$ (phorbazole A)
1282 $X_1 = X_2 = Cl$, $X_3 = H$ (phorbazole B)
1283 $X_2 = Cl$, $X_1 = X_3 = H$ (phorbazole C)
1284 $X_1 = X_2 = X_3 = H$ (phorbazole D)

an Okinawan *Agelas* sponge (*1018*). Likewise, the Okinawan *Agelas* cf. *nemoechinata* sponge contains an oxysceptrin (**1264**) which also exhibits actomyosin ATPase activation (*1019*). Sceptrin (**1260**) was also found in this sponge.

The first halogenated pyrrole sponge metabolites to be discovered were dibromophakellin (**1268**) and bromophakellin (**1269**) from the Great Barrier Reef Sponge *Phakellia flabellata* (*1020, 1021*). The related dibromoisophakellin (**1270**) is found in the marine sponge *Acanthella carteri* (*1022*). The novel compound hymenialdisine (**1271**) was isolated at about the same time by two groups from three different sponges: *Axinella verrucosa* and *Acanthella aurantiaca* from the Mediterranean Sea and the Red Sea, respectively (*1023, 1024*), and *Hymeniacidon aldis* from Okinawa (*1025*). This metabolite has also been identified in *Acanthella carteri* (*1026*) and in *Hymeniacidon* sp. and *Axinella* sp. sponges (*1027*). The latter sponge has furnished the new axinohydantoin (**1272**) (*1027*), while the former sponge from Okinawa has yielded hymenin (**1273**) (*1028*). This metabolite is a competitive antagonist of α-adrenoceptors in vascular smooth muscle (*1028, 1029*). The structurally similar stevensine (**1274**) was isolated from an unidentified sponge (along with sceptrin (**1260**) (*1030*). Stevensine (**1274**) is also found in the sponges *Teichaxinella morchella* and *Ptilocaulis walpersi* in a study that also identified the probable oroidin (**1242**) precursor, 3-amino-1-(2-aminoimidazolyl)prop-1-ene (**1242a**) (*1031*). This amine is also a possible precursor to hymenidin (**1275**) (monodebromooroidin) which is found in an Okinawan *Hymeniacidon* sp. sponge (*1032*). Hymenidin is an antagonist of serotonergic receptors. This sponge also is a source of the novel manzacidins A (**1276**), B (**1277**), and C (**1278**) which are the first examples of tetrahydropyrimidines from a marine source (*1033*). The simple 2-bromoaldisin (**1279**) is found in *H. aldis* and *Lissodendoryx* sp. from Guam and Sri Lanka, respectively; the latter sponge also contains the simple pyrroles **1248** and **1280** (*1034*). The sponge *Phacellia fusca* from the South China Sea also produces 2-bromoaldisin (**1279**) along with the known aldisin (*1035*). The sponge *Phorbas aff. clathrata* has been found to contain the novel chlorinated phorbazoles A (**1281**), B (**1282**), C (**1283**), and D (**1284**) (*1036*). The major pyrrolyloxazole is **1281**, which is optically inactive.

3.14.2. Indoles

Since tryptophan is an essential amino acid and is found in all living things, it is not surprising that halogenated indoles are likewise present in many natural systems, particularly marine organisms (*1037*).

The indole ring system has the distinction of being embodied in the indigo derivative Tyrian Purple (**1285**), which is the ancient Egyptian dye extracted from Mediterranean molluscs to dye clothing used by royalty and the focus of a significant industry (*1038*). The structure of this compound was determined in 1909 by FRIEDLÄNDER who isolated 1.4 grams of Tyrian Purple (6,6′-dibromoindigotin) (*1285*) from 12,000 *Murex brandaris* animals (*1039*) and, later, from the sea snails *Purpura aperta* and *P. lapillus* (*1040*). The biogenesis of this compound has

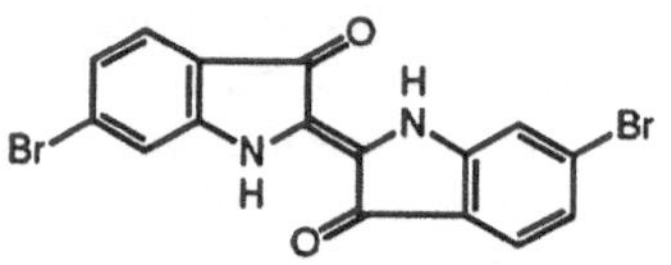

1285 (Tyrian Purple)

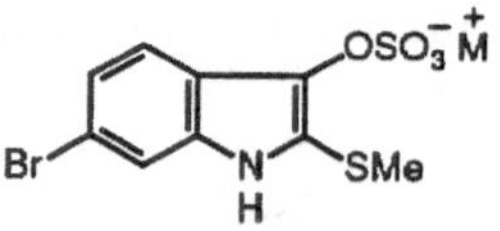

1286
1286a M = $Me_3\overset{+}{N}CH_2CH_2OH$
1286b M = $Me_3\overset{+}{N}CH_2CH_2OCOCH{=}CMe_2$

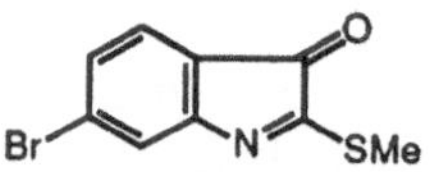

1287

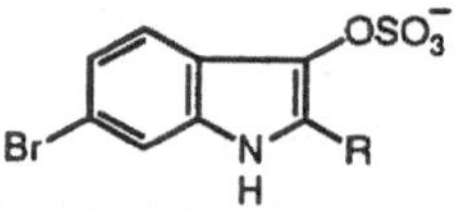

1288 R = H
1289 R = SO_2Me

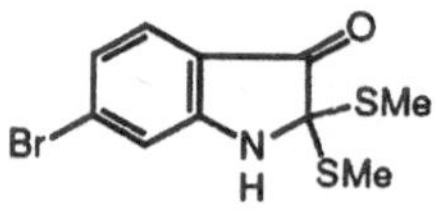

1290

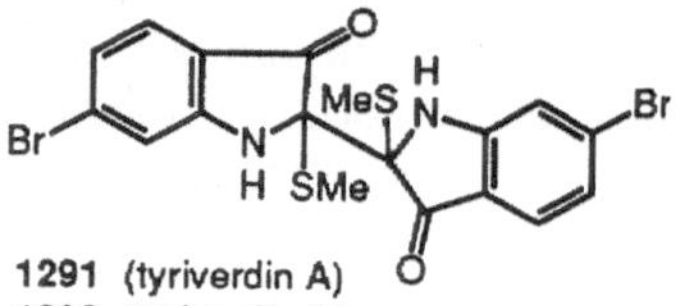

1291 (tyriverdin A)
1292 (tyriverdin B)

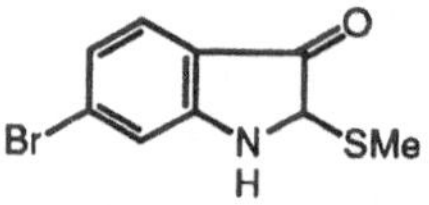

1293 (tyrindoxyl)

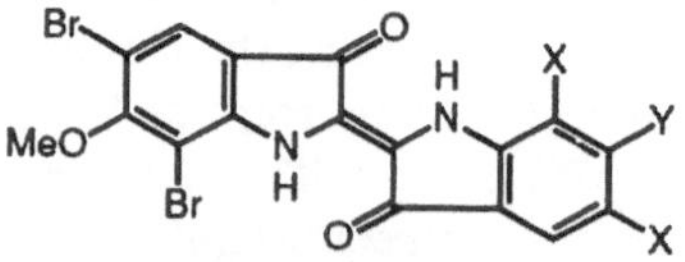

1294 X = Br, Y = OMe
1295 X = H, Y = Br

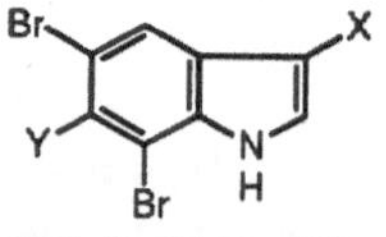

1296 X = H, Y = OMe
1297 X = Br, Y = H
1301 X = Br, Y = OMe

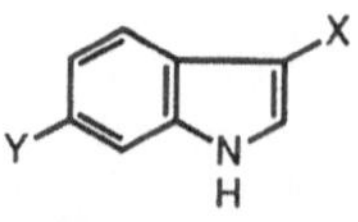

1298 X = Cl, Y = H
1299 X = Br, Y = H
1300 X = Cl, Y = Br
1302 X = Y = Br

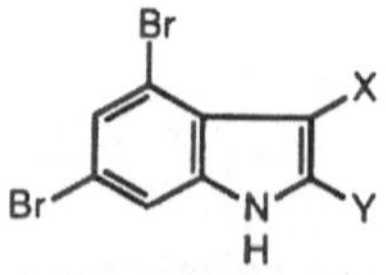

1303 X = Br, Y = H
1304 X = Y = H
1305 X = H, Y = Me

attracted considerable attention and several precursors have been isolated from molluscs, including the indoxylsulfates **1286** and **1287** from *Dicathais orbita* (*1041*), and **1288** and **1289** from *Murex trunculus*, *M. brandis*, *M. erinaceus*, and *Purpura haemastoma* (*1042*). Both **1287** and its possible precursor thioacetal **1290** have been isolated from *Dicathais orbita*, *Mancinella bufo*, *M. keineri*, *M. distinguenda* (*M. ancinella*) (*1043, 1044*). Heating **1290** in toluene converts it to **1287** (*1043*). The choline (**1286a**) and choline ester (**1286b**) salts of **1286** have been isolated from *D. orbita* and *M. keineri*, respectively (*1045*). Model studies suggest that tyriverdins A (**1291**) and B (**1292**), which have been isolated from *Nucella lapilus* (*1037*) and *Thais clavigera* (*1046*), are the immediate precursors of Tyrian Purple (*1046–1048*). These isomers are believed to be formed from tyrindoxyl sulfate (**1286**) via tyrindoxyl (**1293**) and the subsequent reaction of the latter with **1287** (*1037*). The acorn worm *Ptychodera flava laysanica* contains Tyrian Purple (**1285**), the two related new 6,6′-dibromoindigotins (**1294, 1295**) and the dibromoindole **1296** (*1049*). Other studies with this marine worm revealed several other simple halogenated indoles **1297** (*1050*), 3-chloro- (**1298**), 3-bromo- (**1299**), and 6-bromo-3-chloroindole (**1300**) (*1051*), **1301** and **1302** (*1052*). The characteristic odor of this animal is due to 3-chloroindole (**1298**). Another *Ptychodera* sp. from a deep Maui cave has yielded **1303** (*194*), a compound that was previously isolated from the acorn worm *Balanoglossus carnosus* (*1052*). This study also reported 3-bromoindole (**1299**) as the major odorous component of the acorn worm *Glossobalanus* sp. (*1052*). Another study of the latter animal identified new indoles **1304** and **1305** (*1053*).

The New Zealand alga *Rhodophyllis membranacea* produces no fewer than ten novel halogenated indoles **1306–1315**, although the exact structures of **1311** and **1315** have not yet been established (*1054*). In addition, numerous other polyhaloindoles were present in the mass spectra. The Caribbean and red alga *Laurencia brongniartii* contains the four additional new polybromoindoles **1316–1319** (*1055*). Tetrabromoindole **1318** is the only one showing antimicrobial activity. The sea hare *Aplysia dactylomela* produces **1317** and **1319** (*433*), and the sponge *Oceanapia bartschi* produces 3-bromoindole (*1299*), 6-bromo-3-indolecarbaldehyde (**1320**) and **1321** (*1056*). Aldehyde **1320** had been discovered earlier in a marine *Pseudomonas* sp. (*1057*) and is also found in the coral *Dendrophyllia* sp. (*1058*), the sponges *Pleroma menoui* (*1059*), *Plocamissma igzo* (*1060*), and *Pseudosuberites hyalinus* (*1061*). The latter deep water sponge (400 m) also contains 6-bromoindole derivatives **1322–1325** (*1061*) and *Pleroma menoui* from the Coral Sea produces **1326** and **1327** (*1059*). The marine bryozoan *Zoobotryon verticillatum* produces the

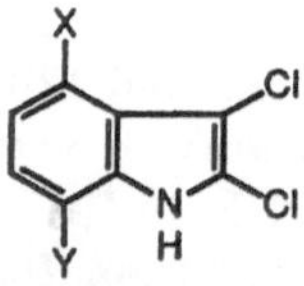

1306 X = H, Y = Br
1307 X = Cl, Y = H
1308 X = H, Y = Cl
1309 X = Y = Br
1310 X = Y = Cl
1311 X, Y = Cl, Br

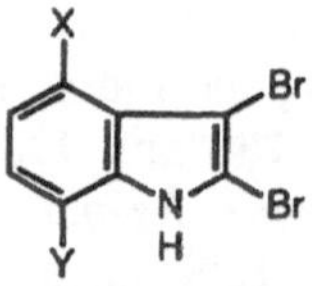

1312 X = H, Y = Br
1313 X = Y = Br

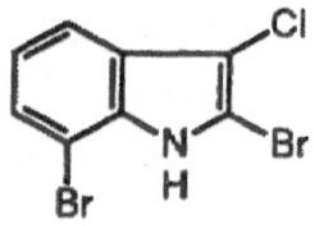

1314

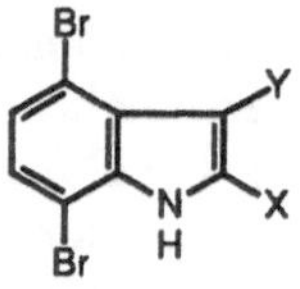

1315 X, Y = Cl, Br

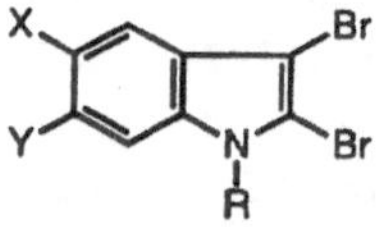

1316 R = Me, X = H, Y = Br
1317 R = Me, X = Br, Y = H
1318 R = H, X = Y = Br
1319 R = Me, X = Y = Br

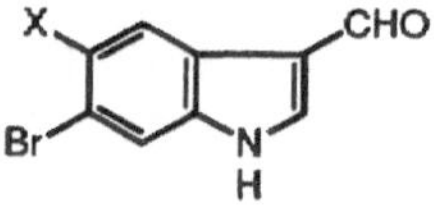

1320 X = H
1321 X = OH

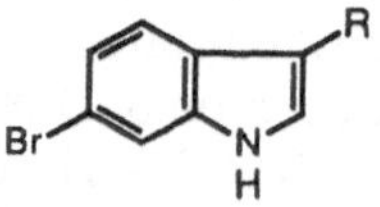

1322 R = CO_2H
1323 R = CH_2CN
1324 R = CH_2CONH_2
1325 R = CH_2CO_2Me
1326 R = CO_2Et
1327 R = $COCH_2OH$

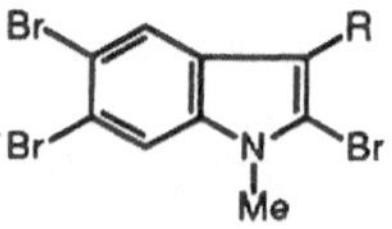

1328 R = CH_2NMe_2
1329 R = $CH_2N(O)Me_2$
1330 R = CHO

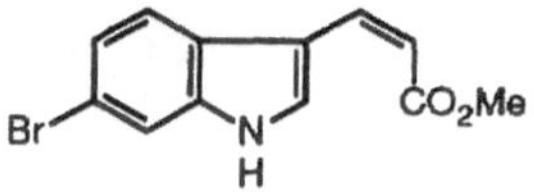

1333

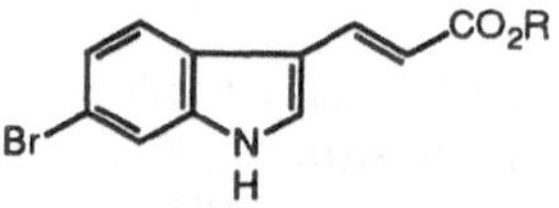

1331 R = Me
1332 R = H
(penaresin)

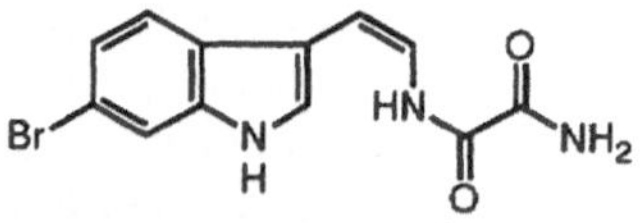

1334 (igzamide)

novel gramine (**1328**) and gramine *N*-oxide (**1329**) derivatives, the latter being the first marine alkaloid *N*-oxide (*1062*). A later study of this organism yielded the new aldehyde **1330** which delays metamorphosis in fertilized sea urchin eggs (*1063*). The marine sponge *Iotrochota* sp. produces indoleacrylate **1331** (*1064*), and the Okinawan sponge *Penares* sp. contains the corresponding acid, penaresin (**1332**), which is ten times better than caffeine as a calcium inducer (*1065*). The Coral Sea sponge *Corallistes undulatus* produces **1331** and the *Z* isomer **1333** in lesser amounts (*1066*). The sponge *Plocamissma igzo* has yielded the novel igzamide (**1334**) (*1060*).

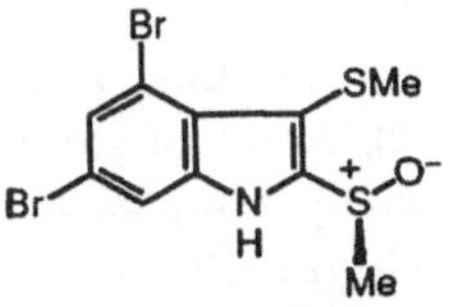

1335 (itomanindole A)

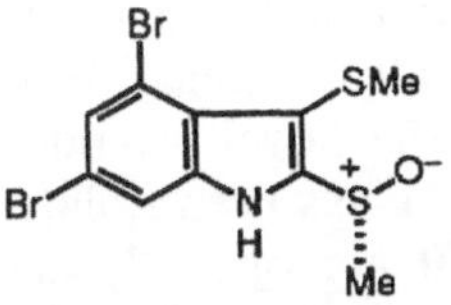

1336 (itomanindole A)

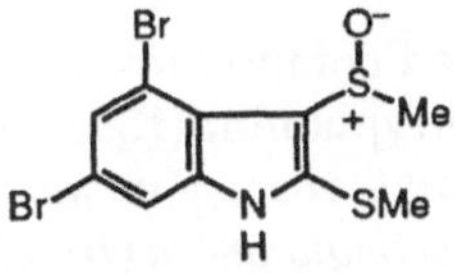

1337 (itomanindole B)

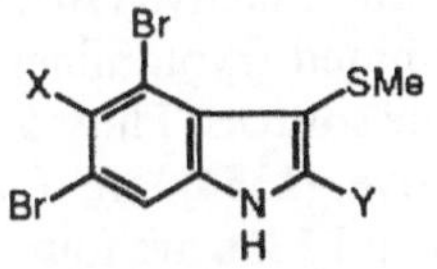

1338 X = H, Y = Br
1339 X = Y = Br
1340 X = H, Y = SMe
1341 X = Br, Y = SMe

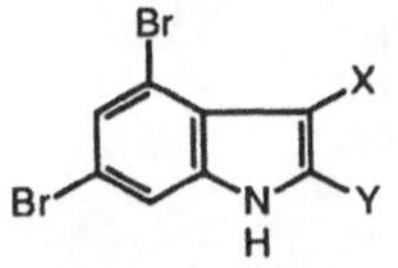

1342 X = H, Y = Br
1343 X = Y = Br
1344 X = H, Y = SMe

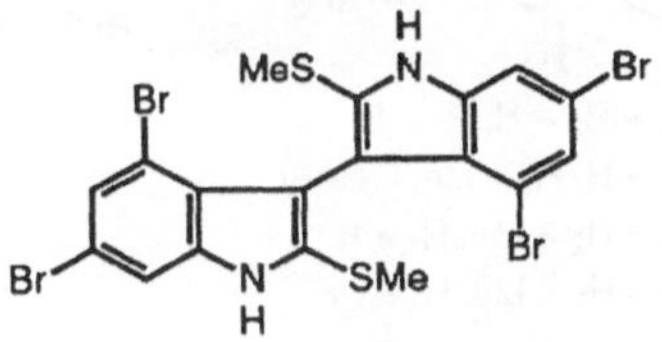

1345

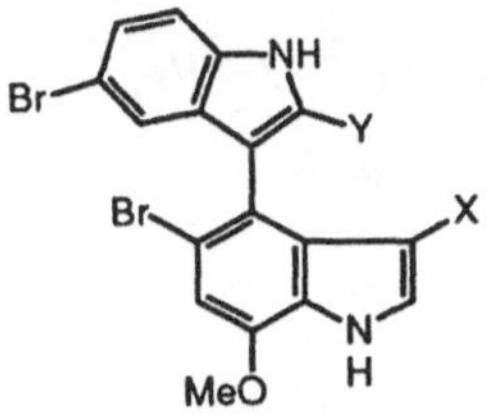

1346 X = Y = Br
1347 X = Br, Y = H
1348 X = H, Y = Br

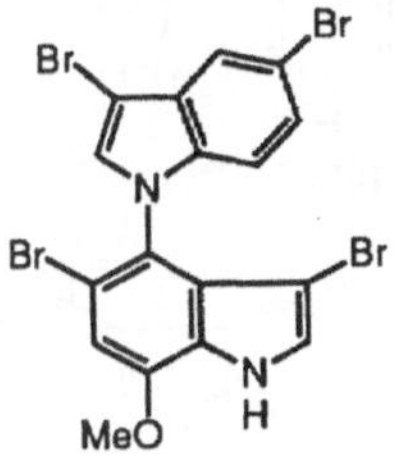

1349

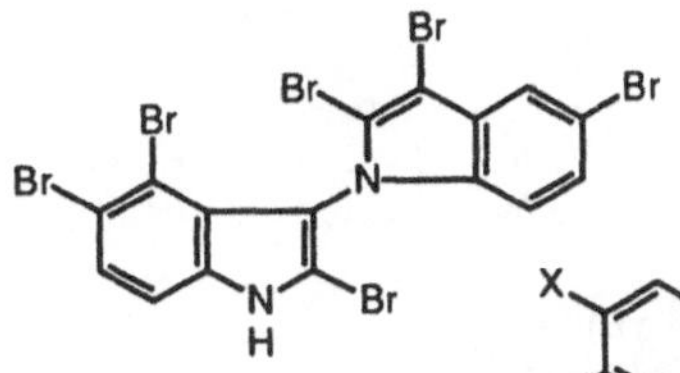

1350

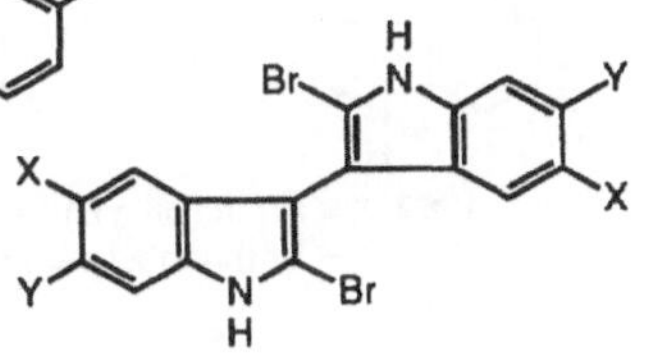

1351 X = Br, Y = H
1352 X = H, Y = Br

The novel sulfur-containing itomanindoles A (**1335, 1336**) and B (**1337**) are found in *Laurencia brongniartii* (*1067*). A more extensive study of this red alga revealed the presence of simple indoles **1338–1344** (*1068, 344*) and the novel *bis*-indole **1345** (*1068*). A set of similar *bis*-indoles **1346–1351** has been isolated from the blue-green alga *Rivularia firma*; **1346–1350** are optically active (biphenyl-type chirality) (*1069*). A subsequent examination of this cyanobacterium identified **1352**, which, like **1351**, is optically inactive (*1070*).

Several brominated tryptamines and derived compounds have been identified in marine sources. Thus, 5,6-dibromotryptamine (**1353**) and the *N*-methyl derivative (**1354**) are found in *Polyfibrospongia maynardii* (*1071*) while **1355** and **1356** are found in *Smenospongia aurea* (*1072, 1073*)

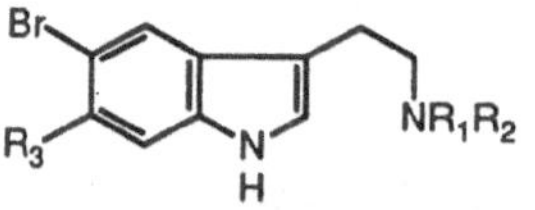

1353 $R_1 = R_2 = H$, $R_3 = Br$
1354 $R_1 = H$, $R_2 = Me$, $R_3 = Br$
1355 $R_1 = R_2 = Me$, $R_3 = H$
1356 $R_1 = R_2 = Me$, $R_3 = Br$

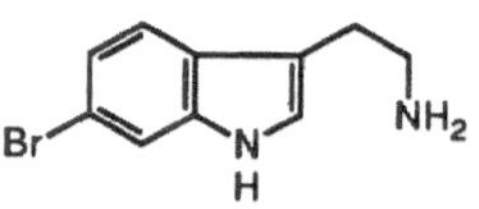

1357

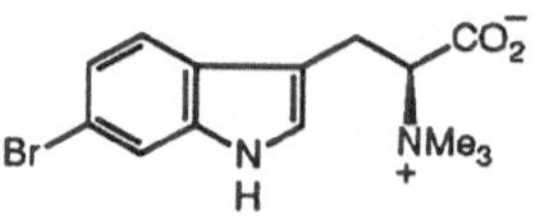

1358

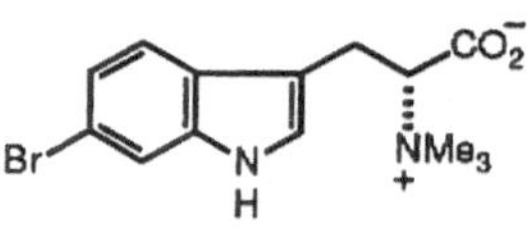

1359

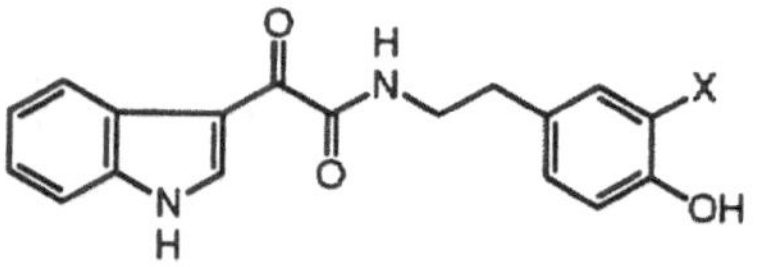

1360 X = Br (polyandrocarpamide A)
1361 X = I (polyandrocarpamide B)

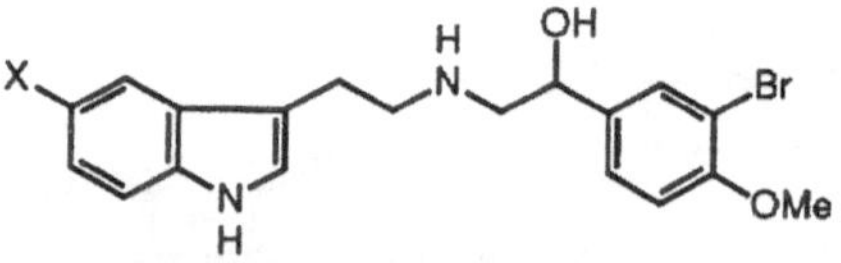

1362 X = H (chelonin B)
1363 X = Br (bromochelonin B)

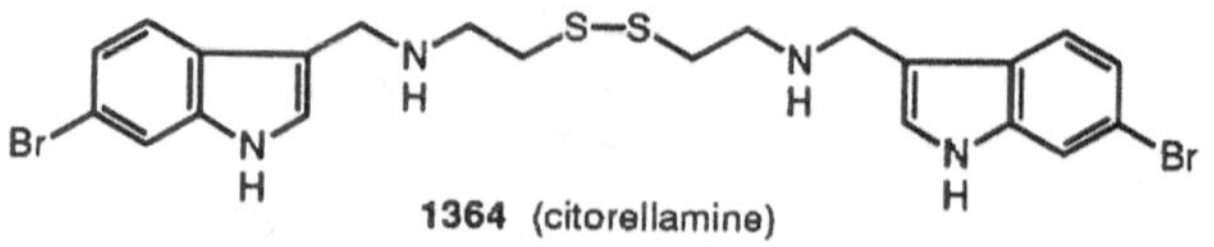

1364 (citorellamine)

and **1356** is found in *S.* (= *Polyfibrospongia*) *echina* (*1072*). The ascidian *Eudistoma fragum* contains **1355** (*1074*), and the tunicate *Didemnum candidum* produces 6-bromotryptamine (**1357**) (*1075*). The sponge *Pachymatisma johnstoni* contains L-6-bromohypaphorine (**1358**) (*1076*) while epimer **1359** is found in an Okinawan *Aplysina* sp. sponge (*1077*). The Philippine ascidian *Polyandrocarpa* sp. produces polyandrocarpamides A (**1360**) and B (**1361**) (*1078*) and the novel chelonin B (**1362**) and bromochelonin B (**1363**) have been identified in the Palau sponge *Chelonaplysilla* sp. (*1079*). A Fijian tunicate *Polycitorella mariae* contains citorellamine (**1364**) (*1080*) (structure revised in (*1081*)) which shows significant cytotoxicity and antibacterial activity (*1080*).

A large number of brominated imidazolinyl-indoles have been discovered in ocean life. The first such metabolite was found in a *Dercitus* sp. sponge and is the 6-bromoaplysinopsin derivative **1365** (*1085*). The analog **1366** was isolated from *Smenospongia aurea* (*1072*) as were **1367** and **1368** (*1073*). For convenience, these compounds are all drawn in the form of the same exocyclic imine tautomer. The Mediterranean anthozoan *Astroides calycularis* also contains **1367** and the new *N*-propionyl

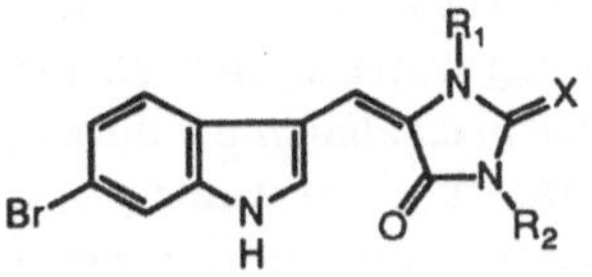

1365 X = NH, R_1 = H, R_2 = Me
1366 X = O, R_1 = R_2 = H
1367 X = NH, R_1 = R_2 = Me
1368 X = NH, R_1 = Me, R_2 = H
1370 X = O, R_1 = R_2 = Me

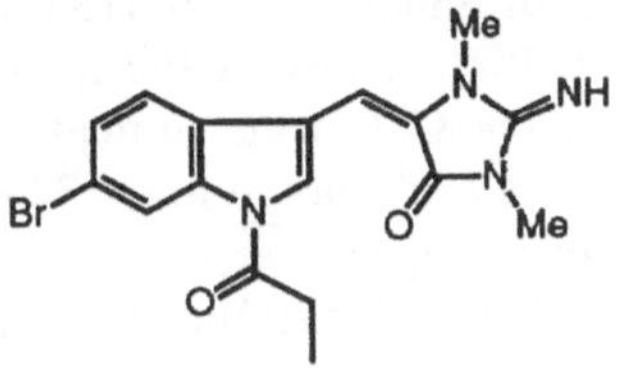

1369

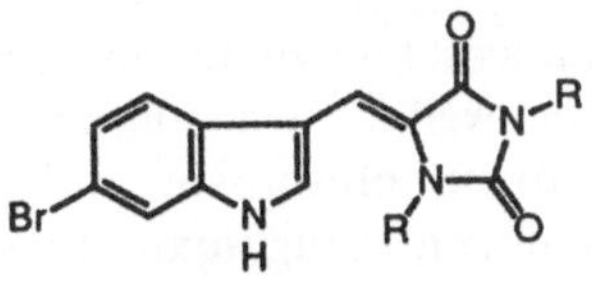

1371 R = Me
1372 R = H

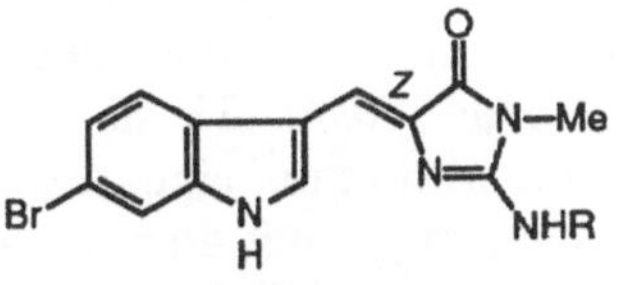

1373 R = H
1374 R = Me

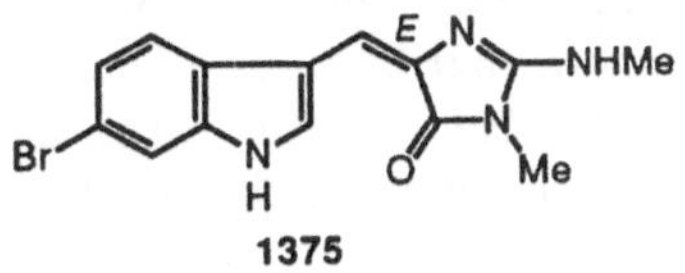

1375

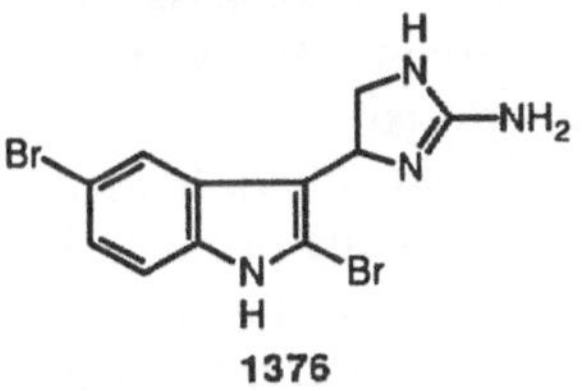

1376

derivative **1369** (*1086*). The coral *Tubastraea* sp. contains an *E*/*Z* mixture of the novel bromoaplysinopsins **1370** and **1371**, while the same study uncovered **1366** and the *Z*-isomer **1372** in the coral *Leptopsammia pruvoti* (*1087*). The coral *Dendrophyllia* sp. contains *E*/*Z* mixtures of **1365**/**1373** and **1374**/**1375** which undergo facile photochemical isomerization to favor the *E*-isomer and thermal reversal to a ground state equilibrium mixture of $> 95:5$ *Z*/*E* (*1058*). The sponge *Discodermia polydiscus*, collected in deep waters in the Bahamas, was found to contain the cytotoxic **1376** (*1088*). This compound showed activity against P388 ($IC_{50} = 1.8\ \mu g/ml$) and A549 ($IC_{50} = 4.6\ \mu g/ml$) cells.

A series of structurally similar brominated metabolites containing two indole rings and possessing potent biological activity have been isolated from several sponges. These compounds appear to play a chemical defense role. For example, topsentin B2 (**1377**) is produced as a defensive agent by the Mediterranean sponge *Topsentia genitrix* together with two similar non-brominated compounds (*1089*). Topsentin B2 (= bromotopsentin) (**1377**) was also isolated from a deep water Caribbean sponge tentatively identified as *Spongosorites* sp. along with the new metabolite **1378** (*1090*). The British Columbian sponge *Hexadella* sp. has yielded in addition to topsentin B2 (**1377**) the new topsentin C (**1379**), dragmacidons A (**1380**) and B (**1381**) (*1091*). Earlier, dragmacidin (**1382**) was extracted from the deep water sponge *Dragmacidon* sp. (*1092*). The California tunicate *Didemnum candidum* produces **1383** as well as the simple *bis*-indole **1384** (*1075*). The deep water (460 meters) Caribbean sponge *Spongosorites ruetzleri* contains nortopsentins A–C (**1385**–**1387**), which exhibit both antifungal and cytotoxic activity (*1093*). This animal also contains topsentin B2 (**1377**). Another study of *Spongosorites* sp. discovered dragmacidin d (**1388**) which is active against feline leukemia (*1094*). The ascidian *Eusynstyela misakiensis* produces the novel eusynstyelamide (**1389**) (*1095*).

Some of the most poisonous substances known to mankind are marine toxins (*1096*). These include the well-known compounds tetrodotoxin, saxitoxin, ciguatoxin, palytoxin, brevetoxin and others. While these compounds do not contain halogen, surugatoxin (**1390**) from the Japanese Ivory shell *Babylonia japonica* is a 6-bromooxindole derivative (*1097*). This toxin is linked with human illness in Suruga Bay. Neosurugatoxin (**1391**) which has 100 times greater antinicotinic activity than surugatoxin and prosurugatoxin (**1392**) are also found in this organism (*1098–1100*). Evidence indicates that bacteria associated with *B. japonica* actually produce **1391** and **1392** (*1101*) and that surugatoxin may actually be an isolation artifact since it is formed when prosurugatoxin is treated with acetic acid (*1100*).

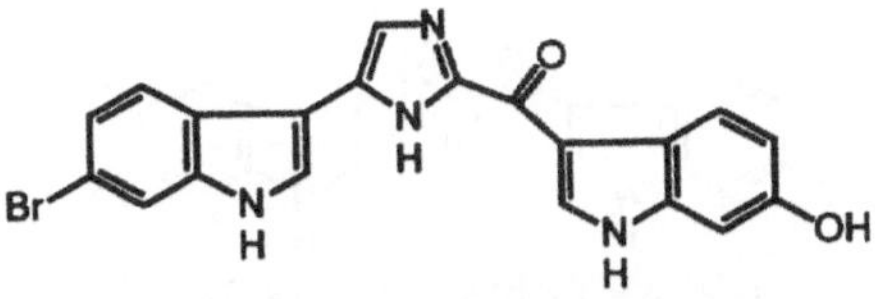

1377 (topsentin B2)

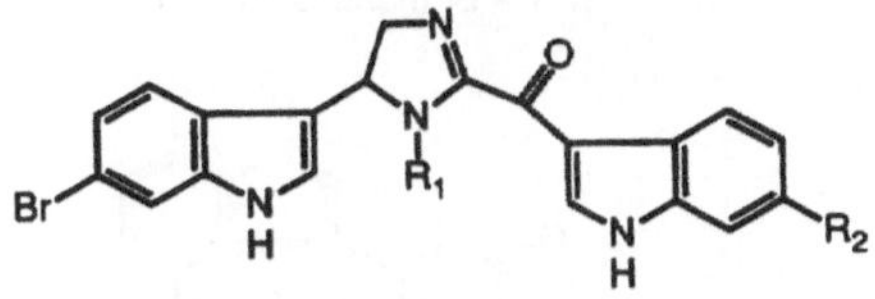

1378 $R_1 = R_2 = H$
1379 $R_1 = Me$, $R_2 = Br$ (topsentin C)

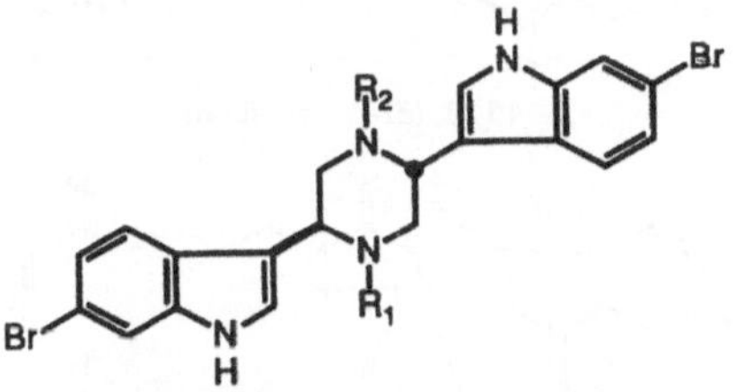

1380 $R_1 = H$, $R_2 = Me$ (dragmacidon A)
1381 $R_1 = R_2 = Me$ (dragmacidon B)
1383 $R_1 = R_2 = H$

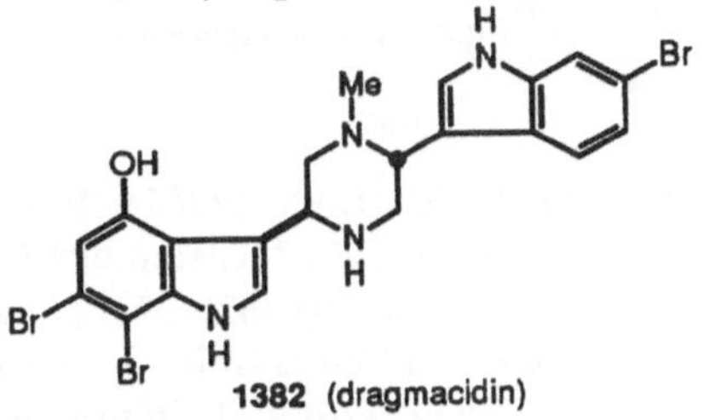

1382 (dragmacidin)

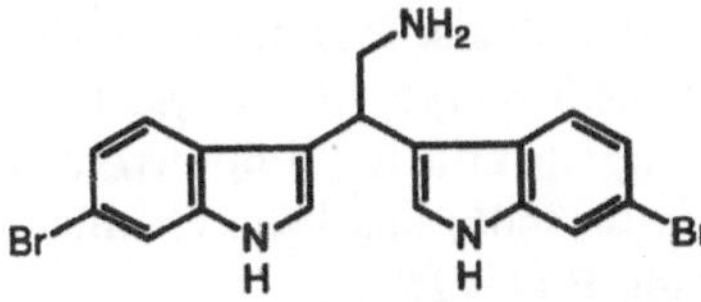

1384

Considering their unpretentious appearance, marine bryozoans, or "moss animals," are unrivaled in their ability to construct complex halogenated indoles (*1102, 1103*). Several bromophysostigmines are produced by *Flustra foliacea.* These include flustramines A (**1393**), B (**1394**)

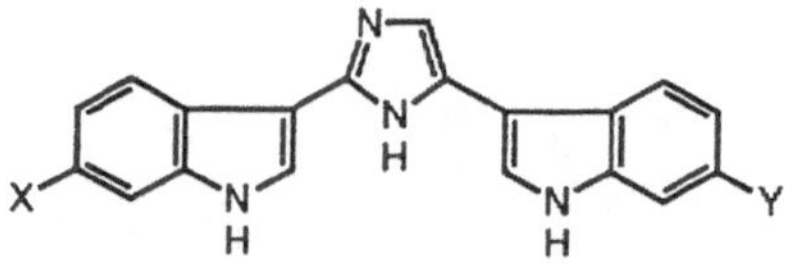

1385 X = Y = Br (nortopsentin A)
1386 X = Br, Y = H (nortopsentin B)
1387 X = H, Y = Br (nortopsentin C)

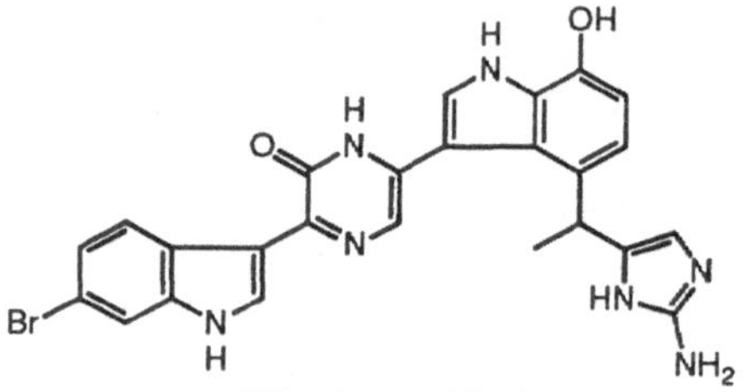

1388 (dragmacidin d)

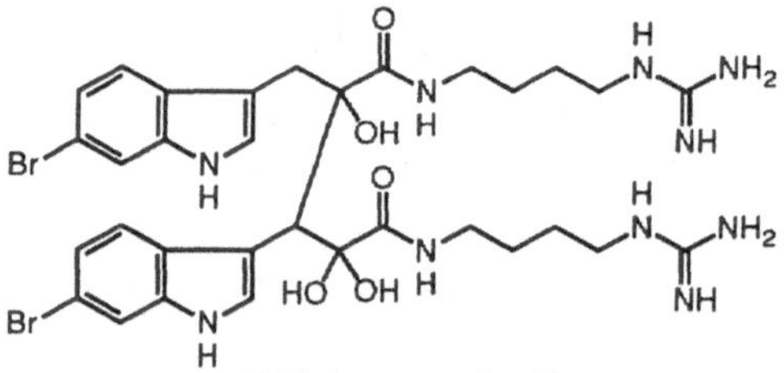

1389 (eusynstyelamide)

(*1104, 1105*), C (**1395**) (*1106*), D (**1396**) (*1107*), E (**1397**) (*1108*), flustraminols A (**1398**) and B (**1399**) (*1106*), flustramides A (**1400**) (*1109*) and B (**1401**) (*1110*), dihydroflustramine C (**1402**) (*1111*) and its N-oxide **1403** (*1107*), isoflustramine D (**1404**) and flustramine D *N*-oxide (**1405**) (*1107*). The related flustrarine B (**1406**) (*1110*) and ring-opened flustrabromine (**1407**) (*1112*) and **1408** (*1109*) are also isolated from this animal. Although the biological profile of these metabolites is yet to be established flustramine E (**1397**) is active against the fungi *Botrytis cinera* and *Rhizotonia solani* (*1108*). Flustrarine B (**1406**) has been chemically correlated with the *N*-oxide of flustramine B (*1591*).

The tunicate *Ciona savignyi* produces the bromophysostigminepteridine alkaloids urochordamine A (**1409**) and B (**1410**), the former of which has potent larval settlement and metamorphosis-promoting activity at 2 ng/ml (*1113*). The bryozoan *Hincksinoflustra denticulata* contains hinckdentine A (**1411**) which embraces the novel pyrimidino[3′,4′:1,2]-pyrrolo[2,3-*d*] azepine ring system (*1114*). Not to be outdone, the

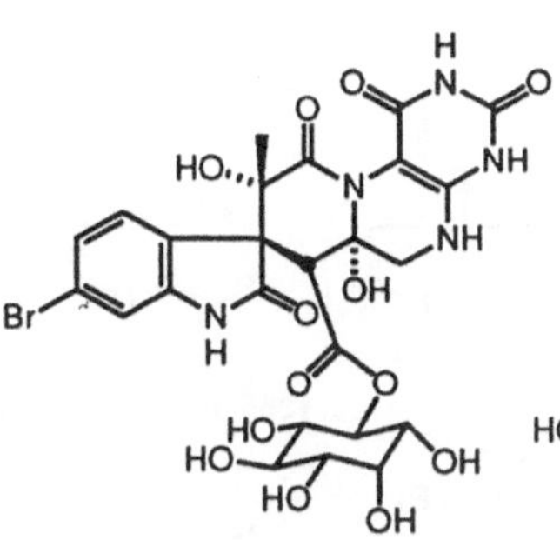

1390 (surugatoxin)

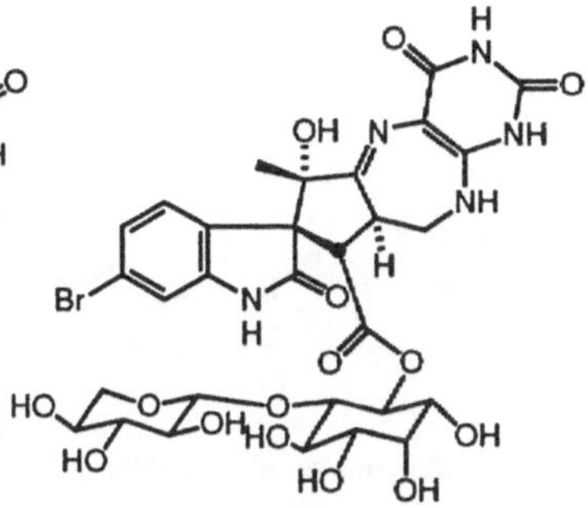

1391 (neosurugatoxin)

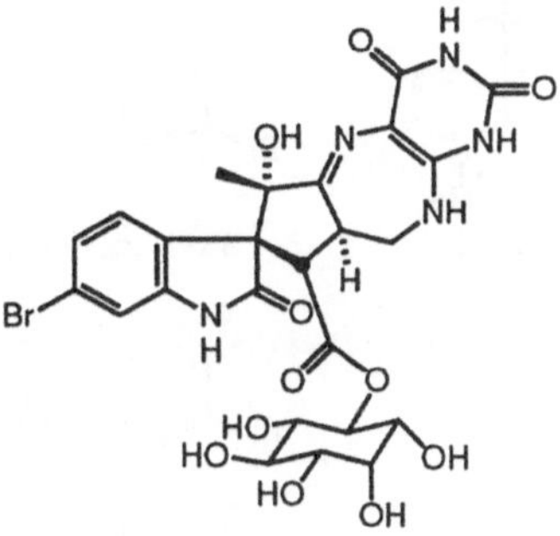

1392 (prosurugatoxin)

bryozoan *Chartella papyracea* produces a complement of stunningly complex indolo-β-lactams, chartellines A (**1412**), B (**1413**), C (**1414**), (*1115, 1116*), and chartellamides A (**1415**) and B (**1416**) (*1117*). Since fewer than a dozen of the estimated 4,000 extant bryozoans have been investigated for their chemical content (*1115*), it seems a certainty that many additional examples of nature's virtuosity in organic synthesis will be revealed by these amazing moss animals.

A series of chlorinated indoloisonitriles and related indoles has been isolated from blue-green algae. The blue-green alga *Hapalosiphon fontinalis* is the source of twenty "hapalindoles," twelve of which contain chlorine: hapalindoles A (**1417**), B (**1418**), L (**1419**), G (**1420**), V (**1421**), N (**1422**), P (**1423**), I (**1424**), K (**1425**), T (**1426**) and the tricyclic E (**1427**) and F (**1428**) (*1118, 1119*). The absolute configuration of hapalindole Q which does not contain chlorine has been determined by synthesis (*1120*) and supports the proposed assignments (*1121, 1122*). The two minor compounds fontonamide (**1429**) and anhydrohapaloxindole A (**1430**) are also present in cultures of *H. fontinalis* (*1122*). These two metabolites may be products of the reaction of singlet oxygen with

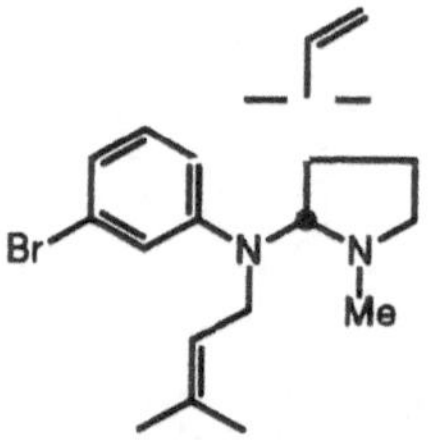

1393 (flustramine A)

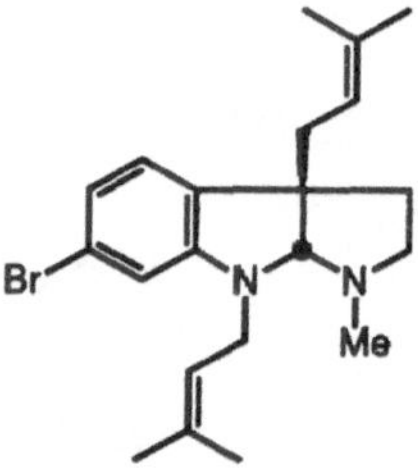

1394 (flustramine B)

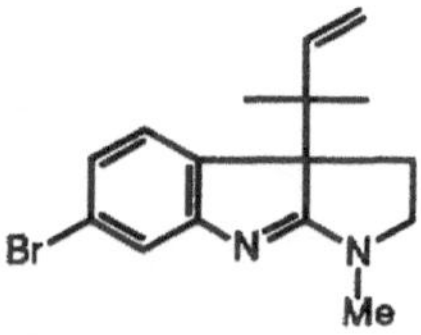

1395 (flustramine C)

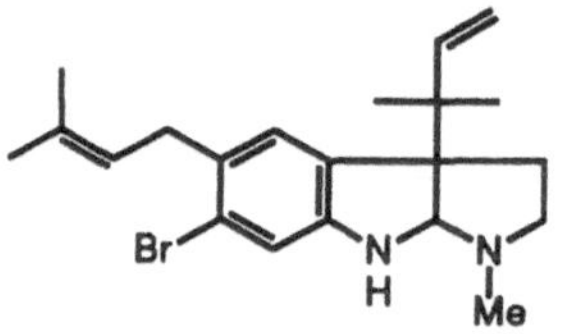

1396 (flustramine D)

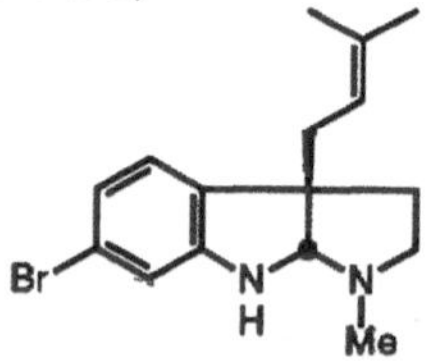

1397 (flustramine E)

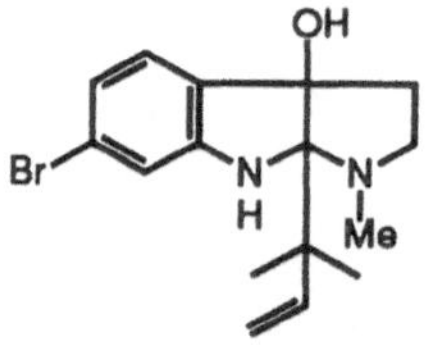

1398 (flustraminol A)

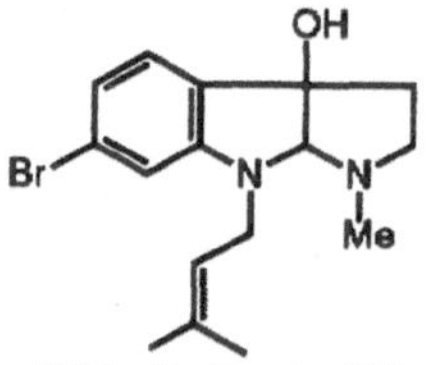

1399 (flustraminol B)

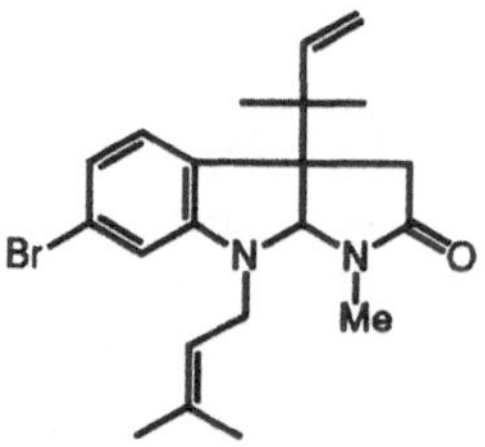

1400 (flustramide A)

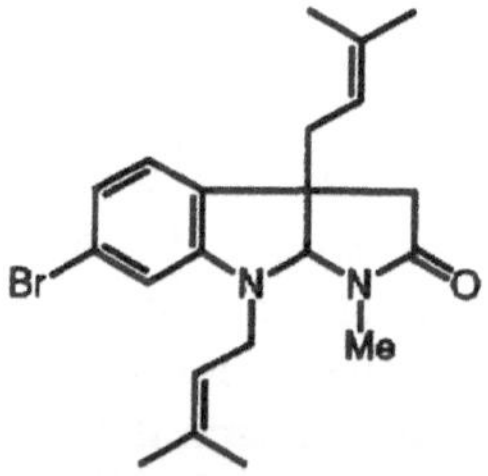

1401 (flustramide B)

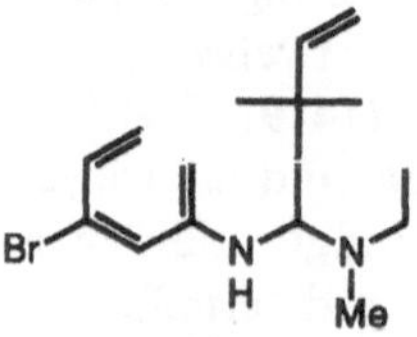

1402 (dihydroflustramine C)

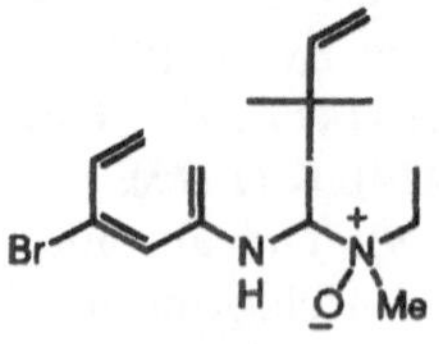

1403

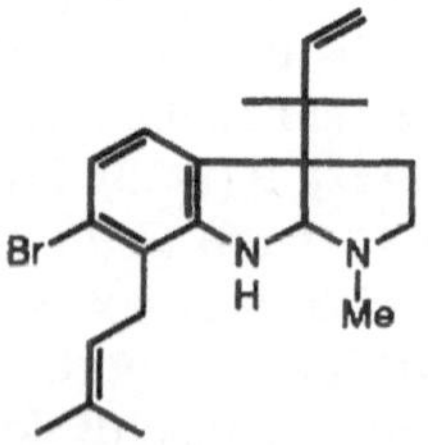

1404 (isoflustramine D)

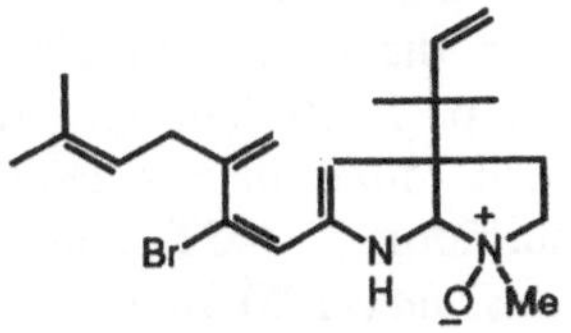

1405

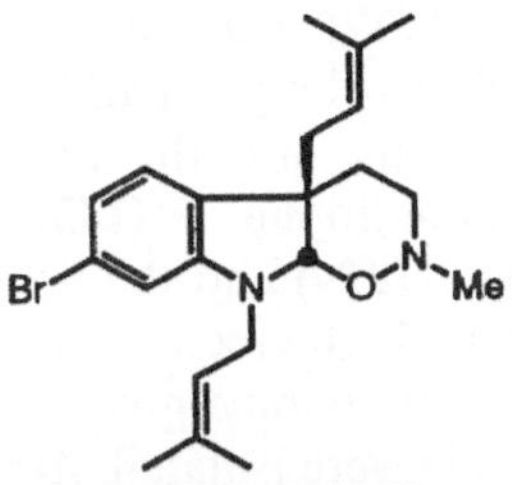

1406 (flustrarine B)

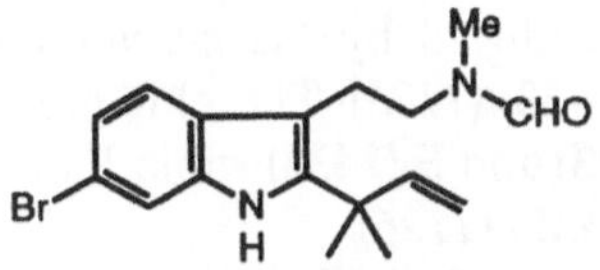

1407 (flustrabromine)

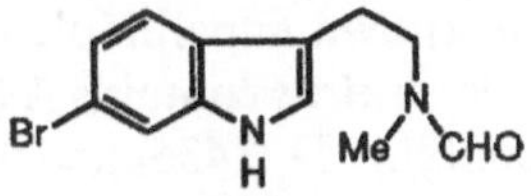

1408

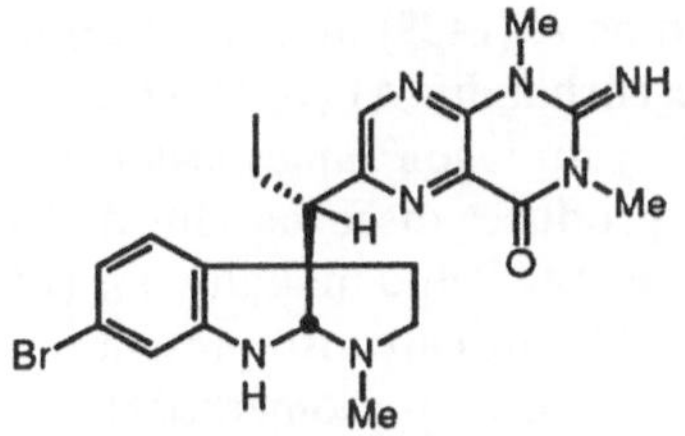

1409 (urochordamine A)

1410 (urochordamine B)
(epimer of 1409)

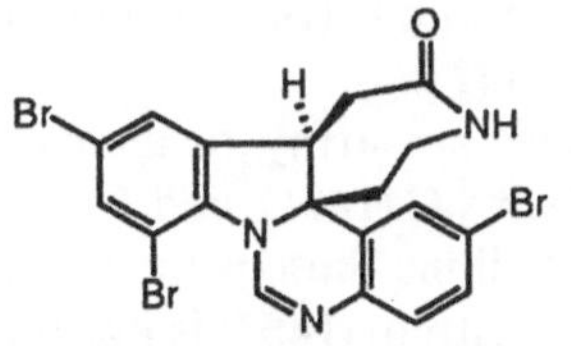

1411 (hinckdentine A)

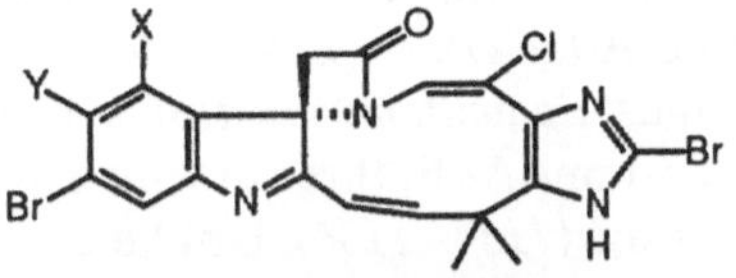

1412 X = Y = Br (chartelline A)
1413 X = Br, Y = H (chartelline B)
1414 X = Y = H (chartelline C)

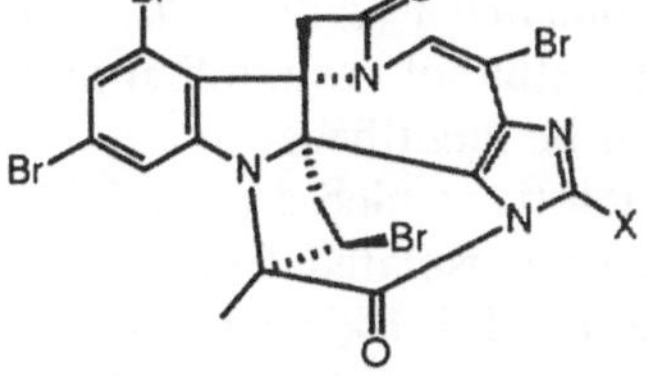

1415 X = H (chartellamide A)
1416 X = Br (chartellamide B)

hapalindole A (**1417**). The biosynthesis of the major metabolite hapalindole A (**1417**) involves a pathway wherein the C(2)-N of glycine is incorporated intact into the isonitrile group (*1123*). The structurally similar hapalindolinone A (**1431**) has been found in *Fischerella* sp. blue-green alga (*1124*) and the novel fischerindole L (**1432**) is produced by the terrestrial blue-green *Fischerella muscicola* (*1125*). In an extensive investigation of three blue-green algae species, several new hapalindole-type compounds were isolated. Ambiguine isonitriles A (**1433**), B (**1434**), D (**1435**), E (**1436**), and F (**1437**) are produced by *Fischerella ambigua* along with the known hapalindole G (**1420**) (*1126*). The blue-green alga *Hapalosiphon hibernicus* contains A (**1433**) and E (**1436**) while *Westiellopsis prolifica* contains D (**1435**) and E (**1436**) (*1126*).

Several marine sponges are the source of highly cytotoxic indoloquinones and related compounds of similar structure. Discorhabdin C (**1438**), which is cytotoxic against the L-1210 mouse leukemia cell line ($IC_{50} < 0.1$ μg/ml), is found in the sponge *Latrunculia* sp. and is the first example of a pyrrolo[1,7]phenanthroline (*1127*). This discovery was followed by the isolation of discorhabdins A (**1439**) and B (**1440**) from this sponge (*1128*). Prianosins A (= discorhabdin A) (**1439**) (*1129*) and B (**1442**) (*1130*) are found in the Okinawan sponge *Prianos melanos*. The Fijian sponge *Zyzzya* cf. *marsailis* also produces discorhabdin A (**1439**) together with the related makaluvamine F (**1442**) and makaluvone (**1443**) (*1131*). Several other nonhalogenated metabolites are also present in this sponge; they display potent cytotoxicity and are topoisomerase II inhibitors (*1131*). The deep water Bahamas sponge *Batzella* sp. produces the pyrroloquinolines batzellines A (**1444**), B (**1445**), and C (**1446**) (*1132*), and isobatzellines A (**1447**), C (**1448**), and D (**1449**) (*1133*).

The fungus *Penicillium crustosum* produces an array of amazingly complex penitrems A–F, three of which, A (**1450**), C (**1451**), and F (**1452**), contain chlorine (*1134–1136*). Unlike the penicillins these penitrems are highly toxic, especially to farm animals. Pennigritrem (**1453**) is produced by *P. nigricans* and is a cyclization product of penitrem A (**1450**) (*1137*). Penitrem A (**1450**) has also been isolated from *P. verrucosum* var. *cyclopium* (*1138*). Years earlier the simpler 8-chlororugulovasines A (**1454**) and B (**1455**) were found in *P. islandicum* (*1139, 1140*). These metabolites which interconvert in boiling methanol are the first naturally occurring halogenated ergot alkaloids to be identified.

A series of antitumor antibiotics which are reminiscent of CC-1065 and its analogs (*1141*) have been isolated from several *Streptomyces* species. Pyrindamycins A (**1456**) and B (**1457**) were isolated from *Streptomyces* SF2582 (*1142*), and as duocarmycins C_2 (**1456**) and C_1 (**1457**), respectively, from *Streptomyces* DO-89 (*1143, 1144*). These metabolites

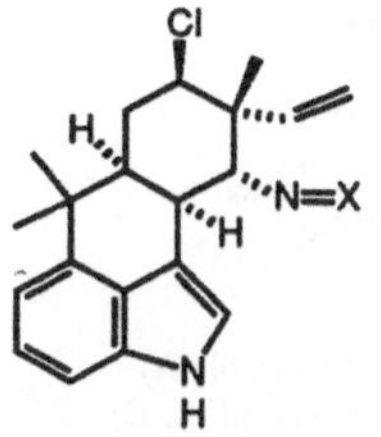

1417 X = C
(hapalindole A)
1418 X = CS
(hapalindole B)

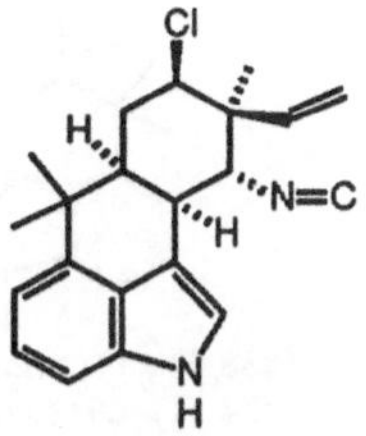

1419
(hapalindole L)

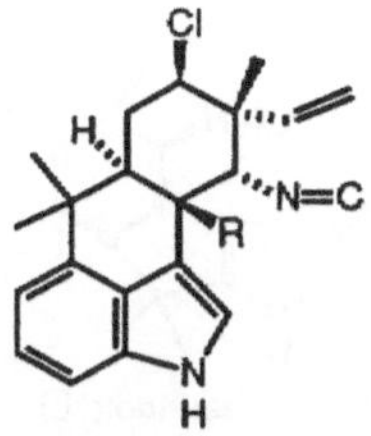

1420 R = H
(hapalindole G)
1421 R = OH
(hapalindole V)

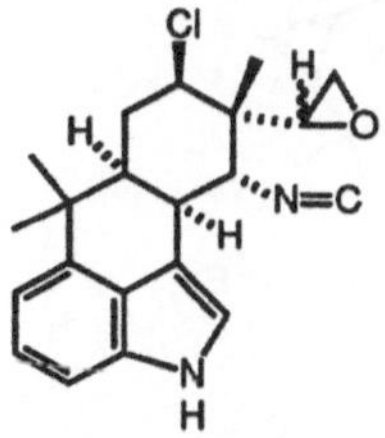

1422 (hapalindole N)
1423 (hapalindole P)

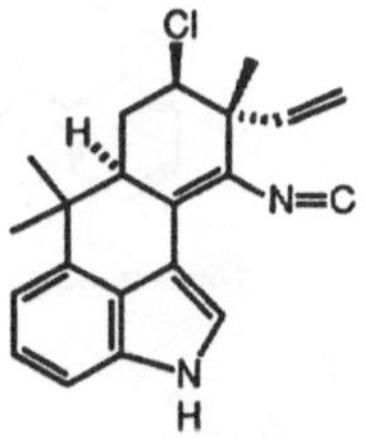

1424
(hapalindole I)

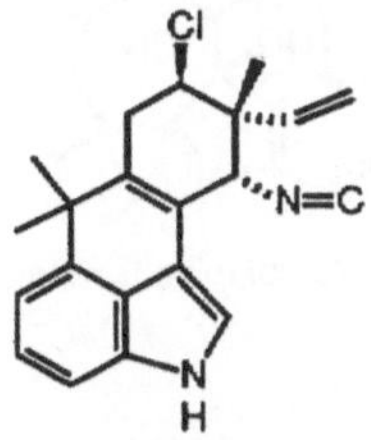

1425
(hapalindole K)

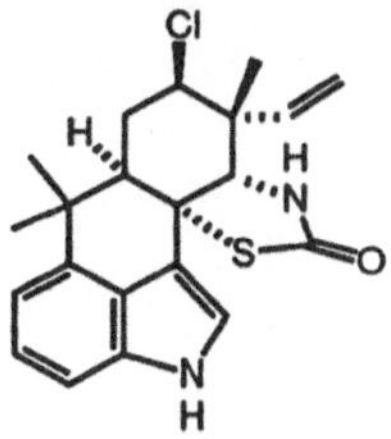

1426
(hapalindole T)

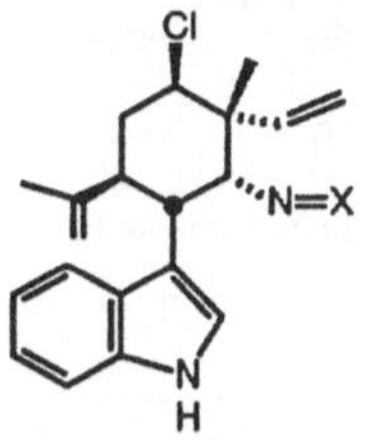

1427 X = C
(hapalindole E)
1428 X = CS
(hapalindole F)

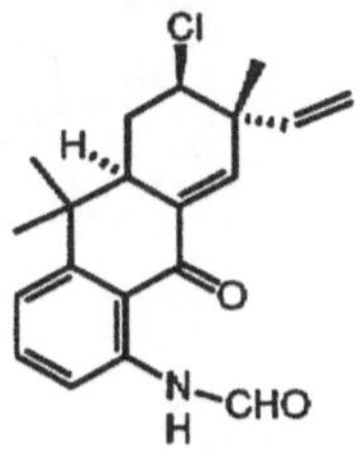

1429
(fontonamide)

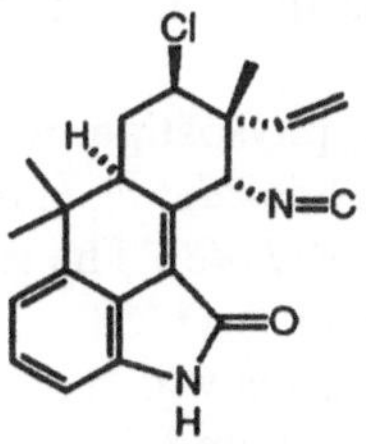

1430
(anhydrohapaloxindole A)

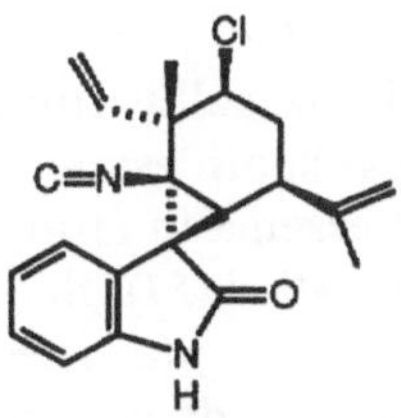

1431
(hapalindolinone A)

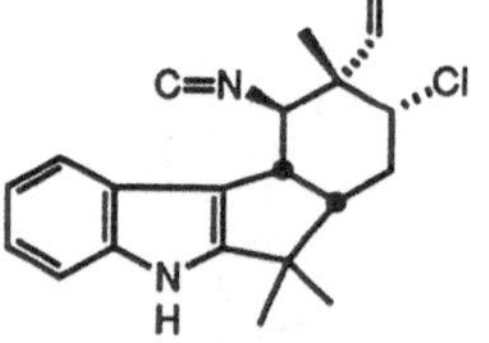

1432 (fischerindole L)

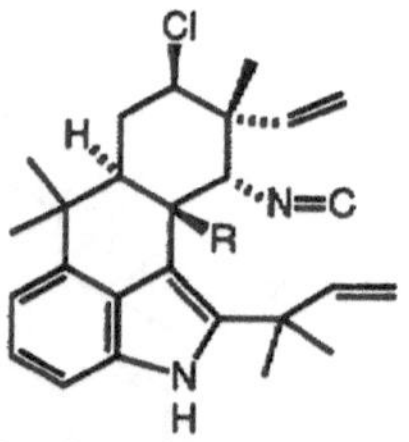

1433 R = H (ambiguine isonitrile A)
1434 R = OH (ambiguine isonitrile B)

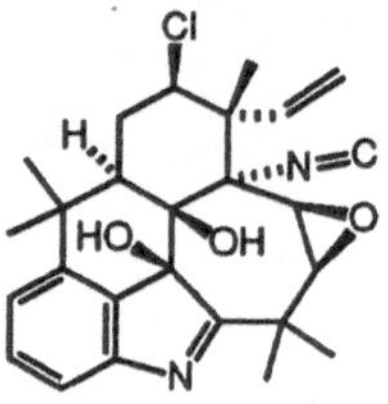

1435 (ambiguine isonitrile D)

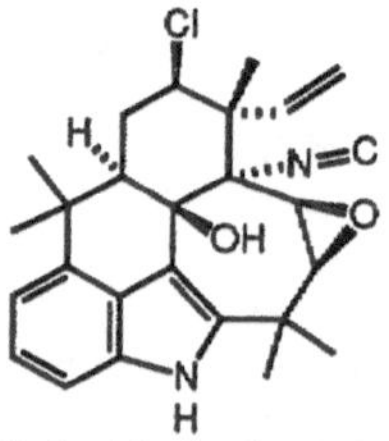

1436 (ambiguine isonitrile E)

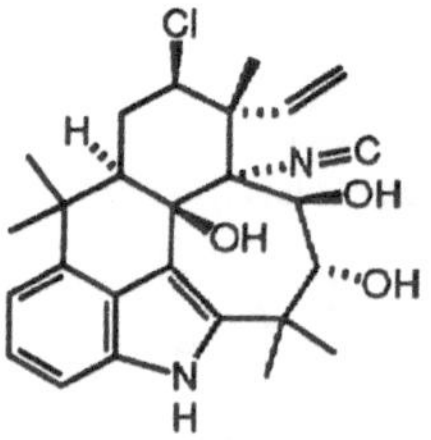

1437 (ambiguine isonitrile F)

exhibit potent cytotoxicity and alkylate DNA via duocarmycin A, which is the corresponding spirocyclopropylhexadienone and also a metabolite (*1145, 1146*). The corresponding bromine compounds are produced in the presence of bromide and are more potent than the chlorine-containing duocarmycins (*1146, 1147*).

The first of a series of sulfur-containing physostigmine-like metabolites to be isolated was sporidesmin A from *Sporidesmium bakeri*, a fungus that causes facial eczema in farm animals (*1148*). The structure was established as **1458** (*1149–1151*). Sporidesmin A (**1458**), B (**1459**), and C (**1460**) were isolated from *Pithomyces chartarum* which is a fungus that also causes eczema and liver damage in New Zealand sheep (*1152–1154*). Later studies identified sporidesmins D (**1461**) (*1155*), E (**1462**) (*1156*), F (**1463**) (*1155*), G (**1464**) (*1157, 1158*), H (**1465**) (*1159*), and J (**1466**) (*1159*) from *P. chartarum*.

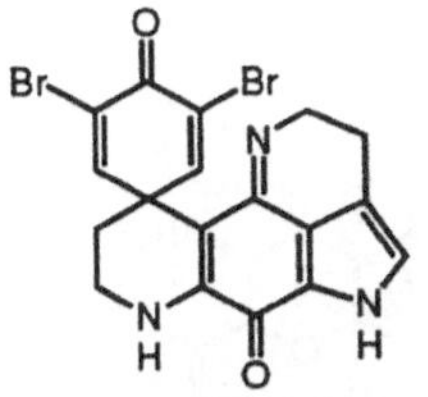

1438 (discorhabdin C)

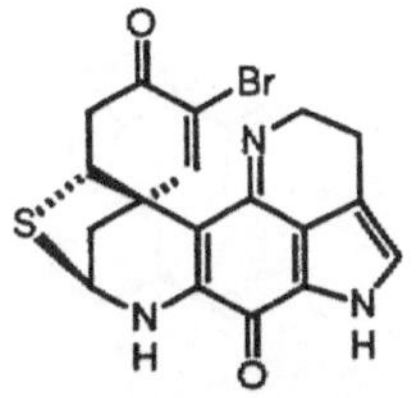

1439 (discorhabdin A = prianosin A)

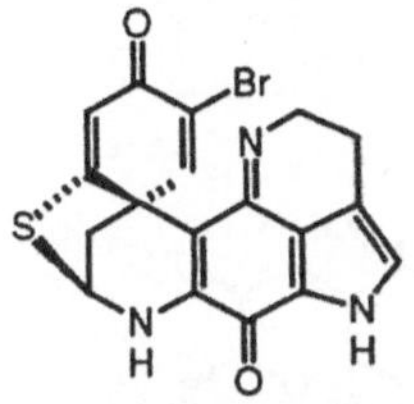

1440 (discorhabdin B)

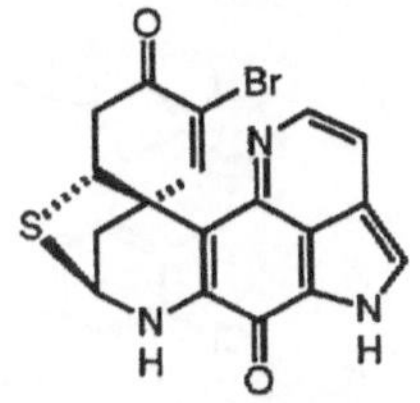

1441 (prianosin B)

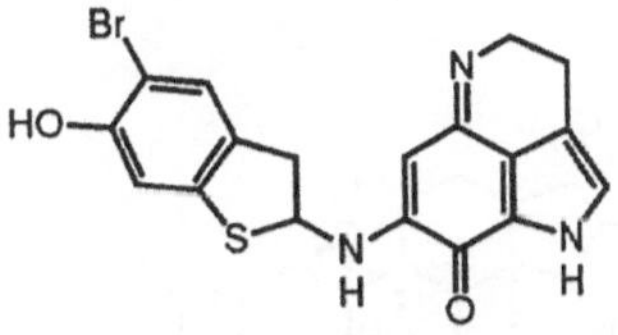

1442 (makaluvamine F)

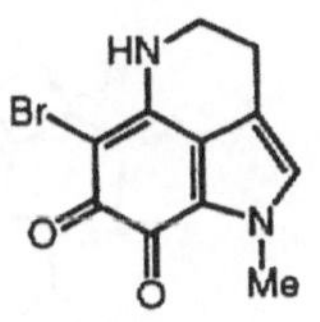

1443 (makaluvone)

1444 R_1 = Me, R_2 = SMe (batzelline A)
1445 R_1 = H, R_2 = SMe (batzelline B)
1446 R_1 = Me, R_2 = H (batzelline C)

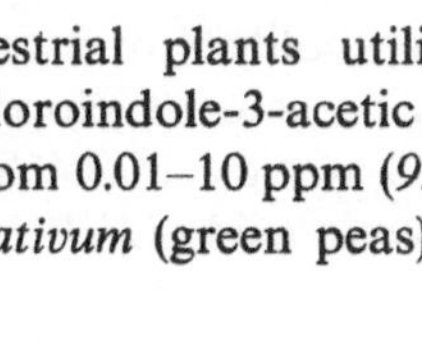

1447 R = SMe (isobatzelline A)
1448 R = H (isobatzelline C)

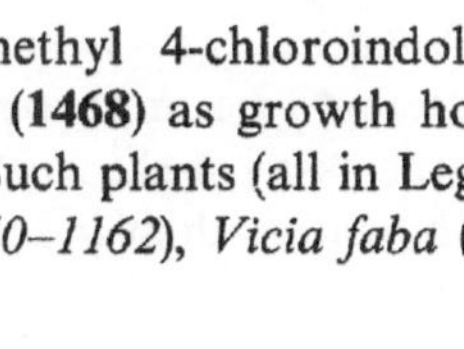

1449 (isobatzelline D)

Several terrestrial plants utilize methyl 4-chloroindole-3-acetate (**1467**) and 4-chloroindole-3-acetic acid (**1468**) as growth hormones in levels ranging from 0.01–10 ppm (*931*). Such plants (all in Leguminosae) include *Pisum sativum* (green peas) (*1160–1162*), *Vicia faba* (fava bean)

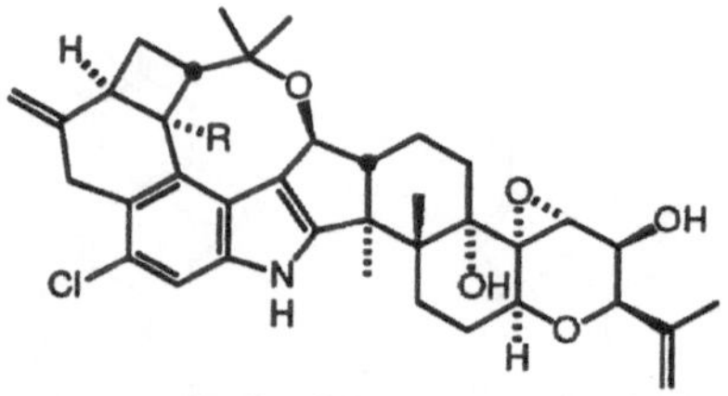

1450 R = OH (penitrem A)
1452 R = H (penitrem F)

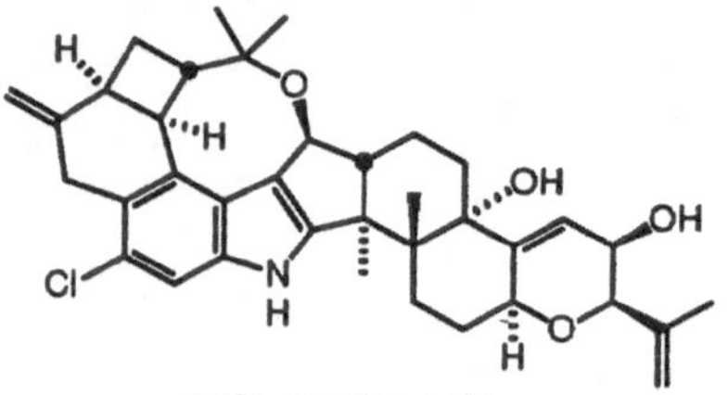

1451 (penitrem C)

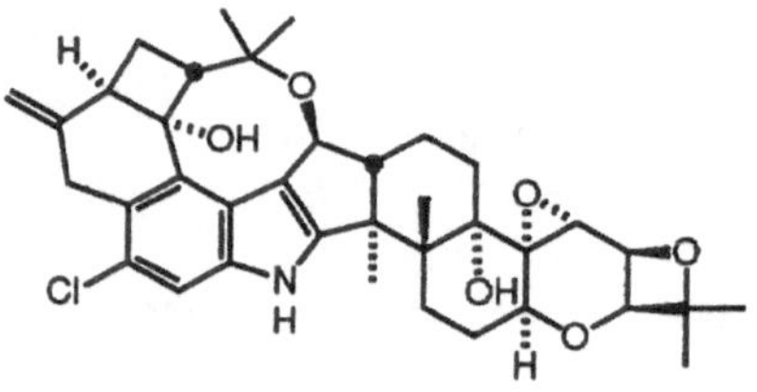

1453 (pennigritrem)

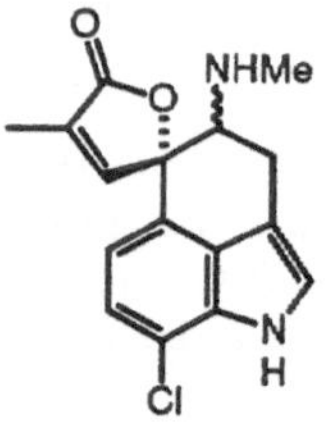

1454 α-epimer (8-chlororugulovasine A)
1455 β-epimer (8-chlororugulovasine B)

(*1162–1164*), *Lathyrus latifolius* (*1162*), *V. amurensis* (*1165*), *L. sativus* (grasspea) (*1166*), *L. maritimus* (sea pea) (*1166*), *L. odoratus* (sweet pea) (*1166*), *V. sativa* (vetch) (*1166*), *Lens culinaris* (lentil) (*1166*), and *Pinus sylvestris* (Pinaceae) (*1167*). These metabolites are usually found in the seeds and seem to be important during seed development (*1168*). Depending on the assay, 4-chloroindole-3-acetic acid (**1468**) is 10–100 times better as a hormone than indole-3-acetic acid (*1163, 1165, 1168, 1169*).

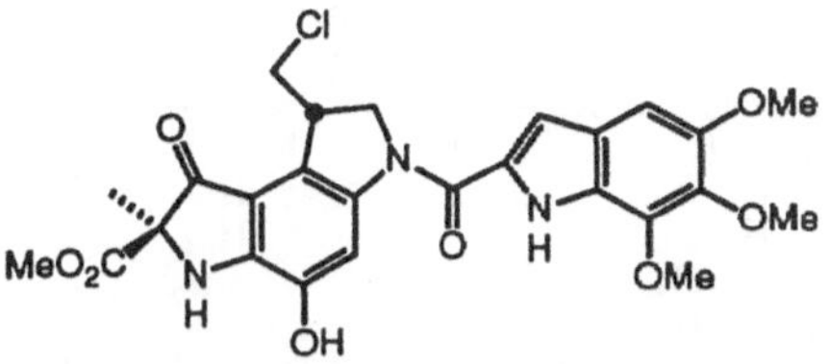

1456 (pyrindamycin A = duocarmycin C_2)

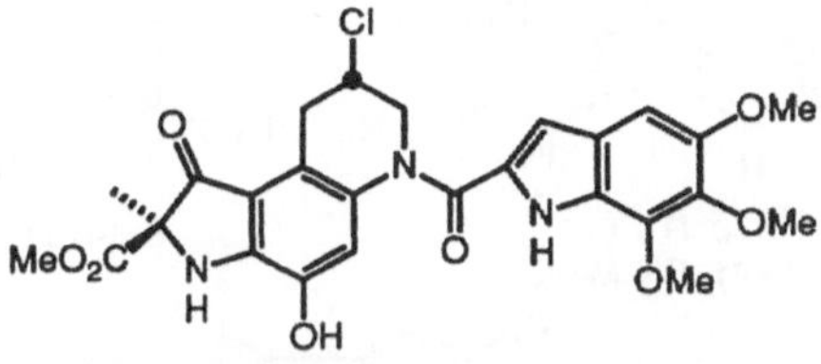

1457 (pyrindamycin B = duocarmycin C_1)

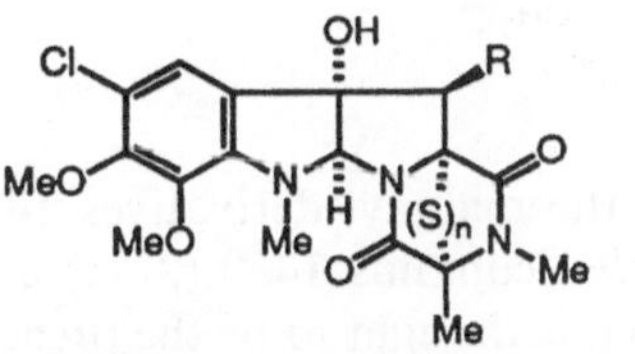

1458 R = OH, n = 2 (sporidesmin A)
1459 R = H, n = 2 (sporidesmin B)
1462 R = OH, n = 3 (sporidesmin E)
1464 R = OH, n = 4 (sporidesmin G)

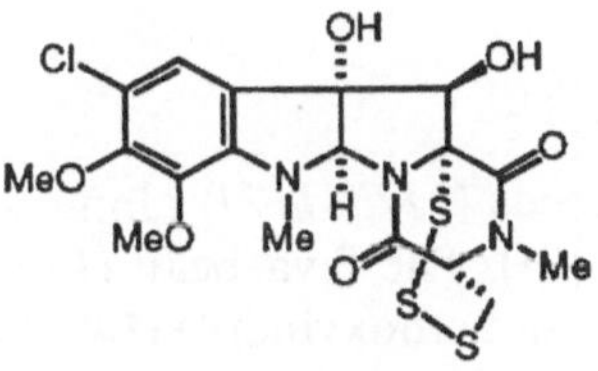

1460 (sporidesmin C)

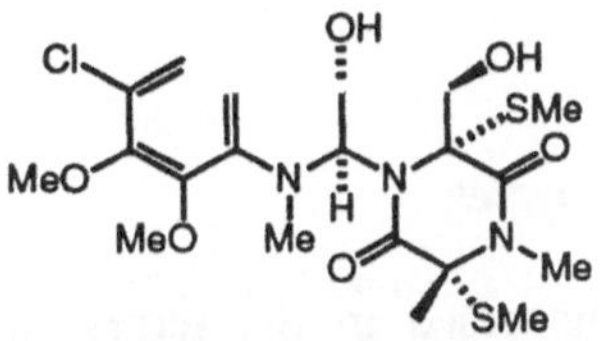

1461 (sporidesmin D)

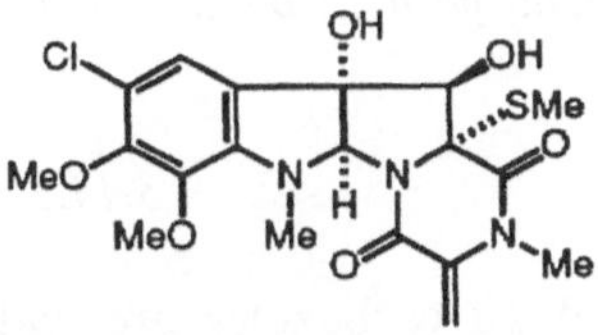

1463 (sporidesmin F)

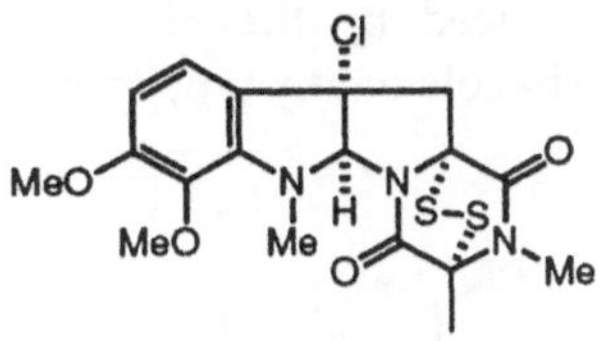

1465 (sporidesmin H)

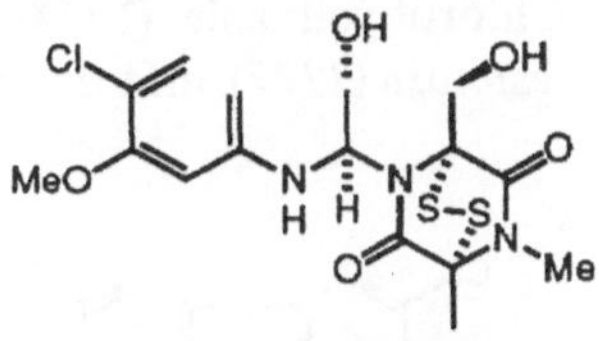

1466 (sporidesmin J)

Within Leguminosae these chlorinated hormones seem limited to the tribus *Vicieae*, since they are not found in *Phaseolus vulgaris* (*1168*). *Pisum sativum* also contains (*S*)-4-chlorotryptophan (**1469**) in the protein

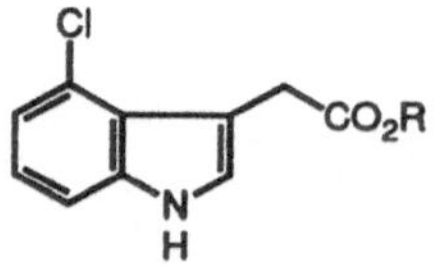

1467 R = Me
1468 R = H

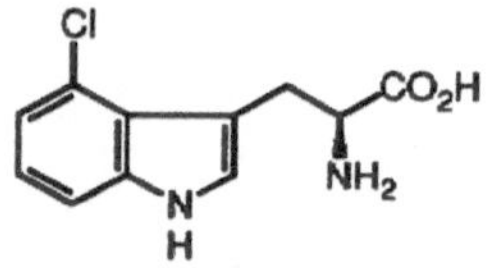

1469

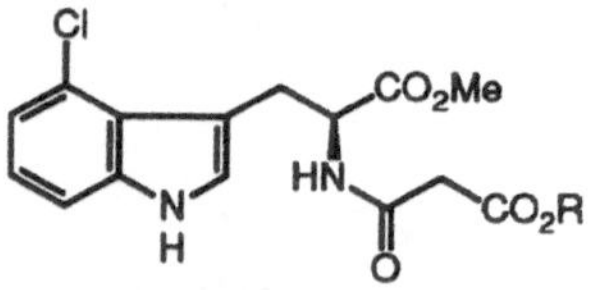

1470 R = Et
1471 R = Me

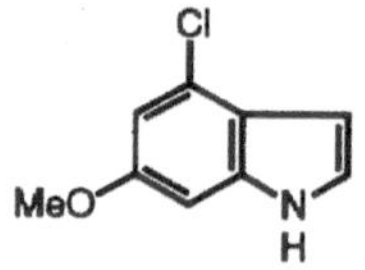

1472

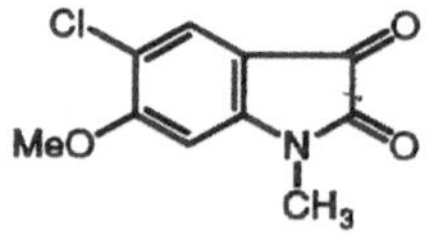

1473

of the seeds (*1170, 1171*) along with the malonyl derivatives **1470** and **1471** (*1171*). The fava bean (*Vicia faba*) contains **1469** (*2216*) and also 4-chloro-6-methoxyindole (**1472**) which is thought to be the precursor of a potent mutagen that forms during intragastric nitrosation. This latter compound may be the cause of the high incidence of gastric cancer in parts of Colombia (*1172–1175*). The chloroisatin **1473** is produced by *Micromonospora carbonacea* (*1176*).

3.14.3. Carbazoles

Although carbazoles are widely dispersed in the terrestrial plant kingdom, naturally occurring halogenated examples are essentially unknown. Chlorohyellazole (**1474**) is produced by the blue-green alga *Hyella caespitosa* (*1177*), and the first carbazole of any type to be isolated

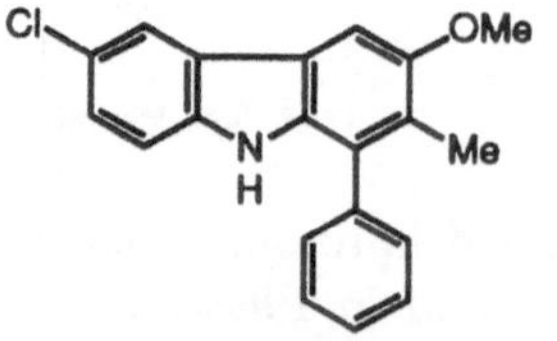

1474 (chlorohyellazole)

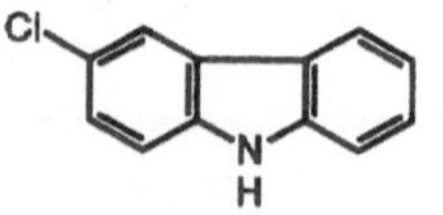

1475

from mammals is 3-chlorocarbazole (**1475**) a compound that exhibits potent monoamine oxidase inhibition and is present in bovine urine (*1178*, *1179*).

3.14.4. Indolocarbazoles

In contrast to halogenated carbazoles, there are several examples of halogenated indolo[2,3-*a*]carbazoles and closely related compounds which occur naturally, some with powerful biological activity.

The Panamanian soil microbe *Nocardia aerocolonigenes* (renamed *Saccharothrix aerocolonigenes*) produces rebeccamycin (**1476**) (*1180–1182*) and 4′-deschlororebeccamycin (**1477**) (*1184*). The corresponding bromorebeccamycin is also produced in the presence of 0.05% bromide (*1185*). These compounds show strong antitumor and antibiotic activity (*1181*) and a clinical candidate from this group seems possible. The biosynthesis of rebeccamycin has been studied; glucose, methionine, and tryptophan are each incorporated, but tryptophan does not furnish the phthalimide nitrogen (*1186*). The microbe *Actinomadura melliaura* produces AT2433-A1 (**1478**) and -A2 (**1479**) which, like the closely related rebeccamycin, also have strong antitumor and antibiotic activity (*1187*, *1188*). The blue-green alga *Tolypothrix tjipanasensis* is the source of 15 new indolo[2,3-*a*]carbazoles, 13 of which contain chlorine (*1189*). These include tjipanazoles A1 (**1480**), A2 (**1481**), C1 (**1482**), C2 (**1483**), C3 (**1484**), C4 (**1485**), B (**1486**), E (**1487**), F1 (**1488**), F2 (**1489**), D (**1490**), I (**1491**), and J (**1492**).

3.14.5. Carbolines

Carbolines, especially β-carbolines, are very abundant as plant alkaloids. In recent years several halogenated β-carbolines have been found in marine life and ascidians dominate these studies (*1191*).

The Caribbean tunicate *Eudistoma olivaceum* produces an array of brominated β-carbolines of several types. The initial study identified eudistomins A (**1493**), C (**1494**), D (**1495**), E (**1496**), F (**1497**), G (**1498**), H (**1499**), J (**1500**), K (**1501**), L (**1502**), N (**1503**), O (**1504**), and P (**1505**) (*1192*). Another study uncovered eudistomins R (**1506**) and S (**1507**) in the Bermudan *E. olivaceum* (*1193*). These compounds have high antiviral, antimicrobial, and calcium-release induction activity (*1192*). The New Zealand ascidian *Ritterella signillinoides* contains the new eudistomin K sulfoxide (**1508**) as well as the known eudistomins K (stereochemistry of

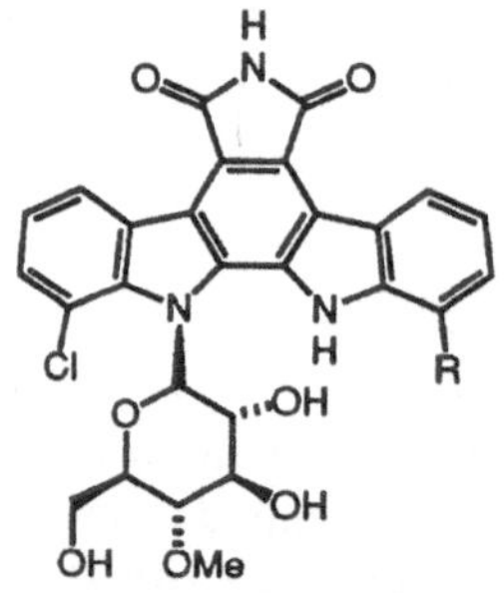

1476 R = Cl (rebeccamycin)
1477 R = H

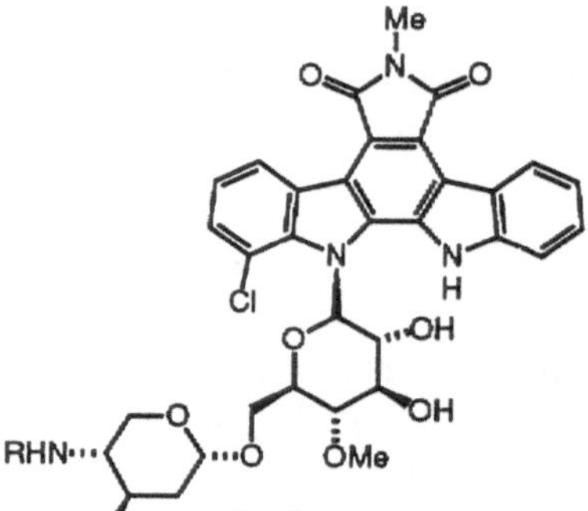

1478 R = Me (AT2433-A1)
1479 R = H (AT2433-A2)

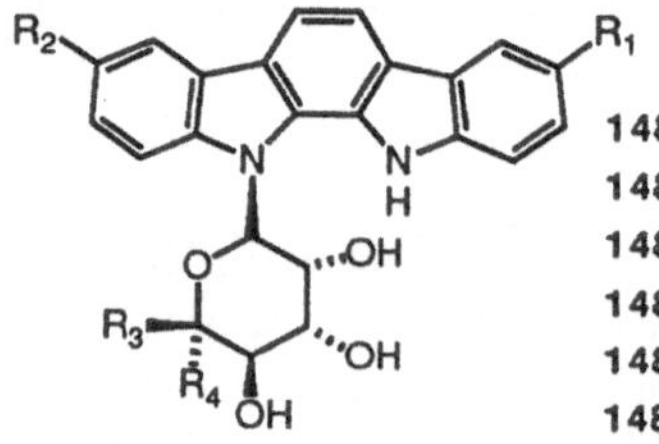

	R_1	R_2	R_3	R_4	
1480	Cl	Cl	Me	H	(tjipanazole A1)
1481	Cl	Cl	H	Me	(tjipanazole A2)
1482	Cl	H	Me	H	(tjipanazole C1)
1483	H	Cl	Me	H	(tjipanazole C2)
1484	Cl	H	H	Me	(tjipanazole C3)
1485	H	Cl	H	Me	(tjipanazole C4)

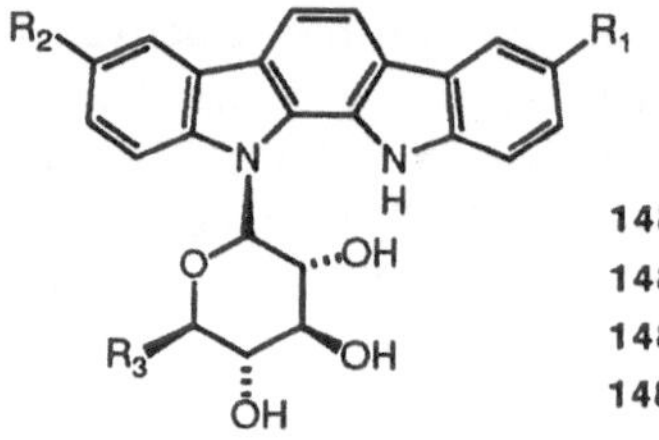

	R_1	R_2	R_3	
1486	Cl	Cl	H	(tjipanazole B)
1487	Cl	Cl	CH_2OH	(tjipanazole E)
1488	Cl	H	H	(tjipanazole F1)
1489	H	Cl	H	(tjipanazole F2)

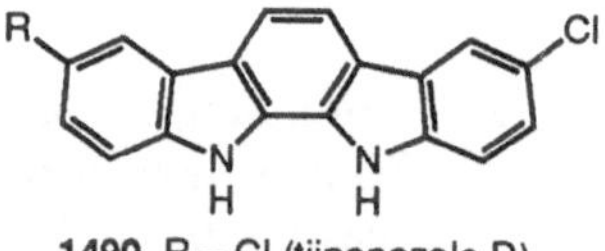

1490 R = Cl (tjipanazole D)
1491 R = H (tjipanazole I)

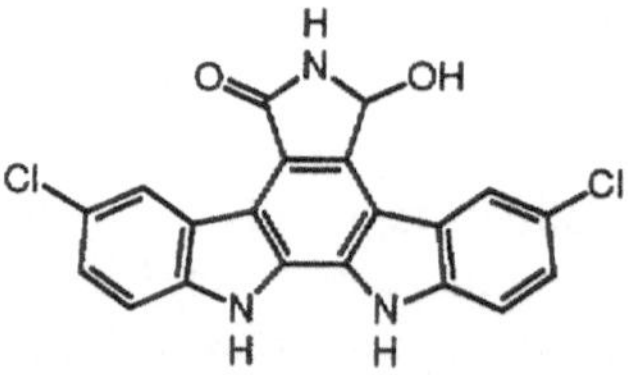

1492 (tjipanazole J)

the favored invertomer revised in (*1194*, *1195*)), C, and O (*1196*, *1197*). Eudistomin K sulfoxide displays antiviral activity against *Polio* and *Herpes* simplex (*1196*), while the synthetic analog 9-methyl-7-bromoeudistomin D is 1000 times more potent than caffeine in stimulating calcium release in skeletal muscle (*1198*). Moreover, several other synthetic eudistomin D analogs are more active than eudistomin D (**1495**) as

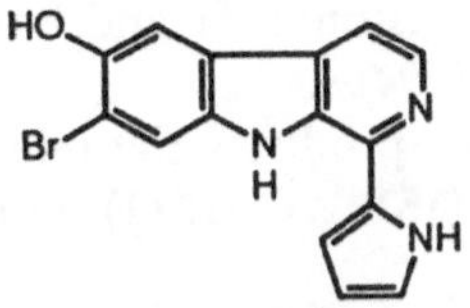

1493 (eudistomin A)

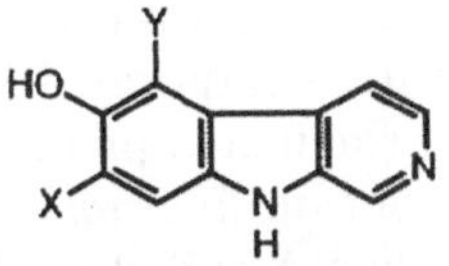

1495 X = H, Y = Br (eudistomin D)
1500 X = Br, Y = H (eudistomin J)

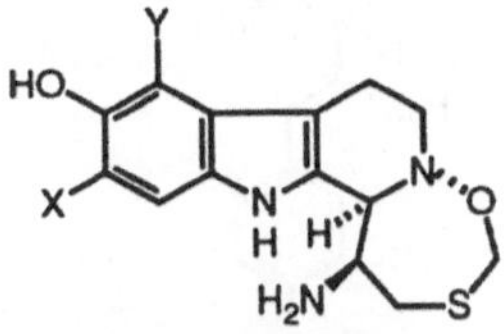

1494 X = Br, Y = H (eudistomin C)
1496 X = H, Y = Br (eudistomin E)

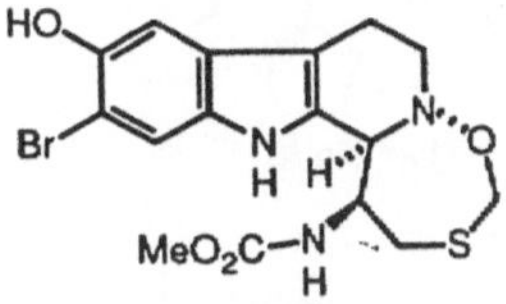

1497 (eudistomin F)

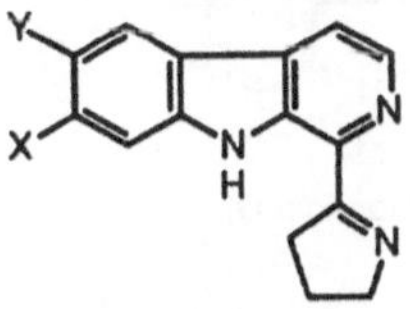

1498 X = Br, Y = H (eudistomin G)
1499 X = H, Y = Br (eudistomin H)
1505 X = Br, Y = OH (eudistomin P)

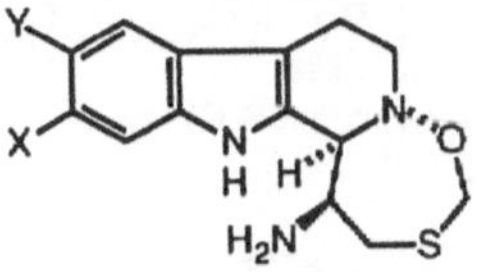

1501 X = Br, Y = H (eudistomin K)
1502 X = H, Y = Br (eudistomin L)

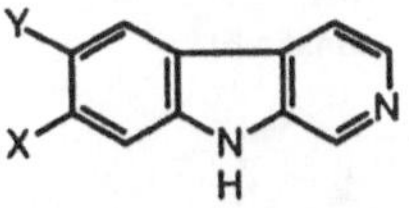

1503 X = H, Y = Br (eudistomin N)
1504 X = Br, Y = H (eudistomin O)

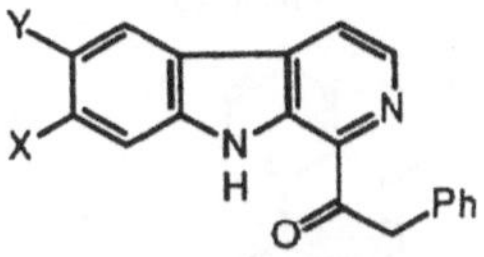

1506 X = Br, Y = H (eudistomin R)
1507 X = H, Y = Br (eudistomin S)

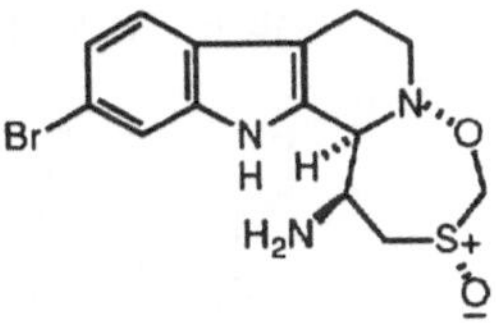

1508 (eudistomin K sulfoxide)

phosphodiesterase inhibitors (*1199*). Biosynthetic studies of eudistomin H (**1499**) in *E. olivaceum* indicate that both 6-bromotryptamine and 6-bromotryptophan are incorporated into eudistomin H and both are better than the non-brominated precursors (*1204*).

The Okinawan tunicate *Eudistoma glaucus* produces the related β-carbolines eudistomidins A (**1509**) (*1200*), B (**1510**), C (**1511**), D (**1512**)

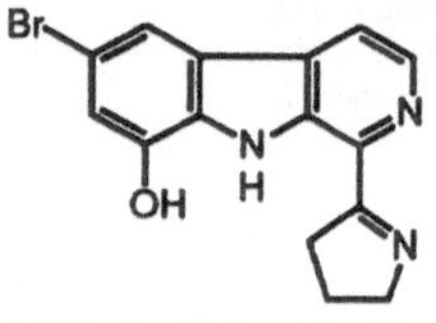

1509 (eudistomidin A)

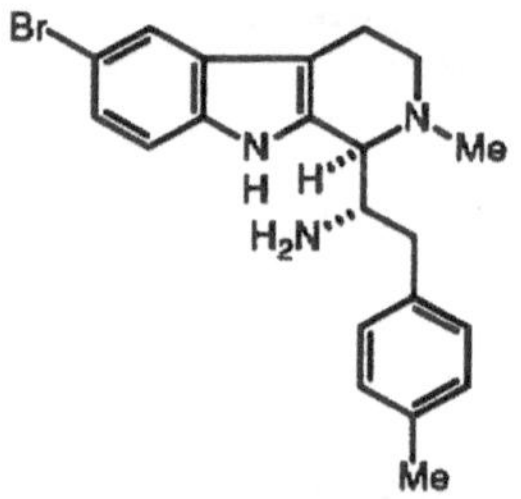

1510 (eudistomidin B)

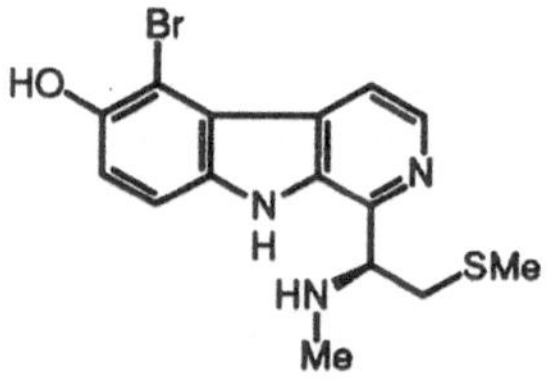

1511 (eudistomidin C)

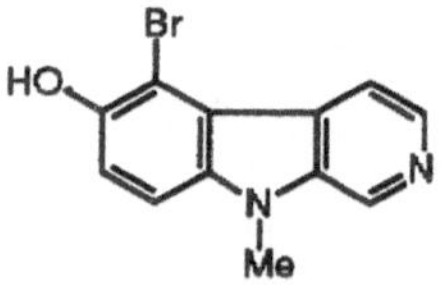

1512 (eudistomidin D)

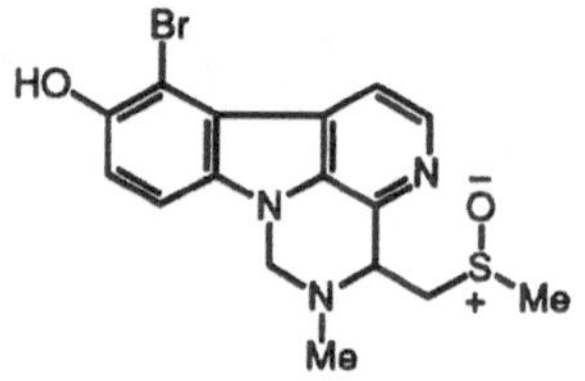

1513 (eudistomidin E)

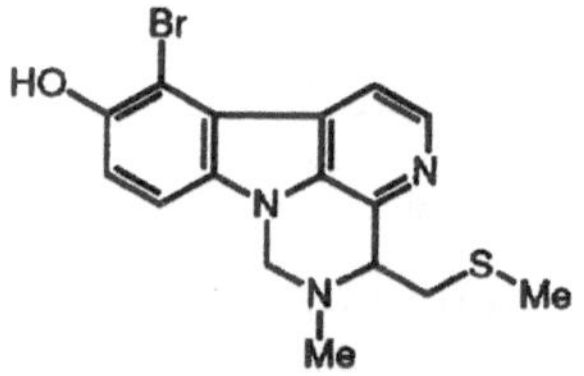

1514 (eudistomidin F)

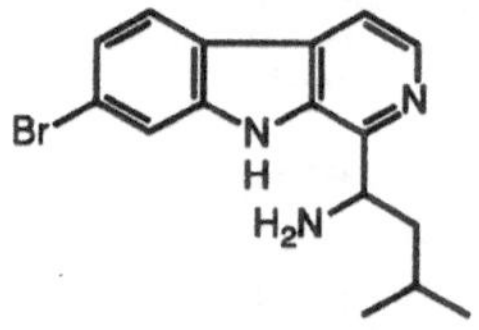

1515 (eudistalbin A)

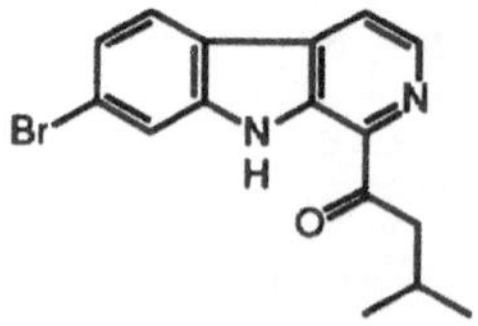

1516 (eudistalbin B)

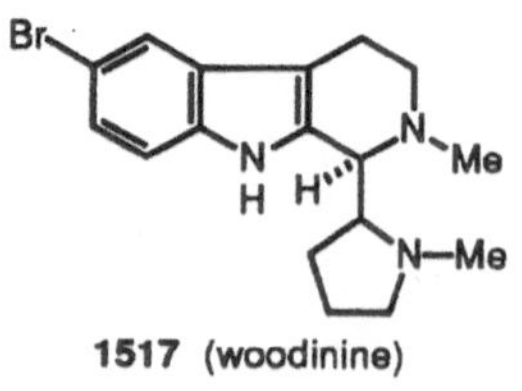

1517 (woodinine)

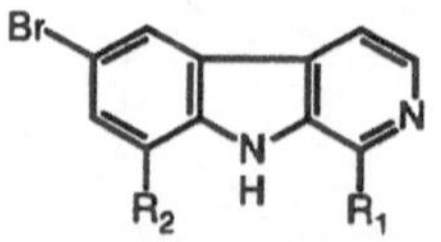

1518 R_1 = Et, R_2 = H
1519 R_1 = Me, R_2 = H
1520 R_1 = Et, R_2 = Br

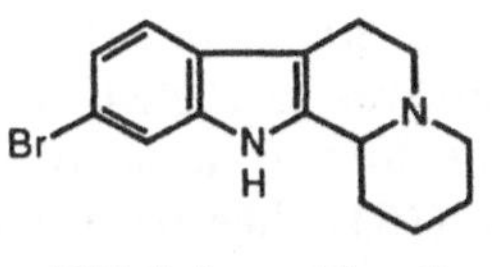

1521 (arborescidine A) 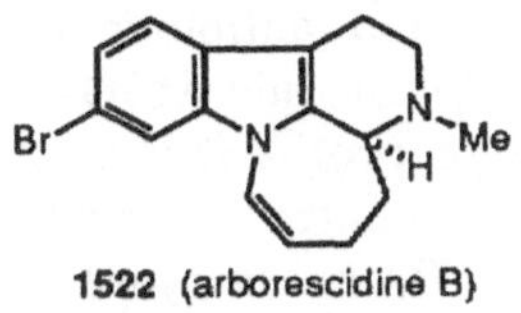 **1522** (arborescidine B)

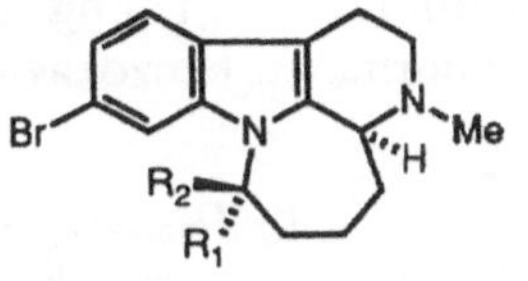

1523 R_1 = OH, R_2 = H (arborescidine C)
1524 R_1 = H, R_2 = OH (arborescidine D)

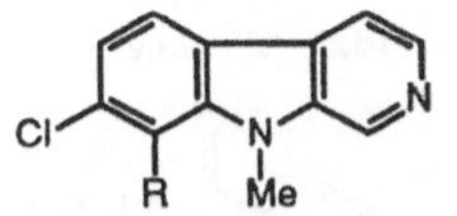

1525 R = H (bauerine A)
1526 R = Cl (bauerine B)
1527 (bauerine C) 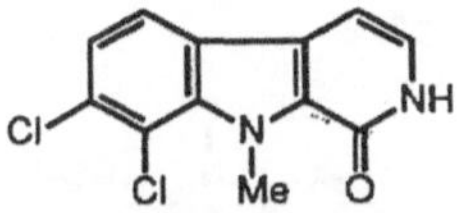

(*1201*), E (**1513**), and F (**1514**) (*1202*) while *E. album* contains the new eudistalbins A (**1515**) and B (**1516**) in addition to eudistomin E (*1203*). Eudistalbin A shows some cytotoxicity against the KB cell line (IC_{50} = 3.2 μg/ml) whereas eudistalbin B is inactive. The New Caledonian ascidian *E. fragum* not only contains bromotryptamine **1355** but also the new woodinine (**1517**) (*1074*). The hydroid *Aglaophenia pluma* produces the three β-carbolines **1518–1520** (*1205*), and the New Caledonian tunicate *Pseudodistoma arborescens* has yielded arborescidines A (**1521**), B (**1522**), C (**1523**), and D (**1524**) (*1206*). Of these latter metabolites only arborescidine D exhibits activity against KB cells (IC_{50} = 3 μg/ml). The terrestrial blue-green alga *Dichothrix baueriana* which was collected from the Na Pali coast on Kauai produces bauerines A–C (**1525–1527**), which are active against *Herpes* simplex virus (*1207*).

3.14.6. Quinolines and Other Nitrogen Heterocycles

Compared with the abundance of natural halogenated π-excessive heterocycles (pyrroles, indoles, *etc.*) only few examples of halogenated π-deficient heterocycles (pyridines, quinolines, etc.) exist. In fact, the only known example of a simple natural halogenated quinoline is 7-bromoquinoline **1528** from the marine bryozoan *Flustra foliacea* (*1271*). The tetrahydroquinoline virantmycin (**1529**) from *Streptomyces nitrosporeus* has antiviral and antibiotic properties (*1208–1211*). The interesting quinoxaline 1,4-dioxide MSD-819 (**1530**) was extracted from *S. ambofaciens* culture broths (*1212, 1213*). The first example of a naturally occurring brominated quinazolinedione (**1531**) has been isolated from the tunicate *Pyura sacciformis* together with the known 6-bromo-3-indolecarbaldehyde (**1320**) (*1214*). Young corn roots (*Zea mays*, Gramineae) produce the chlorinated benzoxazinone **1532** (*1215*).

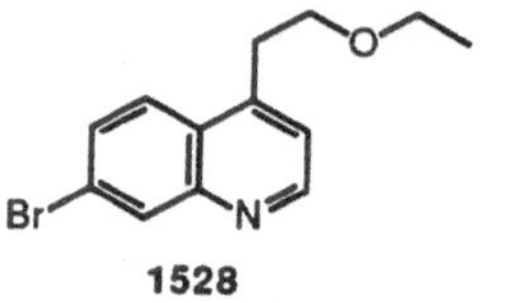

1528

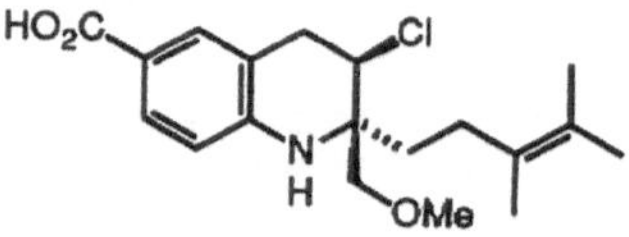

1529 (virantmycin)

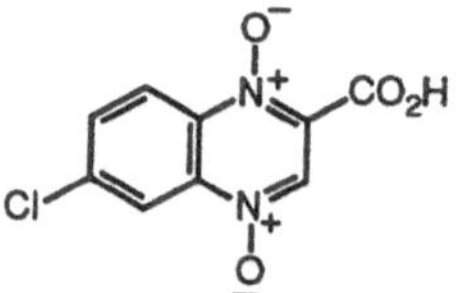

1530 (MSD-819)

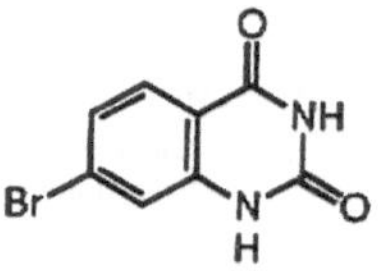

1531

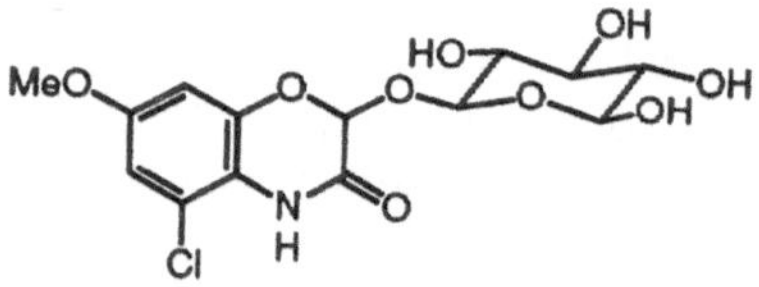

1532

The fungus *Streptosporangium* sp. contains the broad spectrum antifungal chlorophenazine **1533** (*1216*), and chloralbofungin (**1534**) is produced by both *Actinomyces albus* and *A. tumemacerans* (*1217*). The novel chlorinated acridones gravacridonchlorin (**1535**), gravacridonolchlorin (**1536**), and isogravacridonchlorin (**1537**) are present in the roots of *Ruta graveolens* (Rutaceae) (*1218, 1219*). The simple histidine antitumor metabolite girolline (**1538**) is produced by the sponge *Pseudaxinyssa cantharella* (*1220*); the *threo* stereochemistry has been confirmed both by synthesis (*1221, 1222*) and by X-ray crystallography (*1223*).

If the discovery of a benzodiazepine derivative in bovine brain was not enough of a surprise (*1224*), then the isolation of several chlorinated benzodiazepines (**1539–1545**) in wheat grains and potato tubers surely must be (*1225*). These metabolites are produced by the plant in the low ppb range and not by accompanying microorganisms. Compounds **1541** and **1544** are found in wheat only whereas **1542** is found only in potato. For a review of these metabolites see (*2241*).

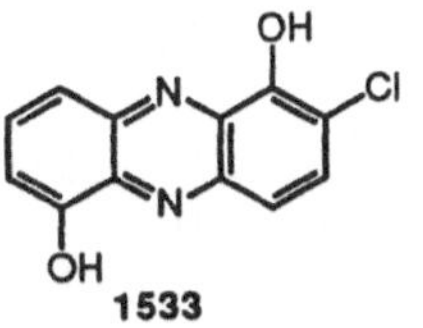

1533

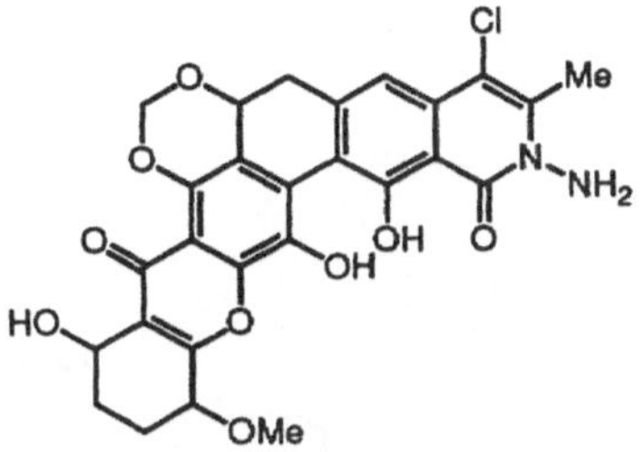

1534 (chloroalbofungin)

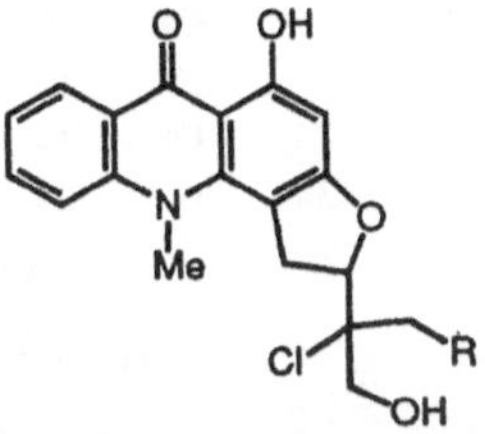

1535 R = H (gravacridonchlorin)
1536 R = OH (gravacridonolchlorin)

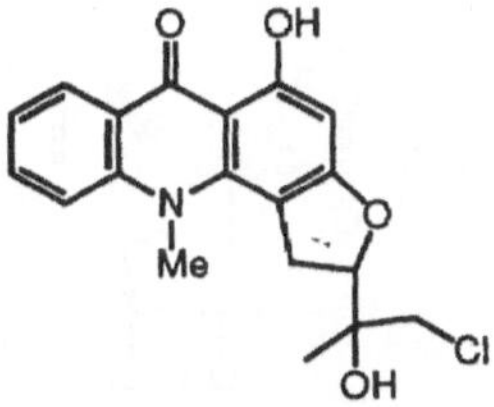

1537 (isogravacridonchlorin)

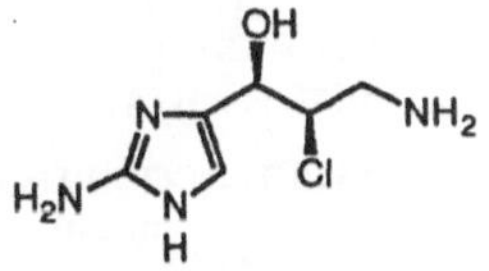

1538 (girolline)

A few nucleic acid bases contain halogen. The novel fluorine-containing nucleocidin which was isolated from the soil microbe *Streptomyces calvus* was found to have the novel structure **1546** (*1226, 1227*). This metabolite which is particularly effective against trypanosomes is the first natural fluoro-sugar. The related chlorosulfonamide AT-625 (**1547**) is produced by *S. rishiriensis* (*1228*). Another study of *Streptomyces* sp. found AT-625 and the alanyl derivative ascamycin **1548** (*1229*). The marine sponge *Echinodictyum* sp. produces the novel brominated pyrrolopyridimine **1549** and the red alga *Hypnea valendiae* contains the iodine-containing tubercidin nucleoside **1550** (*1230*). The latter metabolite is the first example of a naturally occurring 5′-deoxyribosyl nucleoside. A novel inhibitor of adenosine deaminase, 2′-chloropentostatin (**1551**), was isolated simultaneously by three groups from an unidentified actinomycete (ATCC 39365) (*1231*) and from *Actinomadura* sp. ("adechlorin") (*1232, 1233*).

Although Woodward long ago demonstrated that chlorins undergo facile electrophilic chlorination in the laboratory (*1234*), it was not until recently that a chlorinated porphyrin (**1552**) was found in a natural source, the microorganism *Pseudomonas stutzeri* (*1235*). The antitumor, antibacterial compound **1553** was isolated from cultures of *Streptomyces griseolutus* (*1236, 1237*).

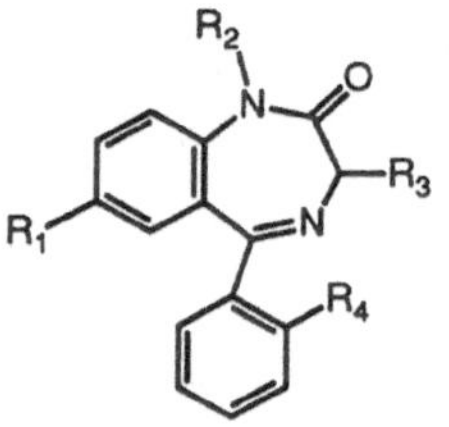

	R_1	R_2	R_3	R_4
1539	Cl	H	H	H
1540	Cl	Me	H	H
1541	H	Me	H	Cl
1542	Cl	H	OH	Cl
1543	Cl	H	H	Cl
1544	Cl	Me	OH	Cl
1545	Cl	Me	H	Cl

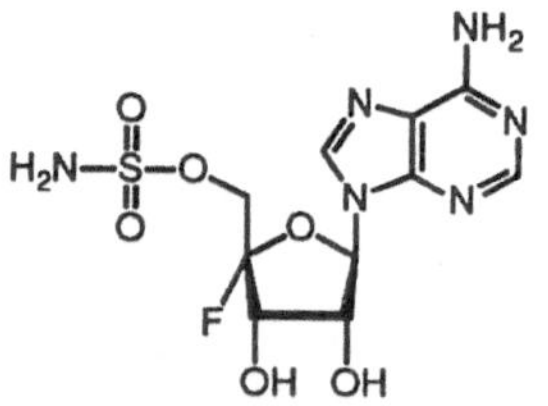

1546 (nucleocidin)

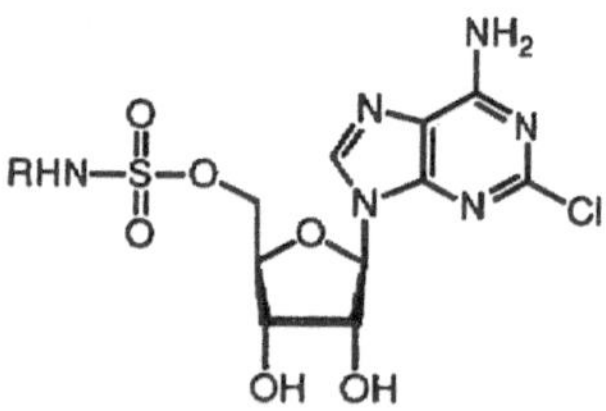

1547 R = H (AT-625)

1548 R = COCH(Me)NH$_2$ (S) (ascamycin)

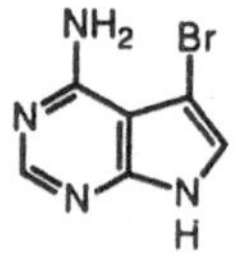

1549

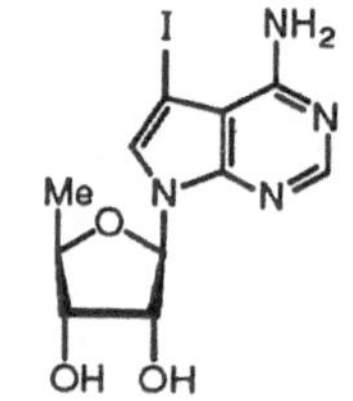

1550

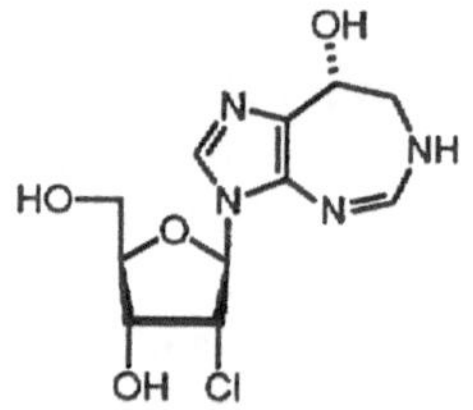

1551 (2'-chloropentostatin= adechlorin)

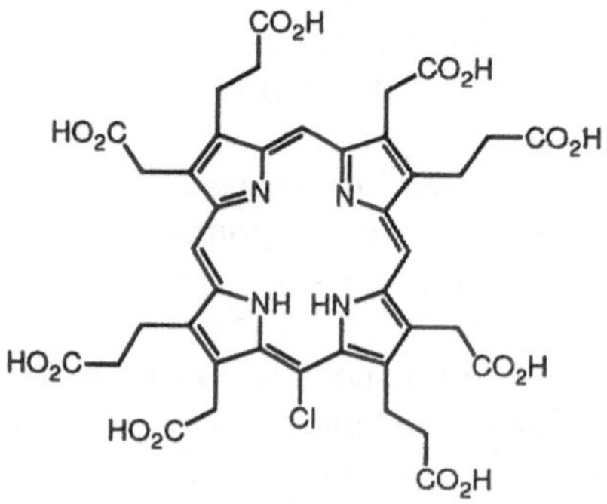

1552

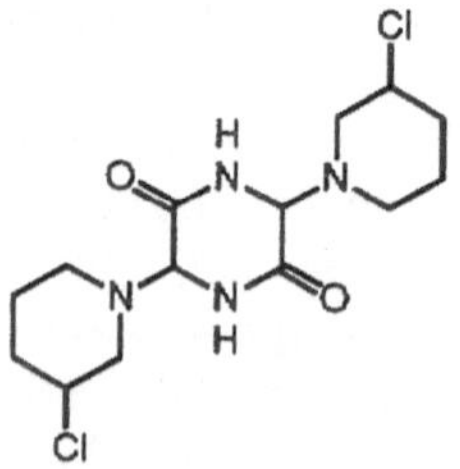

1553

3.14.7. *Benzofurans and Related Compounds*

A few fungal metabolites have been discovered that have chlorinated benzo[*b*]furan structures. The relatively simple furasterin (**1554**) from the fungus *Phialophora asteris* may be related biogenetically to cryptosporiopsin (**196**) (*1238*). Mycorrhizinol (**1555**) which is really a furochroman is produced by the mushroom *Gilmaniella humicola* (*200*), and chloromycorrhizinol A (**1556**) is found in the roots of *Monotropa hypopitys* (Ericaceae) (*1240*). The African plant *Psorospermum febrifugum* (Guttiferae) produces the chlorinated dihydrofuranoxanthone **1557** which is a derivative of psorospermin (*1241*). Other chlorinated xanthones are discussed in a later section. Although it is not a benzofuran itself, the precursor of the *Desmia hornemanni* red alga metabolite 4,5-dimethylbenzo[*b*]furan is the dienone **1558** which has been isolated from fresh alga (*1242*). Mild acid treatment of **1558** rapidly transforms it into 4,5-dimethylbenzo[*b*]furan.

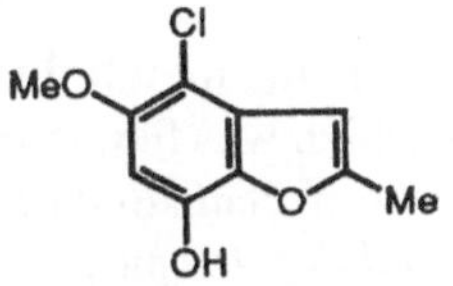

1554 (furasterin)

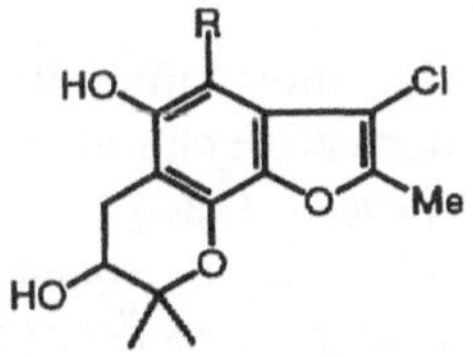

1555 R = H (mycorrhizinol)
1556 R = Cl (chloromycorrhizinol)

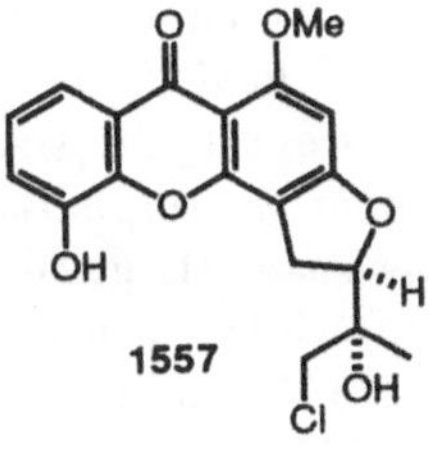

1557

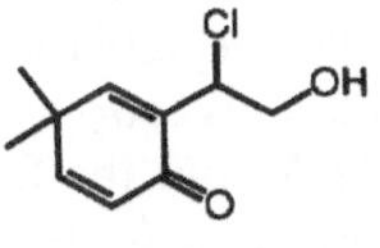

1558

3.14.8. *Pyrones*

The red alga *Ptilonia australasica* which furnished brominated enones **176–178** also contains the novel γ-pyrones **1559** and **1560** (*162*). The Tasmanian red alga *Phacelocarpus labillardieri* has yielded the three novel macrocyclic pyrones **1561–1563** (*1243–1245*).

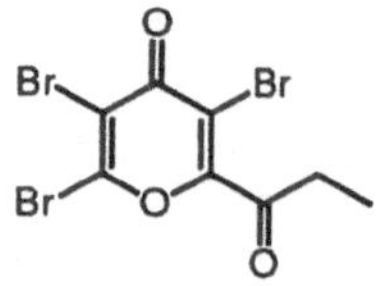

3.14.9. Coumarins and Isocoumarins

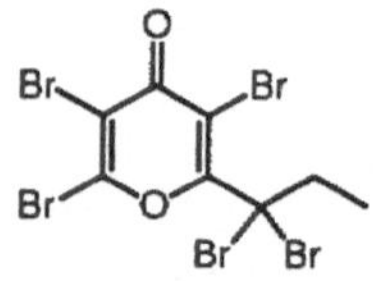

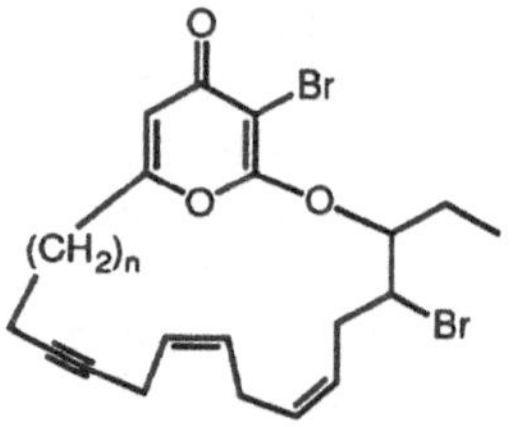

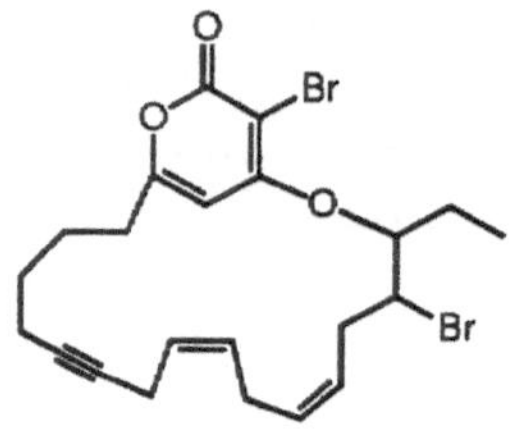

One of the most ubiquitous of all fungal metabolites is the isocoumarin derivative ochratoxin A (**1564**), which was first isolated from *Aspergillus ochraceus* (*1246, 1247*) and later from *Penicillium viridicatum* (*1248*), *P. cyclopium, P. commune, P. variabile, P. purpurescens* (*9, 1249–1251*), *Aspergillus melleus* and *A. sulphureus* (*1252*). This compound has the same order of acute toxicity as aflatoxin B_1 (*1246*) and its structure was established by synthesis (*1247*). The related ochratoxin C (**1565**) is produced by *Aspergillus ochraceus* (*1253*) and 4-hydroxyochratoxin A (**1566**) was isolated from cultures of *Penicillium viridicatum* (*1254*). The carrot fungus *Sporomia affinis* produces dihydroisocoumarins **1567** and **1568** (*1239*) and the fungus *Periconia macrospinosa* contains **1569** (*180*). The Brazilian plant *Swartzia laevicarpa* (Leguminosae) has yielded the isocoumarins **1570** and **1571** (*1255*) while *Tovomita brasiliensis* (Guttiferae) has **1570** as a constituent (*1256*). An unidentified fungus was found to contain LL-Z1640-5 (**1572**) (*1257*). The marine sponge *Mycale adhaerens* produces the novel brominated dihydroisocoumarin hiburipyranone (**1573**) together with known indole ester **1331** (*1258*).

Chlorocoumarin **1574** has been isolated from *Toddalia asiatica* (*1259*), and chloticol (**1575**) is a constituent of *Murraya exotica* (*1260*), both in Rutaceae. The closely related chloculol (**1576**) is found in *M. paniculata* (*1261*). That chloculol is not an artifact produced from the corresponding epoxide during extraction and workup was confirmed by subjecting the epoxide to the complete isolation and purification proto-

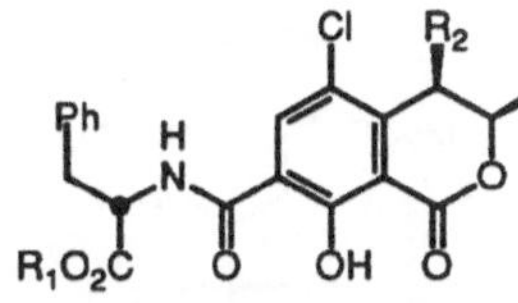

1564 $R_1 = R_2 = H$ (ochratoxin A)
1565 $R_1 = Et, R_2 = H$ (ochratoxin C)
1566 $R_1 = H, R_2 = OH$

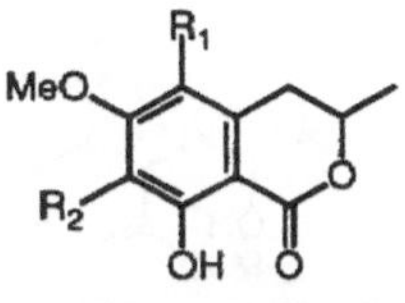

1567 $R_1 = H, R_2 = Cl$
1568 $R_1 = R_2 = Cl$
1569 $R_1 = Cl, R_2 = H$

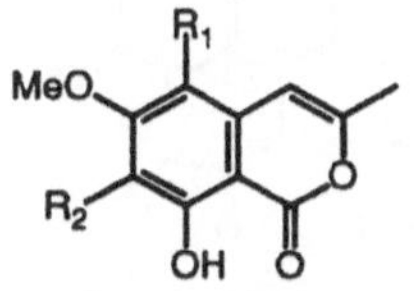

1570 $R_1 = H, R_2 = Cl$
1571 $R_1 = Cl, R_2 = H$

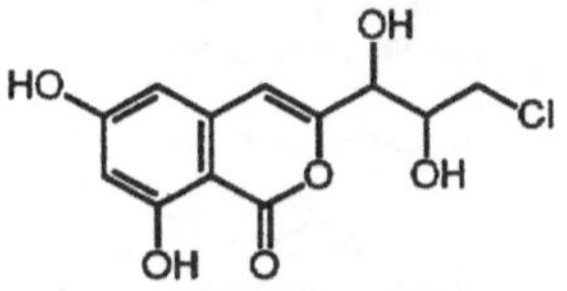

1572 (LL-Z1640-5)

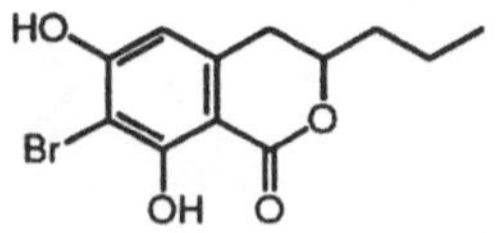

1573 (hiburipyranone)

col (*1261*). Recently, chlorinated coumarin **1577** was characterized from *Triphasia trifolia*, also in Rutaceae (*1262*). Several chlorinated furocoumarins (psoralens) have been discovered, including saxalin (**1578**) from the roots of *Angelica saxatilis* (*1263*), the roots of *Cachrys pubescens* (*2215*), the seeds of Texas *Ammi majus* (Bishop's Weed) (*1264*), and from common parsley (*Petroselinium sativum*) (*1265*). Thin layer chromatographic examination of *Ammi majus* seeds reveals the presence of saxalin, and it is not considered to be an artifact (*1264*, *1265*). The isomeric psoralens **1579** and **1580** are found in *Heracleum granatense* (*1266*) and *H. pyrenaicum* (*1267*), respectively. The latter psoralen **1580** has also been isolated from *Prangos pabularia* (*1268*). Roots of *Peucedanum arenarium* contain peuchlorin (**1581**), peuchlorinin (**1582**), and peuchloridin (**1583**) (*1269*). All these species are in Umbelliferae. The novel dimeric oxaphenalenone gilmaniellin (**1584**) is found in *Gilmaniella humicola* (*1270*).

3.14.10. Flavones and Isoflavones

Like the coumarins, several examples of chlorine-containing flavones and related metabolites are known.

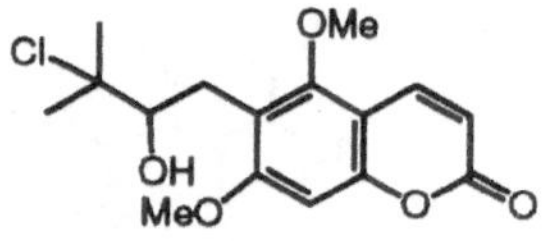

1574

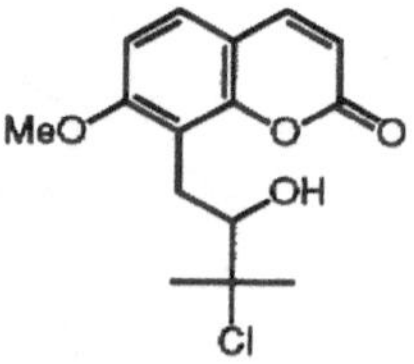

1575 (chloticol)

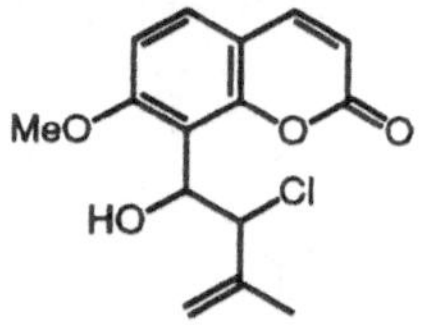

1576 (chloculol)

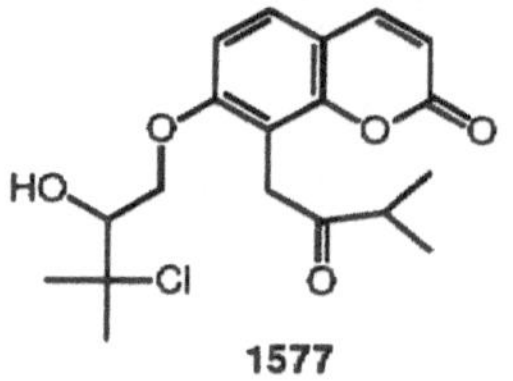

1577

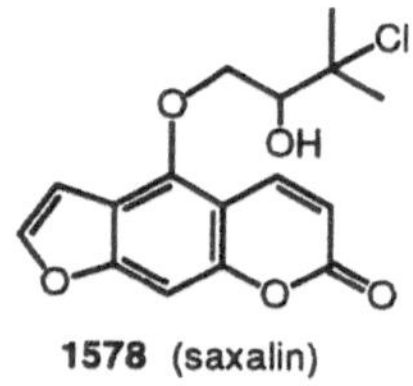

1578 (saxalin)

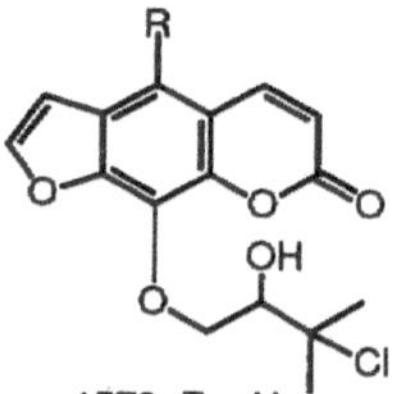

1579 R = H
1580 R = OMe

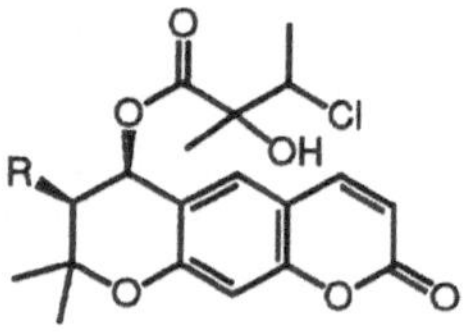

1581 R = OCOC(Me) $\overset{Z}{=}$ CHMe (peuchlorin)

1582 R = OCO— (peuchlorinin)

1583 R = OCOC(Me)OHCH(Cl)Me (peuchloridin)

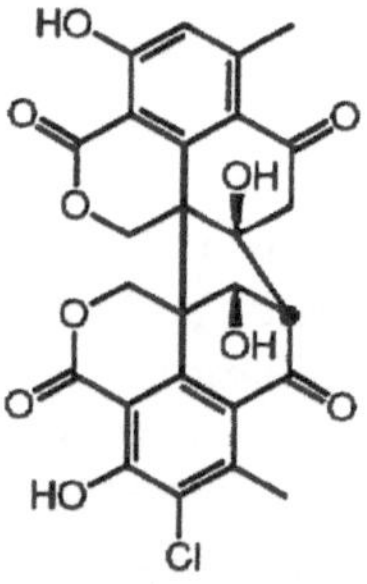

1584 (gilmaniellin)

The lichens *Lecanora sordida* (*1272*), *L. rupicola* (*1273, 1274, 2223*), and *L. carpinea* (*2223*) contain sordidone (rupicolon) (**1585**). Chlorflavonin (**1586**) was discovered in the culture broth of *Aspergillus candidus* (*1275–1277*). A biosynthetic study of this fungal metabolite indicates that it is a true metabolite and is synthesized *de novo* by this microbe (*1278, 1279*). Several chlorogenisteins, **1587**, **1588** (*1280*), **1589**, and **1590** (*1281*), are produced by *Streptomyces griseus*, and **1591** which has antioxidant activity *in vitro* has been isolated from a *Streptomyces* sp.

(*1282*). The isomeric 6-chloroapigenin (**1592**) is found in the "field horsehair" *Equisetum arvense* (Equisetaceae) (*1283*). The estrogen-receptor binding inhibitors BE-14348D (**1593**) and E (**1594**, **1595**) are found in the culture broth of *Streptomyces graminofaciens* (*1284*). The fungus *Monilinia fructicola* produces chloromonilinic acids A (**1596**) and B (**1597**), which are derived from chloromonilicin (*1285*), which is discussed later in Sect. 3.22.6.

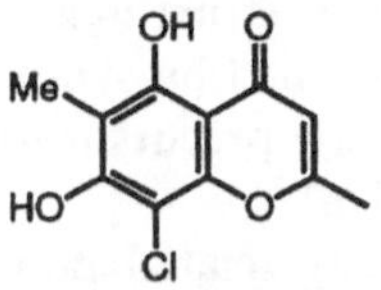

1585 (sordidone)

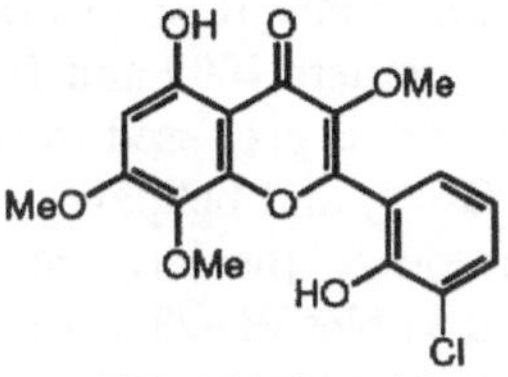

1586 (chlorflavonin)

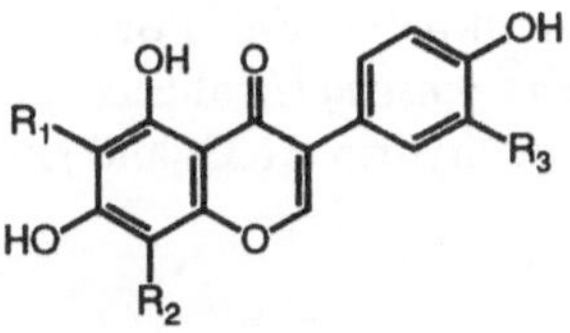

1587 R_1 = Cl, R_2 = R_3 = H
1588 R_1 = R_3 = Cl, R_2 = H
1589 R_1 = R_3 = H, R_2 = Cl
1590 R_1 = R_2 = Cl, R_3 = H
1591 R_1 = H, R_2 = Cl, R_3 = OH

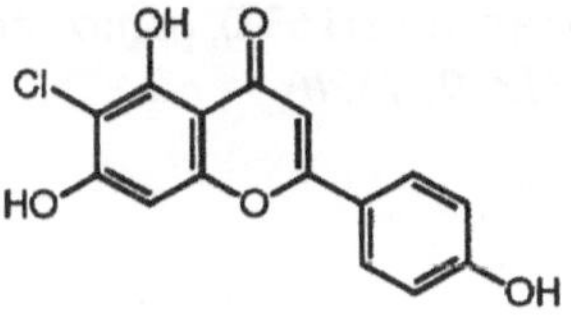

1592

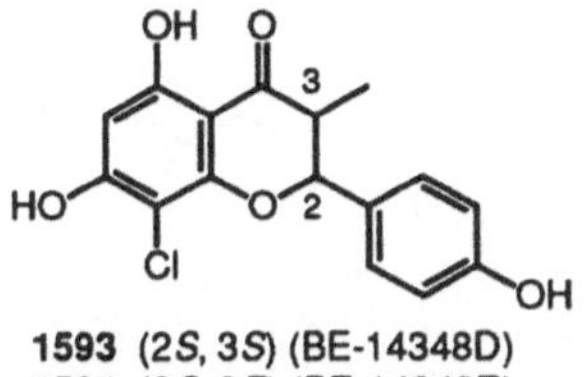

1593 (2*S*, 3*S*) (BE-14348D)
1594 (2*S*, 3*R*) (BE-14348E)
1595 (2*R*, 3*S*) (BE-14348E)

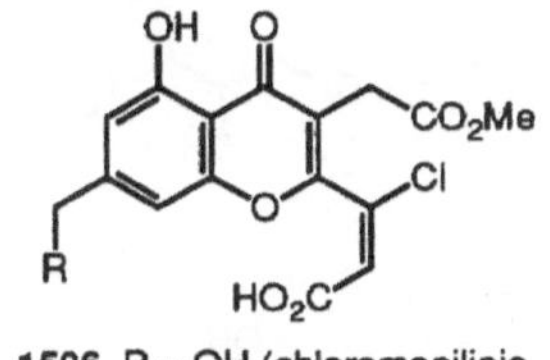

1596 R = OH (chloromonilinic acid A)
1597 R = H (chloromonilinic acid B)

3.15. Polyacetylenes

3.15.1. Terrestrial Polyacetylenes and Derived Thiophenes

A large number of polyacetylenes and derived thiophenes is known from the plant kingdom, mainly due to the work of BOHLMANN and his students and a significant number of these compounds contain chlorine (*1286*, *1287*, *1294*, *1295*).

The majority of these metabolites (**1598**–**1631**) and their plant sources is tabulated in Tables 21–23. In some cases the stereochemistry was established subsequent to the initial isolation and characterization. The phototoxicity properties of **1598** have been investigated and this compound is a cytotoxic photosensitizer (*1299*, *1300*). The biogenesis of **1598** and **1599** has been shown to involve epoxidation of the corresponding polyeneyne and subsequent ring opening (*1301*). Photolysis of **1599** gives rise to the cyclopropane derivative **1605**; both compounds are found in *Centaurea ruthenica* (Compositae) (*1298*). The structures of the two isomeric enol ethers **1609** and **1610** were confirmed by synthesis (*1304*). Enol ether **1606** is proposed to be the biogenetic precursor to the novel helitenuin (**1629**) and helipandurin (**1630**) (*1318*).

In addition to the large number of closely related polyacetylenes presented in Tables 21–23, several other terrestrial examples are known which are the result of an apparent different biogenesis. For example, the important oriental medicinal plant *Panax ginseng* (Araliaceae) produces chloropanaxydiol (**1632**), panaxydol chlorohydrin (**1633**) and ginsenoyne B (**1634**) (*1319*, *1320*).

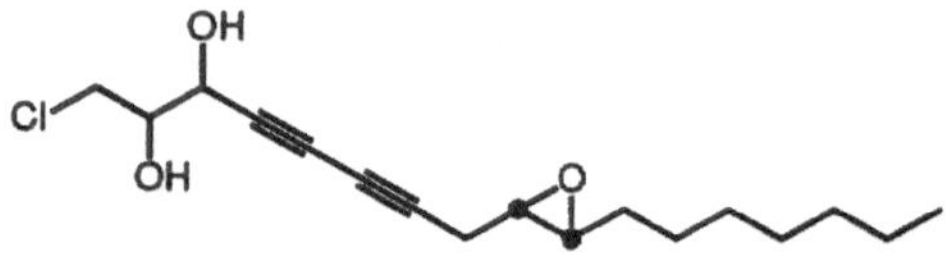

OH Cl OH O

1632 (chloropanaxydiol)

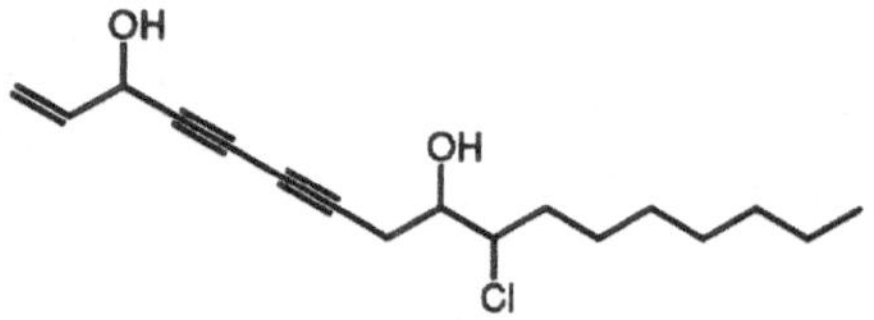

OH OH Cl

1633 (panaxydol chlorohydrin)

OH OH Cl

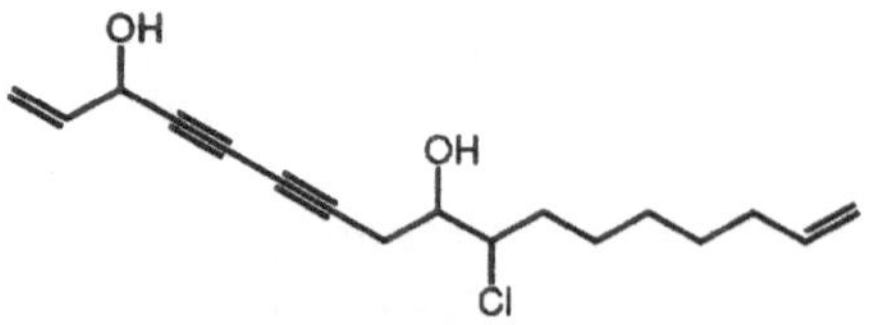

1634 (ginsenoyne B)

Table 21. *Naturally Occurring Plant Chlorine-Containing Polyacetylenes*

Compound	Source	Ref.
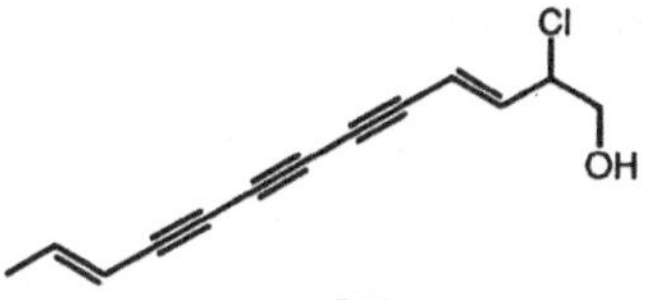 1598	*Centaurea ruthenica*	*1288*
	C. scabiosa	*1289*
	C. alpina	*1286*
	C. tagana	*1286*
	Carthamus tinctorius	*1290*
	C. coeruleus	*1290*
	C. glaucus	*1286*
	C. lanatus	*1294*
	Carduncellus coeruleus	*1286*
	Dicoma zeyheri	*1291*
	D. argyrophylla	*1292*
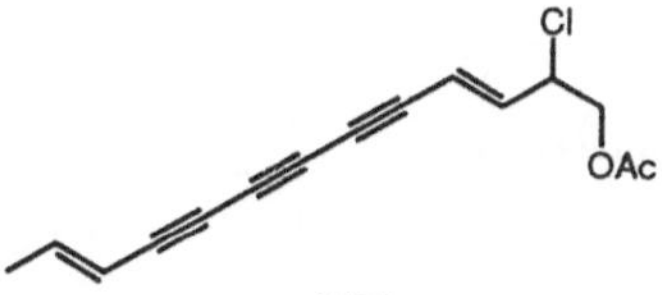 1599	*Centaurea ruthenica*	*1288*
	C. alpina	*1286*
	C. tagana	*1286*
	Carthamus tinctorius	*1290*
	C. coeruleus	*1290*
	C. lanatus	*1290*
	Dicoma zeyheri	*1291*
	D. argyrophylla	*1292*
	Coreopsis nodosa	*1293*
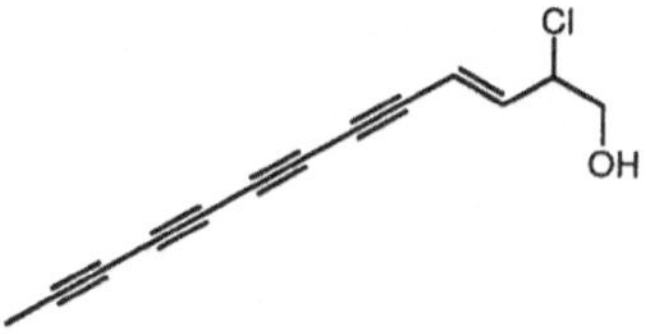 1600	*Centaurea ruthenica*	*1296*
	C. scabiosa	*1289*
	Carthamus tinctorius	*1290*
	Carduncellus coeruleus	*1286*
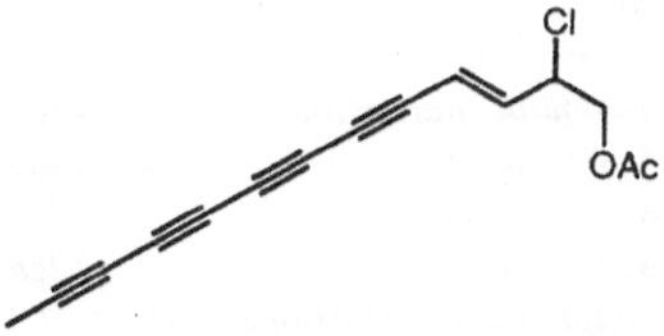 1601	*Centaurea ruthenica*	*1296*
 1602	*Centaurea ruthenica*	*1297*
	C. scabiosa	*1289*
	Carthamus tinctorius	*1290*
	C. lanatus	*1294*
	Carduncellus coeruleus	*1286*

Table 21 (*continued*)

Compound	Source	Ref.
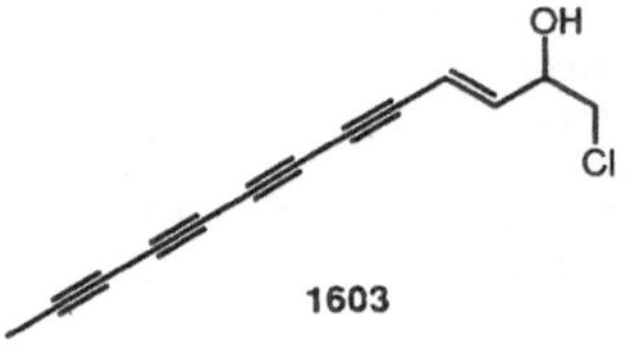 1603	*Centaurea ruthenica*	*1297*
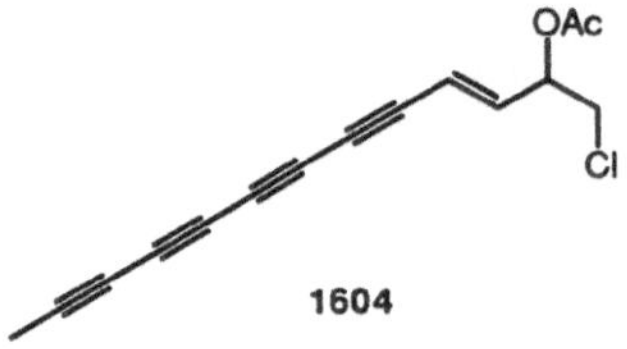 1604	*Centaurea ruthenica*	*1297*
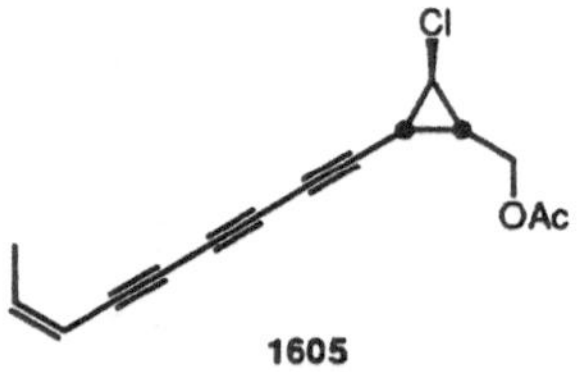 1605	*Centaurea ruthenica*	*1298*

Table 22. *Naturally Occurring Plant Chlorine-Containing Enol Ether Polyacetylenes*

Compound	Source	Ref.
 1606	*Anaphalis margaritacea*	*1302*
	A. triplinervis	*1303*
	A. cinnamomea	*1286*
	A. yeodensis	*1286*
	Gnaphalium obstusifolium	*1303*
	G. sprengelii	*1286*
	Helichrysum allioides	*1286*
	H. arenarium	*1286*
	H. diosmaefolium	*1286*
	H. lanatum	*1286*
	H. latifolium	*1286*
	H. serotinum	*1286*
	H. stoechas	*1286*
	H. transchanicum	*1286*

Table 22 (*continued*)

Compound	Source	Ref.
	H. argenteum	*1286*
	H. argyrophyllum	*1286*
	H. nudiflolium	*1286*
	H. odoratissimum	*1286*
	H. paniculatum	*1286*
	H. petiolatum	*1286*
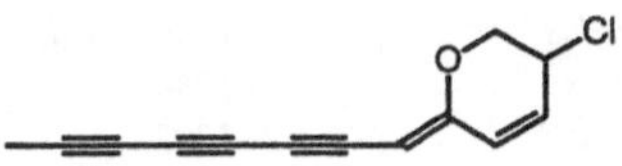 1607	*Anaphalis triplinervis*	*1303*
 1608	*Anaphalis triplinervis*	*1303*
	Gnaphalium sprengelii	*1286*
	Helichrysum diosmaefolium	*1286*
	H. lanatum	*1286*
	H. latifolium	*1286*
	H. serotinum	*1286*
	H. stoechas	*1286*
	H. transchanicum	*1286*
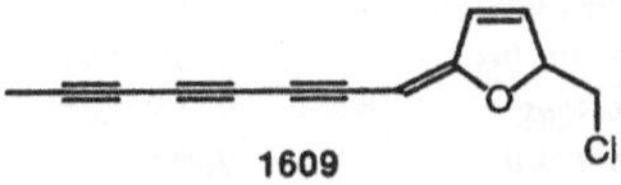 1609	*Achyrocline satureioides*	*1286*
	Gnaphalium obstusifolium	*1303*
	Helichrysum allioides	*1286*
 1610	*Gnaphalium obstusifolium*	*1303*
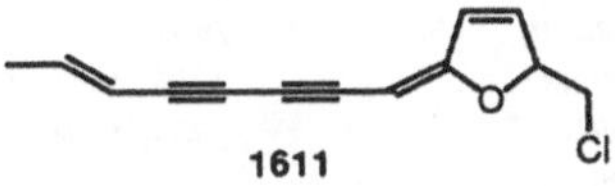 1611	*Achyrocline satureioides*	*1286*

Table 23. *Naturally Occurring Plant Chlorine-Containing Thiophene Polyacetylenes*

Compound	Source	Ref.
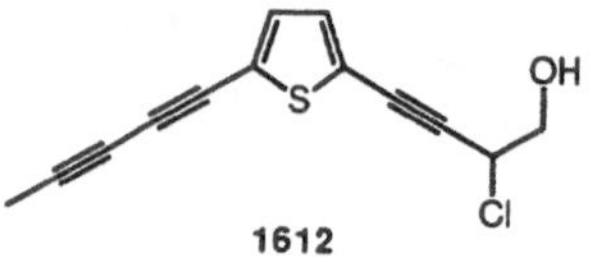 **1612**	*Echinops champtavicus*	*1286*
	E. commutatus	*1286*
	E. dahuricus	*1286, 1305*
	E. ellenbeckii	*1307*
	E. exaltus	*1286*
	E. giganteus	*1307*
	E. hispidus	*1307*
	E. horridus	*1286, 1305, 1294*
	E. humilis	*1286*
	E. longisetus	*1307*
	E. macrochaetus	*1307*
	E. persicus	*1286, 1305, 1294*
	E. ritro	*1286, 1305, 1294*
	E. sphaerocephalus	*1294*
	E. strigosus	*1286, 1305*
	E. viscosus	*1286*
	Pluchea dioscorides	*1286*
	P. indica	*1286*
	Rudbeckia fulgida	*1286*
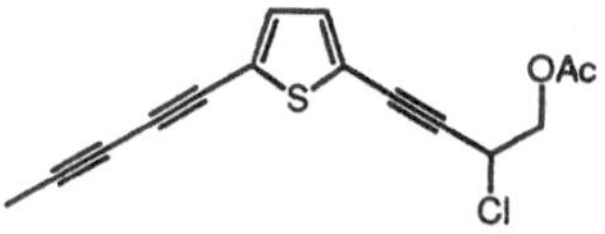 **1613**	*Centaurea cristata*	*1306*
	Echinops champtavicus	*1286*
	E. commutatus	*1286, 1308*
	E. dahuricus	*1286, 1305*
	E. ellenbeckii	*1307*
	E. exaltatus	*1308*
	E. exaltus	*1286*
	E. giganteum	*1308*
	E. giganteus	*1307*
	E. gmelini	*1308*
	E. hispidus	*1307*
	E. hoehnellii	*1307*
	E. horridus	*1286, 1305, 1294*
	E. humilis	*1286, 1308*
	E. longisetus	*1307*
	E. macrochaetus	*1307*
	E. niveus	*1308*
	E. orientalis	*1308*
	E. persicus	*1286, 1305, 1294*
	E. ritro	*1286, 1305, 1294*
	E. sphaerocephalus	*1308*
	E. strigosus	*1286, 1305*
	E. tschimganicus	*1308*
	E. viscosus	*1286*
	Pluchea dioscorides	*1286, 1309*
	P. indica	*1286*

Table 23 (continued)

Compound	Source	Ref.
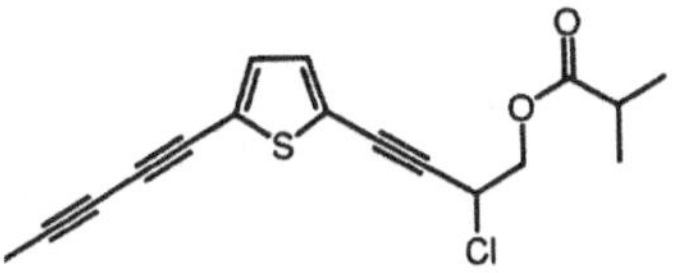 1614	*Pluchea dioscorides*	*1309*
1615	*Eclipta erecta*	*1310*
	E. prostrata	*1295*
	Epaltes brasiliensis	*1312*
	Pterocaulon virgatum	*1311*
1616	*Eclipta erecta*	*1310*
	E. prostrata	*1295*
1617	*Eclipta erecta*	*1310*
	E. prostrata	*1295*
1618	*Ambrosia chamissonis*	*1313*
	Centaurea repens	*299*
	Pterocaulon virgatum	*1311*
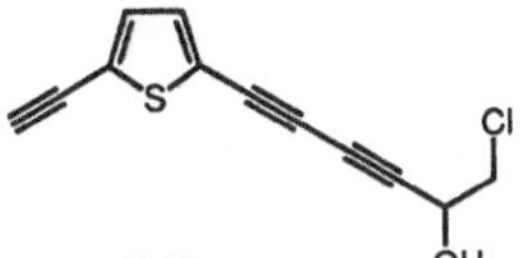 1619	*Pterocaulon virgatum*	*1311*
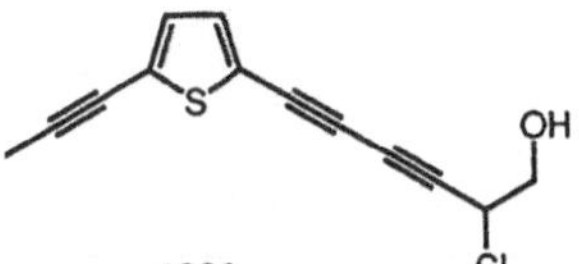 1620	*Ambrosia chamissonis*	*1313*

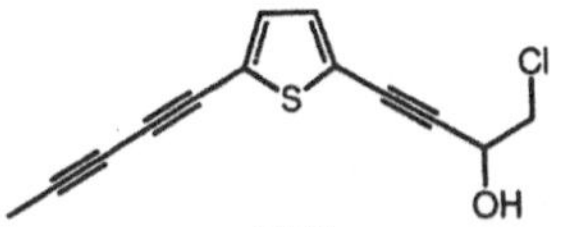

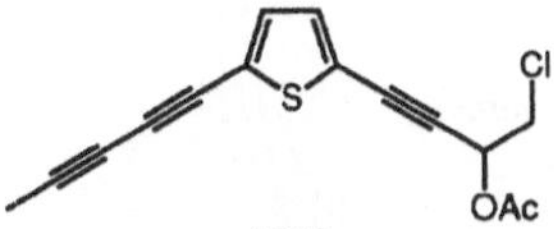

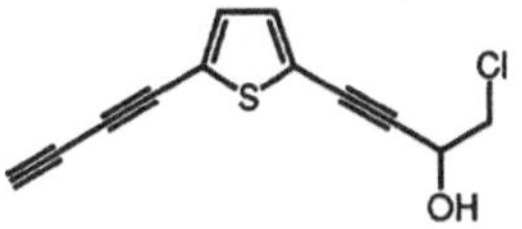

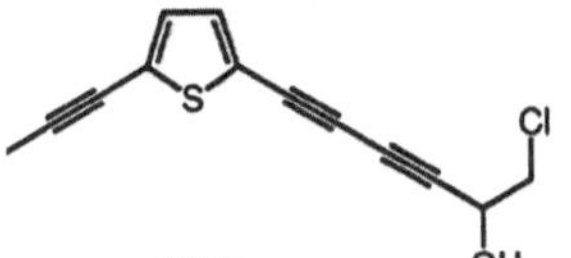

Table 23 (*continued*)

Compound	Source	Ref.
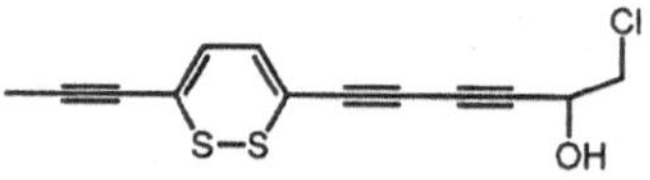 1621	*Ambrosia chamissonis*	*1313*
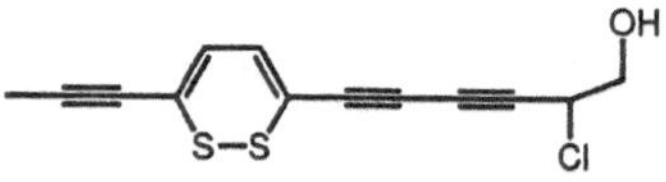 1622	*Ambrosia chamissonis*	*1313*
1623	*Eclipta erecta*	*1310*
	E. prostrata	*1295*
	Pterocaulon virgatum	*1311*
	Tagetes minuta	*1314*
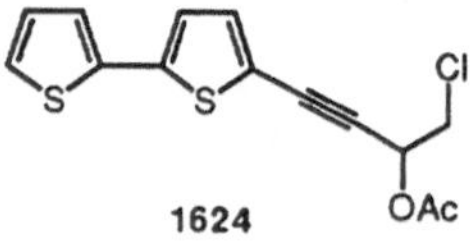 1624	*Porophyllum scoparia*	*1315*
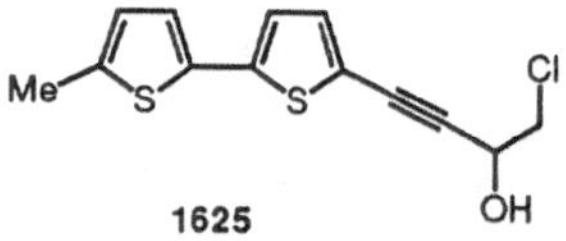 1625	*Pterocaulon virgatum*	*1311*
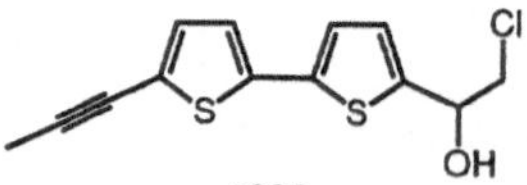 1626	*Epaltes brasiliensis*	*1312*
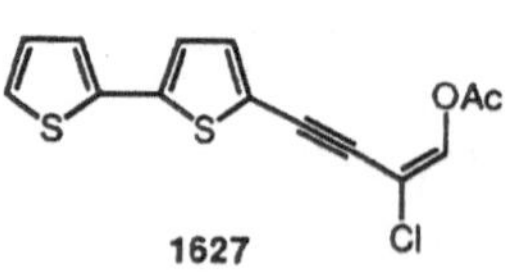 1627	*Berkheya adlamii*	*1316*
	B. bergiana	*1317*
	B. echinata	*1317*
	B. maritima	*1317*
	B. rhapontica	*1317*
	B. speciosa	*1317*

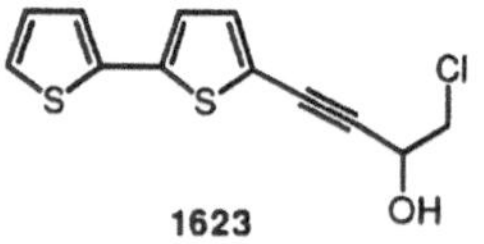

Table 23 (*continued*)

Compound	Source	Ref.
1628	*Berkheya bergiana*	*1317*
	B. echinata	*1317*
	B. macrocephala	*1286*
	B. maritima	*1317*
	B. onopordifolia	*1286*
	B. robusta	*1317*
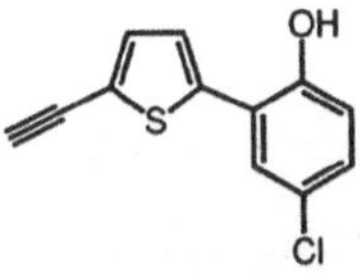 1629 (helitenuin)	*Helichrysum tenuifolium*	*1318*
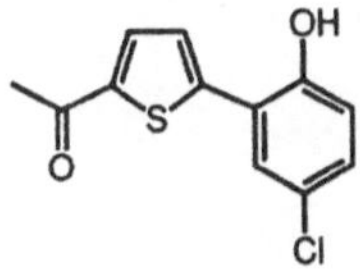 1630 (helitenuon)	*Helichrysum tenuifolium*	*1318*
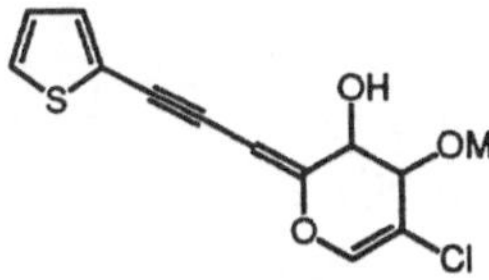 1631 (helipandurin)	*Helichrysum panduratum*	*1318*

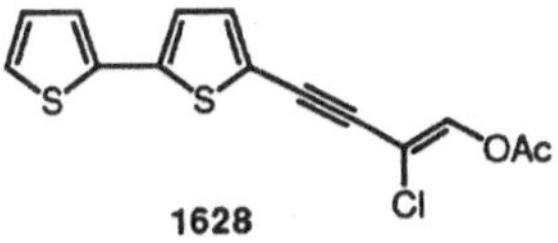

3.15.2. Marine Polyacetylenes

A large group of marine polyacetylenes have been isolated, mainly from sponges. A sponge from the Bay of Naples, *Reniera fulva*, produces renierin-1 (**1635**) and dihydro derivative **1636** (*1321*). The sponge *Xestospongia testudinaria* contains the two similar 18-carbon bromocarboxylic acids **1637** and **1638**, which were isolated as methyl esters (*1322*). An independent study of this sponge revealed the presence of xestospongic acid (**1639**) and its ethyl ester (**1640**) (*1323*). These metabolites are anti-

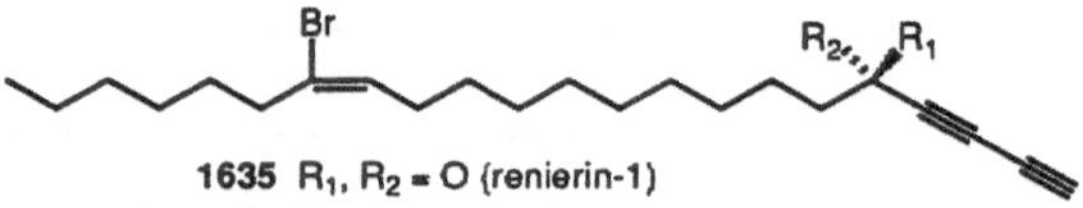

Br

R_2 R_1

1635 $R_1, R_2 = O$ (renierin-1)
1636 $R_1 = OH, R_2 = H$

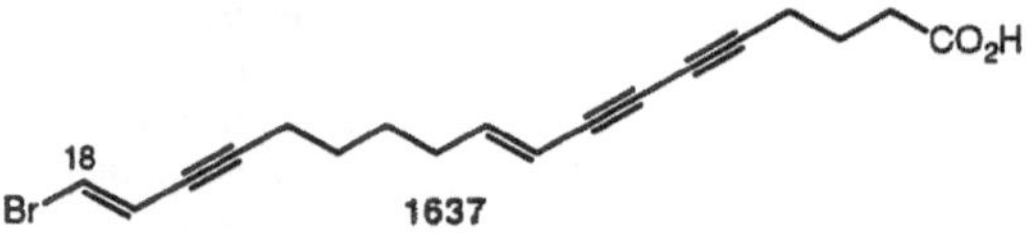

CO_2H

18
Br

1637

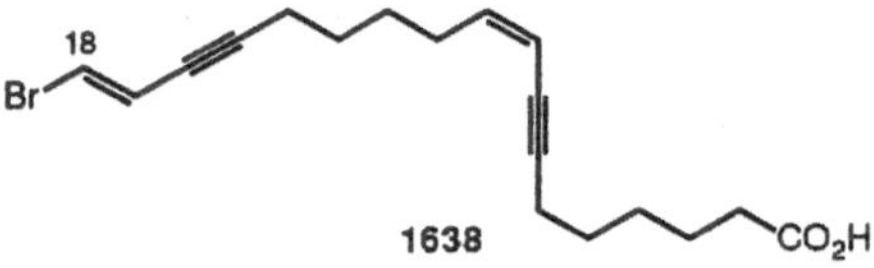

18
Br

1638 CO_2H

CO_2R

18
Br

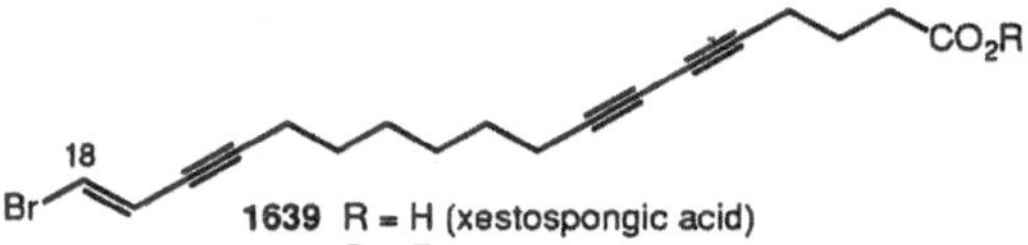

1639 R = H (xestospongic acid)
1640 R = Et

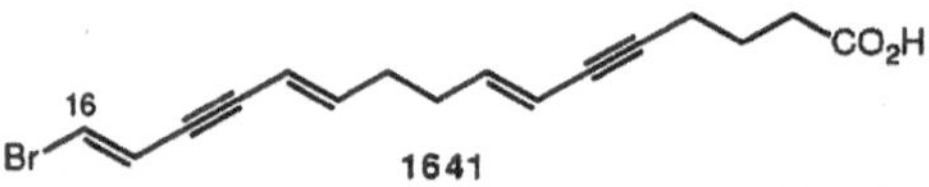

CO_2H

16
Br

1641

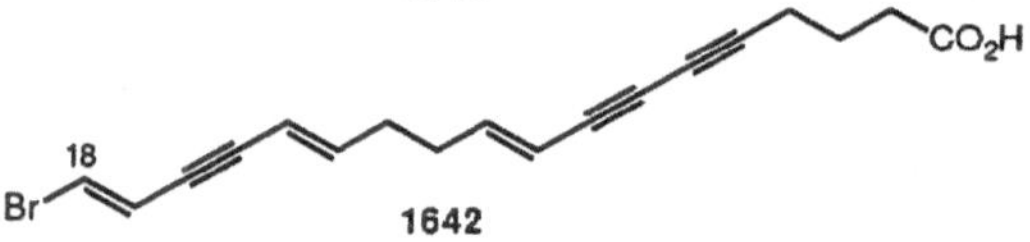

CO_2H

18
Br

1642

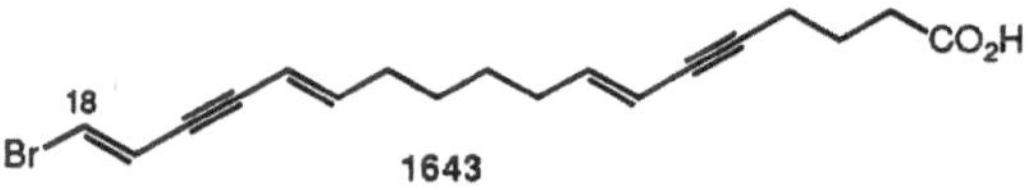

CO_2H

18
Br

1643

CO_2H

16
Br

1644 (major)

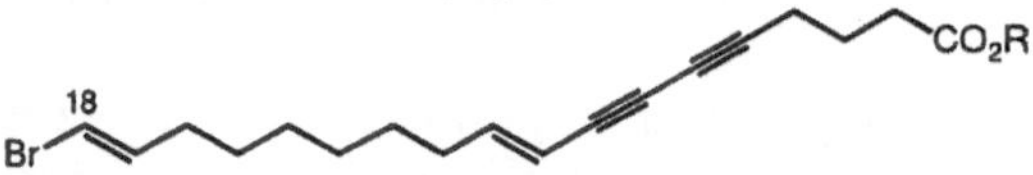

CO_2R

18
Br

1645 R = H
1646 R = Me

CO_2H

18
Br

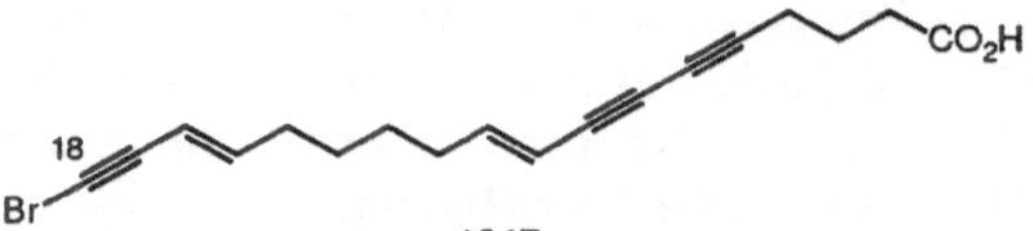

1647

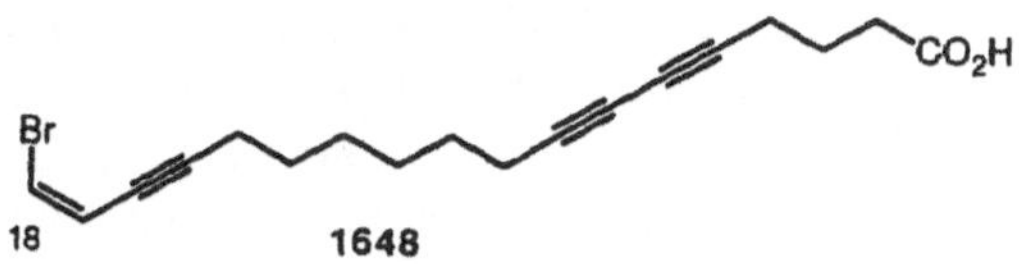

1648

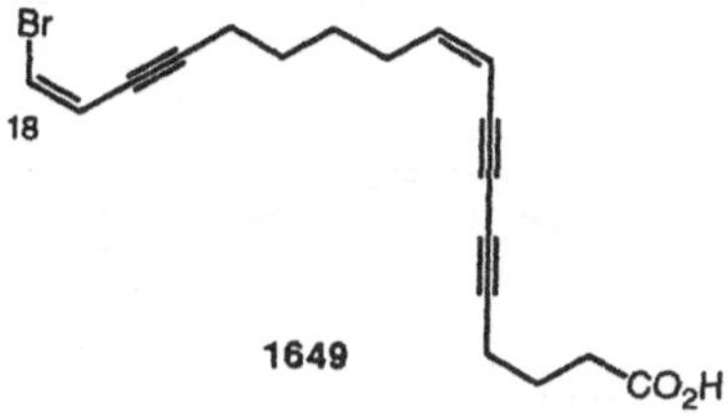

1649

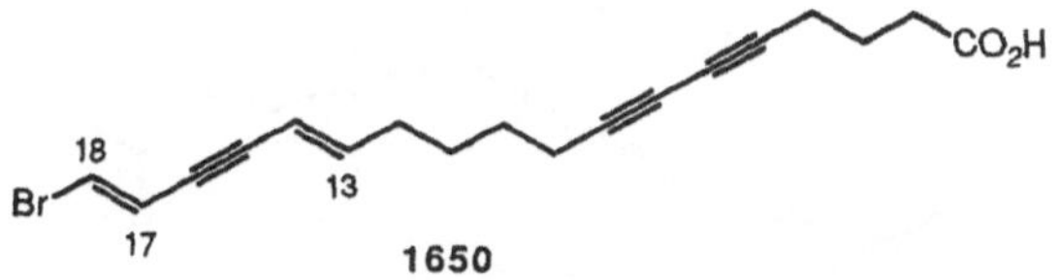

1650
1651 (13*Z*)
1652 (17*Z*)

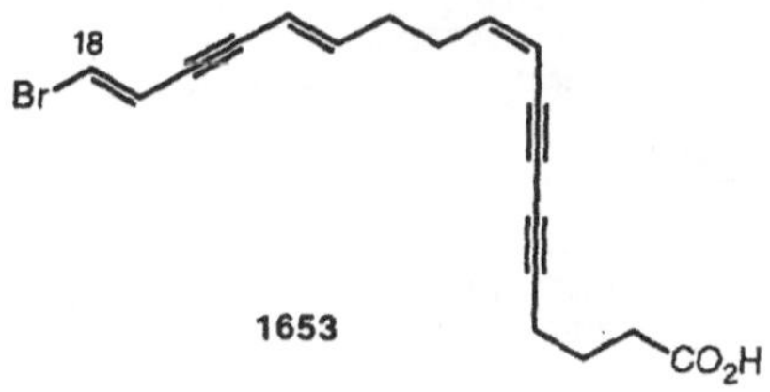

1653

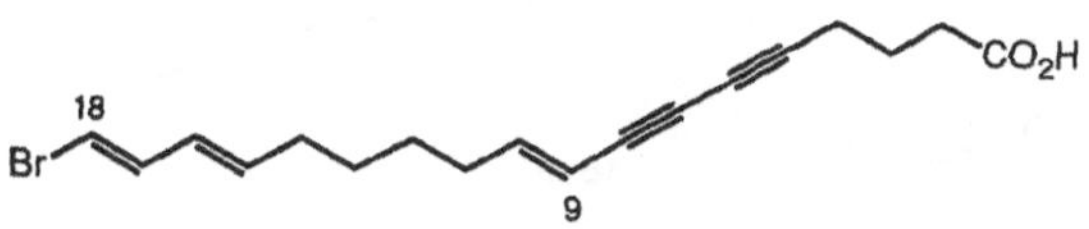

1654
1655 (9*Z*)

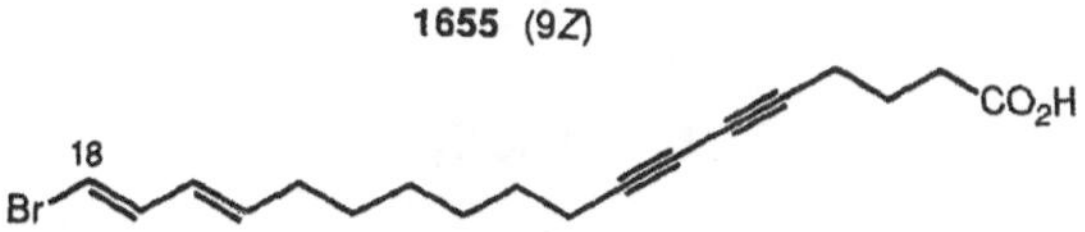

1656

microbial and inhibit Na^+/K^+ ATPase. The sponge *X. muta* from the Bahamas contains a series of brominated polyacetylenes **1637**, **1641**–**1647** which are active against HIV protease, especially **1647**, although this particular compound is very unstable (*1324*).

The sponge *Petrosia volcano* produces several new antifungal brominated polyacetylenic acids **1648**–**1656** in addition to the known xestospongic acid (**1639**) and **1637** (*1325*). Several methyl esters were also isolated but these were shown to be artifacts of the isolation protocol. The nudibranch *Diaulula sandiegensis* employs a set of chloroacetylenes as chemical defense agents (*1326*). Of these compounds, **1657**–**1665**, which appear to be stored in the skin, **1657** is the most abundant comprising 0.11% dry weight.

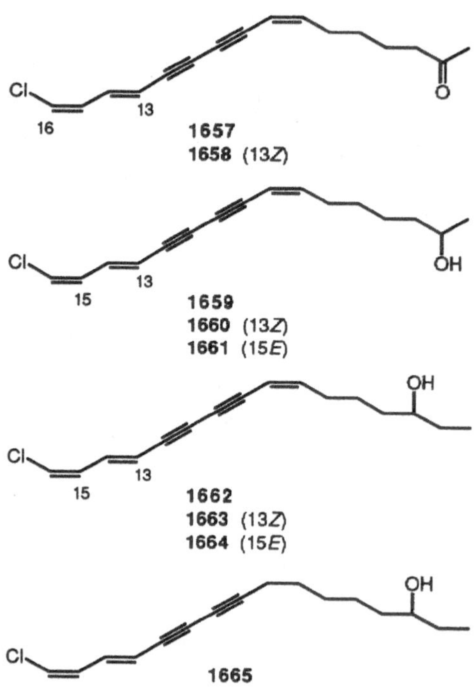

3.16. Enediynes

Although they are few in number, the extraordinary class of natural products containing the enediyne functional unit has captured the imagination of chemists since the discovery of these compounds less than ten years ago. A few of these mechanistically fantastic metabolites contain halogen.

The first such enediynes found to contain halogen are the calicheamicins **1666**–**1672** from *Micromonospora echinospora* ssp. *calichensis* (*1327*–*1330*). Not only do these molecules possess beautifully complex

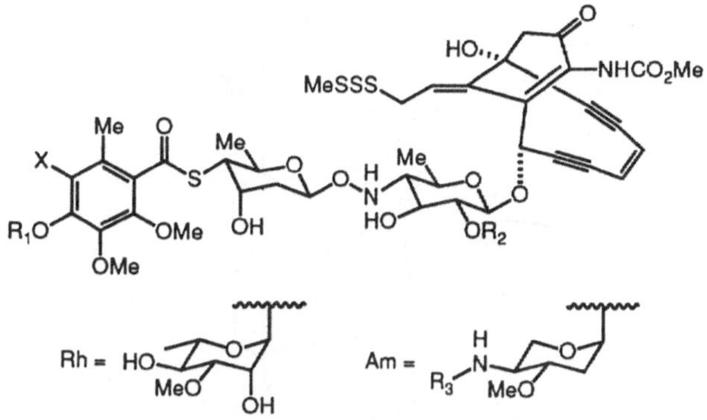

Rh = ; Am =

	X	R_1	R_2	R_3	
1666	Br	Rh	Am	*i*-Pr	(calicheamicin β_1^{Br})
1667	Br	Rh	Am	Et	(calicheamicin γ_1^{Br})
1668	I	H	Am	Et	(calicheamicin α_2^{I})
1669	I	Rh	H	-	(calicheamicin α_3^{I})
1670	I	Rh	Am	*i*-Pr	(calicheamicin β_1^{I})
1671	I	Rh	Am	Et	(calicheamicin γ_1^{I})
1672	I	Rh	Am	Me	(calicheamicin δ_1^{I})

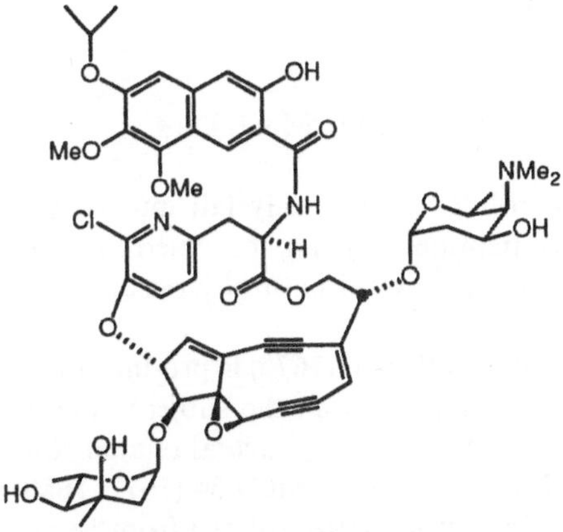

1673 (kedarcidin chromophore)

mechanisms for DNA cleavage, but they are 1000–4000 times more potent than adriamycin as anticancer agents. Kedarcidin (**1673**) is produced by an unidentified actinomycete and contains a relatively rare chloropyridine unit as part of its chromophore (*1331–1334*). Kedarcidin

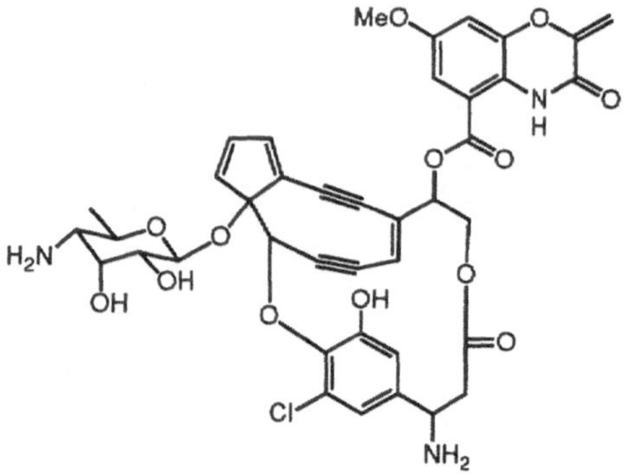

1674 (C-1027 chromophore)

is associated with a protein with a known sequence (*1332*). This metabolite exhibits potent *in vivo* antitumor activity and pronounced antibacterial activity against Gram-positive bacteria, and cleaves DNA site-specifically in a single-stranded manner (*1335*). The soil microbe *Streptomyces globisporus* produces the antitumor antibiotic C-1027 (**1674**), which is also associated with a protein of molecular weight ca. 15,000 (*1336–1341*), and which has been confirmed by synthesis (*1342*). This compound has potent antitumor activity against KB cells ($IC_{50} = 0.1$ ng/ml).

3.17. Macrolides

Although macrolides could easily fall into several other categories (complex phenols, peptides, sugars), it seemed important to group them together separately and several naturally occurring macrolides contain halogen.

The antibiotic chlorothricin (**1675**) is produced by *Streptomyces antibioticus* (*1343*, *1344*) and has been the subject of extensive biosynthetic studies (*1345–1347*). More recently several related compounds, hydroxychlorothricin (**1676**) and MC-031–MC-034 (**1677–1680**) which are cholesterol biosynthesis inhibitors, were isolated from *Streptomyces* sp. (*1348*, *1359*).

A large group of macrolides mainly from the plant family Celastraceae, which initially had very promising clinical anticancer activity are the maytansinoids. These include maytansine (**1681**), maytanprine (**1682**), maytanbutine (**1683**), maytanvaline (**1684**), maytanbutacine (**1685**), maytanacine (**1686**), maytansinol (**1687**), maysine (**1688**), normaysine

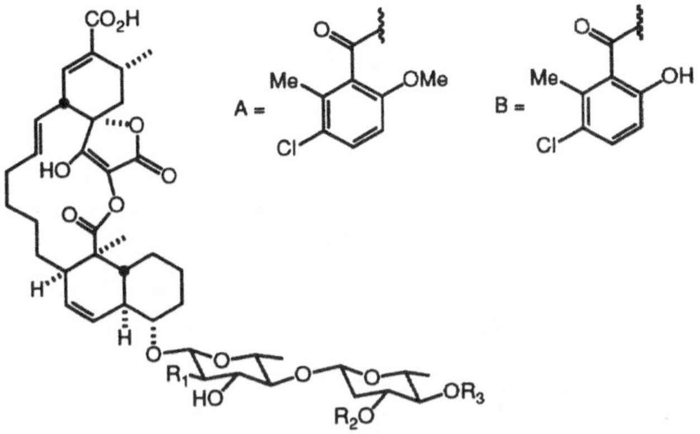

	R_1	R_2	R_3	
1675	H	A	H	(chlorothricin)
1676	OH	A	H	(hydroxychlorothricin)
1677	H	B	H	(MC-031)
1678	OH	B	H	(MC-032)
1679	H	H	B	(MC-033)
1680	OH	H	B	(MC-034)

(**1689**), and maysenine (**1690**) were all isolated from *Maytenus serrata*, *M. buchananii*, and *Putterlickia verrucosa* (*1349*). The related colubrinol acetate (**1691**) and colubrinol (**1692**) are found in *Colubrina texensis*, which also contains maytanbutine (**1683**) (*1350*). Normaytansine (**1693**) is produced by *M. buchananii* (*1351*), and normaytancyprine (**1694**) has been isolated from *P. verrucosa* (*1431*). A *Nocardia* sp. produces the novel ansamitocins P-2 (**1695**), P-3 (**1696**), P-3′ (**1697**), and P-4 (**1698**), as well as the known maytanacine (**1686**) (*1352, 1353*). Ansamitocin P-3 (**1696**) is also found in Oregon mosses and in the associated actinomycetes *Claopodium crispifolium* and *Anomodon attenuatus* (*1354*). This compound has potent antitumor activity against human solid tumor cell lines. Seeds of *Trewia nudiflora* (Euphorbiaceae) produce the six new maytansinoids trewiasine (**1699**), dehydrotrewiasine (**1700**), demethyltrewiasine (**1701**), trenudine (**1702**), treflorine (**1703**), and *N*-methyltrenudone (**1704**) (*1355, 1356*). The Japanese mosses *Isothecium subdiversiforme* and *Thamnobryum sandei* produce the new 15-methoxyansamitocin P-3 (**1705**), as well as the known maytanbutine (**1683**), ansamitocin P-3 (**1696**), and trewiasine (**1699**) (*1357*). All four metabolites have potent antitumor activity.

Monorden (= radiciol) (**1706**) was isolated simultaneously by two groups from *Nectria radicicola* (*1360, 2217*) and *Monosporium bonorden*

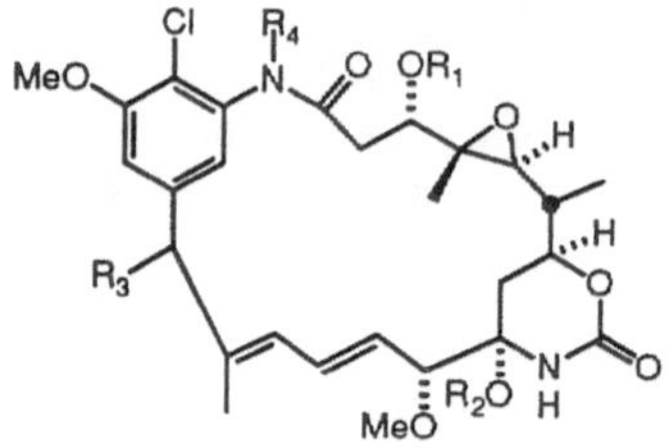

	R_1	R_2	R_3	R_4	
1681	COCH(Me)–N(Me)Ac	H	H	Me	(maytansine)
1682	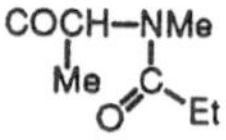 COCH(Me)–N(Me)–C(=O)Et	H	H	Me	(maytanprine)
1683	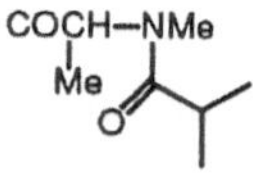 COCH(Me)–N(Me)–C(=O)CHMe$_2$	H	H	Me	(maytanbutine)
1684	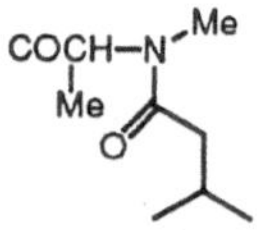 COCH(Me)–N(Me)–C(=O)CH$_2$CHMe$_2$	H	H	Me	(maytanvaline)
1685	COCHMe$_2$	H	OAc	Me	(maytanbutacine)
1686	COMe	H	H	Me	(maytanacine)
1687	H	H	H	Me	(maytansinol)
1691	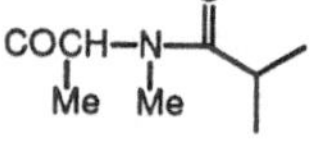 COCH(Me)–N(Me)–C(=O)CHMe$_2$	H	OH	Me	(colubrinol)
1692	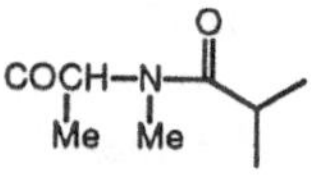 COCH(Me)–N(Me)–C(=O)CHMe$_2$	H	OAc	Me	
1693	COCH(Me)–N(Me)Ac	H	H	H	(normaytansine)
1694	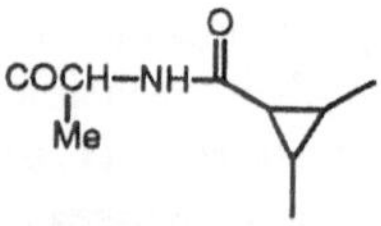 COCH(Me)–NH–C(=O)–(dimethylcyclopropyl)	H	H	H	(normaytancyprine)
1695	COEt	H	H	Me	(ansamitocin P-2)
1696	COCHMe$_2$	H	H	Me	(ansamitocin P-3)
1697	CO*n*-Pr	H	H	Me	(ansamitocin P-3')
1698	COCH$_2$*i*-Pr	H	H	Me	(ansamitocin P-4)

	R_1	R_2	R_3	R_4	
1699	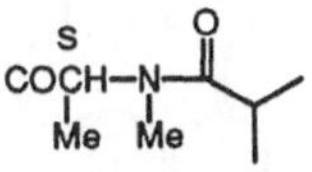 $COCH(Me)-N(Me)-CO-CHMe_2$ (S)	H	OMe	Me	(trewiasine)
1700	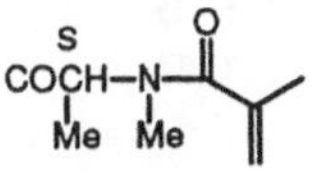 $COCH(Me)-N(Me)-CO-C(Me)=CH_2$ (S)	H	OMe	Me	(dehydrotrewiasine)
1701	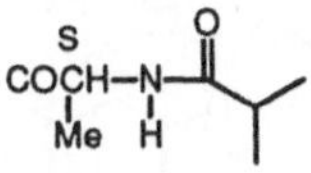 $COCH(Me)-N(H)-CO-CHMe_2$ (S)	H	OMe	Me	(demethyltrewiasine)
1705	$COCHMe_2$	H	OMe	Me	

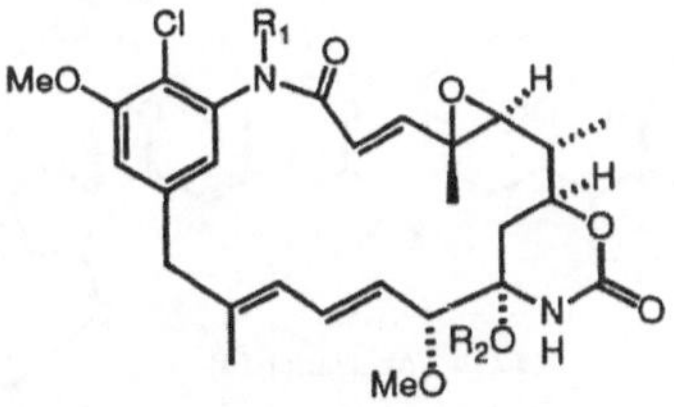

	R_1	R_2	
1688	Me	H	(maysine)
1689	H	H	(normaysine)

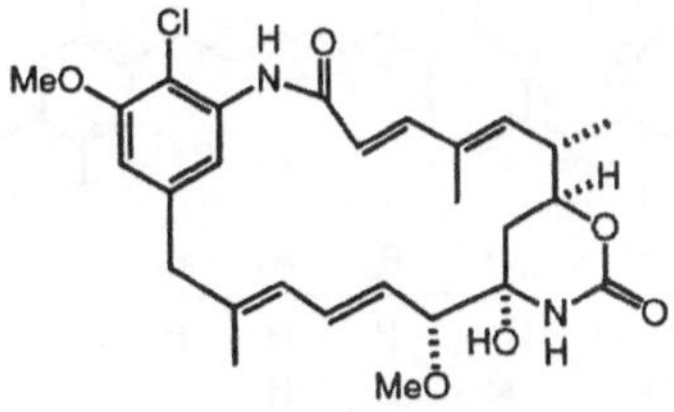

1690 (maysenine)

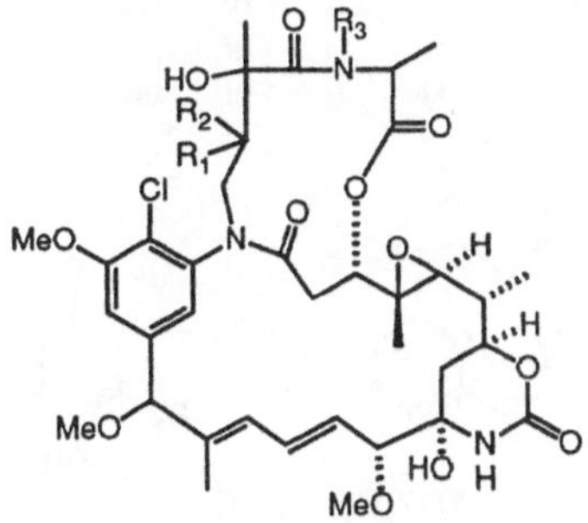

1702 $R_1 = OH, R_2 = R_3 = H$ (trenudine)
1703 $R_1 = R_2 = R_3 = H$ (treflorine)
1704 $R_1, R_2 = O, R_3 = Me$ (*N*-methyltrenudone)

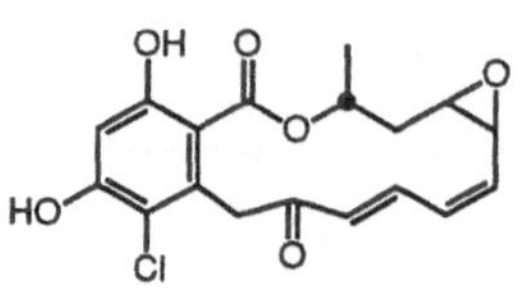

1706
(monordern = radiciol)

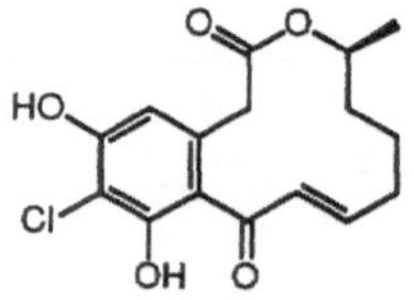

1707

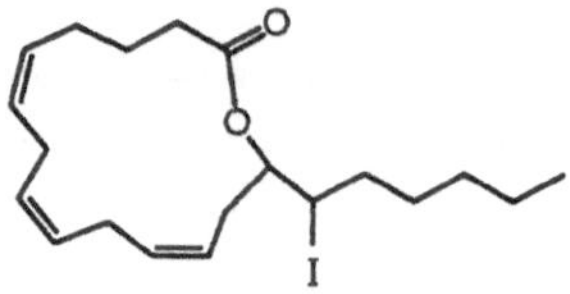

1708

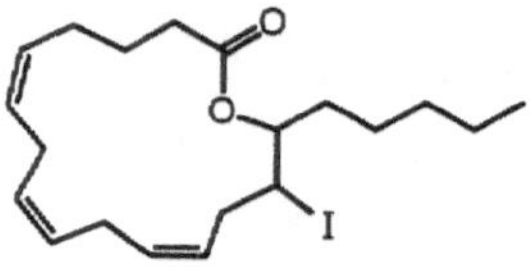

1709

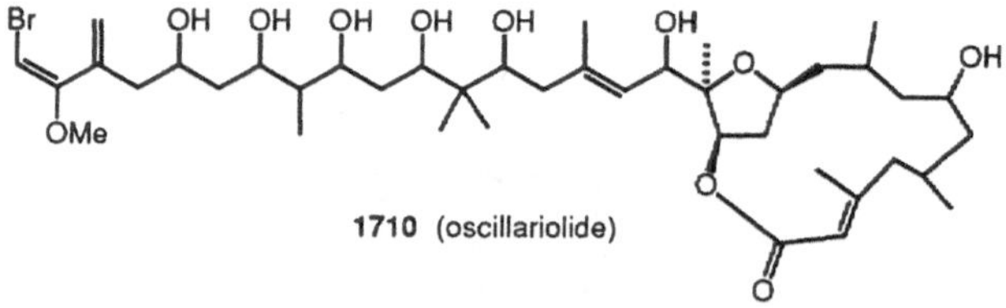

1710 (oscillariolide)

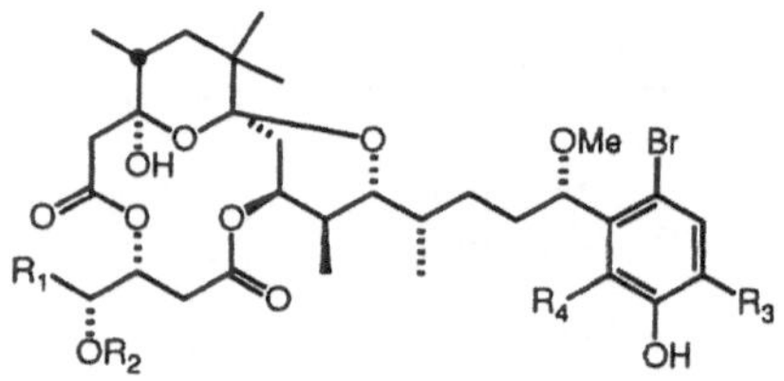

	R_1	R_2	R_3	R_4	
1711	Me	H	H	H	(aplysiatoxin)
1712	Me	Ac	H	H	
1713	Me	H	Br	H	
1714	H	H	H	H	
1715	H	H	Br	H	
1716	Me	H	Br	Br	

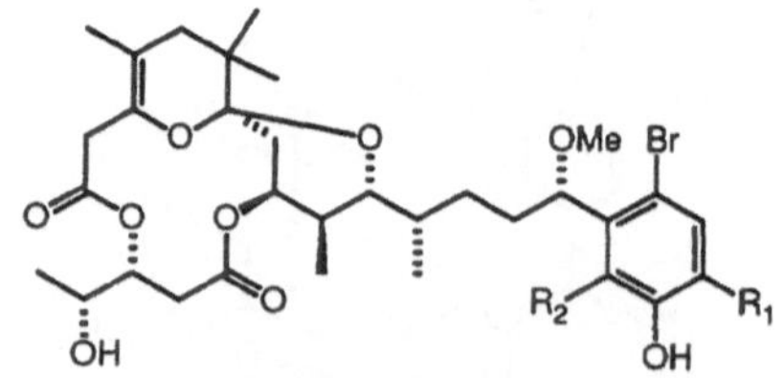

1717 $R_1 = R_2 = H$
1718 $R_1 = Br, R_2 = H$
1719 $R_1 = R_2 = Br$

(*1361, 2219*). Later, this antifungal metabolite was found in *Monocillium nordinii* (*1362*) and *Neocosmospora tenuicristata* (*1363*). The latter study established the stereochemistry of this compound. The related 6-chlorodehydrocurvularin (**1707**) was extracted from the fungus *Cochliobolus spicifer* (*1364*). The two iodinated macrolides **1708** and **1709** are minor products in the reaction of arachidonic acid, lactoperoxidase, iodide, and hydrogen peroxide (*1365*) that was discussed earlier. The blue-green alga *Oscillatoria* sp. produces the novel oscillariolide (**1710**) (*1366*). The sea hare *Stylocheilus longicauda* produces the highly toxic (LD_{100} = 0.3 mg/kg) aplysiatoxin (**1711**), the acetate **1712**, and debromoaplysiatoxin (*1367, 1368*). These compounds may cause "swimmer's itch" in

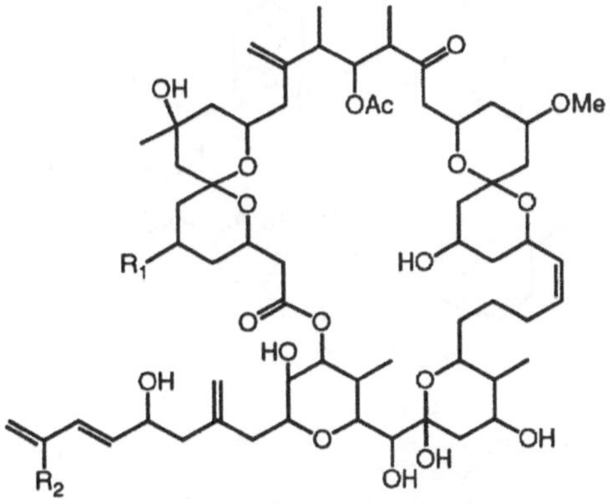

1720 R_1 = OAc, R_2 = Cl (altohyrtin A)
1721 R_1 = OAc, R_2 = Br (altohyrtin B)
1722 R_1 = OH, R_2 = Cl

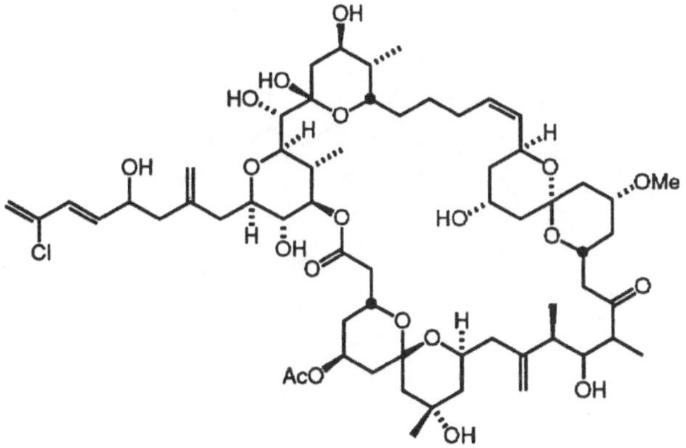

1723 (cinachyrolide A)

Hawaii (*1369*). Research revealed that the origins of these metabolites and others (**1713–1719**) are actually the blue-green algae *Lyngbya majuscula*, *Oscillatoria nigroviridis*, and *Schizothrix calcicola* (*1370, 1371*).

The marine sponge *Hyrtios altum* produces the very potent antitumor agent altohyrtin A (**1720**) (*1372*), as well as altohyrtin B (**1721**) and

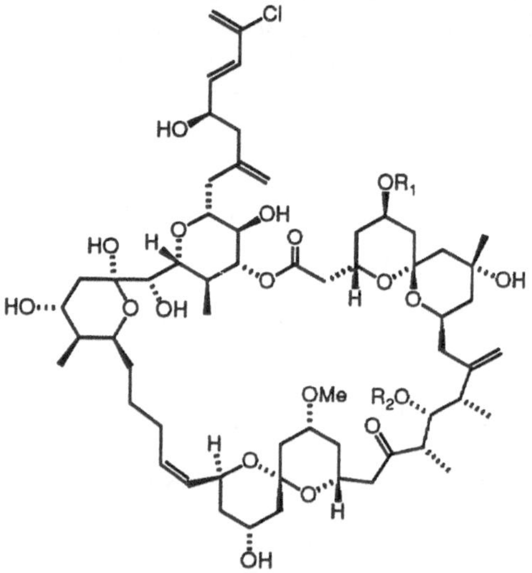

1724 $R_1 = R_2 = COMe$ (spongistatin 1)
1725 $R_1 = H$, $R_2 = COMe$ (spongistatin 3)
1726 $R_1 = COMe$, $R_2 = H$ (spongistatin 4)

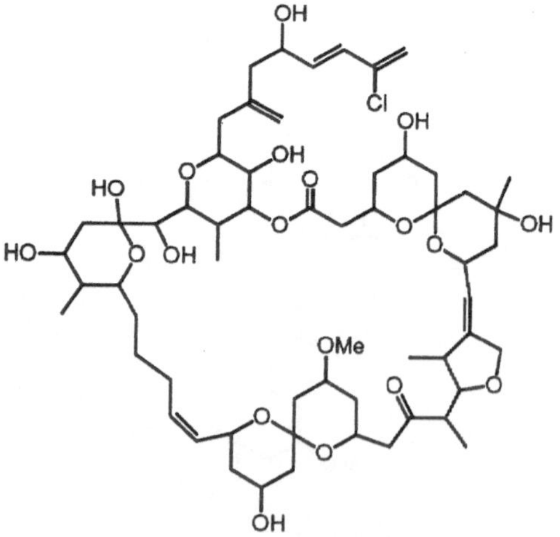

1727 (spongistatin 5)

5-desacetylaltohyrtin A (**1722**) (*1373*). Altohyrtin A exhibits IC_{50} = 0.01 ng/ml against KB cells (*1372*). The sponge *Cinachyra* sp. produces cinachyrolide A (**1723**) (*1374*), which is closely related to altohyrtin A. A set of remarkably active antitumor metabolites, spongistatins 1 (**1724**) and 3 (**1725**), has been isolated from an Indian Ocean sponge, *Spongia* sp. (*1375, 1376*), and spongistatins 4 (**1726**) and 5 (**1727**) were found in the African sponge *Spirastrella spinispirulifera* (*1377*). A more recent study of this marine sponge revealed the presence of spongistatin 9 (**1728**) (*1378*). As a group, these metabolites are among the most potent of all substances screened in the National Cancer Institute (U.S.) panel of 60 human cancer cell lines. Two other spongistatins do not contain chlorine.

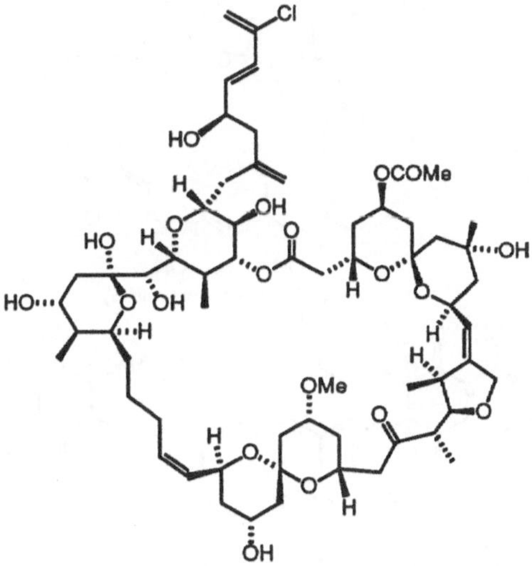

1728 (spongistatin 9)

3.18. Naphthoquinones and Higher Quinones

Several types of natural aromatic quinones other than benzoquinones (Section 3.3.2) contain halogen. Anthraquinones, because of their close biogenetic relationship to depsidones and xanthones, will be covered under the rubric of "Complex Phenols" (3.22).

The fungi *Mollisia caesia* and *M. fallens* produce mollisin (**1729**) (*1379, 1380*) (structure revised in *1381*). The biosynthesis of this antibiotic

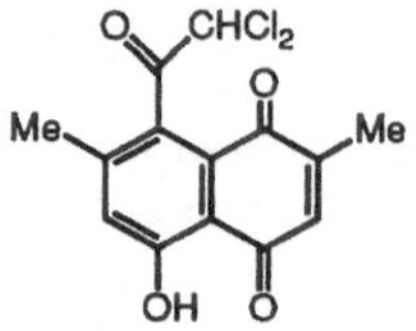

1729 (mollisin)

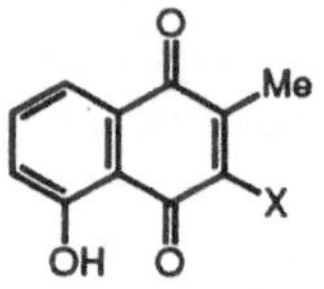

1730 X = Cl (3-chloroplumbagin)
1731 X = Br (3-bromoplumbagin)

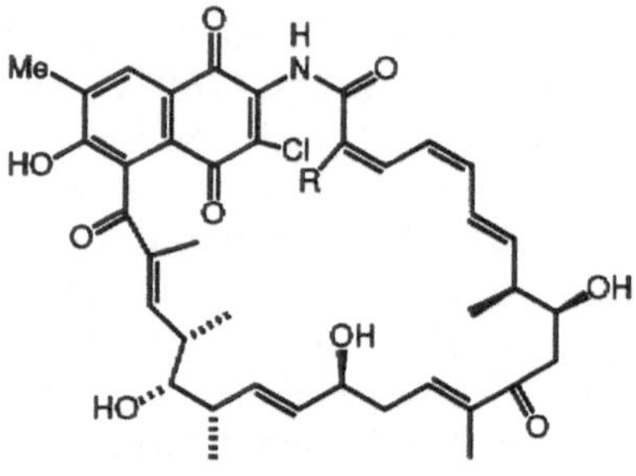

1732 R = Me (naphthomycin A)
1733 R = H (naphthomycin H)

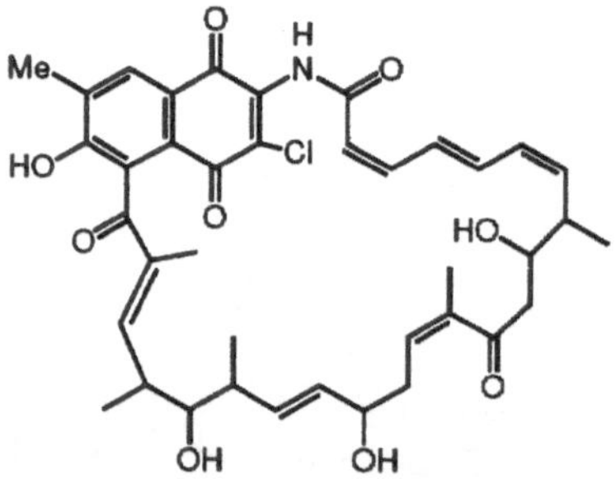

1734 (naphthomycin B)

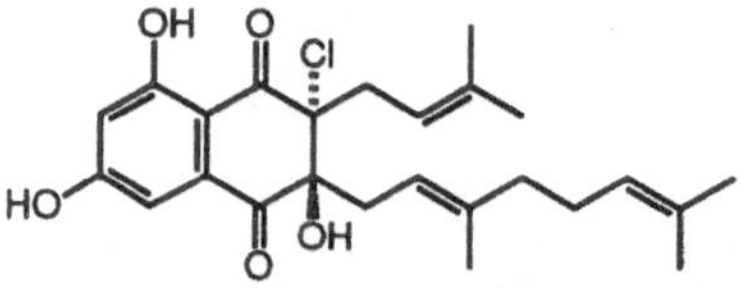

1735 (naphthomevalin)

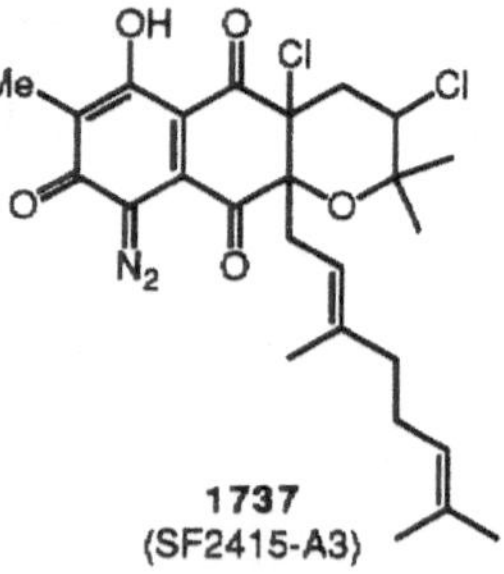

1737 (SF2415-A3)

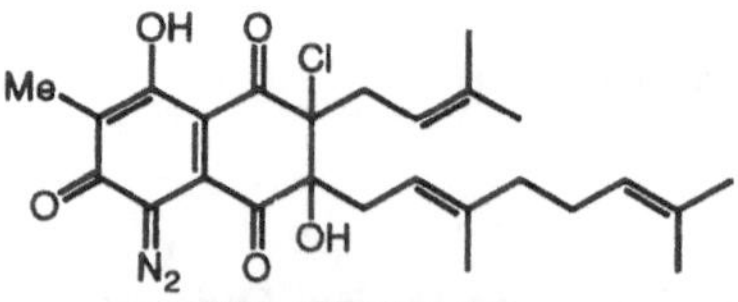

1736 (SF2415-A1)

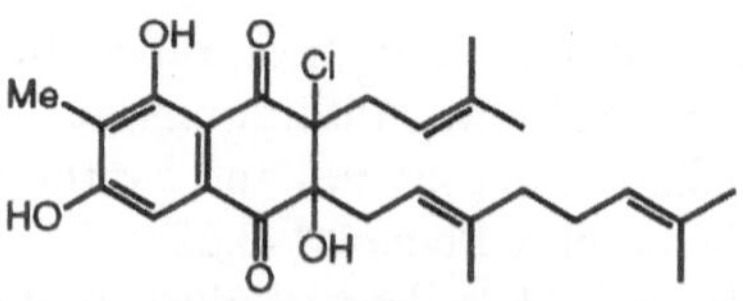

1738 (SF2415-B1)

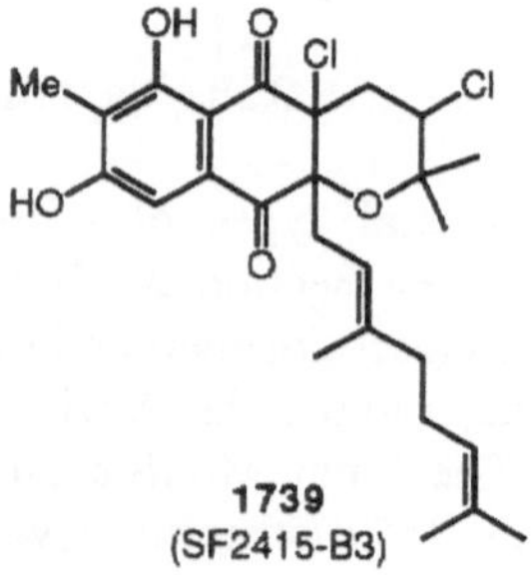

1739 (SF2415-B3)

involves the incorporation of acetate and malonate and chlorination of a β-ketoacid is proposed to occur at an early stage. Interestingly, and in contrast to many fungal metabolites, bromide ion is not incorporated by

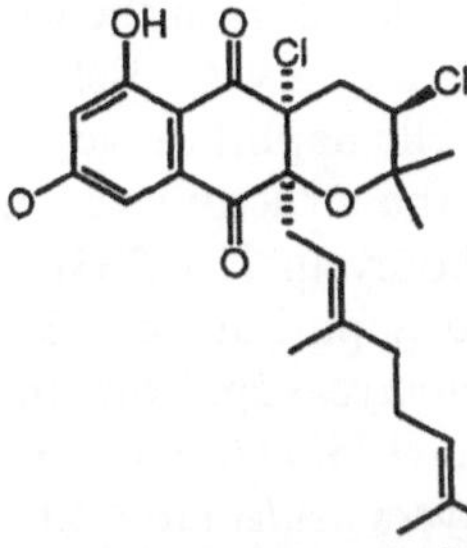

1740 (napyradiomycin A)

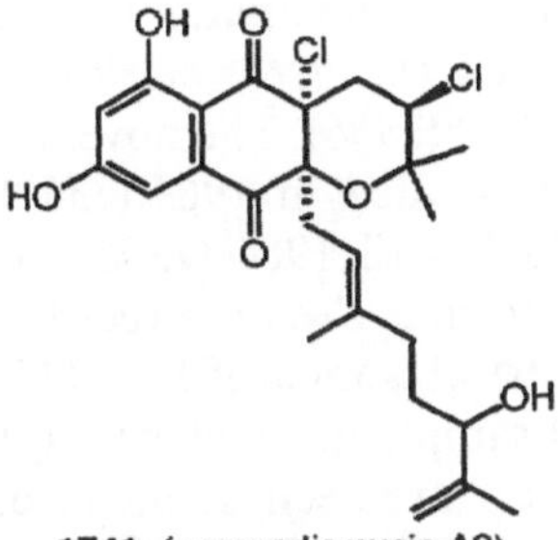

1741 (napyradiomycin A2)

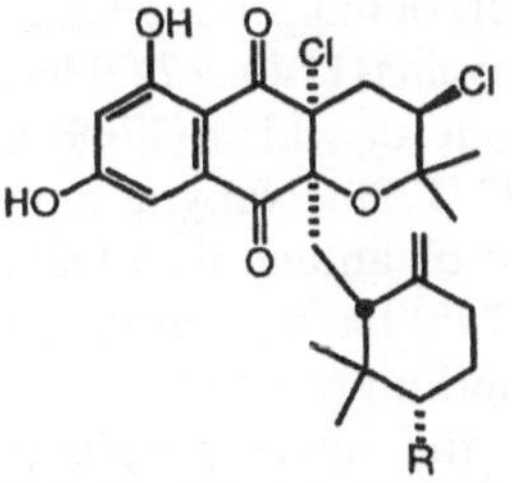

1742 R = Cl (napyradiomycin B1)
1743 R = Br (napyradiomycin B3)

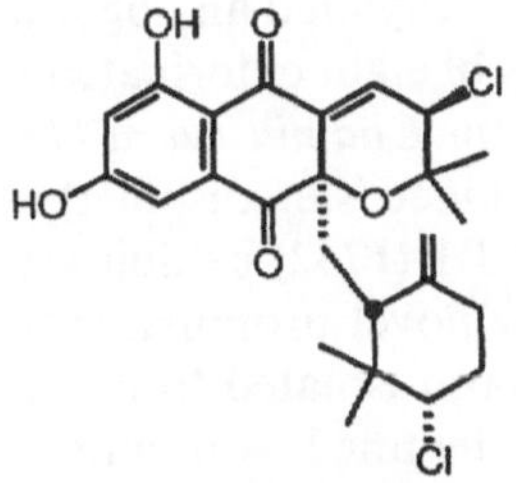

1744 (napyradiomycin B2)

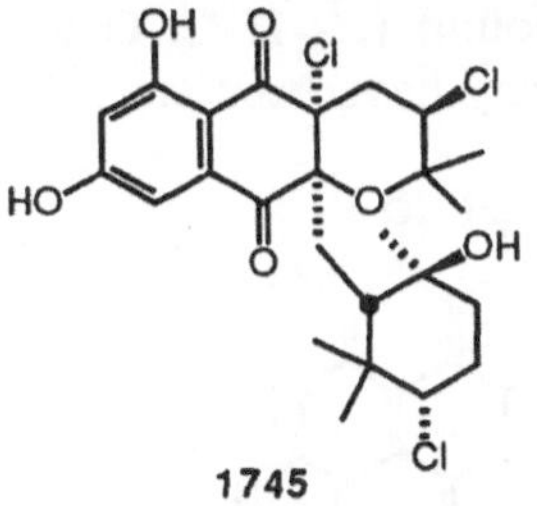

1745 (napyradiomycin B4)

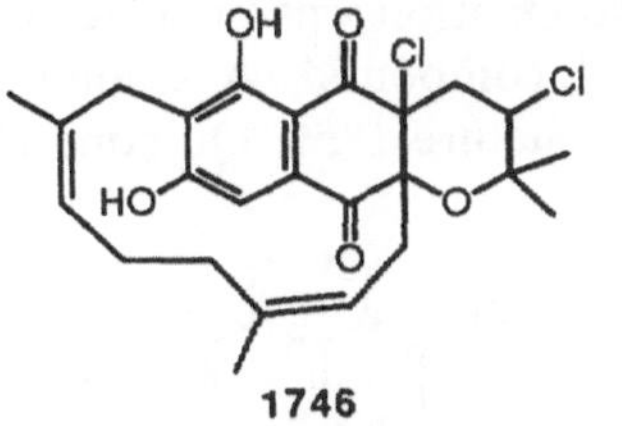

1746 (napyradiomycin C1)

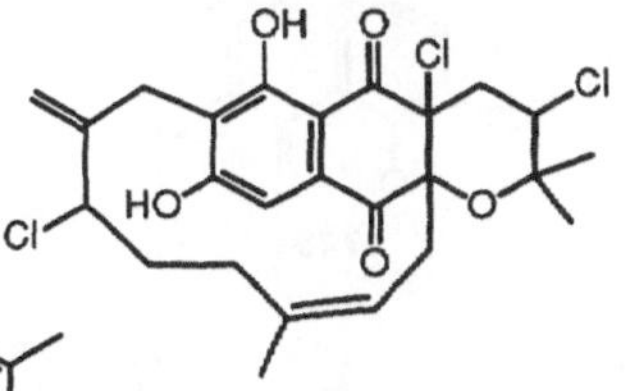

1747 (napyradiomycin C2)

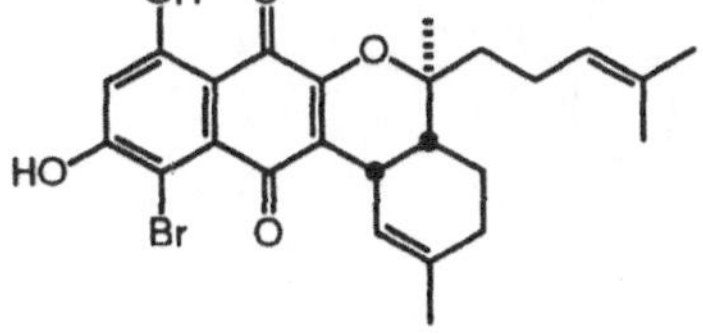

1748 (marinone)

this organism and even blocks the production of mollisin (*1382*). The simple chloronaphthoquinone 3-chloroplumbagin (**1730**) has been discovered in *Drosera intermedia, D. anglica* (Droseraceae) (*1383*), *Plumbago zeylanica* (Plumbaginaceae) (*1384*), and *Dionaea muscipula* (Droseraceae) (*1385*). The bromo analogue **1731** has been isolated from the fruit of the Asian shrub *Diospyros maritima* (Ebenaceae) along with 3-chloroplumbagin (**1730**) (*1386*). The novel ansa macrocyclic naphthomycin A (**1732**) was first isolated in 1969 (*1387*), although the structure was not fully established until 1984 (*1388–1390*). Naphthomycin H (**1733**) has been isolated from a *Streptomyces* sp. (*1391*), and naphthomycin B (**1734**) is a geometrical isomer of **1733** (*1392*). A *Streptomyces* sp. from an Australian soil sample has yielded naphthomevalin (**1735**) (*1393*). Several new antibiotics have been isolated from *Streptomyces aculeolatus*, SF2415-A1 (**1736**), -A3 (**1737**), -B1 (**1738**), and -B3 (**1739**) (*1394, 1395*), one of which (**1738**) is a methylated analogue (stereoisomer) of naphthomevalin (**1735**).

A series of eight chlorinated napyradiomycins (**1740–1747**) have been isolated from *Chainia rubra* (*1396–1398*), one of which (**1740**) is a demethyl analogue (stereoisomer) of SF2415-B3 (**1739**). Napyradiomycins A (**1740**) and B1 (**1742**) exhibit estrogen-receptor antagonism (*1399*). The structurally novel brominated marinone (**1748**) having antibacterial activity has been isolated from a marine bacterium (*1400*).

An unidentified soil fungus produces the novel purple pigment dinaphthofurandione **1749** (*1401, 1402*) which represents a new ring system. No mention was made regarding the possible optical activity of this helical compound. A second compound ($C_{21}H_7Cl_5O_6$) has the tentative structure **1750**. Quinone **1749** may have been the compound

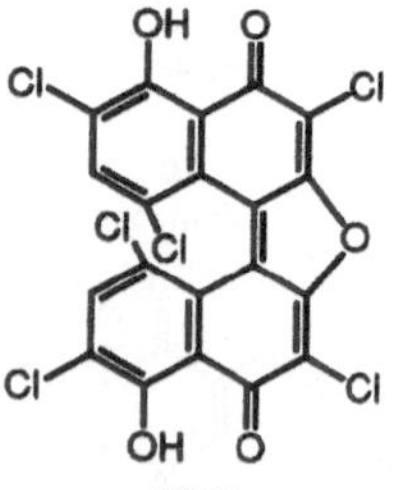

1749

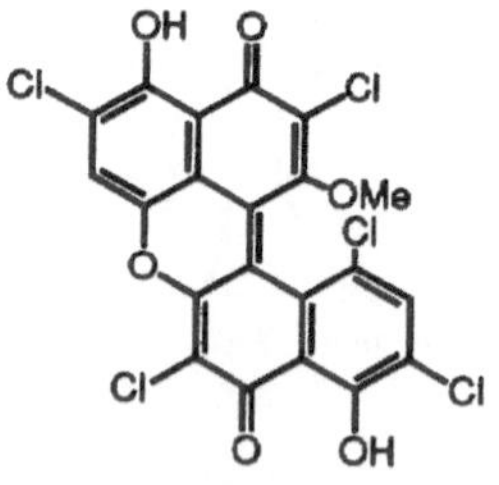

1750

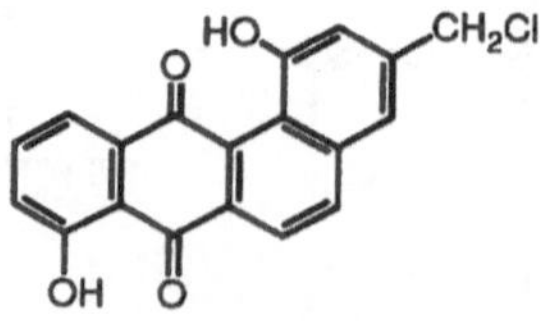

1751 (chlorotetrangulol)

isolated in 1964 from a fungus found growing on decomposed roots of *Eucalyptus obliqua* (*1403*). The chlorinated benz[*a*]anthraquinone chlorotetrangulol (**1751**) is produced by an unidentified actinomycete (*1404*).

3.19. Tetracyclines

Although the tetracycline group of antibiotics could have been more logically included in the "Complex Phenols" section, the importance of these compounds in medicine and in the history of naturally occurring chlorine compounds suggests their special treatment.

Aureomycin (= chlorotetracycline) is produced by *Streptomyces aureofaciens* (*1405*, *1406*) and was determined to have structure **1752**

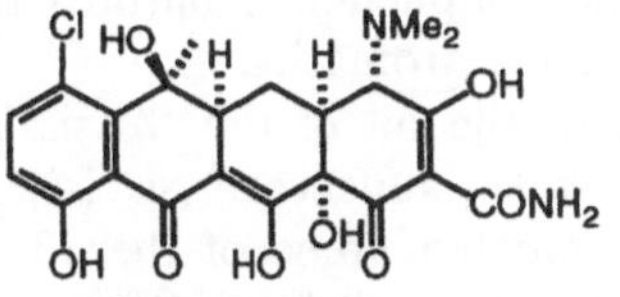

1752 (aureomycin)

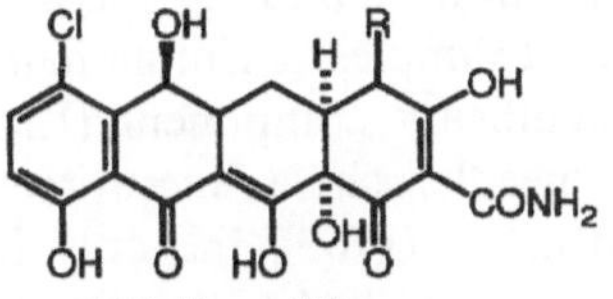

1753 R = α-NMe_2
1754 R = β-NMe_2

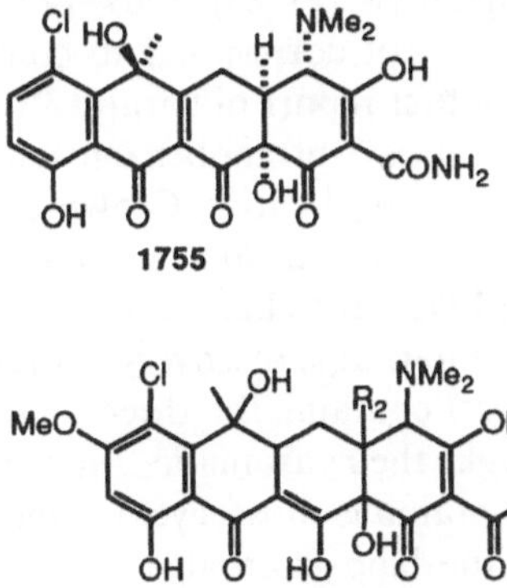

1755

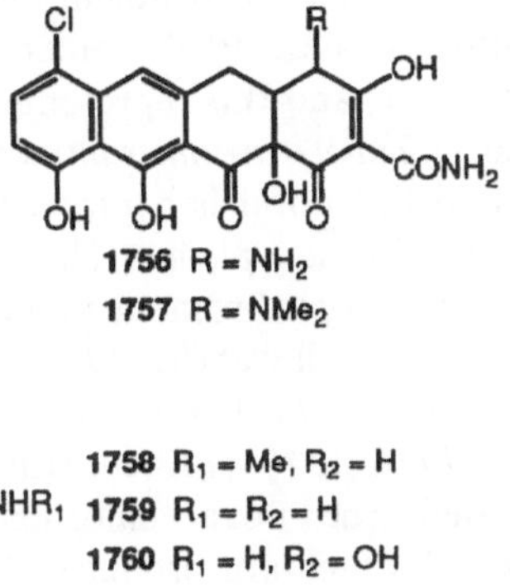

1756 R = NH_2
1757 R = NMe_2

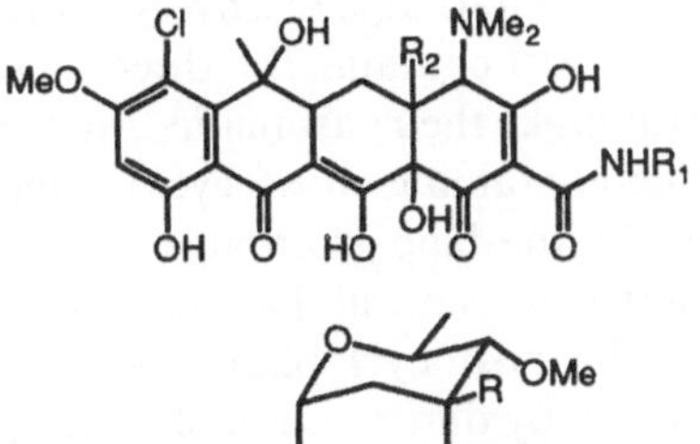

1758 R_1 = Me, R_2 = H
1759 R_1 = R_2 = H
1760 R_1 = H, R_2 = OH

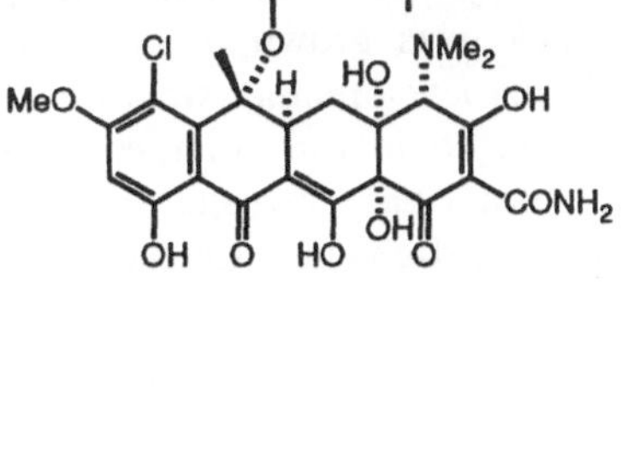

1761 R = NHOH (dactylocycline A)
1762 R = NO_2 (dactylocycline B)
1763 R = NHOAc (dactylocycline D)
1764 R = OH (dactylocycline E)

(*1407–1410*). Several derivatives (**1753**–**1757**) have also been found in various strains of *S. aureofaciens* (*1411–1414*). Brominated derivatives are formed in the presence of bromide ions (*1415*). More recently several new tetracyclines have been discovered in other microorganisms. *Actinomadura brunnea* produces **1758** (*1416*) and **1759** (*1417*) while *Dactylosporangium* sp. has furnished **1760** (*1418*). In separate studies, this latter organism was found to yield dactylocyclines A (**1761**), B (**1762**), D (**1763**), and E (**1764**) which are active against tetracycline-resistant bacteria (*1419–1422*).

3.20. Aromatics

In contrast to the vast majority of natural halogenated aromatic rings that are phenolic – due largely to their enormous reactivity in electrophilic substitution and the pervasiveness of phenols in nature – there are but a few examples of simple halogenated aromatics.

A significant component (1.2%) of the oil of the Mississippi salt marsh "needlerush" (*Juncus roemerianus*, Juncaceae) is 1,2,3,4-tetrachlorobenzene (**1765**), the only halogenated arene of the 13 benzene derivatives identified (*1423*). A study of the ash from 1980 Mount St. Helens eruption has uncovered three simple aromatics, **1766**–**1768**, along with three isomers of pentachlorobiphenyls (PCBs), **1769**–**1771** (*1424*). The nature of these PCB isomers does not correspond to commercial PCB mixtures, and this represents the first report of natural PCBs. The stony coral *Tubastraea micrantha* produces 3-bromobenzoic acid (**1772**) (*1425*), and this coral is avoided by the coral-eating Crown-of-Thorns seastar (*Acanthaster planci*). The Thai plant *Arundo donax* (Gramineae) produces the simple bromoacetamide **1773** which has antifeedant activity against cotton boll weevils (*1426*). A remarkable discovery is that a deep sea gorgonian (collected at 350 meters) contains the three halogenated azulenes **1774**–**1776** (*1427*). Although their aromatic rings are not halogenated, the novel chlorine-containing nostocyclophanes A–D (**1777**–**1780**) which are produced by the blue-green alga *Nostoc linckia* (*1428*) are unprecedented examples of natural [7.7]paracyclophanes. Biosynthetic studies reveal that the nostocyclophanes are polyketide derived and that they are constructed by dimerization of acetate-derived nonaketides (*1429*). When *N. linckia* was grown on a bromide-rich medium, the bromo analogues of nostocyclophanes were not detected. Bromobenzene (**1781**) has been detected in the volatiles of oakmoss (*Evernia prunastri*) (*1618*). Several studies have revealed the presence of chlorinated polycyclic aromatic hydrocarbons in various environmental

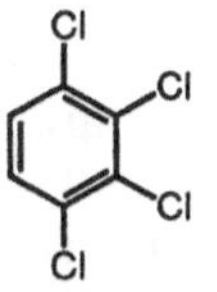

1765 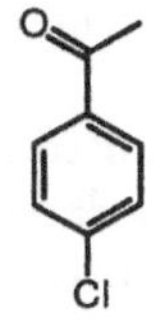 **1766** 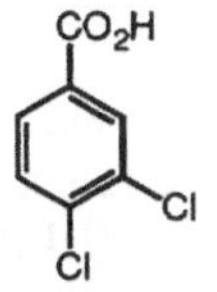 **1767** 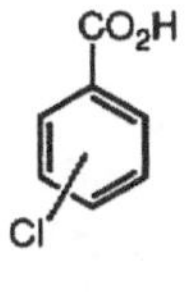 **1768**

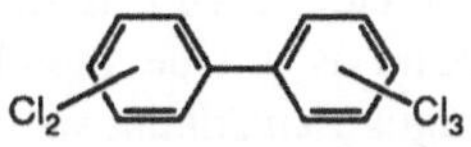

1769 - 1771 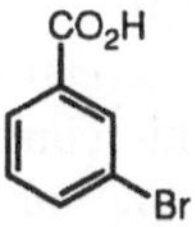 **1772** 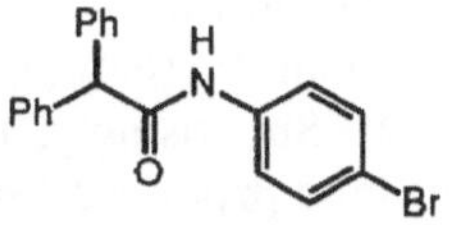 **1773**

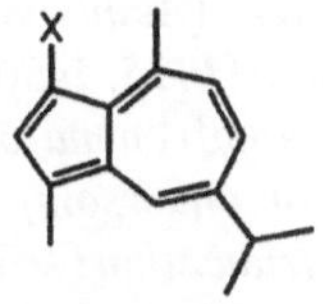

1774 X = Cl
1775 X = Br

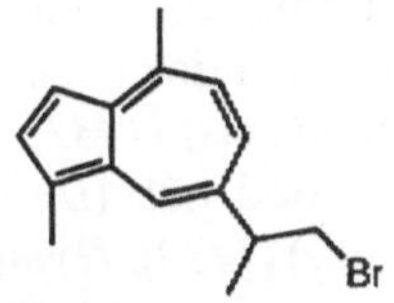

1776 (ehuazulene)

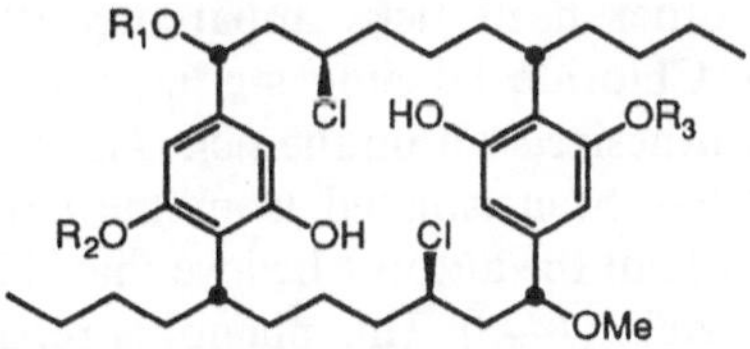

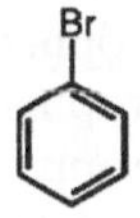

	R_1	R_2	R_3	
1777	Me	β-D-glu	β-D-glu	(nostocyclophane A)
1778	Me	H	β-D-glu	(nostocyclophane B)
1779	H	H	H	(nostocyclophane C)
1780	Me	H	H	(nostocyclophane D)

1781

samples such as urban air (*1430*). The authors of these studies invariably attribute the origin of these chlorinated PAHs to anthropogenic sources, but the possibility exists that these compounds could be products of combustion in nature.

3.21. Simple Phenols

The great reactivity of the phenolic ring toward electrophilic halogenation has been extensively exploited by nature as a myriad of halogenated phenols – simple and complex – have been characterized from natural sources, both terrestrial and marine.

3.21.1. *Terrestrial*

Because chlorophenols are used industrially on a large scale and because several chlorophenols have been classified by the U.S. Environmental Protection Agency (EPA) as Priority Pollutants (*1432*), it was a shock to realize that nature produces numerous chlorophenols, including at least two on the EPA list!

The precursor to the commercial herbicide, 2,4-D, is 2,4-dichlorophenol (**1782**). Surprisingly, this compound is a true metabolite and growth hormone produced by the soil fungus *Penicillium* sp. (*1433*). The isomeric 2,6-dichlorophenol (**1783**) is the sex pheromone of several species of tick, including the lone star tick (*Amblyomma americanum*) (*1434–1436*), Gulf Coast tick (*A. maculatum*) (*1435, 1436*), brown dog tick (*Rhipicephalus sanguineus*) (*1437, 1440*), *Amblyomma variegatum* (*1438*), Rocky Mountain wood tick (*Dermacentor andersoni*) (*1439*), American dog tick (*D. variabilis*) (*1439*), *Hyalomma truncatum* (*1438*), tropical cattle tick (*Boophilus microplus*) (*1440*), camel tick (*Hyalomma dromedarii*) (*1441*), *Dermacentor albipictus* (*1442*), and *Haemaphysalis leporispalustris* (*1442*). However, several other hard ticks apparently do not utilize 2,6-dichlorophenol (*1442*). Chloride labeling studies indicate that the 2,6-dichlorophenol is biosynthesized within the tick (*1435*). The isomeric 2,5-dichlorophenol (**1784**) has been isolated from the common grasshopper (*Romalea microptera*) but the authors believe that this compound has a dietary (herbicide?) origin (*1443*). This phenol is repellant to ants and seems to be part of the defensive secretion. The earliest terrestrial natural chlorophenol to be identified is drosophilin A (**1785**) isolated

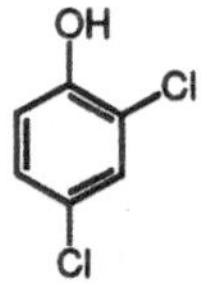

1782 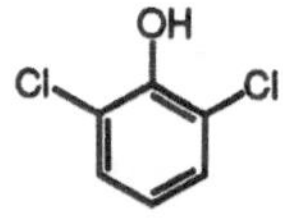 **1783** 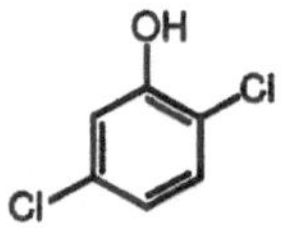 **1784** 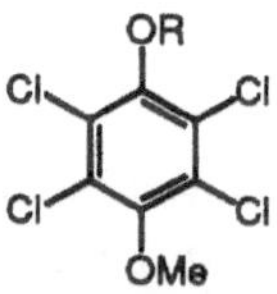 **1785** R = H (drosophilin A) **1786** R = Me

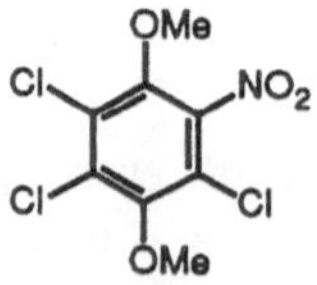

1787 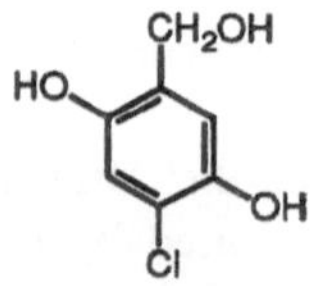 **1788** (amudol) 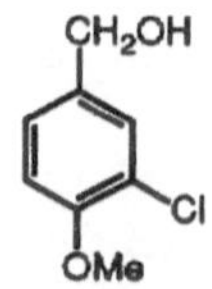 **1789** 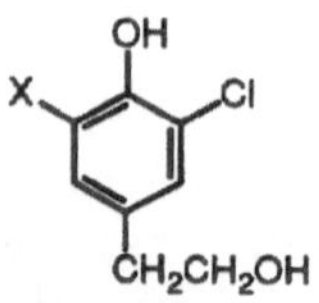 **1790** X = H **1791** X = Cl

from *Drosophila subatrata* (*1444, 1445*). The methyl ether **1786** is produced by *Fomes fastuosus* (*1446*), *F. robiniae* (*1447*), *Phellinus yucatensis* (*1448*), and the mushroom *Agaricus bisporus* (*1449*). This latter study demonstrated that drosophilin A methyl ether (**1786**) is a true metabolite and not a degradation product of pentachlorophenol (*1449*). The fungus *Fomes robiniae* also produces nitro derivative **1787** (*1447*), one of nature's rare nitro-containing metabolites. Amudol (**1788**), which has antibacterial properties, has been isolated from cultures of *Penicillium martinsii* (*1450*). As mentioned in Section 3.3.2, the isomeric 3-chlorogentisyl alcohol (**204**) has been isolated from *Phyllosticta* sp. (*188*), *Phoma* sp. (*Fungi imperfecti*) (*1451*), and *Penicillium canadense* (*1452*). The biosynthesis of this metabolite in *Phyllosticta* sp. involves chloride-induced ring opening of the epoxide in epoxydon and eventual nonenzymatic aromatization to **204** (*189*). The white-rot fungi *Bjerkandera* sp. and *B. adusta* produce benzyl alcohol **1789**, a new natural product, and the previously known 3-chloroanisaldehyde (**1792**) (*vide infra*) is produced by the latter fungus (*2218*). The fungus *Caldariomyces fumago* produces the simple **1790** and **1791**, which appear to form from tyrosol and chloroperoxidase (*166*).

The fungus *Lepista diemii* produces the isomeric chlorobenzaldehydes **1792** and **1793** (*1453*) while acid **1794** is a metabolite of *Maras-*

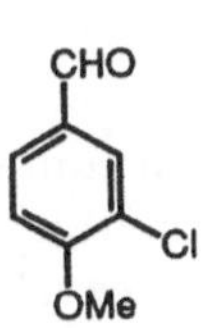

1792 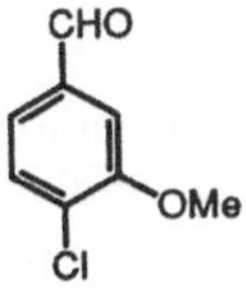 **1793** 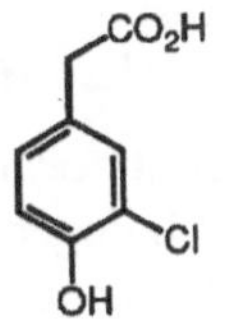 **1794** 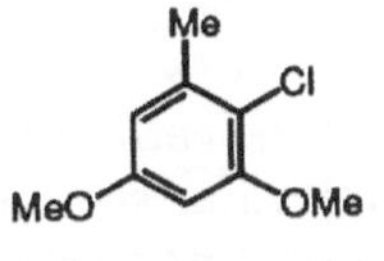 **1795**

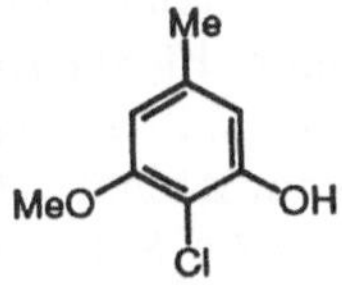

1796 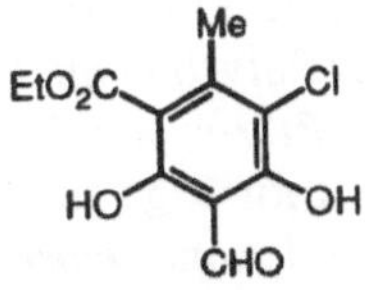 **1797** 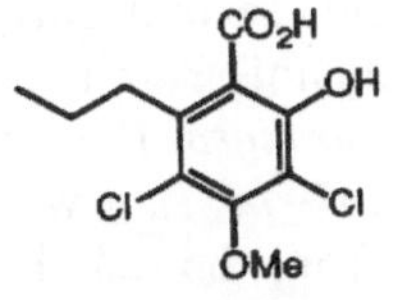 **1798** (differanisole A)

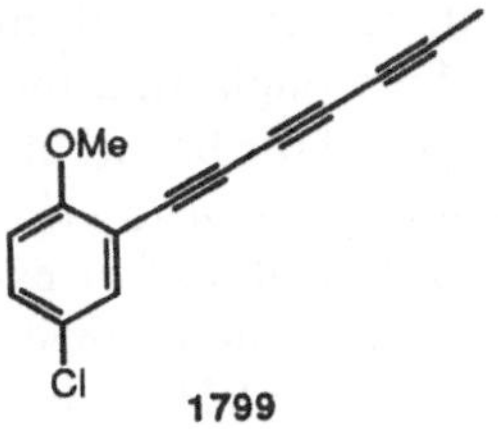

1799 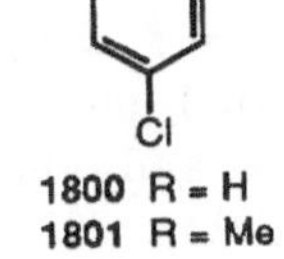 **1800** R = H **1801** R = Me

mius palmivorus (*9*). Oakmoss (*Evernia prunastri*) exudes several volatiles including **1795**, **1796** (*1618*) and ethyl hematommate (**1797**) which is active against the dog roundworm (*Toxocaria canis*) (*1454*). The soil microbe *Chaetomium* sp. produces differanisole A (**1798**) which induces cell differentiation (*1455*). The polyacetylene **1799** is found in the plant *Helichrysum coriaceum* (Compositae) (*1456*).

There is mounting evidence that 2,4,6-trichlorophenol (**1800**) is a natural product of soil microbes. Thus, several studies have found both **1800** and 2,4,6-trichloroanisole (**1801**) in lakes, rivers, and soils in amounts and locations that are consistent with natural sources (*1457–1460*). The formation of **1801** probably involves the biomethylation of **1800** (*1459*). Although **1800** and other chlorinated phenols have been detected in 13th century lake sediments (*1461*), only **1800** is considered to have a natural origin given the present evidence. Most of the others, such as chloroguaiacols and chlorocatechols, have their origin in the bleaching of paper pulp by chlorine. It is interesting to note that 2,4,6-trichloroanisole (**1801**) is responsible for "corked" wines and has been implicated in many "off" flavors in foods (*1462*). Similarly, bromo- and iodophenols and anisoles have very objectionable odors and have ruined many types of foods (*1462–1464*). It remains to be seen if some of these halogenated phenols and anisoles have a natural origin in these incidents.

3.21.2. Marine

Some of the earliest naturally occurring organohalogen compounds to be identified as such were halogenated phenols from marine organisms. For excellent reviews of marine halogenated phenols, see ref. (*1465, 1466*).

Although several early reports appeared on the isolation of brominated phenols from red algae (*Polysiphonia fastigiata*, *Vidalia volubilis*, *Polysiphonia nigra*, *P. elongata*, *Pterosiphonia complanata*, *Halopitys incurvus*) (*1467–1471*), it was not until 1955 that bromocatechol **1802** was identified from the red alga *Polysiphonia morrowii* (*1472*). Subsequently, a number of brominated phenols have been isolated from red algae. These are listed in Table 24. Lanosol (**1803**) and **1802** are highly toxic to unicellular marine algae (*1477*). In those cases where methyl or ethyl ethers have been isolated, it is generally thought that these are artifacts arising by methanol or ethanol solvolysis of the benzyl alcohol or sulfate ester (e.g., **1804**) during extraction/isolation (*1481, 1482*). These isolations are simply listed under the benzyl alcohol. For example, lanosol methyl ether (**1803a**) and others have been isolated several times (*1475, 1480, 1481, 1483–1485, 1487, 1489–1491, 1496*). However, in some cases, lanosol methyl ether has been isolated even when methanol is not used at any point in the isolation (*1496*). It is also claimed that lanosol *n*-propyl ether (**1812**) is not an artifact (*1482*). There is some controversy as to whether or not the benzyl alcohols themselves are artifacts. Since the highly unstable disulfate esters have been isolated in a few cases (e.g., **1804**, **1817**), it is thought that these are hydrolyzed to the corresponding alcohols during extraction/isolation. However, there is evidence to the contrary (*1489*).

The red pigment floridorubin present in the red alga *Lenormandia prolifera* is hydrolyzed to lanosol (**1803**), **1805**, and several other bromophenols (*1497*). Although the function of these red algal bromophenols is unknown, they could be waste products, antibiotics against marine bacteria, or growth regulators (*1498*).

The blue-green alga *Calothrix brevissima* produces lanosol (**1803**), **1805** and the new phenols **1824** and **1825** (*1499*). The tropical green algae *Avrainvillea nigricans* and *A. rawsonii* contain **1810** (*1516, 1518*). The brown alga *Ascophyllum nodosum* produces in culture lanosol (**1803**) and **1805** (*1500*), while the brown alga *Eisenia arborea* produces the two halogenated phloroglucinol triacetates **1826** and **1827** (after acetylation) (*1501*). Marine worms have proved to be a rich source of halogenated phenols as well as closely related metabolites. The acorn worm *Balanoglossus biminiensis* produces 2,6-dibromophenol, possibly as a defensive secretion (*1502*). Up to 15 mg of this compound is present in a single animal. *Phoronopsis viridis* produces **1828** and 2,4,6-tribromophenol (**1829**) (*1503*), but *Saccoglossus kowalewskii* from the coast of Maine produces **1828** as does *Nereis succinea* (*1504, 1505*). It is suggested that the worm uses **1828** as an antibacterial agent and as a defensive substance against higher organisms that are competitors (*1504*). In addition to producing complex phenols the segmented worm *Thelepus setosus* contains **1805** and **1806** (*1506, 1507*). The acorn worm *Ptychodera flava* contains **1829** and **1830a**, while *Glossobalanus* sp. has **1829, 1830**, and **1833**, *Balanoglossus carnosus* has **1829, 1830a**, and **1832**, and *B. misakiensis* has **1828** and **1830** (*1052*). As previously mentioned, a deep water acorn worm *Ptychodera* sp. from Maui produces a variety of brominated compounds including the phenols, resorcinols, and hydroquinones **1828–1832** (*194*). The Florida acorn worm *Ptychodera bahamensis* contains an impressive array of halogenated phenols including the known **1828–1830** and the new compounds **1833–1841** (*195*). The latter compound is either 3,4-dichloro- or 3,5-dichlorophenol and the structure of dibromocatechol **1837** is not yet secured. The marine polychaete *Lanice conchilega* produces 2,4,6-tribromophenol (**1829**) and the new dibromocresol **1842** (*1508*). The sponge *Aplysina aerophoba* from the Canary

Islands contains the dibromohydroquinone methyl ether **1843** (*1509*), whereas an unidentified sponge produces the bromocatechol **1844** and moloka'iamine (**1845**) (*1510*). The Tasmanian bryozoan *Amathia wilsoni* produces the simple phenethylamine (**1846**) which is a possible biogenetic precursor of the amathamides (*1511*). An unidentified sponge has yielded the simple phenylacetonitrile **1847** (*1512*).

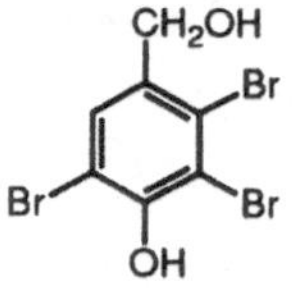

1824 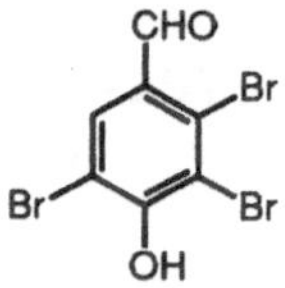 **1825** 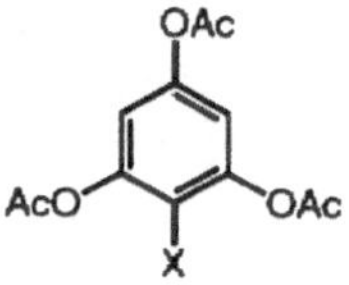 **1826** X = Br **1827** X = I **1828** X = H **1829** X = Br **1830** X = OH

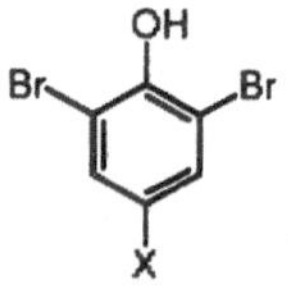

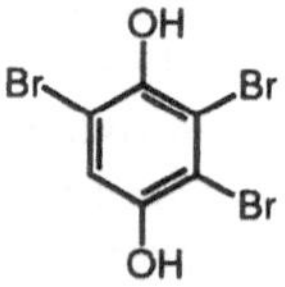

1830a 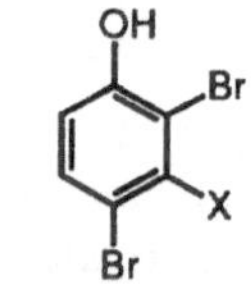 **1831** X = OH **1832** X = H **1833** X = H **1834** X = H, acetate **1835** X = Cl **1836** X = Cl, acetate **1837**

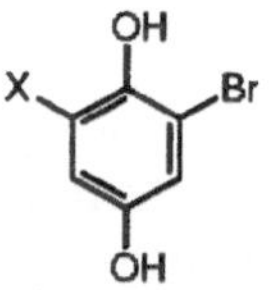

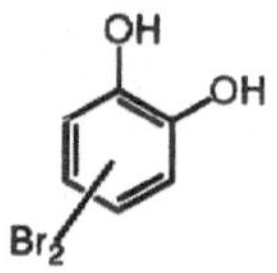

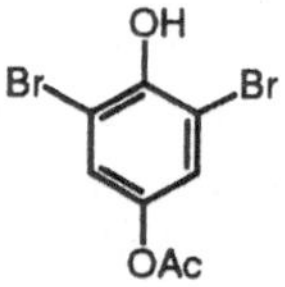

1838 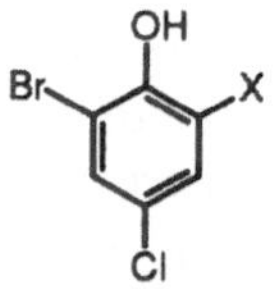 **1839** X = Br **1840** X = Cl **1841** 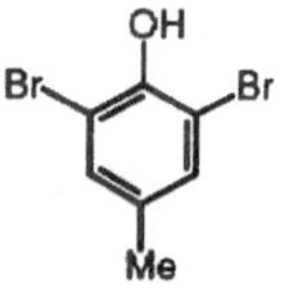 **1842**

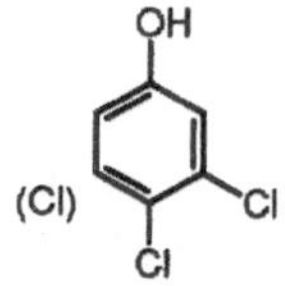

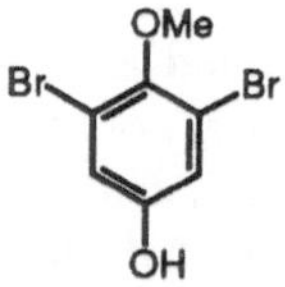

1843 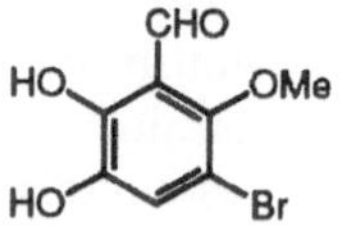 **1844** 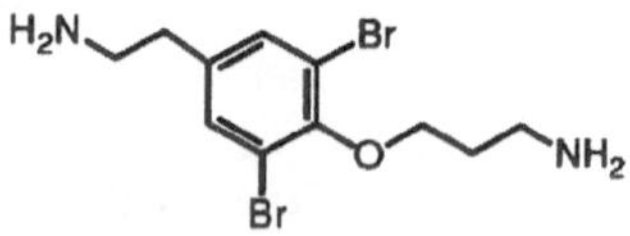 **1845** (moloka'iamine)

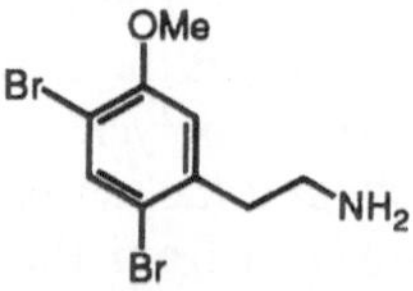

1846 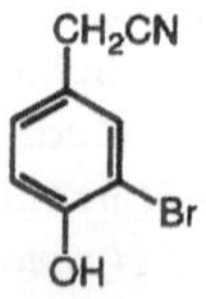 **1847**

Table 24. *Brominated Phenols in Red Algae*

Compound	Source	Ref.
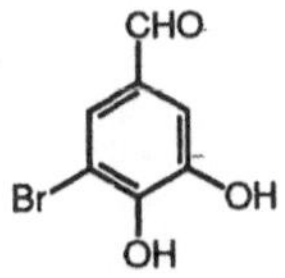 1802	*Odonthalia dentata*	*1483*
	Polysiphonia brodiaei	*1483*
	P. elongata	*1482*
	P. lanosa	*1479, 1482*
	P. morrowii	*1472*
	P. nigrescens	*1483*
	P. urceolata	*1482, 1483, 1485, 1490, 1495*
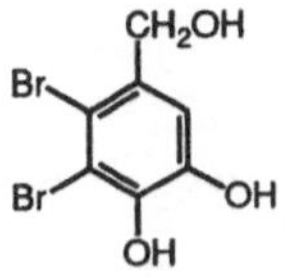 1803 (lanosol)	*Antithamnion plumula*	*1483*
	Ceramium rubrum	*1483*
	Corallina officinalis	*1483*
	Fucus vesiculosus	*1484*
	Odonthalia dentata	*1474, 1483*
	Phycodrys rubens	*1483*
	Polysiphonia brodiaei	*1482, 1483, 1489*
	P. elongata	*1482*
	P. fruticulosa	*1482*
	P. lanosa	*1479, 1482*
	P. nigra	*1482*
	P. nigrescens	*1482, 1483, 1488*
	P. thuyoides	*1482*
	P. urceolata	*1483*
	P. violacea	*1482*
	Rhodomela confervoides	*1474, 1483, 1488*
	R. larix	*1481, 1487*
	R. subfusca	*1482, 1491*
	Rytiphlea tinctoria	*1486*
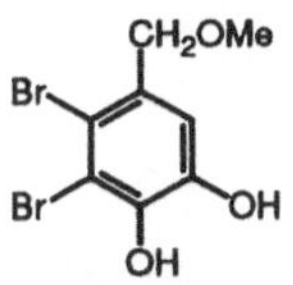 1803a	*Antithamnion plumula*	*1483*
	Ceramium rubrum	*1483*
	Odonthalia corymbifera	*1480*
	O. dentata	*1483*
	O. floccosa	*1496*
	O. washingtoniensis	*1496*
	Phycodrys rubens	*1483*
	Polysiphonia brodiaei	*1483*
	P. nigrescens	*1483*
	P. urceolata	*1483*
	Rhodomela confervoides	*1483*
	R. larix	*1475, 1481*
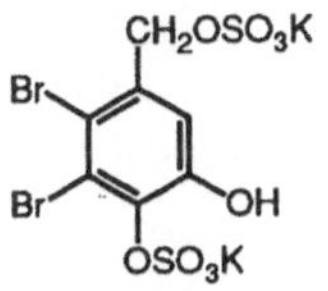 1804	*Polysiphonia brodiaei*	*1476, 1482*
	P. elongata	*1476, 1482*
	P. fruticulosa	*1476, 1482*
	P. lanosa	*1473, 1476, 1482*
	P. nigra	*1476, 1482*

Table 24 (*continued*)

Compound	Source	Ref.
	P. nigrescens	*1476, 1482*
	P. thuyoides	*1482*
	P. violacea	*1482*
	Rhodomela larix	*1481*
	R. subfusca	*1476, 1482*
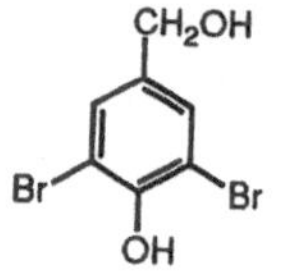 **1805**	*Fucus vesiculosus*	*1484*
	Odonthalia dentata	*1474, 1483*
	Polysiphonia brodiaei	*1482, 1483*
	P. lanosa	*1482*
	P. nigrescens	*1482, 1483, 1488*
	P. urceolata	*1482, 1485, 1490, 1495*
	Rhodomela confervoides	*1474, 1483, 1488*
	R. larix	*1491*
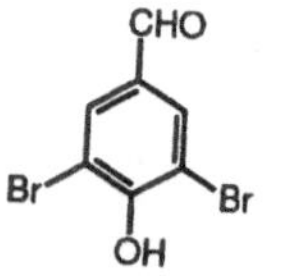 **1806**	*Polysiphonia urceolata*	*1490*
	Rhodomela larix	*1491*
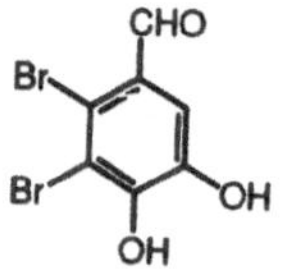 **1807**	*Cystoclonium purpureum*	*1483*
	Odonthalia dentata	*1483*
	Polysiphonia brodiaei	*1483, 1489*
	P. elongata	*1482*
	P. fruticulosa	*1482*
	P. lanosa	*1479, 1482*
	P. nigra	*1482*
	P. nigrescens	*1482*
	P. thuyoides	*1482*
	P. urceolata	*1483*
	P. violacea	*1482*
	Rhodomela confervoides	*1483*
	R. larix	*1475, 1487*
	R. subfusca	*1482*
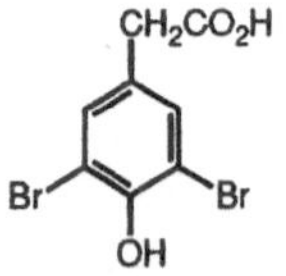 **1808**	*Halopytis incurvus*	*1478*
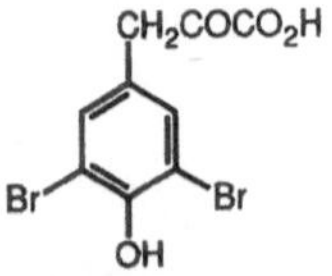 **1809**	*Halopytis incurvus*	*1478*

Table 24 (*continued*)

Compound	Source	Ref.
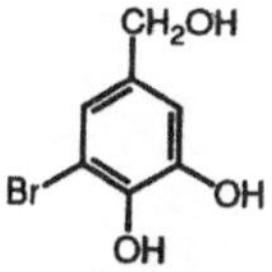 1810	*Halopytis incurvus*	*1482*
	Odonthalia dentata	*1483*
	Polysiphonia brodiaei	*1483*
	P. lanosa	*1479, 1482*
	P. nigrescens	*1482*
	P. urceolata	*1482, 1483, 1485, 1490, 1495*
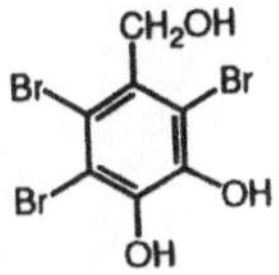 1811	*Polysiphonia lanosa*	*1482*
	P. nigrescens	*1488*
	Rhodomela confervoides	*1488*
	R. subfusca	*1482*
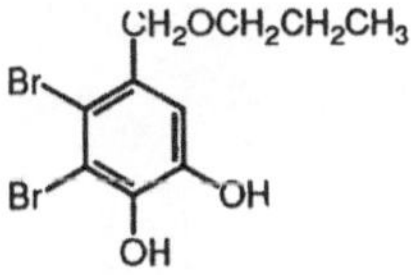 1812	*Polysiphonia lanosa*	*1482*
	P. nigrescens	*1482*
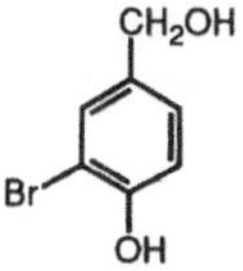 1813	*Halopytis incurvus*	*1482*
	Odonthalia dentata	*1483*
	Polysiphonia brodiaei	*1482*
	P. fruticulosa	*1482*
	P. nigra	*1482*
	P. urceolata	*1482, 1495*
	Rhodomela larix	*1491*
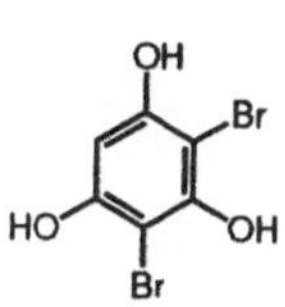 1814	*Rhabdonia verticillata*	*1494*
	Rytiphlea tinctoria	*1486*
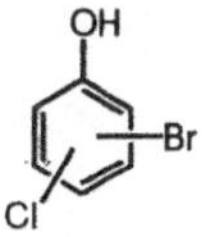 1815	*Polysiphonia nigrescens*	*1488*
	Rhodomela confervoides	*1488*

Table 24 (*continued*)

Compound	Source	Ref.
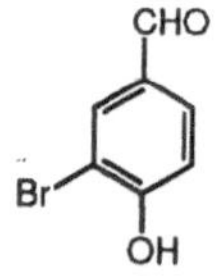 **1816** (CHO; Br; OH)	*Rhodomela larix*	*1491*
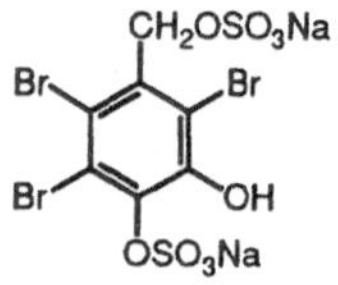 **1817** (CH_2OSO_3Na; Br; Br; Br; OH; OSO_3Na)	*Symphyocladia latiuscula*	*1492*
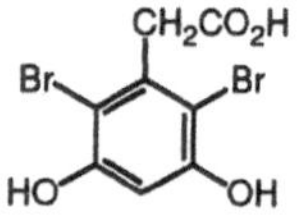 **1818** (CH_2CO_2H; Br; Br; HO; OH)	*Halopitys incurvus*	*1493*
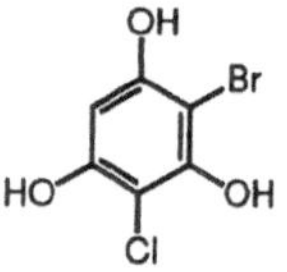 **1819** (OH; Br; HO; OH; Cl)	*Rhabdonia verticillata*	*1494*
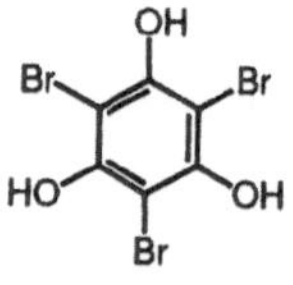 **1820** (OH; Br; Br; HO; OH; Br)	*Rhabdonia verticillata*	*1494*
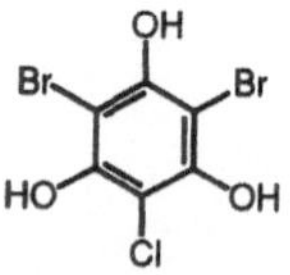 **1821** (OH; Br; Br; HO; OH; Cl)	*Rhabdonia verticillata*	*1494*

Table 24 (*continued*)

Compound	Source	Ref.
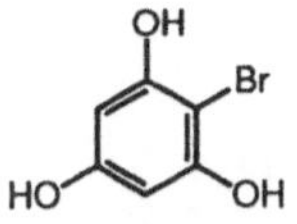 1822	*Rhabdonia verticillata*	*1494*
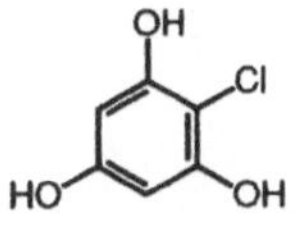 1823	*Rhabdonia verticillata*	*1494*

Brominated phenols and anisoles such as 2,4,6-tribromophenol (**1829**), 2,4,6-tribromoanisole, pentachloroanisole, and drosophilin A methyl ether (**1786**) have been identified in river and marine sediments (*1513*) and in the air over oceans (*1514*), but in these particular studies the possibility of anthropogenic sources could not be precluded.

3.22. Complex Phenols

The number and variety of halogenated phenolic rings in nature is astounding. In addition to the simple examples presented in the previous section, there is a multitude of more complex natural halophenols.

3.22.1. Diphenylmethanes and Related Compounds

Several marine metabolites are halogenated phenolic diphenylmethane ring systems. For example, diphenylmethanes thelephenol (**1848**) and **1849** are two compounds produced by the segmented worm *Thelepus setosus* and **1848** may be the biogenic precursor to thelepin (**2215**) (*vide infra*) which has antifungal activity comparable to griseofulvin (*1506, 1507*). The red alga *Rytiphlea tinctoria* contains the diphenylmethane **1850** (isolated as the pentamethyl ether) (*1486*), while *Rhodomela larix* has yielded diphenylmethanes **1851** and **1852** (*1487*). The latter

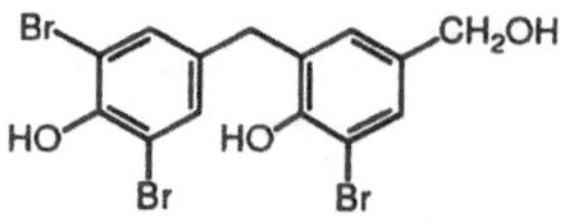

1848 (thelephenol)

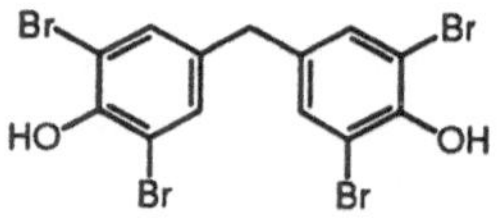

1849

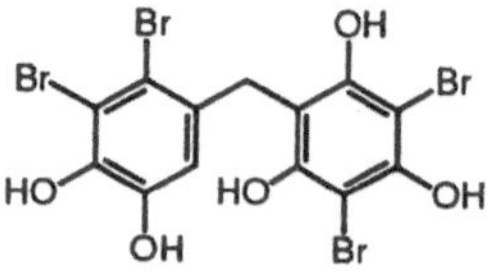

1850

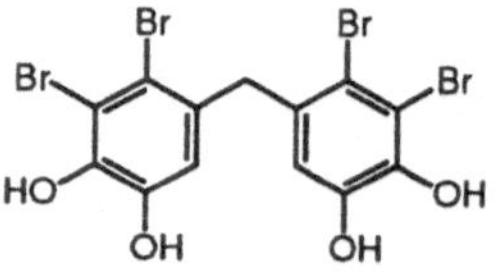

1851

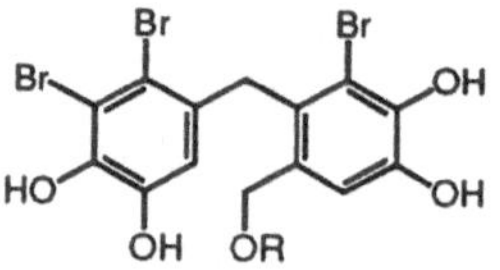

1852 R = H
1852a R = Me

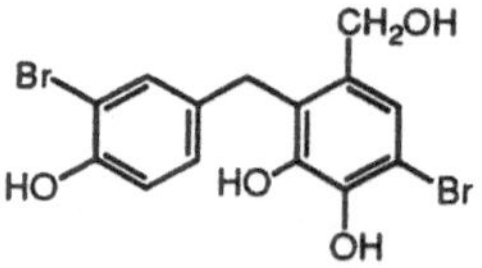

1853 (avrainvilleol)

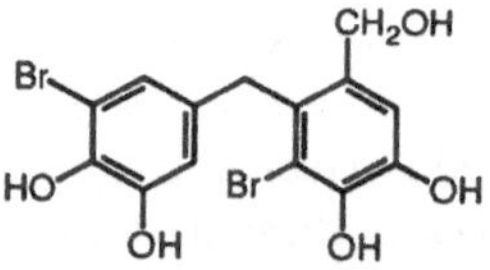

1854 (5'-hydroxyiso avrainvilleol)

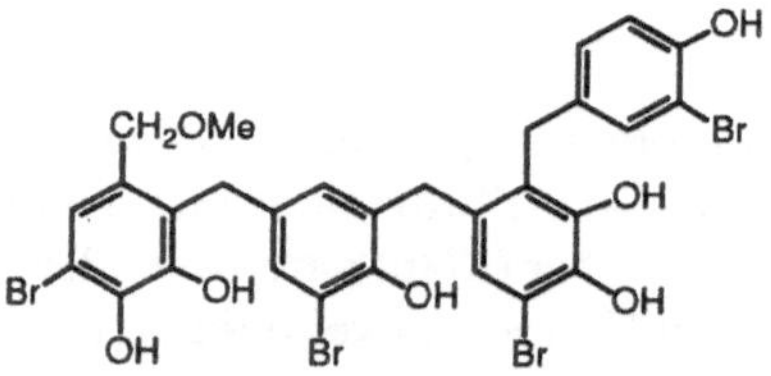

1855 (rawsonol)

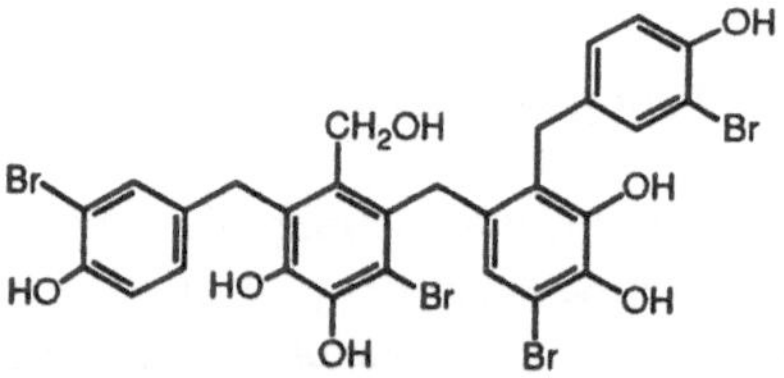

1856 (isorawsonol)

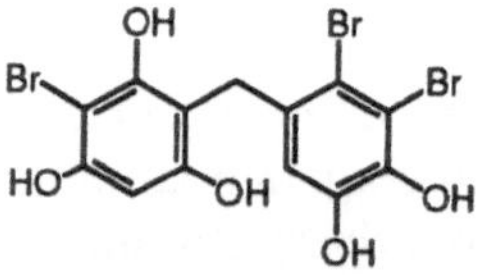

1857 (vidalol A)

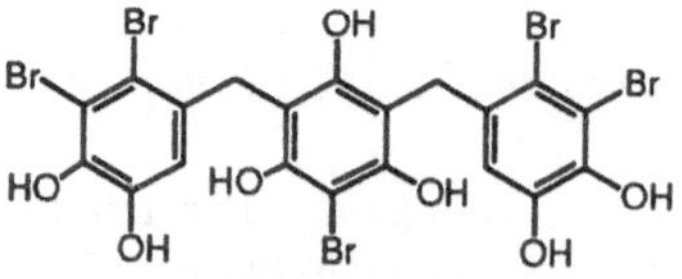

1858 (vidalol B)

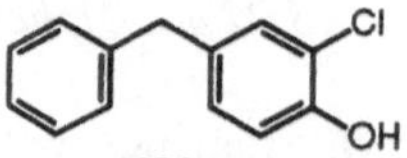

1859

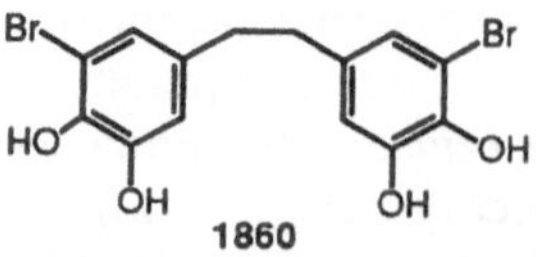

1860

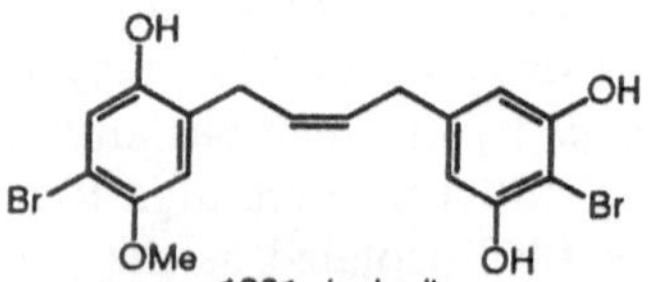

1861 (colpol)

was actually isolated as the benzyl methyl ether (**1852a**) because methanol was used in the extraction protocol. The former metabolite was also found in the red algae *Polysiphonia nigrescens*, *Rhodomela confervoides* (*1488*), and *P. brodiaei* (*1489*). The benzyl methyl ether **1852a** was isolated from *Rhodomela larix* (*1491*). The tropical green alga *Avrainvillea longicaulis* contains the feeding deterrent avrainvilleol (**1853**) (1% of dry weight), which is not only toxic to herbivorous reef fish at 10 ppm, but is antibacterial to the marine bacterium *Vibrio anguillarum* (*1515*). The related alga *Avrainvillea nigricans* from Puerto Rico produces avrainvilleol (**1853**) and the new metabolite 5′-hydroxyisoavrainvilleol (**1854**), which has antibacterial properties (*1516*). Interestingly this same algal species from the Western Caribbean is devoid of halogen metabolites. Rawsonol (**1855**) is a modest HMG-CoA reductase inhibitor isolated from *Avrainvillea rawsoni* (*1517*). The actual metabolite may be the corresponding benzyl alcohol since methanol was used in the isolation procedure. A later study of this green alga identified avrainvilleol (**1853**), rawsonol (**1855**) and the new isorawsonol (**1856**) which is an inhibitor of IMP dehydrogenase (*1518*). The authors propose that **1856** is formed from avrainvilleol (**1853**) *via* electrophilic coupling. The red alga *Vidalia obtusaloba* produces vidalols A (**1857**) and B (**1858**) which are potent inhibitors of phospholipase A_2 (*1519*). Vidalol A is also a feeding deterrent to reef fish. The blue green alga *Anacystis marina* produces the simple chlorinated diphenylmethane **1859** (*62*). Although not diphenylmethanes, the closely analogous diphenylethane **1860** is found in the red alga *Polysiphonia urceolata* (*1485*, *1490*), and colpol, the novel diphenylbutene derivative **1861**, is produced by the alga *Colpomenia sinuosa* and displays cytotoxicity (*1520*). The interesting compound tris(4-chlorophenyl)methanol (**1862**) for which no clear anthropogenic (or natural!) source is evident is found in high levels in marine animals and birds (*2226*).

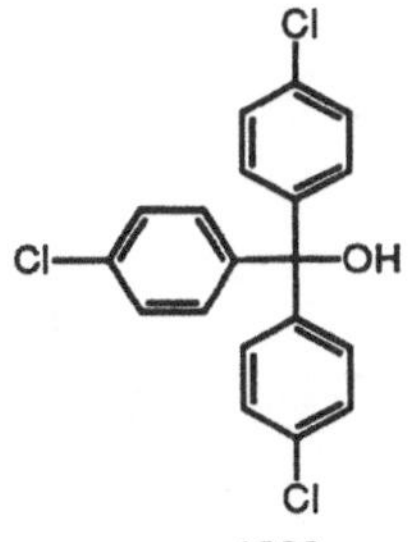

Cl
Cl
OH
Cl

1862

3.22.2. Diphenyl Ethers and Related Compounds

In contrast to the halogenated diphenylmethanes, most naturally occurring diphenyl ethers are brominated metabolites produced by marine sponges rather than marine plants. The first such examples were diphenyl ethers **1863–1867** which were identified in the marine sponge *Dysidea herbacea* (*1522, 1523*). Subsequent studies of this sponge have identified several other diphenyl ethers including **1868** (*1524*), **1869** (*1525*), and **1871–1875** (*1526*). *Dysidea chlorea* contains **1865** (*1525*), and *D. fragilis* contains the new **1870** as well as the known **1866** and **1875**

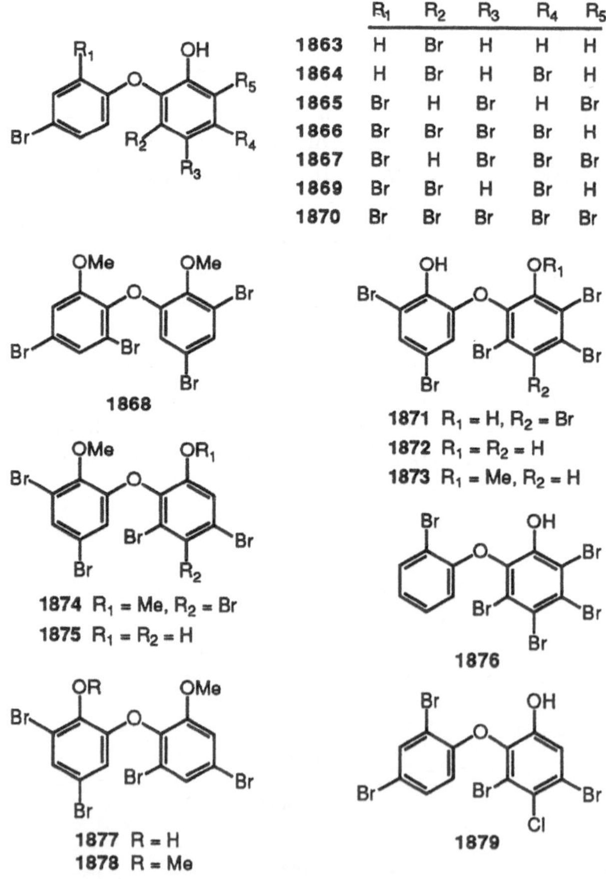

	R_1	R_2	R_3	R_4	R_5
1863	H	Br	H	H	H
1864	H	Br	H	Br	H
1865	Br	H	Br	H	Br
1866	Br	Br	Br	Br	H
1867	Br	H	Br	Br	Br
1869	Br	Br	H	Br	H
1870	Br	Br	Br	Br	Br

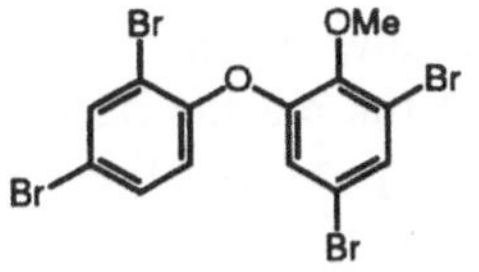

1880

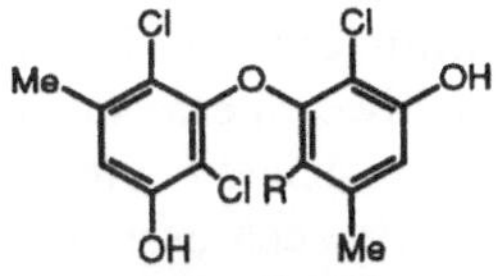

1881 R = H
1882 R = Cl

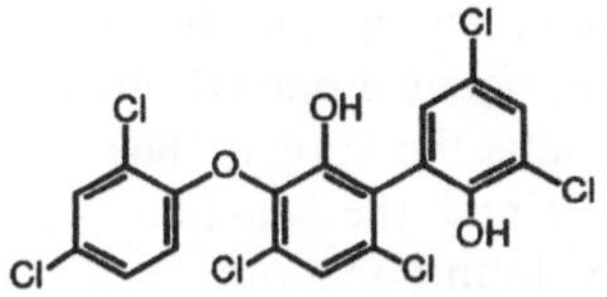

1883 (ambigol A)

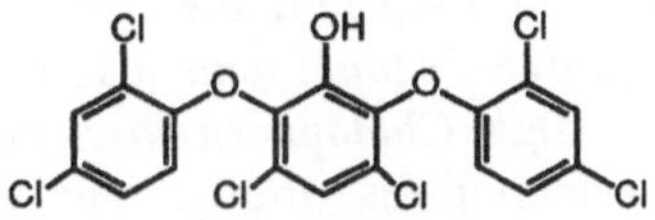

1884 (ambigol B)

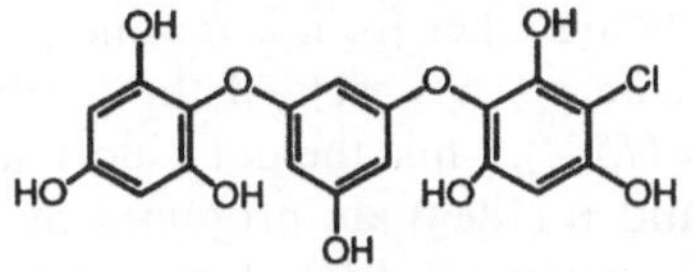

1885

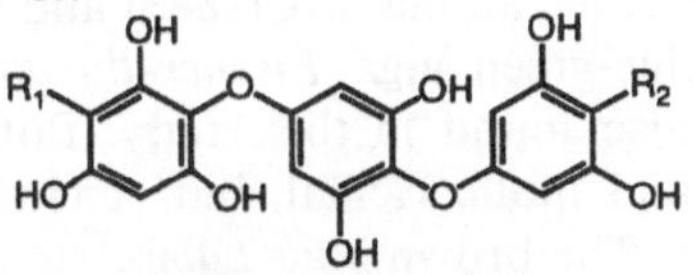

1886 R_1 = H, R_2 = Br
1887 R_1 = Br, R_2 = H

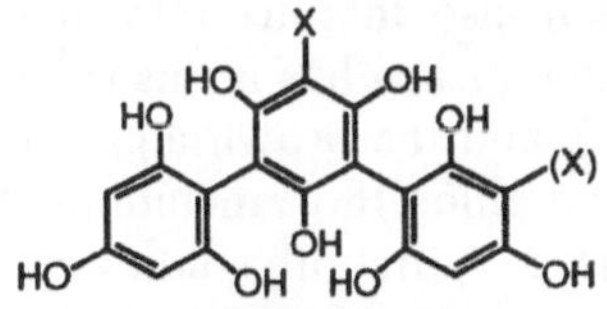

1888 X = Cl
1889 X = Br

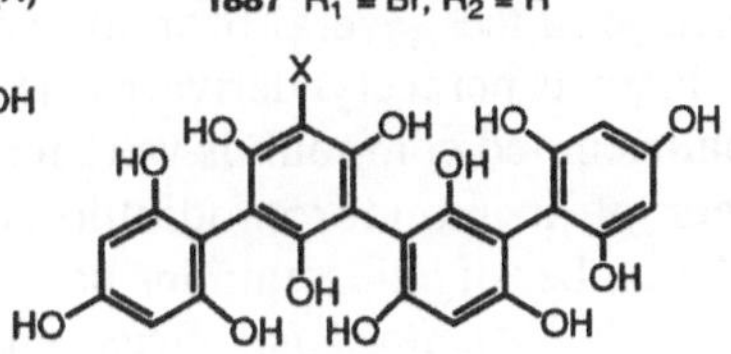

1890 X = Cl
1891 X = Br

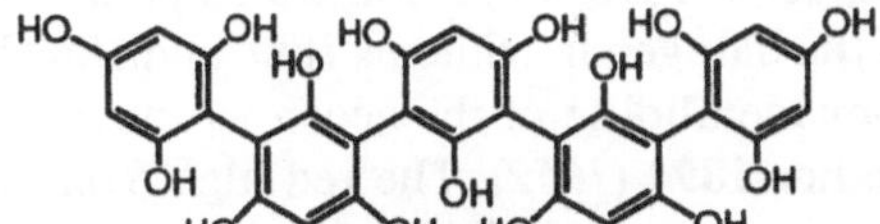

1892 X = Cl
1893 X = Br

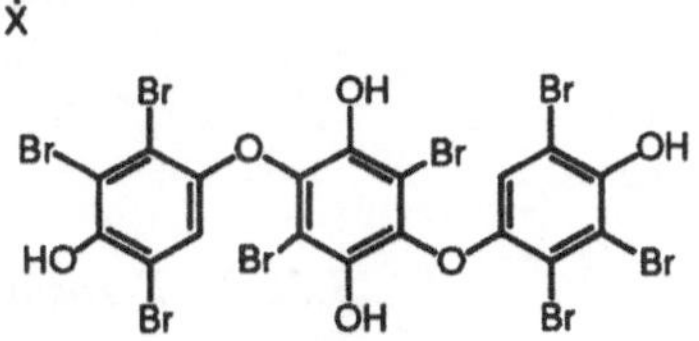

1894

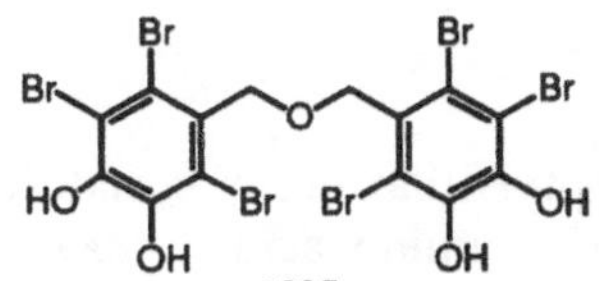

1895

(*1527*). A Philippine *Dysidea* sp. was found to contain **1870** and the new **1876** which was antimicrobial (*1528*). Interestingly, the methyl ether derivatives were inactive. There is evidence that the marine bacterium *Vibrio* sp. which is associated with *Dysidea* sp. produces these brominated diphenyl ethers such as **1875** and **1877** in culture (*1529, 1530*). The Palau sponge *Phyllospongia foliascens* contains **1871** and **1872** and the new metabolite **1878** (*1525*). An unidentified sponge has been reported to contain **1869** and the new chlorinated diphenyl ether **1879** (*1531*). The sponge *Tedania ignis* also contains the known **1869** (*1532*).

The green alga *Cladophora fascicularis* and the sea hare *Aplysia dactylomela* which feeds on the alga are found to contain the new brominated diphenyl ether **1880** (*1533*), the methyl ether of **1865**. This is the first example of a brominated diphenyl ether from a marine plant. The freshwater fungus *Kirschsteiniothelia* sp. has yielded the antibacterial chlorinated ethers **1881** and **1882** (*1534*), while the polychlorinated biologically active ambigol A (**1883**) and B (**1884**) are produced by the terrestrial blue-green alga *Fischerella ambigua* (*1535*). Tjipanazole D (**1490**) was also found in this study. Both ambigols are antibacterial, antifungal, and molluscicidal, but A (**1883**) also inhibits HIV reverse transcriptase. The brown alga *Laminaria ochroleuca* has furnished **1885** (isolated as the peracetyl derivative) (*1536*), and the brown alga *Cystophora congesta* produces several brominated phlorethols including **1886** and **1887** (isolated as peracetyl derivatives) (*1537*). In both studies several additional halogenated compounds were tentatively identified, including three isomers of a monobromodiphlorethol and a monochlorodiphlorethol. A number of phlorotannins have also been found in the North Pacific brown alga *Analipus japonicus*, including several halogenated fucols in which the phloroglucinol units are connected by biphenyl bonds (*1538*). All of these compounds, **1888–1893**, were isolated as peracetates by acetylation. The position of the halogen in trifucols **1888** and **1889** has not been determined. The major metabolite of the acorn worm *Ptychodera flava* is octabrominated ether **1894** (*1052*). The red alga *Symphyocladia latiuscula* produces the interesting antibacterial dibenzyl ether **1895** (*1539*).

3.22.3. Tyrosines

Because of the myriad halogenated phenols that occur in nature it comes as no surprise that the one phenolic amino acid, tyrosine, is frequently found in halogenated form.

3.22.3.1. Simple Tyrosines, Thyroxine, and Related Compounds

Although nature has chosen to use iodine much less frequently than chlorine or bromine in constructing natural products, iodine-containing natural products do have the distinction of being the first organohalides to be discovered and studied in depth. Iodinated tyrosines have been

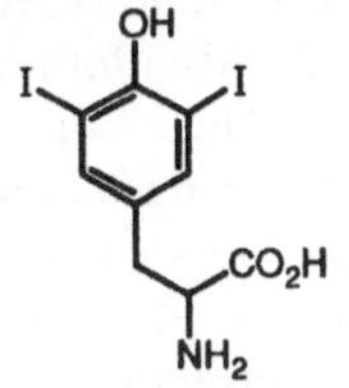

1896

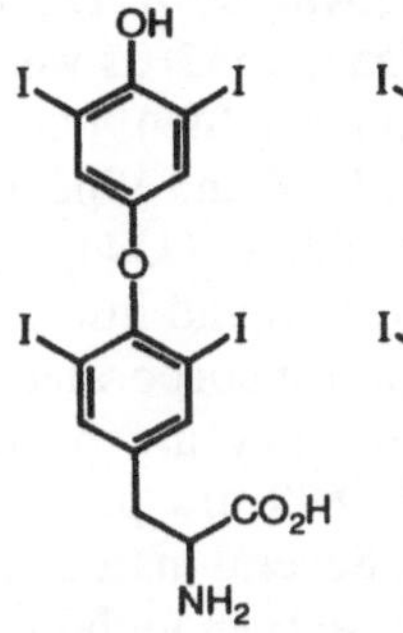

1897 (thyroxine)

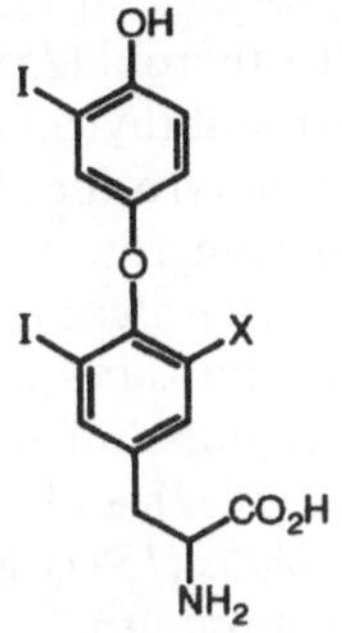

1898 X = H
1899 X = I

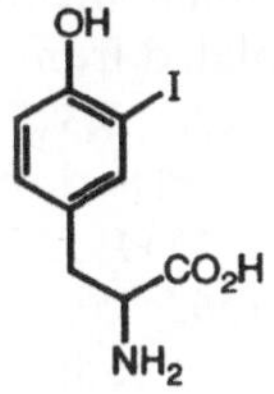

1900

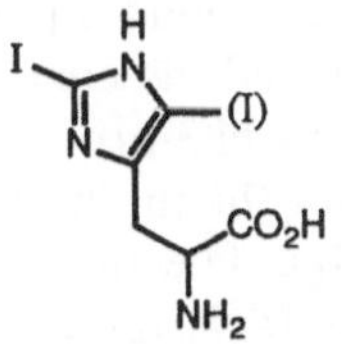

1901

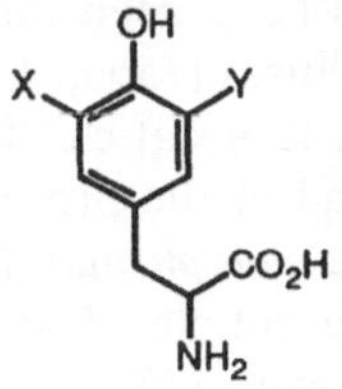

1902 X = Br, Y = H
1903 X = Y = Br
1904 X = Br, Y = Cl
1905 X = Cl, Y = H
1906 X = Y = Cl

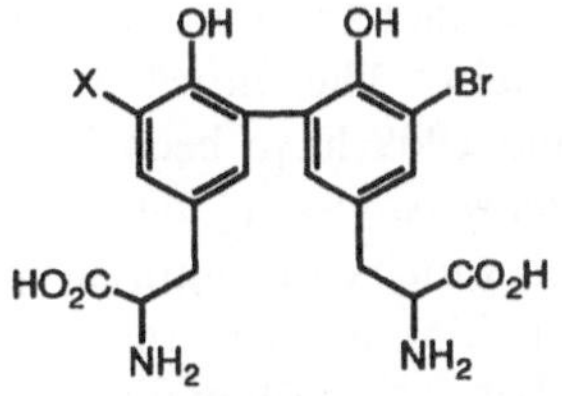

1907 X = H
1908 X = Br

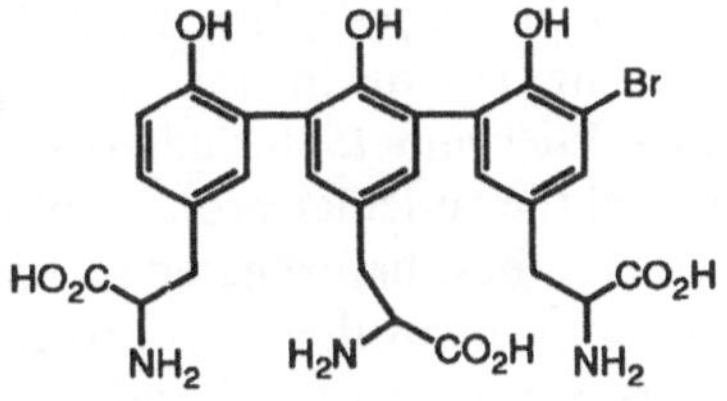

1909

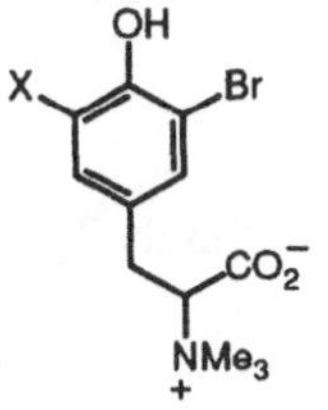

1910 X = H
1911 X = Br

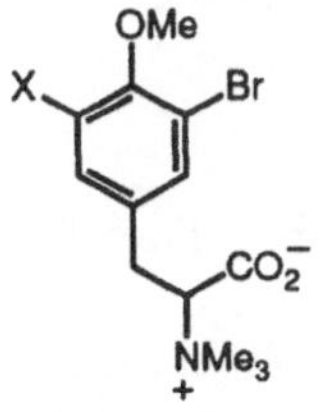

1912 X = H
1913 X = Br

known for nearly 100 years, when "iodogorgoric acid" (3,5-diiodotyrosine (**1896**)) was first isolated from the marine gorgonian *Gorgonia cavolinii* (*1540*) and identified as such (*1541, 1542*). Many other earlier papers described the isolation of "iodine" and "bromine" from numerous marine and insect organisms (*1543–1547*). A major development during this period was the isolation and identification of thyroxine (**1897**) from the mammalian thyroid gland (*1548–1550*). Diiodotyrosine (**1896**) was also isolated from the thyroid (*1551, 1552*) as were 3,3′-diiodothyronine (**1898**) and 3,3′,5′-triiodothyronine (**1899**) (*1553*). These thyroxines (**1897**–**1899**) and/or iodotyrosines **1896** and **1900** have been isolated from sponges (*1554–1556*), numerous ascidians (*1546, 1557–1560*), gorgonians (*1556, 1561, 1562*), marine algae (*1563*) and insects (*1564, 1565*). In some cases, these iodinated tyrosines are incorporated in proteins (as in the thyroid protein thyroglobulin) or they are present in free form. For examples, thyroxine (**1897**) and **1899** are present in free form in the tunicate *Ciona intestinalis* (*1557*). Several insects (dragonfly, squash bug, house fly, cockroach, mosquito) have been found to contain 2-(or 4-)iodohistidine (**1901**) in addition to iodotyrosines (*1565*). Both mono- (**1902**) and dibromotyrosines (**1903**) have been found in sponges (*1554–1556*), gorgonians (*1556, 1561*), molluscs (*1566, 1567*), and insects (*1568*), usually as part of the scleroprotein or cuticle. The novel 3-bromo-5-chlorotyrosine (**1904**) has been found in the proteins of the mollusc *Buccinum undatum* (*1566*) and *Limulus polyphemus* (*1567*). The novel 3-chlorotyrosine (**1905**) occurs in the cuticle of locusts (*Schistocerca gregaria*) (*1569*) and in *Limulus polyphemus* (*1567*). Studies of *Buccinum undatum* (*1570*) and *Limulus polyphemus* (*1567*) revealed the presence of 3,5-dichlorotyrosine (**1906**) in the scleroprotein. The novel brominated dityrosines **1907** and **1908** and trityrosine **1909** have been isolated by hydrolysis of the cuticular proteins of *Cancer pagurus* (*1568*). It has been suggested that these halogenated tyrosines, which can comprise up to 3% of the amino acids in the scleroproteins (*1566*), increase the stability of these proteins by increasing the adhesion between protein fibers or sheets (*1566, 1568*). The marine sponge *Pseudoceratina crassa* produces the novel brominated tyrosines **1910–1913** (*1571*).

3.22.3.2. Transformed Single Tyrosines

Although the simple halogenated tyrosines were the first to be discovered in nature, the transformed tyrosines are indisputably more interesting from a biogenic and structural standpoint.

The first such example was **1914** which was isolated from the sponges *Aplysina* (formerly *Verongia*) *fistularis* and *A. cauliformis* (*1572, 1573*) and later named dibromoverongiaquinol (*1574*). This compound has also been found in *A. cavernicola* (*1574*), *A. aerophoba* (*1575*), *A. laevis* (*1577*), *A. thiona* (*1578*), *Aplysina* sp. (*1576, 1579*), *Verongula rigida* (*1576*), *V. gigantea* (*1576*), and *Aiolochroia crassa* (*1576*). The dichloro analogue **1915** is produced by *Aplysina cavernicola* and was found to inhibit both gram-positive and gram-negative bacteria (*1580*). This sponge also contains bromochloro **1916** and monobromo **1917** (racemic) derivatives (*1574*). *Aplysina fistularis* also contains **1915** and **1916** (*1581*) and *A. thiona* possesses ester **1918** and acid **1919** (*1578*) in addition to the novel ketals, aplysinketals A (**1920**) and B (**1921**), even though no butanol or pentanol was used in the isolation procedure (*1578*). This observation supports the previous control experiments which indicated that ketals **1922** (*A. fistularis*) (*1582, 1583*), **1923** and **1924** (*Aplysina* sp.) (*1579*), (*A. fistularis*) (*1581*) are true natural products. These ketals are also found

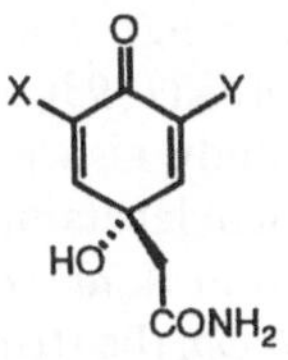

1914 X = Y = Br
1915 X = Y = Cl
1916 X = Br, Y = Cl
1917 X = Br, Y = H

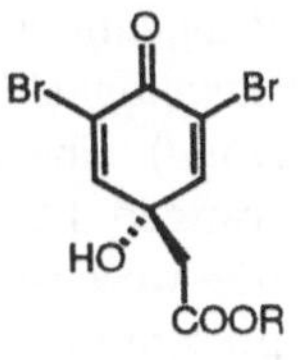

1918 R = Me
1919 R = H

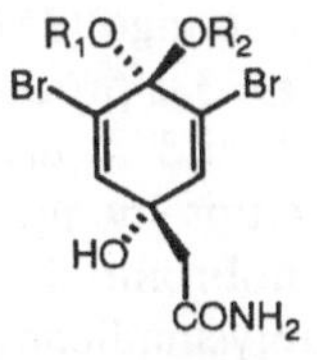

1920 R_1 = *n*-Bu, R_2 = Me (aplysinketal A)
1921 R_1 = *n*-Pen, R_2 = Me (aplysinketal B)
1922 R_1 = R_2 = Me
1923 R_1 = Me, R_2 = Et
1924 R_1 = Et, R_2 = Me

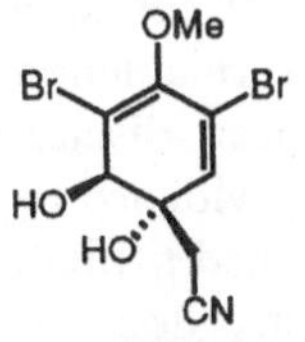

1925 ((+)-aeroplysinin 1)

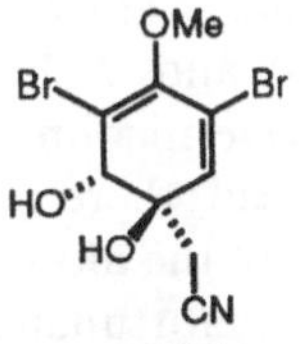

1926 ((−)-aeroplysinin 1)

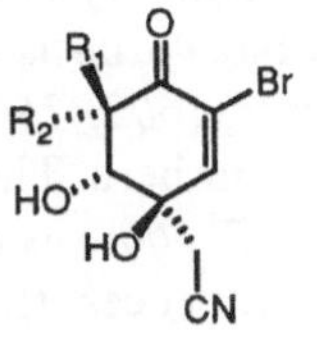

1927 R_1 = H, R_2 = Br
1928 R_1 = Br, R_2 = H

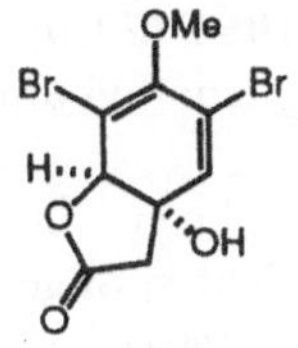

1929 ((+)-aeroplysinin 2)

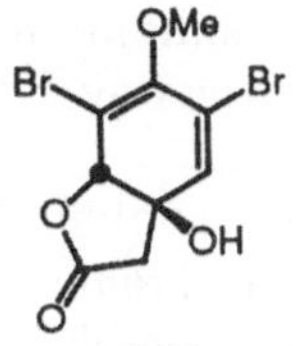

1930 ((−)-aeroplysinin 2)

in all Cuban *Aplysina* and other sponges examined for bromotyrosines (*1576*). *Aplysina aerophoba* was the first sponge found to contain (+)-aeroplysinin-1 (**1925**) (*1575*, *1584*, *1585*). At about the same time the enantiomer **1926** was found in the sponge *Ianthella ardis* (*1586*, *1587*). Interestingly, a study of Cuban sponges revealed the presence of the racemate **1925/1926** in some specimens of *Aiolochroia crassa* and *Verongula gigantea* and single enantiomers in other specimens (*1576*). The Australian sponge *Aplysina laevis* produces **1925** as well as the novel enones **1927** and **1928** (*1577*) and *A. thiona* also contains **1925** (*1578*). The lactone derivative, (+)-aeroplysinin-2 (**1929**), was isolated as the (+)-enantiomer from *Aplysina aerophoba* and as the racemate **1929/1930** from *Ianthella* sp. (*1588*). On standing, **1929** is converted into **1919**. Aeroplysinin-2 has been described in other sponges (*1576*, *1578*, *1579*, *1581*).

Several aromatic brominated tyrosine-derived compounds are also known from this same group of sponges. The dibromophenylacetamide **1931** was found in the sponge *Aplysina archeri* (*1589*), *A. fistularis* (*1581*); the monobromo **1932** is present in *A. aurea* (*1590*), an unidentified Guamanian sponge (*1512*) and *Psammaplysilla* sp. (*1592*). The related compound **1933** is produced by *Psammoposilla purpurea* (*1593*), *Aplysina thiona* (*1578*) and *A. aerophoba* (*1594*). The latter study also identified ester **1934** in this sponge. The rearranged dibromophenylacetamide **1935** and the monobromo derivative **1936** were isolated from *A. aurea* (*1595*). The novel aplysinadiene (**1937**) is present in *A. aerophoba*, the structure of which was confirmed by synthesis (*1596*). The related aplysinolide (**1938**) and aplysinimine (**1939**) are found in *A. thiona* (*1578*). The interesting set of cavernicolins (**1940–1949**) have been isolated from *A. cavernicola* (*1574*, *1597*, *1598*) and *A. fistularis* (*1518*). The former sponge also contains 7-bromo- (**1950**) (*1599*) and 7-chlorocavernicolenone (**1951**) (*1600*). The biogenesis of the cavernicolins has been discussed, with supporting chemistry (*1601*). Indeed, there is good evidence to support a proposed biosynthesis of many of the aforementioned bromotyrosines from tyrosine (*1602*, *1603*), but, unfortunately, space does not permit a full discussion here. The phenylacetonitrile **1952** is found in *Psammaplysilla* sp. (*1592*, *1604*) as is aldehyde **1953** (*1604*). A Caribbean *Verongula* sp. sponge produces the new ammonium salt **1954** (*2220*), and the tunicate *Didemnum* sp. produces the novel diiodophenethyl amine **1955** and urea **1956** (*1605*).

The novel *bis*-oxazolidone metabolite **1957** has been isolated from both *Aplysina lacunosa* (*1606*) and *A. thiona* (*1578*). The sponge *Ianthella ardis* from the Bahamas produces ianthelline (**1958**) (*1607*); verongamine (**1959**) is a very similar metabolite with histamine H_3-antagonist activity

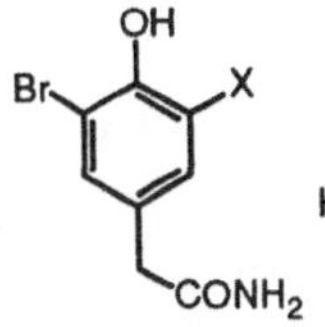

1931 X = Br
1932 X = H

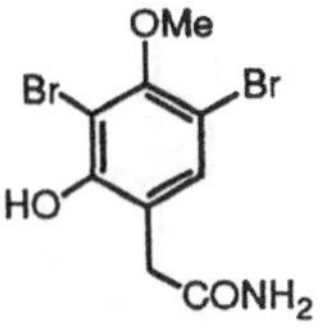

1933

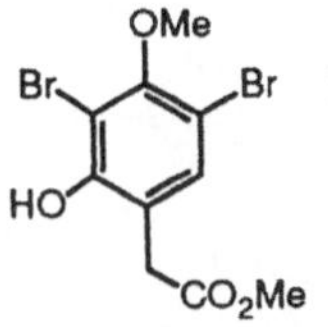

1934

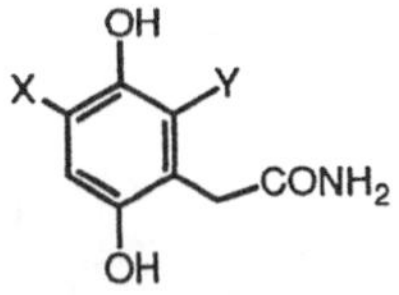

1935 X = Y = Br
1936 X = H, Y = Br
(or X = Br, Y = H)

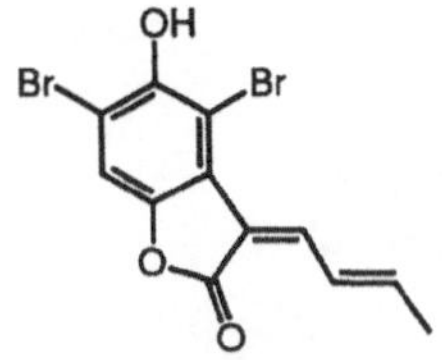

1937
(aplysinadiene)

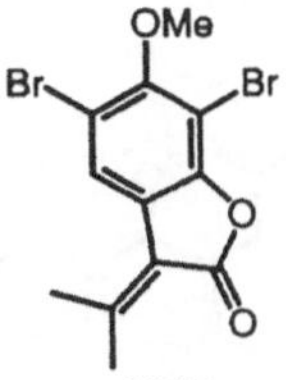

1938
(aplysinolide)

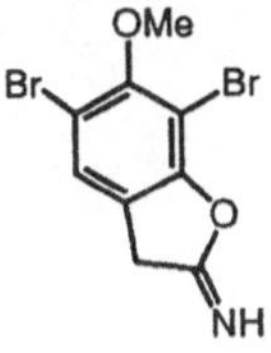

1939
(aplysinimine)

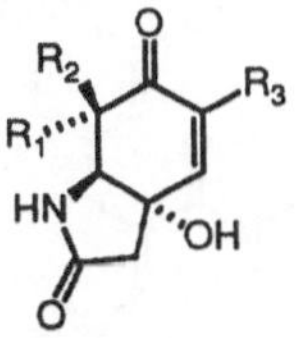

1940 $R_1 = R_2 = H, R_3 = Cl$
1941 $R_1 = H, R_2 = Br, R_3 = Cl$
1942 $R_1 = Br, R_2 = H, R_3 = Cl$
1943 $R_1 = Cl, R_2 = H, R_3 = Br$
1944 $R_1 = H, R_2 = Cl, R_3 = Br$
1945 $R_1 = R_3 = Br, R_2 = H$
1946 $R_1 = H, R_2 = R_3 = Br$
1947 $R_1 = R_2 = H, R_3 = Br$
1948 $R_1 = H, R_2 = R_3 = Cl$
1949 $R_1 = R_3 = Cl, R_2 = H$

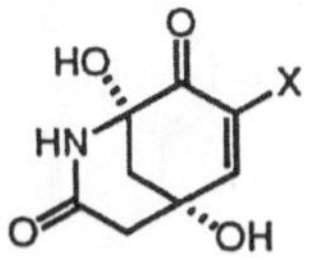

1950 X = Br
1951 X = Cl

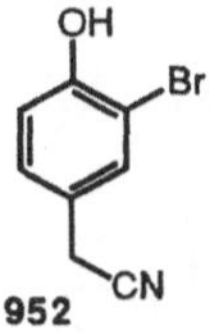

1952

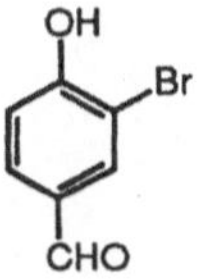

1953

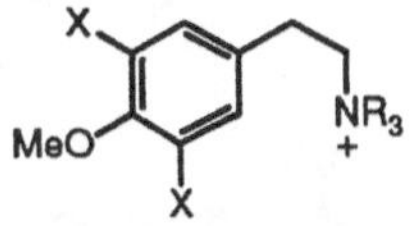

1954 X = Br, R = Me
1955 X = I, R = H

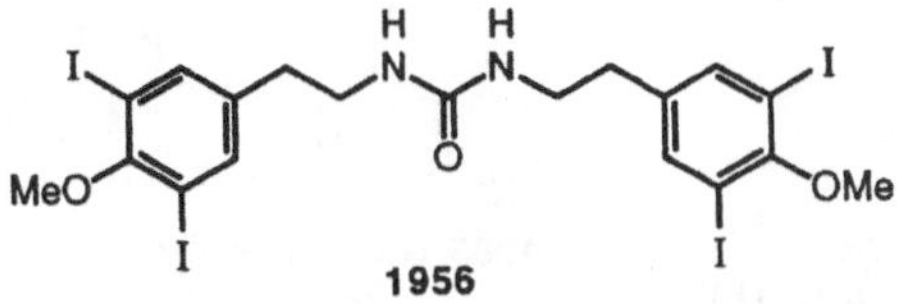

1956

which is found in *Verongula gigantea* (*1608*). The oxime configuration was determined to be *E*. The novel disulfide oxime psammaplin A (**1960**) was isolated nearly simultaneously from two sponges *Psammaplysilla* sp.

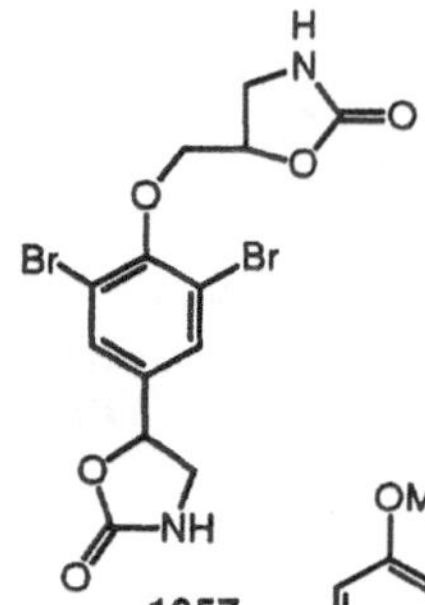

1957

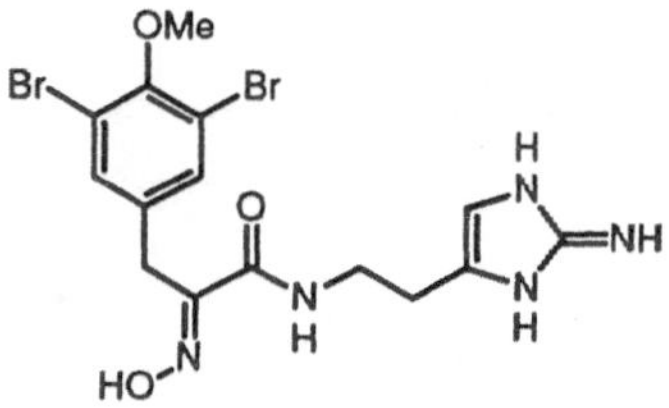

1958 (ianthelline)

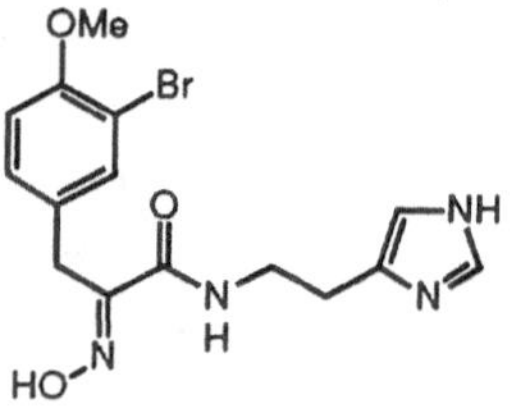

1959 (verongamine)

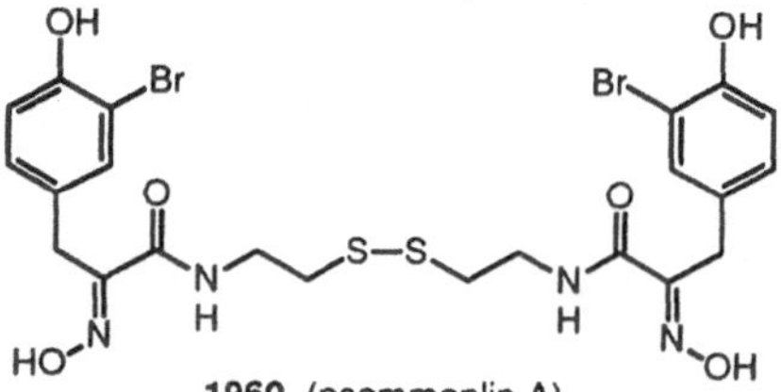

1960 (psammaplin A)

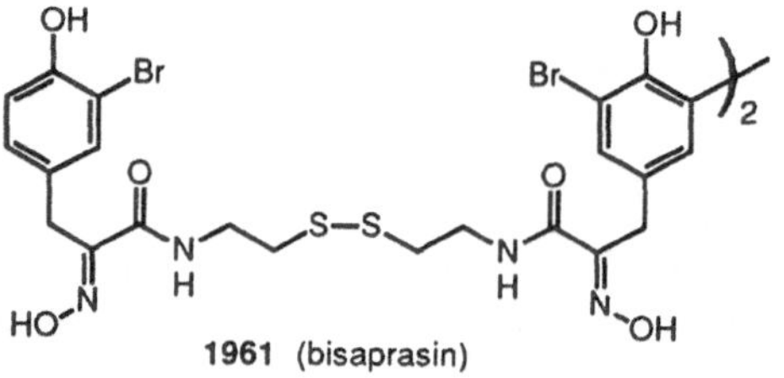

1961 (bisaprasin)

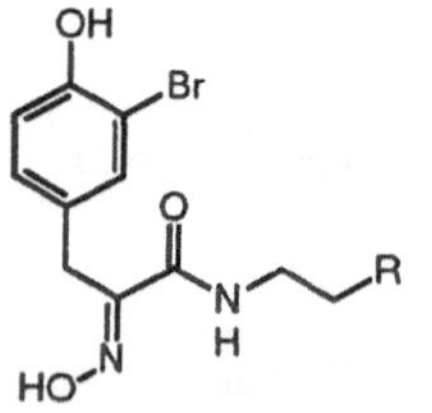

1962 R = SCN (psammaplin B)
1963 R = SO_2NH_2 (psammaplin C)
1964 R = $SSCH_2CH_2CO_2Me$ (psammaplin D)

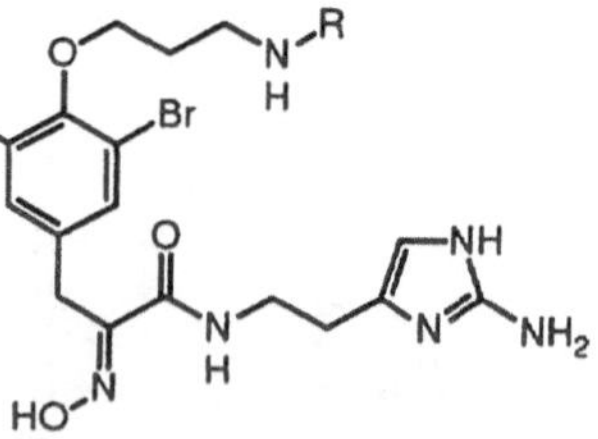

1965 R = $CO(CH_2)_{12}CH_3$ (lipopurealin A)
1966 R = $CO(CH_2)_{11}CHMe_2$ (lipopurealin B)
1967 R = $CO(CH_2)_{14}CH_3$ (lipopurealin C)
1968 R = H (purealidin A)

(*1592*) and *Thorectopsamma xana* (*1609*). The latter study also found bisaprasin (**1961**) and discovered that **1957** and **1958** promote the growth of *Pseudomonas aeruginosa* bacteria. The additional psammaplins B (**1962**), C (**1963**), and D (**1964**) were identified in *Psammaplysilla purpurea* (*1604*). Sponges of this genus have proved to be a rich source of bromotyrosine-derived metabolites. The Okinawan sponge *Psammaplysilla purea* has yielded lipopurealins A (**1965**), B (**1966**) and C (**1967**), compounds that inhibit sodium/potassium-ATPase (*1610*), as well as purealidin A (**1968**) which has antileukemic properties (*1611*). Further study of this sponge discovered purealidins D (**1969**), E (**1970**), F (**1971**), and G (**1972**) which were isolated as salts (*1612, 1613*). Earlier, the related aplysamine-1 (**1973**) was isolated from the Australian sponge *Aplysina* sp.

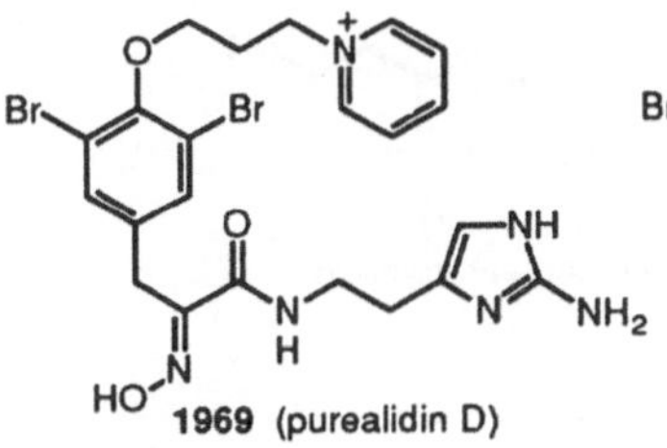

1969 (purealidin D)

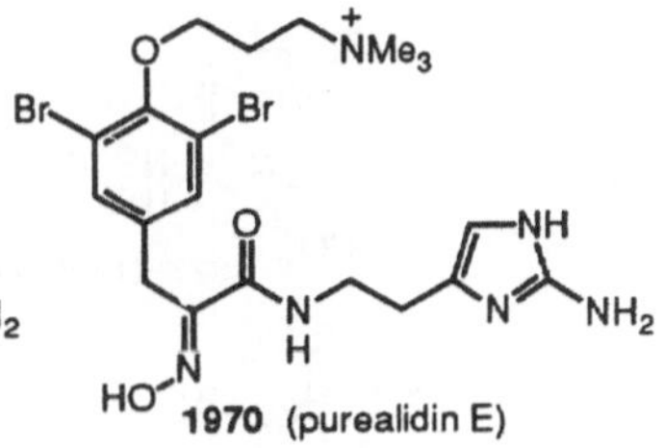

1970 (purealidin E)

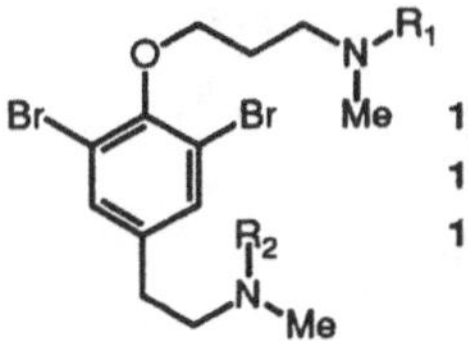

1971 R_1 = H, R_2 = Me (purealidin F)
1972 R_1 = Me, R_2 = H (purealidin G)
1973 R_1 = R_2 = Me (aplysamine-1)

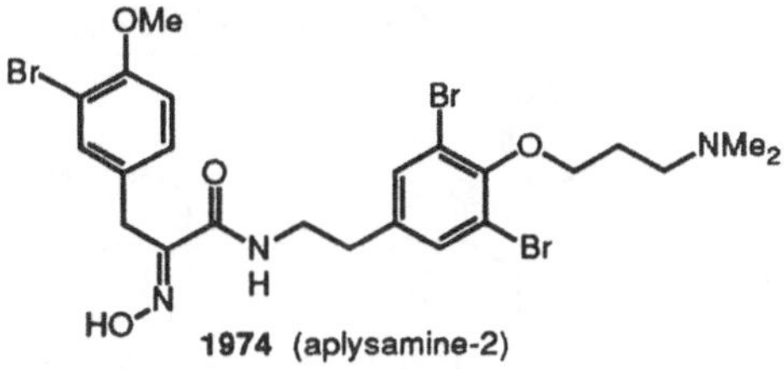

1974 (aplysamine-2)

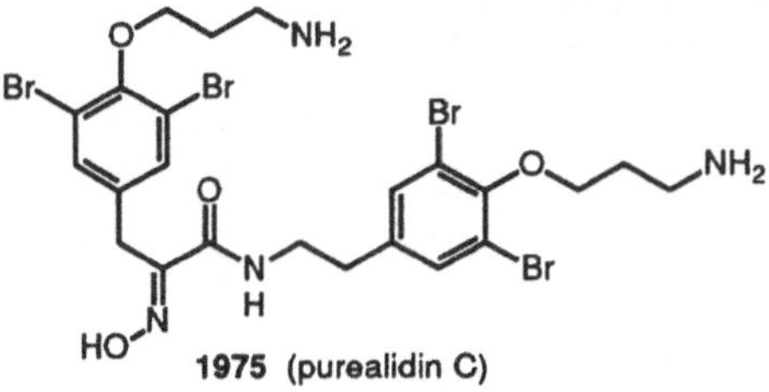

1975 (purealidin C)

as a minor metabolite. The major compound was the *bis*-tyrosine metabolite aplysamine-2 (**1974**) (*1614*). Purealidin C (**1975**), also found in the Okinawan *P. purea*, is a *bis*-dibromo compound (*1615*).

The potent antifungal and antimicrobial anomoian A (**1976**) is produced by the marine sponge *Anomoianthella popeae* (*1616*) and the

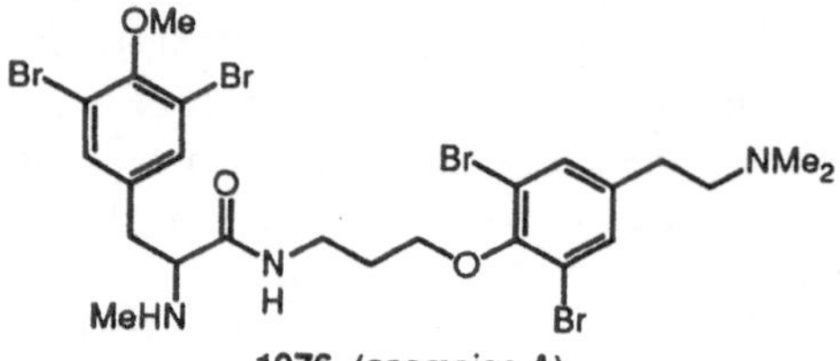

1976 (anomoian A)

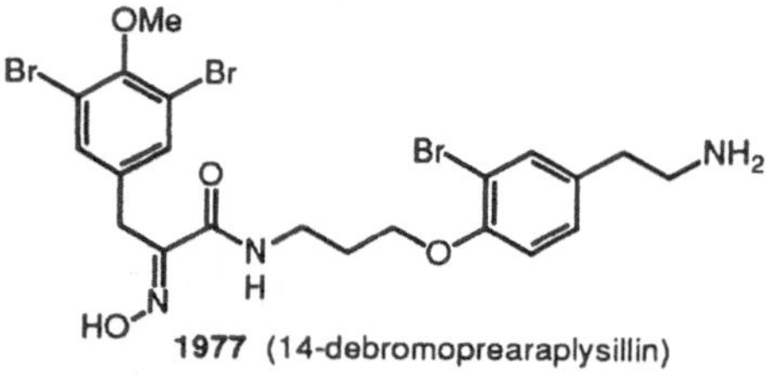

1977 (14-debromoprearaplysillin)

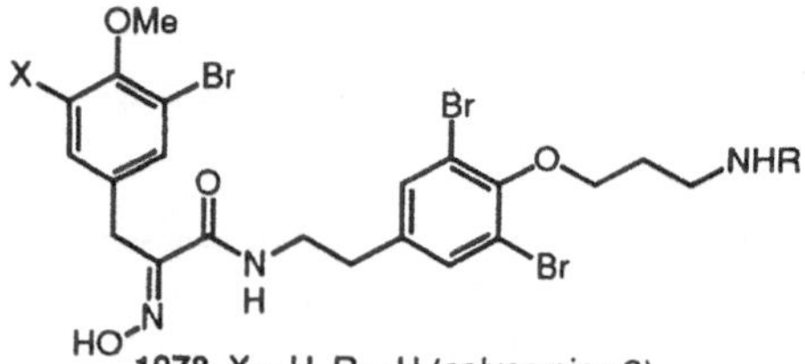

1978 X = H, R = H (aplysamine 3)
1979 X = Br, R = H (aplysamine 4)
1980 X = H, R = $CO(CH_2)_{11}CH(CH_3)_2$ (aplysamine 5)

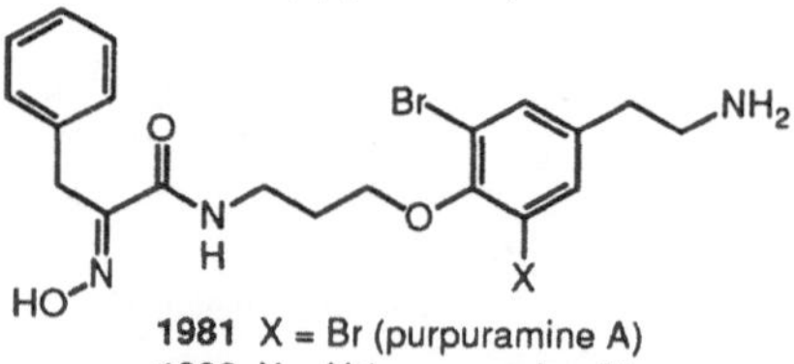

1981 X = Br (purpuramine A)
1982 X = H (purpuramine B)

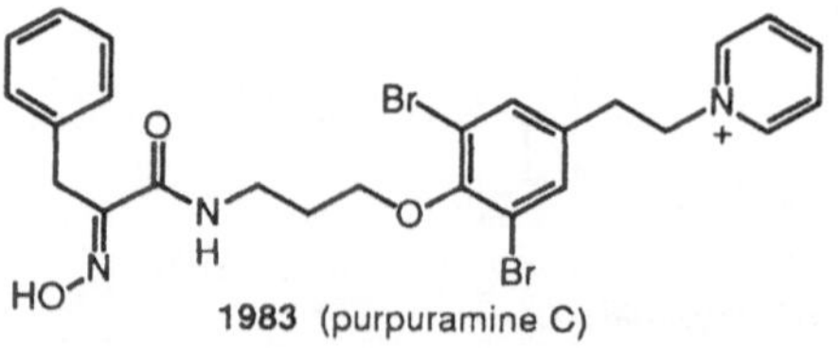

1983 (purpuramine C)

related 14-debromoprearaplysillin (**1977**) is found in *Psammaplysilla purpurea* (*1617*). This compound is the presumed precursor to 14-debromoaraplysillin I (**2020**) to be discussed in the next section and a metabolite also found in this sponge (*1617*). Another collection of *P. purpurea* from Maui has yielded aplysamines-3 (**1978**), -4 (**1979**), and -5 (**1980**) (*1619*). Aplysamine-3 (**1978**) (= purpuramine H) was also found in a Japanese *P. purpurea* together with eight new bromotyrosine-metabolites, purpuramines A (**1981**), B (**1982**), C (**1983**), D (**1984**), E (**1985**), F (**1986**), G (**1987**), and I (**1988**) (*1620*). The *E*-oxime configuration is observed in all of these metabolites, and in fact, this is usually the case in tyrosine oxime metabolites. A version of this sponge from the Bay of Bengal contains in addition to aplysamine-3 and -4 the two new compounds **1989** and **1990** (*1621*). The Caribbean sponge *Pseudoceratina crassa* produces the novel cinnamic acid derivatives **1991** and **1992**, the structures of which were confirmed by synthesis (*1622*).

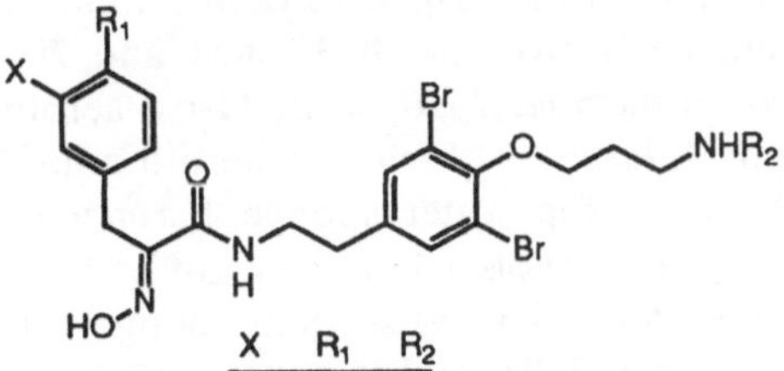

	X	R_1	R_2	
1984	H	H	H	(purpuramine D)
1985	H	H	Me	(purpuramine E)
1986	Br	OH	H	(purpuramine F)
1987	Br	OH	Me	(purpuramine G)
1988	Br	OMe	Me	(purpuramine I)

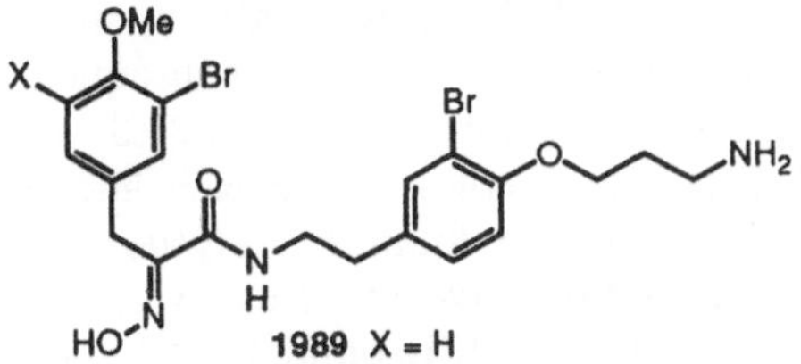

1989 X = H
1990 X = Br

1991 R = H
1992 R = Et

3.22.3.3. Transformed Multiple Tyrosines

A significant number of natural brominated tyrosines contain a novel spirocyclohexadiene isoxazoline ring and would appear to be derived biosynthetically from their aromatic counterparts discussed in the preceding section.

The first such examples are aerothionin (**1993**) and homoaerothionin (**1994**) which were isolated from *Aplysina aerophoba* and *A. thiona* (*1623, 1624*). They were shown to have the absolute configurations indicated from a study of aerothionin from *A. fistularis* (*1625*). Another examination of *A. fistularis* forma *fulva* discovered the existence of fistularin-1 (**1995**), -2(**1996**), and -3 (**1997**), as well as aerothionin (*1626*). An isomer of fistularin-3, isofistularin-3 (**1998**), was found in *A. aerophoba* along with aerophobin-1 (**1999**) and aerophobin-2 (**2000**) (*1627*). Fistularin-3 was also isolated from *A. archeri* together with the new 11-ketofistularin-3 (**2001**) (*1628*). The sponge *Pseudoceratina durissima* produces the known aerothionin and homoaerothionin as well as the new compounds 11,19-dideoxyfistularin-3 (**2002**) and 11-hydroxyaerothionin (**2003**) (*1629*). The two isomeric keto derivatives of **2003**, **2004** and **2005**, are found in *Aplysina fistularis* forma *fulva* (*1630*) while 11-oxoaerothionin (**2006**) has been isolated from *A. lacunosa* along with aerothionin, **2003**, fistularin-3 and **2002** (*1631*). The deep water sponge *Verongula rigida* from the Bahamas contains aerophobin-1 (**1999**) and the new dihydroxyaerothionin (**2007**) (*1632*). The Australian sponge *Agelas oroides* produces 11-*epi*-fistularin-3 (**2008**) and the novel enones agelorins A (**2009**) and B (**2010**) (*1633*).

The Red Sea sponge *Psammaplysilla purpurea* was originally thought to contain the novel spirocyclohexadienyloxazolines psammaplysin-A and -B (*1634*) but in fact these metabolites were later shown to have the oxepin structures **2011** and **2012**, respectively, in samples found in a Palau specimen of this sponge (*1635*). A subsequent investigation of *P. purpurea* also revealed psammaplysin-C (**2013**) which is cytotoxic against the HCT 116 human colon cancer cell line (*1636*). Psammaplysins D (**2014**) and E (**2015**) are found in *Aplysinella* sp., a Micronesian sponge (*1637*). The former is active against HIV. The sponge *Hexadella* sp. has the new hexadellins A (**2016**) and B (**2017**) (*1638*). The former metabolite was also isolated (as araplysillin I) from *Psammaplysilla arabica* together with the new araplysillin II (**2018**) (*1639*). Purealidin B (**2019**), the trimethylammonium derivative of hexadellin A, is found in *P. purea* (*1615*) and 14-debromohexadellin A (14-debromoaraplysillin) (**2020**) was also isolated from this sponge (*1617*). The sponge *Aplysina cauliformis* produces the three novel aplysinamisines I (**2021**), II (**2022**),

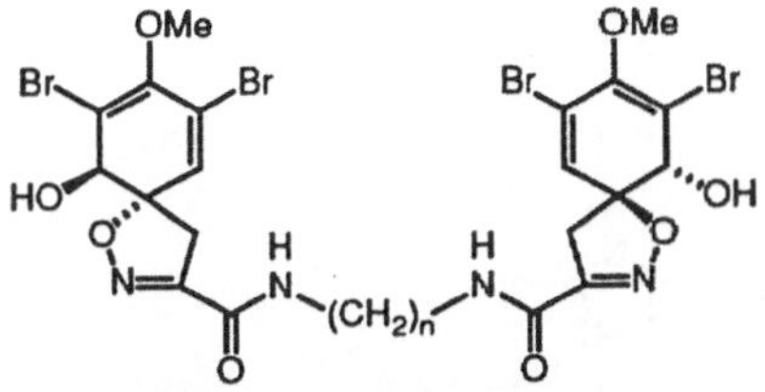

1993 n = 4 (aerothionin)
1994 n = 5 (homoaerothionin)

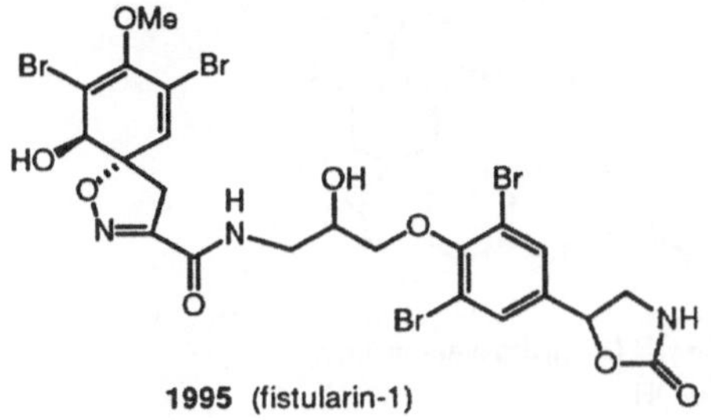

1995 (fistularin-1)

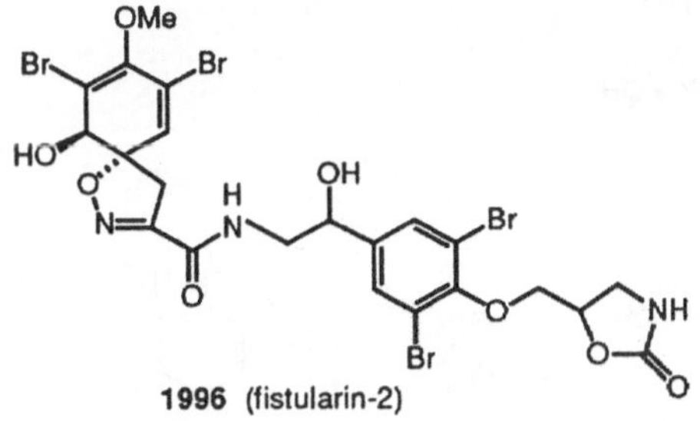

1996 (fistularin-2)

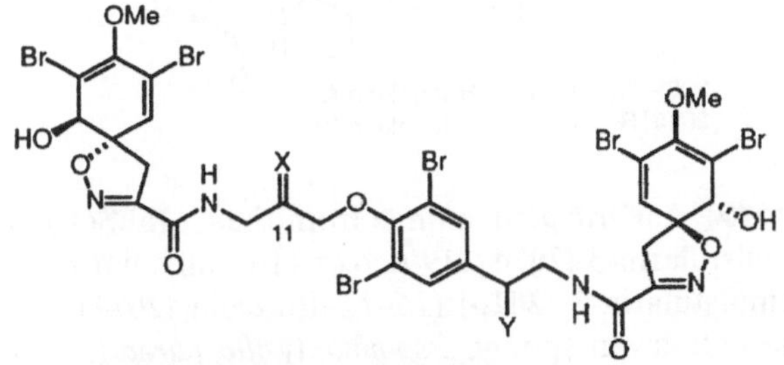

1997 X = H,OH, Y = OH (fistularin-3)
1998 X = H,OH, Y = OH (isofistularin-3)
2001 X = O, Y = OH (11-ketofistularin-3)
2002 X = H,H, Y = H
2008 X = H,OH, Y = OH (11-*epi*-fistularin-3)

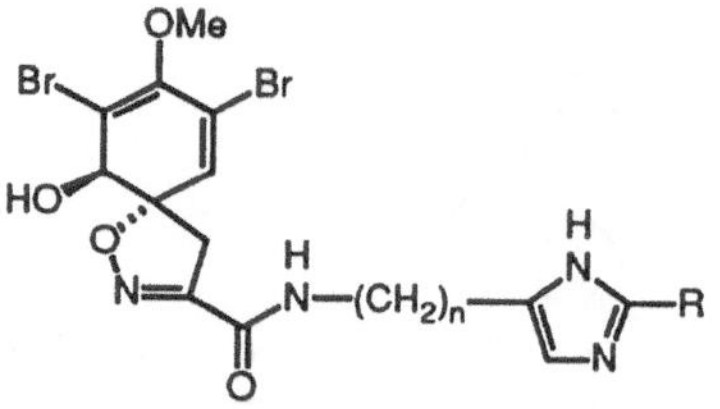

1999 n = 2, R = H (aerophobin-1)
2000 n = 3, R = NH_2 (aerophobin-2)

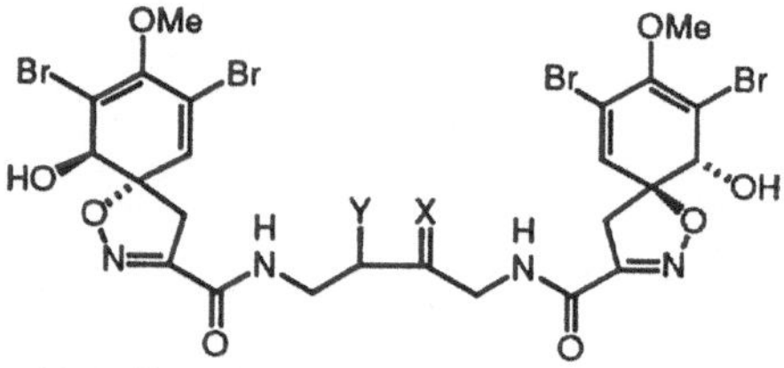

2003 X = H,H, Y = OH (11-hydroxyaerothionin)
2004 X = O, Y = α-OH
2005 X = O, Y = β-OH
2006 X = O, Y = H
2007 X = H,OH, Y = OH

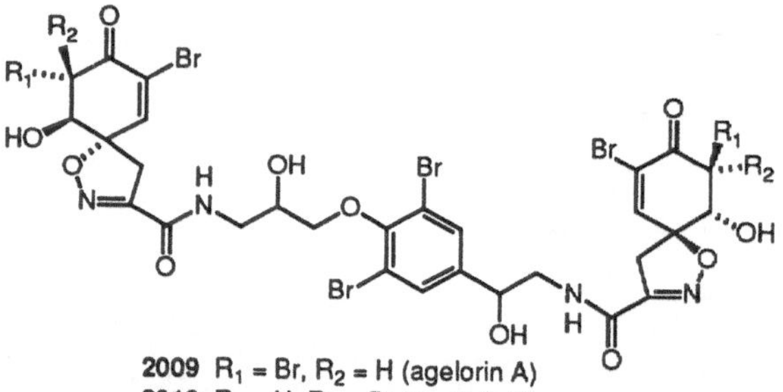

2009 R_1 = Br, R_2 = H (agelorin A)
2010 R_1 = H, R_2 = Br (agelorin B)

and III (**2023**) (*1640*). A *Verongida* sponge from the Coral Sea contains the new 19-deoxyfistularin-3 (**2024**), 19-deoxy-11-oxofistularin-3 (**2025**) and the new hemifistularin-3 (**2026**) (*1641*). Purealin (**2027**) has been isolated from the Okinawan sponge *Psammaplysilla purea* (*1642*). This metabolite modulates enzymatic reactions of ATPase and has been extensively studied (*1642–1646*).

3.22.3.4. Bastadins

The Queensland sponge *Ianthella basta* produces a series of tetrameric brominated tyrosine metabolites known as bastadins. The first two

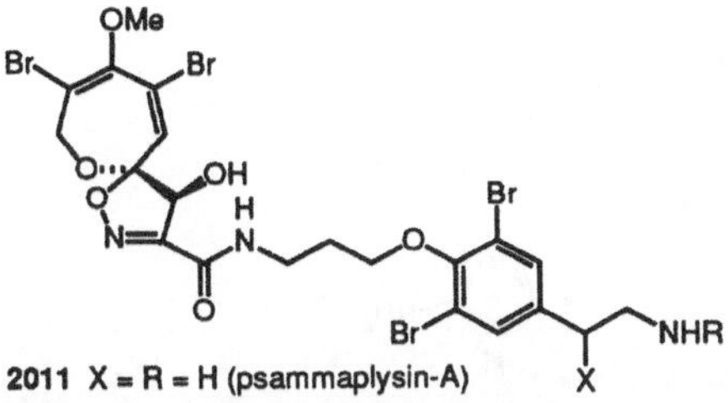

2011 X = R = H (psammaplysin-A)
2012 X = OH, R = H (psammaplysin-B)
2013 X = OH, R = Me (psammaplysin-C)
2014 X = OH, R = $CO(CH_2)_{11}CH(CH_3)_2$ (psammaplysin-D)
2015 X = H, R = CH= (psammaplysin-E)

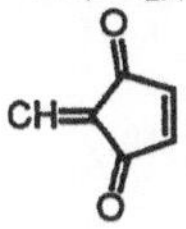

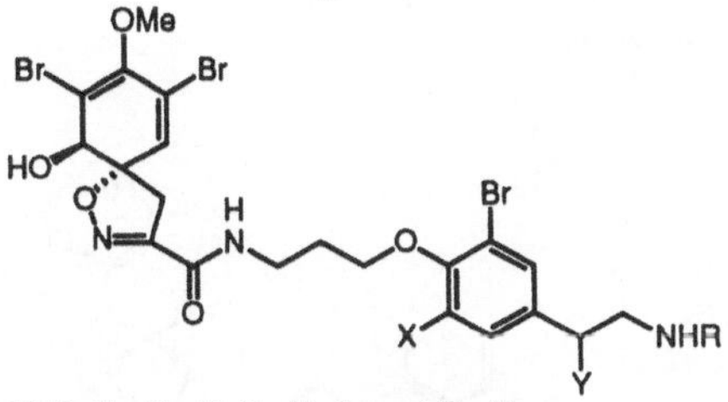

2016 R = Y = H, X = Br (hexadellin A)
2018 R = $CO(CH_2)_{11}CH(CH_3)_2$, X = Br, Y = H (araplysillin II)
2020 R = X = Y = H
2023 R = Ac, X = Br, Y = OH (aplysinamisine III)

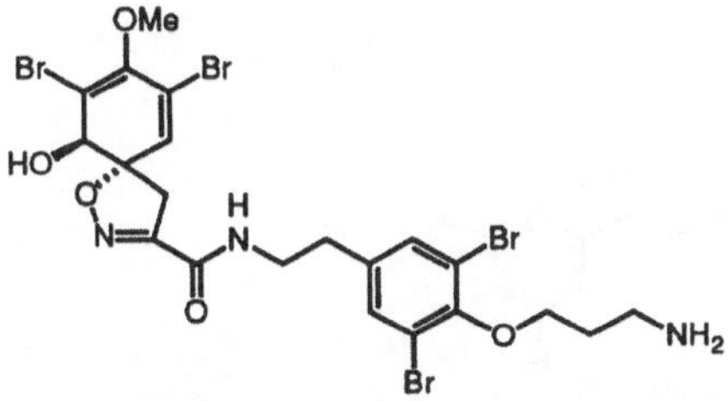

2017 (hexadellin B)

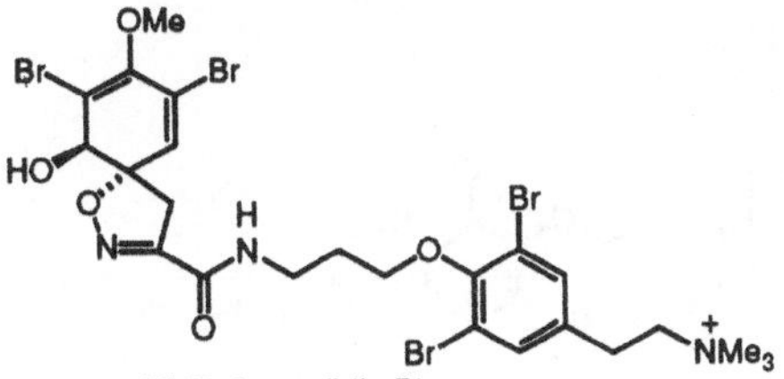

2019 (purealidin B)

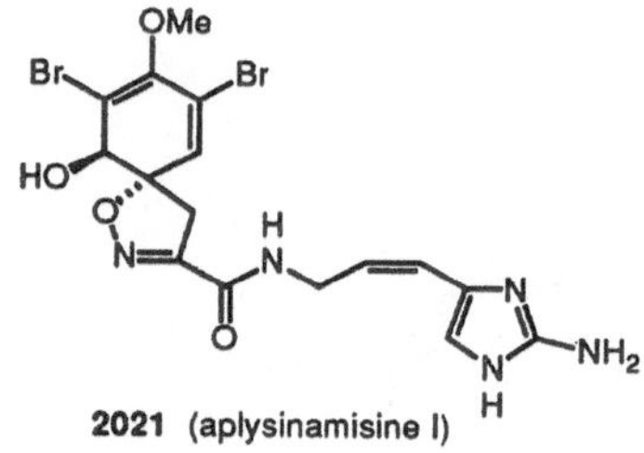

2021 (aplysinamisine I)

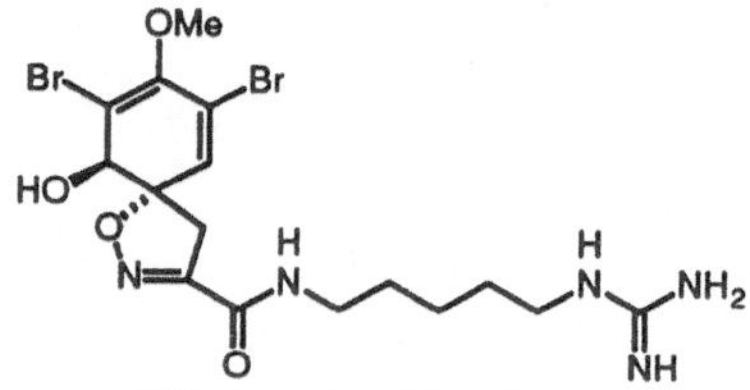

2022 (aplysinamisine II)

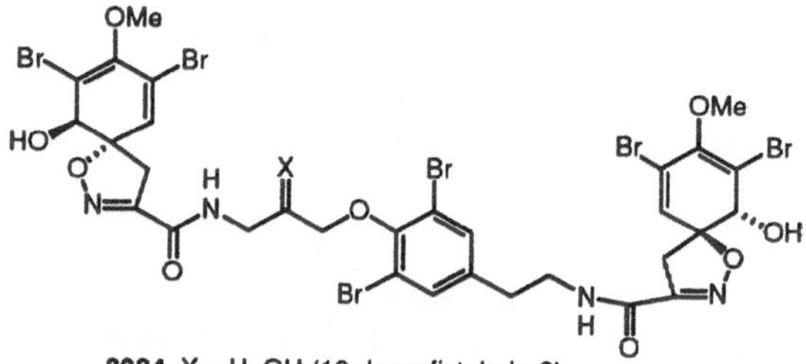

2024 X = H, OH (19-deoxyfistularin-3)
2025 X = O

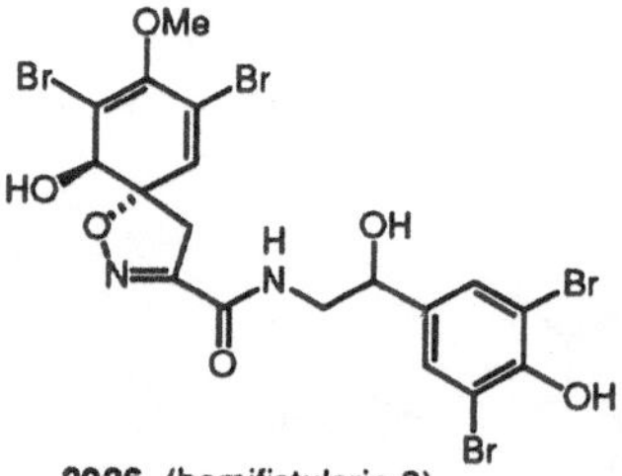

2026 (hemifistularin-3)

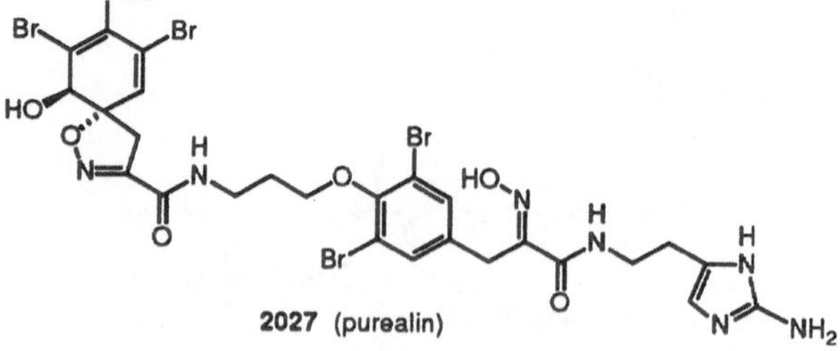

2027 (purealin)

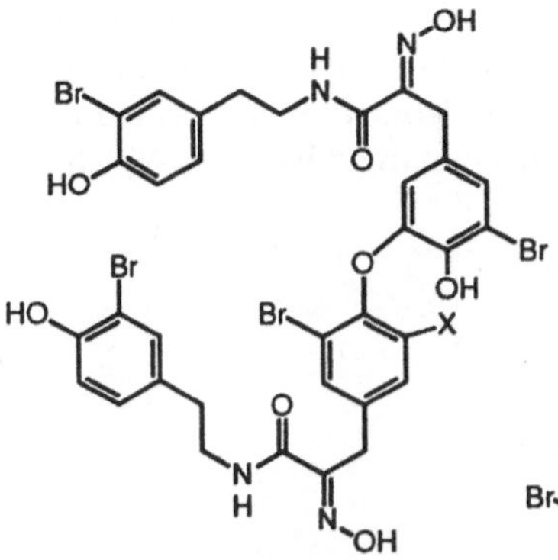

2028 X = H (bastadin-1)
2029 X = Br (bastadin-2)

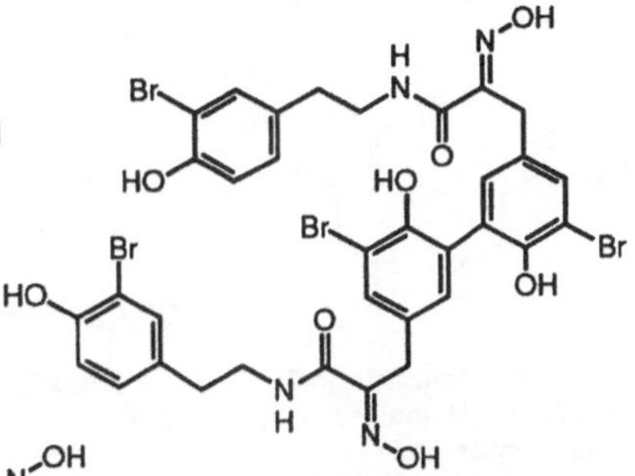

2030 (bastadin-3)

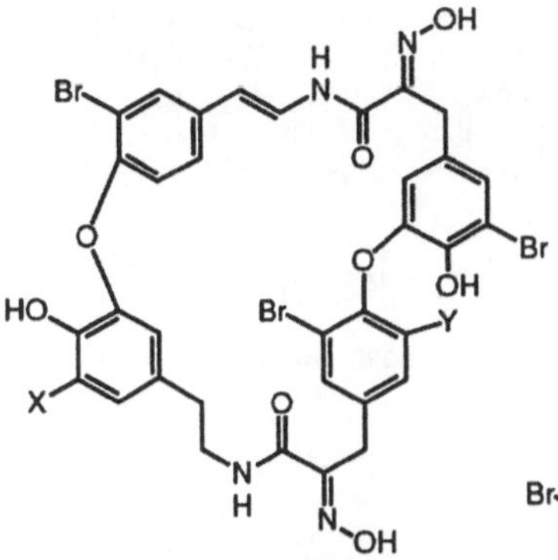

2031 X = Y = Br (bastadin-4)
2034 X = Br, Y = H (bastadin-7)
2038 X = H, Y = Br (bastadin-11)

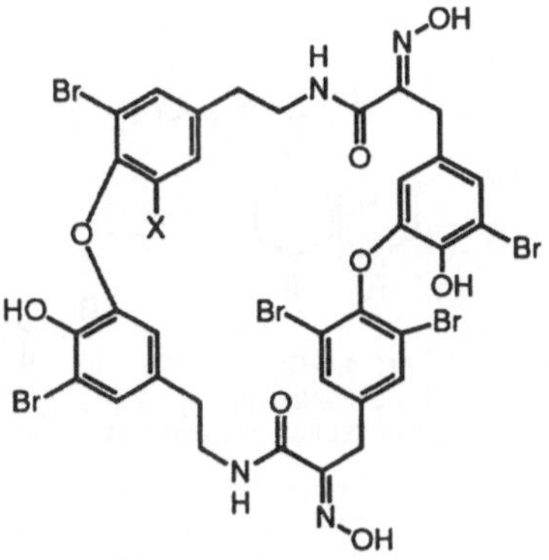

2032 X = H (bastadin-5)
2033 X = Br (bastadin-6)

were bastadin-1 (**2028**) and -2 (**2029**) and these were soon followed by bastadins 3–7 (**2030–2034**) (*1647, 1648*). Total syntheses of bastadins 1–3 support the fact that the *E*-oxime geometry generally obtains in these metabolites (*1649*). A collection of *I. basta* from Papua revealed the new bastadins-8 (**2035**) and -12 (**2036**) (formerly named bastadin-9) which are

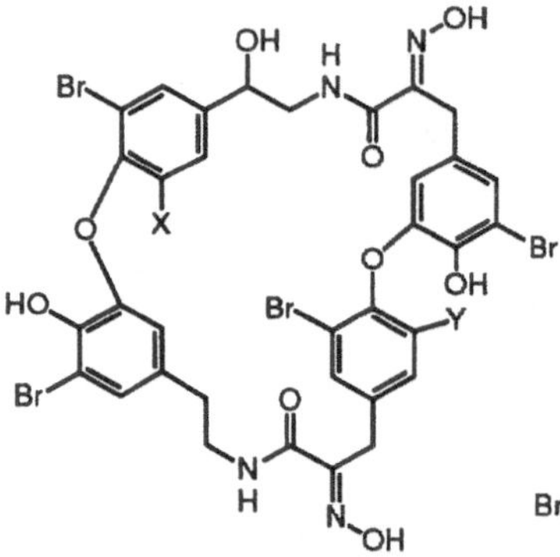

2035 X = H, Y = Br (bastadin-8)
2036 X = Br, Y = H (bastadin-12)
(formerly bastadin-9)
2037 X = Y = H (bastadin-10)

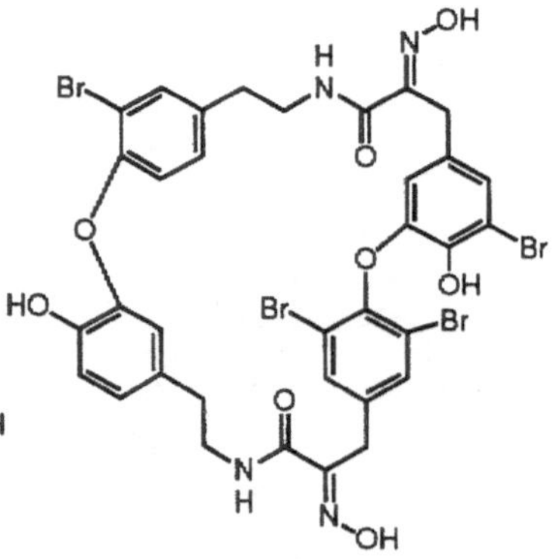

2039 (bastadin-9)

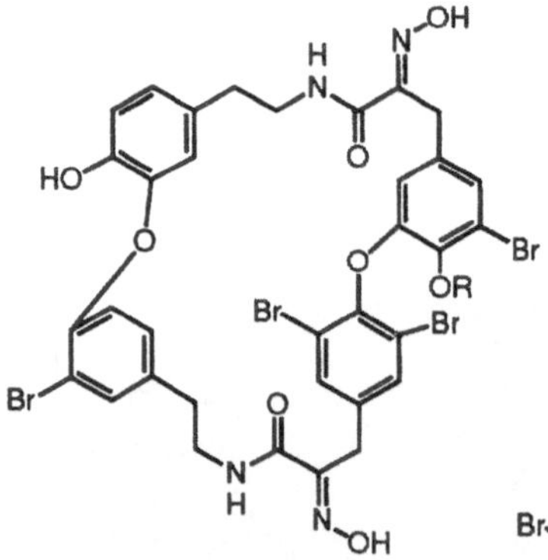

2040 R = H (bastadin-13)
(formerly bastadin-12)
2045 R = SO_3Na (sulfatobastadin-13)

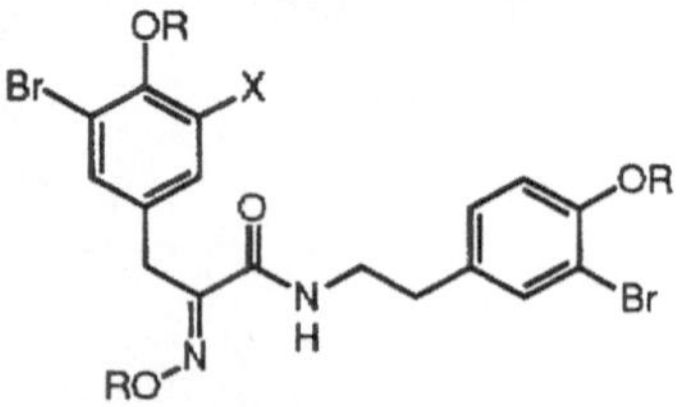

2041 X = R = H (hemibastadin-1)
2042 X = Br, R = H (hemibastadin-2)
2041a X = H, R = Me
2042b X = Br, R = Me

hydroxylated at C-6, in addition to bastadins 4–7 (*1650*). Discovered in a Guamanian collection of this sponge was the presence of bastadins-10 (**2037**) and -11 (**2038**), and -9 (**2039**) (*1651*). Bastadin-13 (**2040**) (formerly named bastadin-12) was isolated from *I. basta* along with the novel hemibastadins-1 (**2041**) and -2 (**2042**) which were isolated as the methyl ethers (*1652*). The Pohnpeian sponge *Psammaplysilla purpurea* has yielded bastadin-14 (**2043**) (*1654*), while bastadin-15 (**2044**) was identified in *Ianthella* sp. together with several other bastadins (*1655*). This sponge also contains 34-sulfatobastadin-13 (**2045**) which is an inhibitor of the

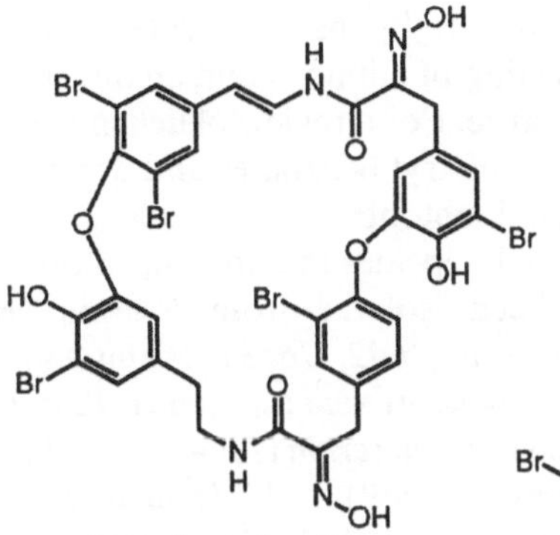

2043 (bastadin-14)

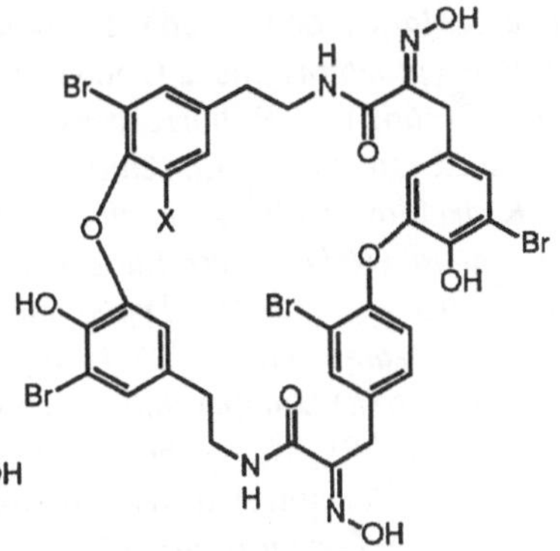

2044 X = Br (bastadin-15)
2048 X = H (bastadin-18)

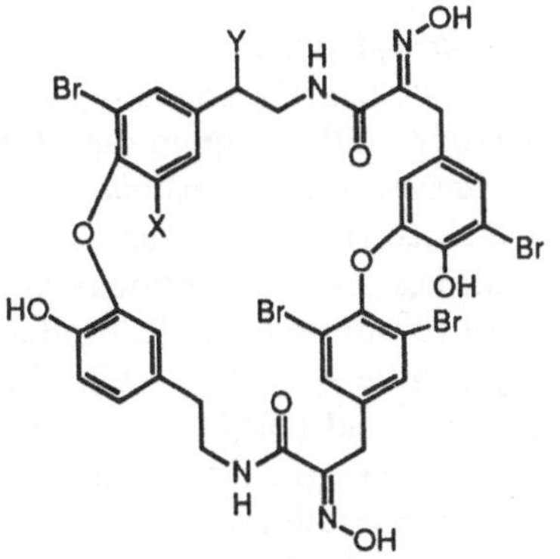

2046 X = Br, Y = H (bastadin-16)
2047 X = H, Y = OH (bastadin-17)

endothelin A receptor (*1656*). The Indonesian *Ianthella basta* produces bastadins-16 (**2046**) and -17 (**2047**); the oxime geometry was established as *E* by ^{13}C NMR chemical shifts (*1657*). Bastadin-18 (**2048**) was also found in a different collection of *I. basta* along with bastadins-1,-2,-5,-6,-8, and -10, and the sponge *I. flabelliformis* yielded bastadins-2 and -5 (*2221*).

3.22.4. Depsides

Depsides and depsidones (next section) are biogenetically related phenolic compounds of polyketide origin that are widely found in lichens. These are symbionts consisting of a fungus and an alga living and growing together in a united structure. For a review of lichen metabolites see (*1658*). Depsides are derivatives of aryl benzoates and depsidones are the corresponding seven-membered lactones.

Although only a few chlorinated depsides are known, chloratranorin (**2049**) is ubiquitous and has been isolated from *Evernia prunastri* (*1659–1661*), *Pseudevernia furfuracea* (*1662, 1663*), *P. intensa* (*1664*), *Parmelia perlata* (*1665, 1666*), *P. pseudoreticulata* (*1666*), *P. olivetorum* (*1667*), *P. cryptochlorophaea* (*1668*), *P. tinctorum* (*1664, 1675*), *P. pseudofatiscens* (*1669*), *P. horrescens* (*1669*), *P. damaziana* (*1670*), *P. furfuracea* (*1671*), *Buellia canescens* (*1672, 1673*), *Ramalina siliquosa* (*1674*), *R. druidarum* (*1664*), *Lecidea carpathica* (*1676*), *Lecidea* sp. (*1703*), *Physicia picta* (*1677*), *Cetrelia cetrarioides* (*1664*), *C. japonica* (*1664*), *Usnea canariensis* (*1664*), *Hypogymnia physodes* (*1664*), *H. billardieri* (*1678*), *H. enteromorpha* (*1678*), *H. lugubris* (1678), *H. mundata* (*1678*), *H. subphysodes* (*1678*), *Anaptychia neoleucomelaena* (*1664, 2222*), *Menegazzia asahinae* (*1679*), *M. terebrara* (*1679*), *M. dispora* (*1680*), *Platismatia glauca* (*1681*), *Parmotrema demethylmicrophyllinicum* (*1682*), *P. praesorediosum* (*1683*), *Lecanora broccha* (*1684*), *L. epibryon* (*1684*), *L. rupicola* (*1664, 2223*), and *L. gangaleoides* (*1664, 1701*). Tumidulin (**2050**) was isolated from *Ramalina ceruchis*, *R. flaccescens*, *R. tumidula*, *R. inanis*, *R. cactacearum*, *R. peruviana*, and *R. chilensis* (*1664, 1685–1688, 2222*). Since the discovery of these two early examples several additional chlorodepsides have been identified. The lichen *Lecanora sulphurella* produces **2051** whose structure was confirmed by synthesis (*1689*), while *Thelomma mammosum* contains 3-chlorodivaricatic acid (**2052**) (*1690*). *Erioderma wrightii* has yielded the novel wrightiin (**2053**) (methyl 3-chloroevernate) whose structure was confirmed by X-ray crystallography (*1691*). Another study of *Erioderma* lichens has uncovered seven new depsides (**2054–2060**), all of which were confirmed by synthesis (*1692, 1693*). The

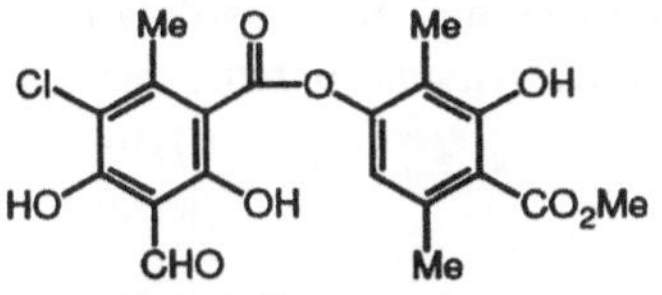

Me O Me
Cl O OH
HO OH CO$_2$Me
CHO Me
2049 (chloratranorin)

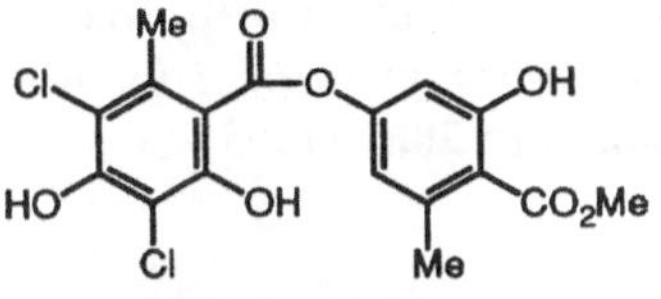

Me O
Cl O OH
HO OH CO$_2$Me
Cl Me
2050 (tumidulin)

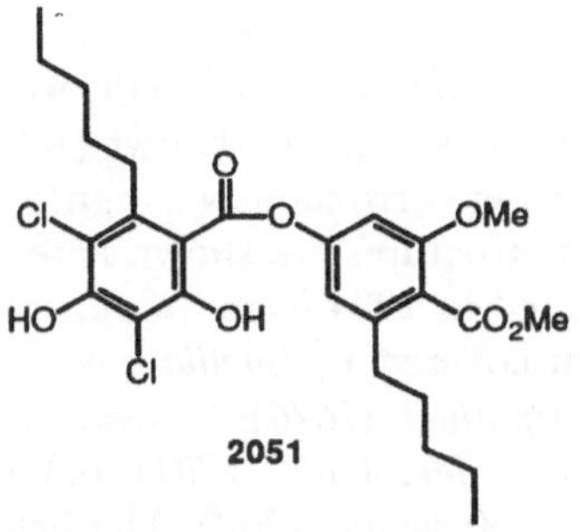

O
Cl O OMe
HO OH CO$_2$Me
Cl
2051

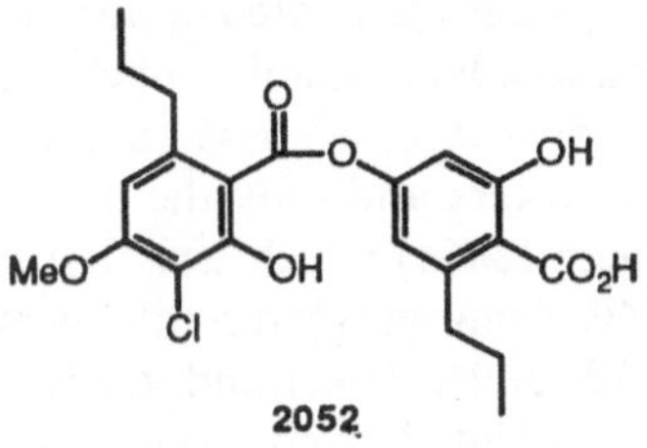

O
O OH
MeO OH CO$_2$H
Cl
2052

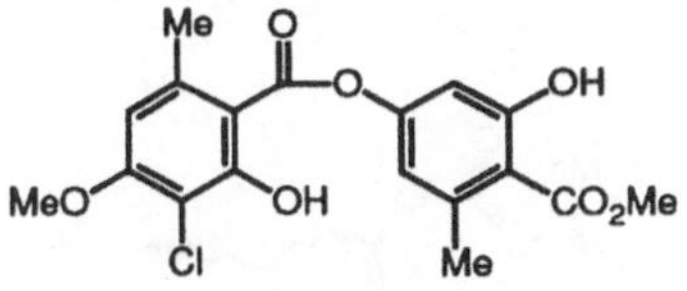

Me O
O OH
MeO OH CO$_2$Me
Cl Me
2053 (wrightiin)

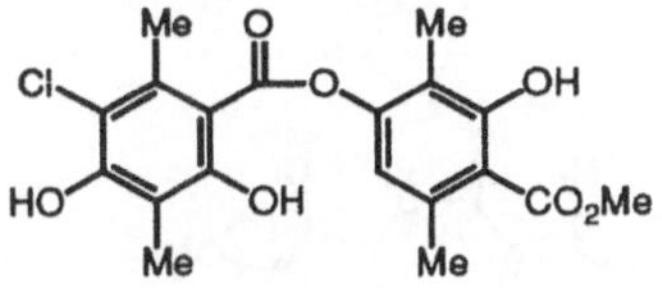

Me O Me
Cl O OH
HO OH CO$_2$Me
Me Me
2054

Me O Me
Cl O OR$_3$
R$_2$O OR$_1$ Me CO$_2$Me
Me Me

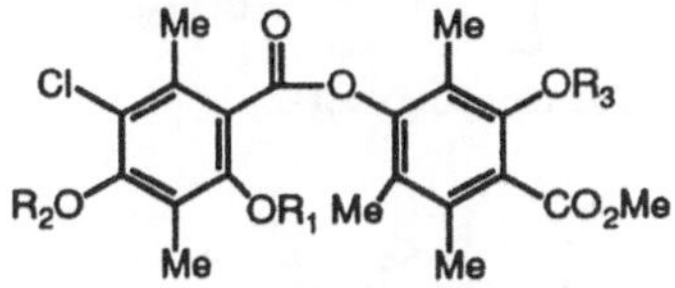

	R$_1$	R$_2$	R$_3$
2055	H	H	H
2056	H	H	Me
2057	Me	H	H
2058	H	Me	H
2059	Me	H	Me

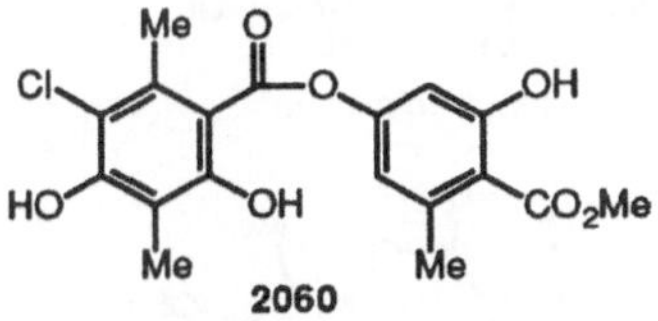

Me O
Cl O OH
HO OH CO$_2$Me
Me Me
2060

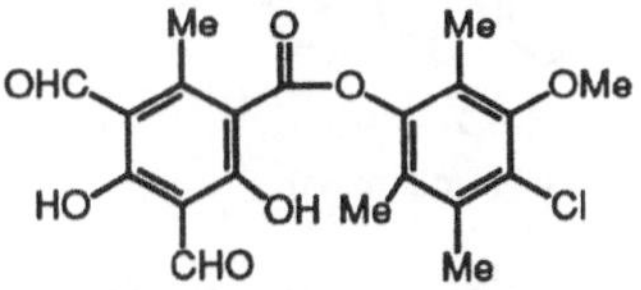

Me O Me
OHC O OMe
HO OH Me Cl
CHO Me
2061 (1'-chloronephroarctin)

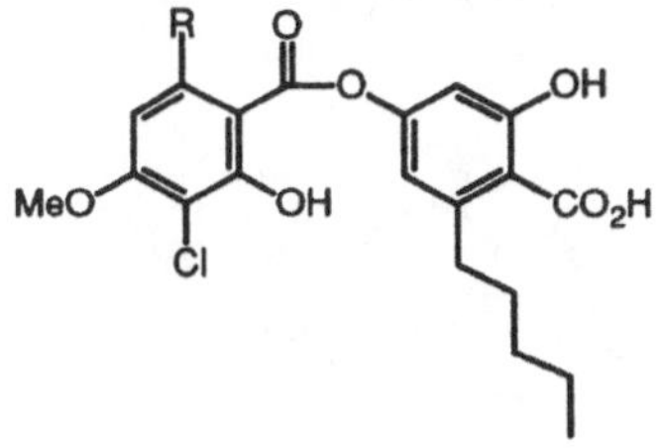

R O
O OH
MeO OH CO$_2$H
Cl

2062 R = n-C_3H_7 (3-chlorostenosporic acid)
2063 R = n-C_5H_{11} (3-chloroperlatolic acid)

new metabolite 1′-chloronephroarctin (**2061**) has been isolated from *Pseudocyphellaria pickeringii* (*1694*) and *Dimelaena* lichens have furnished **2062** and **2063** (*1695*).

3.22.5. Depsidones

Like depsides the related depsidones seem to be confined to the world of lichens and a large number of chlorinated examples are known (*1658*). Structures of several of these compounds were erroneously assigned by the early workers and only the corrected structures are shown here.

Diploicin (**2064**) which was first isolated in 1904 from an unnamed lichen (*1696*) has more recently been discovered in *Buellia canescens* (*1672*, *1673*, *1697*, *1698*) and *Lecidea carpathica* (*1676*). Gangaleoidin (**2065**) was isolated from *Lecanora gangaleoides* (*1699–1701*), but the correct structure was not determined for 30 years (*1702*). The lichen

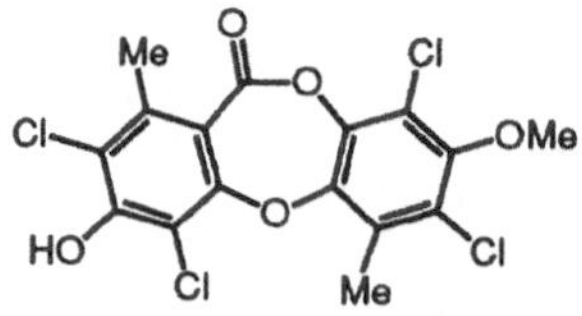

2064 (diploicin)

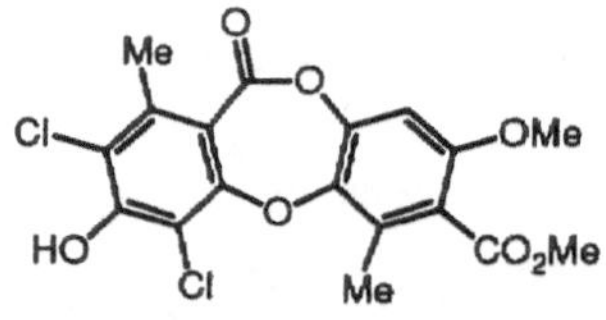

2065 (gangaleoidin)

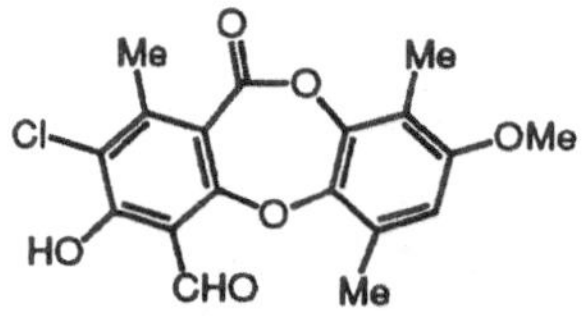

2066 (pannarin)

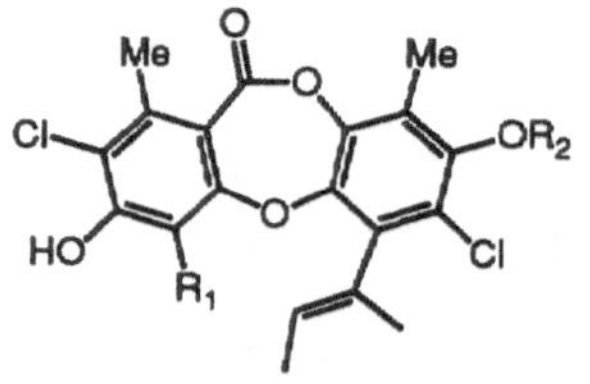

2067 R_1 = Cl, R_2 = Me (nidulin)
2068 R_1 = Cl, R_2 = H (nornidulin)
2069 R_1 = R_2 = H (dechloronornidulin)

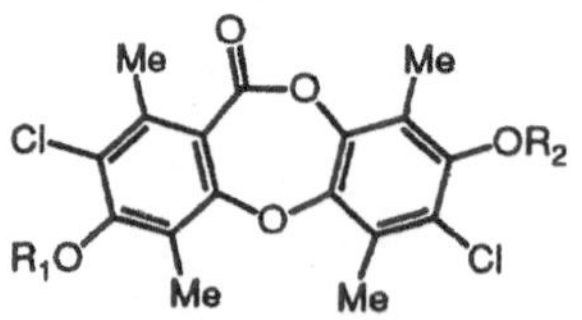

2070 R_1 = H, R_2 = Me (vicanicin)
2071 R_1 = R_2 = H (norvicanicin)
2072 R_1 = Me, R_2 = H (isovicanicin)
2073 R_1 = R_2 = Me (*O*-methylvicanicin)

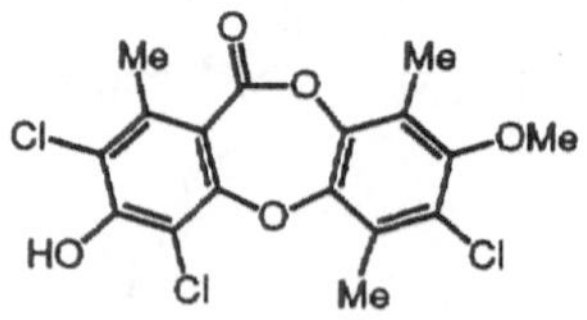

2074 (caloploicin)

pannarin (**2066**), first isolated from *Pannaria lanuginosa*, *P. fulvescens*, and *P. lurida* (*1704*), was later found in *P. pityrea*, *P. rubiginosa*, *Lecanora hercynica*, and *Bombyliospora japonica* (*1664*, *1705*, *2222*) and *Erioderma chilense* (*1706*). The structure of pannarin (**2066**) was revised to that shown by two groups (*1707*, *1708*). Nidulin (**2067**), nornidulin (**2068**) and dechloronornidulin (**2069**) were all originally isolated from *Aspergillus nidulans* (*1709–1711*) and *A. ustus* (*1712*, *2224*), *A. unguis* (*1713*) and nidulin from *Emericella unguis* (*1714*, *1715*). Much work was devoted to solving the structures of these metabolites (*1716–1719*) and the biosynthesis of nidulin has been probed (*1712*, *1720*). Another early lichen depsidone to be discovered is vicanicin (**2070**), first isolated from *Teloschistes flavicans* (*1721*, *1722*) and later from *Caloplaca* sp. (*1723*, *1724*), *Psoroma sphinctrinum* (*1725*), *P. allorhizum* (*1726*), *Erioderma chilense* (*1727*), *E. phaeorhizum* and *E. tomentosum* (*1728*). These latter lichen studies have also yielded norvicanicin (**2071**) (*1725*, *1728*), isovicanicin (**2072**) (*1728*); the latter compound is also present in *Psoroma athrophyllum* (*1726*). The *O*-methylvicanicin (**2073**) is present in *Erioderma* sp. (*1728*, *1729*). *Caloplaca* sp. yielded caloploicin (**2074**) (*1723*, *1724*) the structure of which was confirmed by synthesis (*1730*).

In the years following these early studies numerous new chlorinated depsidones have been discovered. The fungus *Aspergillus unguis* also contains haiderin (**2075**), rubinin (**2076**), shirin (**2077**) and nasrin (**2078**) (*1731*). The related 2-chlorounguinol (**2079**) and emeguisins A–C (**2080–2082**) are found in *Emericella unguis* (*1714*, *1715*). The lichen *Agropsis megalospora* produces argopsin (**2083**) the structure of which was established by Clemmensen reduction to vicanicin (**2070**) and synthesis from pannarin (**2066**) by chlorination (*1732*). This metabolite has also been isolated from *A. friesiana* (= *A. megalospora*) (*1708*), *Erioderma chilense* (*1706*, *1727*), *E. phaeorhizum* and *E. tomentosum* (*1728*). The related eriodermin (**2084**) is found in *E. physcioides* (*1733*) and *E. phaeorhizum* (*1728*) and *E. chilense* produces norpannarin (**2085**) and norargopsin (**2086**) (*1706*). Physciosporin (**2087**) was discovered by three groups in *Pseudocyphellaria physciospora* (*1734*, *1735*), *P. granulata* (*1735*, *1736*), and *P. flaveolata* (*1736*). It was later found in *Erioderma phaeorhizum* together with hypophysciosporin (**2088**) and **2089** (*1728*). Lecideoidin (**2090**) and the dechloro **2091** are produced by *Lecidea* sp. (*1703*). The two dechlorodiploicin derivatives, **2092** and **2093**, are found in *Buellia canescens* (*1737*, *1738*) as is the related scensidin (**2094**) (*1738*). *Diploicia canescens* contains 3-*O*-demethylscensidin (**2095**) and *O*-methyldiploicin (**2096**) (*1729*).

Allorhizin (**2097**) is a novel lichen metabolite isolated from New Zealand *Psoroma allorhizum* (*1726*), while the four novel depsidones

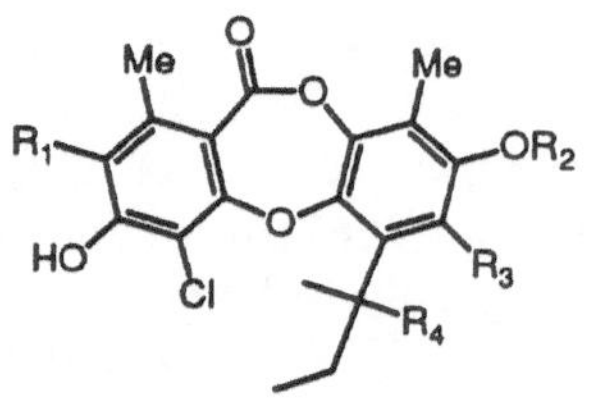

2075 $R_1 = R_2 = R_3 = H$, $R_4 = OH$ (haiderin)
2076 $R_1 = R_3 = H$, $R_2 = Me$, $R_4 = OH$ (rubinin)
2077 $R_1 = R_3 = Cl$, $R_2 = R_4 = H$ (shirin)

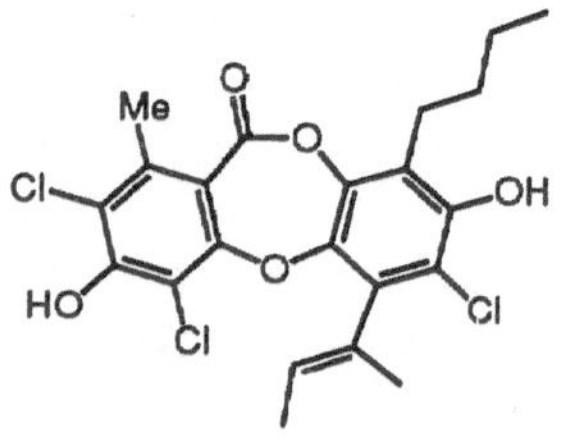

2078 (nasrin)

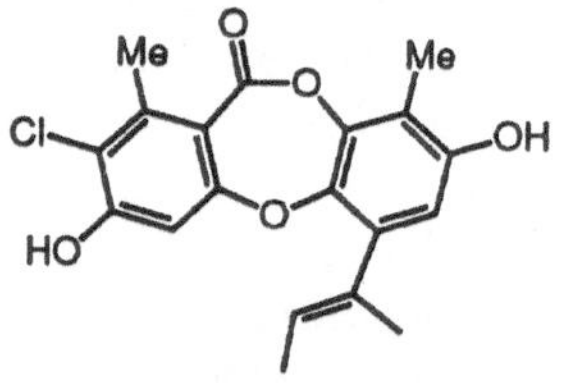

2079 (2-chlorounguinol)

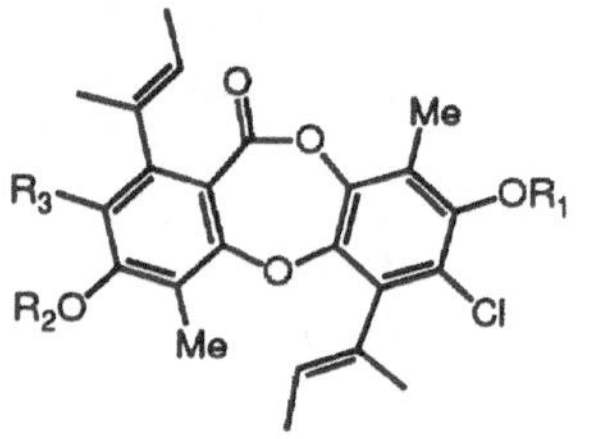

2080 $R_1 = R_2 = R_3 = H$ (emeguisin A)
2081 $R_1 = R_3 = H$, $R_2 = Me$ (emeguisin B)
2082 $R_1 = Me$, $R_2 = H$, $R_3 = Cl$ (emeguisin C)

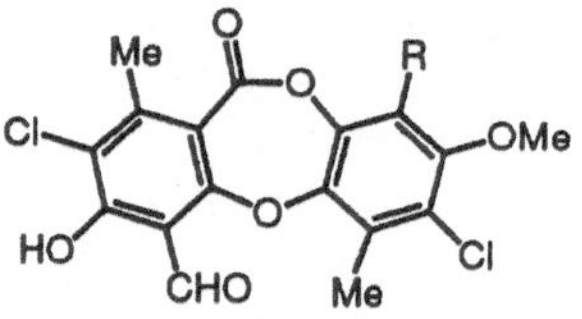

2083 R = H (argopsin, 1-chloropannarin)
2084 R = Me (eriodermin)

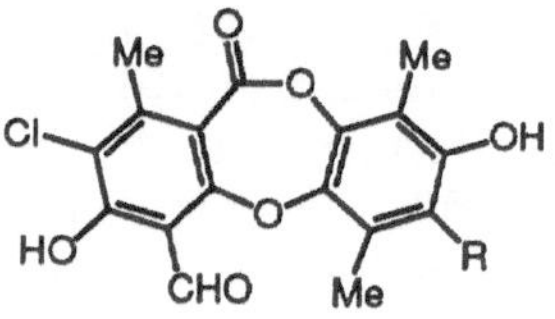

2085 R = H (norpannarin)
2086 R = Cl (norargopsin)

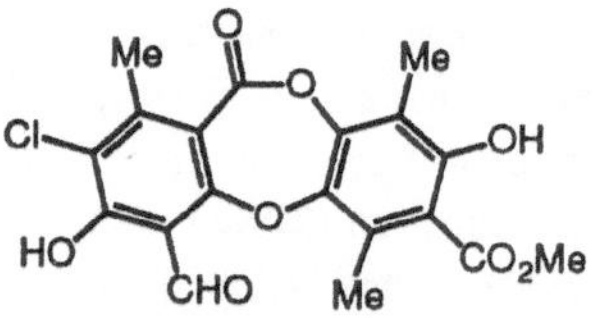

2087 (physciosporin)

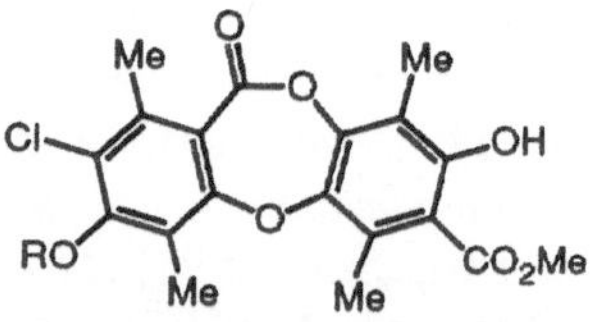

2088 R = H (hypophysciosporin)
2089 R = Me (3-*O*-methylhypophysciosporin)

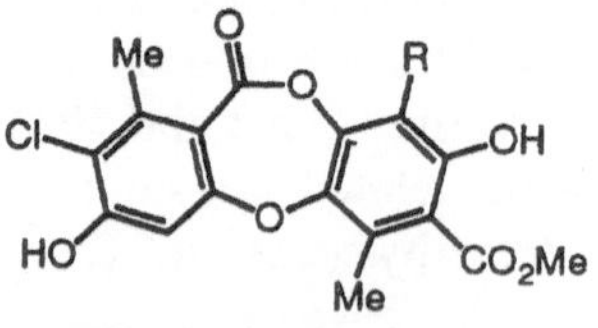

2090 R = Cl (lecideoidin)
2091 R = H (3'-dechlorolecideoidin)

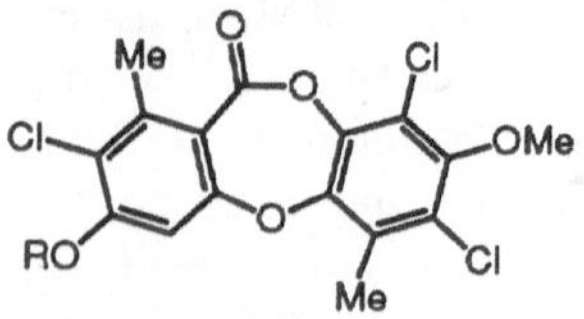

2092 R = H (dechlorodiploicin)
2093 R = Me

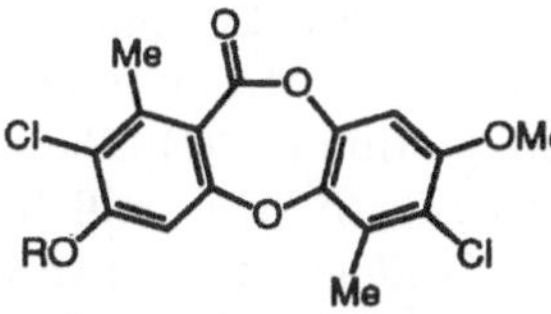

2094 R = Me (scensidin)
2095 R = H

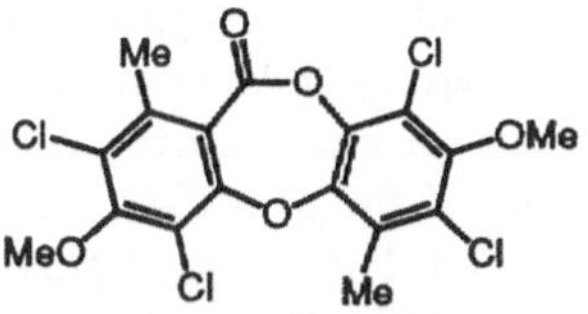

2096 (*O*-methyldiploicin)

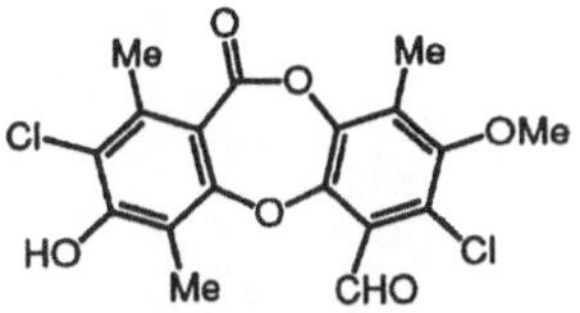

2097 (allorhizin)

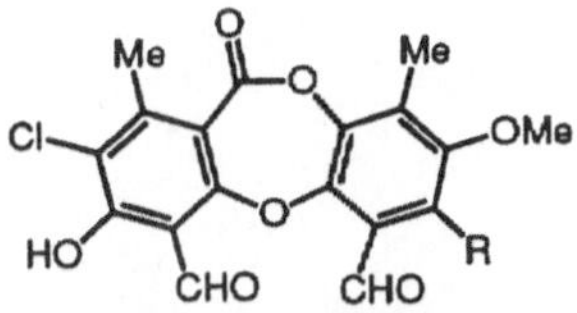

2098 R = H (phyllopsorin)
2099 R = Cl (chlorophyllopsorin)

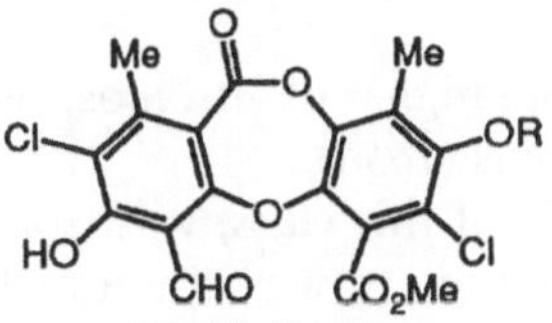

2100 R = H
2101 R = Me

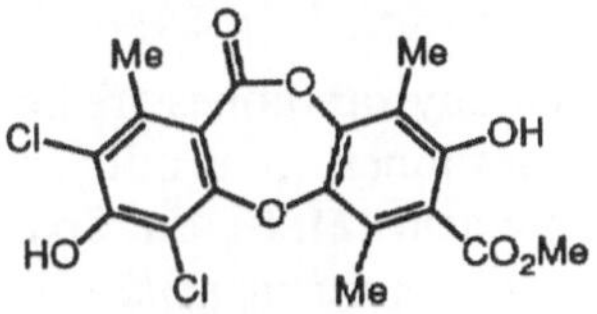

2102 (leoidin)

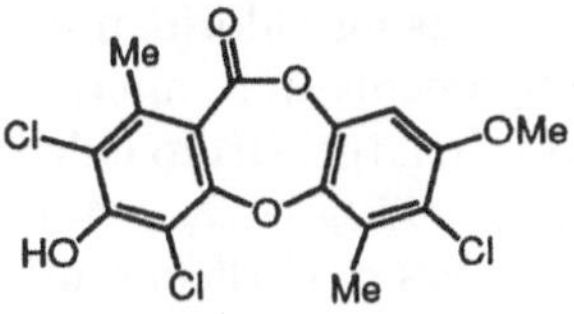

2103 (fulgidin)

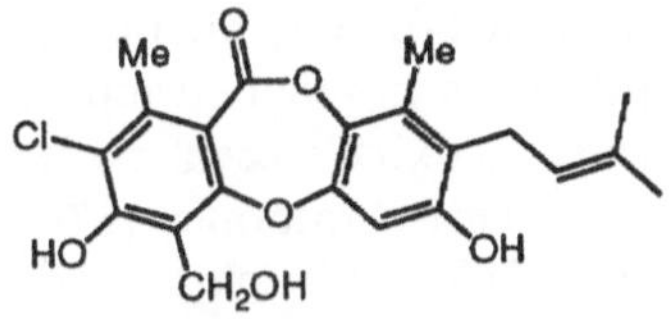

2104 (mollicellin D)

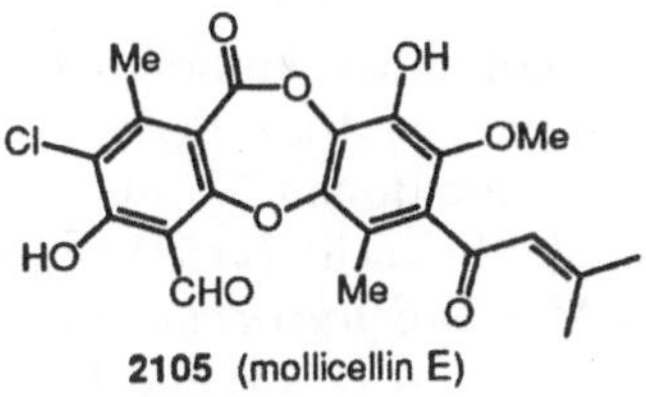

2105 (mollicellin E)

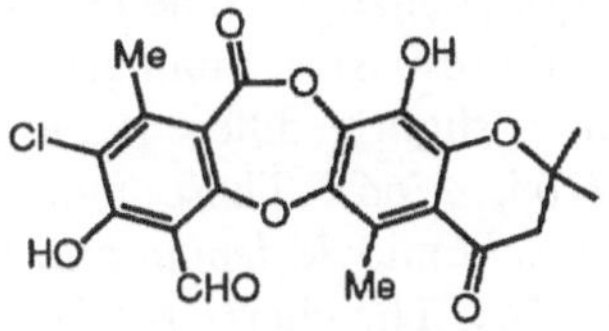

2106 (mollicellin F)

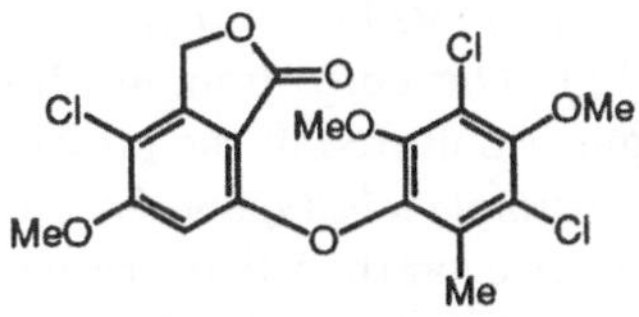

2107 (buellolide)

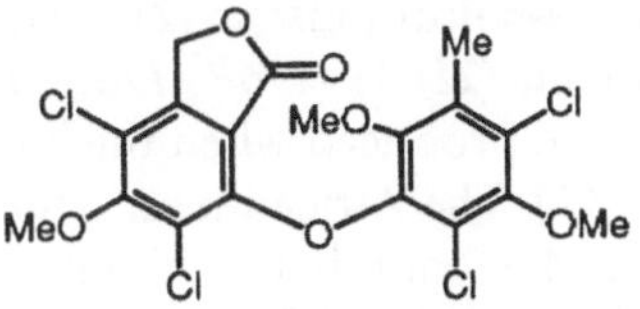

2108 (canesolide)

2098–2101 are found in *Phyllopsora corallina* (*1739*). Leoidin (**2102**) which is produced by *Lecanora gangaleoides* has the structure shown after many years of uncertainty about its constitution (*1701*). The proposed structure of fulgidin (**2103**), from *Fulgensia* sp., may still be in doubt (*1740, 1658*). The fungus *Chaetomium mollicellum* produces several new depsidones three of which are chlorinated, mollicellins D (**2104**), E (**2105**), and F (**2106**) (*1741*). The lactones buellolide (**2107**) and canesolide (**2108**) which may arise from depsidones by catabolism are also found in *Buellia canescens* (*1737*).

3.22.6. Xanthones

Although many xanthones are known from higher plants, most of the chlorinated xanthones are produced by lichens (*1658*).

Thiophanic acid (**2150**), the first example of this class, was isolated initially from *Lecanora rupicola* (*1742*); its structure was confirmed by synthesis (*1743*). It was subsequently isolated from other lichens as summarized in Table 25 which tabulates the other known chlorinated xanthone lichen metabolites (in order of increasing chlorination). Although several of the original structural assignments were in error, the situation now seems to have been resolved, due in large part to independent synthesis of these xanthones (*1747, 1748, 1751, 1752, 1764*). The xanthones (**2109–2152**) in Table 25 are derivatives of norlichexanthone, and are derived biosynthetically from a polyketide *via* the intermediacy of a polyhydroxybenzophenone (*1658*).

However, there exists a small group of chlorinated xanthones that are derived from a slightly different biogenic path and are derivatives of ravenelin (*1765, 1766*). These metabolites are found in the lichens *Rinodina thiomela* and *R. lepida* and include thiomelin (**2153**) and analogues **2154–2161**. The cherry rot fungus *Monilinia fructicola* which was earlier shown to produce chloromonilinic acids A (**1596**) and B (**1597**) also contains 4-chloropinselin (**2162**) (*1767*) and its biosynthetic product chloromonilicin (**2163**) (*1768, 1769, 1783*). The corresponding bromo metabolites are produced when this fungus is cultured in the presence of bromide (*1767*). The African plant *Psorospermum febrifugum* (Guttiferae) has been found to contain two highly cytotoxic xanthones, psorospermin chlorohydrin (**2164**) and **2165** (*1770, 1771*). For example, **2165** is very

Table 25. *Naturally Occurring Lichen Chlorine-Containing Xanthones of the Norlichexanthone Type*

Compound	Source	Ref.
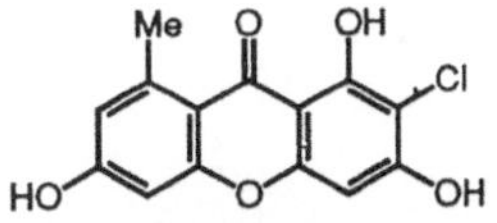 Me, O, OH, Cl, HO, O, OH **2109** (2-chloronorlichexanthone)	*Lecanora* sp. *L. populicola* *L. salina* *Lecidella vorax*	*1764* *1764* *1764* *1764*
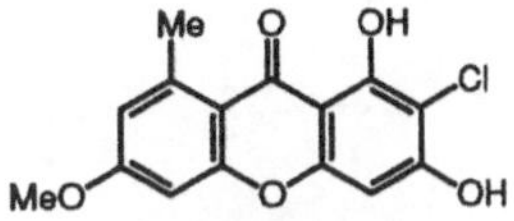 Me, O, OH, Cl, MeO, O, OH **2110** (6-*O*-methyl-2-chloro-norlichexanthone)	*Lecanora salina* *Pertusaria cicatricosa* *P. sulphurata*	*1764* *1764* *1758*
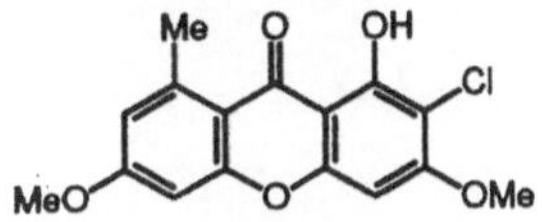 Me, O, OH, Cl, MeO, O, OMe **2111** (2-chlorolichexanthone)	*Lecanora* sp. *Pertusaria cicatricosa* *P. sulphurata*	*1764* *1764* *1758*
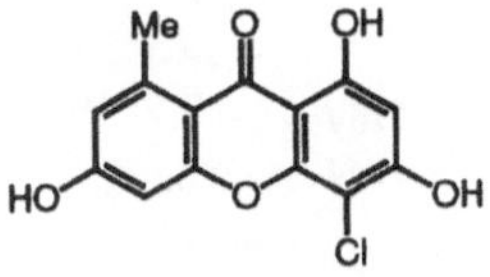 Me, O, OH, HO, O, OH, Cl **2112** (4-chloronorlichexanthone)	*Lecanora straminea*	*1747, 1748, 1752, 1755*
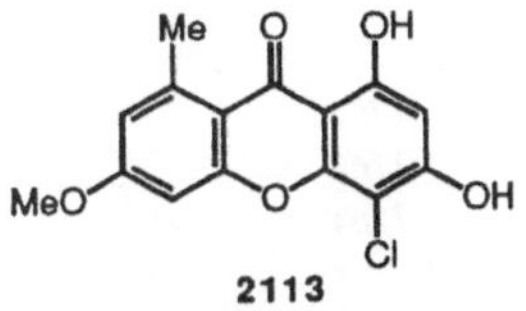 Me, O, OH, MeO, O, OH, Cl **2113** (6-*O*-methyl-4-chloro-norlichexanthone)	*Pertusaria sulphurata*	*1758*
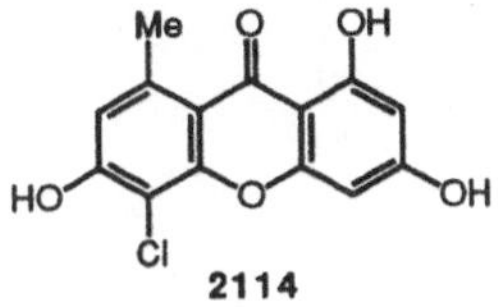 Me, O, OH, HO, O, OH, Cl **2114** (5-chloronorlichexanthone)	*Lecanora straminea*	*1747, 1748, 1752, 1755*

Table 25 (*continued*)

Compound	Source	Ref.
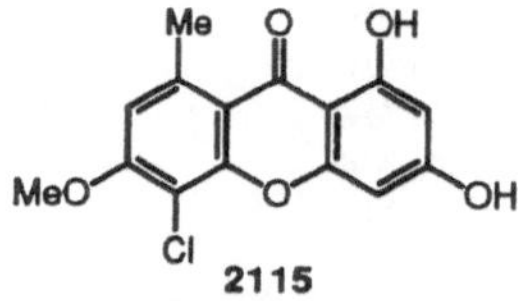 2115 (5-chloro-6-*O*-methylnorlichexanthone)	*Lecanora contractula*	*1764*
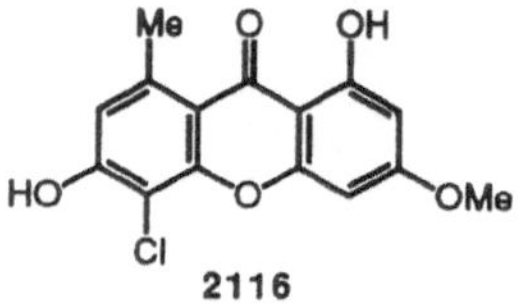 2116 (vinetorin)	*Lecanora vinetorum*	*1764, 2225*
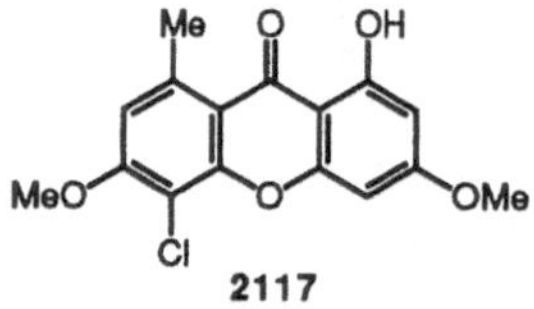 2117 (5-chlorolichexanthone)	*Lecanora contractula*	*1764*
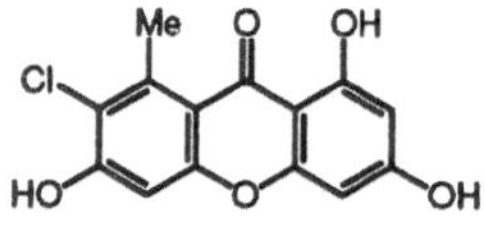 2118 (7-chloronorlichexanthone)	*Lecanora* sp. *L. populicola*	*1764* *1764*
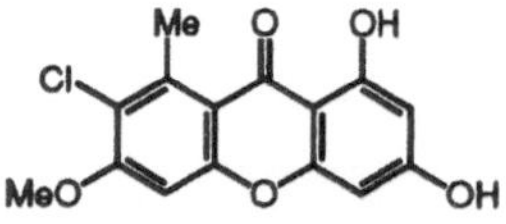 2119 (7-chloro-6-*O*-methylnorlichexanthone)	*Lecanora* sp. *L. populicola* *L. salina*	*1764* *1764* *1764*
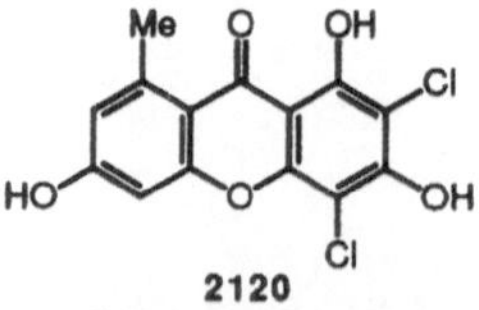 2120 (2,4-dichloronorlichexanthone)	*Lecanora straminea* *Lecidella vorax*	*1755* *1764*

Table 25 (*continued*)

Compound	Source	Ref.
2121 (thiophaninic acid)	*Dimelaena* sp.	*1762*
	D. cf. *australiensis*	*1764*
	Pertusaria sp.	*1753*
	P. flavicans	*1756*
	P. flavicunda	*1664*
	P. sulphurata	*1758*
2122 (2,4-dichlorolichexanthone)	*Dimelaena* cf. *australiensis*	*1764*
	Pertusaria sp.	*1751*
	P. cicatricosa	*1764*
2123 (2,5-dichloronorlichexanthone)	*Buellia* sp.	*1763*
	Lecanora broccha	*1764*
	Lecidella meiococca	*1764*
	L. vorax	*1764*
2124 (2,5-dichloro-6-*O*-methyl-norlichexanthone)	*Dimelaena* sp.	*1762*
	D. cf. *australiensis*	*1764*
	Lecanora contractula	*1764*
	Pertusaria cicatricosa	*1764*
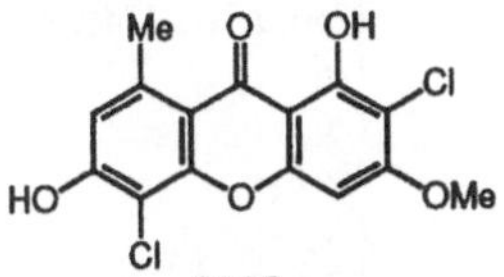 **2125** (3-*O*-methyl-2,5-dichloro-norlichexanthone)	*Lecanora contractula*	*1754*
2126 (2,5-dichlorolichexanthone)	*Dimelaena* cf. *australiensis*	*1764*
	Lecanora sp.	*1754*
	Pertusaria sp.	*1751*
	P. aleianata	*1751*
	P. cicatricosa	*1764*

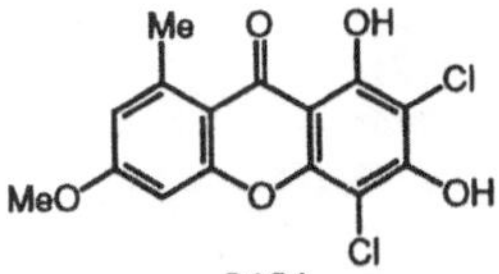

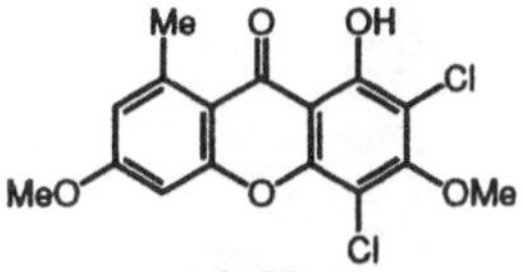

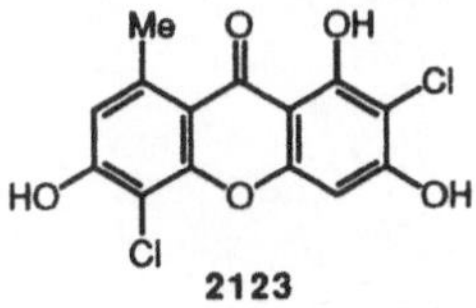

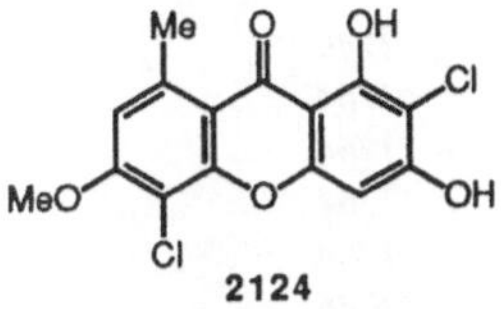

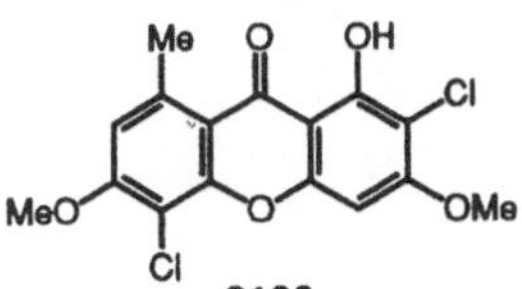

Table 25 (*continued*)

Compound	Source	Ref.
2127 (2,7-dichloronorlichexanthone)	*Buellia* sp.	*1763*
	Lecanora sp.	*1764*
	L. behringii	*1764*
	L. broccha	*1763, 1764*
	L. populicola	*1764*
	L. salina	*1764*
	Lecidella meiococca	*1764*
2128 (2,7-dichloro-6-*O*-methylnorlichexanthone)	*Lecanora* sp.	*1764*
	L. behringii	*1764*
	L. populicola	*1764*
	L. salina	*1764*
2129 (2,7-dichloro-3-*O*-methylnorlichexanthone)	*Lecanora* sp.	*1764*
	L. behringii	*1764*
	L. salina	*1764*
2130 (2,7-dichlorolichexanthone)	*Buellia glaziouana*	*1749*
	Lecanora sp.	*1764*
	L. behringii	*1764*
	L. populicola	*1764*
	L. salina	*1764*
	Lopadium sp.	*1746*
	Pertusaria sp.	*1749*
2131 (4,5-dichloronorlichexanthone)	*Lecanora flavo-pallescens*	*1761*
	L. straminea	*1750, 1747, 1748, 1752*
	Lecidella asema	*1763*
	L. vorax	*1764*
	Micarea austroternaria	*1761*
	M. isabellina	*1764*
	Pertusaria pycnothelia	*1764*
2132 (4,5-dichloro-6-*O*-methylnorlichexanthone)	*Dimelaena* sp.	*1762*
	D. cf. *australiensis*	*1764*

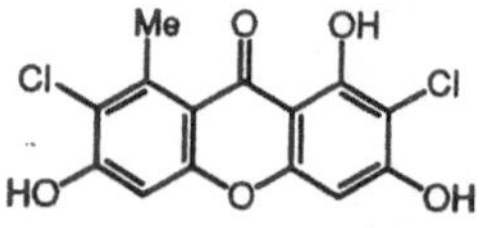

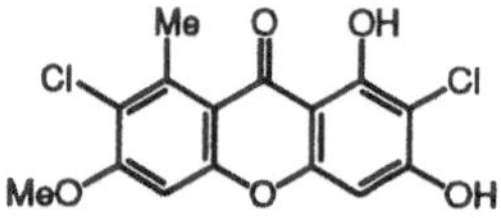

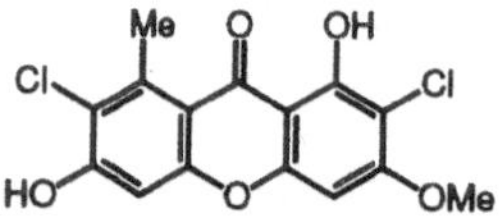

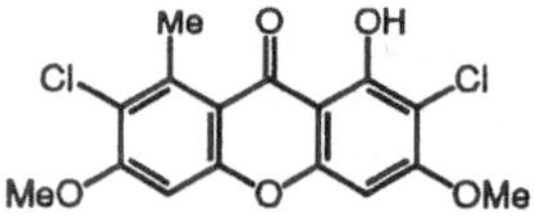

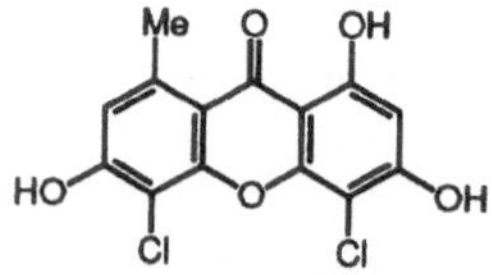

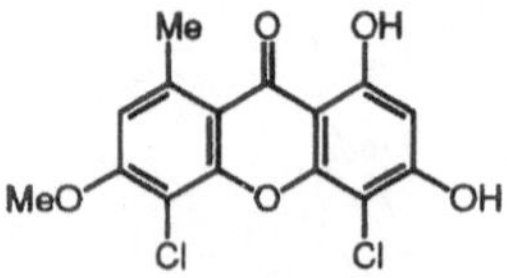

Table 25 (*continued*)

Compound	Source	Ref.
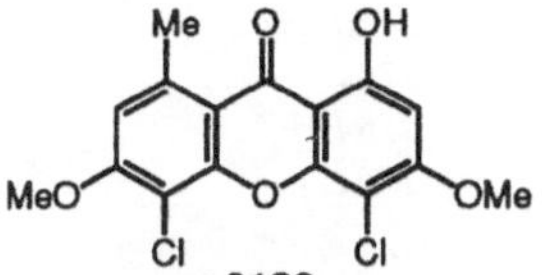 **2133** (4,5-dichlorolichexanthone)	*Buellia glazionana*	*1747, 1748, 1750, 1752*
	Dimelaena cf. *australiensis*	*1764*
	Lecanora straminea	*1747, 1748, 1750, 1752*
	Pertusaria cicatricosa	*1764*
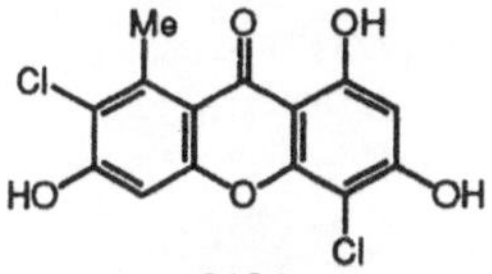 **2134** (4,7-dichloronorlichexanthone)	*Lecidella asema*	*1763*
	L. meiococca	*1764*
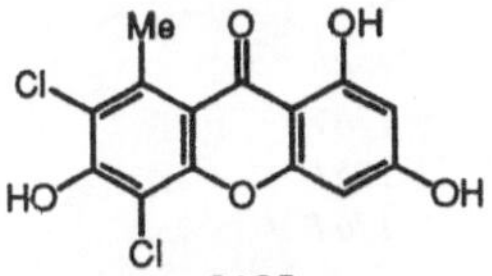 **2135** (5,7-dichloronorlichexanthone)	*Buellia* sp.	*1763*
	Lecanora broccha	*1763, 1764*
	Lecidella asema	*1763*
	L. subalpicida	*1763*
	L. vorax	*1764*
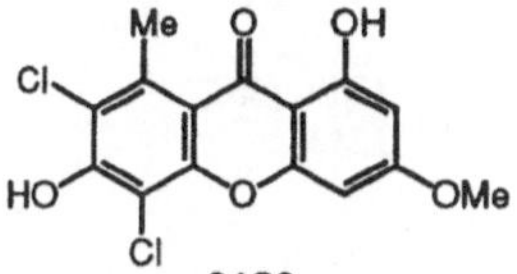 **2136** (5,7-dichloro-3-*O*-methylnorlichexanthone)	*Lecanora broccha*	*1764*
	L. vinetorum	*1764*
	Lecidella meiococca	*1764*
	L. vorax	*1764*
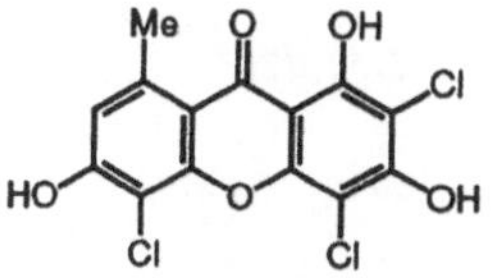 **2137** (arthothelin)	*Arthothelium pacificum*	*1664*
	Buellia sp.	*1759, 1763*
	Dimelaena cf. *australiensis*	*1764*
	Lecanora broccha	*1759*
	L. flavo-pallescens	*1761*
	L. pinguis	*1664*
	L. reuteri	*1744*
	L. straminea	*1745*
	L. subalpicida	*1763*
	L. sulphurata	*1761, 1764*
	Lecidella meiococca	*1764*
	L. quernea	*1753*
	L. vorax	*1764*
	Micarea austroternaria	*1761*
	M. isabellina	*1764*
	Pertusaria pycnothelia	*1764*
	Tapellaria epiphylla	*1746*

Table 25 (*continued*)

Compound	Source	Ref.
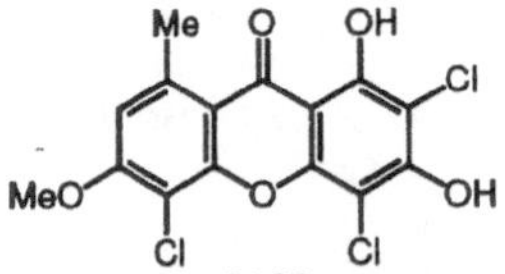 **2138** (6-*O*-methylarthothelin)	*Dimelaena* sp. *D.* cf. *australiensis* *Micarea isabellina* *Pertusaria pycnothelia*	*1760, 1762* *1764* *1764* *1764*
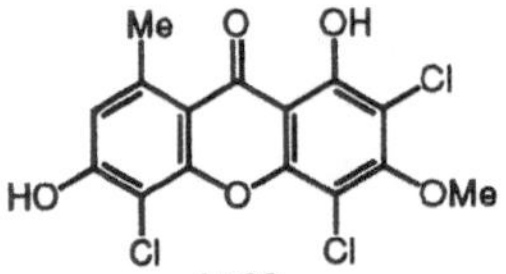 **2139** (thuringion)	*Lecidea pinguis* *L. carpathica*	*1753* *1676*
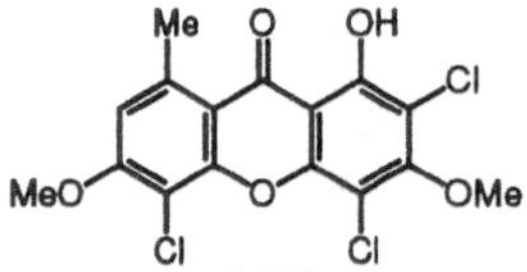 **2140** (2,4,5-trichlorolichexanthone)	*Dimelaena* sp. *D.* cf. *australiensis* *Pertusaria* sp. *P. cicatricosa*	*1760, 1762* *1764* *1751* *1764*
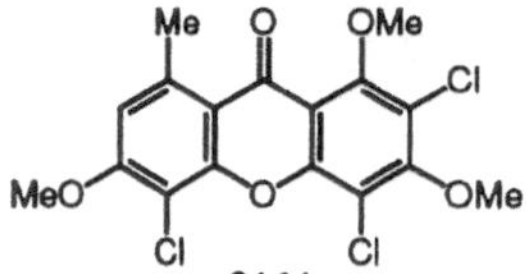 **2141** (1,3,6-tri-*O*-methylarthothelin)	*Dimelaena sp.* *D.* cf. *australiensis*	*1760, 1762* *1764*
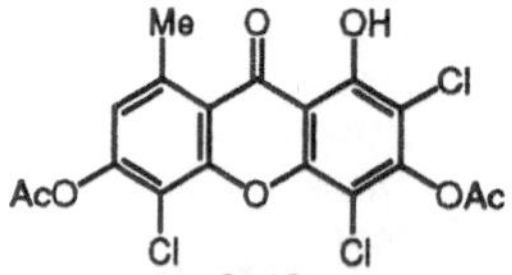 **2142** (erythrommone)	*Haematomma erythromma*	*1751, 1757*
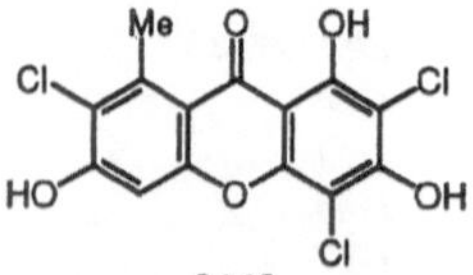 **2143** (2,4,7-trichloronorlichex-anthone)	*Lecanora flavo-pallescens* *L. sulphurata*	*1749, 1761* *1749, 1761, 1764*

Table 25 (*continued*)

Compound	Source	Ref.
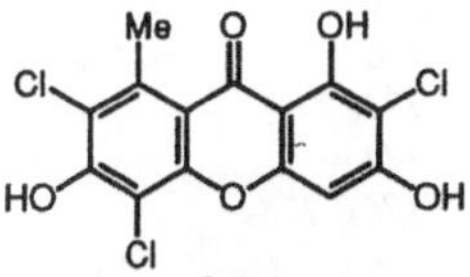 2144 (isoarthothelin)	*Buellia* sp.	*1759, 1763*
	Lecanora broccha	*1759, 1763, 1764*
	L. sulphurata	*1761, 1764*
	Lecidella meiococca	*1764*
	L. subalpicida	*1763*
	L. vorax	*1764*
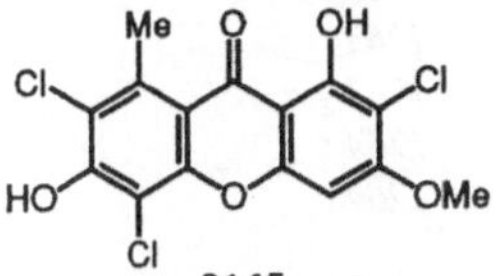 2145 (3-*O*-methyl-2,5,7-trichloronorlichexanthone)	*Lecanora broccha*	*1763, 1764*
	L. capistrata	*1749*
	Lecidella meiococca	*1764*
	L. subalpicida	*1763*
	L. vorax	*1764*
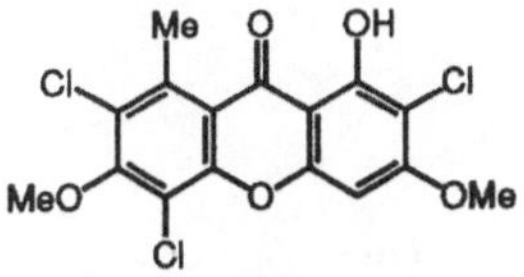 2146 (2,5,7-trichlorolichexanthone)	*Dimelaena* cf. *australiensis*	*1764*
	Lecanora broccha	*1764*
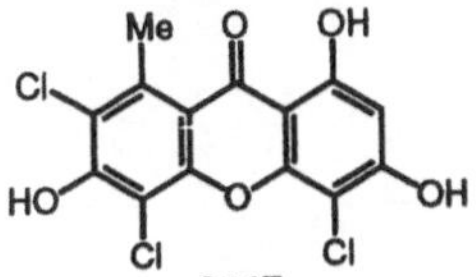 2147 (asemone)	*Lecanora broccha*	*1764*
	Lecidella asema	*1763*
	Micarea isabellina	*1761, 1764*
	Pertusaria pycnothelia	*1764*
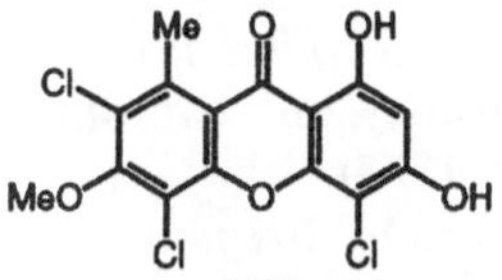 2148 (6-*O*-methylasemone)	*Pertusaria pycnothelia*	*1764*
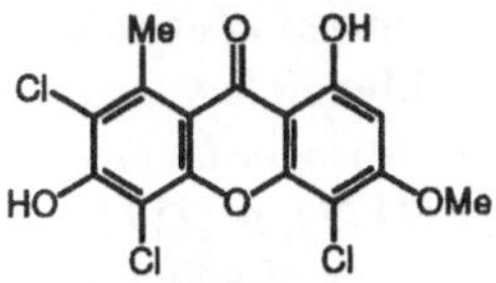 2149 (3-*O*-methylasemone)	*Lecanora broccha*	*1764*
	Lecidella meiococca	*1764*

Table 25 (*continued*)

Compound	Source	Ref.
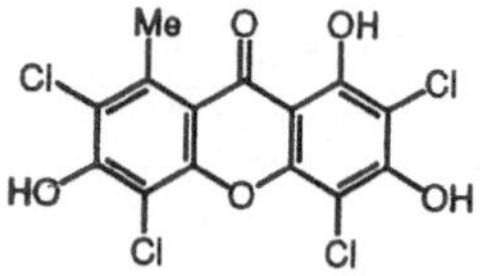 2150 (thiophanic acid)	*Buellia* sp.	*1763*
	Lecanora flavo-pallescens	*1761*
	L. rupicola	*1742*
	L. straminea	*1745*
	L. sulphurata	*1761, 1764*
	Lecidella asema	*1763*
	L. meiococca	*1764*
	L. quernea	*1753*
	L. vorax	*1764*
	Micarea austroternaria	*1761*
	M. isabellina	*1764*
	Pertusaria pycnothelia	*1764*
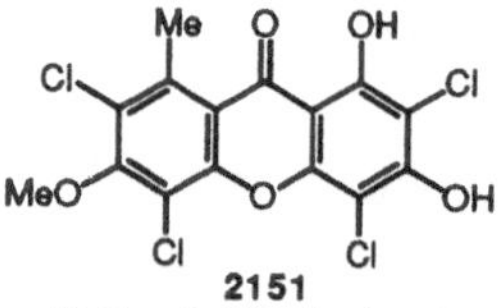 2151 (6-*O*-methylthiophanic acid)	*Micarea isabellina*	*1764*
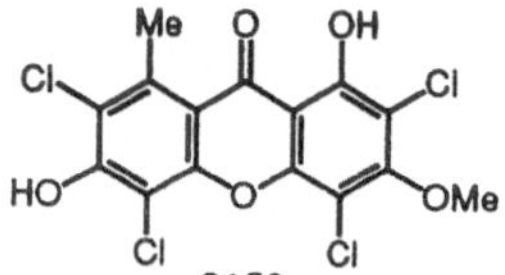 2152 (3-*O*-methylthiophanic acid)	*Lecidella meiococca*	*1764*

active against 9PS cells (ED_{50} < 0.01 ng/ml) (*1771*). *Aspergillus ustus* produces several novel pentacyclic metabolites, the austocystins, two of which, A (**2166**) and C (**2167**), contain chlorine (*1772, 1773*).

The structurally complex lysolipin I (**2168**) is produced by *Streptomyces violaceoniger* and appears to be derived from an unstable precursor lysolipin X (**2169**) (*1774, 1775*). The biosynthesis of these interesting metabolites has been investigated (*1776*). The fungus *Cercospora beticola* which is a highly destructive disease of sugar beets worldwide has been found by several groups to produce a series of highly intricate metabolites, beticolins 1 (**2170**) (= cebetin A), 2 (**2171**), 3 (**2172**), 4 (**2173**), 6 (**2174**), 8 (**2175**), and cebetin B (**2176**) (= CBT), the latter of which is a *bis*-Mg complex of cebetin A (not shown) (*1777–1782*). Following some

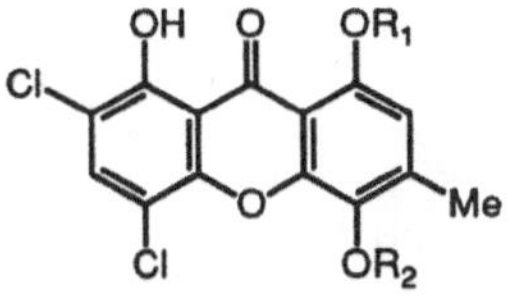

2153 R_1 = H, R_2 = Me (thiomelin)
2154 R_1 = R_2 = Me
2155 R_1 = R_2 = H (TH3)

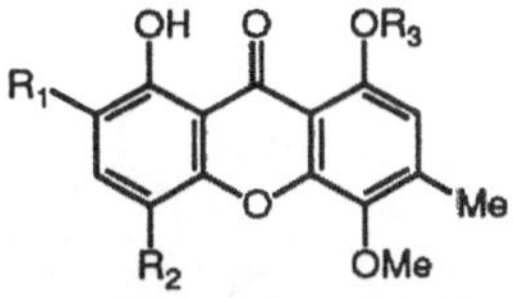

2156 R_1 = Cl, R_2 = R_3 = H
2157 R_1 = Cl, R_2 = H, R_3 = Me
2158 R_1 = H, R_2 = Cl, R_3 = Me
2159 R_1 = R_3 = H, R_2 = Cl

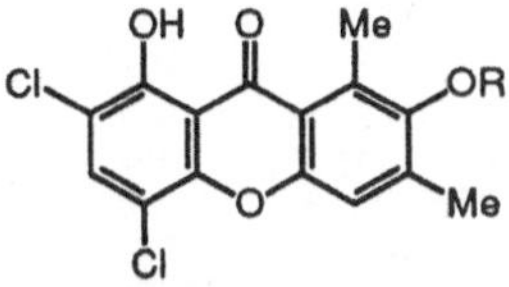

2160 R = H
2161 R = Me (TH2)

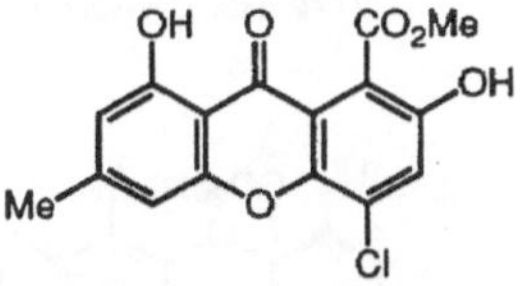

2162 (4-chloropinselin)

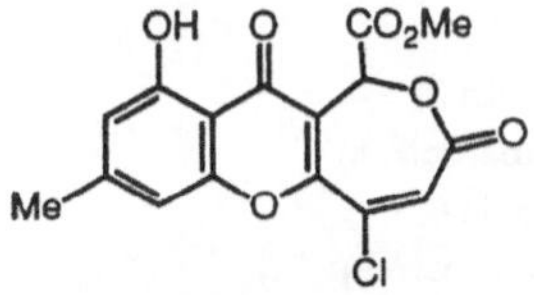

2163 (chloromonilicin)

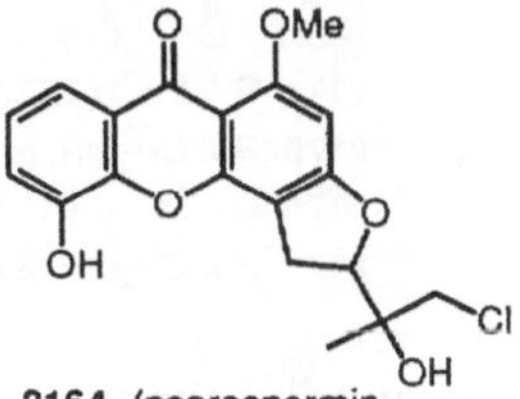

2164 (psorospermin chlorohydrin)

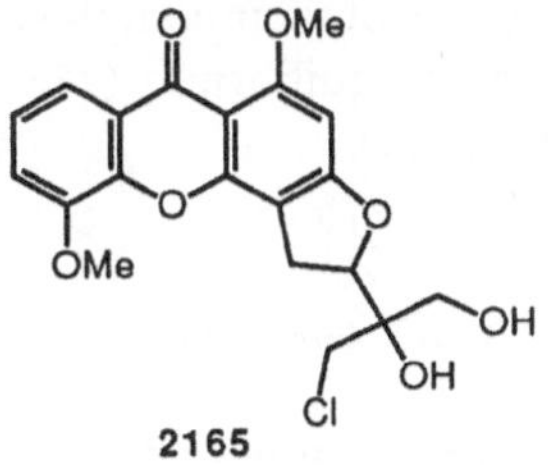

2165

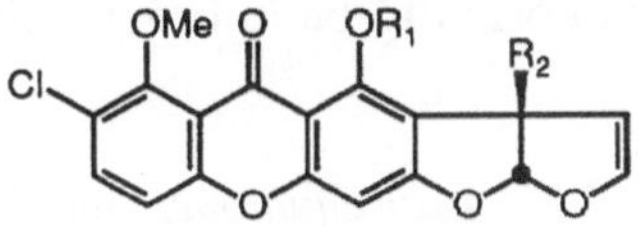

2166 R_1 = Me, R_2 = H (austocystin A)
2167 R_1 = H, R_2 = OH (austocystin G)

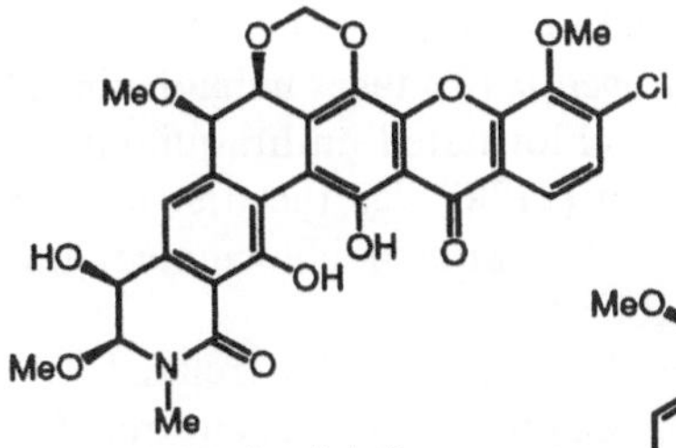

2168 (lysolipin I)

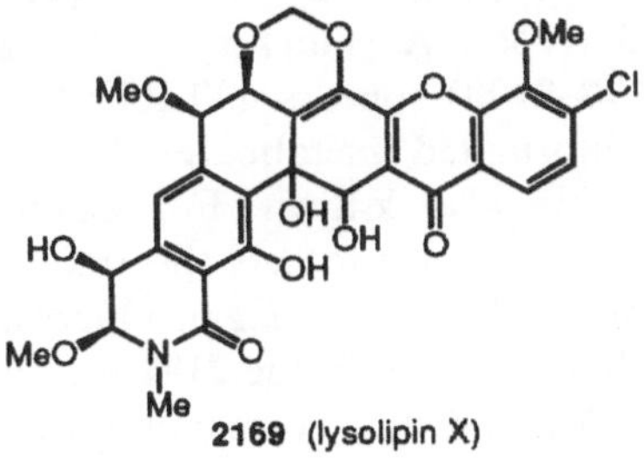

2169 (lysolipin X)

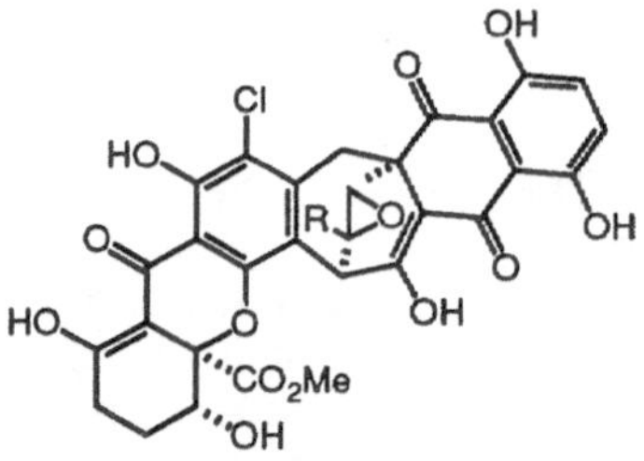

2170 R = Me (beticolin 1 = cebetin A)
2172 R = CH_2OH (beticolin 3)

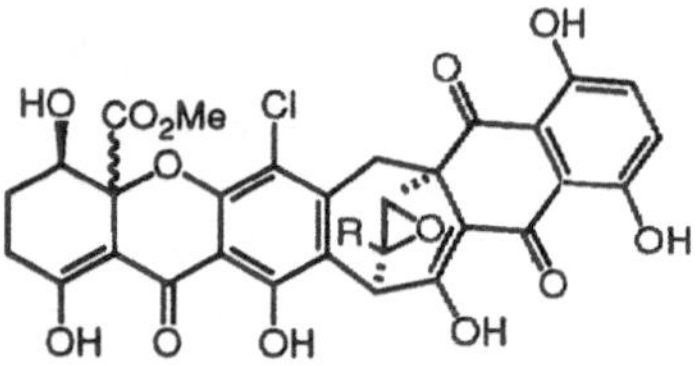

2171 R = Me; α-CO_2Me (beticolin 2)
2173 R = CH_2OH; α-CO_2Me (beticolin 4)
2174 R = Me; β-CO_2Me (beticolin 6)
2175 R = CH_2OH; β-CO_2Me (beticolin 8)

initial confusion regarding the structures because of their complexity and the fact that beticolin 2 and cebetin A are in equilibrium (*1781*), the situation now appears to be in order (*1781, 1782*).

3.22.7. *Anthraquinones and Related Compounds*

Like the fungal and lichen xanthones, anthraquinones which are also produced by both lichens and fungi are derived from extended polyketides by cyclization. Several chlorinated examples exist.

The cheese mold *Penicillium nalgiovensis* produces nalgiolaxin (**2177**) which appears to be the first natural chlorinated anthraquinone to be isolated (*1784, 1785*), although fragilin (**2178**) was the first to be fully characterized (*1786*). A summary of the known chlorinated anthraquinones (**2177**–**2190**) appears in Table 26.

Several chlorinated metabolites that are closely related to anthraquinones are also known. For example, the aspen tree fungus *Phialophora alba* which protects the tree against attack by the decay-causing fungus *Phellinus tremulae* produces anthrone **2191** in addition to **2180** and **2190** (*1799*). Anthrone **2192** is found in cultures of *Aspergillus fumigatus* (*1788*) and the corresponding bromo compound **2193** is

Table 26. *Naturally Occurring Lichen and Fungi Chlorine-Containing Anthraquinones*

Compound	Source	Ref.
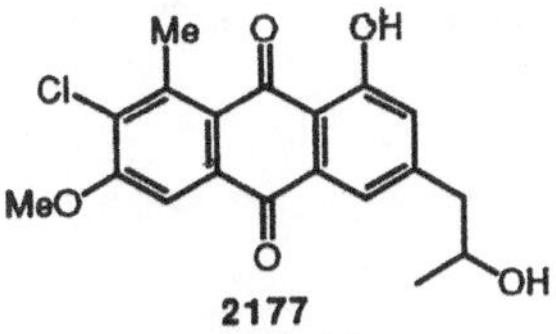 **2177** (nalgiolaxin)	*Penicillium nalgiovensis*	*1784, 1785*
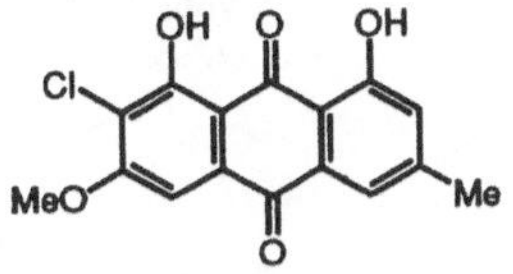 **2178** (fragilin)	*Byssoloma tricholomum*	*1746*
	Caloplaca ssp.	*1794, 1796*
	C. arenaria	*1793*
	C. percrocata	*1793*
	Nephroma laevigatum	*1787*
	Sphaerophorus coralloïdes	*1786*
	S. fragilis	*1786*
	S. globulus	*1786*
	S. melanocarpus	*1664*
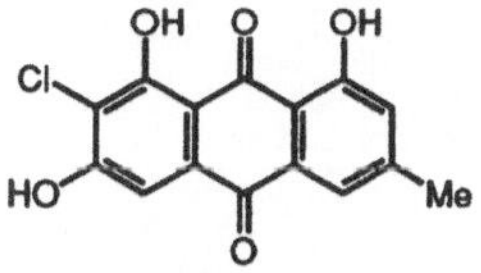 **2179** (7-chloroemodin)	*Anaptychia obscurata*	*1789*
	Aspergillus fumigatus	*1788*
	Byssoloma tricholomum	*1746*
	Caloplaca ssp.	*1794, 1796*
	C. arenaria	*1793*
	C. percrocata	*1793*
	Heterodermia obscurata	*1664*
	Lasallia papulosa	*1790, 1791*
	Lecidea quernea	*1753*
	Nephroma laevigatum	*1787*
	Penicillium sp.	*1664*
	Valsaria rubricosa	*1795*
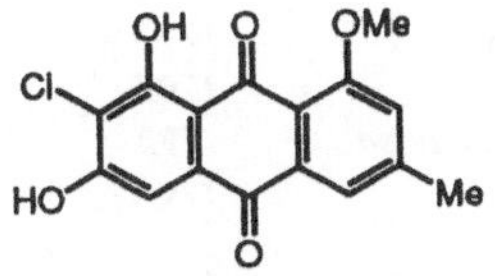 **2180**	*Nephroma laevigatum*	*1787*
	Phialophora alba	*1799*
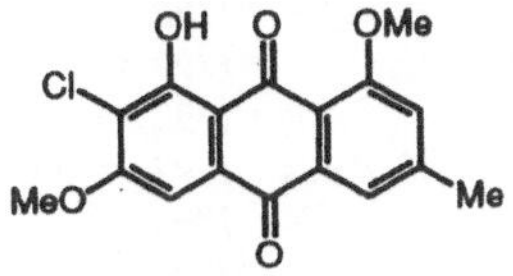 **2181**	*Caloplaca* spp.	*1794*
	Nephroma laevigatum	*1787*

Table 26 (*continued*)

Compound	Source	Ref.
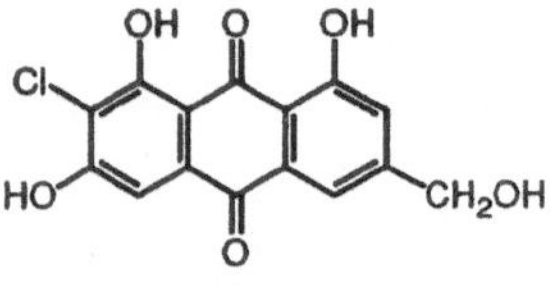 **2182**	*Aspergillus fumigatus*	*1788*
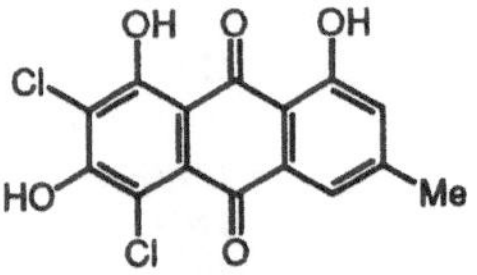 **2183** (5,7-dichloroemodin)	*Anaptychia obscurata*	*1789*
	Heterodermia obscurata	*1664*
	Valsaria rubricosa	*1795*
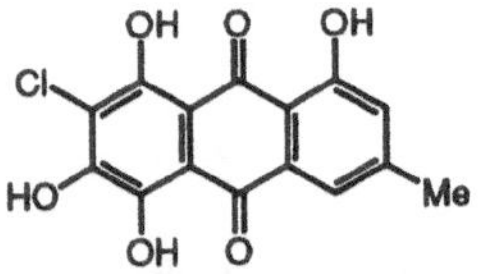 **2184** (papulosin; valsarin)	*Lasallia papulosa*	*1790, 1791*
	Valsaria rubricosa	*1791, 1795*
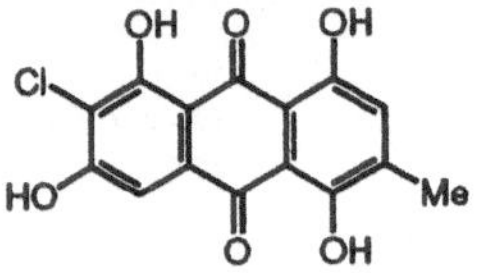 **2185**	*Lasallia papulosa*	*1791*
	Valsaria rubricosa	*1791*
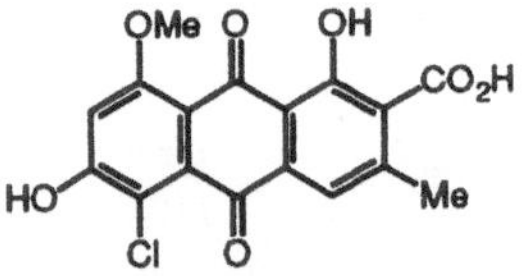 **2186** (5-chlorodermolutein)	*Dermocybe* ssp.	*1797*
	D. alnophila	*1792*
	D. malicoria	*1792*
	D. phoenicea	*1792*
	D. sanguinea	*1792*
	D. sanguinea var. *vitiosa*	*1792*
	D. sommerfeltii	*1792*
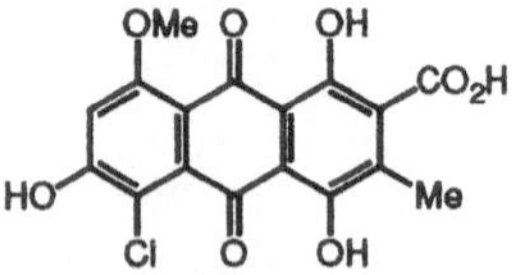 **2187** (5-chlorodermorubin)	*Cortinarius subcroceofolius*	*1792*
	Dermocybe ssp.	*1797, 1798*
	D. crocea	*1792*
	D. marylandensis	*1792*
	D. phoenicea	*1792*
	D. phoenicea var. *occidentalis*	*1792*
	D. punicea	*1792*
	D. sanguinea	*1792*
	D. sanguinea var. *sierraensis*	*1792*

Table 26 (*continued*)

Compound	Source	Ref.
	D. sanguinea var. *vitiosa*	*1792*
	D. semisanguinea	*1792*
	D. sommerfeltii	*1792*
	D. uliginosa	*1792*
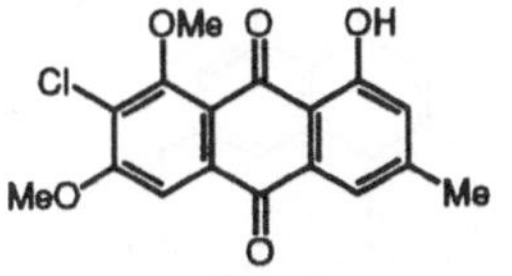 **2188**	*Caloplaca xanthapsis*	*1794*
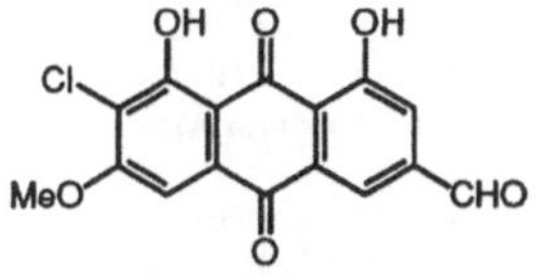 **2189**	*Caloplaca* sp.	*1796*
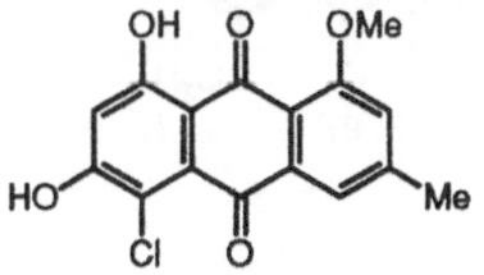 **2190**	*Phialophora alba*	*1799*

formed in the presence of bromide (*1788*). The novel *bis*-anthrones, flavo-obscurin A (**2194**), B_1 (**2195**), and B_2 (**2196**) (the latter two are rotational isomers), are produced by *Anaptychia obscurata* (*1800*). Only a few naturally occurring brominated anthraquinones are known and these from marine sources. The bromide **2197** is produced by the stony coral *Tubastraea micrantha* which is avoided by the predatory seastar *Acanthaster planci* ("crown-of-thorns") (*1425*). The deep water (520 meters) crinoid *Gymnocrinus richeri* contains gymnochromes A–D (**2198–2201**) and isogymnochrome (**2202**) (*1801*).

3.22.8. *Griseofulvin and Related Compounds*

Arguably, the most widely recognized chlorine-containing natural compound is griseofulvin which is still a widely used antibiotic antifungal agent (*1802*).

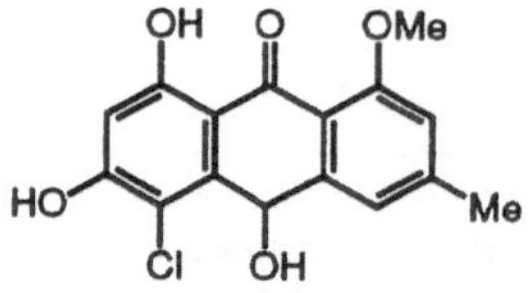

2191

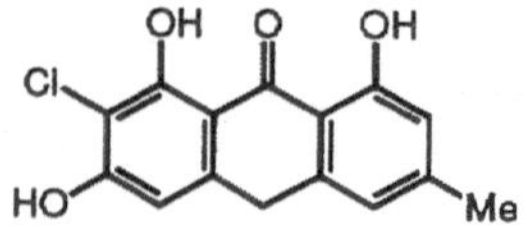

2192

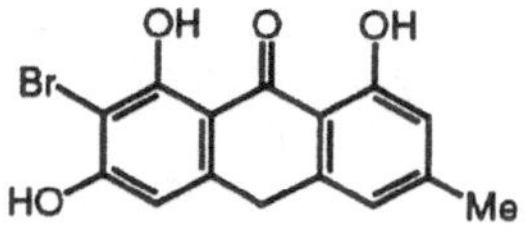

2193

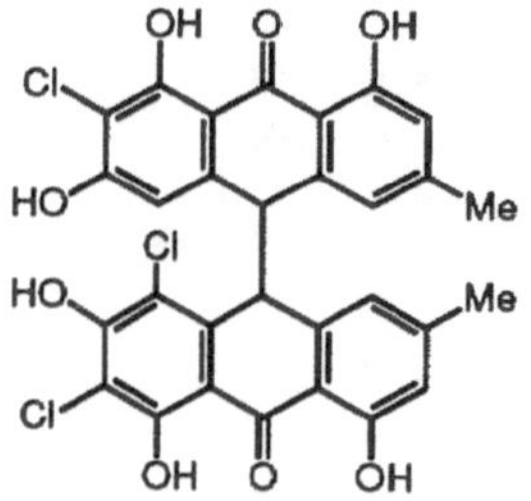

2194 (flavoobscurin A)

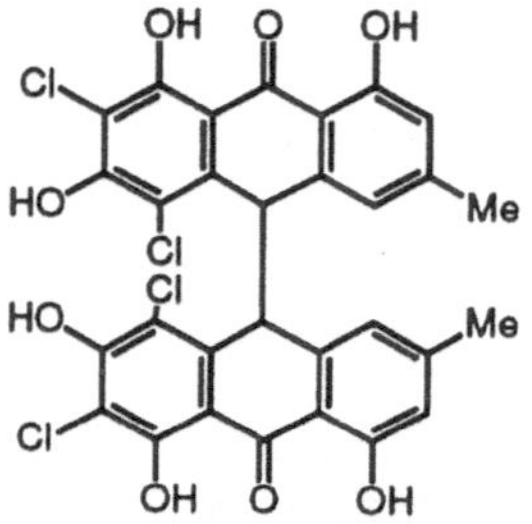

2195 (flavoobscurin B_1)
2196 (flavoobscurin B_2)

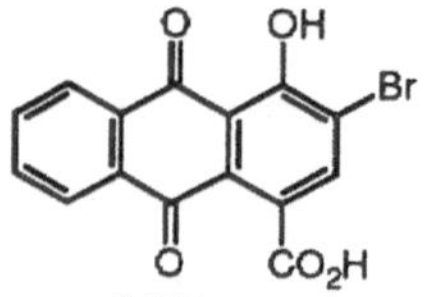

2197

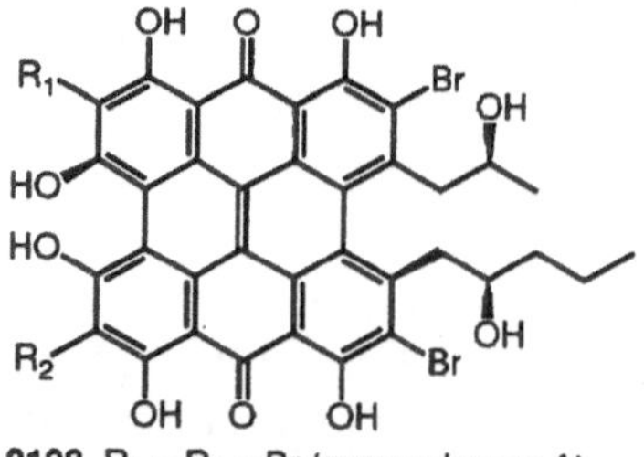

2198 $R_1 = R_2 = Br$ (gymnochrome A)
2199 $R_1 = H$, $R_2 = Br$ (or $R_1 = Br$, $R_2 = H$) (gymnochrome B)

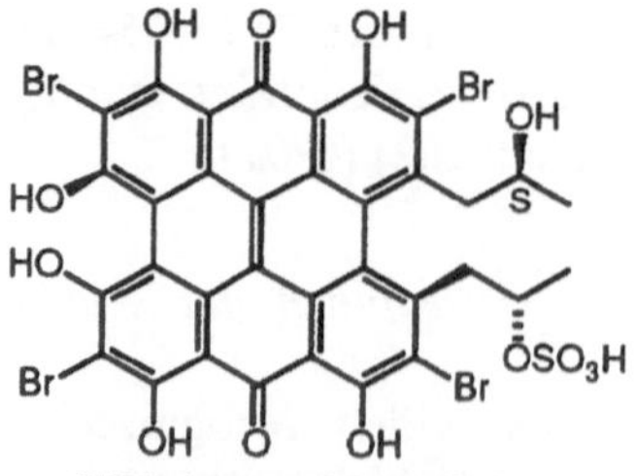

2200 (gymnochrome C)

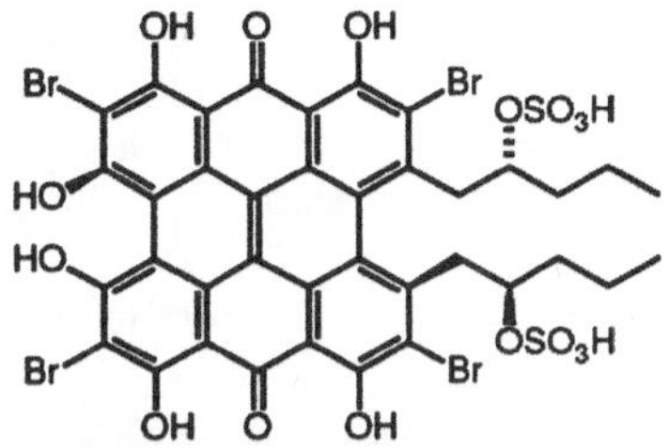

2201 (gymnochrome D)

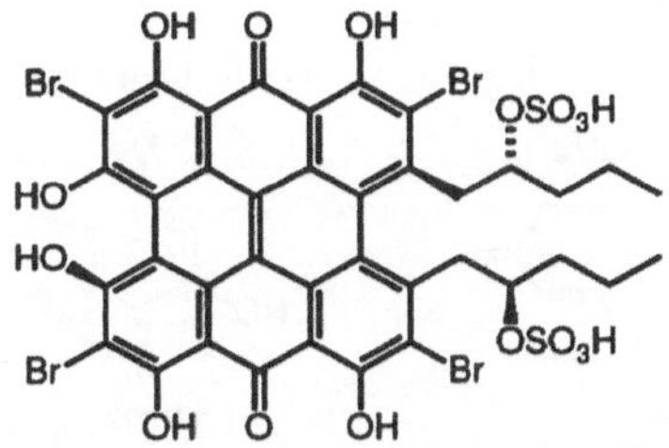

2202 (isogymnochrome D)

However, the related metabolites geodin (**2203**) and erdin (**2204**) were actually isolated prior to griseofulvin from *Aspergillus terreus* (*1803, 1804, 1521*) and later from *Penicillium* sp. (*2227*). Griseofulvin (**2205**) was first isolated in 1939 from *Penicillium griseo-fulvum* (*1805*) and later from many other *Penicillium* species (*1806–1809*). Its absolute configuration was finally determined in 1959 (*1810, 1811*). Subsequently, several related metabolites have been discovered, including dehydrogriseofulvin (**2206**) from *P. patulum* (*1812*) and *P. martinsii* (*1809*). The latter species also produces dihydrogriseofulvin (**2207**) (*1809*). Geodoxin (**2208**) has been isolated from *Aspergillus terreus* (*1813*); more recently, this organism has yielded gillusdin (**2209**) (*1814*). These fungi also produce the presumed biosynthetic precursors griseophenones A–B (**2210, 2211**), dihydrogeodin (**2212**), **2213**, and **2214**, (*1812, 1815, 1817*). The biosynthesis of griseofulvin is consistent with an acetic acid pathway (*1808, 1818*). Interestingly, the marine annelid *Thelepus setosus* produces thelepin (**2215**), a novel brominated spiro compound which has antifungal activity comparable to that of griseofulvin (*1506, 1507*).

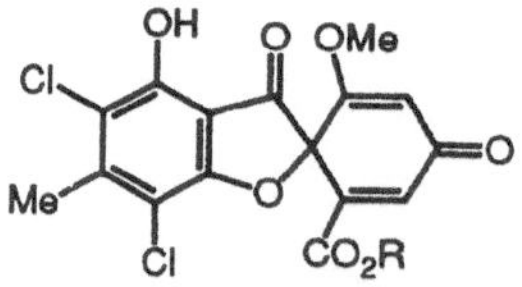

2203 R = Me (geodin)
2204 R = H (erdin)

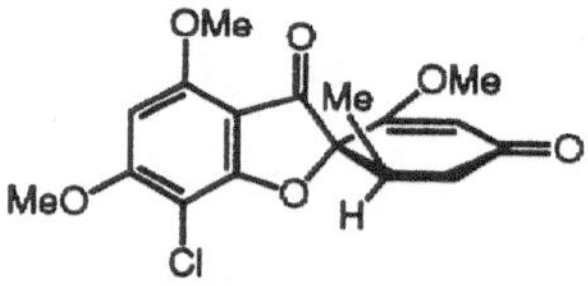

2205 (griseofulvin)

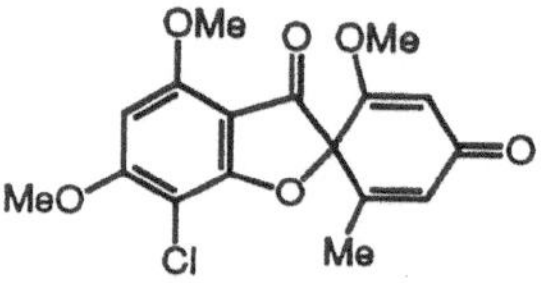

2206 (dehydrogriseofulvin)

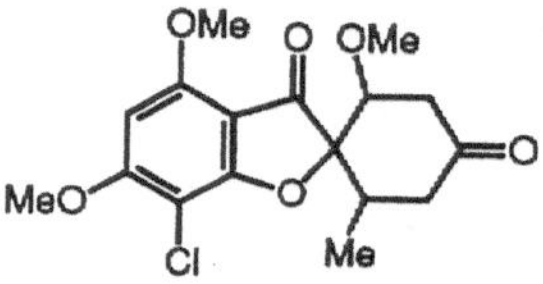

2207 (dihydrogriseofulvin)

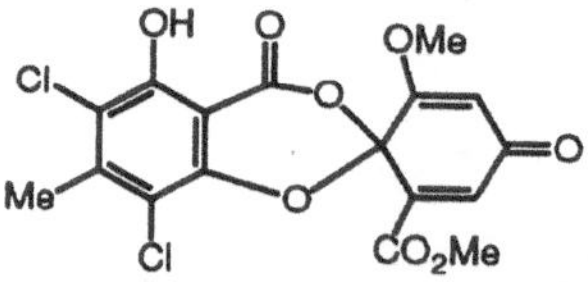

2208 (geodoxin)

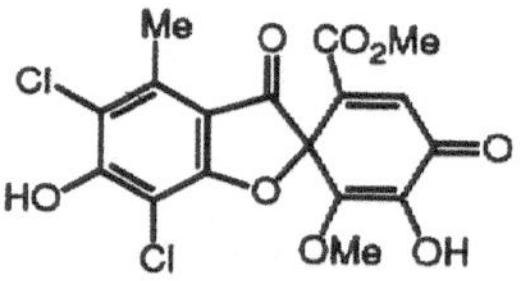

2209 (gillusdin)

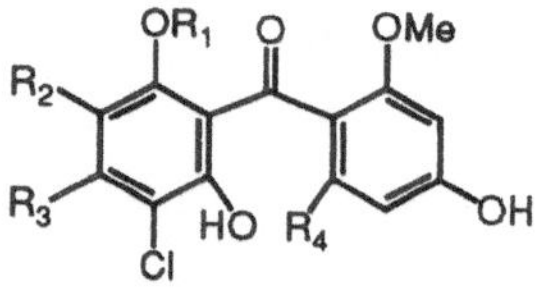

2210 $R_1 = R_4 = Me, R_2 = H, R_3 = OMe$ (griseophenone A)
2211 $R_1 = R_2 = H, R_3 = OMe, R_4 = Me$ (griseophenone B)
2212 $R_1 = H, R_2 = Cl, R_3 = Me, R_4 = CO_2Me$ (dihydrogeodin)
2213 $R_1 = R_2 = H, R_3 = Me, R_4 = CO_2Me$
2214 $R_1 = R_3 = R_4 = Me, R_2 = H$

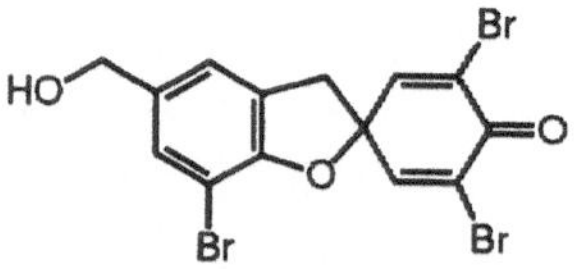

2215 (thelepin)

3.22.9. Miscellaneous Fungal Metabolites and Other Complex Phenols

In addition to the numerous fungal and lichen metabolites discussed in the preceding sections there are several others that do not easily fit into well-defined structural categories.

The fungal metabolite ascochlorin (**2216**) which is produced by *Ascochyta viciae* (*1819–1821*) was simultaneously isolated as "LL-Z1272γ" from *Fusarium* sp. (*1822*) and as "ilicicolin D" from *Cylindrocladium*

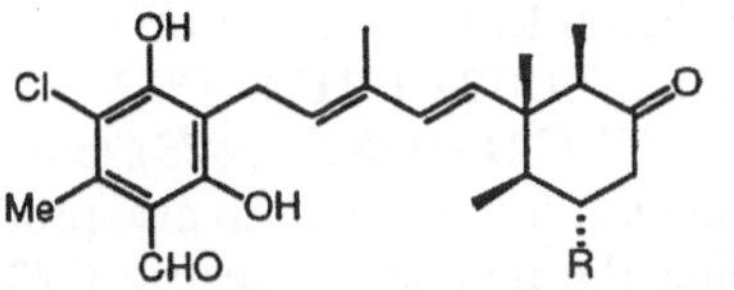

OH Cl Me OH CHO R O

2216 R = H (ascochlorin = LL-Z1272γ = ilicicolin D)
2219 R = OAc (LL-Z1272λ = ilicicolin F)

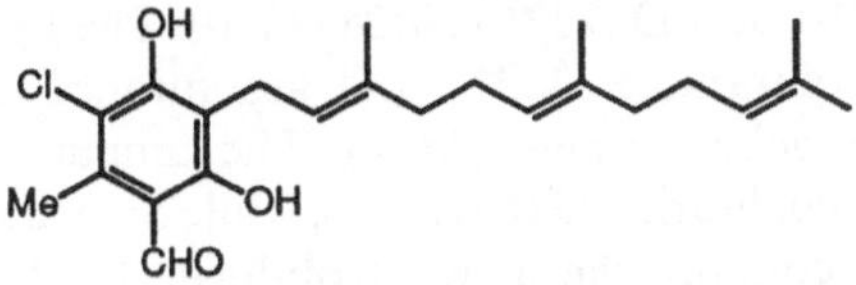

OH Cl Me OH CHO

2217 (LL-Z1272α = ilicicolin A)

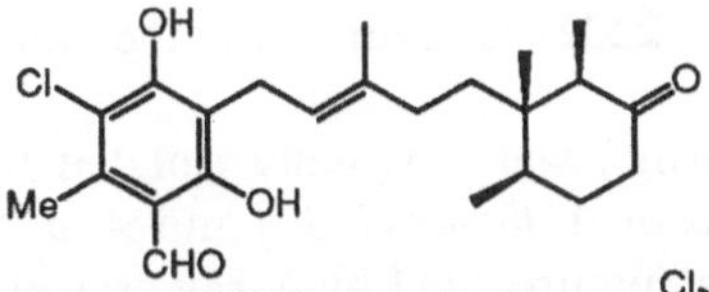

OH Cl Me OH CHO O

2218 (LL-Z1272δ = ilicicolin C)

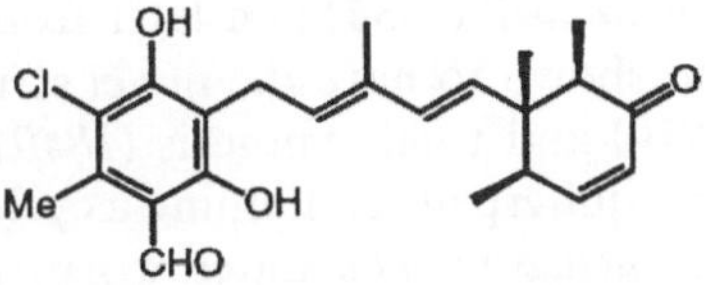

OH Cl Me OH CHO O

2220 (cylindrochlorin = ilicicolin E)

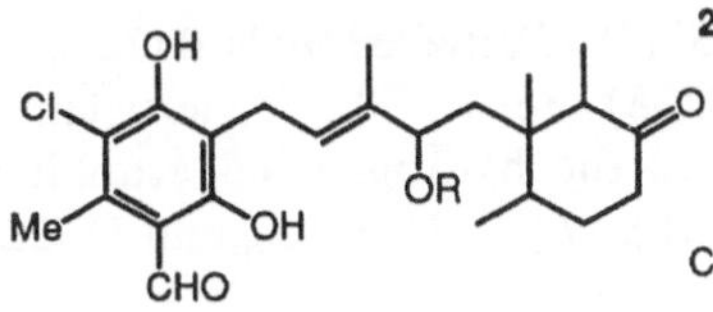

OH Cl Me OH CHO OR O

2221 R = Ac (chloronectrin)
2222 R = H

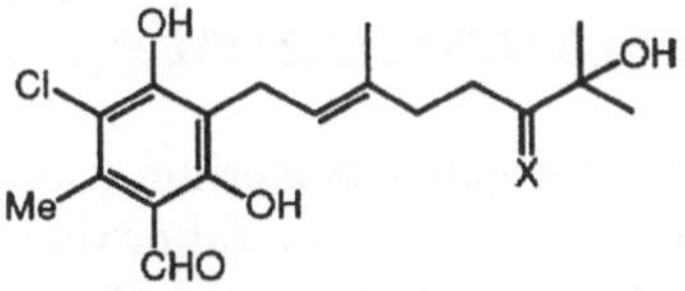

OH Cl Me OH CHO X OH

2223 X = H, OH (colletochlorin A)
2224 X = O (colletochlorin C)

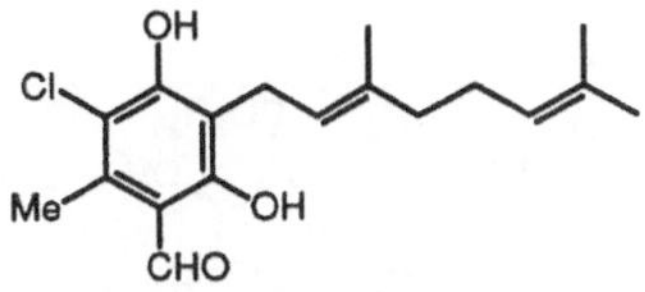

OH Cl Me OH CHO

2225 (colletochlorin B)

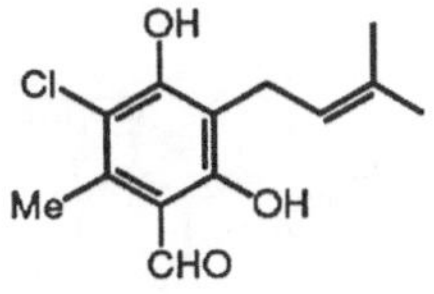

OH Cl Me OH CHO

2226 (colletochlorin D)

OH Cl Me OH CHO O X

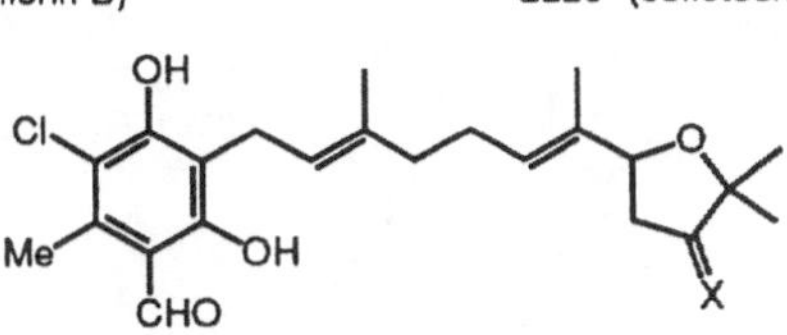

2227 X = O (ascofuranone)
2228 X = H, OH (ascofuranol)

ilicicola, a fungus of dead beech leaves (*1823, 1825*). Other metabolites from these fungi include LL-Z1272α (**2217**) (*1822, 1823*), LL-Z1272δ (**2218**) (*1822, 1823*), LL-Z1272λ (**2219**) (*1822, 1823*), and cylindrochlorin (**2220**) (*1823, 1824*). The fungus *Nectria coccinea* also produces several of these metabolites, including the new chloronectrin (**2221**) (*1826*). The hydroxy analogue **2222** is produced by *Ascochyta viciae* (*1827*). The tobacco pathogen *Colletotrichum nicotianae* produces colletochlorin A (**2223**), C (**2224**), B (**2225**), and D (**2226**) (*1828–1830*). The hypolipidemic active metabolites, ascofuranone (**2227**) and ascofuranol (**2228**), have been isolated from *Ascochyta viciae* (*1831*). The fungus *Acremonium luzulae* also produces ascochlorin (**2216**) (*1832*), while *Strobilurus tenacellus* and *Mycena* ssp. contain the new strobilurin B (**2229**) (*1833*). The simple naphthalene metabolite **2230** (and/or **2231**) is produced by *Verticillium lamellicola* (*1834*) and **2232** is found in the fungus *Scolecobasidiella avellanea* (*9*).

Antibiotic A30641 (= aspirochlorine) was originally isolated from *Aspergillus tamarii* (*1836*) and later from *A. flavus* (*1837*) and *A. oryzae* (*1838*) and shown to have the novel structure **2233** by X-ray crystallography (*1839*) and total synthesis (*1840*). Armillaridin (**2234**) is a novel phenolic sesquiterpene containing a cyclobutane ring that is produced by *Armillaria mellea* (*1841*). Later work with this organism revealed the related metabolites melleolide D (**2235**) (*1842*), melledonals B (**2236**) and C (**2237**) (*1843*), and armillaricin (**2238**) (*1844*). The Venezuelan soil fungus *Emericella falconensis* produces the hydrogenated azaphilones falconensins A (**2239**), C (**2240**), B (**2241**), D (**2242**) (*1845*), and H (**2243**) (*1846*).

The gram-negative bacterium *Xanthomonas juglandis* which is a walnut pathogen produces several novel brominated aryl polyene esters, an *iso*-butoxide cleavage fragment of which, xanthomonadin I (**2244**), has now been identified (*1847–1849*). The first of a series of novel chlorinated

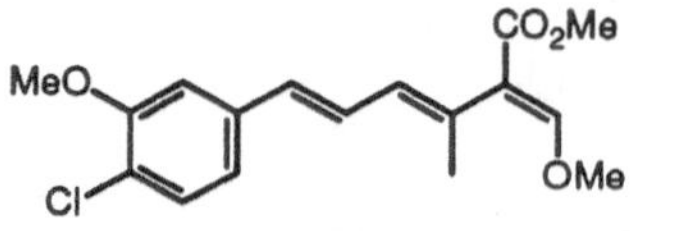

MeO, Cl, CO_2Me, OMe

2229 (strobilurin B)

R_1O_2C, CO_2R_2, HO, OH, Cl, Me, OH

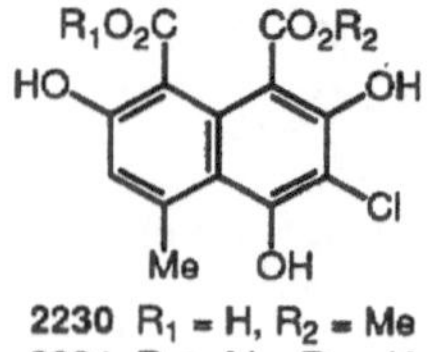

2230 R_1 = H, R_2 = Me
2231 R_1 = Me, R_2 = H

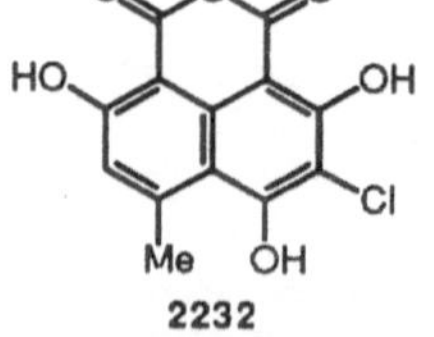

O, O, O, HO, OH, Cl, Me, OH

2232

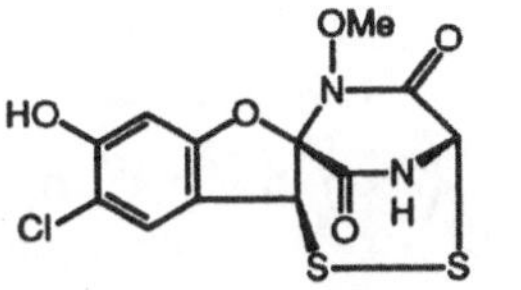

2233 (A30641 = aspirochlorine)

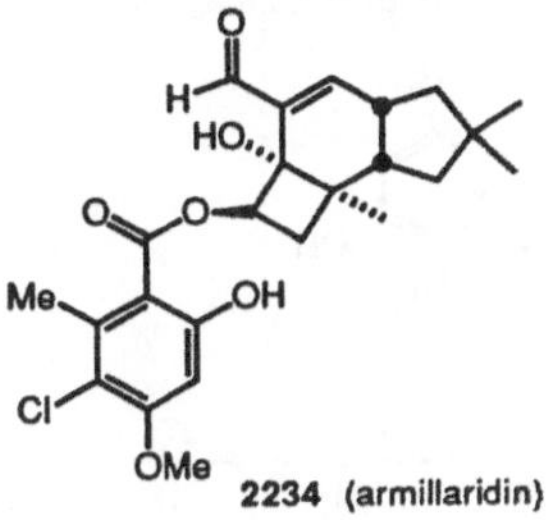

2234 (armillaridin)

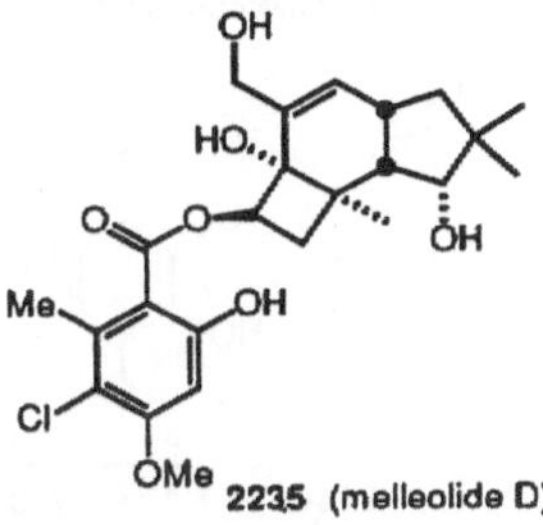

2235 (melleolide D)

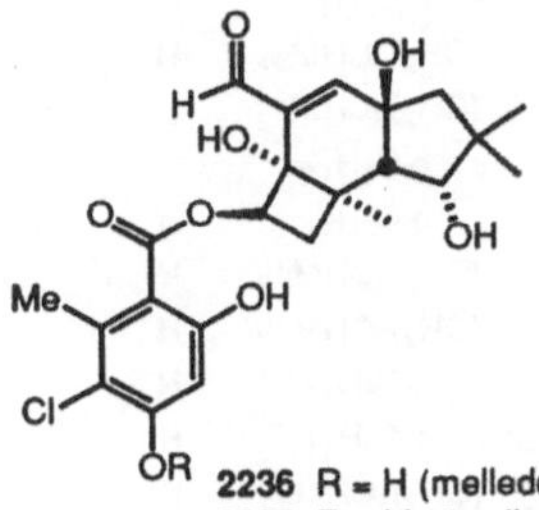

2236 R = H (melledonal B)
2237 R = Me (melledonal C)

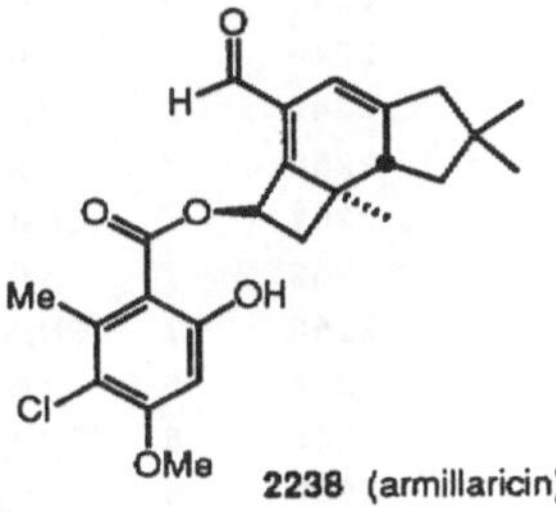

2238 (armillaricin)

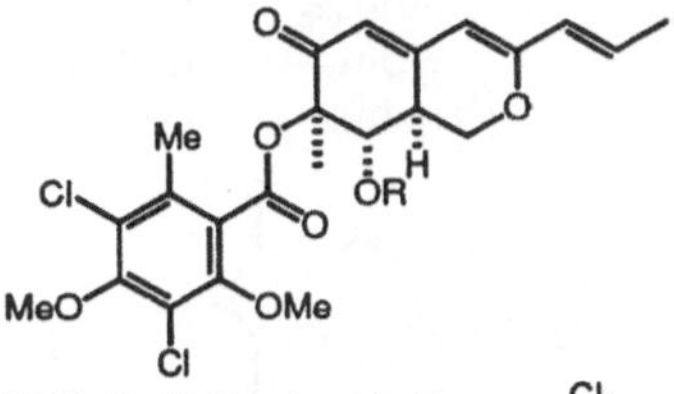

2239 R = H (falconensin A)
2240 R = Ac (falconensin C)

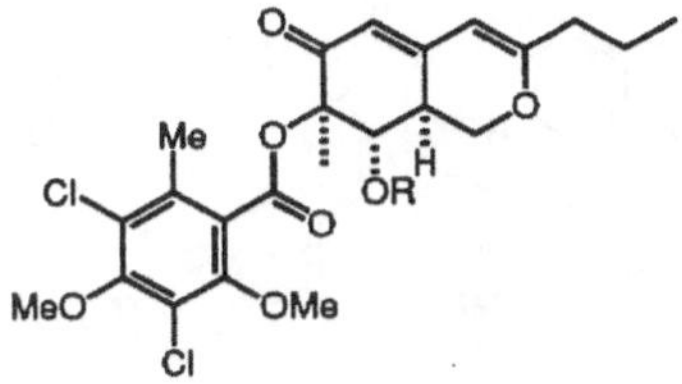

2241 R = H (falconensin B)
2242 R = Ac (falconensin D)

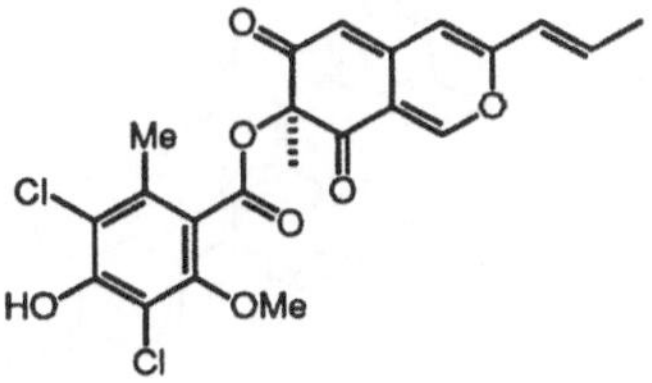

2243 (falconensin H)

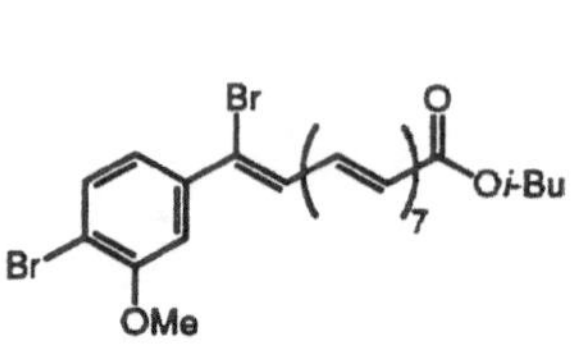

2244 (xanthomonadin I)

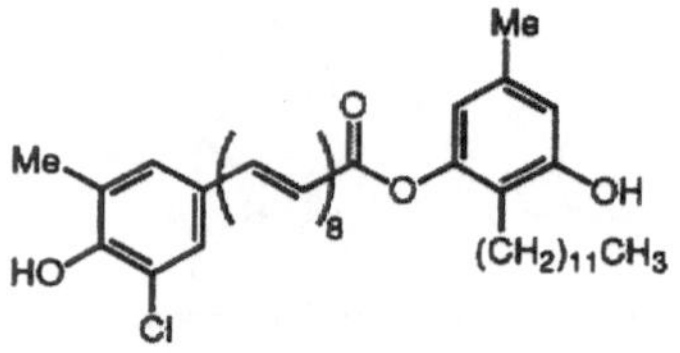

2245 (5-chloroflexirubin)

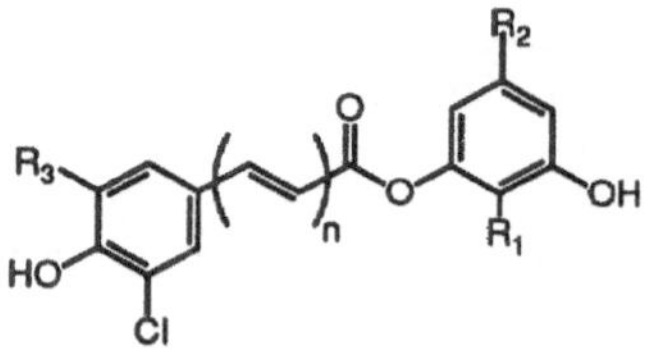

	n	R_1	R_2	R_3
2246	8	n-$C_{10}H_{21}$	n-C_5H_{11}	H
2247	8	$(CH_2)_8CHMe_2$	n-C_5H_{11}	H
2248	8	n-$C_{10}H_{21}$	$(CH_2)_3CHMe_2$	H
2249	8	$(CH_2)_8CHMe_2$	$(CH_2)_3CHMe_2$	H
2250	7	n-$C_{10}H_{21}$	n-C_5H_{11}	H
2251	7	$(CH_2)_8CHMe_2$	n-C_5H_{11}	H
2252	7	n-$C_{10}H_{21}$	$(CH_2)_3CHMe_2$	H
2253	7	$(CH_2)_8CHMe_2$	$(CH_2)_3CHMe_2$	H
2254	6	n-$C_{10}H_{21}$	n-C_5H_{11}	H
2255	6	$(CH_2)_8CHMe_2$	n-C_5H_{11}	H
2256	6	n-$C_{10}H_{21}$	$(CH_2)_3CHMe_2$	H
2257	6	$(CH_2)_8CHMe_2$	$(CH_2)_3CHMe_2$	H
2258	8	n-$C_{12}H_{25}$	CH_2CHMe_2	H (Fla-P1)
2259	8	n-$C_{12}H_{25}$	n-C_3H_7	H (Fla-P2)
2260	7	$(CH_2)_8CHMe_2$	$(CH_2)_3CHMe_2$	Cl

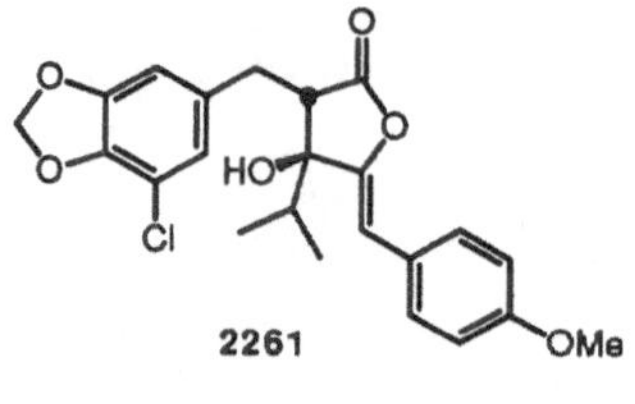

2261

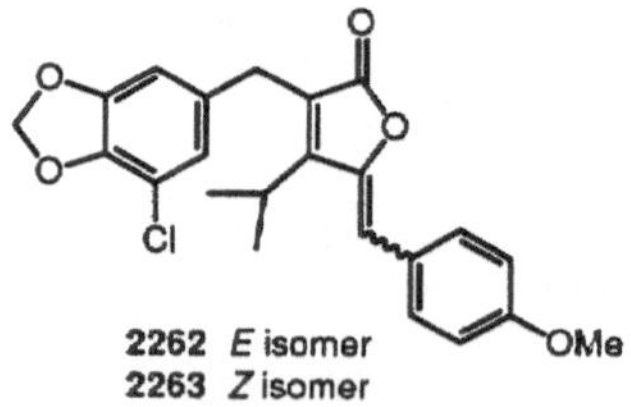

2262 *E* isomer
2263 *Z* isomer

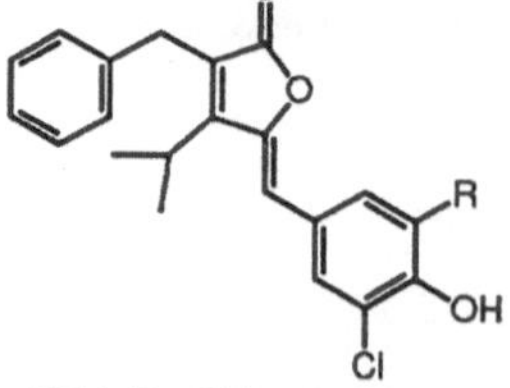

2264 R = Cl (nostoclide I)
2265 R = H (nostoclide II)

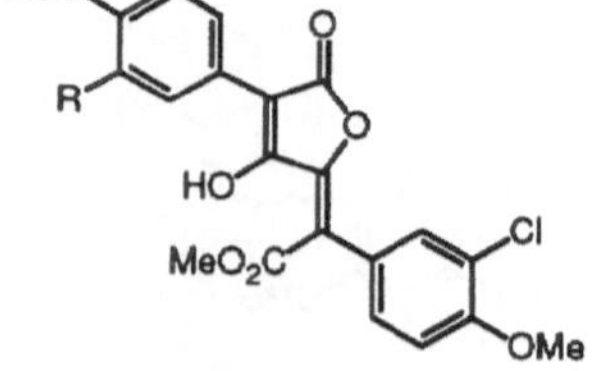

2266 R = H
2267 R = OMe

pigments, 5-chloroflexirubin, was isolated from the "gliding bacterium" *Flexibacter elegans* and shown to have structure **2245** (*1850*). Additional compounds of this type were isolated from *Cytophaga* sp. (*1851*), *C. johnsonae* (*1852*) and *Flavobacterium* sp. (*1853*) and were found to have closely related structures **2246–2260** (*1851–1854*). The freshwater cyanobacterium *Scytonema hofmanni* exudes the algicides **2261–2263** (*1855, 1856*), while the blue-green alga *Nostoc* sp., a cyanobacterium that is associated with the lichen *Peltigera canina* secretes nostoclides I (**2264**) and II (**2265**) into the culture medium in large quantities (*1857*). The two pulvinic acid derivatives **2266** and **2267** are produced by *Pulveroboletus auriflammeus* (*1858*).

The complex polyether phenol X-14766A (**2268**) is produced by *Streptomyces malachitofuscus* (*1859, 1860*). This metabolite is the first halogen-containing polyether antibiotic. The second such example was isolated from cultures of *Actinomadura routienii* and shown to have structure **2269** (*1861, 1862*).

The marine environment has yielded several novel brominated phenolic alkaloids. The mollusc *Aplysia kurodai* produces aplaminone (**2270**) neoaplaminone (**2271**) and sulfate **2272** (*1835*). The asmanian bryozoan *Amathia wilsoni* which was earlier shown to contain amathamides A (**1193**) and B (**1194**) has now been found to contain amathamides C–F (**2273–2276**) (*1863*). The Floridian *Amathia convoluta* produces convolutamides A–F (**2277–2282**). Some of these possess antitumor activity and all contain the novel *N*-acyl-γ-lactam grouping attached to a dibromophenol (*1864*). The red alga *Halopytis pinastroides* produces the novel orthocyclophane cyclotribromoveratrylene (**2283**) (*1865*) whose conformation has been studied by NMR (*1866*).

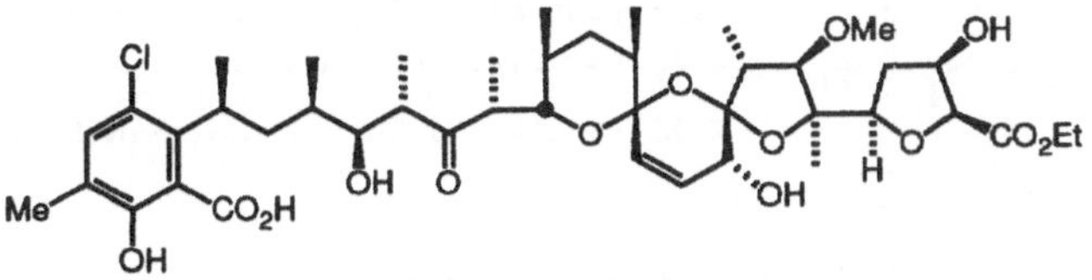

2268 (X-14766A)

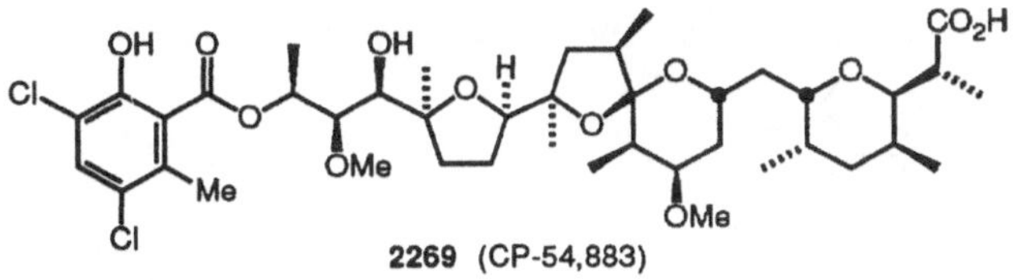

2269 (CP-54,883)

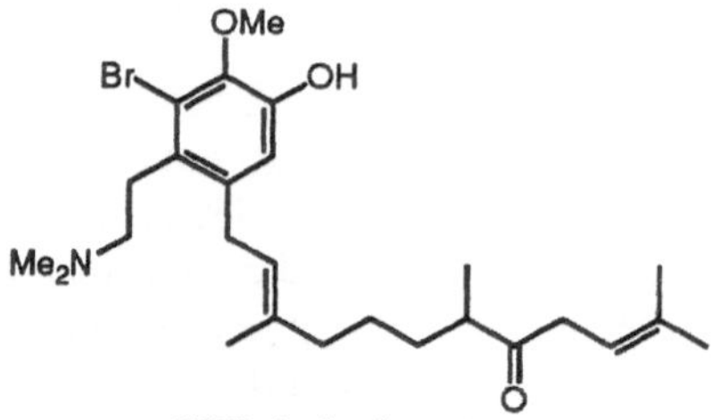

2270 (aplaminone)

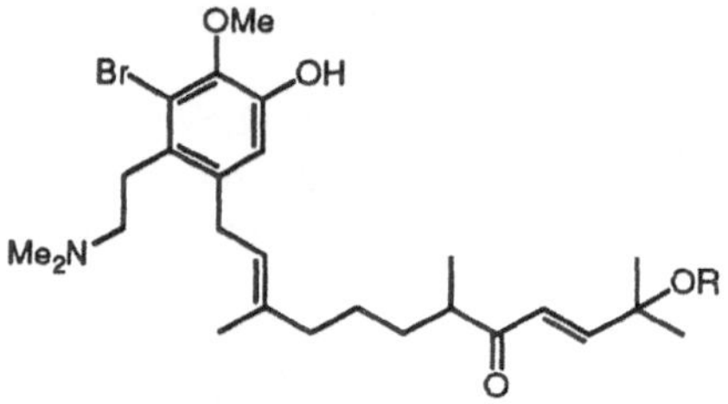

2271 R = H (neoaplaminone)
2272 R = SO_3H

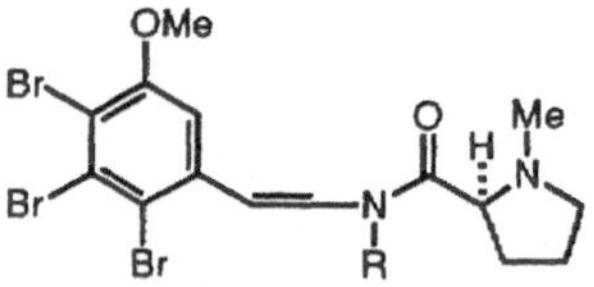

2273 *E*-isomer R = Me (amathamide C)
2274 *E*-isomer R = H (amathamide E)
2275 *Z*-isomer R = H (amathamide F)

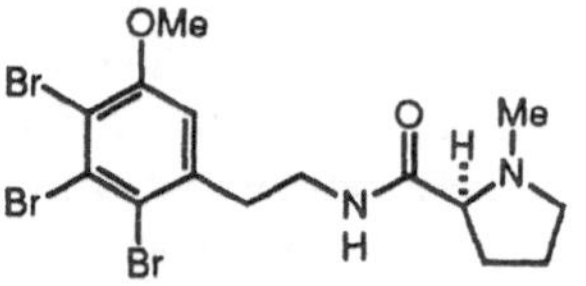

2276 (amathamide D)

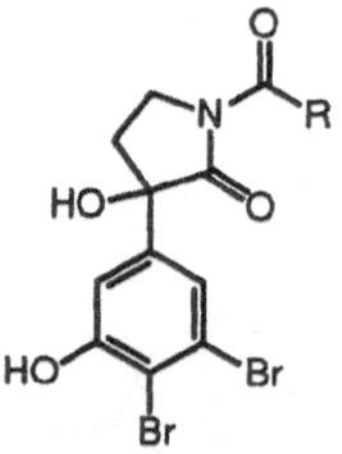

2277 R = n-$C_{13}H_{27}$ (convolutamide A)
2278 R = $(CH_2)_7CH\overset{Z}{=}CH(CH_2)_5CH_3$ (convolutamide B)
2279 R = n-$C_{15}H_{31}$ (convolutamide C)
2280 R = $(CH_2)_7CH\overset{Z}{=}CH(CH_2)_7CH_3$ (convolutamide D)
2281 R = n-$C_{17}H_{35}$ (convolutamide E)
2282 R = $(CH_2)_6CH\overset{?}{=}CH(CH_2)_{10}CH_3$ (convolutamide F)

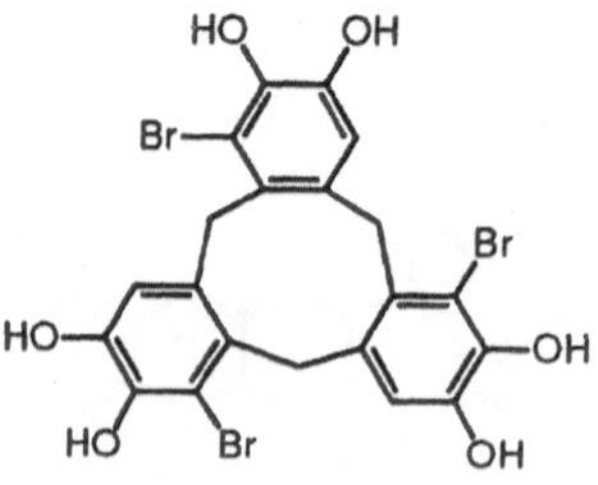

2283 (cyclotribromoveratrylene)

3.23. Glycopeptides

Although the large and growing class of metabolites known as glycopeptides could have been included under "Complex Phenols," their extreme importance as clinical antibiotics suggested that they be treated separately.

For example, the prototypical member of this group of compounds is vancomycin which was isolated in 1956 from *Amycolatopsis orientalis* (*1867*) and has been used for 35 years to treat deep-seated gram positive bacterial infections. Vancomycin is the drug of choice for methicillin-resistant *Staphylococcus aureus* infections, particularly those that occur in hospitals (*1868–1870*), although clinical resistance to vancomycin is increasing (*1871*, *1872*). Of the more than 200 identified and unidentified glycopeptides most contain phenolic chlorine as part of the core heptapeptide unit (*1870*). The role of the chlorine is apparently to contribute both to the stability and specificity of the peptide binding site, since the didechloro and monochloro vancomycin analogues show decreased peptide binding and decreased activity (*1873*).

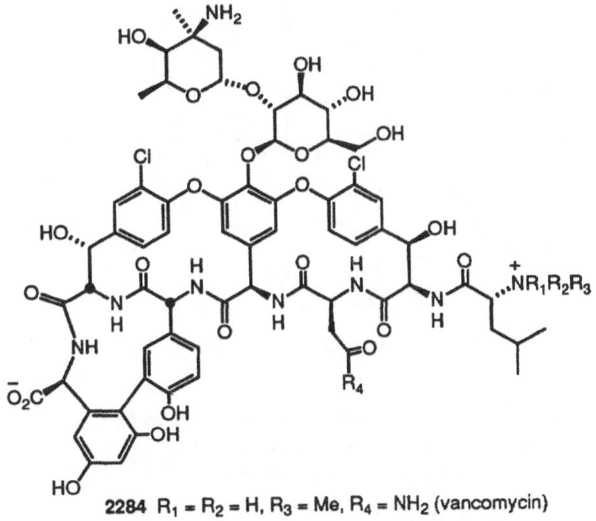

2284 $R_1 = R_2 = H$, $R_3 = Me$, $R_4 = NH_2$ (vancomycin)
2285 $R_1 = H$, $R_2 = R_3 = Me$, $R_4 = NH_2$ (M43D)
2286 $R_1 = R_2 = R_3 = Me$, $R_4 = NH_2$ (M43A)
2287 $R_1 = R_2 = R_3 = Me$, $R_4 = OH$ (M43B)
2288 $R_1 = R_2 = H$, $R_3 = Me$, $R_4 = OH$ (M43F)
2289 $R_1 = R_2 = R_3 = H$, $R_4 = NH_2$ (A51568A)

After many years of effort the structure of vancomycin (**2284**) was finally elucidated (*1874–1880*) and studies of the conformation and peptide binding of vancomycin and other glycopeptides have been reported (*1873*, *1881–1887*). Other vancomycin analogues, **2285–2289**, have been isolated from *Amycolatopsis orientalis* (*1888*, *1889*, *1870*). The glutamine analogues, M43G (**2290**) and A51568NB (**2291**), of vancomycin and A51568A, respectively, where a glutamine amino acid has replaced asparagine are also known (*1870*, *1889*) (not shown). Moreover, M43E (**2292**) is a desoxyvancomycin where the right-hand benzylic hydroxyl is replaced by hydrogen (*1870*).

The organism *Actinoplanes teichomyceticus* produces a set of teicoplanins (**2293–2301**) (renamed from teichomycins) whose core differs slightly from that of vancomycin and which contain a unique long acyl chain attached to a sugar residue (*1890–1897*). The third major group of glycopeptides to be discovered were the avoparcins ("LL-AV290")

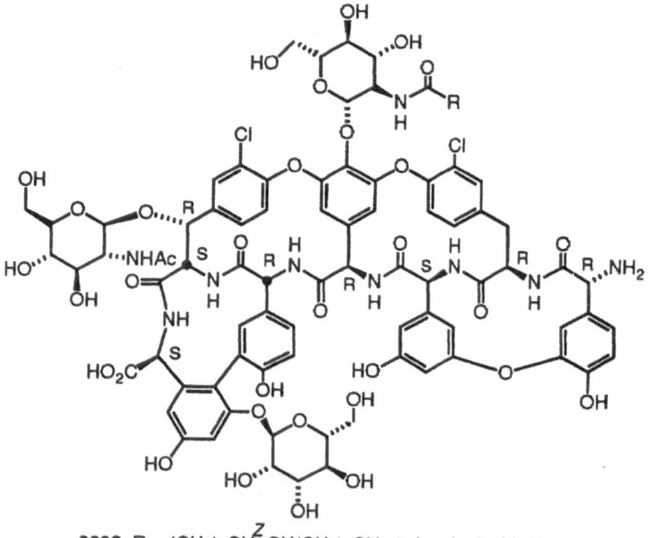

2293 R = $(CH_2)_2CH\overset{Z}{=}CH(CH_2)_4CH_3$ (teicoplanin A2-1)
2294 R = $(CH_2)_6CH(CH_3)_2$ (teicoplanin A2-2)
2295 R = n-C_9H_{19} (teicoplanin A2-3)
2296 R = $(CH_2)_6CH(CH_3)CH_2CH_3$ (teicoplanin A2-4)
2297 R = $(CH_2)_7CH(CH_3)_2$ (teicoplanin A2-5)
2298 R = $(CH_2)_8CH(CH_3)_2$ (teicoplanin RS-1)
2299 R = n-$C_{11}H_{23}$ (teicoplanin RS-2)
2300 R = $(CH_2)_4CH(CH_3)CH_2CH_3$ (teicoplanin RS-3)
2301 R = n-C_8H_{17} (teicoplanin RS-4)

(**2302–2304**), which were isolated from cultures of *Streptomyces candidus* (*1898–1902, 2228*). Helvecardins A (**2305**) and B (**2306**) were isolated from *Pseudonocardia compacta* subsp. *helvetica* and found to be similar to the avoparcins except for possessing the novel sugar 2-*O*-methylrhamnose (*1903, 1904*).

The chloropolysporins A–C from *Faenia interjecta* have structures very similar to those of the avoparcins but with an extra chlorine atom (**2307, 2308**) (*1905–1907*). Chloropolysporin A has an extra galactose but has not been fully characterized (*1907*). The family of actaplanin antibiotics **2309–2315** has been isolated from *Actinoplanes missouriensis* (*1908–1910*). In the presence of bromide ion the organism produces two brominated analogues (*1911*). A series of related antibiotics, the parvodicins (**2316–2322**), are produced by *Actinomadura parvosata* (*1912*). Like the teicoplanins, they feature a long acyl side chain.

An examination of *Kibdelosporangium aridum* from an Arizona desert soil sample has yielded several novel glycopeptides including the aridicins A–C (**2323–2325**) (*1913–1917*) and, from *K. aridum* subsp. *largum*, the kibdelins A–D (**2326–2330**) (*1918, 1919*). The closely related

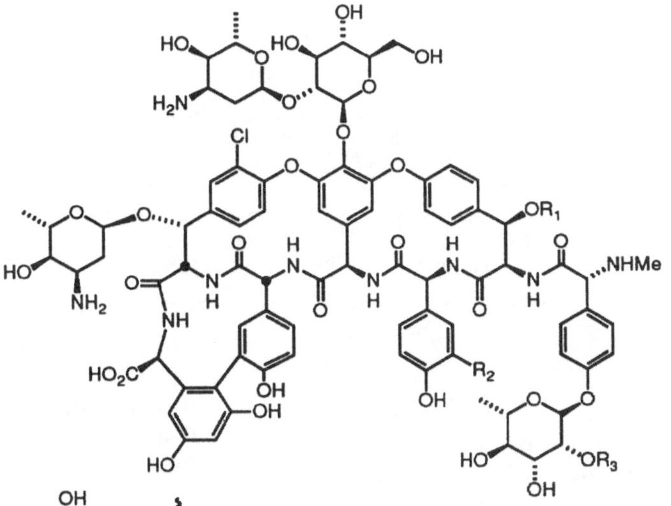

A = 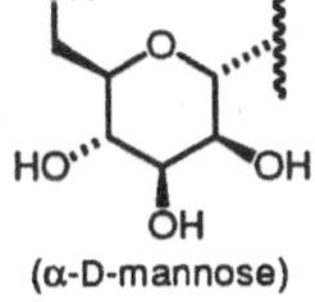 (α-D-mannose)

2302 R_1 = A, R_2 = Cl, R_3 = H (β-avoparcin)
2303 R_1 = A, R_2 = R_3 = H (α-avoparcin)
2304 R_1 = R_3 = H, R_2 = Cl (ε-avoparcin)
2305 R_1 = A, R_2 = Cl, R_3 = Me (helvecardin A)
2306 R_1 = H, R_2 = Cl, R_3 = Me (helvecardin B)

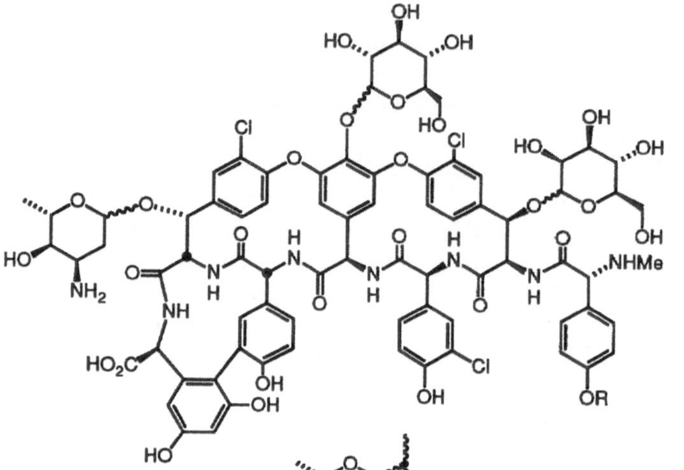

2307 R = (sugar) (chloropolysporin B)

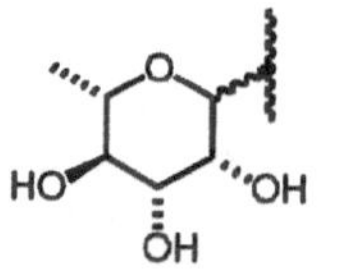

2308 R = H (chloropolysporin C)

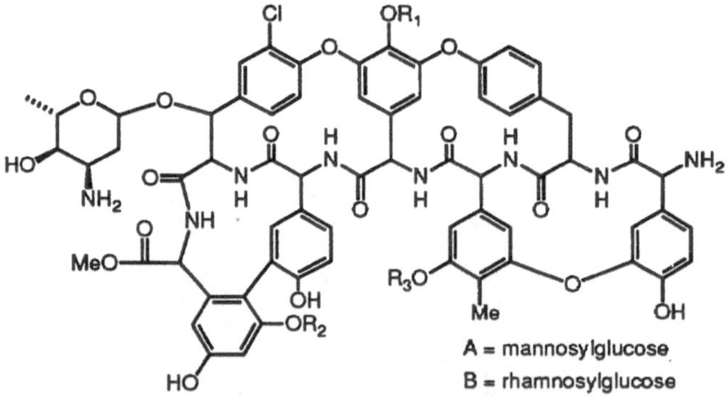

	R_1	R_2	R_3	
2309	A	D	D	(actaplanin A)
2310	B	D	D	(actaplanin B_1)
2311	C	D	D	(actaplanin B_2)
2312	A	D	H	(actaplanin B_3)
2313	B	D	H	(actaplanin C_1)
2314	C	D	H	(actaplanin G)
2315	H	H	H	(actaplanin ϕ)

A = mannosylglucose
B = rhamnosylglucose
C = glucose
D = mannose

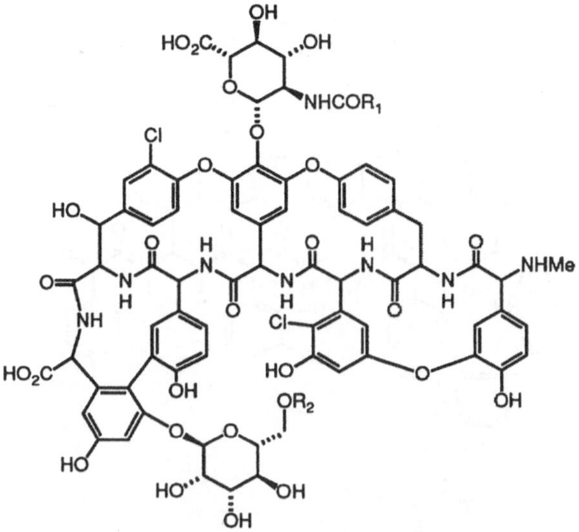

2316 $R_1 = n\text{-}C_9H_{19}$, R_2 = H (parvodicin A)
2317 $R_1 = (CH_2)_7CH(CH_3)_2$, R_2 = H (parvodicin B_1)
2318 $R_1 = n\text{-}C_{10}H_{21}$, R_2 = H (parvodicin B_2)
2319 $R_1 = (CH_2)_8CH(CH_3)_2$, R_2 = H (parvodicin C_1)
2320 $R_1 = n\text{-}C_{11}H_{23}$, R_2 = H (parvodicin C_2)
2321 $R_1 = (CH_2)_8CH(CH_3)_2$, R_2 = Ac (parvodicin C_3)
2322 $R_1 = n\text{-}C_{11}H_{23}$, R_2 = Ac (parvodicin C_4)

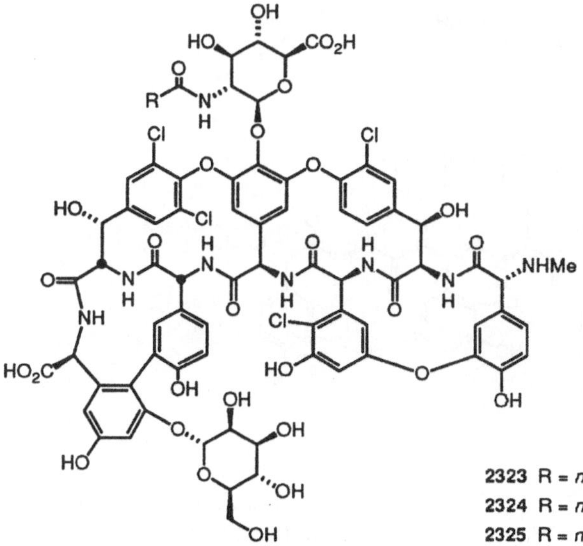

2323 R = $n\text{-}C_9H_{19}$ (aridicin A)
2324 R = $n\text{-}C_{10}H_{21}$ (aridicin B)
2325 R = $n\text{-}C_{11}H_{23}$ (aridicin C)

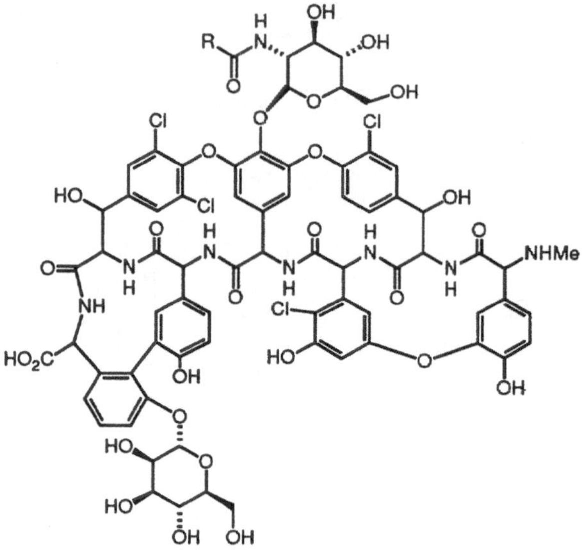

2326 R = n-C_9H_{19} (kibdelin A)
2327 R = $(CH_2)_7CH(CH_3)_2$ (kibdelin B)
2328 R = $(CH_2)_8CH(CH_3)_2$ (kibdelin C_1)
2329 R = n-$C_{11}H_{23}$ (kibdelin C_2)
2330 R = $(CH_2)_2CH \overset{Z}{=} CH(CH_2)_4CH_3$ (kibdelin D)

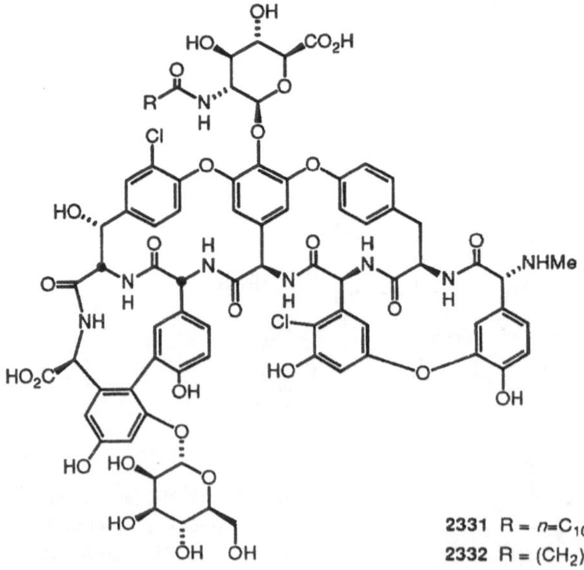

2331 R = n-$C_{10}H_{21}$ (A-40926-A)
2332 R = $(CH_2)_8CH(CH_3)_2$ (A-40926-B)

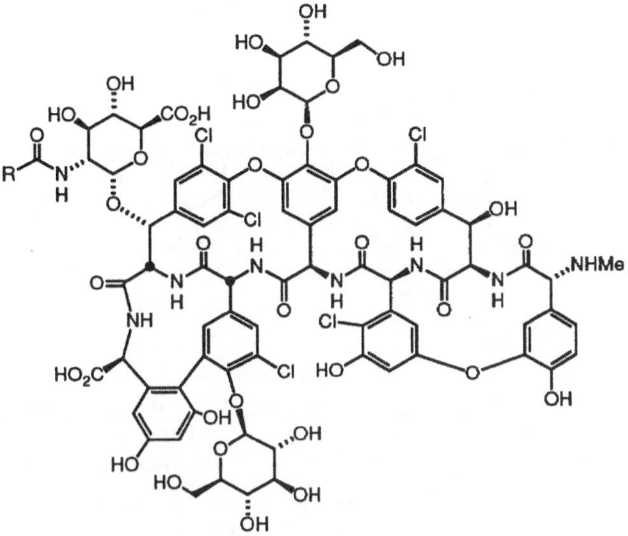

2333 R = $(CH_2)_5CH(CH_3)_2$ (MM 55266)
2334 R = n-C_9H_{19} (MM 55268)

A-40926-A (**2331**) and -B (**2332**) are produced by an *Actinomadura* sp. (*1920*), and MM-55266 (**2333**) and MM-55268 (**2334**) are found in cultures of an *Amycolatopsis* sp. (*1921*).

A set of vancomycin-related glycopeptides has been discovered in *Nocardia orientalis* (*1922*) and *Amycolatopsis orientalis* (*1923*), and designated as orienticins and chloroorienticins (**2335–2342**). An independent study also identified A82846B as identical with chloroorienticin A (*1924*). This latter study also described A82846A (**2343**) which is the same as eremomycin isolated from the same organism (*1925*, *1926*). This organism also produces A42867 (**2344**) (*1927*), MM-47761 (**2345**) (*1928*), and UK-72,051 (= orienticin A) (*1934*).

A series of glycopeptide antibiotics, the A-41030 complex, has been isolated from *Streptomyces virginiae* (**2346–2350**) although not all of the compounds have been identified (*1929*). For example, in A41030-D and -G butyl groups are attached somewhere in the molecule. Fermentation of *Actinoplanes* sp. produces UK-68,597 (**2351**) (*1930*) and a *Saccharothrix* sp. produces galacardins A (**2352**) and B (**2353**) (*1931*). Balhimycin (**2354**) which was isolated from *Amycolatopsis* sp. has a novel keto sugar moiety (*1932*, *1933*).

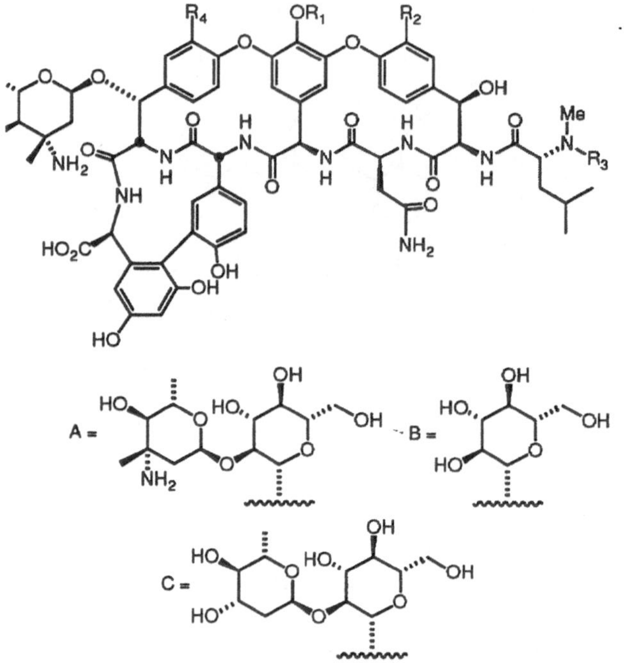

	R_1	R_2	R_3	R_4	
2335	A	Cl	H	Cl	(chloroorienticin A)
2336	B	Cl	H	Cl	(chloroorienticin B)
2337	H	Cl	H	Cl	(chloroorienticin C)
2338	A	Cl	Me	Cl	(chloroorienticin D)
2339	B	Cl	Me	Cl	(chloroorienticin E)
2340	A	H	H	Cl	(orienticin A)
2341	C	H	H	Cl	(orienticin B)
2342	A	H	Me	Cl	(orienticin D)
2343	A	Cl	H	H	(A82846A)

The microbe *Streptomyces hygroscopicus* subsp. *hiwasaensis* produces OA-7653-A (**2355**) and -B (**2356**) which contain the novel dimethylamino terminal amino acid (*1935, 1936*). Complestatin (**2357**) which is produced by *Streptomyces lavendulae* (*1937, 2233*), and kistamicins A (**2358**) and B (**2359**) which are found in cultures of *Microtetraspora parvosata* (*1938, 1939*) are related to the known glycopeptides but with the presence of an indole ring and six atoms of chlorine. Although complestatin has no antibacterial activity it is a potent inhibitor of the complement system. The kistamicins are active against the influenza Type A virus.

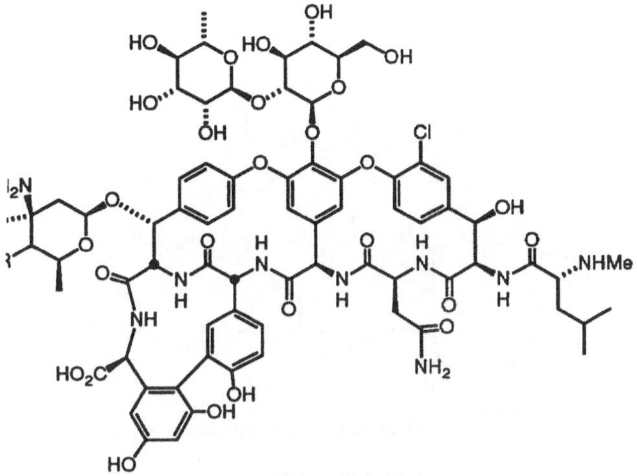

2344 R = β-OH (A42867)
2345 R = α-OH (MM47761)

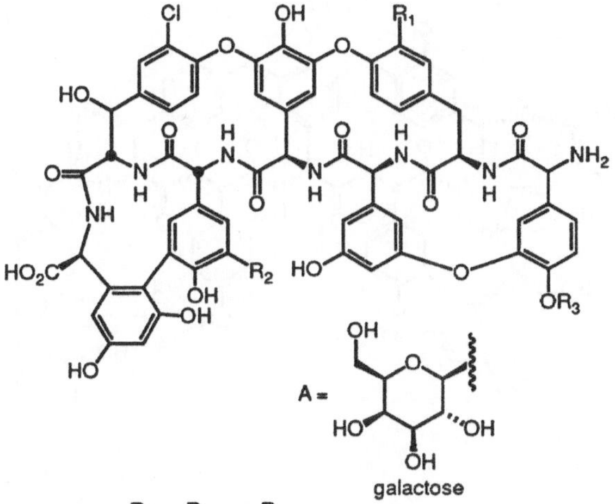

A = galactose

	R_1	R_2	R_3	
2346	Cl	Cl	H	(A41030-A)
2347	Cl	H	H	(A41030-B)
2348	Cl	Cl	A	(A41030-C)
2349	H	H	H	(A41030-E)
2350	Cl	Cl	gal-gal	(A41030-F)

R = α-L-vancosaminyl(1-2)β-D-glucosyl

2351 (UK-68,597)

2352 R = galactose (galacardin A)
2353 R = H (galacardin B)

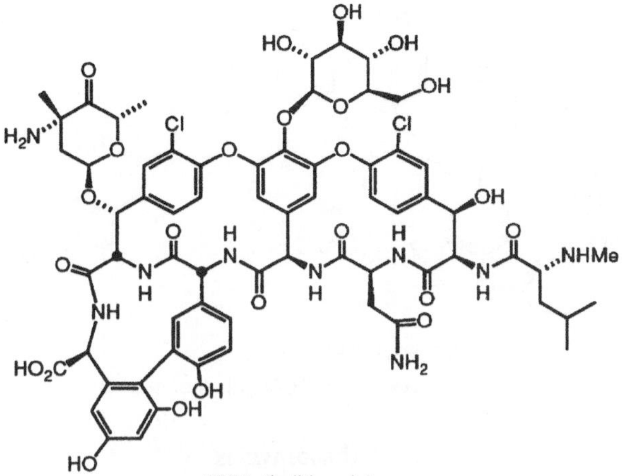

2354 (balhimycin)

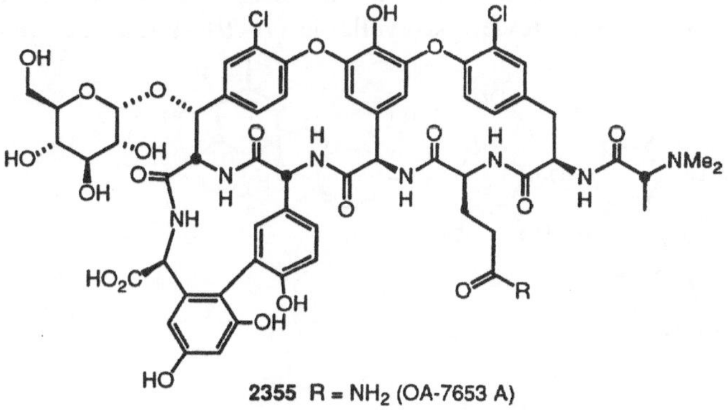

2355 R = NH_2 (OA-7653 A)
2356 R = OH (OA-7653 B)

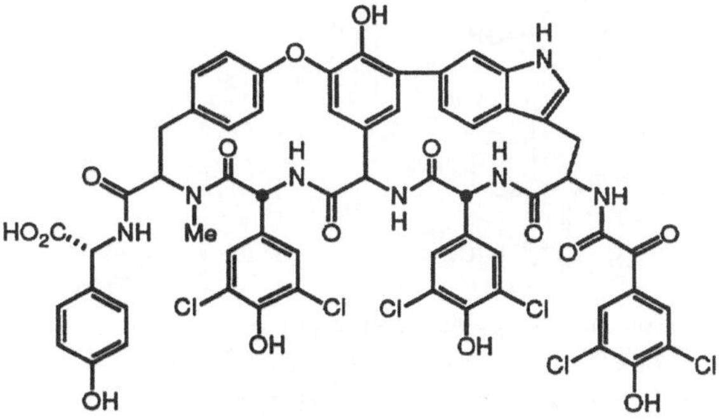

2357 (complestatin)

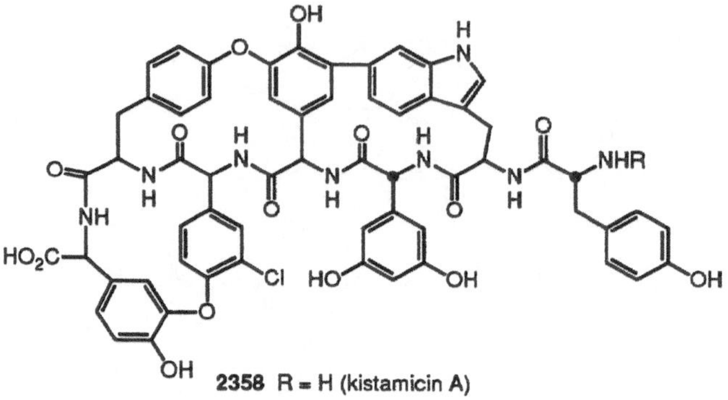

2358 R = H (kistamicin A)
2359 R = $CONHCH_2CH_2Ph$ (kistamicin B)

3.24. Orthosomycins

A relatively small family of novel chlorophenol-oligosaccharide antibiotics known as the "orthosomycins" has been discovered in recent years, and an extensive review is available (*1940*). These metabolites

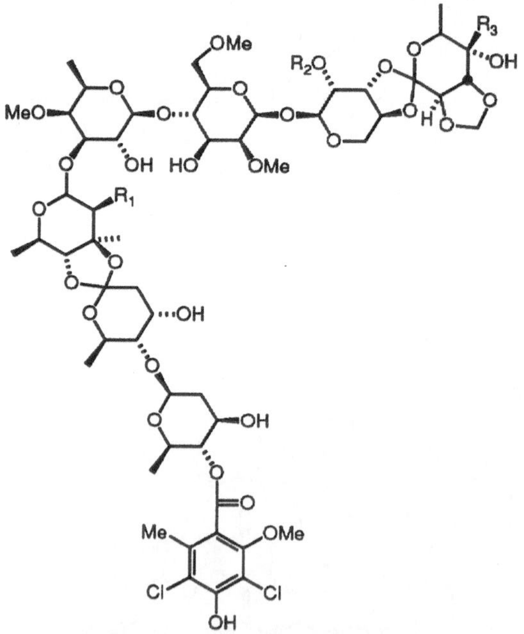

2360 R_1 = H, R_2 = R_3 = Ac (curamycin A)
2361 R_1 = OH, R_2 = CO*i*-Pr, R_3 = Ac (flambamycin)
2362 R_1 = H, R_2 = CO*i*-Pr, R_3 = Ac (avilamycin B)
2363 R_1 = H, R_2 = CO*i*-Pr, R_3 = $CH(OH)CH_3$ (avilamycin C)

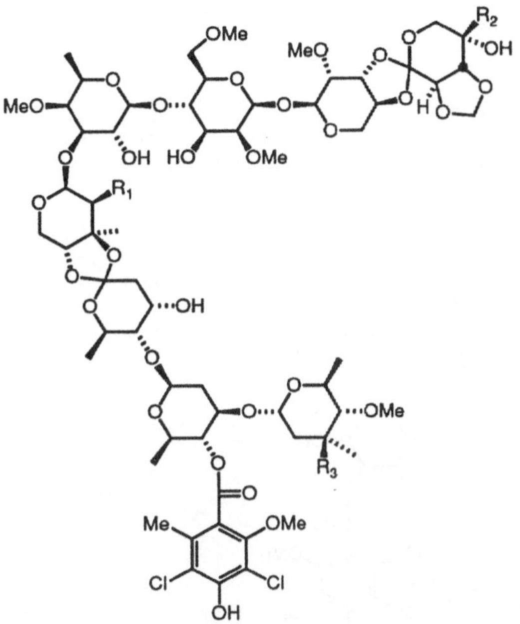

2364 R_1 = OH, R_2 = CH(OMe)Me, R_3 = NO_2 (everninomicin B)
2365 R_1 = R_2 = H, R_3 = NO_2 (everninomicin C)
2366 R_1 = H, R_2 = CH(OMe)Me, R_3 = NO_2 (everninomicin D)

include curamycin A (**2360**) from *Streptomyces cura-coi* (*1941*), flambamycin (**2361**) from *S. hygroscopicus* (*1942–1944*), avilamycins B (**2362**) and C (**2363**) from *S. viridochromogenes* (*1945–1947*) and everninomicins B (**2364**), C (**2365**), D (**2366**) and 2 (**2367**) from *Micromonospora carbonacea* (*1948–1954*). The structure elucidation of these compounds prior to modern NMR spectroscopy represents a significant achievement; an X-ray crystallographic study of the major sugar fragment olgose provided the crowning information (*1954*).

3.25. Dioxins

"Dioxin" – no chemical is more feared by the general public, and no chemical has been more scrutinized by policy regulators and environmental scientists (*1955–1959*). Although the family of chemicals known as dioxins (polychlorodibenzo-*p*-dioxins) (PCDDs) (**2368**) and the related polychlorodibenzofurans (PCDFs) (**2369**) are highly toxic to some

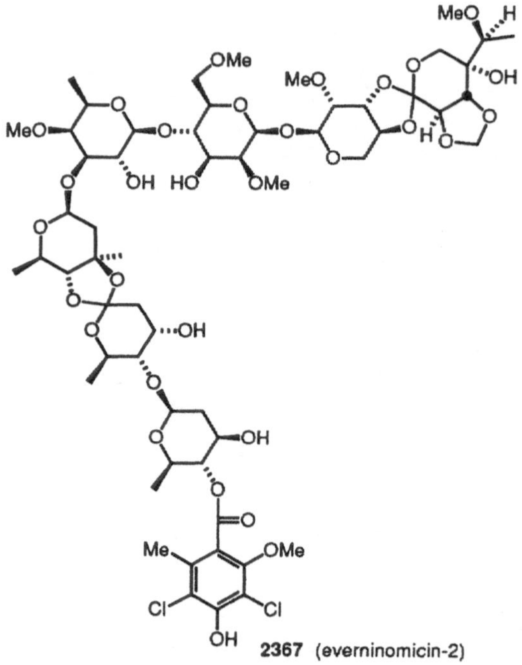

2367 (everninomicin-2)

animal species, the evidence of lasting human toxicity, apart from the skin disease chloracne (*1971*), is much less convincing (*1955*). Studies of industrial plant workers (*1960–1962*), Vietnam Veterans (*1963–1966*), and Seveso inhabitants (*1967–1969*) clearly show that dioxin is not as toxic to humans as was feared by many scientists twenty years ago.

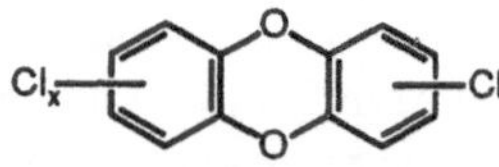

2368 (PCDDs) 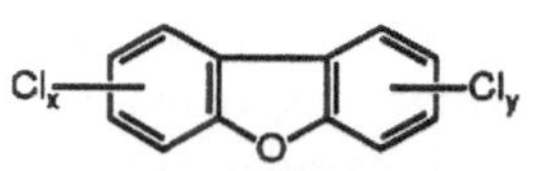 **2369** (PCDFs)

Although dioxin was first identified as an anthropogenic contaminant of chlorinated phenoxy herbicides, such as 2,4,5-T and silvex (*1955*, *1961*, *1970*) it is now recognized that dioxins and furans form during most if not all combustion processes such as waste incineration, coal burning, automobile exhaust, tobacco smoke, power plants and others (*1955–1958*, *1961*, *1972–1978*, *2235*). Under proper operating conditions such as sufficiently high temperatures, dioxins cannot be detected in waste incinerators (*1979*).

Studies of ancient soil samples and lake sediments reveal the presence of PCDDs and PCDFs (*1980*, *1981*). This suggests that these chlorinated compounds are the products of natural combustion processes (e.g., forest fires) and have been on earth for eons. Deep sediments dating back 8,100 years show the presence of PCDDs in amounts well above background levels (*1994*). Since wood burning produces PCDDs and PCDFs it has been suggested that forest fires are the major source of these environmental chemicals (*1982–1984*). Conditions conducive to dioxin formation are the relatively poor efficiency and incomplete oxidation when damp vegetation and wood are burned in the presence of chloride (70–2,100 ppm in wood (*43*) and 200–10,000 ppm in plants (*70*)). It is estimated that 60 kg of dioxins are produced annually in Canadian forest fires (*1984*). Since most forest and brush fires are caused by lightning and there are 200,000 forest fires annually worldwide burning 27,000 square miles (*2229*), it is obvious that dioxins are and have been naturally present in the environment for countless centuries.

Even more surprising are the observations of the biogenic formation of dioxins. In a series of papers a Swedish team of researchers has demonstrated that natural enzymes (horseradish peroxidase, lactoperoxidase) oxidize chlorophenols to PCDDs and PCDFs in the ppm range (*1985–1988*). A French group had earlier noticed the same transformation (*1989*). Nearly 40 different PCDDs and PCDFs have been identified with tetra-, penta-, and hexachloro derivatives predominating (*1986–1988*). Moreover, PCDDs and PCDFs are also formed enzymatically in sewage sludge and in garden compost piles (*1990–1991*, *2016*). Because the three necessary ingredients – chlorophenols, hydrogen peroxide, and peroxidase enzymes – are ubiquitous in living systems (*1987*, *1992*), it is not unreasonable to propose that a major source of "background" PCDDs and PCDFs in the environment is their natural production by biological systems in the soil and water. For example, there is evidence for the natural production of 2,4,6-trichlorophenol from organic matter (humic acid) in surface waters by the action of natural chloroperoxidase (*1458*).

Although many different PCDDs and PCDFs have been identified in both combustion and biogenic studies, for the purpose of counting these

2370  **2371**

compounds as natural products, we have limited the number to two (**2368, 2369**).

The only known naturally occurring brominated dioxins are **2370** and **2371** isolated from the sponge *Tedania ignis* (*1532*).

The photochemical conversion of chlorinated grisadiendiones into polychlorodibenzofurans has been reported Eq. (1) (*1993*), but the relevance of this observation to the natural occurrence of PCDFs is unknown.

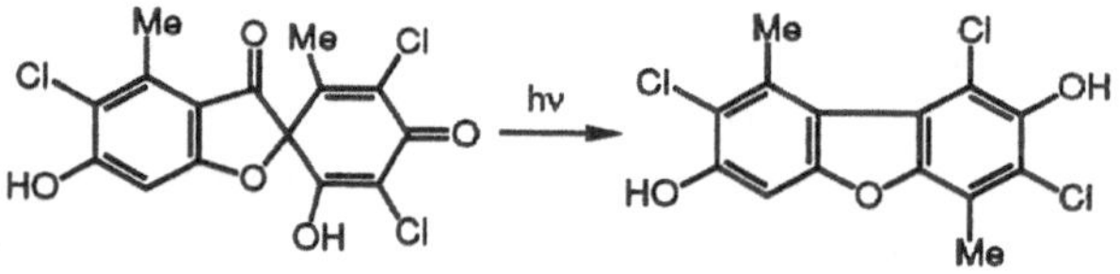

(1)

3.26. Humic Acids

Organic plant matter decays to humic acid substances at a rate of 63 billion tons/year. The total estimated global soil humic acid is 1.0–1.5 trillion tons and another trillion tons of humic acids are in the oceans. Rivers and lakes are also repositories of these highly condensed aromatic phenolic compounds (*1995*).

Numerous studies over the past few years have provided evidence that a large source of environmental organohalogen compounds, particularly organochlorines, may originate from the natural chlorination of humic and fulvic acids and their subsequent breakdown to chlorophenols, chloroform, and other one- and two-carbon chlorinated compounds. Moreover, as we have seen in the previous section, there is evidence that chlorophenols can be enzymatically transformed into dioxins (Scheme II).

Several studies have revealed that a large source of unidentified naturally occurring organohalogen compounds in surface waters (lakes and rivers) that do not receive industrial discharge (*1457*, *1996–2001*). For example, it is estimated that 50–55% of the organohalogens in the River Rhine are of natural origin (*2002*, *2003*) and humus samples from forest and bog drainage water show high concentrations of organochlorines (200–400 ppm) and lesser amounts of organobromines (7–16 ppm) (*2004*). There is also evidence that sea water iodine is incorporated into humic acid deposits (*2005*). Plants, living or dead, can discharge significant quantities of humic acid substances into rivers,

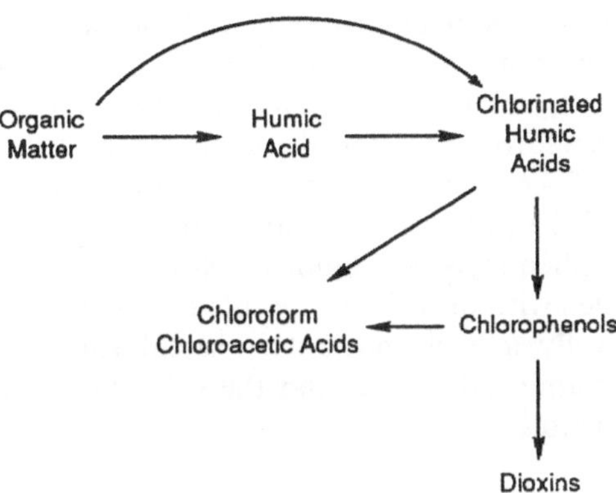

Scheme 11. The possible natural formation of organochlorines from humic acid substances

lakes and the ocean. In one study of cordgrass (*Spartina alterniflora*), it was found that 330 kg of humic substances per hectare of salt marsh were released annually (*2006*, *2007*). Organochlorine compounds in fulvic acids from natural waters in unpolluted environments have been described (*2001*), including high molecular weight aromatic organohalogenated compounds in organic material from both limnic and marine environments (*2008*). The limnic samples contained only organochlorines whereas the marine sediment samples contained organochlorine, organobromine and organoiodine compounds. A natural halogenated soil humic acid containing 0.24% halogen has been isolated (*2009*). Organohalogens have been detected in fulvic acids isolated from groundwater dating back 1300, 4600, and 5200 years (*1997*) and aquatic sediments deposited around the year 1900 contain a remarkably high concentration (30–50 ppm) of organohalogens (*22*).

Model reactions between chlorine and aquatic humic substances from peat and soil produce chloroform, trichloroacetic acid, dichloroacetic acid, and 2,4,6-trichlorophenol (*2011–2013*, *2232*). Moreover, the enzymatic chlorination of fulvic acid by chloroperoxidase in the presence of chloride and hydrogen peroxide has been demonstrated to occur under conditions and concentrations that are typical in the natural environment. Furthermore, an extract from spruce forest soil has chlorinating ability (*2014*), and substantial amounts of cell-free peroxidases are present in natural soils (*2017*). This natural chlorination of humic acids may well represent the source of the large unidentified quantity of organohalogen material found in soil and water. The reaction of humic acid substances with chloride, hydrogen peroxide, and chloroperoxidase yields chloroform, several chloroacetones, chloroacetic acids, chloropyruvic acids, chloromaleic acids and chlorofumaric acids (*2015, 2231*). The similar natural chlorinating ability of surface waters was found to produce 2,4,6-trichlorophenol (*1458*). Even more remarkable was the observation that 2,4,6-trichlorophenol on standing in aqueous solution (pH 8.7, 3 months) produced small amounts of the coupling product, 4-(2,4,6-trichlorophenoxy)-2,6-trichlorophenol (or an isomer) (*1458*). This result coupled with the others presented above suggests that dioxins may form naturally in soil and water via chloroperoxidase-mediated chlorination of humic substances and the subsequent coupling/cyclization of chlorophenols.

4. Biohalogenation

4.1. Introduction and Early Examples

The enzymatic halogenation of organic substrates mentioned briefly in the preceding section has been studied in great depth. Since a number of excellent reviews on various aspects of biohalogenation are available (*2018–2024*, *2065*), the discussion of mechanisms will be limited here.

It was recognized early that many species of marine algae contained enzymes (peroxidases) capable of oxidizing halide (iodide, bromide) to active halogen which was then incorporated as organohalogen (*60, 2025–2034*). For example, examination of 33 species of Phaeophyceae algae from the Atlantic Coast revealed that 22 possessed peroxidase activity (*2032*). Unidentified peroxidases which are capable of halogenating substrates have also been found in at least one sponge (*Iotrochota birotulata*) (*2035*). Other early examples of halogenating peroxidases are mammalian thyroid peroxidase which oxidizes iodide to iodine for incorporation into tyrosine (*2036–2040*), and mammalian lactoperoxidase which is also capable of effecting halogenation, including chlorination (*2040–2042*).

4.2. Chloroperoxidase

Chloroperoxidase (CPO) which was briefly discussed in the previous section on humic acids, was the first haloperoxidase to be isolated, purified and studied in depth.

The fungus *Caldariomyces fumago* proves to be an excellent source of CPO (*2043–2046*) and its biological chlorination ability has been probed

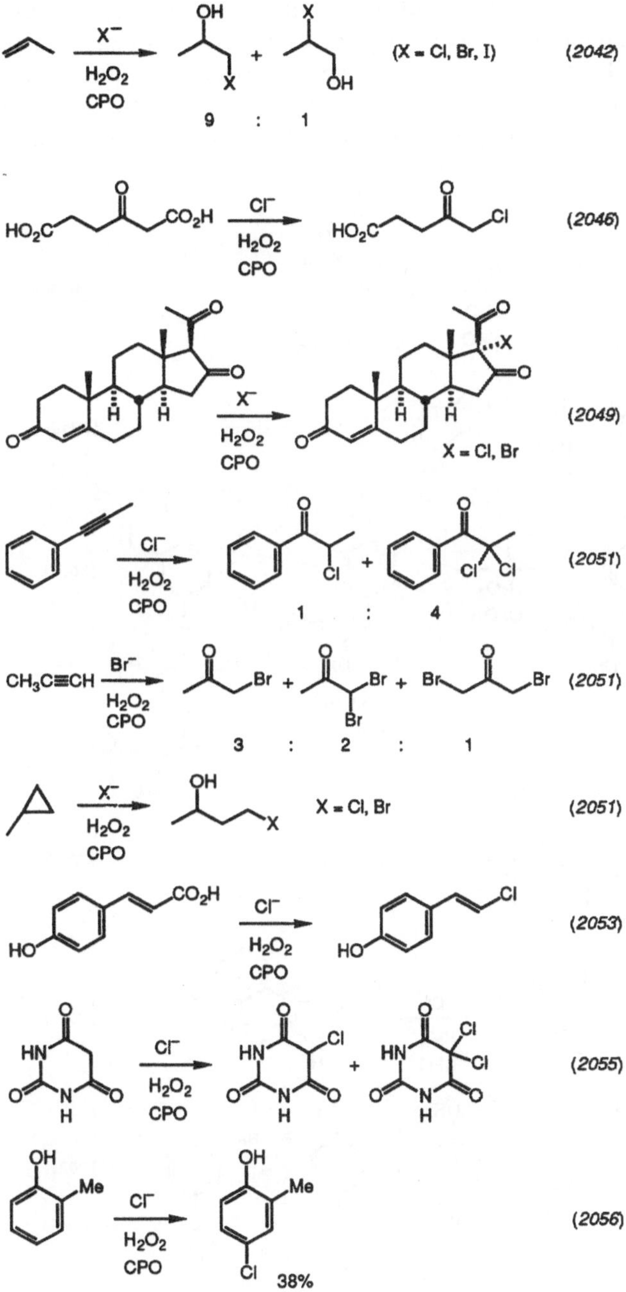

X^-, H_2O_2, CPO (X = Cl, Br, I) 9 : 1 (2042)

Cl^-, H_2O_2, CPO (2046)

X^-, H_2O_2, CPO; X = Cl, Br (2049)

Cl^-, H_2O_2, CPO 1 : 4 (2051)

$CH_3C{\equiv}CH$ Br^-, H_2O_2, CPO 3 : 2 : 1 (2051)

X^-, H_2O_2, CPO; X = Cl, Br (2051)

Cl^-, H_2O_2, CPO (2053)

Cl^-, H_2O_2, CPO (2055)

Cl^-, H_2O_2, CPO 38% (2056)

Scheme IIIa. Representative reactions catalyzed by chloroperoxidase

extensively in a variety of substrates, including 1,3-diones, tyrosine, β-keto acids, phenols, anisoles, alkynes, thiols, cyclopropanes, barbituric acids, NADH, thiazole, cinnamic acids, nitrogen heterocycles, alkenes (enantioselective epoxidation) and fulvic acids (*2046–2059*, *2014*). A sampling of these CPO reactions is shown in Scheme III. Bromide and

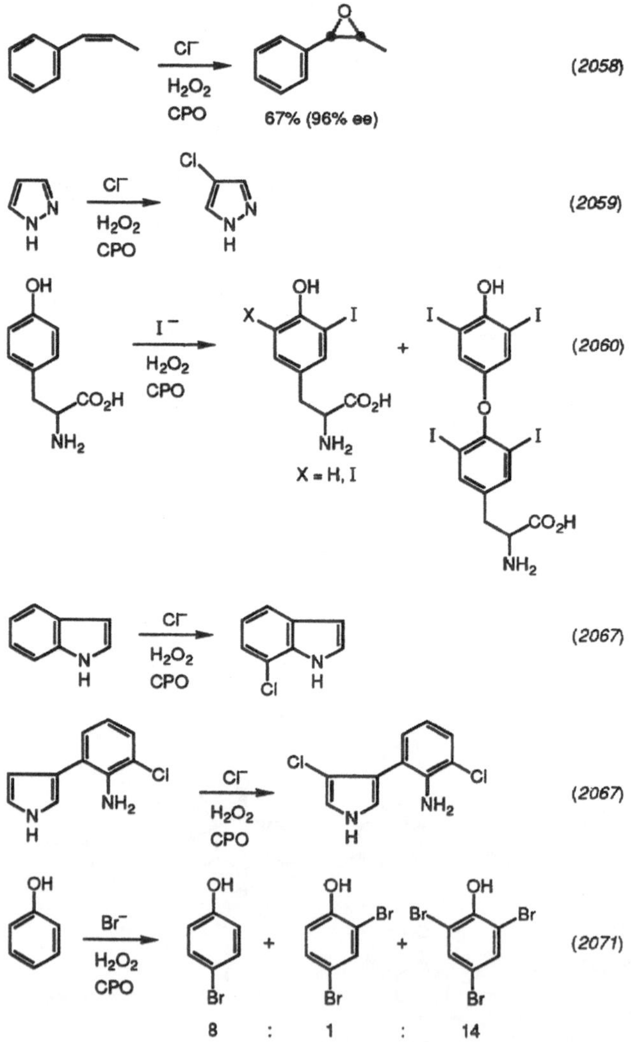

Scheme IIIb. Representative reactions catalyzed by chloroperoxidase

iodide, but not fluoride, are also oxidized by CPO (*2048*, *2051*, *2053*, *2059–2061*); the mechanism seems to involve the generation of free hypohalous acid (HOX) in an ionic mechanism (*2050*, *2054*, *2062–2064*). The major product of the reaction between acetone, Cl^-, H_2O_2, and CPO is chloroform with lesser amounts of other products (*2061*). Bromide ion in place of chloride in this reaction yields bromoform. Some of the other identified products in these reactions are $CHBr_2Cl$, CH_2Br_2, $BrCH_2CH_2Br$, $CHBrCl_2$, $Cl_2CHCHCl_2$, $Cl_2C=CCl_2$, and CCl_4.

CPO has been isolated from several other organisms, including the bacterium *Pseudomonas pyrrocinia* (*2066–2068*), the organism that produces pyrrolnitrin (**1210**), the terrestrial fungus *Curvularia inaequalis* (*2069*, *2070*) and the acorn worm *Notomastus lobatus* (*2071*). The latter organism contains several brominated phenols which are apparently formed enzymatically from phenol, Br^-, and CPO, as demonstrated experimentally (*2071*). CPO activity has been found in 8 of 9 soil extracts, as determined by the catalytic chlorination of monochlorodimedone (*2072*). The properties of this crude enzyme are very similar to those of purified CPO. More than 80 species of Death Valley fungi display CPO activity (*2020*).

4.3 Bromoperoxidase

Many natural sources of bromoperoxidase (BPO) have been discovered, mainly from marine algae (*2023*, *2024*). For example, an examination of the Western Caribbean coast of Central America found BPO activity in 55 out of 72 marine algae species (*2073*). As will be seen in the following discussion, BPO generally effects the oxidation of bromide and iodide, but not chloride or fluoride.

The first BPO was isolated from the red alga *Cystoclonium purpureum* and was found to be a heme protein (*2074*). This enzyme catalyzes the bromination of phenol. The red alga *Bonnemaisonia hamifera* which is a rich source of bromoalkanes (Section 3.2) also contains a bromoperoxidase that converts 3-ketooctanoic acid into the bromoalkanes shown in Scheme IV (*142*). Other red algae that have been found to yield BPO are *Rhodomela larix* (*2075*), *Corallina pilulifera* (*2076–2081*, *2059*), *C. officinalis* (*2077*, *2082*), *C. vancouveriensis* (*2083*, *2084*), *Amphiroa zonata* (*2077*) and *Ceramium rubrum* (*2085*). Several other red algae contain BPOs, but the enzyme activity is much lower (*2077*). Many of the BPOs isolated in the above studies contain vanadium, an element essential for enzymatic activity (*2081*, *2085–2088*). The green alga *Pencillus capitatus* contains one or more BPOs (*143, 2089–2092*) and, at lower pH, chloride can be oxidized to free hypochlorite (*2092*). One study has found that amino acids are converted to nitriles and aldehydes with BPO (*2091*), in a mechanism proposed to involve initial *N*-bromination of the amino acid such as might occur in the biosynthesis of aeroplysinin-1. The green algae *P. lamourouxii* and *Rhipocephalus phoenix* also display

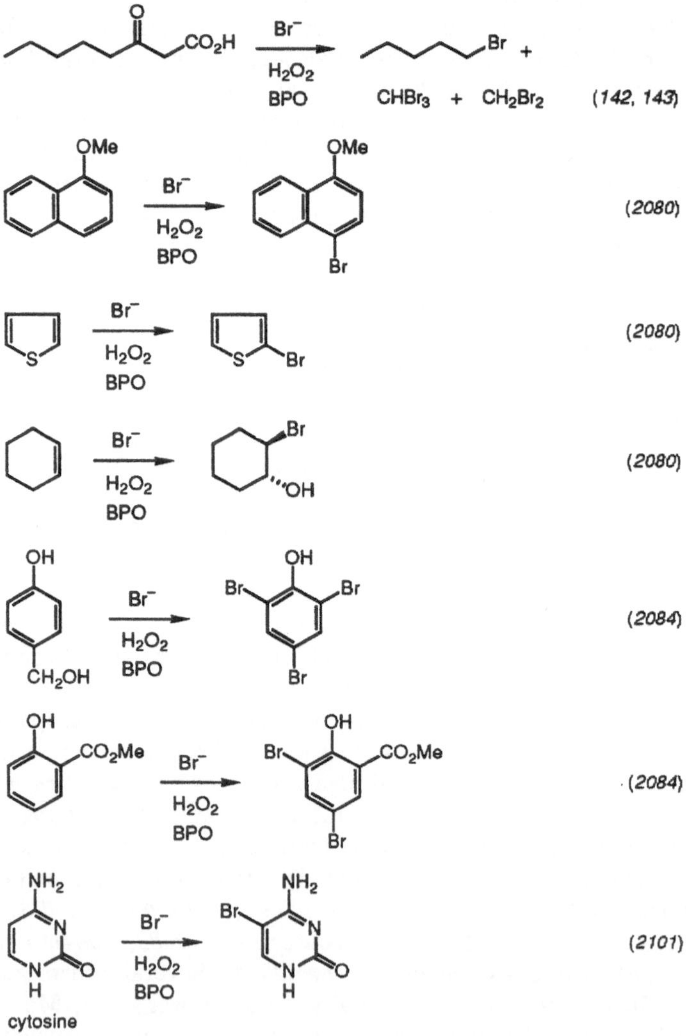

Scheme IV. Representative reactions catalyzed by bromoperoxidase

BPO activity as measured by the catalytic bromination of monochlorodimedone (*2089*). Bromoperoxidases from the brown alga *Ascophyllum nodosum* have been extensively studied; they also require vanadium for activity (*2093–2097*, *2083*). The mechanism seems to involve the generation of free bromine (or HOBr). Other brown algae such as *Laminaria saccharina* (*2098–2100*), *L. digitata* (*2099*, *2100*), *Fucus* sp. (*2098*), *Fucus distichus* (*2101*), *Macrocystis pyrifera* (*2101*, *2102*) and others (*2098*) contain BPOs and require vanadium.

The terrestrial lichen *Xanthoria parientina* contains a vanadium BPO (*2103*) and a BPO has been isolated from *Streptomyces phaeochromogenes* which catalyzes the bromination of pyrrolnitrin (**1210**) in the 2-position of the pyrrole ring (*2104*, *2105*). The bacterium that produces aureomycin (**1752**), *S. aureofaciens*, contains at least two BPOs (*2106–2109*), and *S. griseus* contains at least four distinct BPOs (*2110*). This bacterium produces 6,3′-dichlorogenistein. The strain *S. venezuelae* which produces chloramphenicol (**1084**) has been found to contain a BPO and it is proposed that this enzyme is involved in the chlorination step during the biosynthesis of chloramphenicol (*2111*). The pyrrolnitrin-producing bacterium *Pseudomonas pyrrocinia* contains a BPO that utilizes bromide but not chloride (*2112*, *2066*). *Pseudomonas aureofaciens* (*2113*), *P. putida*, *P. pyrrolnitrica* and *P. aeruginosa* (*2114*) all possess one or more BPOs. The marine annelid *Thelepus setosus* and acorn worm *Ptychodera flava laysanica*, both of which produce and secrete multiple organobromine compounds (*vide supra*), each contain a BPO (*2075*).

4.4. Other Peroxidases

In what must stand as one of the most important aspects of this review, there is growing evidence that mammals – including humans – utilize biohalogenation to fight infection by killing invading microorganisms! Thus, numerous studies have shown that natural blood halide can be oxidized to active halogen by a white blood cell peroxidase in the presence of hydrogen peroxide, and the resulting halogenation is lethal to the foreign invader. For a recent summary of the literature see (*2065*).

The combination myeloperoxidase (MPO), halide (Cl^-, Br^-, I^-) and hydrogen peroxide in white blood cells kills bacteria (*2115–2119*, *2121*), fungi (*2116–2122*) and tumor cells (*2118*, *2120*). For example, the combination of NaCl, H_2O_2, and human MPO is rapidly lethal to *Candida* spp. and KI/H_2O_2/MPO kills *Saccharomyces cerevisiae*, *Aspergillus fumigatus*, *A. nigei*, *Geotrichum candidum* and *Rhodotorula* sp. (*2122*). In these studies, hydrogen peroxide can be replaced by D-amino acid oxidase. A patient who was deficient in MPO died of fungal pneumonia (*Aspergillus fumigatus* and *Candida albicans*) (*2116*). Human leukocytes (neutrophils) which contain MPO can kill the bacteria *E. coli* and *Staphylococcus aureus* in the presence of chloride and hydrogen peroxide (*2117*). Without MPO human neutrophils have a greatly impaired ability to kill fungi and bacteria (*2116*). This enzyme has been isolated from human leucocytes, pus, sputum, bone marrow and tumors (*2123*) and purified (*2124*, *2125*). The related eosinophil peroxidase has also been isolated from human blood and purified (*2121*, *2125–2127*) and a MPO is present in human saliva which may be part of the oral antimicrobial system (*2128*). The halogenating ability of these enzymes has been discussed (*2121*, *2122*, *2129*, *2130*, *2054*) and there is some evidence that a target of a chlorination reaction by the MPO/chloride neutrophil system is NADH, leading to the death of the microorganism since NADH is essential for metabolism (*2054*, *2130*).

Ovoperoxidase has been isolated from sea urchin eggs (*Strongylocentrotus purpuratus*) and is believed to be responsible for the hardening of the fertilization membrane (*2131*). In addition, it is proposed that this enzyme which oxidizes iodide and bromide may kill excess sperm by halogenation after fertilization.

Although the present review documents more than 2400 known natural organohalogen compounds, the number of organisms shown to contain haloperoxidases is comparatively small. Nevertheless from these studies of haloperoxidases it is clear that nature have evolved efficient enzymatic systems for the generation of active halogen from halide and for the subsequent biosynthesis of specific organohalogen metabolites. As one example of the genetic investment made by an organism, it takes 11 enzymes and 72 intermediate compounds for *Streptomyces aureofaciens* to biosynthesize aureomycin (*2179*).

5. Biodegradation

In the previous section, the enzymatic formation of organohalogens in living organisms was discussed. Similarly, enzymes are required to degrade organohalogens so that a recycling of the elements can occur when an organism dies. Organohalogens, whether natural or anthropogenic, are more or less readily metabolized and biodegraded to halide ion by numerous microorganisms. Since several excellent reviews are available (*2132–2138*, *2150*) coverage of this important area will be brief.

One of the earliest examples of halobiodegradation involved conversion of 1-chloro- and 1-bromonaphthalene by *Pseudomonas desmolyticum* to 3-chloro- and 3-bromosalicylic acid and 1-chloro-7,8-dihydroxy-7,8-dihydronaphthalene (*2139*). Several strains of *Pseudomonas*, especially *P. putida*, biodegrade halobenzenes, sometimes with complete loss of halogen as halide, including fluoride (*2140–2148*). The oxidation of aromatics using *P. putida* has been elegantly exploited in organic synthesis, especially enantioselective synthesis (*2147, 2148, 2230*). *Nocardia erythropolis* also metabolizes halogenated aromatic acids (*2149*). The biodegradation of polychlorinated biphenyls (PCBs) by *Pseudomonas* sp. (*2151*), *Alcaligenes* sp., *Acinetobacter* sp. (*2152*) and other soil bacteria (*2153, 2154*) and by anaerobic microorganisms in aquatic sediments (*2155–2158*) is well known. PCBs are also effectively degraded by the white rot fungus *Phanerochaete chrysosporium* (*2159, 2160*). For a review of PCB biodegradation see (*2161*). This fungus has an unusually hearty appetite for aromatic compounds containing chlorine, including chlorophenols (*2162–2165*), DDT (*2166*) and dioxin (*2160*). The natural substrate for this fungus appears to be wood lignin (*2167*).

Soil bacteria are able also to degrade alkyl halides such as $BrCH_2CH_2Br$, $BrCH_2CHBrCH_2Cl$, $CH_3CHBrCHBrCH_3$, $CHCl_3$, CCl_4, CH_2Cl_2, $ClCH_2CH_2Cl$, Br_2ClCH, Br_3CH, and others (*2168–2170*). For example ethylene dibromide is completely degraded to ethylene in two months by soil bacteria (*2132, 2168*). *Pseudomonas* sp. is also able to metabolize (dehalogenate) simple alkyl halides, such 1-chloro-, 1-bromo-, and 1-iodoheptane, 1,6-dichlorohexane, 1,5-dichloropentane, and 1,9-dichloro- and 1,9-dibromononane (*2171*). A *Pseudomonas* sp. has been reported as being able to dehalogenate haloacetic acids in this order of decreasing rates of dehalogenation: fluoride > chloride > bromide > iodide (*2172, 2173*). The dehalogenation is limited to haloacetic acids. *Fusarium solani* also contains such a haloacetate halidohydrolase (*2173*). A fluoroacetate metabolizing enzyme has been isolated from *Pseudomonas cepacia*, which is associated with *Dichapetlum cymosum*, a producer of fluoroacetate (*2174*). Not all biotransformations involve the breakdown of organohalogens. The soil fungus *Rhizoctonia praticola* has the ability to polymerize chloro- and bromophenols into PCBs and humic-like products (*2175, 2176*) and soil bacteria can convert chloroaromatics into chlorophenols by hydroxylation (*2177*). Because the latter compounds are classified by the EPA as Priority Pollutants (*1432*) their biodegradation has been extensively studied (*2178*).

6. Natural Function

The question often asked of natural products chemists is: "Why does this organism produce such a compound?" This question is especially appropriate in the context of naturally occurring organohalogen compounds which are still considered by many to be artifacts or "chemical freaks" of nature. A number of careful studies have provided strong evidence for the role that particular organohalogen compounds serve in their natural setting. For two excellent reviews on the general subject of the role of metabolites in nature see (*2180*, *2181*).

Insect pheromones have been extensively studied for many years as a means with which to monitor and control insect pests. The few examples of chlorinated compounds so far isolated from insects, such as 2,6-dichlorophenol (**1783**) and the cockroach compounds **819** and **820**, clearly serve as sex and aggregation pheromones, respectively. The chlorinated and other halogenated tyrosine-containing proteins which are present in molluscs and locusts as well as horse-shoe crabs are proposed to strengthen cuticle and improve the adhesion between protein sheets (*1566*, *1568*, *2182*).

Several studies have indicated a chemical defensive role for the organohalogen (and other) metabolites of marine invertebrates (*2183*) such as sponges (*2184*, *2185*), sea hares (*2186*), ascidians (*1190*), molluscs (*714*, *721*), and nudibranchs (*2187*, *2188*). These metabolites can be quite toxic to potential predators or harmful bacteria since marine invertebrates lack the mobility enjoyed by fish. The toxicity of sea hare toxins was known since pre-Christian times when they were used as poisons (*2189*). Some organohalogens may function in larval settlement in the life cycle of marine invertebrates (*2190*).

Marine algae produce a myriad of organohalogen compounds, many of which have antimicrobial activity such as HOBr (*139*), perhaps to prevent fouling by harmful bacteria or simply to prevent feeding by reef fish (*2191–2193*). For example, laurinterol (**589**) and related algal metabolites have an antimicrobial activity approaching that of streptomycin (*2192*). Several examples of algal metabolites that are highly toxic to reef fish or that prevent their feeding (*2193*) were cited earlier, such as vidalol A (**1857**) (1519), avrainvilleol (**1853**) (*1515*), and debromoisocymobarbatol (**633**) (*498*). It is interesting that red algae contain more halometabolites than brown or green algae and that blue-green algae contain mainly organochlorine metabolites (*2194*).

Terrestrial plants also produce chemically defensive substances, such as the tafricanins A (**640**) and B (**641**) which are insect antifeedants (*505*) and **1773** from the Thai plant *Arundo donax*, which repels weevils (*1426*).

Chloromethane has the distinction of being the most abundant natural organochlorine compound with an estimated 5 million tons being produced annually from marine and terrestrial biogeneic and combustion sources. There is compelling evidence that chloromethane serves a key role in the degradation of lignin by wood-rotting fungi such as *Phellinus pomaceus* (*2195–2197*) and *Phanerochaete chrysosporium* (*2198*). Thus, a function of chloromethane apparently is to regenerate veratryl alcohol degraded by the attack of lignin peroxidase (*2198*). Since chloromethane is also emitted by the potato, cypress, cedar, and other higher plants it may play a biosynthetic role here as well.

4-Chloro-3-indoleacetic acid (**1468**) is unquestionably a natural plant growth hormone in peas and several types of beans, and 3,5-dichlorohexanophenone **2372**, which is produced by *Dictyostelum discoideum*, is a signal molecule that triggers the transformation of undifferentiated cells of this slime mold into fruiting bodies (*2199*, *2200*). The chlorine atoms in the powerful clinical antibiotic vancomycin (**2284**) are a crucial structural element enforcing the necessary conformation for receptor binding. It is interesting to note that the chlorinated version of the anticancer drug tamoxifen, which is called toremifene, is not carcinogenic in rats, whereas tamoxifen is strongly hepatocarcinogenic (*2201*, *2207*, *2208*). On the other hand, the unnatural α-chloro- and α-bromocinnamaldehyde are strongly mutagenic, whereas cinnamaldehyde is nonmutagenic (*2209*).

The extraordinary discovery of the bromo ester **186** in human cerebrospinal fluid (*168*) asks the question, "What is the function, if any, of this human organohalogen compound?" This compound which has also been isolated from other mammals (cat, rat), is a very effective inducer of REM (rapid-eye-movement) sleep and may play an important role in inducing the sleep phenomenon (*2202–2205*). The concentration of this compound is 100–200 times higher in the retina, hypophysis and cerebral cortex than it is in the blood (*2204*). This organic bromo derivative is also a potent inhibitor of acetylcholinesterase (*2206*).

Several authors have proposed that the role of the halomethanes might be to recycle halogen/halide between oceans, the atmosphere, and land. Thus, much as dimethylsulfide is a natural carrier of sulfur,

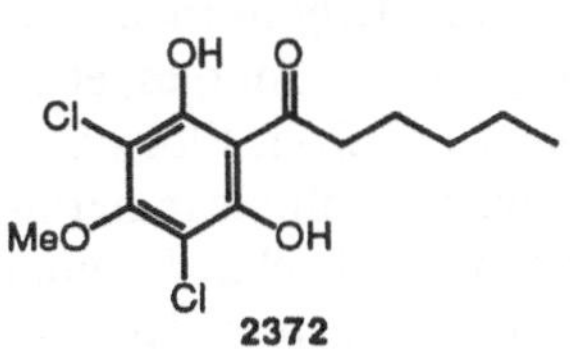

2372

iodomethane may be a natural carrier of iodine, cycling it between land and sea (*105*, *2234*). Similarly, bromoform and/or bromomethane may be carriers of bromide (*123*, *150*, *2234*) and chloromethane may be a natural regulator of the ozone layer.

7. Significance

Despite the existence of at least 2450 naturally occurring organohalogen compounds in our biosphere, it might be argued that the *quantities* of these compounds present in the environment are small compared with the levels of anthropogenic organohalogen compounds, which amount to 20 million tons/year of 150 industrial organohalogen compounds (*68*, *2020*). While this may be true for many of the biogenic organohalogen defensive substances, it is not the case for many simple naturally occurring halogenated alkanes which are released into the atmosphere in large quantities. Some of the global estimates for these compounds are shown in Table 27. For an excellent summary see (*82*). It is clear that naturally produced chloromethane, iodomethane, bromomethane and bromoform make enormous contributions to the global concentrations of these chemicals. For example, both oceanic sources (primarily macro- and microalgae) and biomass burning produce massive quantities of chloromethane and bromomethane. Both sources are important. Natural fires have occurred on earth since land plants evolved 350–400 million years ago (88) and marine algae are found in all the world's oceans (tropical, temperature, and polar) (*62*).

Obviously, due to variability between samples of the same species, global estimates of the chemical content of any organism is very difficult. For example, the tropical green alga *Avrainvillea nigricans* from Puerto Rico contains organobromines whereas the same species from the western Caribbean is apparently devoid of organohalogen (*1516*). However, a few studies have been carried out in an attempt to quantify the organohalogen production of a particular species in a defined area. Thus, a detailed examination of the acorn worm *Ptychodera flava* has determined that the 64 million animals living in a one-square kilometer habitat on Okinawa excrete daily in their feces 43 kilograms of organic matter (primarily bromophenols, indoles, etc.) (*2240*). A similar study of the acorn worm *P. bahamensis* from Florida estimated an annual output of 0.5–1.3 tons of ethyl acetate-soluble material per kilometer of coastline (*195*). Halogenated phenols and related compounds constitute much of this material. The brown alga *Ascophyllum nodosum* which exudes HOBr

Table 27. *Estimates of Naturally Occurring Halomethanes from Various Sources*[a]

Compound	Source	Tons per Year	Ref(s).
CH_3Cl	macroalgae	2,000	*79*
CH_3Cl	oceans	2,500,000–5,000,000	*69, 82, 86, 2234, 2237, 2239*
CH_3Cl	biomass burning	460,000–2,000,000	*69, 88, 2234, 2238*
CH_3Cl	anthropogenic	30,000–300,000	*82*
CH_3Br	macroalgae	100	*79*
CH_3Br	oceans	40,000–300,000	*82, 86, 128, 129, 2234*
CH_3Br	anthropogenic	50,000–80,000	*82, 2234*
CH_3I	macroalgae	200	*79*
CH_3I	all biogenic sources	1,300,000	*86*
CH_3I	oceans	300,000–500,000	*86*
CH_3I	all sources	40,000,000	*105*
CH_3I	oceans	320,000–2,000,000	*82, 124, 125, 2236*
CCl_4	all sources	1,900,000	*105*
$CHBr_3$	Arctic Ocean	1,000	*141*
$CHBr_3$	oceans	1,000,000–2,000,000	*82, 2234*
$CHBr_3$	macroalgae	10,000	*77, 82*
$CHBr_3$	macroalgae	4,000–40,000	*150*
$CHBr_3$	Arctic ice microalgage	4,700–70,000	*150*
$CHBr_3$	Antarctic ice microalgae	53,000–80,000	*150*
$CHBr_3$	macroalgae	200,000	*123*
$CHBr_3$	anthropogenic	600–6,000	*77, 82*

[a] Since the literature values are usually given in grams, for convenience, we used the conversion factor of 1 ton $\cong 10^6$ grams. Thus, the estimates given may be slightly low

when exposed to light produces annually 2 tons of HOBr along a 30-kilometer stretch of dike in the Netherlands (*139*). Many more such studies and careful extrapolations are needed to glean information about the quantities of organohalogens that are produced by particular species.

With the discovery of halogenated humic acids in pristine areas far from industrial sources (Section 3.26) one can estimate that, if the 0.24% halogen content of one sample of soil humic acid (*2009*) is representative of the estimated 2.5 trillion tons of global humic acid (*1995*), then our terrestrial and aquatic humic acid repository alone contains 6 billion tons of natural halogenated material. In any event, it seems likely that the natural biohalogenation of humic material will prove to be a major source of halogenated compounds in the environment. For example, the chloroform that is frequently found in groundwaters with no clear anthropogenic source is probably due to the degradation of natural chlorinated humic acids.

8. Future Outlook

Two unrelated developments guarantee that research in the area of naturally occurring organohalogen compounds will continue far into the next century. First, with their current campaign to ban chlorine, some environmental groups, marching to the slogan that "organochlorines are unnatural," have unwittingly focused the attention of many scientists including this author on the vast (and now rapidly expanding (Table 1)) field of natural organohalogens. Second, after a period of near abandonment, pharmaceutical companies have begun an extensive revitalization of natural products chemistry, especially marine natural products, in a search for new medicinal compounds (*2242*, *2243*). This renewed activity has led to the discovery of many of the organohalogen metabolites discussed in this survey. Indeed, an extensive review of the area of biologically active marine natural products has recently appeared (*2244*).

Given the plethora of halogenated marine natural products presented herein, it is perhaps surprising that only a relatively small number of marine organisms (invertebrates, algae, bacteria, fungi) have been investigated for their chemical content. Thus, whereas some 12,000 natural products have been isolated from terrestrial plants, only 500 had been isolated from marine algae up to 1987. The fact that there are 500,000 species of marine animals, plants, and bacteria, of which 200,000 are invertebrates (*2185*), means that thousands of new organohalogen compounds are awaiting discovery (*11*). For example, of the 4,000 species of bryozoans (moss animals), fewer than a dozen have been examined (*34*). Similarly, the 3,000 known species of opisthobranchs (*2187*), 5,000 sponges, 4,000 segmented worms, and 80,000 molluscs (*17*) each represent a fantastically rich and largely untapped source of new chemical structures, many of which are certain to contain halogen. The diversity of marine life is illustrated by the Great Barrier Reef in Australia. This 100,000 square-mile area consists of 2,500 separate coral reefs. Around just one of these, of less than 14 square miles, there are 930 fish species, 107 corals, and 154 molluscs (11). Moreover, exploration of marine bacteria and fungi is only just beginning (*2251*) and this is a very promising area for future research.

The realization that mammals exploit halogenation in the immune process may be the explanation for the higher concentration of organochlorines in human urine than in drinking water (*36*). In any event, it seems likely that the organochlorine byproducts produced during this "biodisinfection" will soon be isolated and identified, and this research should provide insight into the human immune process.

The analysis of preindustrial glacial ice from Antarctica and northern Sweden has revealed organohalogen levels of 1–3 ppb, including measurable amounts of trichloroacetic acid (*2247*). Moreover, the latter compound, as well as chloroform, is found to be ubiquitous in soil, results confirmed by $^{37}Cl^-$ labeling experiments (*2248*). The analysis of 35,000-year old organic matter (^{14}C-dating) shows the presence of organochlorine, organobromine, and organoiodine compounds (*2249*). Organohalogens have also been found in 1,000-year old peat and 4,000-year old marine clay (*2249*). Even more remarkable are the discoveries that organohalogens (200–300 ppm) have been detected in several thousand-year old peat from the holocene period, in two lignite samples (107 and 166 ppm) that date from the Tertiary era, 15 million years ago, and in a 300-million year old bituminous coal sample (74 ppm) from the Upper Carboniferous period (*2250*). Whether these organohalogens in sediments result from the deposition of biogenic material from plants already containing organohalogens or arise by the formation of organohalogens within the sediments remains to be established, although most of the evidence to date would support the former theory. These studies of sediments – ancient and modern – should provide information about the global quantities of natural organohalogens. In any event, these studies demonstrate that organohalogen compounds predate the arrival of man on earth!

After a slow start (Table 1), the field of naturally occurring organohalogen compounds is in full bloom. It is beyond dispute that chlorine, bromine, iodine, and even fluorine are as natural to our biosphere as carbon, hydrogen, nitrogen and oxygen.

The following is a breakdown of the natural organohalogen compounds presented in this review:

Compounds containing chlorine	1,452
Compounds containing bromine	1,290
Compounds containing iodine	80
Compounds containing fluorine	20
Compounds containing only chlorine	1,145
Compounds containing only bromine	965
Compounds containing only iodine	42
Compounds containing only fluorine	14

Addendum

Since the completion of this review, a number of additional natural organohalogen compounds have been discovered. In fact, as this review goes to press (February, 1996), the total number of natural organohalogens is about 2,570.

3.1. Simple Alkanes

The following compounds were omitted from Table 3: CH_3CHCl_2 (**2373**), $ClCH_2CH_2Cl$ (**2374**), $CH_2 = CCl_2$ (**2375**) and CHF_3 (**2376**). They are released from halogen-containing minerals during mining operations and in deep natural wells (*103, 105*).

3.2. Simple Functionalized Alkanes

The three chlorinated carboxylic acids, Cl_3CCO_2H (**2377**), Cl_2CHCO_2H (**2378**), and $ClCH_2CO_2H$ (**2379**) have been identified several times in surface waters and soils (*2247, 2248*), presumably as natural humic acid decomposition products (Section 3.26), and are therefore considered to be naturally occurring.

3.3. Simple Functionalized Cyclic Organohalogens

3.3.2. Cyclitols and Benzoquinones

The novel alkyne benzoquinone mycenon (**2380**) is produced by a *Mycena* sp. (*2252*). This metabolite is an inhibitor of isocitrate lyase.

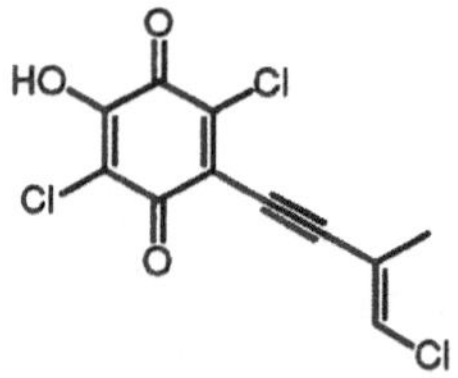

2380 (mycenon)

3.4. Terpenes

3.4.2.4. Marine Sesquiterpenes

In one of the few studies of marine fungi, the novel chloriolins A–C (**2381–2383**) have been isolated from a cultured, unidentified fungus associated with the sponge *Jaspis johnstoni* (*2251*).

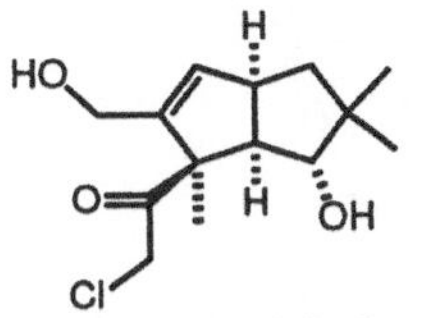

2381 (chloriolin A)

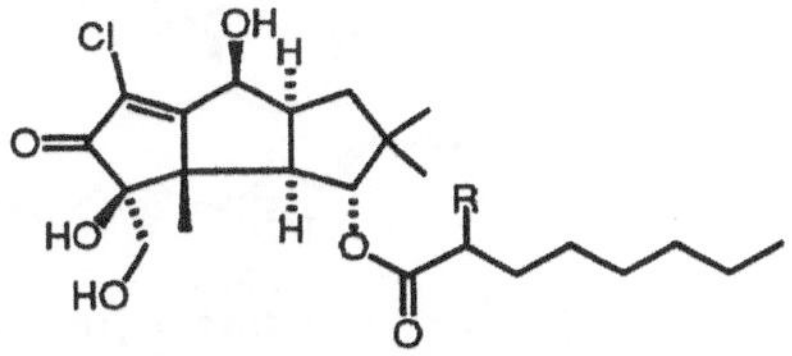

2382 R = OH (chloriolin B)
2383 R = H (chloriolin C)

3.8. Lipids and Fatty Acids

The red alga *Laurencia flexilis* produces aldehyde **2384** which is the first halogen-containing enal from a marine source (*438*). The blue-green alga *Scytonema mirabile* produces mirabimide E (**2385**), an *N*-acylpyrrolinone with a unique tetrachloroethane unit (*2253*).

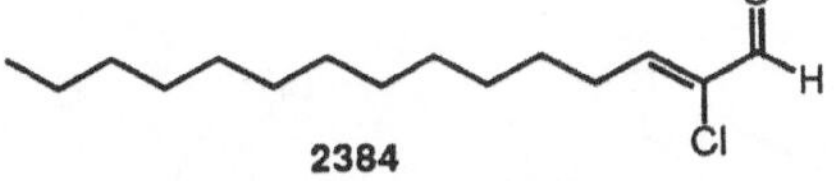

2384

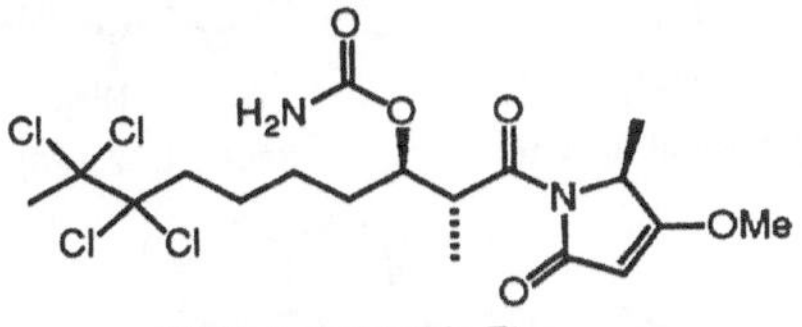

2385 (mirabimide E)

3.10. Prostaglandins

Thirteen new punaglandins (**2386–2398**) have been isolated from the beautiful octocoral *Telesto riisei* (*2254*). These include tertiary acetates **2393–2398** (not shown) of the six known punaglandins 1–4 (**1028–1033**).

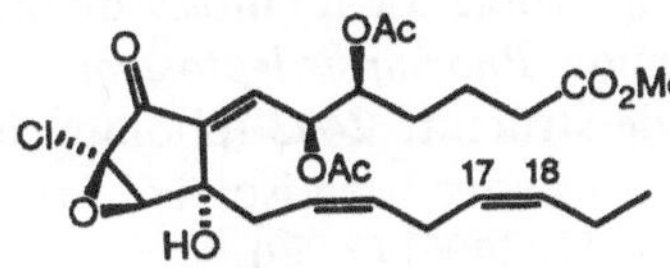

2386
2387 17,18-dihydro

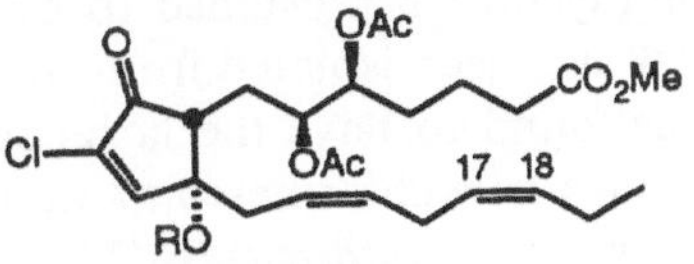

2388 R = H
2389 R = Ac
2390 R = H, 17,18-dihydro

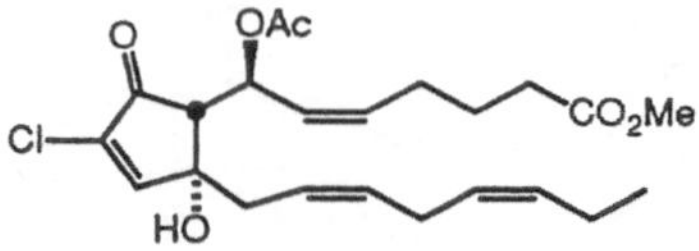

2391
2392 17,18-dihydro

3.12. Amino Acids and Peptides

For the first time, 4-chlorothreonine (**2399**) which is present in several complex cyclic peptides such as **1171–1180** has been found in free form in a *Streptomyces* sp. (*2255*). *S. sviceus* produces the novel amino acid U42,126 (**2400**) (*2256*) which is the deoxy derivative of U43,795 (**1092**). The Japanese sea hare *Dolabella auricularia* contains the cytotoxic cyclodepsipeptide doliculide (**2401**) (*2257*). This iodophenol derivative is very active against HeLa-S_3 cells ($IC_{50} = 0.001$ μg/ml).

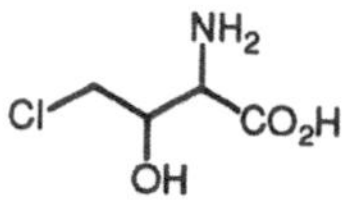

2399 (4-chlorothreonine)

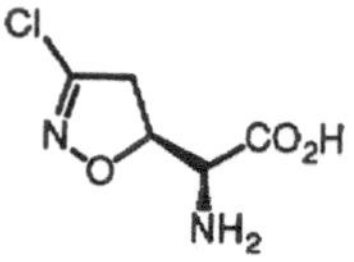

2400 (U42,126)

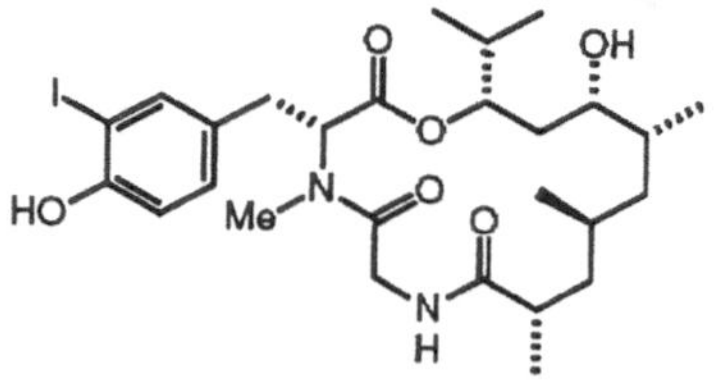

2401 (doliculide)

The Philippine sponge *Theonella swinhoei* produces the antifungal glycopeptide theonegramide (**2402**) which is structurally related to theonellamide F (**1166**) (*2258*).

The mycotoxin presumed to be responsible for lupinosis disease in livestock has been isolated from the fungus *Phomopsis leptostromiformis* and was found to have the hexapeptide structure **2403** (phomopsin A) (*2259*). Two new chlorine-containing astins have been isolated from the Composite *Aster tataricus*, F (**2404**) and H (**2405**) (*2260*).

Ramoplanins (A-16686) A-1 (**2406**), A-2 (**2407**), and A-3 (**2408**) are produced by an *Actinoplanes* sp. (*2261*, *2262*).

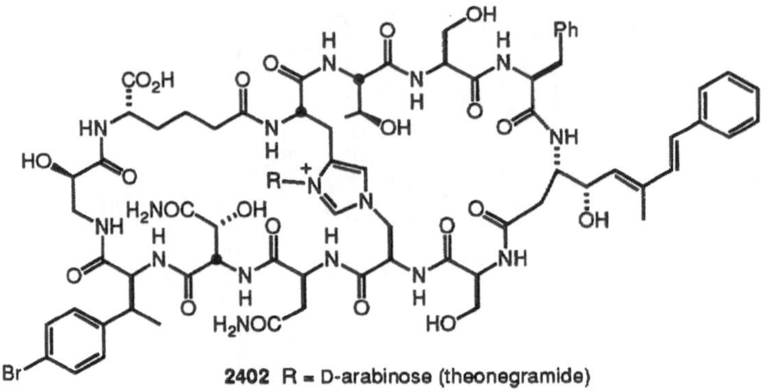

2402 R = D-arabinose (theonegramide)

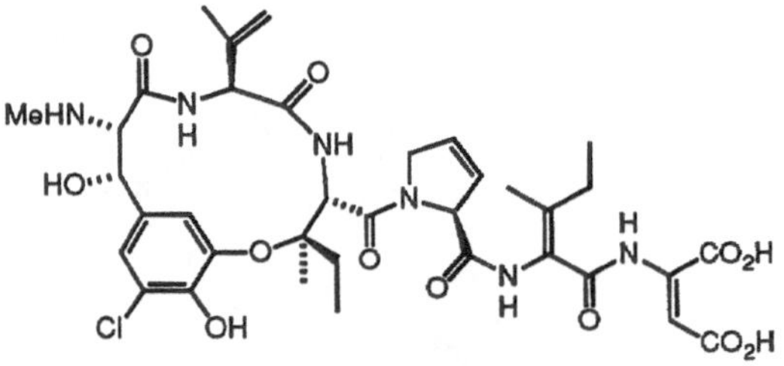

2403 (phomopsin A)

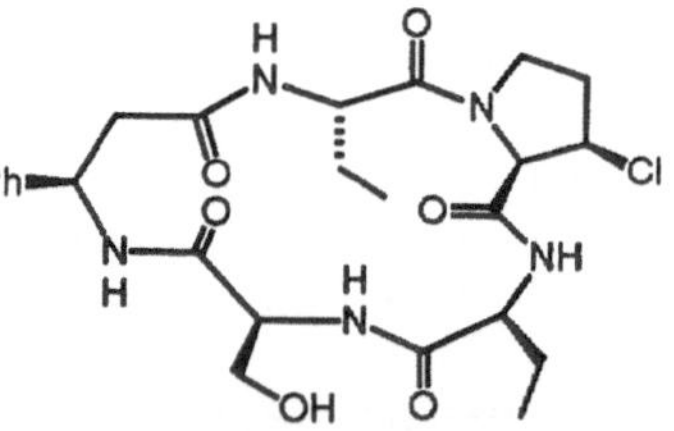

2404 (astin F)

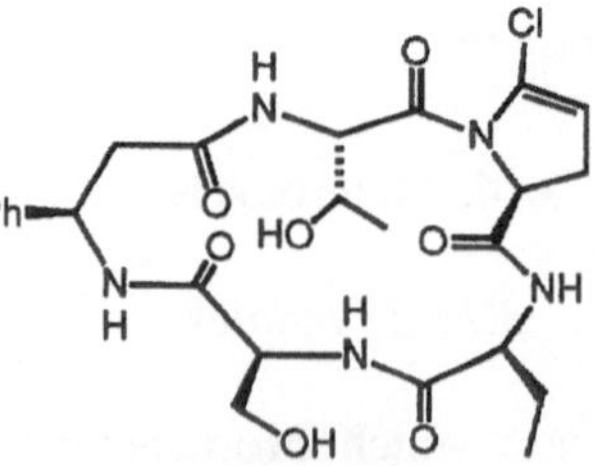

2405 astin H

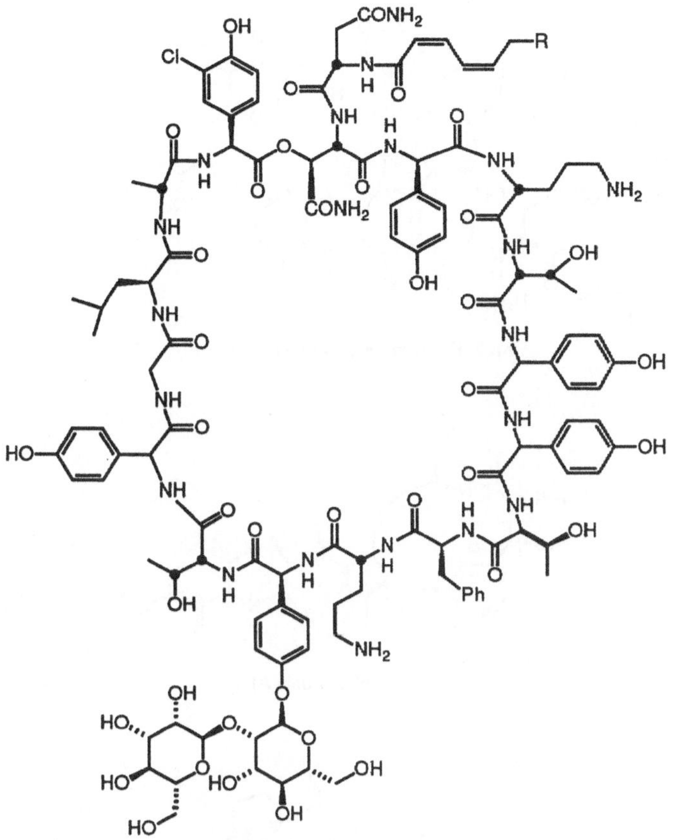

2406 R = CH_2CH_3 (ramoplanin A-1)
2407 R = $CH(CH_3)_2$ (ramoplanin A-2)
2408 R = $CH_2CH(CH_3)_2$ (ramoplanin A-3)

Another family of peptides has been isolated from *Pseudomonas syringae*, the pseudomycins A, B, C, C[1] (**2409–2412**) (not shown) that contain L-4-chlorothreonine (*2268*, *2269*) and that are similar to the syringostatins (**1175–1180**).

3.14. Heterocycles

3.14.2. Indoles

Pseudomonas aureofaciens which produces pyrrolnitrin (**1210**) also produces the new 7-chloro-3-indoleacetic acid (**2413**) (*2263*). The New

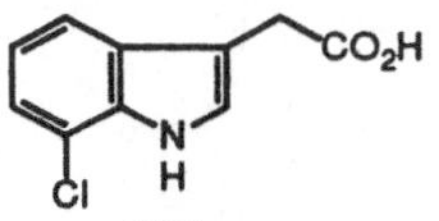

2413

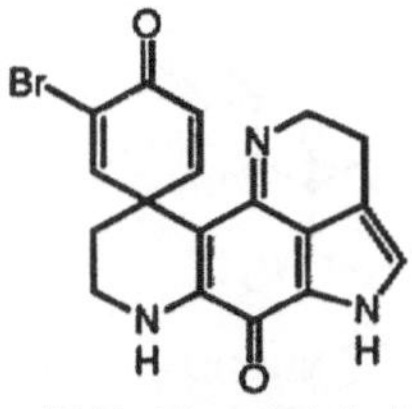

2414 (discorhabdin E)

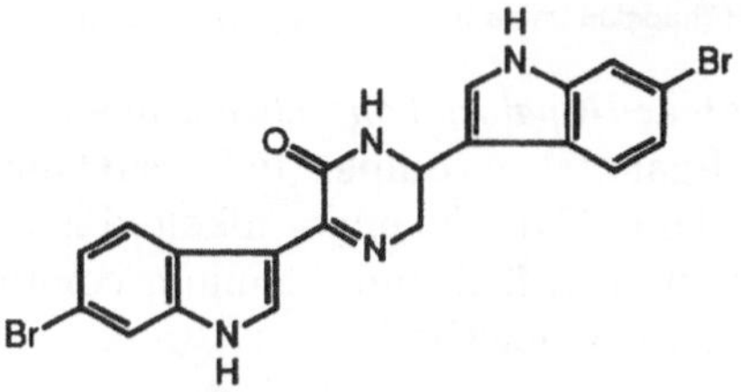

2415 (hamacanthin A)

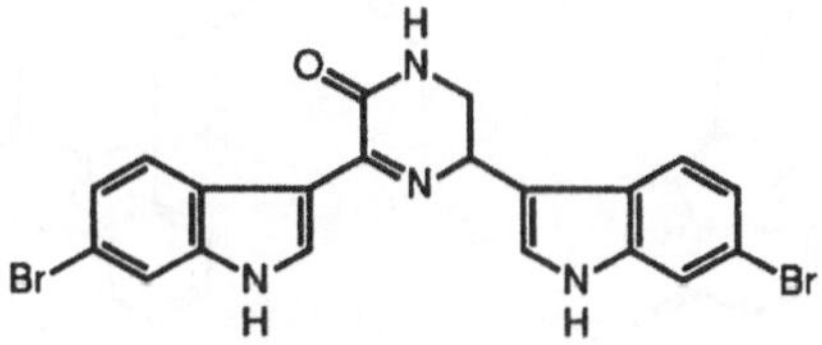

2416 (hamacanthin B)

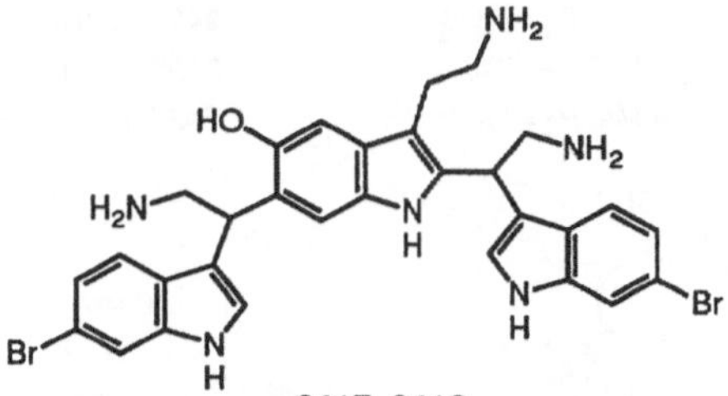

2417, 2418
(diastereomeric gelliusines A and B)

Zealand sponge *Latrunculia* sp. has furnished a discorhabdin (E, **2414**) (*2264*). The deep-water sponge *Hamacantha* sp. produces the novel brominated indoles hamacanthins A (**2415**) and B (**2416**) (*2265*). Both compounds inhibit the fungal pathogens *Candida albicans* and *Cryptococcus neoformans*. A deep-water Caledonian sponge (*Gellius* or *Orina* sp.) produces the novel *tris*-indoles gelliusines A (**2417**) and B (**2418**) (*2266*).

The additional hapalindoles **2419–2421** have been isolated from *Hapalosiphon fontinalis* (*2267*).

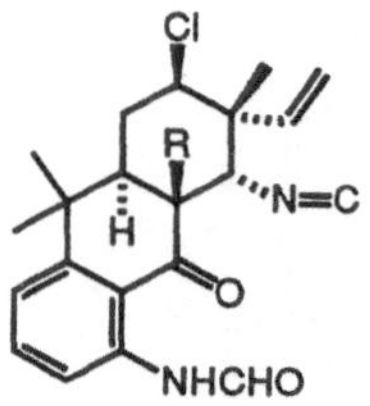

2419 R = H (hapalonamide G)
2420 R = OH (hapalonamide V)

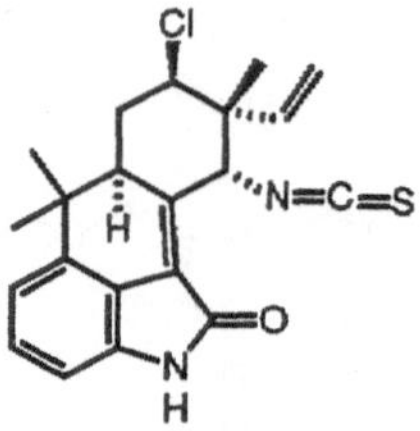

2421
(anhydrohapaloxindole B)

The blue-green algae *Hapalosiphon welwitschii* and *Westiella intricata* produce an extraordinary set of complex indoles related to the fischerindoles and hapalindoles (*2270*). The major alkaloid is *N*-methylwelwitindolinone C isothiocyanate (**2422**), and the minor compounds **2423–2431** were also isolated and characterized.

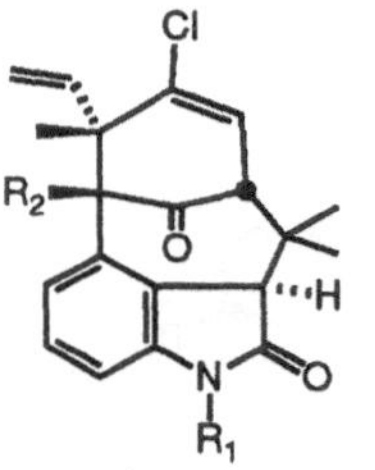

2422 R_1 = Me, R_2 = NCS
2423 R_1 = H, R_2 = NCS
2424 R_1 = Me, R_2 = NC

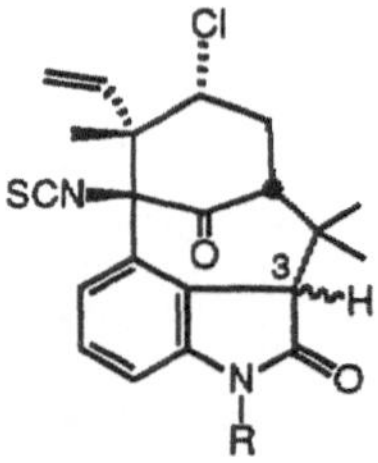

2425 α-3H, R = H
2426 α-3H, R = Me
2427 β-3H, R = Me

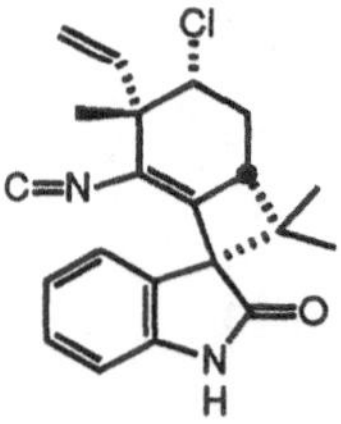

2428

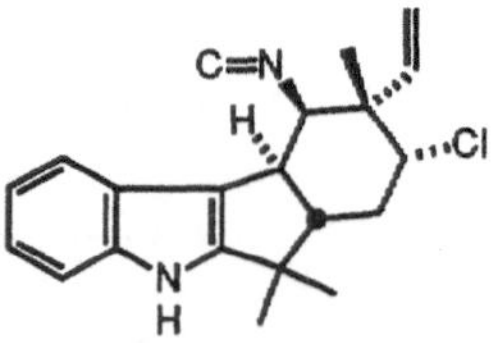

2429

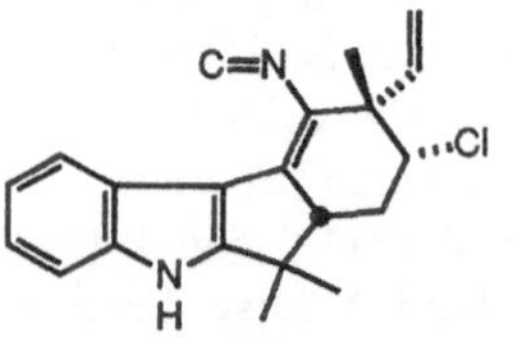

2430

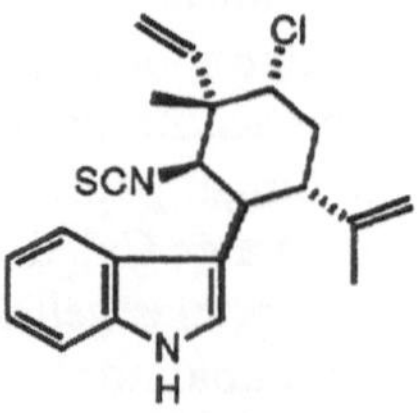

2431

3.14.5. Carbolines

The Great Barrier Reef ascidian *Lissoclinum* sp. produces lissoclin C (**2432**) in addition to the known 6-bromotryptamine (**1357**) (*2271*).

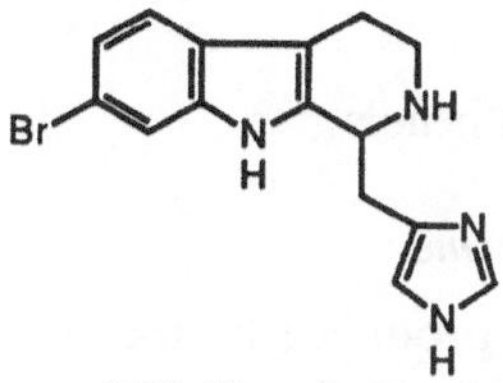

2432 (lissoclin C)

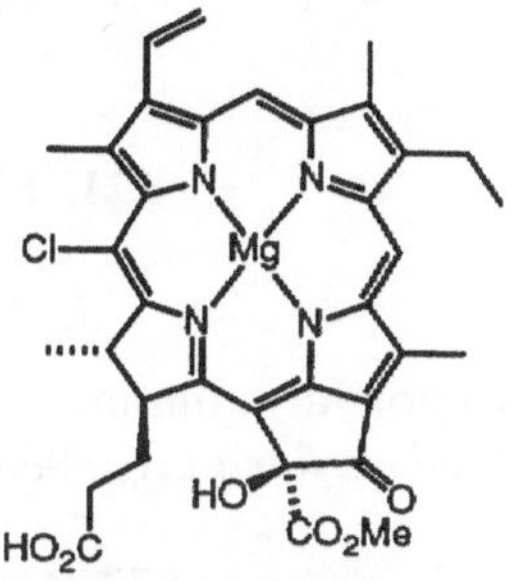

2433 (chlorophyll RC I)

3.14.6. Quinolines and Other Nitrogen Heterocycles

The chlorinated chlorophyll **2433** has been isolated from a wide range of photosynthetic organisms (*2272*).

3.14.9. Coumarins and Isocoumarins

The novel coumarin RP 18,631 (**2434**) is produced by *Streptomyces scopius* (*2273*).

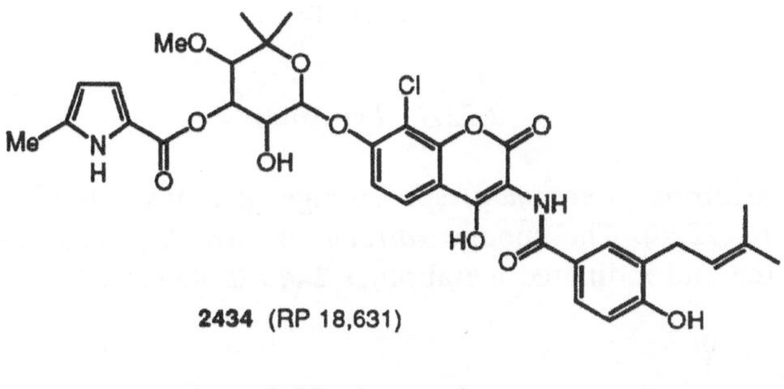

2434 (RP 18,631)

3.20. Aromatics

Both chlorobenzene (**2435**) (*103*) and 3-chloroanthranilic acid (**2436**) (*2263*) are naturally occurring compounds. The former is found in halogen-containing minerals and the latter is produced by *Pseudomonas aureofaciens*.

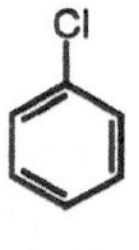

2435 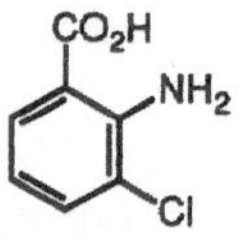 2436  2437

3.21. Simple Phenols

3.21.2. Marine

Tetrabromohydroquinone (**2437**) is produced by the versatile acorn worm *Ptychodera flava laysanica* (*2274*).

3.22. Complex Phenols

3.22.2. Diphenyl Ethers and Related Compounds

Ptychodera flava laysanica also produces the two heavily brominated diphenyl ethers **2438** and **2439** (*2274*).

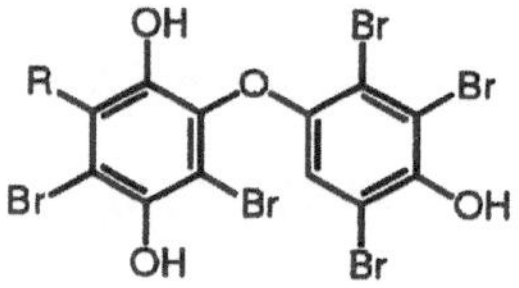

2438 R = H
2439 R = Br

3.22.3. Tyrosines

A Caribbean *Verongula* sp. sponge produces the metabolites **2440–2442** (*2220*). The sponge *Iotrochota birotulata* has afforded the brominated and iodinated metabolites **2443–2445** (*2275*).

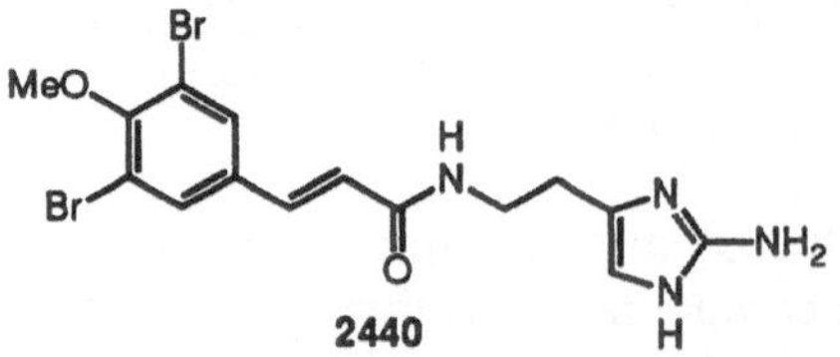

2440

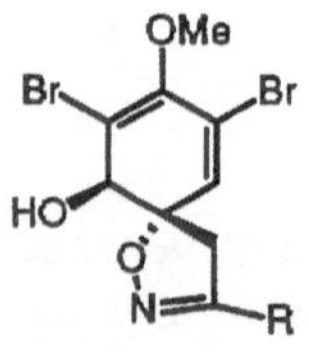

2441 R = CO_2Me
2442 R = $CONH_2$

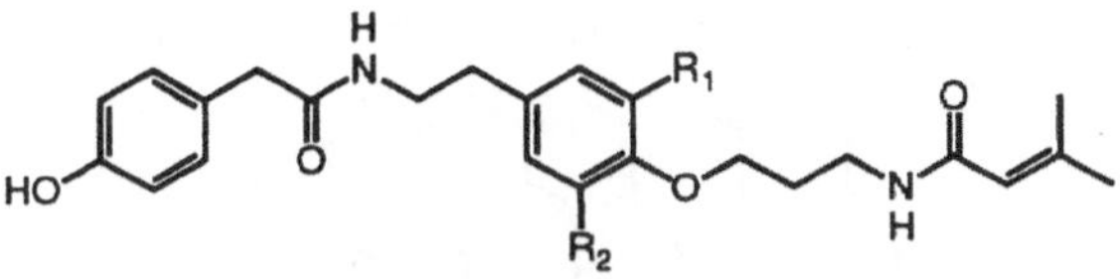

2443 $R_1 = R_2 = Br$

2444 $R_1 = I, R_2 = Br$

2445 $R_1 = R_2 = I$

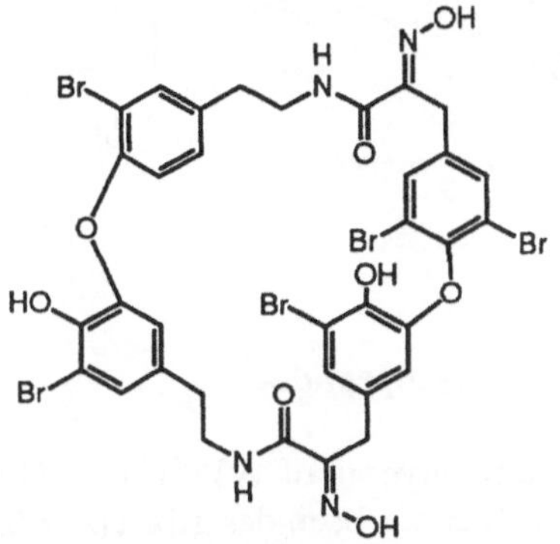

2446 (bastadin-19)

3.22.3.4. Bastadins

The sponge *Ianthella basta* has been found to contain another bastadin, bastadin-19 (**2446**), which is a modulator of skeletal muscle FKBP12/calcium channel complex (*2276*).

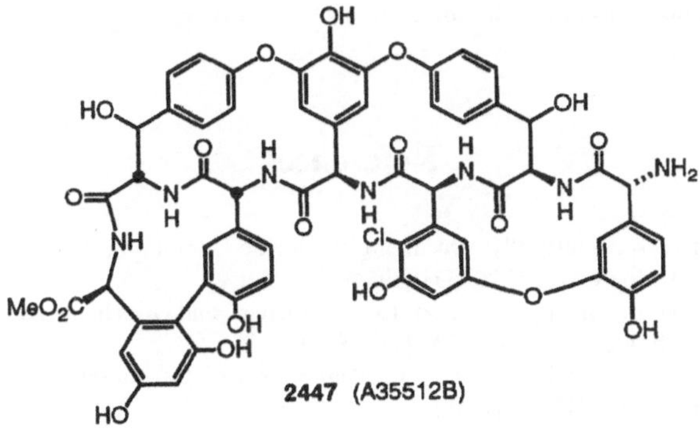

2447 (A35512B)

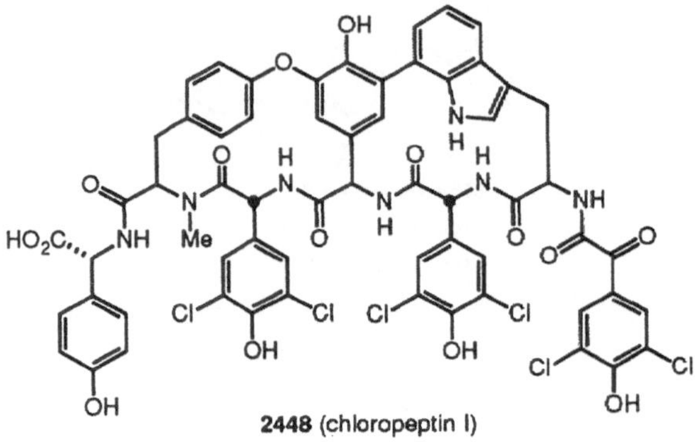

2448 (chloropeptin I)

3.23. Glycopeptides

The structure of the pseudoaglycon of A35512B (**2447**) which is produced by *Streptomyces candidus* has been described (*2277*). A *Streptomyces* sp. produces chloropeptin I (**2448**) which is identical to complestatin except for the attachment of the indole ring (*2278*). Complestatin was also isolated ("chloropeptin II").

Acknowledgements

The author is deeply appreciative to his wife, Tippy, for organizing the reference material contained in this review, and for typing the entire manuscript and drawing all of the structures. Without her monumental effort, this work almost certainly would not have been completed in the lifetime of the author! Special thanks go to the authors' research group for their patience and, especially, to Professor Richard E. Moore for his extraordinary forbearance in awaiting the completion of this review.

References

1. BRACKEN, A.: Naturally Occurring Chlorine-Containing Organic Substances. Manufacturing Chemist **25**, 533 (1954).
2. PETTY, M.A.: An Introduction to the Origin and Biochemistry of Microbial Halometabolites. Bact. Rev. **25**, 111 (1961).
3. FOWDEN, L.: The Occurrence and Metabolism of Carbon-Halogen Compounds. Proc. Roy. Soc. **B171**, 5 (1968).

4. TURNER, W.B.: Fungal Metabolites. New York: Academic Press. 1971.
5. SIUDA, J.F., and J.F. DEBERNARDIS: Naturally Occurring Halogenated Organic Compounds. Lloydia **36**, 107 (1973).
6. MINALE, L.: Natural Product Chemistry of the Marine Sponges. Pure Appl. Chem. **48**, 7 (1976).
7. THOMSON, R.H.: Halogenated Metabolites from Marine Animals and Plants. J. Indian Chem. Soc. **55**, 1209 (1978).
8. FENICAL, W.: Natural Halogenated Organics. In: Marine Organic Chemistry, Chap. 12 (E.K. Duursma and R. Dawson, eds.). Amsterdam: Elsevier. 1981.
9. TURNER, W.B., and D.C. ALDRIDGE: Fungal Metabolites, 2nd ed. New York: Academic Press. 1983.
10. ENGVILD, K.C.: Chlorine-Containing Natural Compounds in Higher Plants. Phytochem. **25**, 781 (1986).
11. GRIBBLE, G.W.: Naturally Occurring Organohalogen Compounds – A Survey. J. Nat. Prod. **55**, 1353 (1992).
12. NAUMANN, K.: Chlorchemie der Natur. Chem. Zeit. **27**, 33 (1993).
13. SCHEUER, P.J.: Chemistry of Marine Natural Products. New York: Academic Press. 1973.
14. FENICAL, W.: Halogenation in the Rhodophyta – A Review. J. Phycol. **11**, 245 (1975).
15. MINALE, L., G. CIMINO, S. DE STEFANO, and G. SODANO: Natural Products from Porifera. Progr. Chem. Org. Nat. Prod. **33**, 1 (1976).
16. BAKER, J.T.: Some Metabolites from Australian Marine Organisms. Pure Appl. Chem. **48**, 35 (1976).
17. SCHEUER, P.J.: The Varied and Fascinating Chemistry of Marine Mollusks. Is. J. Chem. **16**, 52 (1977).
18. FAULKNER, D.J.: Interesting Aspects of Marine Natural Products Chemistry. Tetrahedron **33**, 1421 (1977).
19. Marine Natural Products, Vols. I–V (P.J. Scheuer, ed.). New York: Academic Press. 1981.
20. KREBS, H.C.: Recent Developments in the Field of Marine Natural Products with Emphasis on Biologically Active Compounds. Progr. Chem. Org. Nat. Prod. **49**, 151 (1986).
21. RINEHART, JR., K.L., P.D. SHAW, L.S. SHIELD, J.B. GLOER, G.C. HARBOUR, M.E.S. KOKER, D. SAMAIN, R.E. SCHWARTZ, A.A. TYMIAK, D.L. WELLER, G.T. CARTER, M.H.G. MUNRO, R.G. HUGHES, JR., H.E. RENIS, E.B. SWYNENBERG, D.A. STRINGFELLOW, J.J. VAVRA, J.H. COATS, G.E. ZURENKO, S.L. KUENTZEL, L.H. LI, G.J. BAKUS, R.C. BRUSCA, L.L. CRAFT, D.N. YOUNG, and J.L. CONNOR: Marine Natural Products as Sources of Antiviral, Antimicrobial, and Antineoplastic Agents. Pure Appl. Chem. **53**, 795 (1981).
22. MÜLLER, G., and W. SCHMITZ: Halogenorganische Verbindungen in aquatischen Sedimenten: Antropogene und Biogene. Chem. Zeit. **109**, 415 (1985).
23. FAULKNER, D.J.: Marine Natural Products: Metabolites of Marine Algae and Herbivorous Marine Molluscs. Nat. Prod. Rep. **1**, 251 (1984).
24. FAULKNER, D.J.: Marine Natural Products: Metabolites of Marine Invertebrates. Nat. Prod. Rep. **1**, 551 (1984).
25. FAULKNER, D.J.: Marine Natural Products. Nat. Prod. Rep. **3**, 1 (1986).
26. FAULKNER, D.J.: Marine Natural Products. Nat. Prod. Rep. **4**, 539 (1987).
27. FAULKNER, D.J.: Marine Natural Products. Nat. Prod. Rep. **5**, 613 (1988).
28. FAULKNER, D.J.: Marine Natural Products. Nat. Prod. Rep. **7**, 269 (1990).
29. FAULKNER, D.J.: Marine Natural Products. Nat. Prod. Rep. **8**, 97 (1991).

30. FAULKNER, D.J.: Marine Natural Products. Nat. Prod. Rep. **9**, 323 (1992).
31. FAULKNER, D.J.: Marine Natural Products. Nat. Prod. Rep. **10**, 497 (1993).
32. MOORE, R.E.: Marine Aliphatic Natural Products. Aliphatic and Related Natural Prod. Chem. **1**, 20 (1979).
33. THOMSON, R.H.: Marine Natural Products. Chem. Brit. **14**, 133 (1978).
34. CHRISTOPHERSEN, C.: Secondary Metabolites from Marine Bryozoans. A Review. Acta Chem. Scand. **B39**, 517 (1985).
35. ASPLUND, G., and A. GRIMVALL: Organohalogens in Nature. Environ. Sci. Technol. **25**, 1346 (1991).
36. FLEMING, B.: Chlorinated Organics in Perspective: From Drinking Water to Mill Effluent. Pulp & Paper April, 115 (1991).
37. PREMUZIC, E.: Chemistry of Natural Products Derived from Marine Sources. Progr. Chem. Org. Nat. Prod. **29**, 417 (1971).
38. HOPP, V.: Chlor und seine Verbindungen—ihr Kreislauf in Natur und Technik. Chem. Zeit. **115**, 341 (1991).
39. HARPER, D.B., and D. O'HAGAN: The Fluorinated Natural Products. Nat. Prod. Rep. **11**, 123 (1994).
40. GRIBBLE, G.W.: The Natural Production of Chlorinated Compounds. Environ. Sci. Technol. **28**, 310A (1994).
41. GOLDBERG, E.D.: The Oceans as a Chemical System. In: The Sea, 2 (M.N. Hill, ed.), pp. 3–25. New York: Wiley-Interscience. 1963.
42. STIJVE, T.: Inorganic Bromide in Higher Fungi. Z. Naturforsch. **39C**, 863 (1984).
43. ISIDOROV, V.A.: Organic Chemistry of the Earth's Atmosphere, p. 107. Berlin, Heidelberg: Springer. 1990.
44. KEENE, W.C., A.A.P. PSZENNY, D.J. JACOB, R.A. DUCE, J.N. GALLOWAY, J.J. SCHULTZ-TOKOS, H. SIEVERING, and J.F. BOATMAN: The Geochemical Cycling of Reactive Chlorine Through the Marine Troposphere. Global Biogeochem. Cycles **4**, 407 (1990).
45. STOLARSKI, R.S., and R.J. CICERONE: Stratospheric Chlorine: A Possible Sink for Ozone. Can. J. Chem. **52**, 1610 (1974).
46. SYMONDS, R.B., W.I. ROSE, and M.H. REED: Contribution of Cl- and F-Bearing Gases to the Atmosphere by Volcanoes. Nature **334**, 415 (1988).
47. TABAZADEH, A., and R.P. TURCO: Stratospheric Chlorine Injection by Volcanic Eruptions: HCl Scavenging and Implications for Ozone. Science **260**, 1082 (1993).
48. WESTRICH, H.R., and T.M. GERLACH: Magmatic Gas Source for the Stratospheric SO_2 Cloud from the June 15, 1991, Eruption of Mount Pinatubo. Geology **20**, 867 (1992).
49. JOHNSTON, D.A.: Volcanic Contribution of Chlorine to the Stratosphere: More Significant to Ozone than Previously Estimated? Science **209**, 491 (1980).
50. SYMONDS, R.B., W.I. ROSE, T.M. GERLACH, P.H. BRIGGS, and R.S. HARMON: Evaluation of Gases, Condensates, and SO_2 Emissions from Augustine Volcano, Alaska: The Degassing of a Cl-Rich Volcanic System. Bull. Volcanol. **52**, 355 (1990).
51. SYMONDS, R.B., M.H. REED, and W.I. ROSE: Origin, Speciation, and Fluxes of Trace-Element Gases at Augustine Volcano, Alaska: Insights into Magma Degassing and Fumarolic Processes. Geochim. Cosmochim. Acta **56**, 633 (1992).
52. OSKARSSON, N.: The Interaction Between Volcanic Gases and Tephra. Fluorine Adhering to Tephra of the 1970 Hekla Eruption. J. Volcanol. Geother. Res. **8**, 251 (1980).
53. WOODS, D.C., R.L. CHUAN, and W.I. ROSE: Halite Particles Injected into the Stratosphere by the 1982 El Chichón Eruption. Science **230**, 170 (1985).

54. MANKIN, W.G., and M.T. COFFEY: Increased Stratospheric Hydrogen Chloride in the El Chichón Cloud. Science **226**, 170 (1983).
55. CADLE, R.D., A.L. LAZRUS, B.J. HUEBERT, L.E. HEIDT, W.I. ROSE, D.C. WOODS, R.L. CHUAN, R.E. STOIBER, D.B. SMITH, and R.A. ZIELINSKI: Atmospheric Implications of Studies of Central American Volcanic Eruption Clouds. J. Geophys. Res. **84**, 6961 (1979).
56. OLMEZ, I., D.L. FINNEGAN, and W.H. ZOLLER: Iridium Emissions from Kilauea Volcano. J. Geophys. Res. **91**, 653 (1986).
57. BLAKE, G.A., J. KEENE, and T.G. PHILLIPS: Chlorine in Dense Interstellar Clouds: The Abundance of HCl in OMC-1. Astrophys. J. **295**, 501 (1985).
58. WINNEWISSER, G., and E. HERBST: Organic Molecules in Space. Top. Curr. Chem. **139**, 119 (1987).
59. SAUVAGEAU, C.: Algae Containing Free Iodine. Rev. Bot. App. Agr. Col. **6**, 169 (1926); Chem. Abstr. **20**, 3485 (1926).
60. KYLIN, H.: The Occurrence of Iodides, Bromides and Iodide-Oxidases in Marine Algae. Z. Physiol. Chem. **186**, 50 (1929).
61. LOW, E.M.: Iodine and Bromine in Sponges. J. Mar. Res. **8**, 97 (1949).
62. FAULKNER, D.J.: Natural Organohalogen Compounds. In: The Handbook of Environmental Chemistry, Vol. 1, Part A (O. Hutzinger, ed.), p. 229. Berlin: Springer. 1980.
63. ISIDOROV, V.A.: Organic Chemistry of the Earth's Atmosphere, p. 49. Berlin, Heidelberg: Springer. 1990.
64. GREGSON, R.P., B.A. BALDO, P.G. THOMAS, R.J. QUINN, P.R. BERGQUIST, J.F. STEPHENS, and A.R. HORNE: Fluorine Is a Major Constituent of the Marine Sponge *Halichondria moorei*. Science **206**, 1108 (1979).
65. RODRIGUEZ, J.M.: Probing Stratospheric Ozone. Science **261**, 1128 (1993).
66. CICERONE, R.J.: Fires, Atmospheric Chemistry, and the Ozone Layer. Science **263**, 1243 (1994).
67. PEARSON, C.R.: C_1 and C_2 Halocarbons. In: The Handbook of Environmental Chemistry, Vol. 3/Part B (O. Hutzinger, ed.). Berlin, Heidelberg: Springer. 1982.
68. LEISINGER, T.: Microorganisms and Xenobiotic Compounds. Experientia **39**, 1183 (1983).
69. RASMUSSEN, R.A., L.E. RASMUSSEN, M.A.K. KHALIL, and R.W. DALLUGE: Concentration Distribution of Methyl Chloride in the Atmosphere. J. Geophys. Res. **85**, 7350 (1980).
70. HARPER, D.B.: Halomethane from Halide Ion – A Highly Efficient Fungal Conversion of Environmental Significance. Nature **315**, 55 (1985).
71. EDWARDS, P.R., I. CAMPBELL, and G.S. MILNE: The Impact of Chloromethanes on the Environment, Part 2: Methyl Chloride and Methylene Chloride. Chem. Ind., 619 (1982).
72. COWAN, M.I., A.T. GLEN, S.A. HUTCHINSON, M.E. MACCARTNEY, J.M. MACKINTOSH, and A.M. MOSS: Production of Volatile Metabolites by Species of Fomes. Trans. Br. Mycol. Soc. **60**, 347 (1973).
73. WHITE, R.H.: Biosynthesis of Methyl Chloride in the Fungus *Phellinus pomaceus*. Arch. Microbiol. **132**, 100 (1982).
74. HARPER, D.B., J.T. KENNEDY, and J.T.G. HAMILTON: Chloromethane Biosynthesis in Poroid Fungi. Phytochem. **27**, 3147 (1988).
75. HARPER, D.B., and J.T. KENNEDY: Effect of Growth Conditions on Halomethane Production by *Phellinus* Species: Biological and Environmental Implications. J. Gen. Microbiol. **132**, 1231 (1986).

76. TURNER, E.M., M. WRIGHT, T. WARD, D.J. OSBORNE, and R. SELF: Production of Ethylene and Other Volatiles and Changes in Cellulase and Laccase Activities During the Life Cycle of the Cultivated Mushroom, *Agaricus bisporus*. J. Gen. Microbiol. **91**, 167 (1975).
77. GSCHWEND, P.M., J.K. MACFARLAND, and K.A. NEWMAN: Volatile Halogenated Organic Compounds Released to Seawater from Temperate Marine Macroalgae. Science **227**, 1033 (1985).
78. WUOSMAA, A.M., and L.P. HAGER: Methyl Chloride Transferase: A Carbocation Route for Biosynthesis of Halometabolites. Science **249**, 160 (1990).
79. MANLEY, S.L., and M.N. DASTOOR: Methyl Halide (CH_3X) Production from the Giant Kelp, *Macrocystis*, and Estimates of Global CH_3X Production by Kelp. Limnol. Oceanogr. **32**, 709 (1987).
80. BLACKMAN, A.J., N.W. DAVIES, and C.E. RALPH: Volatile and Odorous Compounds from the Bryozoan *Biflustra perfragilis*. Biochem. Syst. Ecol. **20**, 339 (1992).
81. ISIDOROV, V.A., I.G. ZENKEVICH, and B.V. IOFFE: Volatile Organic Compounds in the Atmosphere of Forests. Atmos. Environ. **19**, 1 (1985).
82. WEVER, R.: Formation of Halogenated Gases by Natural Sources. In: Microbial Production and Consumption of Greenhouse Gases: Methane, Nitrogen Oxides, and Halomethanes, Chap. 15 (J.E. Rogers and W.B. Whitman, eds.). Washington, D.C.: American Society for Microbiology. 1991.
83. HARPER, D.B., J.T.G. HAMILTON, J.T. KENNEDY, and K.J. MCNALLY: Chloromethane, a Novel Methyl Donor for Biosynthesis of Esters and Anisoles in *Phellinus pomaceus*. Appl. Environ. Microbiol. **55**, 1981 (1989).
84. HARPER, D.B., J.A. BUSWELL, J.T. KENNEDY, and J.T.G. HAMILTON: Chloromethane, Methyl Donor in Veratryl Alcohol Biosynthesis in *Phanerochaete chrysosporium* and Other Lignin-Degrading Fungi. Appl. Environ. Microbiol. **56**, 3450 (1990).
85. HARPER, D.B., J.A. BUSWELL, and J.T. KENNEDY: Effect of Chloromethane on Veratryl Alcohol and Lignin Peroxidase Production by the Fungus *Phanerochaete chrysosporium*. J. Gen. Microbiol. **137**, 2867 (1991).
86. ISIDOROV, V.A.: Natural Sources of Organic Components of the Atmosphere. In: Organic Chemistry of the Earth's Atmosphere, Chap. 3. Berlin, Heidelberg: Springer. 1990.
87. PALMER, T.Y.: Combustion Sources of Atmospheric Chlorine. Nature **263**, 44 (1976).
88. CRUTZEN, P.J., and M.O. ANDREAE: Biomass Burning in the Tropics: Impact on Atmospheric Chemistry and Biogeochemical Cycles. Science **250**, 1669 (1990).
89. LOVELOCK, J.E.: Natural Halocarbons in the Air and in the Sea. Nature **256**, 193 (1975).
90. STEPHENS, E.R., and F.R. BURLESON: Distribution of Light Hydrocarbons in Ambient Air. J. Air Pollut. Control Assoc. **19**, 929 (1969).
91. TASSIOS, S., and D.R. PACKHAM: The Release of Methyl Chloride from Biomass Burning in Australia. J. Air Pollut. Control Assoc. **35**, 41 (1985).
92. GUERIN, M.R.: Organic Chemistry of the Atmosphere, Chap. 3 (L.D. Hansen and D.J. Eatough, eds.). Boca Raton, FL: CRC Press. 1991.
93. STOIBER, R.E., D.C. LEGGETT, T.F. JENKINS, R.P. MURRMANN, and W.I. ROSE, JR.: Organic Compounds in Volcanic Gas from Santiaguito Volcano, Guatemala. Geol. Soc. Amer. Bull. **82**, 2299 (1971).
94. RASMUSSEN, R.A., M.A.K. KHALIL, R.W. DALLUGE, S.A. PENKETT, and B. JONES: Carbonyl Sulfide and Carbon Disulfide from the Eruptions of Mount St. Helens. Science **215**, 665 (1982).
95. CADLE, R.D.: A Comparison of Volcanic with Other Fluxes of Atmospheric Trace Gas Constituents. Rev. Geophys. Space Phys. **18**, 746 (1980).

96. GERLACH, T.M.: Evaluation of Volcanic Gas Analyses from Kilauea Volcano. J. Volcan. Geotherm. Res. **7**, 295 (1980).
97. INN, E.C.Y., J.F. VEDDER, E.P. CONDON, and D. O'HARA: Gaseous Constituents in the Plume from Eruptions of Mount St. Helens. Science **211**, 821 (1981).
98. DEGROOT, W.F.: Methyl Chloride as a Gaseous Tracer for Wood Burning? Letter to the Editor. Environ. Sci. Technol. **23**, 252 (1989).
99. EDGERTON, S.A., M.A.K. KHALIL, and R.A. RASMUSSEN: Emissions from Wood Burning. Letter to the Editor. Environ. Sci. Technol. **23**, 906 (1989).
100. VARNES, J.L.: The Release of Methyl Chloride from Potato Tubers. Am. Potato J. **59**, 593 (1982).
101. GRIMSRUD, E.P., and R.A. RASMUSSEN: Survey and Analysis of Halocarbons in the Atmosphere by Gas Chromatography-Mass Spectrometry. Atmos. Environ. **9**, 1014 (1975).
102. MCCONNELL, O., and W. FENICAL: Halogen Chemistry of the Red Alga *Asparagopsis*. Phytochem. **16**, 367 (1977).
103. ISIDOROV, V.A., E.B. PRILEPSKY, and V.G. POVAROV: Photochemically and Optically Active Components of Minerals and Gas Emissions of Mining Plants. J. Ecol. Chem. N2–3, 201 (1993).
104. EDWARDS, P.R., I. CAMPBELL, and G.S. MILNE: The Impact of Chloromethanes on the Environment, Part 1: The Atmospheric Chlorine Cycle. Chem. Ind. (London), 574 (1982).
105. LOVELOCK, J.E., R.J. MAGGS, and R.J. WADE: Halogenated Hydrocarbons in and over the Atlantic. Nature **241**, 194 (1973).
106. ISIDOROV, V.A., V.G. POVAROV, and E.B. PRILEPSKY: Geological Sources of Volatile Organic Components in Regions of Seismic and Volcanic Activity. J. Ecol. Chem. N1, 19 (1993).
107. PYYSALO, H.: Identification of Volatile Compounds in Seven Edible Fresh Mushrooms. Acta Chem. Scand. **B30**, 235 (1976).
108. GIL, V., and A.J. MACLEOD: Some Glucosinolates of *Farsetia aegyptia* and *Farsetia ramosissima*. Phytochem. **19**, 227 (1980).
109. MACLEOD, A.J., and N.G. DETROCONIS: Volatile Flavor Components of Sapodilla Fruit. J. Agric. Food Chem. **30**, 515 (1982).
110. CHOI, S.H., and H. KATO: Volatile Components of *Sergia lucens* and Its Fermented Product. Agric. Biol. Chem. **48**, 1479 (1984).
111. KAMEOKA, H., K. KUBO, and M. MIYAZAWA: Volatile Flavor Components of Malabar-Nightshade (*Basella rubra* L.). J. Food Comp. Anal. **4**, 315 (1991).
112. FOGELQVIST, E.: Carbon Tetrachloride, Tetrachloroethylene, 1,1,1-Trichloroethane and Bromoform in Arctic Seawater. J. Geophys. Res. **90**, 9181 (1985).
113. KRYSELL, M., and D.W.R. WALLACE: Arctic Ocean Ventilation Studied with a Suite of Anthropogenic Halocarbon Traces. Science **242**, 746 (1988).
114. LOVELOCK, J.E.: Atmospheric Halocarbons and Stratospheric Ozone. Nature **252**, 292 (1974).
115. YAGI, K., J. WILLIAMS, N.-Y. WANG, and R.J. CICERONE: Agricultural Soil Fumigation as a Source of Atmospheric Methyl Bromide. Proc. Natl. Acad. Sci. USA **90**, 8420 (1993).
116. BERG, W.W., L.E. HEIDT, W. POLLOCK, P.D. SPERRY, R.J. CICERONE, and E.S. GLADNEY: Brominated Organic Species in the Arctic Atmosphere. Geophys. Res. Lett. **11**, 429 (1984).
117. RASMUSSEN, R.A., and M.A.K. KHALIL: Gaseous Bromine in the Arctic Haze. Geophys. Res. Lett. **11**, 433 (1984).

118. CICERONE, R.J., L.E. HEIDT, and W.H. POLLOCK: Measurements of Atmospheric Methyl Bromide and Bromoform. J. Geophys. Res. **93**, 3745 (1988).

119. MANLEY, S.L., and M.N. DASTOOR: Methyl Iodide (CH_3I) Production by Kelp and Associated Microbes. Mar. Biol. **98**, 477 (1988).

120. SHARP, G.J., Y. YOKOUCHI, and H. AKIMOTO: Trace Analysis of Organobromine Compounds in Air by Adsorbent Trapping and Capillary Gas Chromatography/Mass Spectroscopy. Environ. Sci. Technol. **26**, 815 (1992).

121. STURGES, W.T., C.W. SULLIVAN, R.C. SCHNELL, L.E. HEIDT, and W.H. POLLOCK: Bromoalkane Production by Antarctic Ice Algae. Tellus **45B**, 120 (1993).

122. CLASS, TH., R. KOHNLE, and K. BALLSCHMITER: Chemistry of Organic Traces in the Air VII: Bromo- and Bromochloromethanes in Air Over the Atlantic Ocean. Chemosphere **15**, 429 (1986).

123. MANLEY, S.L., K. GOODWIN, and W.J. NORTH: Laboratory Production of Bromoform, Methylene Bromide, and Methyl Iodide by Macroalgae and Distribution in Nearshore Southern California Waters. Limnol. Oceanogr. **37**, 1652 (1992).

124. REIFENHÄUSER, W., and K.G. HEUMANN: Determinations of Methyl Iodide in the Antarctic Atmosphere and the South Polar Sea. Atmos. Environ. **26A**, 2905 (1992).

125. RASMUSSEN, R.A., M.A.K. KHALIL, R. GUNAWARDENA, and S.D. HOYT: Atmospheric Methyl Iodide (CH_3I). J. Geophys. Res. **87**, 3086 (1982).

126. SINGH, H.B., L.J. SALAS, and R.E. STILES: Methyl Halides in and over the Eastern Pacific. J. Geophys. Res. **88**, 3684 (1983).

127. BUTLER, J.H.: The Potential Role of the Ocean in Regulating Atmospheric CH_3Br. Geophys. Res. Lett. **21**, 185 (1994).

128. KHALIL, M.A.K., R.A. RASMUSSEN, and R. GUNAWARDENA: Atmospheric Methyl Bromide: Trends and Global Mass Balance. J. Geophys. Res. **98**, 2887 (1993).

129. SINGH, H.B., and M. KANAKIDOU: An Investigation of the Atmospheric Sources and Sinks of Methyl Bromide. Geophys. Res. Lett. **20**, 133 (1993).

130. BERG, W.W., P.D. SPERRY, K.A. RAHN, and E.S. GLADNEY: Atmospheric Bromine in the Arctic. J. Geophys. Res. **88**, 6719 (1983).

131. STURGES, W.T., and L.A. BARRIE: Chlorine, Bromine and Iodine in Arctic Aerosols. Atmos. Environ. **22**, 1179 (1988).

132. BURRESON, B.J., R.E. MOORE, and P.P. ROLLER: Volatile Halogen Compounds in the Alga *Asparagopsis taxiformis* (Rhodophyta). J. Agric. Food. Chem. **24**, 856 (1976).

133. MOORE, R.E.: Volatile Compounds from Marine Algae. Acct. Chem. Res. **10**, 40 (1977).

134. CLASS, T., and K. BALLSCHMITER: Chemistry of Organic Traces in Air, VIII: Sources and Distribution of Bromo- and Bromochloromethanes in Marine Air and Surface Water of the Atlantic Ocean. J. Atmos. Chem. **6**, 35 (1988).

135. SCHALL, C., and K.G. HEUMANN: GC Determination of Volatile Organoiodine and Organobromine Compounds in Arctic Seawater and Air Samples. Fresenius' J. Anal. Chem. **346**, 717 (1993).

136. TOKARCZYK, R., and R.M. MOORE: Production of Volatile Organohalogens by Phytoplankton Cultures. Geophys. Res. Lett. **21**, 285 (1994).

137. KLICK, S.: The Release of Volatile Halocarbons to Seawater by Untreated and Heavy Metal Exposed Samples of the Brown Seaweed *Fucus vesiculosus*. Marine Chem. **42**, 211 (1993).

138. REIFENHÄUSER, W., and K.G. HEUMANN: Bromo- and Bromochloromethanes in the Antarctic Atmosphere and the South Polar Sea. Chemosphere **24**, 1293 (1992).

139. WEVER, R., M.G.M. TROMP, B.E. KRENN, A. MARJANI, and M. VAL TOL: Brominating Activity of the Seaweed *Ascophyllum nodosum*: Impact on the Biosphere. Environ. Sci. Technol. **25**, 446 (1991).

140. DYRSSEN, D., and E. FOGELQVIST: Bromoform Concentrations of the Arctic Ocean in the Svalbard Area. Oceanol. Acta **4**, 313 (1981).

141. KRYSELL, M.: Bromoform in the Nansen Basin in the Arctic Ocean. Marine Chem. **33**, 187 (1991).

142. THEILER, R., J.C. COOK, L.P. HAGER, and J.F. SIUDA: Halohydrocarbon Synthesis by Bromoperoxidase. Science **202**, 1094 (1978).

143. BEISSNER, R.S., W.J. GUILFORD, R.M. COATES, and L.P. HAGER: Synthesis of Brominated Heptanones and Bromoform by a Bromoperoxidase of Marine Origin. Biochem. **20**, 3724 (1981).

144. KLICK, S., and K. ABRAHAMSSON: Biogenic Volatile Iodated Hydrocarbons in the Ocean. J. Geophys. Res. **97**, 12683 (1992).

145. BURRESON, B.J., R.E. MOORE, and P. ROLLER: Haloforms in the Essential Oil of the Alga *Asparagopsis taxiformis* (Rhodophyta). Tetrahedron Lett., 473 (1975).

146. MOORE, R.M., and R. TOKARCZYK: Chloro-Iodomethane in N. Atlantic Waters: A Potentially Significant Source of Atmospheric Iodine. Geophys. Res. Lett. **19**, 1779 (1992).

147. MOORE, R.M., and R. TOKARCZYK: Volatile Biogenic Halocarbons in the Northwest Atlantic. Global Biogeochem. Cycles **7**, 195 (1993).

148. FABIAN, P., R. BORCHERS, B.C. KRÜGER, and S. LAI: CF_4 and C_2F_6 in the Atmosphere. J. Geophys. Res. **92**, 9831 (1987).

149. CICERONE, R.J.: Atmospheric Carbon Tetrafluoride: A Nearly Inert Gas. Science **206**, 59 (1979).

150. STURGES, W.T., G.F. COTA, and P.T. BUCKLEY: Bromoform Emission from Arctic Ice Algae. Nature **358**, 660 (1992).

151. MANÖ, S., and M.O. ANDREAE: Emission of Methyl Bromide from Biomass Burning. Science **263**, 1255 (1994).

152. WOOLARD, F.X., R.E. MOORE, and P.P. ROLLER: Halogenated Acetamides, But-3-en-2-ols, and Isopropanols from *Asparagopsis taxiformis* (Delile) Trev. Tetrahedron **32**, 2843 (1976).

153. FENICAL, W.: Polyhaloketones from the Red Seaweed *Asparagopsis taxiformis*. Tetrahedron Lett., 4463 (1974).

154. COMBAUT, G., Y. BRUNEAU, J. TESTE, and L. CODOMIER: Composes Halogenes d'une Algue Rouge, *Falkenbergia rufolanosa* Tetrasporophyte d'*Asparagopsis armata*. Phytochem. **17**, 1661 (1978).

155. SIUDA, J.F., G.R. VANBLARICOM, P.D. SHAW, R.D. JOHNSON, R.H. WHITE, L.P. HAGER, and K.L. RINEHART JR.: 1-Iodo-3,3-dibromo-2-heptanone, 1,1,3,3-Tetrabromo-2-heptanone, and Related Compounds from the Red Alga *Bonnemaisonia hamifera*. J. Am. Chem. Soc. **97**, 937 (1975).

156. JACOBSEN, N., and J.O. MADSEN: Halogenated Metabolites Including Brominated 2-Heptanols and 2-Heptyl Acetates from the Tetrasporophyte of the Red Alga *Bonnemaisonia hamifera*. Tetrahedron Lett., 3065 (1978).

157. MCCONNELL, O.J., and W. FENICAL: Polyhalogenated 1-Octene-3-ones, Antibacterial Metabolites from the Red Seaweed *Bonnemaisonia asparagoides*. Tetrahedron Lett., 1851 (1977).

158. MCCONNELL, O.J., and W. FENICAL: Halogen Chemistry of the Red Alga *Bonnemaisonia*. Phytochem. **19**, 233 (1980).

159. MCCONNELL, O.J., and W. FENICAL: Halogenated Metabolites – Including Favorsky

Rearrangement Products – from the Red Seaweed *Bonnemaisonia nootkana*. Tetrahedron Lett., 4159 (1977).

160. ROSE, A.F., J.A. PETTUS JR., and J.J. SIMS: Marine Natural Products, XIII: Isolation and Synthesis of Some Halogenated Ketones from the Red Seaweed *Delisea fimbriata*. Tetrahedron Lett., 1847 (1977).

161. DE NYS, R., J.C. COLL, and B.F. BOWDEN: *Delisea pulchra* (cf. *fimbriata*) Revisited. The Structural Determination of Two New Metabolites from the Red Alga *Delisea pulchra*. Aust. J. Chem. **45**, 1625 (1992).

162. KAZLAUSKAS, R., R.O. LIDGARD, and R.J. WELLS: New Polybrominated Metabolites from the Red Alga *Ptilonia australasica*. Tetrahedron Lett., 3165 (1978).

163. OHTA, K., and M. TAKAGI: Antimicrobial Compounds of the Marine Red Alga *Marginisporum aberrans*. Phytochem. **16**, 1085 (1977).

164. PETERS, R.A., and M. SHORTHOUSE: Identification of a Volatile Constituent Formed by Homogenates of *Acacia georginae* Exposed to Fluoride. Nature **231**, 123 (1971).

165. KUBOTA, K., K. YOKOYAMA, T. YAMANISHI, and S. AKATSUKA: Odor of Dried Shell Powder of Antarctic Krills and Liquid Seasoning of the Hydrolysate. Nippon Nogei Kagaku Kaishi **54**, 727 (1980); Chem. Abstr. **94**, 63937 (1980).

166. FRANSSEN, M.C.R., M.A. POSTHUMUS, and H.C. VAN DER PLAS: New Halometabolites from *Caldariomyces fumago*. Phytochem. **27**, 1093 (1988).

167. NICOD, F., F. TILLEQUIN, and J. VAQUETTE: Métabolite Halogéné Nouveau, Substance Majoritaire de *Ptilonia magellanica*, Algue Rhodophycée. J. Nat. Prod. **50**, 259 (1987).

168. YANAGISAWA, I., and H. YOSHIKAWA: A Bromine Compound Isolated from Human Cerebrospinal Fluid. Biochim. Biophys. Acta **329**, 283 (1973).

169. MYNDERSE, J.S., and R.E. MOORE: The Isolation of (–)-*E*-1-Chlorotridec-1-ene-6,8-diol from a Marine Cyanophyte. Phytochem. **17**, 1325 (1978).

170. NOZOE, S., N. ISHII, G. KUSANO, K. KIKUCHI, and T. OHTA: Neocarzilins A and B, Novel Polyenones from *Streptomyces carzinostaticus*. Tetrahedron Lett. **33**, 7547 (1992).

171. NOZOE, S., K. KIKUCHI, N. ISHII, and T. OHTA: Synthesis of Neocarzilin A: An Absolute Stereochemistry. Tetrahedron Lett. **33**, 7551 (1992).

172. KAZLAUSKAS, R., P.T. MURPHY, and R.J. WELLS: A Brominated Metabolite from the Red Alga *Vidalia spiralis*. Aust. J. Chem. **35**, 219 (1982).

173. CLUTTERBUCK, P.W., S.L. MUKHOPADHYAY, A.E. OXFORD, and H. RAISTRICK: Studies in the Biochemistry of Micro-Organisms 65. (A) A Survey of Chlorine Metabolism by Moulds. (B) Caldariomycin, $C_5H_8O_2Cl_2$, a Metabolic Product of *Caldariomyces fumago* Woronichin. Biochem. J. **34**, 664 (1940).

174. BECKWITH, J.R., and L.P. HAGER: Synthesis of D,L-Caldariomycin. J. Org. Chem. **26**, 5206 (1961).

175. JOHNSON, S.M., I.C. PAUL, K.L. RINEHART, JR., and R. SRINIVASAN: The Absolute Configuration of Caldariomycin. J. Am. Chem. Soc. **90**, 136 (1968).

176. PATTERSON, E.L., W.W. ANDRES, and L.A. MITSCHER: Isolation of the Bromo Analogue of Caldariomycin from *Caldariomyces fumago*. Appl. Microb. **15**, 528 (1967).

177. NAKANISHI, S., K. ANDO, I. KAWAMOTO, T. YASUZAWA, H. SANO, and H. KASE: KS-504 Compounds, Novel Inhibitors of Ca^{+2} and Calmodulin-Dependent Cyclic Nucleotide Phosphodiesterase from *Mollisia ventosa*. J. Antibiot. **42**, 1775 (1989).

178. HIRAYAMA, N., and E. SHIMIZU: Structures of Novel Calmodulin Inhibitors KS504a, KS504b and KS504e. Acta Cryst. **C46**, 1515 (1990).

179. MCGAHREN, W.J., J.H. VAN DEN HENDE, and L.A. MITSCHER: Chlorinated Cyclo-

pentenone Fungiotoxic Metabolites from the Fungus *Sporomia affinis.* J. Am. Chem. Soc. **91**, 157 (1969).

180. GILES, D., and W.B. TURNER: Chlorine-Containing Metabolites of *Periconia macrospinosa.* J. Chem. Soc. (C), 2187 (1969).

181. STRUNZ, G.M., A.S. COURT, J. KOMLOSSY, and M.A. STILLWELL: Structures of Cryptosporiopsin: A New Antibiotic Substance Produced by a Species of *Cryptosporiopsis.* Can. J. Chem. **47**, 2087 (1969).

182. STRUNZ, G.M., A.S. COURT, J. KOMLOSSY, and M.A. STILLWELL: Addendum: Cryptosporiopsin, an Amended Structure. Can. J. Chem. **47**, 3700 (1969).

183. STRUNZ, G.M., P.I. KAZINOTI, and M.A. STILLWELL: A New Chlorinated Cyclopentenone Produced by a *Cryptosporiopsis* sp. Can. J. Chem. **52**, 3623 (1974).

184. LOUSBERG, R.J.J.CH., Y. TIRILLY, and M. MOREAU: Isolation of (−)-Cryptosporiopsin, a Chlorinated Cyclopentenone Fungitoxic Metabolite from *Phialophora asteris* f. sp. *helianthi.* Experientia **32**, 331 (1976).

185. SINGH, J., K.L. DHAR, and C.K. ATAL: Studies on the Genus Piper, Part XI: Occurrence of Pipoxide Chlorohydrin from *Piper hookeri.* Indian J. Pharm. **33**, 50 (1971).

186. JOSHI, B.S., D.H. GAWAD, and H. FUHLER: Revised Structures of Pipoxide and Pipoxide Chlorohydrin. Tetrahedron Lett., 2427 (1979).

187. SAKAMURA, S., K. NABETA, S. YAMADA, and A. ICHIHARA: Minor Constituents from *Phyllosticta* sp. and Their Correlation with Epoxydon (Phyllosinol). Agric. Biol. Chem. **39**, 403 (1975).

188. SAKAMURA, S., J. ITO, and R. SAKAI: Phytotoxic Metabolites of *Phyllosticta* sp. Agric. Biol. Chem. **35**, 105 (1971).

189. NABETA, K., A. ICHIHARA, and S. SAKAMURA: Biosynthesis of Epoxydon and Related Compounds by *Phyllosticta* sp. Agric. Biol. Chem. **39**, 409 (1975).

190. KIRIYAMA, N., Y. HIGUCHI, and Y. YAMAMOTO: Studies on the Metabolic Products of *Aspergillus terreus,* II: Structure and Biosynthesis of the Metabolites of the Strain ATCC 12238. Chem. Pharm. Bull. (Japan) **25**, 1265 (1977).

191. STADLER, M., H. ANKE, W.-R. ARENDHOLZ, F. HANSSKE, U. ANDERS, O. STERNER, and K.-E. BERGQUIST: Lachnumon and Lachnumol A, New Metabolites with Nematicidal and Antimicrobial Activities from the Ascomycete *Lachnum papyraceum* (Karst.) Karst, I: Producing Organism, Fermentation, Isolation and Biological Activities. J. Antibiot. **46**, 961 (1993).

192. STADLER, M., H. ANKE, K.-E. BERGQUIST, and O. STERNER: Lachnumon and Lachnumol A, New Metabolites with Nematicidal and Antimicrobial Activities from the Ascomycete *Lachnum papyraceum* (Karst.) Karst, II: Structural Elucidation. J. Antibiot. **46**, 968 (1993).

193. HIGA, T., and P.J. SCHEUER: Constituents of the Hemichordate *Ptychodera flava laysanica.* Mar. Nat. Prod. NATO Conf., 35 (1977).

194. HIGA, T., R.K. OKUDA, R.M. SEVERNS, P.J. SCHEUER, C.-H. HE, X. CHANGFU, and J. CLARDY: Unprecedented Constituents of a New Species of Acorn Worm. Tetrahedron **43**, 1063 (1987).

195. CORGIAT, J.M., F.C. DOBBS, M.W. BURGER, and P.J. SCHEUER: Organohalogen Constituents of the Acorn Worm *Ptychodera bahamensis.* Comp. Biochem. Physiol. **106B**, 83 (1993).

196. BOLLINGER, P., and T. ZARDIN-TARTAGLIA: Isolierung und Strukturaufklärung von Mikrolin. Helv. Chim. Acta **59**, 1809 (1976).

197. WEBER, H.P., and T.J. PETCHER: Die Kristallstruktur und absolute Konfiguration von Mikrolin. Helv. Chim. Acta **59**, 1821 (1976).

198. Trofast, J., and B. Wickberg: Mycorrhizin A and Chloromycorrhizin A, Two Antibiotics from a Mycorrhizal Fungus of *Monotropa hypopitys* L. Tetrahedron **33**, 875 (1977).

199. Stalhandske, C., C. Svensson, and C. Särnstrand: Chloromycorrhizin A. Acta Cryst. **B33**, 870 (1977).

200. Chexal, K.K., Ch. Tamm, J. Clardy, and K. Hirotsu: Gilmicolin and Mycorrhizinol, Two New Metabolites of *Gilmaniella humicola* Barron. Helv. Chim. Acta **62**, 1129 (1979).

201. Kitamura, E., A. Hirota, M. Nakagawa, M. Nakayama, H. Nozaki, T. Tada, M. Nukina, and H. Hirota: (1*R*, 6*R*, 9*S*, 10*S*)-9-Chloro-10-hydroxy-8-methoxycarbonyl-4-methylene-2,5-dioxabicyclo[4.4.0]dec-3-one-7-ene, A First Chlorine-Containing Shikimate-Related Metabolite from Fungi. Tetrahedron Lett. **31**, 4605 (1990).

202. Curtin, T.P., and J. Reilly: Sclerotiorin, $C_{20}H_{21}O_5Cl$, a Chlorine-Containing Metabolic Product of *Penicillium sclerotiorin*. Biochem. J. **34**, 1419 (1940).

203. Birkinshaw, J.H.: Studies in the Biochemistry of Micro-Organisms, 89: Metabolic Products of *Penicillium multicolor* G.-M. and P. with Special Reference to Sclerotiorin. Biochem. J. **52**, 283 (1952).

204. Udagawa, S.: (–)-Sclerotiorin, a Major Metabolite of *Penicillium hirayamae*. Chem. Pharm. Bull. (Japan) **11**, 366 (1963).

205. Gregory, E.M., and W.B. Turner: 7-epi-Sclerotiorin. Chem. Ind., 1625 (1963).

206. Ellestad, G.A., and W.B. Whalley: The Chemistry of Fungi, Part LII: (–)-Sclerotiorin. J. Chem. Soc., 7260 (1965).

207. Dean, F.M., J. Staunton, and W.B. Whalley: The Chemistry of Fungi, Part XXXVI: A Revised Structure for Sclerotiorin. J. Chem. Soc., 3004 (1959).

208. Whalley, W.B., G. Ferguson, W.C. Marsh, and R.J. Restivo: The Chemistry of Fungi, Part LXVIII: The Absolute Configuration of (+)-Sclerotiorin and of the Azaphilones. J. Chem. Soc., Perkin Trans. 1, 1366 (1976).

209. Gray, R.W., and W.B. Whalley: The Chemistry of Fungi, Part LXIII: Rubrorotiorin, a Metabolite of *Penicillium hirayamae* Udagawa. J. Chem. Soc. C, 3575 (1971).

210. Gray, R.W., and W.B. Whalley: (–) -7-Epi-5-Chloroisorotiorin, a Novel Metabolite. J. Chem. Soc., Chem. Commun., 762 (1970).

211. Takahashi, M., K. Koyama, and S. Natori: Four New Azaphilones from *Chaetomium globosum* var. *flavo-viridae*. Chem. Pharm. Bull. (Japan) **38**, 625 (1990).

212. Omura, S., H. Tanaka, K. Matsuzaki, H. Ikeda, and R. Masuma: Isochromophilones I and II, Novel Inhibitors Against gp120-CD4 Binding from *Penicillium* sp. J. Antibiot. **46**, 1908 (1993).

213. Rose, A.F., P.J. Scheuer, J.P. Springer, and J. Clardy: Stylocheilamide, an Unusual Constituent of the Sea Hare *Stylocheilus longicauda*. J. Am. Chem. Soc. **100**, 7665 (1978).

214. Naylor, S., F.J. Hanke, L.V. Manes, and P. Crews: Chemical and Biological Aspects of Marine Monoterpenes. Progr. Chem. Org. Nat. Prod. **44**, 189 (1983).

215. Faulkner, D.J., and M.O. Stallard: 7-Chloro-3,7-dimethyl-1,4,6-tribromo-1-octen-3-ol, a Novel Monoterpene Alcohol from *Aplysia californica*. Tetrahedron Lett., 1171 (1973).

216. Faulkner, D.J., and M.O. Stallard, J. Fayos, and J. Clardy: (3*R*, 4*S*, 7*S*)-*trans*, *trans*-3,7-Dimethyl-1,8,8-tribromo-3,4,7-trichloro-1,5-octadiene, a Novel Monoterpene from the Sea Hare, *Aplysia californica*. J. Am. Chem. Soc. **95**, 3413 (1973).

217. STALLARD, M.O., and D.J. FAULKNER: Chemical Constituents of the Digestive Gland of the Sea Hare *Aplysia californica* – I. Comp. Biochem. Physiol. **49B**, 25 (1974).

218. WILLCOTT, M.R., R.E. DAVIS, D.J. FAULKNER, and M.O. STALLARD: The Configuration and Conformation of 7-Chloro-1,6-dibromo-3,7-dimethyl-3,4-epoxy-1-octene. Tetrahedron Lett., 3967 (1973).

219. CREWS, P., and E. KHO: Cartilagineal, an Unusual Monoterpene Aldehyde from Marine Alga. J. Org. Chem. **39**, 3303 (1974).

220. ICHIKAWA, N., Y. NAYA, and S. ENOMOTO: New Halogenated Monoterpenes from *Desmia* (*Chondrococcus*) *hornemanni.* Chem. Lett., 1333 (1974).

221. NAYA, Y., Y. HIROSE, and N. ICHIKAWA: Labile Halogenated Monoterpenes from *Desmia* (*Chondrococcus*) *japonicus* Harvey. Chem. Lett., 839 (1976).

222. MYNDERSE, J.S., and D.J. FAULKNER: Polyhalogenated Monoterpenes from the Red Alga *Plocamium cartilagineum.* Tetrahedron **31**, 1963 (1975).

223. IMPERATO, F., L. MINALE, and R. RICCIO: Constituents of the Digestive Gland of Molluscs of the Genus *Aplysia*, II: Halogenated Monoterpenes from *Aplysia limacina.* Experientia **33**, 1273 (1977).

224. CREWS, P., S. NAYLOR, F.J. HANKE, E.R. HOGUE, E. KHO, and R. BRASLAU: Halogen Regiochemistry and Substituent Stereochemistry Determination in Marine Monoterpenes by ^{13}C NMR. J. Org. Chem. **49**, 1371 (1984).

225. KÖNIG, G.M., A.D. WRIGHT, and O. STICHER: A New Polyhalogenated Monoterpene from the Red Alga *Plocamium cartilagineum.* J. Nat. Prod. **53**, 1615 (1990).

226. STIERLE, D.B., and J.J. SIMS: Marine Natural Products, XV: Polyhalogenated Cyclic Monoterpenes from the Red Alga *Plocamium cartilagineum* of Antarctica. Tetrahedron **35**, 1261 (1979).

227. STIERLE, D.B., R.M. WING, and J.J. SIMS: Marine Natural Products, XVI: Polyhalogenated Acyclic Monoterpenes from the Red Alga *Plocamium* of Antarctica. Tetrahedron **35**, 2855 (1979).

228. IRELAND, C., M.O. STALLARD, D.J. FAULKNER, J. FINER, and J. CLARDY: Some Chemical Constituents of the Digestive Gland of the Sea Hare *Aplysia californica.* J. Org. Chem. **41**, 2461 (1976).

229. BLUNT, J.W., N.J. BOWMAN, M.H.G. MUNRO, M.J. PARSONS, G.J. WRIGHT, and Y.K. KON: Polyhalogenated Monoterpenes of the New Zealand Marine Red Alga *Plocamium cartilagineum.* Aust. J. Chem. **38**, 519 (1985).

230. KAZLAUSKAS, R., P.T. MURPHY, R.J. QUINN, and R.J. WELLS: Two Polyhalogenated Monoterpenes from the Red Alga *Plocamium costatum.* Tetrahedron Lett., 4451 (1976).

231. CREWS, P.: Monoterpene Halogenation by the Red Alga *Plocamium oregonum.* J. Org. Chem. **42**, 2634 (1977).

232. CREWS, P., and E. KHO-WISEMAN: Acyclic Polyhalogenated Monoterpenes from the Red Alga *Plocamium violaceum.* J. Org. Chem. **42**, 2812 (1977).

233. BLUNT, J.W., M.P. HARTSHORN, M.H.G. MUNRO, and S.C. YORKE: A Novel, C_8 Dichlorodienol Metabolite of the Red Alga *Plocamium cruciferum.* Tetrahedron Lett., 4417 (1978).

234. BATES, P., J.W. BLUNT, M.P. HARTSHORN, A.J. JONES, M.H.G. MUNRO, W.T. ROBINSON, and S.C. YORKE: Halogenated Metabolites of the Red Alga *Plocamium cruciferum.* Aust. J. Chem. **32**, 2545 (1979).

235. STIERLE, D.B., and J.J. SIMS: Plocamenone, A Unique Halogenated Monoterpene from the Red Alga, *Plocamium.* Tetrahedron Lett. **25**, 153 (1984).

236. COLL, J.C., B.W. SKELTON, A.H. WHITE, and A.D. WRIGHT: Tropical Marine Algae,

II: The Structure Determination of New Halogenated Monoterpenes from *Plocamium hamatum* (Rhodophyta, Gigartinales, Plocamiaceae). Aust. J. Chem. **41**, 1743 (1988).

237. DUNLOP, R.W., P.T. MURPHY, and R.J. WELLS: A New Polyhalogenated Monoterpene from the Red Alga *Plocamium angustum*. Aust. J. Chem. **32**, 2735 (1979).

238. SIMS, J.J., A.F. ROSE, and R.R. IZAC: Applications of ^{13}C-NMR to Marine Natural Products. In: Marine Natural Products, Vol. 2, Chap. 5 (P.J. Scheuer, ed.). New York: Academic Press. 1978.

239. LEARY, J.V., R. KFIR, J.J. SIMS, and D.W. FULBRIGHT: The Mutagenicity of Natural Products from Marine Algae. Mutation Res. **68**, 301 (1979).

240. BURRESON, B.J., F.X. WOOLARD, and R.E. MOORE: Evidence for the Biogenesis of Halogenated Myrcenes from the Red Alga *Chondrococcus hornemanni*. Chem. Lett., 1111 (1975).

241. BURRESON, B.J., F.X. WOOLARD, and R.E. MOORE: Chondrocole A and B, Two Halogenated Dimethylhexahydrobenzofurans from the Red Alga *Chondrococcus hornemanni*. Tetrahedron Lett., 2155 (1975).

242. WOOLARD, F.X., R.E. MOORE, M. MAHENDRAN, and A. SIVAPALAN: (−)-3-Bromomethyl-3-chloro-7-methyl-1,6-octadiene from Sri Lankan *Chondrococcus hornemanni*. Phytochem. **15**, 1069 (1976).

243. COLL, J.C., and A.D. WRIGHT: Tropical Marine Algae, I: New Halogenated Monoterpenes from *Chondrococcus hornemannii* (Rhodophyta, Gigartinales, Rhizophyllidaceae). Aust. J. Chem. **40**, 1893 (1987).

244. COLL, J.C., and A.D. WRIGHT: Tropical Marine Algae, VI: New Monoterpenes from Several Collections of *Chondrococcus hornemannii* (Rhodophyta, Gigartinales, Rhizophyllidaceae). Aust. J. Chem. **42**, 1983 (1989).

245. WRIGHT, A.D., G.M. KÖNIG, O. STICHER, and R. DE NYS: Five New Monoterpenes from the Marine Red Alga *Portieria hornemannii*. Tetrahedron **47**, 5717 (1991).

246. KATAYAMA, A., K. INA, H. NOZAKI, and M. NAKAYAMA: Structure Elucidation of Kurodainol, a Novel Halogenated Monoterpene from Sea Hare (*Aplysia kurodai*). Agric. Biol. Chem. **46**, 859 (1982).

247. MIYAMOTO, T., R. HIGUCHI, T. KOMORI, T. FUJIOKA, and K. MIHASHI: Isolation and Structures of New Isoprenoids, Aplykurodin A and B, and Some Halogenated Terpenoids from the Marine Mollusk, *Aplysia kurodai*, Collected Along the Coast of Fukuoka. Chem. Abstr. **104**, 183529n (1986).

248. MIYAMOTO, T., R. HIGUCHI, N. MARUBAYASHI, and T. KOMORI: Studies on the Constituents of Marine Opisthobranchia, IV: Two New Polyhalogenated Monoterpenes from the Sea Hare *Aplysia kurodai*. Liebigs Ann. Chem., 1191 (1988).

249. DE NOPOLI, L., E. FATTORUSSO, S. MAGNO, and L. MAYOL: Acyclic Polyhalogenated Monoterpenes from Four Marine Hydroids. Biochem. Sys. Ecol. **12**, 321 (1984).

250. CREWS, P., L. CAMPBELL, and E. HERON: Different Chemical Types of *Plocamium violaceum* (Rhodophyta) from the Monterey Bay Region, California. J. Phycol. **13**, 297 (1977).

251. MYNDERSE, J.S., and D.J. FAULKNER: Variations in the Halogenated Monoterpene Metabolites of *Plocamium cartilagineum* and *P. violaceum*. Phytochem. **17**, 237 (1978).

252. MYNDERSE, J.S., and D.J. FAULKNER: Violacene, a Polyhalogenated Monocyclic Monoterpene from the Red Alga *Plocamium violaceum*. J. Am. Chem. Soc. **96**, 6771 (1974).

253. ENGEN, D.V., J. CLARDY, E. KHO-WISEMAN, P. CREWS, M.D. HIGGS, and D.J. FAULKNER: Violacene: A Reassignment of Structure. Tetrahedron Lett., 29 (1978).

254. Crews, P., E. Kho-Wiseman, and P. Montana: Halogenated Alicyclic Monoterpenes from the Red Algae *Plocamium.* J. Org. Chem. **43**, 116 (1978).

255. Mynderse, J.S., D.J. Faulkner, J. Finer, and J. Clardy: (1*R*, 2*S*, 4*S*, 5*R*)-1-Bromo-*trans*-2-chlorovinyl-4,5-dichloro-1,5-dimethylcyclohexane, a New Monoterpene Skeletal Type from the Red Alga *Plocamium violaceum.* Tetrahedron Lett., 2175 (1975).

256. Crews, P., and E. Kho: Plocamene B, a New Cyclic Monoterpene Skeleton from a Red Marine Alga. J. Org. Chem. **40**, 2568 (1975).

257. Higgs, M.D., D.J. Vanderah, and D.J. Faulkner: Polyhalogenated Monoterpenes from *Plocamium cartilagineum* from the British Coast. Tetrahedron **33**, 2775 (1977).

258. Norton, R.S., R.G. Warren, and R.J. Wells: Three New Polyhalogenated Monoterpenes from *Plocamium* Species. Tetrahedron Lett., 3905 (1977).

259. González, A.G., J.M. Arteaga, J.D. Martín, M.L. Rodríguez, J. Fayos, and M. Martínez-Ripolls: Two New Polyhalogenated Monoterpenes from the Red Alga *Plocamium cartilagineum.* Phytochem. **17**, 947 (1978).

260. Rivera, P., L. Astudillo, J. Rovirosa, and A. San-Martín: Halogenated Monoterpenes of the Red Alga *Shottera nicaensis.* Biochem. Sys. Ecol. **15**, 3 (1987).

261. San-Martín, A., R. Negrete, and J. Rovirosa: Insecticide and Acaricide Activities of Polyhalogenated Monoterpenes from Chilean *Plocamium cartilagineum.* Phytochem. **30**, 2165 (1991).

262. Capon, R.J., L.M. Engelhardt, E.L. Ghisalberti, P.R. Jefferies, V.A. Patrick, and A.H. White: Structural Studies of Polyhalogenated Monoterpenes from *Plocamium* Species. Aust. J. Chem. **37**, 537 (1984).

263. Watanabe, K., M. Miyakado, N. Ohno, A. Okada, K. Yanagi, and K. Moriguchi: A Polyhalogenated Insecticidal Monoterpene from the Red Alga *Plocamium telfairiae.* Phytochem. **28**, 77 (1989).

264. Combaut, G., J.-M. Kornprobst, and J. Mollion: Chemistry of Seaweeds from Senegal. Chem. Abstr. **97**, 69273m (1982).

265. Castedo, L., M.L. Garcia, E. Quinoa, and R. Riguera: Marine Natural Products from the Galician Coast, Part II: Halogenated Monoterpenes from *Plocamium coccineum* of Northwest Spain. J. Nat. Prod. **47**, 724 (1984).

266. Sardina, F.J., E. Quiñoá, L. Castedo, and R. Riguera: Structural Elucidation of Marine Halogenated Monoterpenes by 2D-NMR and NOE Difference Spectroscopy. A Stereochemical Correction. Chem. Lett., 697 (1985).

267. Barrow, K.D., and C.A. Temple: Biosynthesis of Halogenated Monoterpenes in *Plocamium cartilagineum.* Phytochem. **24**, 1697 (1985).

268. Crews, P., P. Ng, E. Kho-Wiseman, and C. Pace: Halogenated Monoterpenes of the Red Alga *Microcladia.* Phytochem. **15**, 1707 (1976).

269. Sakata, K., Y. Iwase, K. Ina, and D. Fujita: Halogenated Terpenes Isolated from the Red Alga *Plocamium leptophyllum* as Feeding Inhibitors for Marine Herbivores. Nippon Suisan Gakkaishi **57**, 743 (1991).

270. Aazizi, M.A., G.M. Assef, and R. Faure: Gelidene, a New Polyhalogenated Monocyclic Monoterpene from the Red Marine Alga *Gelidium sesquipedale.* J. Nat. Prod. **52**, 829 (1989).

271. Woolard, F.X., R.E. Moore, D. Van Engen, and J. Clardy: The Structure and Absolute Configuration of Chondrocolactone, A Halogenated Monoterpene from the Red Alga *Chondrococcus hornemanni,* and a Revised Structure for Chondrocole A. Tetrahedron Lett., 2367 (1978).

272. Crews, P., B.L. Myers, S. Naylor, E.L. Clason, R.S. Jacobs, and G.B. Staal: Bio-Active Monoterpenes from Red Seaweeds. Phytochem. **23**, 1449 (1984).

273. McCONNELL, O.J., and W. FENICAL: Ochtodene and Ochtodiol: Novel Polyhalogenated Cyclic Monoterpenes from the Red Seaweed *Ochtodes secundiramea.* J. Org. Chem. **43**, 4238 (1978).
274. GERWICK, W.H.: 2-Chloro-1,6(*S**),8-tribromo-3-(8)(*Z*)-ochtodene: A Metabolite of the Tropical Red Seaweed *Ochtodes secundiramea.* Phytochem. **23**, 1323 (1984).
275. PAUL, V.J., O.J. McCONNELL, and W. FENICAL: Cyclic Monoterpenoid Feeding Deterrents from the Red Marine Alga *Ochtodes crockeri.* J. Org. Chem. **45**, 3401 (1980).
276. SUMATHYKUTTY, M.A., and J.M. RAO: 8-Hentriacontanol and Other Constituents from *Piper attenuatum.* Phytochem. **30**, 2075 (1991).
277. STIERLE, D.B., R.M. WING, and J.J. SIMS: Marine Natural Products, XI: Costatone and Costatolide, New Halogenated Monoterpenes from the Red Seaweed, *Plocamium costatum.* Tetrahedron Lett., 4455 (1976).
278. KUSUMI, T., H. UCHIDA, Y. INOUYE, M. ISHITSUKA, H. YAMAMOTO, and H. KAKISAWA: Novel Cytotoxic Monoterpenes Having a Halogenated Tetrahydropyran from *Aplysia kurodai.* J. Org. Chem. **52**, 4597 (1987).
279. WATANABE, K., K. UMEDA, Y. KURITA, C. TAKAYAMA, and M. MIYAKADO: Two Insecticidal Monoterpenes, Telfairine and Aplysiaterpenoid A, from the Red Alga *Plocamium telfairiae*: Structure Elucidation, Biological Activity, and Molecular Topographical Consideration by a Semiempirical Molecular Orbital Study. Pestic. Biochem. Physiol. **37**, 275 (1990).
280. KUPCHAN, S.M., J.E. KELSEY, M. MARUYAMA, and J.M. CASSADY: Eupachlorin Acetate, A Novel Chloro-Sesquiterpenoid Lactone Tumor Inhibitor from *Eupatorium rotundifolium.* Tetrahedron Lett., 3517 (1968).
281. KUPCHAN, S.M., J.E. KELSEY, M. MARUYAMA, J.M. CASSADY, J.C. HEMINGWAY, and J.R. KNOX: Tumor Inhibitors, XLI: Structural Elucidation of Tumor-Inhibitory Sesquiterpene Lactones from *Eupatorium rotundifolium.* J. Org. Chem. **34**, 3876 (1969).
282. HARLEY-MASON, J., A.T. HEWSON, O. KENNARD, and R.C. PETTERSEN: Isolation of *Centaurea repens* Centaurepensin, a Guaianolide Sesquiterpene Lactone Ester Containing Two Chlorine Atoms; Determination of Structure and Absolute Configuration by X-Ray Crystallography. J. Chem. Soc., Chem. Commun., 460 (1972).
283. LOPEZ DE LERMA, J., J. FAYOS, S. GARCÍA-BLANCO, and M. MARTÍNEZ-RIPOLL: Centaurepensin. A Redetermination of Its Absolute Configuration by X-Ray Crystallography. Acta Cryst. **B34**, 2669 (1978).
284. CASSADY, J.M., D. ABRAMSON, P. COWALL, C. CHANG, J.L. McLAUGHLIN, and Y. AYNEHCHI: Centaurepensin: A Cytotoxic Constituent of *Centaurea solstitialis* and *C. repens* (Asteraceae). J. Nat. Prod. **42**, 427 (1979).
285. STEVENS, K.L., and R.Y. WONG: Structure of Chlororepdiolide, a New Sesquiterpene Lactone from *Centaurea repens.* J. Nat. Prod. **49**, 833 (1986).
286. EVSTRATOVA, R.I., V.I. SCHEICHENKO, and K.S. RYBALKO: The Structure of Acroptilin – A Sesquiterpene Lactone from *Acroptilon repens.* Khim. Prir. Soedin., 161 (1973).
287. STEVENS, K.L., and R.Y. WONG: Acroptilin $C_{19}H_{23}ClO_7$. Cryst. Struct. Comm. **11**, 949 (1982).
288. GONZÁLEZ, A.G., J. BERMEJO, J.L. BRETÓN, and J. TRIANA: Constituents of Compositae, XV: Chlorohyssopifolin A and B, Two New Sesquiterpene Lactones Isolated from *Centaurea hyssopifolia* Vahl. Tetrahedron Lett., 2017 (1972).
289. GONZÁLEZ, A.G., J. BERMEJO, J.L. BRETÓN, G.M. MASSANET, and J. TRIANA: Chloro-

hyssopifolin C, D, E and Vahlenin, Four New Sesquiterpene Lactones from *Centaurea hyssopifolia*. Phytochem. **13**, 1193 (1974).

290. GONZÁLEZ, A.G., J. BERMEJO, J.L. BRETON, G.M. MASSANET, B. DOMINGUEZ, and J.M. AMARO: The Chemistry of the Compositae, Part XXXI: Absolute Configuration of the Sesquiterpene Lactones Centaurepensin (Chlorohyssopifolin A), Acroptilin (Chlorohyssopifolin C), and Repin. J. Chem. Soc., Perkin Trans. 1, 1663 (1976).

291. GONZÁLEZ, A.G., J. BERMEJO, J.M. AMARO, G.M. MASSANET, A. GALINDO, and I. CABRERA: Sesquiterpene Lactones from *Centaurea linifolia* Vahl. Can. J. Chem. **56**, 491 (1978).

292. RUSTAIYAN, A., L. NAZARIANS, and F. BOHLMANN: Guaianolides from *Acroptilon repens*. Phytochem. **20**, 1152 (1981).

293. SHAM'YANOV, I.D., A. MALLABAEV, and G.P. SIDYAKIN: Structure of Sesquiterpene Lactone Elegin. Khim. Prir. Soedin., 442 (1978).

294. SHAM'YANOV, I.D., A. MALLABAEV, and G.P. SIDYAKIN: Salegin – A New Sesquiterpene Lactone from *Saussurea elegans*. Khim. Prir. Soedin., 865 (1979).

295. MERRILL, G.B., and K.L. STEVENS: Sesquiterpene Lactones from *Centaurea solstitialis*. Phytochem. **24**, 2013 (1985).

296. GONZÁLEZ, A.G., J. BERMEJO, and G.M. MASSANET: Aportacion al Estudio Quimiotaxonomico del Genero Centaurea: Determinacion Estructural de las Lactonas Sesquiterpenicas Presentes en Centaureas de Canarias y de la Peninsula Iberica. Rev. Latinoamer. Quim. **8**, 176 (1977).

297. EL-DAHMY, S., F. BOHLMANN, T.M. SARG, A. ATEYA, and N. FARREG: New Guaianolides from *Centaurea aegyptica*. Planta Med. **51**, 176 (1985).

298. SARG, T.M., M. EL-DOMIATY, and S. EL-DAHMY: Further Guaianolides from *Centaurea aegyptica*. Sci. Pharm. **55**, 107 (1987).

299. JAKUPOVIC, J., Y. JIA, V.P. PATHAK, F. BOHLMANN, and R.M. KING: Bisabolone Derivatives and Sesquiterpene Lactones from *Centaurea* Species. Planta Med. **52**, 399 (1986).

300. DAWIDAR, A.M., M.A. METWALLY, M. ABOU-ELZAHAB, and M. ABDEL-MOGIB: Chemical Constituents of Two *Centaurea* Species. Pharmazie **44**, 735 (1989).

301. AL-EASA, H.S., J. MANN, and A.-F. RIZK: Guaianolides from *Centaurea sinaica*. Phytochem. **29**, 1324 (1990).

302. DANIEWSKI, W.M., and G. NOWAK: Further Sesquiterpene Lactones of *Centaurea bella*. Phytochem. **32**, 204 (1993).

303. ÖKSÜZ, S., S. SERIN, and G. TOPCU: Sesquiterpene Lactones from *Centaurea hermannii*. Phytochem. **35**, 435 (1994).

304. RUSTAIYAN, A., Z. SHARIF, A. TAJARODI, J. ZIESCHE, and F. BOHLMANN. Neue Guaianolide aus *Centaurea imperalis*. Planta Med. **50**, 193 (1984).

305. NOWAK, G., M. HOLUB, and M. BUDESINSKY: Sesquiterpene Lactones, XXXVI: Sesquiterpene Lactones in Several Subgenera of the Genus *Centaurea* L. Acta Soc. Bot. Pol. **58**, 95 (1989).

306. SINGH, P., and M. BHALA: Guaianolides from *Saussurea candicans*. Phytochem. **27**, 1203 (1988).

307. TODOROVA, M.N., I.V. OGNYANOV, and S. SHATAR: Sesquiterpene Lactones in Mongolian *Saussurea lipshitzii*. Collect. Czech. Chem. Commun. **56**, 1106 (1991).

308. JAKUPOVIC, J., R. BOEKER, A. SCHUSTER, F. BOHLMANN, and S.B. JONES: Further Guaianolides and 5-Alkylcoumarins from *Gutenbergia* and *Bothriocline* Species. Phytochem. **26**, 1069 (1987).

309. MARCO, J.A., J.F. SANZ, R. ALBIACH, A. RUSTAIYAN, and Z. HABIBI: Bisabolene

Derivatives and Sesquiterpene Lactones from *Cousinia* Species. Phytochem. **32**, 395 (1993).

310. YUSUPOV, M.I., A. MALLABAEV, SH.Z. KASYMOV, and G.P. SIDYAKIN: Biebsanin – A New Sesquiterpene Lactone from *Achillea biebersteinii*. Khim. Prir. Soedin. **15**, 580 (1979).

311. BOHLMANN, F., J. JAKUPOVIC, R.M. KING, and H. ROBINSON: New Germacranolides, Guaianolides and Rearranged Guaianolides from *Lasiolaena santosii*. Phytochem. **20**, 1613 (1981).

312. BOHLMANN, F., J. JAKUPOVIC, A. SCHUSTER, R.M. KING, and H. ROBINSON: Guaianolides and Homoditerpenes from *Lasiolaena morii*. Phytochem. **21**, 161 (1982).

313. VICHNEWSKI, W., P. KULANTHAIVEL, V.L. GOEDKEN, and W. HERZ: Two Sesquiterpene Lactones from *Trichogonia gardneri*. Phytochem. **24**, 291 (1985).

314. CASTRO, V., J. JAKUPOVIC, and F. BOHLMANN: Sesquiterpene Lactones from *Mikania* Species. Phytochem. **25**, 1750 (1986).

315. MATA, R., G. DELGADO, and A. ROMO DE VIVAR: Sesquiterpene Lactones of *Artemisia klotzchiana*. Phytochem. **24**, 1515 (1985).

316. WAGNER, H., B. FESSLER, H. LOTTER, and V. WRAY: New Chlorine-Containing Sesquiterpene Lactones from *Chrysanthemum parthenium*. Planta Med. **54**, 171 (1988).

317. ALI, A.A., O.M. ABDALLAH, and W. STEGLICH: Chlorosesquiterpene Lactones from *Ambrosia maritima*. Pharmazie **44**, 800 (1989).

318. GIL, R.R., J.A. PASTORIZA, J.C. OBERTI, A.B. GUTIÉRREZ, and W. HERZ: Guaianolides from *Stevia sanguinea*. Phytochem. **28**, 2841 (1989).

319. DE GUTIERREZ, A.N., E.E. SIGSTAD, C.A.N. CATALÁN, A.B. GUTIÉRREZ, and W. HERZ: Guaianolides from *Kaunia lasiophthalma*. Phytochem. **29**, 1219 (1990).

320. REIS, L.V., M.R. TAVARES, F.M.S.B. PALMA, and M.J. MARCELO-CURTO: Sesquiterpene Lactones from *Cynara humilis*. Phytochem. **31**, 1285 (1992).

321. ALI, A.A., N.A. EL-EMARY, A.A. KHALIFA, and A.W. FRAHM: Guaianolides from *Venidium fastuosum*. Phytochem. **31**, 2781 (1992).

322. BOHLMANN, F., U. FRITZ, R.M. KING, and H. ROBINSON: Fourteen Heliangolides from *Calea* Species. Phytochem. **20**, 743 (1981).

323. BOHLMANN, F., R.K. GUPTA, R.M. KING, and H. ROBINSON: Three Furanoheliangolides from *Calea villosa*. Phytochem. **21**, 2593 (1982).

324. HERZ, W., and P. KULANTHAIVEL: Sesquiterpene Lactones from *Liatris acidota*, *L. aspera* and *L. mucronata*. Phytochem. **22**, 513 (1983).

325. JAKUPOVIC, J., S. BANERJEE, V. CASTRO, F. BOHLMANN, A. SCHUSTER, J.D. MSONTHI, and S. KEELEY: Poskeanolide, A Seco-Germacranolide, and Other Sesquiterpene Lactones from *Vernonia* Species. Phytochem. **25**, 1359 (1986).

326. ARRIAGA-GINER, F.J., J. BORGES-DEL-CASTILLO, M.T. MANRESA-FERRERO, P. VÁSQUEZ-BUENO, F. RODRIGUEZ-LUIS, and S. VALDÉS-IRAHETA: Eudesmane Derivatives from *Pluchea odorata*. Phytochem. **22**, 1767 (1983).

327. YOSHIHIRA, K., M. FUKUOKA, M. KUROYANAGI, and S. NATORI: 1-Indanone Derivatives from Bracken, *Pteridium acquilinum* var. *latiusculum*. Chem. Pharm. Bull. (Japan) **19**, 1491 (1971).

328. HAYASHI, Y., M. NISHIZAWA, S. HARITA, and T. SAKAN: Structures and Syntheses of Hypolepin A, B, and C, Sesquiterpenes from *Hypolepis punctata*. Chem. Lett., 375 (1972).

329. FUKUOKA, M., M. KUROYANAGI, M. TOYAMA, K. YOSHIHIRA, and S. NATORI: Pterosins J, K, and L and Six Acylated Pterosins from Bracken *Pteridium aquilinum* var. *latiusculum*. Chem. Pharm. Bull. (Japan) **20**, 2282 (1972).

330. MURAKAMI, T., K. OWASHI, N. TANAKA, T. SATAKE, and C.-M. CHEN: Chemische Untersuchungen der Inhaltsstoffe von *Dennstaedtia scabra* (Wall.) Moore. Chem. Pharm. Bull. (Japan) **23**, 1630 (1975).

331. MURAKAMI, T., S. TAGUCHI, and C.-M. CHEN: Chemische Untersuchungen der Inhaltsstoffe von *Hypolepis punctata* (Thunb.) Mett. Chem. Pharm. Bull. (Japan) **24**, 2241 (1976).

332. FUKUOKA, M., M. KUROYANAGI, K. YOSHIHIRA, and S. NATORI: Chemical and Toxicological Studies on Bracken Fern, *Pteridium aquilinum* var. *latiusculum*, II: Structures of Pterosins, Sesquiterpenes Having 1-Indanone Skeleton. Chem. Pharm. Bull. (Japan) **26**, 2365 (1978).

333. KUROYANAGI, M., M. FUKUOKA, K. YOSHIHIRA, and S. NATORI: Chemical and Toxicological Studies on Bracken Fern, *Pteridium aquilinum* var. *latiusculum*, III: Further Characterization of Pterosins and Pterosides, Sesquiterpenes and the Glucosides Having 1-Indanone Skeleton, from the Rhizomes. Chem. Pharm. Bull. (Japan) **27**, 592 (1979).

334. KOBAYASHI, A., and K. KOSHIMIZU: Cytotoxic Effects of Bracken Fern Constituents, Pterosins, on Sea Urchin Embryos and a Ciliate. Agric. Biol. Chem. **44**, 393 (1980).

335. MURAKAMI, T., T. SATAKE, K. NINOMIYA, H. IIDA, K. YAMAUCHI, N. TANAKA, Y. SAIKI, and C.-M. CHEN: Pterosin-Derivate aus der Familie Pteridaceae. Phytochem. **19**, 1743 (1980).

336. TAKANA, N., T. MURAKAMI, Y. SAIKI, C.-M. CHEN, and L.D. GOMEZ P: Chemical and Chemotaxonomical Studies of Ferns, XXXVII: Chemical Studies on the Constituents of Costa Rican Ferns (2). Chem. Pharm. Bull. (Japan) **29**, 3455 (1981).

337. TANAKA, N., T. SATAKE, A. TAKAHASHI, M. MOCHIZUKI, T. MURAKAMI, Y. SAIKI, J.-Z. YANG, and C.-M. CHEN: Chemical and Chemotaxonomical Studies of Ferns, XXXIX: Chemical Studies on the Constituents of *Pteris bella* Tagawa and *Pteridium aquilinum* subsp. *wightianum* (Wall) Shich. Chem. Pharm. Bull. (Japan) **30**, 3640 (1982).

338. KURAISHI, T., T. MURAKAMI, T. TANIGUCHI, Y. KOBUKI, H. MAEHASHI, N. TANAKA, Y. SAIKI, and C.-M. CHEN: Chemical and Chemotaxonomical Studies of Ferns, LIV: Pterosin Derivatives of the Genus *Microlepia* (Pteridaceae). Chem. Pharm. Bull. (Japan) **33**, 2305 (1985).

339. MURAKAMI, T., H. MAEHASHI, N. TANAKA, T. SATAKE, T. KURAISHI, Y. KOMAZAWA, Y. SAIKI, and C.-M. CHEN: Chemical and Chemotaxonomical Studies of Filices, LV: Studies on the Constituents of Several Species of *Pteris*. J. Pharmacol. Soc. Japan (Yakugaku Zasshi) **105**, 640 (1985).

340. AHMAD, V.U., T.A. FAROOQUI, K. FIZZA, A. SULTANA, and R. KHATOON: Three New Eudesmane Sesquiterpenes from *Pluchea arguta*. J. Nat. Prod. **55**, 730 (1992).

341. HASHIMOTO, T., M. TORI, and Y. ASAKAWA: Drimane-Type Sesquiterpenoids from the Liverwort *Makinoa crispata*. Phytochem. **28**, 3377 (1989).

342. GARG, S.N., S.K. AGARWAL, K. FIDELIS, M.B. HOSSAIN, and D. VAN DER HELM: New Jaeschkeanadiol Derivatives from *Ferula jaeschkeana*. J. Nat. Prod. **56**, 539 (1993).

343. MARTIN, J.D., and J. DARIAS: Algal Sesquiterpenoids. In: Marine Natural Products, Vol. I, Chap. 3 (P.J. Scheuer, ed.). New York: Academic Press. 1978.

344. ERICKSON, K.L.: Constituents of *Laurencia*. In: Marine Natural Products, Vol. V, Chap. 4 (P.J. Scheuer, ed.). New York: Academic Press. 1983.

345. WHITE, R.H., and L.P. HAGER: A Biogenetic Sequence of Halogenated Sesquiterpenes from *Laurencia intricata*. In: Dahlem Workshop on the Nature of Seawater, (E.D. Goldberg, ed.), pp. 633–650. Dahlem Conference: Berlin. 1975.

346. KÖNIG, G.M., and A.D. WRIGHT: New C_{15} Acetogenins and Sesquiterpenes from the Red Alga *Laurencia* sp. cf. *L. gracilis*. J. Nat. Prod. **57**, 477 (1994).

347. VAZQUEZ, J.T., M. CHANG, K. NAKANISHI, J.D. MARTÍN, V.S. MARTÍN, and R. PÉREZ: Puertitols: Novel Sesquiterpenes from *Laurencia obtusa*. Structure Elucidation and Absolute Configuration and Conformation Based on Circular Dichroism. J. Nat. Prod. **51**, 1257 (1988).
348. NORTE, M., J.J. FERNÁNDEZ, and A. PADILLA: Bisabolane Halogenated Sesquiterpenes from *Laurencia*. Phytochem. **31**, 326 (1992).
349. HOWARD, B.M., and W. FENICAL: α- and β-Snyderol; New Bromo-Monocyclic Sesquiterpenes from the Seaweed *Laurencia*. Tetrahedron Lett., 41 (1976).
350. AYYAD, S.-E.N., A.-A.M. DAWIDAR, H.W. DIAS, R.A. HOWIE, J. JAKUPOVIC, and R.H. THOMSON: Three Halogenated Metabolites from *Laurencia obtusa*. Phytochem. **29**, 3193 (1990).
351. PAUL, V.J., and W. FENICAL: Palisadins A, B and Related Monocyclofarnesol-Derived Sesquiterpenoids from the Red Marine Alga *Laurencia* cf. *palisada*. Tetrahedron Lett. **21**, 2787 (1980).
352. NORTE, M., R. GONZÁLEZ, A. PADILLA, J.J. FERNÁNDEZ, and J.T. VÁZQUEZ: New Halogenated Sesquiterpenes from the Red Alga *Laurencia caespitosa*. Can. J. Chem. **69**, 518 (1991).
353. WRATTEN, S.J., and D.J. FAULKNER: Carbonimidic Dichlorides from the Marine Sponge *Pseudaxinyssa pitys*. J. Am. Chem. Soc. **99**, 7367 (1977).
354. IMRE, S., S. ISLIMYELI, A. OZTUNC, and R.H. THOMSON: Obtusenol, a Sesquiterpene from *Laurencia obtusa*. Phytochem. **20**, 833 (1981).
355. VAN TAMELEN, E.E., and E.J. HESSLER: The Direct Brominative Cyclization of Methyl Farnesate. J. Chem. Soc., Chem. Commun., 411 (1966).
356. FAULKNER, D.J.: 3β-Bromo-8-epicaparrapi Oxide, the Major Metabolite of *Laurencia obtusa*. Phytochem. **15**, 1992 (1976).
357. PETTIT, G.R., C.L. HERALD, M.S. ALLEN, R.B. VON DREELE, L.D. VANELL, J.P.Y. KAO, and W. BLAKE: The Isolation and Structure of Aplysistatin. J. Am. Chem. Soc. **99**, 262 (1977).
358. CAPON, R., E.L. GHISALBERTI, P.R. JEFFERIES, B.W. SKELTON, and A.H. WHITE: Sesquiterpene Metabolites from *Laurencia filiformis*. Tetrahedron **37**, 1613 (1981).
359. DE NYS, R., A.D. WRIGHT, G.M. KÖNIG, O. STICHER, and P.M. ALINO: Five New Sesquiterpenes from the Red Alga *Laurencia flexilis*. J. Nat. Prod. **56**, 877 (1993).
360. KÖNIG, G.M., A.D. WRIGHT, and F.R. FRONCZEK: X-Ray Crystal Structure of 3,4-Epoxypalisadin A. J. Nat. Prod. **57**, 151 (1994).
361. CARNEY, J.R., A.T. PHAM, W.Y. YOSHIDA, and P.J. SCHEUER: Napalilactone, a New Halogenated Norsesquiterpenoid from the Soft Coral *Lemnalia africana*. Tetrahedron Lett. **33**, 7115 (1992).
362. GONZÁLEZ, A.G., J. DARIAS, and J.D. MARTÍN: Furocaespitane, a New Furan from *Laurencia caespitosa*. Tetrahedron Lett., 3625 (1973).
363. GONZÁLEZ, A.G., J.D. MARTÍN, V.S. MARTÍN, and M. NORTE: Carbon-13 NMR Application to *Laurencia* Polyhalogenated Sesquiterpenes. Tetrahedron Lett., 2719 (1979).
364. ESTRADA, D.M., J.D. MARTÍN, R. PÉREZ, P. RIVERA, M.L. RODRÍGUEZ, and J.Z. RUANO: Furocaespitane and Related C_{12} Metabolites from *Laurencia caespitosa*. Tetrahedron Lett. **28**, 687 (1987).
365. SUZUKI, M., E. KUROSAWA, and T. IRIE: Spirolaurenone, a New Sesquiterpenoid Containing Bromine from *Laurencia glandulifera* Kützing. Tetrahedron Lett., 4995 (1970).
366. SUZUKI, M., N. KOWATA, and E. KUROSAWA: The Structure of Spirolaurenone, a Halogenated Sesquiterpenoid from the Red Alga *Laurencia glandulifera* Kützing. Tetrahedron **36**, 1551 (1980).

367. SUZUKI, M., E. KUROSAWA, and T. IRIE: Three New Sesquiterpenoids Containing Bromine, Minor Constituents of *Laurencia glandulifera* Kützing. Tetrahedron Lett., 821 (1974).

368. SUZUKI, M., A. FURUSAKI, and E. KUROSAWA: The Absolute Configurations of Halogenated Chamigrene Derivatives from the Marine Alga, *Laurencia glandulifera* Kützing. Tetrahedron **35**, 823 (1979).

369. SUZUKI, M., E. KUROSAWA, and T. IRIE: Glanduliferol, a New Halogenated Sesquiterpenoid from *Laurencia glandulifera* Kützing. Tetrahedron Lett., 1807 (1974).

370. SUZUKI, M., and E. KUROSAWA: Halogenated Chamigrene-Type Sesquiterpenoids from the Red Algae of the Genus *Laurencia*. Chem. Abstr. **92**, 215566z (1980).

371. SIMS, J.J., W. FENICAL, R.M. WING, and P. RADLICK: Marine Natural Products, I: Pacifenol, a Rare Sesquiterpene Containing Bromine and Chlorine from the Red Alga, *Laurencia pacifica*. J. Am. Chem. Soc. **93**, 3774 (1971).

372. SIMS, J.J., W. FENICAL, R.M. WING, and P. RADLICK: Marine Natural Products, IV: Prepacifenol, a Halogenated Epoxy Sesquiterpene and Precursor to Pacifenol from the Red Alga, *Laurencia filiformis*. J. Am. Chem. Soc. **95**, 972 (1973).

373. SIMS, J.J., W. FENICAL, R.M. WING, and P. RADLICK: Marine Natural Products, III: Johnstonol, an Unusual Halogenated Epoxide from the Red Alga *Laurencia johnstonii*. Tetrahedron Lett., 195 (1972).

374. HOWARD, B.M., and W. FENICAL: 10-Bromo-α-chamigrene. Tetrahedron Lett., 2519 (1976).

375. WOLINSKY, L.E., and D.J. FAULKNER: A Biomimetic Approach to the Synthesis of *Laurencia* Metabolites. Synthesis of 10-Bromo-α-chamigrene. J. Org. Chem. **41**, 597 (1976).

376. FENICAL, W.: Chemical Variation in a New Bromochamigrene Derivative from the Red Seaweed *Laurencia pacifica*. Phytochem. **15**, 511 (1976).

377. SUZUKI, M., E. KUROSAWA, and A. FURUSAKI: The Structure and Absolute Stereochemistry of a Halogenated Chamigrene Derivative from the Red Alga *Laurencia* Species. Bull. Chem. Soc. Japan **61**, 3371 (1988).

378. SELOVER, S.J., and P. CREWS: Kylinone, a New Sesquiterpene Skeleton from the Marine Alga *Laurencia pacifica*. J. Org. Chem. **45**, 69 (1980).

379. HOWARD, B.M., and W. FENICAL: Structures and Chemistry of Two New Halogen-Containing Chamigrene Derivatives from *Laurencia*. Tetrahedron Lett., 1687 (1975).

380. ICHINOSE, I., and T. KATO: Biogenetic Type Synthesis of 10-Bromo-α-chamigrene. Chem. Lett., 61 (1979).

381. BITTNER, M.L., M. SILVA, V.J. PAUL, and W. FENICAL: A Rearranged Chamigrene Derivative and Its Potential Biogenetic Precursor from a New Species of the Marine Red Algal Genus *Laurencia* (Rhodomelaceae). Phytochem. **24**, 987 (1985).

382. GONZÁLEZ, A.G., J. DARIAS, and J.D. MARTÍN: Caespitol, a New Halogenated Sesquiterpene from *Laurencia caespitosa*. Tetrahedron Lett., 2381 (1973).

383. MCMILLAN, J.A., I.C. PAUL, R.H. WHITE, and L.P. HAGER: Molecular Structure of Acetoxyintricatol: A New Bromo Compound from *Laurencia intricata*. Tetrahedron Lett., 2039 (1974).

384. SIMS, J.J., G.H.Y. LIN, and R.M. WING: Marine Natural Products, X: Elatol, a Halogenated Sesquiterpene Alcohol from the Red Alga *Laurencia elata*. Tetrahedron Lett., 3487 (1974).

385. WARASZKIEWICZ, S.M., and K.L. ERICKSON: Halogenated Sesquiterpenoids from the Hawaiian Marine Alga *Laurencia nidifica*: Nidificene and Nidifidiene. Tetrahedron Lett., 2003 (1974).

386. WARASZKIEWICZ, S.M., and K.L. ERICKSON: Halogenated Sesquiterpenoids from the Hawaiian Marine Alga *Laurencia nidifica*, II: Nidifidienol. Tetrahedron Lett., 281 (1975).
387. WARASZKIEWICZ, S.M., and K.L. ERICKSON: Halogenated Sesquiterpenoids from the Hawaiian Marine Alga *Laurencia nidifica*, IV: Nidifocene. Tetrahedron Lett., 1443 (1976).
388. WARASZKIEWICZ, S.M., K.L. ERICKSON, J. FINER, and J. CLARDY: Nidifocene: A Reassignment of Structure. Tetrahedron Lett., 2311 (1977).
389. SUZUKI, T.: Two New Sesquiterpene Alcohols Containing Bromine from the Marine Alga, *Laurencia nipponica* Yamada. Chem. Lett., 541 (1980).
390. KURATA, K., A. FURUSAKI, C. KATAYAMA, H. KIKUCHI, and T. SUZUKI: A New Labile Sesquiterpene Diol Having Bromine from the Marine Red Alga, *Laurencia nipponica* Yamada. Chem. Lett., 773 (1981).
391. SUZUKI, T., H. KIKUCHI, and E. KUROSAWA: Six New Sesquiterpenoids from the Red Alga *Laurencia nipponica* Yamada. Bull. Chem. Soc. Japan **55**, 1561 (1982).
392. SUZUKI, M., M. SEGAWA, T. SUZUKI, and E. KUROSAWA: Structures of Halogenated Chamigrene Derivatives, Minor Constituents from the Red Alga *Laurencia nipponica* Yamada. Bull. Chem. Soc. Japan **56**, 3824 (1983).
393. KURATA, K., T. SUZUKI, M. SUZUKI, E. KUROSAWA, A. FURUSAKI, and T. MATSUMOTO: Laureacetal-D and -E, Two New Secochamigrane Derivatives from the Red Alga *Laurencia nipponica* Yamada. Chem. Lett., 557 (1983).
394. KURATA, K., T. SUZUKI, M. SUZUKI, E. KUROSAWA, A. FURUSAKI, K. SUEHIRO, T. MATSUMOTO, and C. KATAYAMA: Structures of Two New Halogenated Chamigrane-Type Sesquiterpenoids from the Red Alga *Laurencia nipponica* Yamada. Chem. Lett., 561 (1983).
395. SUZUKI, M., M. SEGAWA, T. SUZUKI, and E. KUROSAWA: Structures of Two New Halochamigrene Derivatives from the Red Alga *Laurencia nipponica* Yamada. Bull. Chem. Soc. Japan **58**, 2435 (1985).
396. KIKUCHI, H., T. SUZUKI, M. SUZUKI, and E. KUROSAWA: A New Chamigrane-Type Bromo Diether from the Red Alga *Laurencia nipponica* Yamada. Bull. Chem. Soc. Japan **58**, 2437 (1985).
397. WATANABE, K., K. UMEDA, and M. MIYAKADO: Isolation and Identification of Three Insecticidal Principles from the Red Alga *Laurencia nipponica*. Agric. Biol. Chem. **53**, 2513 (1989).
398. KURATA, K., T. SUZUKI, M. SUZUKI, E. KUROSAWA, A. FURUSAKI, and T. MATSUMOTO: Laurencial, a Novel Sesquiterpene, α, β-Unsaturated Aldehyde from the Red Alga *Laurencia nipponica* Yamada. Chem. Lett., 299 (1983).
399. FURUSAKI, A., C. KATAYAMA, T. MATSUMOTO, M. SUZUKI, T. SUZUKI, H. KIKUCHI, and E. KUROSAWA: The Crystal and Molecular Structure of 7,8-Epoxyhalochamigrene. Bull. Chem. Soc. Japan **55**, 3398 (1982).
400. FENICAL, W., and J.N. NORRIS: Chemotaxonomy in Marine Algae: Chemical Separation of Some *Laurencia* Species (*Rhodophyta*) from the Gulf of California. J. Phycol. **11**, 104 (1975).
401. SUZUKI, T., A. FURUSAKI, N. HASHIBA, and E. KUROSAWA: Novel Skeletal Bromo Ether from the Marine Alga, *Laurencia nipponica* Yamada. Tetrahedron Lett., 3731 (1977).
402. SUZUKI, T., and E. KUROSAWA: New Bromo Acetal from the Marine Alga, *Laurencia nipponica* Yamada. Chem. Lett., 301 (1979).
403. SUZUKI, M., and E. KUROSAWA: Two New Halogenated Sesquiterpenes from the Red Alga *Laurencia majuscula* Harvey. Tetrahedron Lett., 4805 (1978).

404. SUZUKI, M., A. FURUSAKI, N. HASHIBA, and E. KUROSAWA: The Structures and Absolute Stereochemistry of Two Halogenated Chamigrenes from the Red Alga *Laurencia majuscula* Harvey. Tetrahedron Lett., 879 (1979).

405. CACCAMESE, S., A. COMPAGNINI, and R.M. TOSCANO: Pacifenol from the Mediterranean Red Alga *Laurencia majuscula*. J. Nat. Prod. **49**, 173 (1986).

406. CACCAMESE, S., A. COMPAGNINI, R.M. TOSCANO, F. NICOLO, and G. CHAPUIS: A New Labile Bromoterpenoid from the Red Alga *Laurencia majuscula*: Dehydrochloroprepacifenol. Tetrahedron **43**, 5393 (1987).

407. COLL, J.C., and A.D. WRIGHT: Tropical Marine Algae, III: New Sesquiterpenes from *Laurencia majuscula* (Rhodophyta, Rhodophyceae, Ceramiales, Rhodomelaceae). Aust. J. Chem. **42**, 1591 (1989).

408. OJIKA, M., Y. SHIZURI, and K. YAMADA: A Halogenated Chamigrene Epoxide and Six Related Halogen-Containing Sesquiterpenes from the Red Alga *Laurencia okamurai*. Phytochem. **21**, 2410 (1982).

409. GUELLA, G., G. CHIASERA, I. MANCINI, and F. PIETRA: Conformational Analysis of Marine Polyhalogenated β-Chamigrenes Through Temperature-Dependent NMR Spectra. Helv. Chim. Acta **74**, 774 (1991).

410. WRIGHT, A.D., J.C. COLL, and I.R. PRICE: Tropical Marine Algae, VII: The Chemical Composition of Marine Algae from North Queensland Waters. J. Nat. Prod. **53**, 845 (1990).

411. CAPON, R.J., E.L. GHISALBERTI, T.A. MORI, and P.R. JEFFERIES: Sesquiterpenes from *Laurencia* Sp. J. Nat. Prod. **51**, 1302 (1988).

412. DE NYS, R., J.C. COLL, and B.F. BOWDEN: Tropical Marine Algae, VIII: The Structural Determination of Novel Sesquiterpenoid Metabolites from the Red Alga *Laurencia majuscula*. Aust. J. Chem. **45**, 1611 (1992).

413. GONZÁLEZ, A.G., J. DARIAS, A. DÍAZ, J.D. FOURNERON, J.D. MARTÍN, and C. PÉREZ: Evidence for the Biogenesis of Halogenated Chamigrenes from the Red Alga *Laurencia obtusa*. Tetrahedron Lett., 3051 (1976).

414. GONZÁLEZ, A.G., J.D. MARTÍN, V.S. MARTÍN, M. NORTE, J. FAYOS, and M. MARTÍNEZ-RIPOLL: A New Polyhalogenated Sesquiterpene from *Laurencia obtusa*. Tetrahedron Lett., 2035 (1978).

415. PERALES, A., M. MARTÍNEZ-RIPOLL, and J. FAYOS: Structure of Obtusol Acetate, a Halogenated Chamigrene-Type Sesquiterpene. Acta Cryst. **B35**, 2771 (1979).

416. GONZÁLEZ, A.G., J.D. MARTÍN, V.S. MARTÍN, M. MARTÍNEZ-RIPOLL, and J. FAYOS: X-Ray Study of Sesquiterpene Constituents of the Alga *L. obtusa* Leads to Structure Revision. Tetrahedron Lett., 2717 (1979).

417. GERWICK, W.H., A. LOPEZ, R. DAVILA, and R. ALBORS: Two New Chamigrene Sesquiterpenoids from the Tropical Red Alga *Laurencia obtusa*. J. Nat. Prod. **50**, 1131 (1987).

418. BRENNAN, M.R., K.L. ERICKSON, D.A. MINOTT, and K.O. PASCOL: Chamigrane Metabolites from a Jamaican Variety of *Laurencia obtusa*. Phytochem. **26**, 1053 (1987).

419. KENNEDY, D.J., I.A. SELBY, and R.H. THOMSON: Chamigrane Metabolites from *Laurencia obtusa* and *L. scoparia*. Phytochem. **27**, 1761 (1988).

420. MARTÍN, J.D., P. CABALLERO, J.J. FERNANDEZ, M. NORTE, R. PERÉZ, and M.L. RODRIGUEZ: Metabolites from *Laurencia obtusa*. Phytochem. **28**, 3365 (1989).

421. GONZÁLEZ, A.G., J. DARIAS, J.D. MARTÍN, V.S. MARTÍN, M. NORTE, and C. PÉREZ: *Laurencia* Sesquiterpene Biogenetic-Type Interconversions. Tetrahedron Lett. **21**, 1151 (1980).

422. ELSWORTH, J.F., and R.H. THOMSON: A New Chamigrane from *Laurencia glomerata*. J. Nat. Prod. **52**, 893 (1989).
423. GONZÁLEZ, A.G., J. DARIAS, J.D. MARTÍN, and C. PÉREZ: Revised Structure of Caespitol and Its Correlation with Isocaespitol. Tetrahedron Lett., 1249 (1974).
424. BANO, S., M.S. ALI, and V.U. AHMAD: Marine Natural Products, VI: A Halogenated Chamigrene Epoxide from the Red Alga *Laurencia pinnatifida*. Planta Med. **53**, 508 (1987).
425. BANO, S., M.S. ALI, and V.U. AHMAD: Marine Natural Products, VIII: Two Minor Halogenated Sesquiterpenoids from the Red Alga *Laurencia pinnatifida*. Sci. Pharm. **56**, 125 (1988).
426. BANO, S., M.S. ALI, and V.U. AHMAD: Marine Natural Products, IX: A New Halogenated Sesquiterpene Pinnatifidone from the Red Alga *Laurencia pinnatifida*. Z. Naturforsch. **43B**, 1347 (1988).
427. AHMAD, V.U., and M.S. ALI: Terpenoids from Marine Red Alga *Laurencia pinnatifida*. Phytochem. **30**, 4172 (1991).
428. ATTA-UR-RAHMAN, V.U. AHMAD, S. BANO, S.A. ABBAS, K.A. ALVI, M.S. ALI, H.S.M. LU, and J. CLARDY: Pinnatazane, a Bridged Cyclic Ether Sesquiterpene from *Laurencia pinnatifida*. Phytochem. **27**, 3879 (1988).
429. STALLARD, M.O., and D.J. FAULKNER: Chemical Constituents of the Digestive Gland of the Sea Hare *Aplysia californica*, II: Chemical Transformations. Comp. Biochem. Physiol. **49B**, 37 (1974).
430. FAULKNER, D.J., M.O. STALLARD, and C. IRELAND: Prepacifenol Epoxide, a Halogenated Sesquiterpene Diepoxide. Tetrahedron Lett., 3571 (1974).
431. GONZÁLEZ, A.G., J.D. MARTÍN, M. NORTE, R. PÉREZ, V. WEYLER, A. PERALES, and J. FAYOS: New Halogenated Constituents of the Digestive Gland of the Sea Hare *Aplysia dactylomela*. Tetrahedron Lett. **24**, 847 (1983).
432. SAKAI, R., T. HIGA, C.W. JEFFORD, and G. BERNARDINELLI: The Absolute Configurations and Biogenesis of Some New Halogenated Chamigrenes from the Sea Hare *Aplysia dactylomela*. Helv. Chim. Acta **69**, 91 (1986).
433. RAO, C.B., C. SATYANARAYANA, D.V. RAO, E. FAHY, and D.J. FAULKNER: Metabolites of *Aplysia dactylomela* from the Indian Ocean. Indian J. Chem. **B28**, 322 (1989).
434. IWATA, C., T. AKIYAMA, and K. MIYASHITA: Synthesis of Four Possible Isomers of 9-(Bromomethylene)-1,2,5-trimethylspiro[5.5]undeca-1,7-dien-3-one: Structure Elucidation of a Brominated Rearranged Chamigrane-Type Sesquiterpene. Chem. Pharm. Bull. (Japan) **36**, 2872 (1988).
435. GUELLA, G., I. MANCINI, G. CHIASERA, and F. PIETRA: Rogiolol Acetate: A Novel β-Chamigrene-Type Sesquiterpene Isolated from a Marine Sponge. Helv. Chim. Acta **73**, 1612 (1990).
436. GUELLA, G., I. MANCINI, and F. PIETRA: C_{15} Acetogenins and Terpenes of the Dictyoceratid Sponge *Spongia zimocca* of Il Rogiolo: A Case of Seaweed-Metabolite Transfer to, and Elaboration Within, a Sponge? Comp. Biochem. Physiol. **B103**, 1019 (1992).
437. COX, P.J., and R.A. HOWIE: Structure of 2,10-Dibromo-3-chloro-7*R*,8*S*-epoxychamigrene. Z. Kristallogr. **188**, 1 (1989).
438. DE NYS, R., G.M. KÖNIG, A.D. WRIGHT, and O. STICHER: Two Metabolites from the Red Alga *Laurencia flexilis*. Phytochem. **34**, 725 (1993).
439. GONZÁLEZ, A.G., J.D. MARTÍN, V.S. MARTÍN, R. PÉREZ, B. TAGLE, and J. CLARDY: Rhodolaureol and Rhodolauradiol, Two New Halogenated Tricyclic Sesquiterpenes from a Marine Alga. J. Chem. Soc., Chem. Commun., 260 (1985).

440. GONZÁLEZ, A.G., J.D. MARTÍN, V.S. MARTÍN, M. NORTE, and R. PÉREZ: Biomimetic Approach to the Syntheses of Rhodolaureol and Rhodolauradiol. Tetrahedron Lett. **23**, 2395 (1982).

441. KAZLAUSKAS, R., P.T. MURPHY, R.J. WELLS, J.J. DALY, and W.E. OBERHÄNSLI: Heterocladol, a Halogenated Selinane Sesquiterpene of Biosynthetic Significance from the Red Alga *Laurencia filiformis*: Its Isolation, Crystal Structure and Absolute Configuration. Aust. J. Chem. **30**, 2679 (1977).

442. HOWARD, B.M., and W. FENICAL: Structure, Chemistry, and Absolute Configuration of 1(*S*)-Bromo-4(*R*)-hydroxy-(−)-selin-7-ene from a Marine Red Alga *Laurencia* sp. J. Org. Chem. **42**, 2518 (1977).

443. ROSE, A.F., and J.J. SIMS: Marine Natural Products, XIV: 1-*S*-Bromo-4-*R*-hydroxy-selin-7-ene, a Metabolite of the Marine Alga *Laurencia* sp. Tetrahedron Lett., 2935 (1977).

444. ROSE, A.F., J.J. SIMS, R.M. WING, and G.M. WIGER: Marine Natural Products, XVII: The Structure of (1*S*, 4*R*, 7*R*)-1-Bromo-4-hydroxy-7-chloroselinane, a Metabolite of the Marine Alga *Laurencia* sp. Tetrahedron Lett., 2533 (1978).

445. DIETER, R.K., R. KINNEL, J. MEINWALD, and T. EISNER: Brasudol and Isobrasudol: Two Bromosesquiterpenes. Tetrahedron Lett., 1645 (1979).

446. BRENNAN, M.R., and K.L. ERICKSON: Austradiol Acetate and Austradiol Diacetate, 4,6-Dihydroxy-(+)-selinane Derivatives from an Australian *Laurencia* sp. J. Org. Chem. **47**, 3917 (1982).

447. BARNEKOW, D.E., J.H. CARDELLINA II, A.S. ZEKTZER, and G.E. MARTIN: Novel Cytotoxic and Phytotoxic Halogenated Sesquiterpenes from the Green Alga *Neomeris annulata*. J. Am. Chem. Soc. **111**, 3511 (1989).

448. BARNEKOW, D.E., and J.H. CARDELLINA II: Determining the Absolute Configuration of Hindered Secondary Alcohols – A Modified Horeau's Method. Tetrahedron Lett. **30**, 3629 (1989).

449. TALPIR, R., A. RUDI, Y. KASHMAN, Y. LOYA, and A. HIZI: Three New Sesquiterpene Hydroquinones from Marine Origin. Tetrahedron **50**, 4179 (1994).

450. BURGOYNE, D.L., E.J. DUMDEI, and R.J. ANDERSEN: Acanthenes A to C: A Chloro, Isothiocyanate, Formamide Sesquiterpene Triad Isolated from the Northeastern Pacific Marine Sponge *Acanthella* sp. and the Dorid Nudibranch *Cadlina luteomarginata*. Tetrahedron **49**, 4503 (1993).

451. WRATTEN, S.J., D.J. FAULKNER, D. VAN ENGEN, and J. CLARDY: A Vinyl Carbonimidic Dichloride from the Marine Sponge *Pseudaxinyssa pitys*. Tetrahedron Lett., 1391 (1978).

452. WRATTEN, S.J., and D.J. FAULKNER: Minor Carbonimidic Dichlorides from the Marine Sponge *Pseudaxinyssa pitys*. Tetrahedron Lett., 1395 (1978).

453. HALL, S.S., D.J. FAULKNER, J. FAYOS, and J. CLARDY: Oppositol, a Brominated Sesquiterpene Alcohol of a New Skeletal Class from the Red Alga, *Laurencia subopposita*. J. Am. Chem. Soc. **95**, 7187 (1973).

454. WRATTEN, S.J., and D.J. FAULKNER: Metabolites of the Red Alga *Laurencia subopposita*. J. Org. Chem. **42**, 3343 (1977).

455. GONZÁLEZ, A.G., J.M. AGUIAR, J.D. MARTÍN, and M. NORTE: Three New Sesquiterpenoids from the Marine Alga *Laurencia perforata*. Tetrahedron Lett., 2499 (1975).

456. GONZÁLEZ, A.G., J.M. AGUIAR, J. DARIAS, E. GONZÁLEZ, J.D. MARTÍN, V.S. MARTÍN, and C. PÉREZ: Perforenol, a New Polyhalogenated Sesquiterpene from *Laurencia perforata*. Tetrahedron Lett., 3931 (1978).

457. COLL, J.C., B.W. SKELTON, A.H. WHITE, and A.D. WRIGHT: Tropical Marine Algae,

V: The Structure Determination of Two Novel Sesquiterpenes from the Red Alga *Laurencia tenera* (Rhodophyceae, Ceramiales, Rhodomelaceae). Aust. J. Chem. **42**, 1695 (1989).

458. González, A.G., J. Darias, J.D. Martín, C. Pérez, J.J. Sims, G.H.Y. Lin, and R.M. Wing: Isocaespitol, a New Halogenated Sesquiterpene from *Laurencia caespitosa*. Tetrahedron **31**, 2449 (1975).

459. González, A.G., J.D. Martín, C. Pérez, M.A. Ramírez, and F. Ravelo: Total Synthesis of 8-Desoxy-isocaespitol, a New Polyhalogenated Sesquiterpene from *Laurencia caespitosa*. Tetrahedron Lett. **21**, 187 (1980).

460. Chang, M., J.T. Vazquez, K. Nakanishi, F. Cataldo, D.M. Estrada, J. Fernandez, A. Gallardo, J.D. Martin, M. Norte, R. Pérez, and M.L. Rodríguez: Regular and Irregular Sesquiterpenes Containing a Halogenated Hydropyran from *Laurencia caespitosa*. Phytochem. **28**, 1417 (1989).

461. González, A.G., J.D. Martín, M. Norte, R. Pérez, P. Rivera, J.Z. Ruano, M.L. Rodríguez, J. Fayos, and A. Perales: X-Ray Structure Determination of New Brominated Metabolites Isolated from the Red Seaweed *Laurencia obtusa*. Tetrahedron Lett. **24**, 4143 (1983).

462. Hollenbeak, K.H., F.J. Schmitz, M.B. Hossain, and D. van der Helm: Marine Natural Products. Deodactol, Antineoplastic Sesquiterpenoid from the Sea Hare *Aplysia dactylomela*. Tetrahedron **35**, 541 (1979).

463. Schmitz, F.J., D.P. Michaud, and K.H. Hollenbeak: Marine Natural Products: Dihydroxydeodactol Monoacetate, a Halogenated Sesquiterpene Ether from the Sea Hare *Aplysia dactylomela*. J. Org. Chem. **45**, 1525 (1980).

464. Gopichand, Y., F.J. Schmitz, J. Shelly, A. Rahman, and D. van der Helm: Marine Natural Products: Halogenated Acetylenic Ethers from the Sea Hare *Aplysia dactylomela*. J. Org. Chem. **46**, 5192 (1981).

465. González, A.G., V. Darias, and E. Estévez: Chemotherapeutic Activity of Polyhalogenated Terpenes from Spanish Algae. Planta Med. **44**, 44 (1982).

466. Murakami, T., and N. Tanaka: Occurrence, Structure and Taxonomic Implications of Fern Constituents. Progr. Chem. Org. Nat. Prod. **54**, 1 (1988).

467. Gonález, A.G., J. Darias, and J.D. Martín: Biomimetic Interconversions of Two New Types of Metabolite from *Laurencia perforata*. Tetrahedron Lett., 3375 (1977).

468. González, A.G., J. Darias, J.D. Martín, and M.A. Melián: Total Synthesis of Racemic Perforenone and 3-Debromo-perforatone. Tetrahedron Lett., 481 (1978).

469. Sham'yanov, I.D., A. Mallabaev, U. Rakhmankulov, and G.P. Sidyakin: Sesquiterpene Lactones of *Saussurea elegans*. Khim. Prir. Soedin. **12**, 819 (1976).

470. Yamamura, S., and Y. Hirata: Structures of Aplysin and Aplysinol, Naturally Occurring Bromo-Compounds. Tetrahedron **19**, 1485 (1963).

471. Irie, T., M. Suzuki, and Y. Hayakawa: Isolation of Aplysin, Debromoaplysin, and Aplysinol from *Laurencia okamurai* Yamada. Bull. Chem. Soc. Japan **42**, 843 (1969).

472. Cameron, A.F., G. Ferguson, and J.M. Robertson: The Crystal Structure and Absolute Stereochemistry of Laurinterol. The Absolute Stereochemistry of Aplysin. J. Chem. Soc., Chem. Commun., 271 (1967).

473. Cameron, A.F., G. Ferguson, and J.M. Robertson: Laurencia Natural Products, Part II: Crystal Structure and Absolute Stereochemistry of Laurinterol Acetate, a Bicyclo[3.1.0]hexane Derivative. J. Chem. Soc. (B), 692 (1969).

474. McMillan, J.A., I.C. Paul, S. Caccamese, and K.L. Rinehart Jr.: Aplysinol from *Laurencia decidua*: Crystal Structure and Absolute Stereochemistry. Tetrahedron Lett., 4219 (1976).

475. IRIE, T., A. FUKUZAWA, M. IZAWA, and E. KUROSAWA: Laurenisol, A New Sesquiterpenoid Containing Bromine from *Laurencia nipponica* Yamada. Tetrahedron Lett., 1343 (1969).

476. IRIE, T., M. SUZUKI, E. KUROSAWA, and T. MASAMUNE: Laurinterol, Debromolaurinterol and Isolaurinterol, Constituents of *Laurencia intermedia* Yamada. Tetrahedron **26**, 3271 (1970).

477. SUZUKI, T., M. SUZUKI, and E. KUROSAWA: α-Bromocuparene and α-Isobromocuparene, New Bromo Compounds from *Laurencia* Species. Tetrahedron Lett., 3057 (1975).

478. KAZLAUSKAS, R., P.T. MURPHY, R.J. QUINN, and R.J. WELLS: New Laurene Derivatives from *Laurencia filiformis*. Aust. J. Chem. **29**, 2533 (1976).

479. SUZUKI, M., and E. KUROSAWA: New Bromo Compounds from *Laurencia glandulifera* Kützing. Tetrahedron Lett., 4817 (1976).

480. IZAK, R.R., and J.J. SIMS: Marine Natural Products, 18: Iodinated Sesquiterpenes from the Red Algae Genus *Laurencia*. J. Am. Chem. Soc. **101**, 6136 (1979).

481. IZAK, R.R., J.S. DRAGE, and J.J. SIMS: Caraibical, a New Aromatic Sesquiterpene from the Marine Alga *Laurencia caraibica*. Tetrahedron Lett. **22**, 1799 (1981).

482. IRIE, T., Y. YASUNARI, T. SUZUKI, N. IMAI, E. KUROSAWA, and T. MASAMUNE: A New Sesquiterpene Hydrocarbon from *Laurencia glandulifera*. Tetrahedron Lett., 3619 (1965).

483. SUZUKI, M., and E. KUROSAWA: New Aromatic Sesquiterpenoids from the Red Alga *Laurencia okamurai* Yamada. Tetrahedron Lett., 2503 (1978).

484. SUZUKI, M., and E. KUROSAWA: Halogenated and Non-Halogenated Aromatic Sesquiterpenes from the Red Algae *Laurencia okamurai* Yamada. Bull. Chem. Soc. Japan **52**, 3352 (1979).

485. CACCAMESE, S., L.P. HAGER, K.L. RINEHART JR., and R.B. SETZER: Characterization of *Laurencia* Species by Gas Chromatography-Mass Spectrometry. Bot. Mar. **22**, 41 (1979).

486. SUZUKI, M., and E. KUROSAWA: Halogenated Sesquiterpene Phenols and Ethers from the Red Alga *Laurencia glandulifera* Kützing. Bull. Chem. Soc. Japan **52**, 3349 (1979).

487. BLUNT, J.W., R.J. LAKE, and M.H.G. MUNRO: Sesquiterpenes from the Marine Red Alga *Laurencia distichophylla*. Phytochem. **23**, 1951 (1984).

488. GONZÁLEZ, A.G., J.M. ARTEAGA, J.J. FERNANDEZ, J.D. MARTÍN, M. NORTE, and J.Z. RUANO: Terpenoids of the Red Alga *Laurencia pinnatifida*. Tetrahedron **40**, 2751 (1984).

489. WRIGHT, A.D., G.M. KÖNIG, R. DE NYS, and O. STICHER: Seven New Metabolites from the Marine Red Alga *Laurencia majuscula*. J. Nat. Prod. **56**, 394 (1993).

490. GONZÁLEZ, A.G., J.M. AGUIAR, J.D. MARTÍN, and M.L. RODRÍGUEZ: Perforene, a New Halogenated Sesquiterpene from the Red Alga *Laurencia perforata*. Tetrahedron Lett., 205 (1976).

491. ICHIBA, T., and T. HIGA: New Cuparene-Derived Sesquiterpenes with Unprecedented Oxygenation Patterns from the Sea Hare *Aplysia dactylomela*. J. Org. Chem. **51**, 3364 (1986).

492. OHTA, K., and M. TAKAGI: Halogenated Sesquiterpenes from the Marine Red Alga *Marginisporum aberrans*. Phytochem. **16**, 1062 (1977).

493. AFAQ-HUSAIN, S., M. SHAMEEL, K. USMANGHANI, M. AHMAD, S. PERVEEN, and V.U. AHMAD: Brominated Sesquiterpene Metabolites of *Hypnea pannosa* Gigartinales. J. Appl. Phycol. **3**, 111 (1991).

494. HÖGBERG, H.-E., R.H. THOMSON, and T.J. KING: The Cymopols, a Group of Prenylated Bromohydroquinones from the Green Calcareous Alga *Cymopolia barbata*. J. Chem. Soc., Perkin Trans. 1, 1696 (1976).

495. MCCONNELL, O.J., P.A. HUGHES, and N.M. TARGETT: Diastereomers of Cyclocymopol and Cyclocymopol Monomethyl Ether from *Cymopolia barbata*. Phytochem. **21**, 2139 (1982).

496. ESTRADA, D.M., J.D. MARTÍN, and C. PÉREZ: A New Brominated Monoterpenoid Quinol from *Cymopolia barbata*. J. Nat. Prod. **50**, 735 (1987).

497. WALL, M.E., M.C. WANI, G. MANIKUMAR, H. TAYLOR, T.J. HUGHES, K. GAETANO, W.H. GERWICK, A.T. MCPHAIL, and D.R. MCPHAIL: Plant Antimutagenic Agents, 7: Structure and Antimutagenic Properties of Cymobarbatol and 4-Isocymobarbatol, New Cymopols from Green Alga (*Cymopolia barbata*). J. Nat. Prod. **52**, 1092 (1989).

498. PARK, M., W. FENICAL, and M.E. HAY: Debromoisocymobarbatol, a New Chromanol Feeding Deterrent from the Marine Alga *Cymopolia barbata*. Phytochem. **31**, 4115 (1992).

499. GARSON, M.J., D.C. MANKER, K.E. MAXWELL, B.W. SKELTON, and A.H. WHITE: Novel Brominated Metabolites from a Dictyocerated Sponge of the *Cacospongia* Genus. Aust. J. Chem. **42**, 611 (1989).

500. RAVI, B.N., H.P. PERZANOWSKI, R.A. ROSS, T.R. ERDMAN, P.J. SCHEUER, J. FINER, and J. CLARDY: Recent Research in Marine Natural Products: The Puupehenones. Pure Appl. Chem. **51**, 1893 (1979).

501. AIELLO, A., E. FATTORUSSO, and M. MENNA: A New Antibiotic Chloro-Sesquiterpene from the Caribbean Sponge *Smenospongia aurea*. Z. Naturforsch. **48B**, 209 (1993).

502. JONES, D.F.: Examination of the Gibberellins of *Zea mays* and *Phaseolus multiflorus* Using Thin-Layer Chromatography. Nature **202**, 1309 (1964).

503. DURLEY, R.C., J. MACMILLAN, and R.J. PRYCE: Investigation of Gibberellins and Other Growth Substances in the Seed of *Phaseolus multiflorus* and of *Phaseolus vulgaris* by Gas Chromatography and by Gas Chromatography-Mass Spectrometry. Phytochem. **10**, 1891 (1971).

504. CRUSE, W.B.T., M.N.G. JAMES, A.A. AL-SHAMMA, J.K. BEAL, and R.W. DOSKOTCH: The Molecular Structure of Gutierolide, a Novel Chloro-diterpenoid Lactone. J. Chem. Soc., Chem. Commun., 1278 (1971).

505. HANSON, J.R., D.E.A. RIVETT, S.V. LEY, and D.J. WILLIAMS: The X-Ray Structure and Absolute Configuration of Insect Antifeedant Clerodane Diterpenoids from *Teucrium africanum*. J. Chem. Soc., Perkin Trans. 1, 1005 (1982).

506. CARREIRAS, M.C., B. RODRÍGUEZ, F. PIOZZI, G. SAVONA, M.R. TORRES, and A. PERALES: A Chlorine-Containing and Two 17β-*neo*-Clerodane Diterpenoids from *Teucrium polium* subsp. *vincentinum*. Phytochem. **28**, 1453 (1989).

507. XIE, N., Z. MIN, S. ZHAO, Y. LU, Q. ZHENG, C. WANG, M. MIZUNO, M. IINUMA, and T. TANAKA: A Chlorine-Containing *neo*-Clerodane Diterpene from *Teucrium pernyi*. Chem. Pharm. Bull. (Japan) **40**, 2193 (1992).

508. SHIMOMURA, H., Y. SASHIDA, K. OGAWA, and Y. IITAKA: The Chemical Constituents of *Ajuga* Plants, I: *neo*-Clerodanes from the Leaves of *Ajuga nipponensis* Makino. Chem. Pharm. Bull. (Japan) **31**, 2192 (1983).

509. LAL, A.R., R.C. CAMBIE, P.S. RUTLEDGE, and P.D. WOODGATE: *Ent*-Atisane Diterpenes from *Euphorbia fidjiana*. Phytochem. **29**, 1925 (1990).

510. PCOLINSKI, M.J., R.W. DOSKOTCH, A.Y. LEE, and J. CLARDY: Chlorosilphanol A and Silphanepoxol, Labdane Diterpenes from *Silphium perfoliatum*. J. Nat. Prod. **57**, 776 (1994).

511. SASSA, T., M. TOGASHI, and T. KITAGUCHI: The Structures of Cotylenins A, B, C, D, and E. Agric. Biol. Chem. **39**, 1735 (1975).

512. MATSUDA, H., Y. TOMIIE, S. YAMAMURA, and Y. HIRATA: The Structure of Aplysin-20. J. Chem. Soc., Chem. Commun., 898 (1967).

513. YAMAMURA, S., and Y. HIRATA: A Naturally-Occurring Bromo-Compound, Aplysin-20 from *Aplysia kurodai.* Bull. Chem. Soc. Japan **44**, 2560 (1971).

514. YAMAMURA, S., and Y. TERADA: Isoaplysin-20, A Natural Bromine-Containing Diterpene, from *Aplysia kurodai.* Tetrahedron Lett., 2171 (1977).

515. NISHIZAWA, M., H. TAKENAKA, K. HIROTSU, T. HIGUCHI, and Y. HAYASHI: Synthesis and Structure Determination of Isoaplysin-20. J. Am. Chem. Soc. **106**, 4290 (1984).

516. IMAMURA, P.M., and E.A. RÚVEDA: The C-13 Configuration of the Bromine-Containing Diterpene Isoaplysin-20. Synthesis of Debromoisoplysin-20 and Its C-13 Epimer. J. Org. Chem. **45**, 510 (1980).

517. OJIKA, M., H. KIGOSHI, K. YOSHIKAWA, Y. NAKAYAMA, and K. YAMADA: A New Bromo Diterpene, *epi*-Aplysin-20, and *ent*-Isoconcinndiol from the Marine Mollusc *Aplysia kurodai.* Bull. Chem. Soc. Japan **65**, 2300 (1992).

518. OJIKA, M., Y. YOSHIDA, M. OKUMURA, S. IEDA, and K. YAMADA: Aplysiadiol, a New Brominated Diterpene from the Marine Mollusk *Aplysia kurodai.* J. Nat. Prod. **53**, 1619 (1990).

519. RODRÍGUEZ, M.L., J.D. MARTÍN, and D. ESTRADA: The Absolute Configuration of (+)-Isoconcinndiol. Acta Cryst. **C45**, 306 (1989).

520. FUJIWARA, S., K. TAKEDA, T. UYEHARA, and T. KATO: Structural Revision of Isoconcinndiol by Its Synthesis. Chem. Lett., 1763 (1986).

521. ESTRADA, D.M., J.L. RAVELO, C. RUIZ-PÉREZ, J.D. MARTÍN, and X. SOLANS: Dactylomelol, a New Class of Diterpene from the Sea Hare *Aplysia dactylomela.* Tetrahedron Lett. **30**, 6219 (1989).

522. SCHMITZ, F.J., K.H. HOLLENBEAK, D.C. CARTER, M.B. HOSSAIN, and D. VAN DER HELM: Marine Natural Products: 14-Bromoobtus-1-ene-3,11-diol, a New Diterpenoid from the Sea Hare *Aplysia dactylomela.* J. Org. Chem. **44**, 2445 (1979).

523. SCHMITZ, F.J., D.P. MICHAUD, and P.G. SCHMIDT: Marine Natural Products: Parguerol, Deoxyparguerol, and Isoparguerol. New Brominated Diterpenes with Modified Pimarane Skeletons from the Sea Hare *Aplysia dactylomela.* J. Am. Chem. Soc. **104**, 6415 (1982).

524. PETTIT, G.R., C.L. HERALD, J.J. EINCK, L.D. VANELL, P. BROWN, and D. GUST: Isolation and Structure of Angasiol. J. Org. Chem. **43**, 4685 (1978).

525. ATTA-UR-RAHMAN, K.A. ALVI, S.A. ABBAS, T. SULTANA, M. SHAMEEL, M.I. CHOUDHARY, and J.C. CLARDY: A Diterpenoid Lactone from *Aplysia juliana.* J. Nat. Prod. **54**, 886 (1991).

526. FENICAL, W.: Diterpenoids. In: Marine Natural Products, Vol. 2, Chap. 3 (P.J. Scheuer, ed.). New York: Academic Press. 1978.

527. SIMS, J.J., G.H.Y. LIN, R.M. WING, and W. FENICAL: Marine Natural Products. Concinndiol, a Bromo-diterpene Alcohol from the Red Alga, *Laurencia concinna.* J. Chem. Soc., Chem. Commun., 470 (1973).

528. HOWARD, B.M., W. FENICAL, J. FINER, K. HIROTSU, and J. CLARDY: Neoconcinndiol Hydroperoxide, a Novel Marine Diterpenoid from the Red Alga *Laurencia.* J. Am. Chem. Soc. **99**, 6440 (1977).

529. HOWARD, B.M., and W. FENICAL: Isoconcinndiol, a Brominated Diterpenoid from *Laurencia snyderae* var. *Guadalupensis.* Phytochem. **19**, 2774 (1980).

530. FUKUZAWA, A., M. MIYAMOTO, Y. KUMAGAI, A. ABIKO, Y. TAKAYA, and T.

MASAMUNE: Structure of New Bromoditerpenes, Pinnatols, from the Marine Red Alga *Laurencia pinnata* Yamada. Chem. Lett., 1259 (1985).

531. SUZUKI, M., E. KUROSAWA, and K. KURATA: Venustanol, a Brominated Labdane Diterpene from the Red Alga *Laurencia venusta*. Phytochem. **27**, 1209 (1988).

532. GONZÁLEZ, A.G., J.F. CICCIO, A.P. RIVERA, and J.D. MARTÍN: New Halogenated Diterpenes from the Red Alga *Laurencia perforata*. J. Org. Chem. **50**, 1261 (1985).

533. FENICAL, W., B. HOWARD, K.B. GIFKINS, and J. CLARDY: Irieol A and Iriediol, Dibromoditerpenes of a New Skeletal Class from *Laurencia*. Tetrahedron Lett., 3983 (1975).

534. HOWARD, B.M., and W. FENICAL: Structures of the Irieols, New Dibromoditerpenoids of a Unique Skeletal Class from a Marine Red Alga *Laurencia irieii*. J. Org. Chem. **43**, 4401 (1978).

535. FUKUZAWA, A., Y. KUMAGAI, T. MASAMUNE, A. FURUSAKI, T. MATSUMOTO, and C. KATAYAMA: Pinnaterpenes A, B, and C, New Dibromoditerpenes from the Red Alga *Laurencia pinnata* Yamada. Chem. Lett., 1389 (1982).

536. FUKUZAWA, A., Y. TAKAYA, H. MATSUE, and T. MASAMUNE: Structure of a New Bromoditerpene, Prepinnaterpene, from the Marine Red Alga *Laurencia pinnata* Yamada. Chem. Lett., 1263 (1985).

537. HOWARD, B.M., W. FENICAL, S.F. DONOVAN, and J. CLARDY: Neoirieone, a Diterpenoid of a New Skeletal Class from the Red Marine Alga *Laurencia* cf. *irieii*. Tetrahedron Lett. **23**, 3847 (1982).

538. HOWARD, B.M., and W. FENICAL: Obtusadiol, a Unique Bromoditerpenoid from the Marine Red Alga *Laurencia obtusa*. Tetrahedron Lett., 2453 (1978).

539. CACCAMESE, S., R.M. TOSCANO, S. CERRINI, and E. GAVUZZO: Laurencianol, a New Halogenated Diterpenoid from the Marine Alga *Laurencia obtusa*. Tetrahedron Lett. **23**, 3415 (1982).

540. HIGGS, M.D., and D.J. FAULKNER: A Diterpene from *Laurencia obtusa*. Phytochem. **21**, 789 (1982).

541. SUZUKI, T., S. TAKEDA, H. HAYAMA, I. TANAKA, and K. KOMIYAMA: The Structure of Brominated Diterpene from the Marine Red Alga *Laurencia obtusa* (Hudson) Lamouroux. Chem. Lett., 969 (1989).

542. TAKEDA, S., E. KUROSAWA, K. KOMIYAMA, and T. SUZUKI: The Structures of Cytotoxic Diterpenes Containing Bromine from the Marine Red Alga *Laurencia obtusa* (Hudson) Lamouroux. Bull. Chem. Soc. Japan **63**, 3066 (1990).

543. TAKEDA, S., T. MATSUMOTO, K. KOMIYAMA, E. KUROSAWA, and T. SUZUKI: A New Cytotoxic Diterpene from the Marine Red Alga *Laurencia obtusa* (Hudson) Lamouroux. Chem. Lett., 277 (1990).

544. BRENNAN, M.R., I.K. KIM, and K.L. ERICKSON: Kahukuenes, New Diterpenoids from the Marine Alga *Laurencia majuscula*. J. Nat. Prod. **56**, 76 (1993).

545. FENICAL, W., J. FINER, and J. CLARDY: Sphaerococcenol A; a New Rearranged Bromo-Diterpene from the Red Alga *Sphaerococcus coronopifolius*. Tetrahedron Lett., 731 (1976).

546. FATTORUSSO, E., S. MAGNO, C. SANTACROCE, D. SICA, B. DI BLASIO, C. PEDONE, G. IMPELLIZZERI, S. MANGIAFICO, and G. ORIENTE: Bromosphaerol, a New Bromine-Containing Diterpenoid from the Red Alga *Sphaerococcus coronopifolius*. Gazz. Chim. Ital. **106**, 779 (1976).

547. CAFIERI, F., L. DE NAPOLI, E. FATTORUSSO, G. IMPELLIZZERI, M. PIATTELLI, and S. SCIUTO: Bromosphaerodiol, a Minor Bromo Compound from the Red Alga *Sphaerococcus coronopifolius*. Experientia **32**, 1549 (1977).

548. CAFIERI, F., P. CIMINIELLO, E. FATTORUSSO, and C. SANTACROCE: 12*S*-Hydroxy-

bromosphaerol, a New Bromoditerpene from the Red Alga *Sphaerococcus coronopifolius*. Experientia **38**, 298 (1982).

549. CAFIERI, F., P. CIMINIELLO, C. SANTACROCE, and E. FATTORUSSO: (1*S*)-1,2-Dihydro-1-hydroxybromosphaerol, a Minor Bromoditerpene from the Red Alga *Sphaerococcus coronopifolius*. Phytochem. **21**, 2412 (1982).

550. CAFIERI, F., P. CIMINIELLO, C. SANTACROCE, and E. FATTORUSSO: Three Diterpenes from the Red Alga *Sphaerococcus coronopifolius*. Phytochem. **22**, 1824 (1983).

551. CAFIERI, F., E. FATTORUSSO, and C. SANTACROCE: Bromocorodienol, a Diterpenoid Based on a Novel Bicyclic Skeleton from the Red Alga *Sphaerococcus coronopifolius*. Tetrahedron Lett. **25**, 3141 (1984).

552. CAFIERI, F., E. FATTORUSSO, L. MAYOL, and C. SANTACROCE: Coronopifoliol, a Diterpene Based on an Unprecedented Tetracyclic Skeleton from the Red Algae *Sphaerococcus coronopifolius*. J. Org. Chem. **50**, 3982 (1985).

553. CAFIERI, F., E. FATTORUSSO, L. MAYOL, and C. SANTACROCE: Structure of Bromotetrasphaerol, a Further Irregular Diterpene from the Red Alga *Sphaerococcus coronopifolius*. Tetrahedron **42**, 4273 (1986).

554. BARTLETT, P.D.: Nonclassical Ions. New York: W.A. Benjamin. 1965.

555. CAFIERI, F., L. DE NAPOLI, E. FATTORUSSO, and C. SANTACROCE: Diterpenes from the Red Alga *Sphaerococcus coronopifolius*. Phytochem. **26**, 471 (1987).

556. CAFIERI, F., P. CIMINIELLO, E. FATTORUSSO, and A. MANGONI: Two Novel Bromoditerpenes from the Red Alga *Sphaerococcus coronopifolius*. Gazz. Chim. Ital. **120**, 139 (1990).

557. DE ROSA, S., S. DE STEFANO, P. SCARPELLI, and N. ZAVODNIK: Terpenes from the Red Alga *Sphaerococcus coronopifolius* of the North Adriatic Sea. Phytochem. **27**, 1875 (1988).

558. CACCAMESE, S., O. CASCIO, and A. COMPAGNINI: Isolation of an Antimicrobial Bromoterpene from a Marine Alga Aided by Improved Bioautography. J. Chromatogr. **478**, 255 (1989).

559. LE NY, G.: Participation Transannulaire de la Double Liaison lors de l'Acetolyse du *p*-Bromobenzène Sulfonate de Cycloheptène-4 yle. Compt. rend. **251**, 1526 (1960).

560. LAWTON, R.G.: 1,5-Participation in the Solvolysis of β-(Δ^3-Cyclopentenyl)ethyl *p*-Nitrobenzenesulfonate. J. Am. Chem. Soc. **83**, 2399 (1961).

561. ÖZTUNC, A., S. IMRE, H. LOTTER, and H. WAGNER: *ent*-13-Epiconcinndiol from the Red Alga *Chondria tenuissima* and Its Absolute Configuration. Phytochem. **28**, 3403 (1989).

562. SULLIVAN, B., and D.J. FAULKNER: Metabolites of the Marine Sponge *Dendrilla* sp. J. Org. Chem. **49**, 3204 (1984).

563. BOBZIN, S.C., and D.J. FAULKNER: Diterpenes from the Pompeian Marine Sponge *Chelonaplysilla* sp. J. Nat. Prod. **54**, 225 (1991).

564. CORRIERO, G., A. MADAIO, L. MAYOL, V. PICCIALLI, and D. SICA: Rotalin A and B, Two Novel Diterpene Metabolites from the Encrusting Mediterranean Sponge *Mycale rotalis* (Bowerbank). Tetrahedron **45**, 277 (1989).

565. CHANG, C.W.J., A. PATRA, J.A. BAKER, and P.J. SCHEUER: Kalihinols, Multifunctional Diterpenoid Antibiotics from Marine Sponges *Acanthella* spp. J. Am. Chem. Soc. **109**, 6119 (1987).

566. FUSETANI, N., K. YASUMURO, H. KAWAI, T. NATORI, L. BRINEN, and J. CLARDY: Kalihinene and Isokalinhinol B, Cytotoxic Diterpene Isonitriles from the Marine Sponge *Acanthella klethra*. Tetrahedron Lett. **31**, 3599 (1990).

567. TRIMURTULU, G., and D.J. FAULKNER: Six New Diterpene Isonitriles from the Sponge *Acanthella cavernosa*. J. Nat. Prod. **57**, 501 (1994).

568. ALVI, K.A., L. TENENBAUM, and P. CREWS: Anthelmintic Polyfunctional Nitrogen-Containing Terpenoids from Marine Sponges. J. Nat. Prod. **54**, 71 (1991).

569. TURSCH, B., J.C. BRAEKMAN, D. DALOZE, and M. KAISIN: Terpenoids from Coelenterates. In: Marine Natural Products, Vol. II, Chap. 4 (P.J. Scheuer, ed.). New York: Academic Press. 1978.

570. FINDLAY, J.A., G. LI, and P.E. PENNER: Novel Metabolites from the Nudibranch *Aplysia punctata*. International Research Congress on Natural Products, Halifax, Nova Scotia, Canada, July–August, 1994.

571. BURKS, J.E., D. VAN DER HELM, C.Y. CHANG, and L.S. CIERESZKO: The Crystal and Molecular Structure of Briarein A, a Diterpenoid from the Gorgonian *Briareum asbestinum*. Acta Cryst. **B33**, 704 (1977).

572. GRODE, S.H., T.R. JAMES, and J.H. CARDELLINA: Brianthein Z, a New Polyfunctional Diterpene from the Gorgonian *Briareum polyanthes*. Tetrahedron Lett. **24**, 691 (1983).

573. GRODE, S.H., T.R. JAMES JR., J.H. CARDELLINA II, and K.D. ONAN: Molecular Structure of the Briantheins, New Insecticidal Diterpenes from *Briareum polyanthes*. J. Org. Chem. **48**, 5203 (1983).

574. VAN DER HELM, D., R.A. LOGHRY, J.A. MATSON, and A.J. WEINHEIMER: Crystal and Molecular Structure of Brianthein X. J. Crystallogr. Spectrosc. Res. **16**, 713 (1986).

575. LINZ, G.S., M.J. MUSMAR, A.J. WEINHEIMER, G.E. MARTIN, and J.A. MATSON: Two-Dimensional NMR Studies of Marine Natural Products, III: Reassignment of the ^{13}C-NMR Spectrum of Brianthein-X Using Heteronuclear Relayed Coherence Transfer. Spectrosc. Lett. **19**, 545 (1986).

576. COVAL, S.J., S. CROSS, G. BERNARDINELLI, and C.W. JEFFORD: Brianthein V, a New Cytotoxic and Antiviral Diterpene Isolated from *Briareum asbestinum*. J. Nat. Prod. **51**, 981 (1988).

577. BOWDEN, B.F., J.C. COLL, I.M. VASILESCU, and P.N. ALDERSLADE: Studies of Australian Soft Corals, XLVII: New Halogenated Briaran Diterpenes from a *Briareum* sp. (Octocorallia, Gorgonacea). Aust. J. Chem. **42**, 1727 (1989).

578. KOBAYASHI, J., J.F. CHENG, N. NAKAMURA, Y. OHIZUMI, Y. TOMOTAKE, T. MATSUZAKI, K.J.S. GRACE, R.S. JACOBS, Y. KATO, L.S. BRINEN, and J. CLARDY: Structure and Stereochemistry of Antiinflammatory Diterpenoid from the Okinawan Gorgonian *Briareum* sp. Experientia **47**, 501 (1991).

579. WRATTEN, S.J., D.J. FAULKNER, K. HIROTSU, and J. CLARDY: Stylatulide, a Sea Pen Toxin. J. Am. Chem. Soc. **99**, 2824 (1977).

580. WRATTEN, S.J., and D.J. FAULKNER: Some Diterpenes from the Sea Pen *Stylatula* sp. Tetrahedron **35**, 1907 (1979).

581. WRATTEN, S.J., W. FENICAL, D.J. FAULKNER, and J.C. WEKELL: Ptilosarcone, the Toxin from the Sea Pen *Ptilosarcus gurneyi*. Tetrahedron Lett., 1559 (1977).

582. HENDRICKSON, R.L., and J.H. CARDELLINA: Structure and Stereochemistry of Insecticidal Diterpenes from the Sea Pen *Ptilosarcus gurneyi*. Tetrahedron **42**, 6565 (1986).

583. WILLIAMS, D.E., and R.J. ANDERSEN: Terpenoid Metabolites from Skin Extracts of the Dendronotid Nudibranch *Tochuina tetraquetra*. Can. J. Chem. **65**, 2244 (1987).

584. CLASTRES, A., A. AHOND, C. POUPAT, P. POTIER, and S.K. KAN: Invertébrés Marins du Lagon Néo-Calédonien, II: Étude Structurale de Trois Nouveaux Diterpénes Isolés du Pennatulaire *Pteroides laboutei*. J. Nat. Prod. **47**, 155 (1984).

585. LOOK, S.A., W. FENICAL, D. VAN ENGEN, and J. CLARDY: Erythrolides: Unique Marine Diterpenoids Interrelated by a Naturally Occurring Di-π-methane Rearrangement. J. Am. Chem. Soc. **106**, 5026 (1984).

586. PORDESIMO, E.O., F.J. SCHMITZ, L.S. CIERESZKO, M.B. HOSSAIN, and D. VAN DER

HELM: New Briarein Diterpenes from the Caribbean Gorgonians *Erythropodium caribaeorum* and *Briareum* sp. J. Org. Chem. **56**, 2344 (1991).

587. DOOKRAN, R., D. MAHARAJ, B.S. MOOTOO, R. RAMSEWAK, S. MCLEAN, W.F. REYNOLDS, and W.F. TINTO: Diterpenes from the Gorgonian Coral *Erythropodium caribaeorum* from the Southern Caribbean. J. Nat. Prod. **56**, 1051 (1993).

588. KSEBATI, M.B., and F.J. SCHMITZ: Diterpenes from a Soft Coral, *Minabea* sp., from Truk Lagoon. Bull. Soc. Chim. Belg. **95**, 835 (1986).

589. GROWEISS, A., S.A. LOOK, and W. FENICAL: Solenolides, New Antiinflammatory and Antiviral Diterpenoids from a Marine Octocoral of the Genus *Solenopodium*. J. Org. Chem. **53**, 2401 (1988).

590. LUO, Y., K. LONG, and Z. FANG: Studies on the Chemical Constituents of the Chinese Gorgonia, III: Isolation and Identification of a New Polyacetoxychlorine Containing Diterpene Lactone (Praelolide). Zhongshan Daxue Xuebao, Ziran Kexueban, 83 (1983); Chem. Abs. **99**, 50572c (1983).

591. YAO, J., J. QIAN, H. FAN, K. SHIH, S. HUANG, Y. LIN, and K. LONG: Crystal and Molecular Structure of Junceellin. Zhongshan Daxue Xuebao, Ziran Kexueban, 83 (1984); Chem. Abs. **101**, 211491c (1984).

592. LONG, K., Y. LIN, and W. HUANG: Chemical Constituents of the Chinese Gorgonia, VI: Junceellin B, a New Chlorine-Containing Diterpenoid from *Junceella squamata*. Zhongshan Daxue Xuebao, Ziran Kexueban, 15 (1987); Chem. Abs. **107**, 215049m (1987).

593. SHIN, J., M. PARK, and W. FENICAL: The Junceelloides, New Anti-Inflammatory Diterpenoids of the Briarane Class from the Chinese Gorgonian *Junceella fragilis*. Tetrahedron **45**, 1633 (1989).

594. ISAACS, S., S. CARMELY, and Y. KASHMAN: Juncins A–F, Six New Briarane Diterpenoids from the Gorgonian *Junceella juncea*. J. Nat. Prod. **53**, 596 (1990).

595. HE, H., and D.J. FAULKNER: New Chlorinated Diterpenes from the Gorgonian *Junceella gemmacea*. Tetrahedron **47**, 3271 (1991).

596. AHOND, A., B.F. BOWDEN, J.C. COLL, J.-D. FOURNERON, and S.J. MITCHELL: Studies of Australian Soft Corals, XXV: Several Caryophyllene-Based Diterpenes from a *Nephthea* Species. Aust. J. Chem. **34**, 2657 (1981).

597. ALMOURABIT, A., A. AHOND, A. CHIARONI, C. POUPAT, C. RICHE, P. POTIER, P. LABOUTE, and J.-L. MENOU: Invertébrés Marins du Lagon Néo-Calédonien, IX: Havannachlorhydrines, Nouveaux Métabolites de *Xenia membranacea*: Étude Structurale et Configuration Absolue. J. Nat. Prod. **51**, 282 (1988).

598. ALMOURABIT, A., A. AHOND, C. POUPAT, and P. POTIER: Invertébrés Marins du Lagon Néo-Calédonien, XII: Isolement et Étude Structurale de Nouveaux Diterpénes Extraits de L'Alcyonaire *Xenia membranacea*. J. Nat. Prod. **53**, 894 (1990).

599. GUERRIERO, A., M. D'AMBROSIO, and F. PIETRA: Slowly Interconverting Conformers of the Briarane Diterpenoids Verecynarmin B, C, and D, Isolated from the Nudibranch Mollusc *Armina maculata* and the Pennatulacean Octocoral *Veretillum cynomorium* of East Pyrenean Waters. Helv. Chim. Acta **71**, 472 (1988).

600. MALOCHET-GRIVOIS, C., P. COTELLE, J.F. BIARD, J.P. HÉNICHART, C. DEBITUS, C. ROUSSAKIS, and J.F. VERBIST: Dichlorolissoclimide, a New Cytotoxic Labdane Derivative from *Lissoclinum voeltzkowi* Michaelson (Urochordata). Tetrahedron Lett. **32**, 6701 (1991).

601. DE GIULIO, A., S. DE ROSA, G. DI VINCENZO, G. STRAZZULLO, and N. ZAVODNIK: Norsesterterpenes from the North Adriatic Sponge *Ircinia oros*. J. Nat. Prod. **53**, 1503 (1990).

602. N'DIAYE, I., G. GUELLA, I. MANCINI, J.-M. KORNPROBST, and F. PIETRA: Konakhin,

a Novel Type of Degraded Sesterterpene; Isolation from a Marine Sponge of Senegal. J. Chem. Soc., Chem. Commun., 97 (1991).

603. Blunt, J.W., M.P. Hartshorn, T.J. McLennan, M.H.G. Munro, W.T. Robinson, and S.C. Yorke: Thyrsiferol: A Squalene-Derived Metabolite of *Laurencia thyrsifera*. Tetrahedron Lett., 69 (1978).

604. Suzuki, T., S. Takeda, M. Suzuki, E. Kurosawa, A. Kato, and Y. Imanaka: Cytotoxic Squalene-Derived Polyethers from the Marine Red Alga *Laurencia obtusa* (Hudson) Lamouroux. Chem. Lett., 361 (1987).

605. Suzuki, T., M. Suzuki, A. Furusaki, T. Matsumoto, A. Kato, Y. Imanaka, and E. Kurosawa: Teurilene and Thyrsiferyl 23-Acetate, *meso* and Remarkably Cytotoxic Compounds from the Marine Red Alga *Laurencia obtusa* (Hudson) Lamouroux. Tetrahedron Lett. **26**, 1329 (1985).

606. Sakemi, S., T. Higa, C.W. Jefford, and G. Bernardinelli: Venustatriol. A New, Anti-Viral, Triterpene Tetracyclic Ether from *Laurencia venusta*. Tetrahedron Lett. **27**, 4287 (1986).

607. Suzuki, M., Y. Matsuo, S. Takeda, and T. Suzuki: Intricatetraol, a Halogenated Triterpene Alcohol from the Red Alga *Laurencia intricata*. Phytochem. **33**, 651 (1993).

608. Kusumi, T., M. Igari, M.O. Ishitsuka, A. Ichikawa, Y. Itezono, N. Nakayama, and H. Kakisawa: A Novel Chlorinated Biscembranoid from the Marine Soft Coral *Sarcophyton glaucum*. J. Org. Chem. **55**, 6286 (1990).

609. Cimino, G., S. De Rosa, S. De Stefano, R. Puliti, G. Strazzullo, C.A. Mattia, and L. Mazzarella: Absolute Stereochemistry of Disidein and of Two New Related Halogenated Sesterterpenoids. Two-Dimensional NMR Studies and X-Ray Crystal Structure. Tetrahedron **43**, 4777 (1987).

610. Tschesche, R., M. Baumgarth, and P. Welzel: Weitere Inhaltsstoffe aus *Jaborosa integrifolia* Lam.-III. Tetrahedron **24**, 5169 (1968).

611. Tschesche, R., K. Annen, and P. Welzel: Zur Konfiguration der Jaborosalactone, ein neuer Abbauweg für Withanolide. Chem. Ber. **104**, 3556 (1971).

612. González, A.G., J.L. Bréton, and J.M. Trujillo: Esteroides de Withanias, III: Lactonas Esteroidales de la *Withania frutescens* Pauq. An. Quim. **70**, 69 (1974).

613. Ray, A.B., M. Sahai, and B.C. Das: Physalolactone: A New Withanolide from *Physalis peruviana*. J. Indian Chem. Soc. **55**, 1175 (1978).

614. Frolow, F., A.B. Ray, M. Sahai, E. Glotter, H.E. Gottlieb, and I. Kirson: Withaperuvin and 4-Deoxyphysalolactone, Two New Ergostane-Type Steroids from *Physalis peruviana* (Solanaceae). J. Chem. Soc., Perkin Trans. 1, 1029 (1981).

615. Ali, A., M. Sahai, A.B. Ray, and D.J. Slatkin: Physalolactone C, a New Withanolide from *Physalis peruviana*. J. Nat. Prod. **47**, 648 (1984).

616. Shingu, K., S. Yahara, T. Nohara, and H. Okabe: Three New Withanolides, Physagulins A, B and D from *Physalis angulata* L. Chem. Pharm. Bull. (Japan) **40**, 2088 (1992).

617. Nittala, S.S., V.V. Velde, F. Frolow, and D. Lavie: Chlorinated Withanolides from *Withania somnifera* and *Acnistus breviflorus*. Phytochem. **20**, 2547 (1981).

618. Nittala, S.S., and D. Lavie: Withanolides and *Acnistus breviflorus*. Phytochem. **20**, 2735 (1981).

619. González, A.G., V. Darias, D.A.M. Herrera, and M.C. Suárez: Cytostatic Activity of Natural Withanolides from Spanish Withanias. Fitoterapia **53**, 85 (1982).

620. Burton, G., A.S. Veleiro, and E.G. Gros: Analytical and Preparative Separation of Withanolides from Crude Extracts of *Acnistus breviflorus* Leaves by High-Performance Liquid Chromatography. J. Chromatog. **248**, 472 (1982).

621. BURTON, G., A.S. VELEIRO, and E.G. GROS: Reversed-Phase Chromatographic Separation of Withanolides from *Acnistus breviflorus*. J. Chromatog. **315**, 435 (1984).

622. VELEIRO, A.S., G. BURTON, and E.G. GROS: 2,3-Dihydrojaborosalactone A, a Withanolide from *Acnistus breviflorus*. Phytochem. **24**, 1799 (1985).

623. BESSALLE, R., and D. LAVIE: Withanolide C, a Chlorinated Withanolide from *Withania somnifera*. Phytochem. **31**, 3648 (1992).

624. KIRSON, I., and E. GLOTTER: Recent Developments in Naturally Occurring Ergostane-Type Steroids. A Review. J. Nat. Prod. **44**, 633 (1981).

625. FAJARDO, V., F. PODESTA, M. SHAMMA, and A.J. FREYER: New Withanolides from *Jaborosa magellanica*. J. Nat. Prod. **54**, 554 (1991).

626. DJERASSI, C., and C.J. SILVA: Sponge Sterols: Origin and Biosynthesis. Acct. Chem. Res. **24**, 371 (1991).

627. CARNEY, J.R., P.J. SCHEUER, and M. KELLY-BORGES: Three Unprecedented Chloro Steroids from the Maui Sponge *Strongylacidon* sp., Kiheisterones C, D, and E. J. Org. Chem. **58**, 3460 (1993).

628. SHIMURA, H., K. IGUCHI, Y. YAMADA, S. NAKAIKE, T. YAMAGISHI, K. MATSUMOTO, and C. YOKOO: Aragusterol C: A Novel Halogenated Marine Steriod from an Okinawan Sponge, *Xestospongia* sp., Possessing Potent Antitumor Activity. Experientia **50**, 134 (1994).

629. SAKUMA, M., and H. FUKAMI: Novel Steroid Glycosides as Aggregation Pheromone of the German Cockroach. Tetrahedron Lett. **34**, 6059 (1993).

630. MOORE, R.E.: Algal Nonisoprenoids. In: Marine Natural Products, Vol. I, Chap. 2 (P.J. Scheuer, ed.). New York: Academic Press. 1978.

631. IRIE, T., M. SUZUKI, and T. MASAMUNE: Laurencin, a Constituent from *Laurencia* Species. Tetrahedron Lett., 1091 (1965).

632. CAMERON, A.F., K.K. CHEUNG, G. FERGUSON, and J.M. ROBERTSON: The Crystal Structure and Stereochemistry of Laurencin. J. Chem. Soc., Chem. Commun., 638 (1965).

633. IRIE, T., M. SUZUKI, and T. MASAMUNE: Laurencin, a Constituent of *Laurencia glandulifera* Kützing. Tetrahedron **24**, 4193 (1968).

634. CAMERON, A.F., K.K. CHEUNG, G. FERGUSON, and J.M. ROBERTSON: Laurencia Natural Products, Part I: Crystal Structure and Absolute Stereochemistry of Laurencin. J. Chem. Soc. (B), 559 (1969).

635. FUKUZAWA, A., M. AYE, and A. MURAI: A Direct Enzymatic Synthesis of Laurencin from Laurediol. Chem. Lett., 1579 (1990).

636. KUROSAWA, E., A. FUKUZAWA, and T. IRIE: *trans*- and *cis*-Laurediol, Unsaturated Glycols from *Laurencia nipponica* Yamada. Tetrahedron Lett., 2121 (1972).

637. IRIE, T., M. IZAWA, and E. KUROSAWA: Laureatin, a Constituent from *Laurencia nipponica* Yamada. Tetrahedron Lett., 2091 (1968).

638. IRIE, T., M. IZAWA, and E. KUROSAWA: Isolaureatin, a Constituent from *Laurencia nipponica* Yamada. Tetrahedron Lett., 2735 (1968).

639. IRIE, T., M. IZAWA, and E. KUROSAWA: Laureatin and Isolaureatin, Constituents of *Laurencia nipponica* Yamada. Tetrahedron **26**, 851 (1970).

640. KUROSAWA, E., A. FURUSAKI, M. IZAWA, A. FUKUZAWA, and T. IRIE: The Absolute Configurations of Laureatin and Isolaureatin. Tetrahedron Lett., 3857 (1973).

641. FUKUZAWA, A., M. AYE, M. NAKAMURA, M. TAMURA, and A. MURAI: Biosynthetic Formation of Cyclic Bromo-Ethers Initiated by Lactoperoxidase. Chem. Lett., 1287 (1990).

642. FUKUZAWA, A., E. KUROSAWA, and T. IRIE: Acid-Catalyzed Rearrangement of Laureatin to Isolaureatin and Related Reactions. J. Org. Chem. **37**, 680 (1972).

643. FUKUZAWA, A., Y. TAKASUGI, and A. MURAI: Prelaureatin, a New Biogenetic Key Intermediate Isolated from *Laurencia nipponica*. Tetrahedron Lett. **32**, 5597 (1991).
644. FUKUZAWA, A., Y. TAKASUGI, A. MURAI, M. NAKAMURA, and M. TAMURA: Enzymatic Single-Step Formation of Laureatin and Its Key Intermediate, Prelaureatin, from (3*Z*,6*S*,7*S*)-Laurediol. Tetrahedron Lett. **33**, 2017 (1992).
645. FUKUZAWA, A., E. KUROSAWA, and T. IRIE: Laurefucin and Acetyllaurefucin, New Bromo Compounds from *Laurencia nipponica* Yamada. Tetrahedron Lett., 3 (1972).
646. FURUSAKI, A., E. KUROSAWA, A. FUKUZAWA, and T. IRIE: The Revised Structure and Absolute Configuration of Laurefucin from *Laurencia nipponica* Yamada. Tetrahedron Lett., 4579 (1973).
647. KUROSAWA, E., A. FUKUZAWA, and T. IRIE: Isoprelaurefucin, New Bromo Compound from *Laurencia nipponica* Yamada. Tetrahedron Lett., 4135 (1973).
648. KAUL, P.N., S.K. KULKARNI, and E. KUROSAWA: Novel Substances of Marine Origin as Drug Metabolism Inhibitors. J. Pharm. Pharmacol. **30**, 589 (1978).
649. FUKUZAWA, A., and E. KUROSAWA: Laurallene, New Bromoallene from the Marine Red Alga *Laurencia nipponica* Yamada. Tetrahedron Lett., 2797 (1979).
650. SUZUKI, M., K. KOIZUMI, H. KIKUCHI, T. SUZUKI, and E. KUROSAWA: Epilaurallene, a New Nonterpenoid C_{15}-Bromoallene from the Red Alga *Laurencia nipponica* Yamada. Bull. Chem. Soc. Japan **56**, 715 (1983).
651. KURATA, K., A. FURUSAKI, K. SUEHIRO, C. KATAYAMA, and T. SUZUKI: Isolaurallene, a New Nonterpenoid C_{15}-Bromoallene from the Red Alga *Laurencia nipponica* Yamada. Chem. Lett., 1031 (1982).
652. FURUSAKI, A., S. KATSURAGI, K. SUEHIRO, and T. MATSUMOTO: The Conformations of (*Z*)-2,3,4,7,8,9-Hexahydrooxonin and (*Z*)-Cyclononene. X-Ray Structure Determinations of Isolaurallene and Neolaurallene, and Force-Field Calculations. Bull. Chem. Soc. Japan **58**, 803 (1985).
653. FUKUZAWA, A., M. AYE, M. NAKAMURA, M. TAMURA, and A. MURAI: Structure Elucidation of Laureoxanyne, a New Nonisoprenoid C_{15} Enyne, Using Lactoperoxidase. Tetrahedron Lett. **31**, 4895 (1990).
654. FUKUZAWA, A., and E. KUROSAWA: Laureepoxide, New Bromo Ether from the Marine Red Alga *Laurencia nipponica* Yamada. Tetrahedron Lett. **21**, 1471 (1980).
655. SUZUKI, T., K. KOIZUMI, M. SUZUKI, and E. KUROSAWA: Kumausallene, a New Bromoallene from the Marine Red Alga *Laurencia nipponica* Yamada. Chem. Lett., 1639 (1983).
656. SUZUKI, T., K. KOIZUMI, M. SUZUKI, and E. KUROSAWA: Kumausynes and Deacetylkumausynes, Four New Halogenated C_{15} Acetylenes from the Red Alga *Laurencia nipponica* Yamada. Chem. Lett., 1643 (1983).
657. FUKUZAWA, A., M. AYE, Y. TAKAYA, H. FUKUI, T. MASAMUNE, and A. MURAI: Laureoxolane, a New Bromo Ether from *Laurencia nipponica*. Tetrahedron Lett. **30**, 3665 (1989).
658. KIKUCHI, H., T. SUZUKI, E. KUROSAWA, and M. SUZUKI: The Structure of Notoryne, a Halogenated C_{15} Nonterpenoid with a Novel Carbon Skeleton from the Red Alga *Laurencia nipponica* Yamada. Bull. Chem. Soc. Japan **64**, 1763 (1991).
659. WARASZKIEWICZ, S.M., H.H. SUN, and K.L. ERICKSON: C_{15}-Halogenated Compounds from the Hawaiian Marine Alga *Laurencia nidifica*, V: The Maneonenes. Tetrahedron Lett., 3021 (1976).
660. SUN, H.H., S.M. WARASZKIEWICZ, and K.L. ERICKSON: C_{15}-Halogenated Compounds from the Hawaiian Marine Alga *Laurencia nidifica*, VI: The Isomaneonenes. Tetrahedron Lett., 4227 (1976).
661. WARASZKIEWICZ, S.M., H.H. SUN, K.L. ERICKSON, J. FINER, and J. CLARDY:

C_{15} Halogenated Compounds from the Hawaiian Marine Alga *Laurencia nidifica*. Maneonenes and Isomaneonenes. J. Org. Chem. **43**, 3194 (1978).

662. WHITE, R.H., and L.P. HAGER: Intricenyne and Related Halogenated Compounds from *Laurencia intricata*. Phytochem. **17**, 939 (1978).

663. CARDELLINA, J.H., S.B. HORSLEY, J. CLARDY, S.R. LEFTOW, and J. MEINWALD: Secondary Metabolites from the Red Alga *Laurencia intricata*: Halogenated Enynes. Can J. Chem. **60**, 2675 (1982).

664. HOWARD, B.M., G.R. SCHULTE, W. FENICAL, B. SOLHEIM, and J. CLARDY: Three New Vinyl Acetylenes from the Marine Red Alga *Laurencia*. Tetrahedron **36**, 1747 (1980).

665. KING, T.J., S. IMRE, A. ÖZTUNC, and R.H. THOMSON: Obtusenyne, a New Acetylenic Nine-Membered Cyclic Ether from *Laurencia obtusa*. Tetrahedron Lett., 1453 (1979).

666. SUZUKI, M., K. KURATA, T. SUZUKI, and E. KUROSAWA: The Absolute Configuration of Isoprelaurefucin. Bull. Chem. Soc. Japan **59**, 2953 (1986).

667. SUZUKI, M., and E. KUROSAWA: (3*E*)-Laureatin and (3*E*)-Isolaureatin, Halogenated C_{15} Non-Terpenoid Compounds from the Red Alga *Laurencia nipponica* Yamada. Bull. Chem. Soc. Japan **60**, 3791 (1987).

668. HOWARD, B.M., W. FENICAL, E.V. ARNOLD, and J. CLARDY: Obtusin, a Unique Bromine-Containing Polycyclic Ketal from the Red Marine Alga *Laurencia obtusa*. Tetrahedron Lett., 2841 (1979).

669. CACCAMESE, S., R. AZZOLINA, E.N. DUESLER, I.C. PAUL, and K.L. RINEHART: Laurencienyne, a New Acetylenic Cyclic Ether from the Marine Red Alga *Laurencia obtusa*. Tetrahedron Lett. **21**, 2299 (1980).

670. DUESLER, E.N., K.L. RINEHART, I.C. PAUL, S. CACCAMESE, and R. AZZOLINA: Laurencienyne, $C_{17}H_{23}BrCl_2O_3$. Cryst. Struct. Comm. **9**, 777 (1980).

671. CACCAMESE, S., R. AZZOLINA, R.M. TOSCANO, and K.L. RINEHART JR.: Variations in the Halogenated Metabolites of *Laurencia obtusa* from Eastern Sicily. Biochem. Sys. Ecol. **9**, 241 (1981).

672. FALSHAW, C.P., T.J. KING, S. IMRE, S. ISLIMYELI, and R.H. THOMSON: Laurenyne, a New Acetylene from *Laurencia obtusa*: Crystal Structure and Absolute Configuration. Tetrahedron Lett. **21**, 4951 (1980).

673. COX, P.J., S. IMRE, S. ISLIMYELI, and R.H. THOMSON: Obtusallene I, a New Halogenated Allene from *Laurencia obtusa*. Tetrahedron Lett. **23**, 579 (1982).

674. COX, P.J., and R.A. HOWIE: X-Ray Structure Analysis of Obtusallene. Acta Cryst. **B38**, 1386 (1982).

675. ÖZTUNC, A., S. IMRE, H. WAGNER, M. NORTE, J.J. FERNÁNDEZ, and R. GONZÁLEZ: A New Haloether from *Laurencia* Possessing a Lauroxacyclododecane Ring. Structural and Conformational Studies. Tetrahedron **47**, 2273 (1991).

676. ÖZTUNC, A., S. IMRE, H. LOTTER, and H. WAGNER: Two C_{15} Bromoallenes from the Red Alga *Laurencia obtusa*. Phytochem. **30**, 255 (1991).

677. ÖZTUNC, A., S. IMRE, H. WAGNER, M. NORTE, J.J. FERNÁNDEZ, and R. GONZÁLEZ: A New and Highly Oxygenated Bromoallene from a Marine Source. Tetrahedron Lett. **32**, 4377 (1991).

678. GONZÁLEZ, A.G., J.D. MARTÍN, M. NORTE, P. RIVERA, and J.Z. RUANO: Two New C_{15} Acetylenes from the Marine Red Alga *Laurencia obtusa*. Tetrahedron **40**, 3443 (1984).

679. NORTE, M., J.J. FERNANDEZ, J.Z. RUANO, M.L. RODRÍGUEZ, and R. PÉREZ: Graciosin and Graciosallene, Two Bromoethers from *Laurencia obtusa*. Phytochem. **27**, 3537 (1988).

680. NORTE, M., J.J. FERNÁNDEZ, and J.Z. RUANO: Three New Bromo Ethers from the Red Alga *Laurencia obtusa*. Tetrahedron **45**, 5987 (1989).

681. CACCAMESE, S., and R.M. TOSCANO: Neoobtusin, a New Brominated Ketal from the Marine Red Alga *Laruencia obtusa*. Gazz. Chim. Ital. **116**, 177 (1986).
682. IMRE, S., H. LOTTER, H. WAGNER, and R.H. THOMSON: Epoxy-*trans*-isodihydrorhodophytin, a New Metabolite from *Laurencia obtusa*. Z. Naturforsch. **42C**, 507 (1987).
683. HOWARD, B.M., W. FENICAL, K. HIROTSU, B. SOLHEIM, and J. CLARDY: The Rhodophytin and Chondriol Natural Products; Structures of Several New Acetylenes from *Laurencia*, and a Reassignment of Structure for *cis*-Rhodophytin. Tetrahedron **36**, 171 (1980).
684. FENICAL, W.: Rhodophytin, a Halogenated Vinyl Peroxide of Marine Origin. J. Am. Chem. Soc. **96**, 5580 (1974).
685. SUZUKI, M., E. KUROSAWA, A. FURUSAKI, S. KATSURAGI, and T. MATSUMOTO: Neolaurallene, a New Halogenated C_{15} Nonterpenoid from the Red Alga *Laurencia okamurai* Yamada. Chem. Lett., 1033 (1984).
686. SUZUKI, M., and E. KUROSAWA: A C_{15} Non-Terpenoid from the Red Alga *Laurencia okamurai*. Phytochem. **24**, 1999 (1985).
687. LOWE, G.: The Absolute Configuration of Allenes. J. Chem. Soc., Chem. Commun., 411 (1965).
688. KIGOSHI, H., Y. SHIZURI, H. NIWA, and K. YAMADA: Laurencenyne, a Plausible Precursor of Various Nonterpenoid C_{15}-Compounds, and Neolaurencenyne from the Red Alga *Laurencia okamurai*. Tetrahedron Lett. **22**, 4729 (1981).
689. KIGOSHI, H., Y. SHIZURI, H. NIWA, and K. YAMADA: Isolation and Structures of *trans*-Laurencenyne, a Possible Cyclic Ether, and *trans*-Neolaurencenyne from *Laurencia okamurai*. Tetrahedron Lett. **23**, 1475 (1982).
690. SUZUKI, M., and E. KUROSAWA: Okamurallene, a Novel Halogenated C_{15} Metabolite from the Red Alga *Laurencia okamurai* Yamada. Tetrahedron Lett. **22**, 3853 (1981).
691. SUZUKI, M., and E. KUROSAWA: Deoxyokamurallene and Isookamurallene, New Halogenated Nonterpenoid C_{15}-Compounds from the Red Alga *Laurencia okamurai* Yamada. Chem. Lett., 289 (1982).
692. SUZUKI, M., Y. SASAGE, M. IKURA, K. HIKICHI, and E. KUROSAWA: Structure Revision of Okamurallene and Structure Elucidation of Further C_{15} Non-Terpenoid Bromoallenes from *Laurencia intricata*. Phytochem. **28**, 2145 (1989).
693. SUZUKI, M., H. KONDO, and I. TANAKA: The Absolute Stereochemistry of Okamurallene and Its Congeners, Halogenated C_{15} Nonterpenoids from the Red Alga *Laurencia intricata*. Chem. Lett., 33 (1991).
694. FUKUZAWA, A., and T. MASAMUNE: Laurepinnacin and Isolaurepinnacin, New Acetylenic Cyclic Ethers from the Marine Red Alga *Laurencia pinnata* Yamada. Tetrahedron Lett. **22**, 4081 (1981).
695. GONZÁLEZ, A.G., J.D. MARTÍN, V.S. MARTÍN, M. NORTE, R. PÉREZ, J.Z. RUANO, S.A. DREXLER, and J. CLARDY: Non-Terpenoid C_{15} Metabolites from the Red Seaweed *Laurencia pinnatifida*. Tetrahedron **38**, 1009 (1982).
696. NORTE, M., A.G. GONZÁLEZ, F. CATALDO, M.L. RODRÍGUEZ, and I. BRITO: New Examples of Acyclic and Cyclic C_{15} Acetogenins from *Laurencia pinnatifida*. Reassignment of the Absolute Configuration for *E* and *Z* Pinnatifidienyne. Tetrahedron **47**, 9411 (1991).
697. SUZUKI, M., and E. KUROSAWA: Venustin A and B, New Halogenated C_{15} Metabolites from the Red Alga *Laurencia venusta* Yamada. Chem. Lett., 1177 (1980).
698. SUZUKI, M., E. KUROSAWA, A. FURUSAKI, and T. MATSUMOTO: The Structures of (3*Z*)-Epoxyvenustin, (3*Z*)-Venustin, and (3*Z*)-Venustinene, New Halogenated C_{15}-

Nonterpenoids from the Red Alga *Laurencia venusta* Yamada. Chem. Lett., 779 (1983).

699. BLUNT, J.W., R.J. LAKE, M.H.G. MUNRO, and S.C. YORKE: A New Vinyl Acetylene from the Red Alga *Laurencia thyrsifera*. Aust. J. Chem. **34**, 2393 (1981).

700. BLUNT, J.W., R.J. LAKE, and M.H.G. MUNRO: Metabolites of the Marine Red Alga *Laurencia thyrsifera*, III. Aust. J. Chem. **37**, 1545 (1984).

701. KENNEDY, D.J., I.A. SELBY, H.J. COWE, P.J. COX, and R.H. THOMSON: Bromoallenes from the Alga *Laurencia microcladia*. J. Chem. Soc., Chem. Commun., 153 (1984).

702. GUELLA, G., and F. PIETRA: Rogiolenyne A, B, and C: The First Branched Marine C_{15} Acetogenins. Isolation from the Red Seaweed *Laurencia microcladia* or the Sponge *Spongia zimocca* of Il Rogiolo. Helv. Chim. Acta **74**, 47 (1991).

703. GUELLA, G., I. MANCINI, G. CHIASERA, and F. PIETRA: Rogiolenyne D, the Likely Immediate Precursor of Rogiolenyne A and B, Branched C_{15} Acetogenins Isolated from the Red Seaweed *Laurencia microcladia* of Il Rogiolo. Conformation and Absolute Configuration in the Whole Series. Helv. Chim. Acta **75**, 303 (1992).

704. GUELLA, G., I. MANCINI, G. CHIASERA, and F. PIETRA: On the Unusual Propensity by the Red Seaweed *Laurencia microcladia* of Il Rogiolo to Form C_{15} Oxepanes: Isolation of Rogioloxepane A, B, C, and Their Likely Biogenetic Acyclic Precursor, Prerogioloxepane. Helv. Chim. Acta **75**, 310 (1992).

705. COLL, J.C., and A.D. WRIGHT: Tropical Marine Algae, IV: Novel Metabolites from a Red Alga *Laurencia implicata* (Rhodophyta, Rhodophyceae, Ceramiales, Rhodomelaceae). Aust. J. Chem. **42**, 1685 (1989).

706. WRIGHT, A.D., G.M. KÖNIG, and O. STICHER: New Sesquiterpenes and C_{15} Acetogenins from the Marine Red Alga *Laurencia implicata*. J. Nat. Prod. **54**, 1025 (1991).

707. KIM, I.K., M.R. BRENNAN, and K.L. ERICKSON: Lauroxolanes from the Marine Alga *Laurencia majuscula*. Tetrahedron Lett. **30**, 1757 (1989).

708. SUZUKI, M., Y. MATSUO, and M. MASUDA: Structures of Laurenenyne-A and -B, Novel Halogenated Acetogenins from a Species of the Red Alga *Laurencia*. Tetrahedron **49**, 2033 (1993).

709. FENICAL, W., J.J. SIMS, and P. RADLICK: Chondriol, A Halogenated Acetylene from the Marine Alga *Chondria oppositiclada*. Tetrahedron Lett., 313 (1973).

710. FENICAL, W., K.B. GIFKINS, and J. CLARDY: X-Ray Determination of Chondriol; a Re-Assignment of Structure. Tetrahedron Lett., 1507 (1974).

711. MCDONALD, F.J., D.C. CAMPBELL, D.J. VANDERAH, F.J. SCHMITZ, D.M. WASHECHECK, J.E. BURKS, and D. VAN DER HELM: Marine Natural Products. Dactylyne, an Acetylenic Dibromochloro Ether from the Sea Hare *Aplysia dactylomela*. J. Org. Chem. **40**, 665 (1975).

712. VANDERAH, D.J., and F.J. SCHMITZ: Marine Natural Products: Isodactylyne; a Halogenated Acetylenic Ether from the Sea Hare *Aplysia dactylomela*. J. Org. Chem. **41**, 3480 (1976).

713. KAUL, P.N., and S.K. KULKARNI: New Drug Metabolism Inhibitor of Marine Origin. J. Pharm. Sci. **67**, 1293 (1978).

714. KINNEL, R., A.J. DUGGAN, T. EISNER, and J. MEINWALD: Panacene, an Aromatic Bromoallene from a Sea Hare (*Aplysia brasiliana*). Tetrahedron Lett., 3913 (1977).

715. KINNEL, R.B., R.K. DIETER, J. MEINWALD, D. VAN ENGEN, J. CLARDY, T. EISNER, M.O. STALLARD, and W. FENICAL: Brasilenyne and *cis*-Dihydrorhodophytin: Antifeedant Medium-Ring Haloethers from a Sea Hare (*Aplysia brasiliana*). Proc. Natl. Acad. Sci. USA **76**, 3576 (1979).

716. FENICAL, W., H.L. SLEEPER, V.J. PAUL, M.O. STALLARD, and H.H. SUN: Defensive

Chemistry of *Navanax* and Related Opisthobranch Molluscs. Pure Appl. Chem. **51**, 1865 (1979).
717. SCHULTE, G.R., M.C.H. CHUNG, and P.J. SCHEUER: Two Bicyclic C_{15} Enynes from the Sea Hare *Aplysia oculifera*. J. Org. Chem. **46**, 3870 (1981).
718. DE SILVA, E.D., R.E. SCHWARTZ, P.J. SCHEUER, and J.N. SHOOLERY: Srilankenyne, a New Metabolite from the Sea Hare *Aplysia oculifera*. J. Org. Chem. **48**, 395 (1983).
719. OJIKA, M., T. NEMOTO, and K. YAMADA: Doliculols A and B, the Non-Halogenated C_{15} Acetogenins with Cyclic Ether from the Sea Hare *Dolabella auricularia*. Tetrahedron Lett. **34**, 3461 (1993).
720. CAPON, R.J., E.L. GHISALBERTI, and P.R. JEFFERIES: New Tetrahydropyrans from a Marine Sponge. Tetrahedron **38**, 1699 (1982).
721. POINER, A., V.J. PAUL, and P.J. SCHEUER: Kumepaloxane, a Rearranged Trisnor Sesquiterpene from the Bubble Shell *Haminoea cymbalum*. Tetrahedron **45**, 617 (1989).
722. GIORDANO, F., L. MAYOL, G. NOTARO, V. PICCIALLI, and D. SICA: Structure and Absolute Configuration of Two New Polybrominated C_{15} Acetogenins from the Sponge *Mycale rotalis*. J. Chem. Soc., Chem. Commun., 1559 (1990).
723. NOTARO, G., V. PICCIALLI, D. SICA, L. MAYOL, and F. GIORDANO: A Further C_{15} Nonterpenoid Polybromoether from the Encrusting Sponge *Mycale rotalis*. J. Nat. Prod. **55**, 626 (1992).
724. GEISSMAN, T.A., and D.H.G. CROUT: Organic Chemistry of Secondary Plant Metabolism. San Francisco: Freeman, Cooper & Co. 1969.
725. KITAGAWA, I., T. TANI, K. AKITA, and I. YOSIOKA: Linarioside, a New Chlorine Containing Iridoid Glucoside, from *Linaria japonica* Miq. Tetrahedron Lett., 419 (1972).
726. KITAGAWA, I., T. TANI, K. AKITA, and I. YOSIOKA: On the Constituents of *Linaria japonica* Miq., I: The Structure of Linarioside, a New Chlorinated Iridoid Glucoside, and Identification of Two Related Glucosides. Chem. Pharm. Bull. (Japan) **21**, 1978 (1973).
727. KAPOOR, S.K., J. REISCH, and K. SZENDREI: Iridoids of *Cymbalaria muralis*. Phytochem. **13**, 1018 (1974).
728. ILIEVA, E., N. HANDJIEVA, S. SPASSOV, and S. POPOV: 5-*O*-Allosylantirrinoside from *Linaria* Species. Phytochem. **32**, 1068 (1993).
729. HANDJIEVA, N.V., E.I. ILIEVA, S.L. SPASSOV, S.S. POPOV, and H. DUDDECK: Iridoid Glycosides from *Linaria* Species. Tetrahedron **49**, 9261 (1993).
730. JENSEN, S.R., and B.J. NIELSEN: Iridoids in *Thunbergia* Species. Phytochem. **28**, 3059 (1989).
731. DAMTOFT, S., H. FRANZYK, S. JENSEN, and B.J. NIELSEN: Iridoids and Verbascosides in *Retzia*. Phytochem. **34**, 239 (1993).
732. DEMUTH, H., S.R. JENSEN, and B.J. NIELSEN: Iridoid Glucosides from *Asystasia bella*. Phytochem. **28**, 3361 (1989).
733. EL-NAGGAR, L.J., J.L. BEAL, and R.W. DOSKOTCH: Iridoid Glycosides from *Mentzelia decapetala*. J. Nat. Prod. **45**, 539 (1982).
734. KOBAYASHI, H., H. KARASAWA, T. MIYASE, and S. FUKUSHIMA: Studies on the Constituents of *Cistanchis* Herba, II: Isolation and Structures of New Iridoids, Cistanin and Cistachlorin. Chem. Pharm. Bull. (Japan) **32**, 1729 (1984).
735. KITAGAWA, I., Y. FUKUDA, T. TANIYAMA, and M. YOSHIKAWA: Absolute Stereostructures of Rehmaglutins A, B, and D. Three New Iridoids Isolated from Chinese Rehmanniae Radix. Chem. Pharm. Bull. (Japan) **34**, 1399 (1986).
736. KITAGAWA, I., Y. FUKUDA, T. TANIYAMA, and M. YOSHIKAWA: Chemical Studies on Crude Drug Processing, VII: On the Constituents of Rehmanniae Radix. (1). Abso-

lute Stereostructures of Rehmaglutins A, B, and D Isolated from Chinese Rehmanniae Radix, the Dried Root of *Rehmannia glutinosa* Libosch. Chem. Pharm. Bull. (Japan) **39**, 1171 (1991).

737. Morota, T., H. Nishimura, H. Sasaki, M. Chin, K. Sugama, T. Katsuhara, and H. Mitsuhashi: Five Cyclopentanoid Monoterpenes from *Rehmannia glutinosa*. Phytochem. **28**, 2385 (1989).

738. Yoshikawa, M., Y. Fukuda, T. Taniyama, and I. Kitagawa: Absolute Stereostructures of Rehmaglutin C and Glutinoside. A New Iridoid Lactone and a New Chlorinated Iridoid Glucoside from Chinese Rehmanniae Radix. Chem. Pharm. Bull. (Japan) **34**, 1403 (1986).

739. Uesato, S., T. Hashimoto, and H. Inouye: Three New Secoiridoid Glucosides from *Eustoma russellianum*. Phytochem. **18**, 1981 (1979).

740. Calis, I., T. Ersöz, A.J. Chulia, and P. Rüedi: Septemfidoside: A New Bis-Iridoid Diglucoside from *Gentiana septemfida*. J. Nat. Prod. **55**, 385 (1992).

741. Popov, S., N.V. Handzhieva, and N. Marekov: Halogen-Containing Valepotriates Isolated from *Valeriana officinalis* Roots. Dokl. Bolg. Akad. Nauk. **26**, 913 (1973); Chem. Abstr. **80**, 6820 (1974).

742. Haines, T.H., M. Pousada, B. Stern, and G.L. Mayers: Microbial Sulpholipids: (*R*)-13-Chloro-1-(*R*)-14-docosanediol Disulphate and Polychlorosulpholipids in *Ochromonas danica*. Biochem. J. **113**, 565 (1969).

743. Elovson, J., and P.R. Vagelos: A New Class of Lipids: Chlorosulfolipids. Proc. Natl. Acad. Sci. **62**, 957 (1969).

744. Elovson, J., and P.R. Vagelos: Structure of the Major Species of Chlorosulfolipid from *Ochromonas danica*. 2,2,11,13,15,16-Hexachloro-*N*-docosane 1,14-Disulfate. Biochem. **9**, 3110 (1970).

745. Haines, T.H.: Progress in the Chemistry of Fats and Other Lipids, Vol. XI (R.T.M. Holman, ed.), pp. 299–345. New York: Pergamon Press. 1971.

746. Haines, T.H.: Halogen- and Sulfur-Containing Lipids of *Ochromonas*. Ann. Rev. Microbiol. **27**, 403 (1973).

747. Mooney, C.L., and T.H. Haines: Chlorination and Sulfation Reactions in the Biosynthesis of Chlorosulfolipids in *Ochromonas danica, in vivo*. Biochem. **12**, 4469 (1973).

748. Elovson, J.: Biosynthesis of Chlorosulfolipids in *Ochromonas danica*. Assembly of the Docosane-1,14-diol Structure *in vivo*. Biochem. **13**, 3483 (1974).

749. Mercer, E.I., and C.L. Davies: Chlorosulpholipids of *Tribonema aequale*. Phytochem. **13**, 1607 (1974).

750. Mercer, E.I., and C.L. Davies: Distribution of Chlorosulpholipids in Algae. Phytochem. **18**, 457 (1979).

751. Chen, J.L., P.J. Proteau, M.A. Roberts, W.H. Gerwick, D.L. Slate, and R.H. Lee: Structure of Malhamensilipin A, an Inhibitor of Protein Tyrosine Kinase, from the Cultured Chrysophyte *Poterioochromonas malhamensis*. J. Nat. Prod. **57**, 524 (1994).

752. Stepanichenko, N.N., A.A. Tyshchenko, S.D. Gusakova, N.Sh. Navrezova, R. Khamidova, S.Z. Mukhamedzhanov, A.U. Umarov, and O.S. Otroshchenko: Metabolites of the Pathogenic Fungus *Verticillium dahliae*, V: 9,10-Dichlorostearic Acid – A Minor Component of the Lipid Fraction. Khim. Prir. Soedin. **13**, 627 (1977).

753. White, R.H., and L.P. Hager: Occurrence of Fatty Acid Chlorohydrins in Jellyfish Lipids. Biochem. **16**, 4944 (1977).

754. Tinsley, I.J., and R.R. Lowry: Bromine Content of Lipids of Marine Organisms. J. Amer. Oil Chem. Soc. **57**, 31 (1980).

755. WESÉN, C., H. MU, A. LUND KVERNHEIM, P. LARSSON: Identification of Chlorinated Fatty Acids in Fish Lipids by Partitioning Studies and by Gas Chromatography with Hall Electrolytic Conductivity Detection. J. Chromatog. **625**, 257 (1992).

756. SUNDIN, P., C. WESÉN, H. MU, and G. ODHAM: Are Chlorinated Fatty Acids in Fish Lipids of Natural Origin? International Conference on Naturally-Produced Organohalogens, Delft, The Netherlands. September, 1993.

757. SONG, W.-C., D.L. HOLLAND, K.H. GIBSON, E. CLAYTON, and A. OLDFIELD: Identification of Novel Hydroxy Fatty Acids in the Barnacle *Balanus balanoides*. Biochim. Biophys. Acta **1047**, 239 (1990).

758. GUSAKOVA, S.D., and A.U. UMAROV: Dibromo- and Tetrabromostearic Acids in the Seed Oil of *Eremostachys molucelloides*. Khim. Prir. Soedin. **12**, 717 (1976).

759. SCHMITZ, F.J., and Y. GOPICHAND: (7*E*, 13?, 15*Z*)-14,16-Dibromo-7,13,15-hexadecatrien-5-ynoic Acid. A Novel Dibromo Acetylenic Acid from the Marine Sponge *Xestospongia muta*. Tetrahedron Lett., 3637 (1978).

760. QUINN, R.J., and D.J. TUCKER: A Brominated Bisacetylenic Acid from the Marine Sponge *Xestospongia testudinaria*. Tetrahedron Lett. **26**, 1671 (1985).

761. HIRSCH, S., S. CARMELY, and Y. KASHMAN: Brominated Unsaturated Acids from the Marine Sponge *Xestospongia* sp. Tetrahedron **43**, 3257 (1987).

762. WIJEKOON, W.M.D., E. AYANOGLU, and C. DJERASSI: Phospholipid Studies of Marine Organisms, 9: New Brominated Demospongic Acids from the Phospholipids of Two *Petrosia* Species. Tetrahedron Lett. **25**, 3285 (1984).

763. CARBALLEIRA, N.M., and F. SHALABI: Novel Brominated Phospholipid Fatty Acids from the Caribbean Sponge *Petrosia* sp. J. Nat. Prod. **56**, 739 (1993).

764. LAM, W., S. HAHN, E. AYANOGLU, and C. DJERASSI: Phospholipid Studies of Marine Organisms, 22: Structure and Biosynthesis of a Novel Brominated Fatty Acid from a Hymeniacidonid Sponge. J. Org. Chem. **54**, 3428 (1989).

765. OHTA, S., H. OKADA, H. KOBAYASHI, J.M. OCLARIT, and S. IKEGAMI: Clathrynamides A, B, and C: Novel Amides from a Marine Sponge *Clathria* sp. that Inhibit Cell Division of Fertilized Starfish Eggs. Tetrahedron Lett. **34**, 5935 (1993).

766. CARDELLINA II, J.H., F.-J. MARNER, and R.E. MOORE: Malyngamide A, a Novel Chlorinated Metabolite of the Marine Cyanophyte *Lyngbya majuscula*. J. Am. Chem. Soc. **101**, 240 (1979).

767. AINSLIE, R.D., J.J. BARCHI JR., M. KUNIYOSHI, R.E. MOORE, and J.S. MYNDERSE: Structure of Malyngamide C. J. Org. Chem. **50**, 2859 (1985).

768. GERWICK, W.H., S. REYES, and B. ALVARADO: Two Malyngamides from the Caribbean Cyanobacterium *Lyngbya majuscula*. Phytochem. **26**, 1701 (1987).

769. PRAUD, A., R. VALLS, L. PIOVETTI, and B. BANAIGS: Malyngamide G: Proposition de Structure pour un Nouvel Amide Chloré d'une Algue Bleuverte Epiphyte de *Cystoseira crinita*. Tetrahedron Lett. **34**, 5437 (1993).

770. HODDER, A.R., and R.J. CAPON: A Novel Brominated Lipid from an Australian Cyanobacterium, *Lyngbya* sp. J. Nat. Prod. **54**, 1668 (1991).

771. MATSUNAGA, S., N. FUSETANI, Y. KATO, and H. HIROTA: Aurantosides A and B: Cytotoxic Tetramic Acid Glycosides from the Marine Sponge *Theonella* sp. J. Am. Chem. Soc. **113**, 9690 (1991).

772. WATANABE, T., T. SUGIYAMA, M. TAKAHASHI, J. SHIMA, K. YAMASHITA, K. IZAKI, K. FURIHATA, and H. SETO: The Structure of Enacyloxin II, a Novel Linear Polyenic Antibiotic Produced by *Gluconobacter* Sp. W-315. Agric. Biol. Chem. **54**, 259 (1990).

773. RABINOWITZ, J.L., M. ZANGER, and V. PODOLSKI: Identification by Nuclear Magnetic Resonance of Iodinated Lipids in the Dog Thyroid. Biochem. Biophys. Res. Commun. **68**, 1161 (1976).

774. Boeynaems, J.-M., and W.C. Hubbard: Transformation of Arachidonic Acid into an Iodolactone by the Rat Thyroid. J. Biol. Chem. **255**, 9001 (1980).
775. Boeynaems, J.M., J.T. Watson, J.A. Oates, and W.C. Hubbard: Iodination of Docosahexaenoic Acid by Lactoperoxidase and Thyroid Gland *in vitro*: Formation of an Iodolactone. Lipids **16**, 323 (1981).
776. Gribble, G.W.: Fluoroacetate Toxicity. J. Chem. Ed. **50**, 460 (1973).
777. Meyer, M., and D. O'Hagan: Rare Fluorinated Natural Products. Chem. Brit., 785 (1992).
778. Marais, J.S.C.: Monofluoroacetic Acid, the Toxic Principle of "gifblaar" *Dichapetalum cymosum* (Hook) Engl. Onderstepoort J. Vet. Sci. **20**, 67 (1944).
779. Hall, R.J., and R.B. Cain: Organic Fluorine in Tropical Soils. New Phytol. **71**, 839 (1972).
780. Vickery, B., and M.L. Vickery: Fluoride Metabolism in *Dichapetalum toxicarium*. Phytochem. **11**, 1905 (1972).
781. Vickery, B., M.L. Vickery, and J.T. Ashu: Analysis of Plants for Fluoroacetic Acids. Phytochem. **12**, 145 (1973).
782. O'Hagan, D., R. Perry, J.M. Lock, J.J.M. Meyer, L. Dasaradhi, J.T.G. Hamilton, and D.B. Harper: High Levels of Monofluoroacetate in *Dichapetalum braunii*. Phytochem. **33**, 1043 (1993).
783. Nwude, N., L.E. Parsons, and A.O. Adaudi: Acute Toxicity of the Leaves and Extracts of *Dichapetalum barteri* (Engl.) in Mice, Rabbits and Goats. Toxicol. **7**, 23 (1977).
784. Saunders, B.C.: Some Aspects of the Chemistry and Toxic Action of Organic Compounds Containing Phosphorus and Fluorine, pp. 114–170. Cambridge: University Press. 1957.
785. Meyer, J.J.M., and N. Grobbelaar: The Determination, Uptake and Transport of Fluoroacetate in *Dichapetalum cymosum*. J. Plant Physiol. **135**, 546 (1990).
786. Baron, M.L., C.M. Bothroyd, G.I. Rogers, A. Staffa, and I.D. Rae: Detection and Measurement of Fluoroacetate in Plant Extracts by ^{19}F NMR. Phytochem. **26**, 2293 (1987).
787. Murray, L.R., J.D. McConnell, and J.H. Whittem: Suspected Presence of Fluoroacetate in *Acacia georginae*. Aust. J. Sci. **24**, 41 (1961).
788. Oelrichs, P.B., and T. McEwan: Isolation of the Toxic Principle in *Acacia georginae*. Nature **190**, 808 (1961).
789. Peters, R.A., and M. Shorthouse: Fluorocitrate in Plants and Food Stuffs. Phytochem. **11**, 1337 (1972).
790. Bennett, L.W., G.W. Miller, M.H. Yu, and R.I. Lynn: Production of Fluoroacetate by Callus Tissue from Leaves of *Acacia georginae*. Fluoride **16**, 111 (1983).
791. De Oliveira, M.M.: Chromatographic Isolation of Monofluoroacetic Acid from *Palicourea marcgravii* St. Hil. Experientia **19**, 586 (1963).
792. McEwan, T.: Isolation and Identification of the Toxic Principle of *Gastrolobium grandiflorum*. Nature **201**, 827 (1964).
793. Hall, R.J.: The Distribution of Organic Fluorine in Some Toxic Tropical Plants. New Phytol. **71**, 855 (1972).
794. Kamgue, R.T., O. Sylla, J.L. Pousset, A. Laurens, J.C. Brunet, and A. Sere: Isolation and Characterization of Toxic Principles from *Spondianthus preussii* var. *glaber* Engler. Plant. Med. Phytother. **13**, 252 (1979).
795. Vartiainen, T., and J. Gynther: Fluoroacetic Acid in Guar Gum. Food Chem. Toxicol. **22**, 307 (1984).

796. Vickery, B., M.L. Vickery, and F. Kaberia: A Possible Biomimetic Synthesis of Fluoroacetic Acid. Experientia **35**, 299 (1979).

797. Mead, R.J., A.J. Oliver, and D.R. King: Metabolism and Defluorination of Fluoroacetate in the Brush-Tailed Possum (*Trichosurus vulpecula*). Aust. J. Biol. Sci. **32**, 15 (1979).

798. Peters, R.A., and M. Shorthouse: Formation of Monofluorocarbon Compounds by Single Cell Cultures of *Glycine max* Growing on Inorganic Fluoride. Phytochem. **11**, 1339 (1972).

799. Peters, R.A., R.W. Wakelin, A.J.P. Martin, J. Webb, and F.T. Birks: Observations Upon the Toxic Principle in the Seeds of *Dichapetalum toxicarium*. Biochem. J. **71**, 245 (1959).

800. Peters, R.A., and R.J. Hall: Further Observations Upon the Toxic Principle of *Dichapetalum toxicarium*. Biochem. Pharmacol. **2**, 25 (1959).

801. Peters, R.A., R.J. Hall, F.F.V. Ward, and N. Sheppard: The Chemical Nature of the Toxic Compounds Containing Fluorine in the Seeds of *Dichapetalum toxicarium*. Biochem. J. **77**, 17 (1960).

802. Peters, R., and R.J. Hall: Fluorine Compounds in Nature. The Distribution of Carbon-Fluorine Compounds in Some Species of *Dichapetalum*. Nature **187**, 573 (1960).

803. Ward, P.F.V., R.J. Hall, and R.A. Peters: Fluoro-Fatty Acids in the Seeds of *Dichapetalum toxicarium*. Nature **201**, 611 (1964).

804. Harper, D.B., J.T.G. Hamilton, and D. O'Hagan: Identification of *Threo*-18-Fluoro-9,10-dihydroxystearic Acid: A Novel ω-Fluorinated Fatty Acid from *Dichapetalum toxicarium*. Tetrahedron Lett. **31**, 7661 (1990).

805. For a color photo of this animal, see the cover of the November 1994 issue of the *Journal of Chemical Education*.

806. Baker, B.J., R.K. Okuda, P.T.K. Yu, and P.J. Scheuer: Punaglandins: Halogenated Antitumor Eicosanoids from the Octocoral *Telesto riisei*. J. Am. Chem. Soc. **107**, 2976 (1985).

807. Nagaoka, H., H. Miyaoka, T. Miyakoshi, and Y. Yamada: Synthesis of Punaglandin 3 and 4. Revision of the Structures. J. Am. Chem. Soc. **108**, 5019 (1986).

808. Suzuki, M., Y. Morita, A. Yanagisawa, R. Noyori, B.J. Baker, and P.J. Scheuer: Synthesis of (7*E*)- and (7*Z*)-Punaglandin 4. Structural Revision. J. Am. Chem. Soc. **108**, 5021 (1986).

809. Suzuki, M., Y. Morita, A. Yanagisawa, B.J. Baker, P.J. Scheuer, and R. Noyori: Synthesis and Structural Revision of (7*E*)- and (7*Z*)-Punaglandin 4. J. Org. Chem. **53**, 286 (1988).

810. Iguchi, K., S. Kaneta, K. Mori, Y. Yamada, A. Honda, and Y. Mori: Chlorovulones, New Halogenated Marine Prostanoids with an Antitumor Activity from the Stolonifer *Clavularia viridis* Quoy and Gaimard. Tetrahedron Lett. **26**, 5787 (1985).

811. Nagaoka, H., K. Iguchi, T. Miyakishi, N. Yamada, and Y. Yamada: Determination of Absolute Configuration of Chlorovulones by CD Measurement and by Enantioselective Synthesis of (−)-Chlorovulone II. Tetrahedron Lett. **27**, 223 (1986).

812. Iguchi, K., S. Kaneta, K. Mori, Y. Yamada, A. Honda, and Y. Mori: Bromovulone I and Iodovulone I, Unprecedented Brominated and Iodinated Marine Prostanoids with Antitumor Activity Isolated from the Japanese Stolonifer *Clavularia viridis* Quoy and Gaimard. J. Chem. Soc., Chem. Commun., 981 (1986).

813. Iguchi, K., S. Kaneta, K. Mori, and Y. Yamada: A New Marine Epoxy Prostanoid with an Antiproliferative Activity from the Stolonifer *Clavularia viridis* Quoy and Gaimard. Chem. Pharm. Bull. (Japan) **35**, 4375 (1987).

814. FUKUSHIMA, M., T. KATO, Y. YAMADA, I. KITAGAWA, S. KUROZUMI, and P.J. SCHEUER: Inhibition of Tumor Growth by Novel Marine Eicosanoids, Clavulones and Punaglandins. Proc. Am. Assoc. Cancer Res. **26**, 980 (1985).

815. FUKUSHIMA, M., and T. KATO: Antitumor Marine Icosanoids: Clavulones and Punaglandins. Adv. in Prostaglandin, Thromboxane, and Leukotriene Research **15**, 415 (1985).

816. NAGAOKA, H., T. MIYAKOSHI, J. KASUGA, and Y. YAMADA: Synthesis of a Halogenated Clavulone Analog. Tetrahedron Lett. **26**, 5053 (1985).

817. TODD, J.S., P.J. PROTEAU, and W.H. GERWICK: Egregiachlorides A–C: New Chlorinated Oxylipins from the Marine Brown Alga *Egregia menziesii*. Tetrahedron Lett. **34**, 7689 (1993).

818. KAZLAUSKAS, R., P.T. MURPHY, R.J. QUINN, and R.J. WELLS: A New Class of Halogenated Lactones from the Red Alga *Delisea fimbriata* (Bonnemaisoniaceae). Tetrahedron Lett., 37 (1977).

819. PETTUS JR., J.A., R.M. WING, and J.J. SIMS: Marine Natural Products, XII: Isolation of a Family of Multihalogenated Gamma-Methylene Lactones from the Red Seaweed *Delisea fimbriata*. Tetrahedron Lett., 41 (1977).

820. OHTA, K.: Antimicrobial Compounds in the Marine Red Alga *Beckerella subcostatum*. Agric. Biol. Chem. **41**, 2105 (1977).

821. MCCOMBS, J.D., J.W. BLUNT, M.V. CHAMBERS, M.H.G. MUNRO, and W.T. ROBINSON: Novel 2(5*H*)-Furanones from the Red Marine Alga *Delisea elegans* (Lamouroux). Tetrahedron **44**, 1489 (1988).

822. DE NYS, R., A.D. WRIGHT, G.M. KÖNIG, and O. STICHER: New Halogenated Furanones from the Marine Alga *Delisea pulchra* (cf. *fimbriata*). Tetrahedron **49**, 11213 (1993).

823. MIAO, S., and R.J. ANDERSON: Rubrolides A–H, Metabolites of the Colonial Tunicate *Ritterella rubra*. J. Org. Chem. **56**, 6275 (1991).

824. KRONBERG, L., and R. FRANZÉN: Determination of Chlorinated Furanones, Hydroxy-Furanones, and Butenedioic Acids in Chlorine-Treated Water and in Pulp Bleaching Liquor. Environ. Sci. Technol. **27**, 1811 (1993).

825. GOTTLIEB, D., P.K. BHATTACHARYYA, H.W. ANDERSON, and H.E. CARTER: Some Properties of an Antibiotic Obtained from a Species of *Streptomyces*. J. Bacteriol. **55**, 409 (1948).

826. BARTZ, Q.R.: Isolation and Characterization of Chloromycetin. J. Biol. Chem. **172**, 445 (1948).

827. MALIK, V.S.: Chloramphenicol. Adv. Appl. Microbiol. **15**, 297 (1972).

828. JONES, A., and L.C. VINING: Biosynthesis of Chloramphenicol in *Streptomyces* sp. 3022a. Identification of *p*-Amino-L-phenylalanine as a Product from the Action of Arylamine Synthetase on Chorismic Acid. Can. J. Microbiol. **22**, 237 (1976).

829. SIMONSEN, J.N., K. PARAMASIGAMANI, L.C. VINING, A.G. MCINNES, J.A. WALTER, and J.L.C. WRIGHT: Biosynthesis of Chloramphenicol. Studies on the Origin of the Dichloroacetyl Moiety. Can. J. Microbiol. **24**, 136 (1978).

830. REBSTOCK, M.C., H.M. CROOKS JR., J. CONTROULIS, and Q.R. BARTZ: Chloramphenicol (Chloromycetin), IV: Chemical Studies. J. Am. Chem. Soc. **71**, 2458 (1949).

831. CONTROULIS, J., M.C. REBSTOCK, and H.M. CROOKS JR.: Chloramphenicol (Chloromycetin), V: Synthesis. J. Am. Chem. Soc. **71**, 2463 (1949).

832. STRATTON, C.D., and M.C. REBSTOCK: A New Metabolite of *Streptomyces venezuelae*: D-*Threo*-1-*p*-aminophenyl-2-dichloroacetamido-1,3-propanediol. Arch. Biochem. Biophys. **103**, 159 (1963).

833. SMITH, C.G.: Effect of Halogens on the Chloramphenicol Fermentation. J. Bacteriol. **75**, 577 (1958).

834. KONDO, S., Y. HORIUCHI, M. HAMADA, T. TAKEUCHI, and H. UMEZAWA: A New Antitumor Antibiotic, Bactobolin Produced by *Pseudomonas*. J. Antibiot. **32**, 1069 (1979).

835. UEDA, I., T. MUNAKATA, and J. SAKAI: Structure of 2-Amino-*N*-(3-dichloromethyl-3,4,4a,5,6,7-hexahydro-5,6,8-trihydroxy-3-methyl-1-oxo-1*H*-2-benzopyran-4-yl)-propanamide Hydrobromide Dihydrate. Acta Cryst. **B36**, 3128 (1980).

836. EZAKI, N., S. MIYADOH, T. HISAMATSU, T. KASAI, and Y. YAMADA: BN-183B, a New Antitumor Antibiotic Produced by *Pseudomonas*. Taxonomy, Isolation, Physico-Chemical and Biological Properties. J. Antibiot. **33**, 213 (1980).

837. MUNAKATA, T.: Bactobolins, Antitumor Antibiotics from Pseudomonas, 2: Synthesis and Antimicrobial Activities of Related Compounds. Yakugaku Zasshi **101**, 138 (1981); Chem. Abstr. **95**, 42826 (1981).

838. ARGOUDELIS, A.D., R.R. HERR, D.J. MASON, T.R. PYKE, and J.F. ZIESERL: New Amino Acids from *Streptomyces*. Biochem. **6**, 165 (1967).

839. NARAYANAN, S., M.R.S. IYENGAR, P.L. GANJU, S. RENGARAJU, T. SHOMURA, T. TSURUOKA, S. INOUYE, and T. NIIDA: γ-Chloronorvaline, a Leucine Analog from *Streptomyces*. J. Antibiot. **33**, 1249 (1980).

840. CHAIET, L., B.H. ARISON, R.L. MONAGHAN, J.P. SPRINGER, J.L. SMITH, and S.B. ZIMMERMAN: *R*-(*Z*)-4-Amino-3-chloro-2-pentenedioic Acid, a New Antibiotic. Fermentation, Isolation, and Characterization. J. Antibiot. **37**, 207 (1984).

841. KURODA, Y., M. OKUHARA, T. GOTO, M. YAMASHITA, E. IGUCHI, M. KOHSAKA, H. AOKI, and H. IMANAKA: J. Antibiot. **33**, 259 (1980).

842. KURODA, Y., M. OKUHARA, T. GOTO, M. OKAMOTO, M. YAMASHITA, M. KOHSAKA, H. AOKI, and H. IMANAKA: FR-900148, a New Antibiotic, II: Structure Determination of FR-900148. J. Antibiot. **33**, 267 (1980).

843. YASUDA, N., and K. SAKANE: Revised Structure and the Chemical Transformations of FR900148. J. Antibiot. **44**, 801 (1991).

844. MARTIN, D.G., C.G. CHIDESTER, S.A. MIZSAK, D.J. DUCHAMP, L. BACZYNSKYJ, W.C. KRUEGER, R.J. WNUK, and P.A. MEULMAN: The Isolation, Structure, and Absolute Configuration of U-43,795, a New Antitumor Agent. J. Antibiot. **28**, 91 (1975).

845. FORTIER, G., and S.L. MACKENZIE: Identification of a New, Naturally Occurring Non-Proteic Amino Acid in Xylem Sap of *Pisum sativum*. Biochem. Biophys. Res. Commun. **139**, 383 (1986).

846. WILD, H.: Enantioselective Total Synthesis of the Antifungal Natural Products Chlorotetaine, Bacilysin, and Anticapsin and of Related Compounds: Revision of the Relative Configuration. J. Org. Chem. **59**, 2748 (1994).

847. SANADA, M., T. MIYANO, S. IWADARE, J.M. WILLIAMSON, B.H. ARISON, J.L. SMITH, A.W. DOUGLAS, J.M. LIESCH, and E. INAMINE: Biosynthesis of Fluorothreonine and Fluoroacetic Acid by the Thienamycin Producer, *Streptomyces cattleya*. J. Antibiot. **39**, 259 (1986).

848. HATANAKA, S.: Amino Acids from Mushrooms. Progr. Chem. Org. Nat. Prod. **59**, 1 (1992).

849. CHILTON, W.S., and G. TSOU: A Chloro Amino Acid from *Amanita solitaria*. Phytochem. **11**, 2853 (1972).

850. HATANAKA, S.-I., S. KANEKO, Y. NIIMURA, F. KINOSHITA, and G. SOMA: L-2-Amino-4-chloro-4-pentenoic Acid, a New Natural Amino Acid from *Amanita pseudoporphyria* Hongo. Tetrahedron Lett., 3931 (1974).

851. MORIGUCHI, M., Y. HARA, and S. HATANAKA: Antibacterial Activity of L-2-Amino-4-chloro-4-pentenoic Acid Isolated from *Amanita pseudoporphyria* Hongo. J. Antibiot. **40**, 904 (1987).

852. OGISHI, H.: 2-Amino-5-chloro-5-hexenoic Acid. Chem. Abstr. **88**, 149206 (1978).

853. OHTA, T., S. NAKAJIMA, S. HATANAKA, M. YAMAMOTO, Y. SHIMMEN, S. NIEHIMURA, Z. YAMAIZUMI, and S. NOZOE: A Chlorohydrin Amino Acid from *Amanita abrupta.* Phytochem. **26**, 565 (1987).

854. HOFHEINZ, W., and W.E. OBERHÄNSLI: Dysidin, ein neuartiger, chlorhaltiger Naturstoff aus dem Schwamm *Dysidea herbacea.* Helv. Chim. Acta **60**, 660 (1977).

855. KAZLAUSKAS, R., R.O. LIDGARD, R.J. WELLS, and W. VETTER: A Novel Hexachloro-Metabolite from the Sponge *Dysidea herbacea.* Tetrahedron Lett., 3183 (1977).

856. CHARLES, C., J.C. BRAEKMAN, D. DALOZE, B. TURSCH, and R. KARLSSON: Chemical Studies of Marine Invertebrates, XXXII: Isodysidenin, a Further Hexachlorinated Metabolite from the Sponge *Dysidea herbacea.* Tetrahedron Lett., 1519 (1978).

857. CHARLES, C., J.C. BRAEKMAN, D. DALOZE, and B. TURSCH: Chemical Studies of Marine Invertebrates, XLII: The Relative and Absolute Configuration of Dysidenin. Tetrahedron **36**, 2133 (1980).

858. KÖHLER, H., and H. GERLACH: Synthese des Dysidins. Helv. Chim. Acta **67**, 1783 (1984).

859. BISKUPIAK, J.E., and C.M. IRELAND: Revised Absolute Configuration of Dysidenin and Isodysidenin. Tetrahedron Lett. **25**, 2935 (1984).

860. DE LASZLO, S.E., and P.G. WILLIARD: Total Synthesis of (+)-Demethyldysidenin and (–)-Demethylisodysidenin, Hexachlorinated Amino Acids from the Marine Sponge *Dysidea herbacea.* Assignment of Absolute Stereochemistry. J. Am. Chem. Soc. **107**, 199 (1985).

861. ERICKSON, K.L., and R.J. WELLS: New Polychlorinated Metabolites from a Barrier Reef Collection of the Sponge *Dysidea herbacea.* Aust. J. Chem. **35**, 31 (1982).

862. KAZLAUSKAS, R., P.T. MURPHY, and R.J. WELLS: A Diketopiperazine Derived from Trichloroleucine from the Sponge *Dysidea herbacea.* Tetrahedron Lett., 4945 (1978).

863. UNSON, M.D., and D.J. FAULKNER: Cyanobacterial Symbiont Biosynthesis of Chlorinated Metabolites from *Dysidea herbacea* (Porifera). Experientia **49**, 349 (1993).

864. SU, J.-Y., Y.-L. ZHONG, L.-M. ZENG, S. WEI, Q.-W. WANG, T.C.W. MAK, and Z.-Y. ZHOU: Three New Diketopiperazines from a Marine Sponge *Dysidea Fragilis.* J. Nat. Prod. **56**, 637 (1993).

865. GEBREYESUS, T., T. YOSIEF, S. CARMELY, and Y. KASHMAN: Dysidamide, a Novel Hexachloro-Metabolite from a Red Sea Sponge *Dysidea* sp. Tetrahedron Lett. **29**, 3863 (1988).

866. CARMELY, S., T. GEBREYESUS, Y. KASHMAN, B.W. SKELTON, A.H. WHITE, and T. YOSIEF: Dysidamide, a Novel Metabolite from a Red Sea Sponge *Dysidea herbacea.* Aust. J. Chem. **43**, 1881 (1990).

867. ISAACS, S., R. BERMAN, Y. KASHMAN, T. GEBREYESUS, and T. YOSIEF: New Polyhydroxy Sterols, Dysidamides, and a Dideoxyhexose from the Sponge *Dysidea herbacea.* J. Nat. Prod. **54**, 83 (1991).

868. LEE, G.M., and T.F. MOLINSKI: Herbaceamide, a Chlorinated *N*-Acyl Amino Ester from the Marine Sponge, *Dysidea herbacea.* Tetrahedron Lett. **33**, 7671 (1992).

869. UNSON, M.D., C.B. ROSE, D.J. FAULKNER, L.S. BRINEN, J.R. STEINER, and J. CLARDY: New Polychlorinated Amino Acid Derivatives from the Marine Sponge *Dysidea herbacea.* J. Org. Chem. **58**, 6336 (1993).

870. RAPP, C., G. JUNG, W. KATZER, and W. LOEFFLER: Chlorotetain from *Bacillus subtilis*, an Antifungal Dipeptide with an Unusual Chlorine-Containing Amino Acid. Angew. Chem. Int. Ed. Engl. **27**, 1733 (1988).

871. MARUMO, S., and Y. SUMIKI: Islanditoxin, a Toxic Metabolite Produced by *Penicillin islandicum.* J. Agric. Chem. Soc. Japan **29**, 305 (1955).

872. MARUMO, S.: Islanditoxin, a Toxic Metabolite Produced by *Pencillium islandicum Sopp.*, Part I. Bull. Agric. Chem. Soc. **19**, 258 (1955).

873. MARUMO, S., K. MIYAO, and A. MATSUYAMA: Islanditoxin, a Toxic Metabolite Produced by *Pencillium islandicum Sopp.*, Part II: Acid Hydrolysis of Islanditoxin. Bull. Agric. Chem. Soc. **19**, 262 (1955).

874. MARUMO, S.: Islanditoxin, a Toxic Metabolite Produced by *Penicillium islandicum*, III: Structure of Islanditoxin. Bull. Agric. Chem. Soc. Japan **23**, 428 (1959).

875. URAGUCHI, K., M. SAITO, Y. NOGUCHI, K. TAKAHASHI, M. ENOMOTO, and T. TATSUNO: Chronic Toxicity and Carcinogenicity in Mice of the Purified Mycotoxins, Luteoskyrin and Cyclochlorotine. Food Cosmet. Toxicol. **10**, 193 (1972).

876. YOSHIOKA, H., K. NAKATSU, M. SATO, and T. TATSUNO: Molecular Structure of Cyclochlorotine, a Toxic Chlorine-Containing Cyclic Pentapeptide. Chem. Lett., 1319 (1973).

877. KOSEMURA, S., T. OGAWA, and K. TOTSUKA: Isolation and Structure of Asterin, a New Halogenated Cyclic Penta-Peptide from *Aster tataricus*. Tetrahedron Lett. **34**, 1291 (1993).

878. MORITA, H., S. NAGASHIMA, K. TAKEYA, and H. ITOKAWA: Astins A and B, Antitumor Cyclic Pentapeptides from *Aster tataricus*. Chem. Pharm. Bull. (Japan) **41**, 992 (1993).

879. MORITA, H., S. NAGASHIMA, O. SHIROTA, K. TAKEYA, and H. ITOKAWA: Two Novel Monochlorinated Cyclic Pentapeptides, Astins D and E from *Aster tataricus*. Chem. Lett., 1877 (1993).

880. SCANNELL, J.P., D.L. PRUESS, J.F. BLOUNT, H.A. AX, M. KELLETT, F. WEISS, T.C. DEMNY, T.H. WILLIAMS, and A. STEMPEL: Antimetabolites Produced by Microorganisms, XII: (*S*)-Alanyl-3-[α-(*S*)-chloro-3-(*S*)-hydroxy-2-oxo-4-azetidinylmethyl]-(*S*)-alanine, a New β-Lactam Containing Natural Product. J. Antibiot. **28**, 1 (1975).

881. NISBET, L.J., R.J. MEHTA, Y. OH, C.H. PAN, C.G. PHELEN, M.J. POLANSKY, M.C. SHEARER, A.J. GIOVENELLA, and S.F. GRAPPEL: Chlorocardicin, a Monocyclic β-Lactam from a *Streptomyces* sp., I. J. Antibiot. **38**, 133 (1985).

882. CHAN, J.A., E.A. SHULTIS, J.J. DINGERDISSEN, C.W. DE BROSSE, G.D. ROBERTS, and K.M. SNADER: Chlorocardicin; a Monocyclic β-Lactam from a *Streptomyces* sp., II. J. Antibiot. **38**, 139 (1985).

883. OKUMURA, Y., R. OKAMOTO, and T. ISHIKURA: Neoviridogrisein-cls, New Analogs of Viridogrisein, Produced Under the Supplementation of 4-Chloro-L-proline. Agric. Biol. Chem. **48**, 543 (1984).

884. GUPTA, S., D.W. ROBERTS, and J.A.A. RENWICK: Insecticidal Cyclodepsipeptides from *Metarhizium anisopliae*. J. Chem. Soc., Perkin Trans. 1, 2347 (1989).

885. FUSETANI, N., and S. MATSUNAGA: Bioactive Sponge Peptides. Chem. Rev. **93**, 1793 (1993).

886. ANDERSEN, R.J., and R.J. STONARD: Clionamide, a Major Metabolite of the Sponge *Cliona celata* Grant. Can. J. Chem. **57**, 2325 (1979).

887. STONARD, R. J., and R.J. ANDERSEN: Celenamides A and B, Linear Peptide Alkaloids from the Sponge *Cliona celata*. J. Org. Chem. **45**, 3687 (1980).

888. STONARD, R.J., and R.J. ANDERSEN: Linear Peptide Alkaloids from the Sponge *Cliona celata* Grant. Celenamides C and D. Can. J. Chem. **58**, 2121 (1980).

889. AZUMI, K., H. YOKOSAWA, and S. ISHII: Halocyamines: Novel Antimicrobial Tetrapeptide-Like Substances Isolated from the Hemocytes of the Solitary Ascidian *Halocynthia roretzi*. Biochem. **29**, 159 (1990).

890. ZABRISKIE, T.M., J.A. KLOCKE, C.M. IRELAND, A.H. MARCUS, T.F. MOLINSKI, D.J. FAULKNER, C. XU, and J.C. CLARDY: Jaspamide, a Modified Peptide from

a *Jaspis* Sponge, with Insecticidal and Antifungal Activity. J. Am. Chem. Soc. **108**, 3123 (1986).

891. CREWS, P., L.V. MANES, and M. BOEHLER: Jasplakinolide, a Cyclodepsipeptide from the Marine Sponge, *Jaspis* sp. Tetrahedron Lett. **27**, 2797 (1986).

892. BRAEKMAN, J.C., D. DALOZE, B. MOUSSIAUX, and R. RICCIO: Jaspamide from the Marine Sponge *Jaspis johnstoni.* J. Nat. Prod. **50**, 994 (1987).

893. INMAN, W., and P. CREWS: Novel Marine Sponge Derived Amino Acids, 8: Conformational Analysis of Jasplakinolide. J. Am. Chem. Soc. **111**, 2822 (1989).

894. INMAN, W., P. CREWS, and R. MCDOWELL: Novel Marine Sponge Derived Amino Acids, 9: Lithium Complexation of Jasplakinolide. J. Org. Chem. **54**, 2523 (1989).

895. FUSETANI, N., T. SUGAWARA, S. MATSUNAGA, and H. HIROTA: Orbiculamide A: A Novel Cytotoxic Cyclic Peptide from a Marine Sponge *Theonella* sp. J. Am. Chem. Soc. **113**, 7811 (1991).

896. KOBAYASHI, J., F. ITAGAKI, H. SHIGEMORI, M. ISHIBASHI, K. TAKAHASHI, M. OGURA, S. NAGASAWA, T. NAKAMURA, H. HIROTA, T. OHTA, and S. NOZOE: Keramamides B–D: Novel Peptides from the Okinawan Marine Sponge *Theonella* sp. J. Am. Chem. Soc. **113**, 7812 (1991).

897. KOBAYASHI, J., M. SATO, T. MURAYAMA, M. ISHIBASHI, M.R. WÄLCHI, M. KANAI, J. SHOJI, and Y. OHIZUMI: Konbamide, a Novel Peptide with Calmodulin Antagonist Activity from the Okinawan Marine Sponge *Theonella* sp. J. Chem. Soc., Chem. Commun., 1050 (1991).

898. KOBAYASHI, J., M. SATO, M. ISHIBASHI, H. SHIGEMORI, T. NAKAMURA, and Y. OHIZUMI: Keramamide A, a Novel Peptide from the Okinawan Marine Sponge *Theonella* sp. J. Chem. Soc., Perkin Trans. 1, 2609 (1991).

899. SHIBA, T., Y. MUKUNOKI, and H. AKIYAMA: Isolation and Characterization of a New Amino Acid, 5-Chloro-D-tryptophan from Antibiotic Longicatenamycin. Tetrahedron Lett., 3085 (1974).

900. GULAVITA, N.K., S.P. GUNASEKERA, S.A. POMPONI, and E.V. ROBINSON: Polydiscamide A: A New Bioactive Depsipeptide from the Marine Sponge *Discodermia* sp. J. Org. Chem. **57**, 1767 (1992).

901. LINDQUIST, N., W. FENICAL, G.D. VAN DUYNE, and J. CLARDY: Isolation and Structure Determination of Diazonamides A and B, Unusual Cytotoxic Metabolites from the Marine Ascidian *Diazona chinensis.* J. Am. Chem. Soc. **113**, 2303 (1991).

902. OMURA, S., D. VAN DER PYL, J. INOKOSHI, Y. TAKAHASHI, and H. TAKESHIMA: Pepticinnamins, New Farnesyl-Protein Transferase Inhibitors Produced by an Actinomycete, I: Producing Strain, Fermentation, Isolation and Biological Activity. J. Antibiot. **46**, 222 (1993).

903. SHIOMI, K., H. YANG, J. INOKOSHI, D. VAN DER PYL, A. NAKAGAWA, H. TAKESHIMA, and S. OMURA: Pepticinnamins, New Farnesyl-Protein Transferase Inhibitors Produced by an Actinomycete, II: Structural Elucidation of Pepticinnamin E. J. Antibiot. **46**, 229 (1993).

904. CHAN, W.R., W.F. TINTO, P.S. MANCHAND, and L.J. TODARO: Stereostructures of Geodiamolides A and B, Novel Cyclodepsipeptides from the Marine Sponge *Geodia* sp. J. Org. Chem. **52**, 3091 (1987).

905. DE SILVA, E.D., R.J. ANDERSEN, and T.M. ALLEN: Geodiamolides C to F, New Cytotoxic Cyclodepsipeptides from the Marine Sponge *Pseudaxinyssa* sp. Tetrahedron Lett. **31**, 489 (1990).

906. ZINK, D., O.D. HENSENS, Y.K.T. LAM, R. REAMER, and J.M. LIESCH: Cochinmicins, Novel and Potent Cyclodepsipeptide Endothelin Antagonists from a *Microbispora* sp. J. Antibiot. **45**, 1717 (1992).

907. Lam, Y.K.T., D.L. Zink, D.L. Williams Jr., and B.W. Burgess: Additional Cochinmicins from a *Microbispora* sp. J. Antibiot. **45**, 1792 (1992).
908. Trischman, J.A., D.M. Tapiolas, P.R. Jensen, R. Dwight, W. Fenical, T.C. McKee, C.M. Ireland, T.J. Stout, and J. Clardy: Salinamides A and B: Anti-Inflammatory Depsipeptides from a Marine Streptomycete. J. Am. Chem. Soc. **116**, 757 (1994).
909. Trimurtulu, G., I. Ohtani, G.M.L. Patterson, R.E. Moore, T.H. Corbett, F.A. Valeriote, and L. Demchik: Total Structures of Cryptophycins, Potent Antitumor Depsipeptides from the Blue-Green Alga *Nostoc* sp. Strain GSV 224. J. Am. Chem. Soc. **116**, 4729 (1994).
910. Matsunaga, S., N. Fusetani, K. Hashimoto, and M. Wälchli: Theonellamide F. A Novel Antifungal Bicyclic Peptide from a Marine Sponge *Theonella* sp. J. Am. Chem. Soc. **111**, 2582 (1989).
911. Moore, R.E., V. Bornemann, W.P. Niemczura, J.M. Gregson, J.-L. Chen, T.R. Norton, G.M.L. Patterson, and G.L. Helms: Puwainaphycin C, a Cardioactive Cyclic Peptide from the Blue-Green Alga *Anabaena* BQ-16-1. Use of Two-Dimensional ^{13}C-^{13}C and ^{13}C-^{15}N Correlation Spectroscopy in Sequencing the Amino Acid Units. J. Am. Chem. Soc. **111**, 6128 (1989).
912. Iwasaki, H., S. Horii, M. Asai, K. Mizumo, J. Ueyanagi, and A. Miyake: Enduracidin, a New Antibiotic, VIII: Structures of Enduracidins A and B. Chem. Pharm. Bull. (Japan) **21**, 1184 (1973).
913. Segre, A., R.C. Bachmann, A. Ballio, F. Bosoa, I. Grgurina, N.S. Iacobellis, G. Marino, P. Pucci, M. Simmaco, and J.Y. Takemoto: The Structures of Syringomycin A, E and G. FEBS Lett. **255**, 27 (1989).
914. Fukuchi, N., A. Isogai, S. Yamashita, K. Suyama, J.Y. Takemoto, and A. Suzuki: Structure of Phytotoxin Syringomycin Produced by a Sugar Cane Isolate of *Pseudomonas syringae* pv. *syringae*. Tetrahedron Lett. **31**, 1589 (1990).
915. Fukuchi, N., A. Isogai, J. Nakayama, S. Takayama, S. Yamashita, K. Suyama, J.Y. Takemoto, and A. Suzuki: Structure and Stereochemistry of Three Phytotoxins, Syringomycin, Syringotoxin, and Syringostatin, Produced by *Pseudomonas syringae* pv. *syringae*. J. Chem. Soc., Perkin Trans. 1, 1149 (1992).
916. Isogai, A., N. Fukuchi, S. Yamashita, K. Suyama, and A. Suzuki: Structures of Syringostatins A and B, Novel Phytotoxins Produced by *Pseudomonas syringae* pv. *syringae* Isolated from Lilac Blights. Tetrahedron Lett. **31**, 695 (1990).
917. Fukuchi, N., A. Isogai, J. Nakayama, S. Takayama, S. Yamashita, K. Suyama, and A. Suzuki: Isolation and Structural Elucidation of Syringostatins, Phytotoxins Produced by *Pseudomonas syringae* pv. *syringae* Lilac Isolate. J. Chem. Soc., Perkin Trans. 1, 875 (1992).
918. Barger, G., and J.J. Blackie: Alkaloids of Senecio, Part III: Jacobine, Jacodine, and Jaconine. J. Chem. Soc., 584 (1937).
919. Bradbury, R.B., and J.B. Willis: The Alkaloids of *Senecio jacobaea* L., II: The Structures of the Acids, and the Relationship Between Jacobine and Jaconine. Aust. J. Chem. **9**, 258 (1956).
920. Geissman, T.A.: The Alkaloids of *Senecio jacobaea* L.: The Structures of the Alkaloids and the Necic Acids. Aust. J. Chem. **12**, 247 (1959).
921. Bradbury, R.B., and S. Masamune: The Alkaloids of *Senecio jacobaea* L., IV: The Structures of Jacobine, Jaconine, Jacoline and Their Constituent Acids. J. Am. Chem. Soc. **81**, 5201 (1959).
922. Johnson, A.E., R.J. Molyneux, and L.D. Stuart: Toxicity of Riddell's Groundsel (*Senecio riddellii*) to Cattle. Am. J. Vet. Res. **46**, 577 (1985). We thank Dr. Molyneux for bringing this paper to our attention.

923. GORDON-GRAY, G.G.: The Senecio Alkaloids, Part XIX: The Isolation of Chlorodeoxysceleratine. J. Chem. Soc. C, 781 (1967).
924. ALIEVA, SH.A., U.A. ABDULLAEV, M.V. TELEZHENETSKAYA, and S.YU. YUNUSOV: Alkaloids of *Doronicum macrophyllum*. Khim. Prir. Soedin. **12**, 194 (1976).
925. MATTOCKS, A.R.: Chemistry and Toxicology of Pyrrolizidine Alkaloids. New York: Academic Press. 1986.
926. LÜTHY, J., U. ZWEIFEL, B. KARLHUBER, and C. SCHLATTER: Pyrrolizidine Alkaloids of *Senecio alpinus* L. and Their Detection in Feedingstuffs. J. Agric. Food Chem. **29**, 302 (1981).
927. WONG, R.Y., and J.N. ROITMAN: Structure and Absolute Configuration of (+)-Doronine-Benzene (1:1), $C_{21}H_{30}ClNO_8 \cdot C_6H_6$. Acta Cryst. **C40**, 163 (1984).
928. RÖDER, E., H. WIEDENFELD, and P. KNÖZINGER-FISCHER: Pyrrolizidinalkaloide aus *Senecio abrotanifolius*, ssp. *abrotanifolius* und ssp. *abrotanifolius*, var. *tiroliensis*. Planta Med. **50**, 203 (1984).
929. BREDENKAMP, M.W., A. WIECHERS, and P.H. VAN ROOYEN: A New Pyrrolizidine Alkaloid from *Senecio latifolius* DC. Tetrahedron Lett. **26**, 929 (1985).
930. STERMITZ, F.R., M.A. PASS, R.B. KELLY, and J.R. LIDDELL: Pyrrolizidine Alkaloids from *Cryptantha* Species. Phytochem. **33**, 383 (1993).
931. ENGVILD, K.C.: The Natural Chlorinated Plant Hormone of Pea, 4-Chloroindole-3-Acetic Acid, an Endogenous Herbicide? International Conference on Naturally-Produced Organohalogens, Delft, The Netherlands. September, 1993.
932. BATIROV, E.K., V.M. MALIKOV, and S.Y. YUNUSOV: Lolidine – A New Chlorine-Containing Alkaloid from the Seeds of *Lolium cuneatum*. Khim. Prir. Soedin., 63 (1976).
933. HORIUCHI, Y., S. KONDO, T. IKEDA, D. IKEDA, K. MIURA, M. HAMADA, T. TAKEUCHI, and H. UMEZAWA: New Antibiotics, Clazamycins A and B. J. Antibiot. **32**, 762 (1979).
934. NAKAMURA, H., Y. IITAKA, and H. UMEZAWA: Crystal and Molecular Structure of Clazamycin A. J. Antibiot. **32**, 765 (1979).
935. DOLAK, L.A., and C. DEBOER: Clazamycin B Is Antibiotic 354. J. Antibiot. **33**, 83 (1980).
936. VERDOORN, G.H., and B.-E. VAN WYCK: Oxypterine, a Chlorinated Alkaloid from *Lotononis* Subsection *Rostrata*. Phytochem. **31**, 1029 (1992).
937. TOMITA, M., Y. OKAMOTO, T. KIKUCHI, K. OSAKI, M. NISHIKAWA, K. KAMIYA, Y. SASAKI, K. MATOBA, and K. GOTO: Acutumine and Acutumidine, Chlorine-Containing Alkaloids with a Novel Skeleton (1): X-Ray Analysis of Acutumine. Tetrahedron Lett., 2421 (1967).
938. TOMITA, M., Y. OKAMOTO, T. KIKUCHI, K. OSAKI, M. NISHIKAWA, K. KAMIYA, Y. SASAKI, K. MATOBA, and K. GOTO: Acutumine and Acutumidine, Chlorine-Containing Alkaloids with a Novel Skeleton (2): Chemical Proof. Tetrahedron Lett. 2425 (1967).
939. NISHIKAWA, M., K. KAMIYA, M. TOMITA, Y. OKAMOTO, T. KIKUCHI, K. OSAKI, Y. TOMIIE, I. NITTA, and K. GOTO: The X-Ray Analyses of Acutumine and Its Acetate; A Trial of a Short Cut in the Structure Elucidation. J. Chem. Soc. B, 652 (1968).
940. BRADBURY, R.B., and C.C.J. CULVENOR: The Alkaloids of *Senecio jacobaea* L. Chem. Ind., 1021 (1954).
941. BRADBURY, R.B., and C.C.J. CULVENOR: The Alkaloids of *Senecio jacobaea* L, I: Isolation of the Alkaloids and Identification of Jacodine as Seneciphylline. Aust. J. Chem. **7**, 378 (1954).
942. OKAMOTO, Y., E. YUGE, Y. NAGAI, R. KATSUTA, A. KISHIMOTO, Y. KOBAYASHI, T. KIKUCHI, and M. TOMITA: Acutuminine, a New Alkaloid from the Leaves of *Menispermum dauricum* DC. Tetrahedron Lett., 1933 (1969).

943. BARTON, D.H.R., A.J. KIRBY, and G.W. KIRBY: Phenol Oxidation and Biosynthesis, Part XVII: Investigations on the Biosynthesis of Sinomenine. J. Chem. Soc. C, 929 (1968).

944. KOBAYASHI, J., and M. ISHIBASHI: Marine Alkaloids. In: The Alkaloids, Vol. 41 (A. Brossi and G.A. Cordell, eds.). San Diego, CA: Academic Press. 1992.

945. MOLINSKI, T.F.: Marine Pyridoacridine Alkaloids: Structure, Synthesis, and Biological Chemistry. Chem. Rev. **93**, 1825 (1993).

946. BLACKMAN, A.J., and D.J. MATTHEWS: Amathamide Alkaloids from the Marine Bryozoan *Amathia wilsoni* Kirkpatrick. Heterocycles **23**, 2829 (1985).

947. ALAM, M., R. SANDUJA, and G.M. WELLINGTON: Tubastraine: Isolation and Structure of a Novel Alkaloid from the Stony Coral *Tubastraea micrantha*. Heterocycles **27**, 719 (1988).

948. BLOOR, S.J., and F.J. SCHMITZ: A Novel Pentacyclic Aromatic Alkaloid from an Ascidian. J. Am. Chem. Soc. **109**, 6134 (1987).

949. DE GUZMAN, F.S., and F.J. SCHMITZ: Chemistry of 2-Bromoleptoclinidinone, Structure Revision. Tetrahedron Lett. **30**, 1069 (1989).

950. GOULLE, V., J.-M. LEHN, B. SCHOENTJES, and F.J. SCHMITZ: Ruthenium(II) Complex of the Alkaloid 2-Bromoleptoclinidinone. Preparation and Interaction with Double-Stranded DNA. Helv. Chim. Acta **74**, 1471 (1991).

951. MOLINSKI, T.F., E. FAHY, D.J. FAULKNER, G.D. VAN DUYNE, and J. CLARDY: Petrosamine, a Novel Pigment from the Marine Sponge *Petrosia* sp. J. Org. Chem. **53**, 1340 (1988).

952. KIM, J., E.O. PORDESIMO, S.I. TOTH, F.J. SCHMITZ, and I.V. ALTENA: Pantherinine, a Cytotoxic Aromatic Alkaloid, and 7-Deazainosine from the Ascidian *Aplidium pantherinum*. J. Nat. Prod. **56**, 1813 (1993).

953. COORAY, N.M., P.J. SCHEUER, L. PARKANYI, and J. CLARDY: Shermilamine A: A Pentacyclic Alkaloid from a Tunicate. J. Org. Chem. **53**, 4619 (1988).

954. KOBAYASHI, J., K. KONDO, H. SHIGEMORI, M. ISHIBASHI, T. SASAKI, and Y. MIKAMI: Theoneberine: The First Brominated Benzyltetrahydroprotoberberine Alkaloid from the Okinawan Marine Sponge *Theonella* sp. J. Org. Chem. **57**, 6680 (1992).

955. BLACKMAN, A.J., C. LI, D.C.R. HOCKLESS, B.W. SKELTON, and A.H. WHITE: Cylindricines A and B, Novel Alkaloids from the Ascidian *Clavelina cylindrica*. Tetrahedron **49**, 8645 (1993).

956. FUSON, R.C., and C.L. ZIRKLE: Ring Enlargement by Rearrangement of the 1,2-Aminochloroalkyl Group; Rearrangement of 1-Ethyl-2-chloromethylpyrrolidine to 1-Ethyl-3-chloropiperidine. J. Am. Chem. Soc. **70**, 2760 (1948).

957. MEHRI, H., S. BAASSOU, and M. PLAT: Methylene-10,10'Bis[(+)Na^-Norvallesamidine], Dichlorure de Methylene N_b, N'_b [Bis (+)meloninium], Chlorures de N'_b-Chloromethyl Celastromelinium et Celastromelidium: Alcaloides Dimeres, Artefacts Possibles d'Extraction de *Melodinus celastroides*. J. Nat. Prod. **54**, 372 (1991).

958. JONG, T.-T., and M.-Y. JEAN: Alkaloids from *Houttuyniae cordata*. J. Chin. Chem. Soc. **40**, 301 (1993).

959. OTSUKA, T., S. TAKASE, H. TERANO, and M. OKUHARA: New Angiogenesis Inhibitors, WF-16775 A_1 and A_2. J. Antibiot. **45**, 1970 (1992).

960. SPANDE, T.F., H.M. GARRAFFO, M.W. EDWARDS, H.J.C. YEH, L. PANNELL, and J.W. DALY: Epibatidine: A Novel (Chloropyridyl) Azabicycloheptane with Potent Analgesic Activity from an Ecuadoran Poison Frog. J. Am. Chem. Soc. **114**, 3475 (1992).

961. For a color photograph of this frog, see the July 1994 cover of *Environmental Science and Technology*.

962. FLETCHER, S.R., R. BAKER, M.S. CHAMBERS, R.H. HERBERT, S.C. HOBBS, S.R. THOMAS, H.M. VERRIER, A.P. WATT, and R.G. BALL: Total Synthesis and Determination of the Absolute Configuration of Epibatidine. J. Org. Chem. **59**, 1771 (1994).

963. LINDA, P., and G. MARINO: Electrophilic Substitution in Five-Membered Heterocyclic Systems, Part VI: Kinetics of the Bromination of the 2-Methoxycarbonyl Derivatives of Furan, Thiophen, and Pyrrole in Acetic Acid Solution. J. Chem. Soc. B, 392 (1968).

964. TAKEDA, R.: Pseudomonas Pigments, I: Pyoluteorin, a New Chlorine-Containing Pigment Produced by *Pseudomonas aeruginosa*. Hakko Kogaku Zasshi **36**, 281 (1958).

965. TAKEDA, R.: Structure of a New Antibiotic, Pyoluteorin. J. Am. Chem. Soc. **80**, 4749 (1958).

966. HOWELL, C.R., and R.D. STIPANOVIC: Suppression of *Pythium ultimum* – Induced Damping-Off of Cotton Seedlings by *Pseudomonas fluorescens* and Its Antibiotic, Pyoluteorin. Phytopath. **70**, 712 (1980).

967. ARIMA, K., H. IMANAKA, M. KOUSAKA, A. FUKUDA, and G. TAMURA: Pyrrolnitrin, a New Antibiotic, I: Isolation and Properties of Pyrrolnitrin. J. Antibiot. (A) **18**, 201 (1965).

968. IMANAKA, H., M. KOUSAKA, G. TAMURA, and K. ARIMA: Pyrrolnitrin, a New Antibiotic, III: Structure of Pyrrolnitrin. J. Antibiot. (A) **18**, 207 (1965).

969. LIVELY, D.H., M. GORMAN, M.E. HANEY, and J.A. MABE: Metabolism of Tryptophans by *Pseudomonas aureofaciens*, I: Biosynthesis of Pyrrolnitrin. Antimicrol. Agents Chemother., 462 (1966).

970. VAN PÉE, K.-H., O. SALCHER, and F. LINGENS: Formation of Pyrrolnitrin and 3-(2-Amino-3-chlorophenyl)pyrrole from 7-Chlorotryptophan. Angew. Chem. Int. Ed. Engl. **19**, 828 (1980).

971. ROITMAN, J.N., N.E. MAHONEY, W.J. JANISIEWICZ, and M. BENSON: A New Chlorinated Phenylpyrrole Antibiotic Produced by the Antifungal Bacterium *Pseudomonas cepacia*. J. Agric. Food Chem. **38**, 538 (1990).

972. HASHIMOTO, M., and K. HATTORI: Isopyrrolnitrin: A Metabolite from *Pseudomonas*. Bull. Chem. Soc. Japan **39**, 410 (1966).

973. HASHIMOTO, M., and K. HATTORI: Oxypyrrolnitrin: A Metabolite of *Pseudomonas*. Chem. Pharm. Bull. (Japan) **14**, 1314 (1966).

974. HASHIMOTO, M., and K. HATTORI: A New Metabolite from *Pseudomonas pyrrolnitrica*. Chem. Pharm. Bull. (Japan) **16**, 1144 (1968).

975. ELANDER, R.P., J.A. MABE, R.L. HAMILL, and M. GORMAN: Biosynthesis of Pyrrolnitrins by Analogue-Resistant Mutants of *Pseudomonas fluorescens*. Folia Microbiol. **16**, 156 (1971).

976. HAMILL, R., R. ELANDER, J. MABE, and M. GORMAN: Metabolism of Tryptophans by *Pseudomonas aureofaciens*. Antimicrob. Agents Chemother., 388 (1967).

977. KIRNER, S., and K.-H. VAN PÉE: The Biosynthesis of Nitro Compounds: The Enzymatic Oxidation to Pyrrolnitrin of Its Amino-Substituted Precursor. Angew. Chem. Int. Ed. Engl. **33**, 352 (1994).

978. HORI, Y., Y. ABE, H. NAKAJIMA, S. TAKASE, T. FUJITA, T. GOTO, M. OKUHARA, and M. KOHSAKA: WB2838[3-Chloro-4-(2-amino-3-chlorophenyl)pyrrole]: Non-Steroidal Androgen-Receptor Antagonist Produced by a *Pseudomonas*. J. Antibiot. **46**, 1327 (1993).

979. EZAKI, N., T. SHOMURA, M. KOYAMA, T. NIWA, M. KOJIMA, S. INOUYE, T. ITO, and T. NIIDA: New Chlorinated Nitro-Pyrrole Antibiotics, Pyrrolomycin A and B (SF-2080 A and B). J. Antibiot. **34**, 1363 (1981).

980. KOYAMA, M., Y. KODAMA, T. TSURUOKA, N. EZAKI, T. NIWA, and S. INOUYE: Structure and Synthesis of Pyrrolomycin A, a Chlorinated Nitro-Pyrrole Antibiotic. J. Antibiot. **34**, 1569 (1981).

981. KANEDA, M., S. NAKAMURA, N. EZAKI, and Y. IITAKA: Structure of Pyrrolomycin B, a Chlorinated Nitro-Pyrrole Antibiotic. J. Antibiot. **34**, 1366 (1981).

982. EZAKI, N., M. KOYAMA, T. SHOMURA, T. TSURUOKA, and S. INOUYE: Pyrrolomycins C, D and E, New Members of Pyrrolomycins. J. Antibiot. **36**, 1263 (1983).

983. KOYAMA, M., N. EZAKI, T. TSURUOKA, and S. INOUYE: Structural Studies on Pyrrolomycins C, D and E. J. Antibiot. **36**, 1483 (1983).

984. EZAKI, N., M. KOYAMA, Y. KODAMA, T. SHOMURA, K. TASHIRO, T. TSURUOKA, S. INOUYE, and S. SAKAI: Pyrrolomycins F_1, F_{2a}, F_{2b} and F_3, New Metabolites Produced by the Addition of Bromide to the Fermentation. J. Antibiot. **36**, 1431 (1983).

985. YANO, K., J. OONO, K. MOGI, T. ASAOKA, and T. NAKASHIMA: Pyrroxamycin, a New Antibiotic. Taxonomy, Fermentation, Isolation, Structure Determination and Biological Properties. J. Antibiot. **40**, 961 (1987).

986. NOGAMI, T., Y. SHIGIHARA, N. MATSUDA, Y. TAKAHASHI, H. NAGANAWA, H. NAKAMURA, M. HAMADA, Y. MURAOKA, T. TAKITA, Y. IITAKA, and T. TAKEUCHI: Neopyrrolomycin, a New Chlorinated Phenylpyrrole Antibiotic. J. Antibiot. **43**, 1192 (1990).

987. HAYAKAWA, Y., K. KAWAKAMI, H. SETO, and K. FURIHATA: Structure of a New Antibiotic, Roseophilin. Tetrahedron Lett. **33**, 2701 (1992).

988. BURKHOLDER, P.R., R.M. PFISTER, and F.H. LEITZ: Production of a Pyrrole Antibiotic by a Marine Bacterium. Appl. Microb. **14**, 649 (1966).

989. LOVELL, F.M.: The Structure of a Bromine-Rich Marine Antibiotic. J. Am. Chem. Soc. **88**, 4510 (1966).

990. ANDERSEN, R.J., M.S. WOLFE, and D.J. FAULKNER: Autotoxic Antibiotic Production by a Marine *Chromobacterium*. Mar. Biol. **27**, 281 (1974).

991. FENICAL, W.: Chemical Studies of Marine Bacteria: Developing a New Resource. Chem. Rev. **93**, 1673 (1993).

992. EMRICH, R., H. WEYLAND, and K. WEBER: 2,3,4-Tribromopyrrole from the Marine Polychaete *Polyphysia crassa*. J. Nat. Prod. **53**, 703 (1990).

993. CARTÉ, B., and D.J. FAULKNER: Defensive Metabolites from Three Nembrothid Nudibranchs. J. Org. Chem. **48**, 2314 (1983).

994. CARTÉ, B., and D.J. FAULKNER: Role of Secondary Metabolites in Feeding Associations Between a Predatory Nudibranch, Two Grazing Nudibranchs, and a Bryozoan. J. Chem. Ecol. **12**, 795 (1986).

995. YOSHIDA, W.Y., K.K. LEE, A.R. CARROLL, and P.J. SCHEUER: A Complex Pyrrolo-Oxazinone and Its Iodo Derivative Isolated from a Tunicate. Helv. Chim. Acta **75**, 1721 (1992).

996. RUDI, A., I. GOLDBERG, Z. STEIN, F. FROLOW, Y. BENAYAHU, M. SCHLEYER, and Y. KASHMAN: Polycitone A and Polycitrins A and B: New Alkaloids from the Marine Ascidian *Polycitor* sp. J. Org. Chem. **59**, 999 (1994).

997. YAMAGISHI, Y., M. MATSUOKA, A. ODAGAWA, S. KATO, K. SHINDO, and J. MOCHIZUKI: Rumbrin, a New Cytoprotective Substance Produced by *Auxarthron umbrinum*, I: Taxonomy, Production, Isolation and Biological Properties. J. Antibiot. **46**, 884 (1993).

998. YAMAGISHI, Y., K. SHINDO, and H. KAWAI: Rumbrin, a New Cytoprotective Substance Produced by *Auxarthron umbrinum*, II: Physico-Chemical Properties and Structure Determination. J. Antibiot. **46**, 888 (1993).

999. YAMAGISHI, Y., M. MATSUOKA, A. ODAGAWA, S. KATO, K. SHINDO, and J. MOCHIZUKI: Thiazohalostatin, a New Cytoprotective Substance Produced by *Actinomadura*, I: Taxonomy, Production, Isolation and Biological Activities. J. Antibiot. **46**, 1633 (1993).

1000. SHINDO, K., Y. YAMAGISHI, and H. KAWAI: Thiazohalostatin, a New Cytoprotective Substance Produced by *Actinomadura*, II: Physico-Chemical Properties and Structure Determination. J. Antibiot. **46**, 1638 (1993).

1001. HÖFLE, G., S. POHLAN, G. UHLIG, K. KABBE, and D. SCHUMACHER: Keronopsins A and B, Chemical Defence Substances of the Marine Ciliate *Pseudokeronopsis rubra* (Protozoa): Identification by Ex Vivo HPLC. Angew. Chem. Int. Ed. Engl. **33**, 1495 (1994).

1002. FORENZA, S., L. MINALE, R. RICCIO, and E. FATTORUSSO: New Bromo-Pyrrole Derivatives from the Sponge *Agelas oroides*. J. Chem. Soc., Chem. Commun., 1129 (1971).

1003. GARCIA, E.E., L.E. BENJAMIN, and R.I. FRYER: Reinvestigation into the Structure of Oroidin, a Bromopyrrole Derivative from Marine Sponge. J. Chem. Soc., Chem. Commun., 78 (1973).

1004. GUNASEKERA, S.P., S. CRANICK, and R.E. LONGLEY: Immunosuppressive Compounds from a Sponge *Agelas flabelliformis*. J. Nat. Prod. **52**, 757 (1989).

1005. TADA, H., and T. TOZKO: Two Bromopyrroles from a Marine Sponge *Agelas* sp. Chem. Lett., 803 (1988).

1006. CHEVOLOT, L., S. PADUA, B.N. RAVI, P.C. BLYTH, and P.J. SCHEUER: Isolation of 1-Methyl-4,5-dibromopyrrole-2-carboxylic Acid and Its 3′-(Hydantoyl)propylamide (Midpacamide) from a Marine Sponge. Het. **7**, 891 (1977).

1007. FATHI-AFSHAR, R., and T.M. ALLEN: Biologically Active Metabolites from *Agelas mauritiana*. Can. J. Chem. **66**, 45 (1988).

1008. STEMPIEN, M.F., R.F. NIGRELLI, and J.S. CHIB: Isolation and Synthesis of Physiologically Active Substances from Sponges of the Genus *Agelas*. 164th ACS Meeting, Abstracts MEDI 21, New York, NY, August 27–September 1, 1972.

1009. NAKAMURA, H., Y. OHIZUMI, J. KOBAYASHI, and Y. HIRATA: Keramadine, a Novel Antagonist of Serotonergic Receptors Isolated from the Okinawan Sea Sponge *Agelas* sp. Tetrahedron Lett. **25**, 2475 (1984).

1010. HANESSIAN, S., and J.S. KALTENBRONN: Synthesis of a Bromine-Rich Marine Antibiotic. J. Am. Chem. Soc. **88**, 4509 (1966).

1011. ISHIDA, K., M. ISHIBASHI, H. SHIGEMORI, T. SASAKI, and J. KOBAYASHI: Agelasine G, a New Antileukemic Alkaloid from the Okinawan Marine Sponge *Agelas* sp. Chem. Pharm. Bull. (Japan) **40**, 766 (1992).

1012. JIMÉNEZ, C., and P. CREWS: Mauritamide A and Accompanying Oroidin Alkaloids from the Sponge *Agelas mauritiana*. Tetrahedron Lett. **35**, 1375 (1994).

1013. FEDOREYEV, S.A., S.G. ILYIN, N.K. UTKINA, O.B. MAXIMOV, M.V. RESHETNYAK, M.YU. ANTIPIN, YU.T. STRUCHKOV: The Structure of Dibromoagelaspongin – A Novel Bromine-Containing Guanidine Derivative from the Marine Sponge *Agelas* sp. Tetrahedron **45**, 3487 (1989).

1014. D'AMBROSIO, M., A. GUERRIERO, C. DEBITUS, O. RIBES, J. PUSSET, S. LEROY, and F. PIETRA: Agelastatin A, A New Skeleton Cytotoxic Alkaloid of the Oroidin Family. Isolation from the Axinellid Sponge *Agelas dendromorpha* of the Coral Sea. J. Chem. Soc., Chem. Commun., 1305 (1993)

1015. WALKER, R.P., D.J. FAULKNER, D.V. ENGEN, and J. CLARDY: Sceptrin, an Antimicrobial Agent from the Sponge *Agelas sceptrum*. J. Am. Chem. Soc. **103**, 6772 (1981).

1016. KEIFER, P.A., R.E. SCHWARTZ, M.E.S. KOKER, R.G. HUGHES, D. RITTSCHOF, and K.L. RINEHART: Bioactive Bromopyrrole Metabolites from the Caribbean Sponge *Agelas conifera*. J. Org. Chem. **56**, 2965 (1991).

1017. RINEHART, K.L.: Biologically Active Marine Natural Products. Pure Appl. Chem. **61**, 525 (1989).

1018. KOBAYASHI, J., M. TSUDA, T. MURAYAMA, H. NAKAMURA, Y. OHIZUMI, M. ISHIBASHI, M. IWAMURA, T. OHTA, and S. NOZOE: Ageliferins, Potent Actomyosin ATPase Activators from the Okinawan Marine Sponge *Agelas* sp. Tetrahedron **46**, 5579 (1990).

1019. KOBAYASHI, J., M. TSUDA, and Y. OHIZUMI: A Potent Actomyosin ATPase Activator from the Okinawan Sponge *Agelas* cf. *nemoechinata*. Experientia **47**, 301 (1991).

1020. SHARMA, G.M., and P.R. BURKHOLDER: Structure of Dibromophakellin, a New Bromine-Containing Alkaloid from the Sponge *Phakellia flabellata*. J. Chem. Soc., Chem. Commun., 151 (1971).

1021. SHARMA, G., and B. MAGDOFF-FAIRCHILD: Natural Products of Marine Sponges, 7: The Constitution of Weakly Basic Guanidine Compounds, Dibromophakellin and Monobromophakellin. J. Org. Chem. **42**, 4118 (1977).

1022. FEDOREYEV, S.A., N.K. UTKINA, S.G. ILYIN, M.V. RESHETNYAK, and O.B. MAXIMOV: The Structure of Dibromoisophakellin from the Marine Sponge *Acanthella carteri*. Tetrahedron Lett. **27**, 3177 (1986).

1023. CIMINO, G., S. DE ROSA, S. DE STEFANO, L. MAZZARELLA, R. PULITI, and G. SODANO: Isolation and X-Ray Crystal Structure of a Novel Bromo-Compound from Two Marine Sponges. Tetrahedron Lett. **23**, 767 (1982).

1024. MATTIA, C.A., L. MAZZARELLA, and R. PULITI: 4-(2-Amino-4-oxo-2-imidazolin-5-ylidene)-2-bromo-4,5,6,7-tetrahydropyrrolo[2,3-*c*]azepin-8-one Methanol Solvate: A New Bromo Compound from the Sponge *Acanthella aurantiaca*. Acta Cryst. **B38**, 2513 (1982).

1025. KITAGAWA, I., M. KOBAYASHI, K. KITANAKA, M. KIDO, and Y. KYOGOKU: Marine Natural Products, XII: On the Chemical Constituents of the Okinawan Marine Sponge *Hymeniacidon aldis*. Chem. Pharm. Bull. (Japan) **31**, 2321 (1983).

1026. UTKINA, N.K., S.A. FEDOREEV, and O.B. MAKSIMOV: Nitrogen-Containing Metabolites of the Marine Sponge *Acanthella carteri*. Khim. Prir. Soedin., 535 (1984).

1027. PETTIT, G.R., C.L. HERALD, J.E. LEET, R. GUPTA, D.E. SCHAUFELBERGER, R.B. BATES, P.J. CLEWLOW, D.L. DOUBEK, K.P. MANFREDI, K. RÜTZLER, J.M. SCHMIDT, L.P. TACKETT, F.B. WARD, M. BRUCK, and F. CAMOU: Antineoplastic Agents, 168: Isolation and Structure of Axinohydantoin. Can. J. Chem. **68**, 1621 (1990).

1028. KOBAYASHI, J., Y. OHIZUMI, H. NAKAMURA, Y. HIRATA, K. WAKAMATSU, and T. MIYAZAWA: Hymenin, a Novel α-Adrenoceptor Blocking Agent from the Okinawan Marine Sponge *Hymeniacidon* sp. Experientia **42**, 1064 (1986).

1029. KOBAYASHI, J., H. NAKAMURA, and Y. OHIZUMI: α-Adrenoceptor Blocking Action of Hymenin, a Novel Marine Alkaloid. Experientia **44**, 86 (1988).

1030. ALBIZATI, K.F., and D.J. FAULKNER: Stevensine, a Novel Alkaloid of an Unidentified Marine Sponge. J. Org. Chem. **50**, 4163 (1985).

1031. WRIGHT, A.E., S.A. CHILES, and S.S. CROSS: 3-Amino-1-(2-aminoimidazolyl)prop-1-ene from the Marine Sponges *Teichaxinella morchella* and *Ptilocaulis walpersi*. J. Nat. Prod. **54**, 1684 (1991).

1032. KOBAYASHI, J., Y. OHIZUMI, H. NAKAMURA, and Y. HIRATA: A Novel Antagonist of Serotonergic Receptors, Hymenidin, Isolated from the Okinawan Marine Sponge *Hymeniacidon* sp. Experientia **42**, 1176 (1986).

1033. Kobayashi, J., F. Kanda, M. Ishibashi, and H. Shigemori: Manzacidins A–C, Novel Tetrahydropyrimidine Alkaloids from the Okinawan Marine Sponge *Hymeniacidon* sp. J. Org. Chem. **56**, 4574 (1991).

1034. Schmitz, F.J., S.P. Gunasekera, V. Lakshmi, and L.M.V. Tillekeratne: Marine Natural Products: Pyrrololactams from Several Sponges. J. Nat. Prod. **48**, 47 (1985).

1035. Zeng, L.-M., X. Fu, J.-Y. Su, F. De Guzman, F.J. Schmitz, M.B. Hossain, and D. van der Helm: Studies on the Chemical Constituents of the South China Sea Sponge *Phacellia fusca*. Chin. J. Chem. **9**, 136 (1991).

1036. Rudi, A., Z. Stein, S. Green, I. Goldberg, Y. Kashman, Y. Benayahu, and M. Schleyer: Phorbazoles A–D, Novel Chlorinated Phenylpyrrolyloxazoles from the Marine Sponge *Phorbas aff. clathrata*. Tetrahedron Lett. **35**, 2589 (1994).

1037. Christophersen, C.: Marine Indoles. In: Marine Natural Products, Vol. V, Chap. 5 (P.J. Scheuer, ed.). New York: Academic Press. 1983.

1038. Baker, J.T.: Tyrian Purple: An Ancient Dye, A Modern Problem. Endeavour **33**, 11 (1974).

1039. Friedländer, P.: Über den Farbstoff des antiken Purpurs aus *Murex brandaris*. Chem. Ber. **42**, 765 (1909).

1040. Friedländer, P.: Über die Farbstoffe aus *Purpura aperta* und *Purpura lapillus*. Chem. Ber. **55**, 1655 (1922).

1041. Baker, J.T., and M.D. Sutherland: Pigments of Marine Animals, VIII: Precursors of 6,6′-Dibromoindigotin (Tyrian Purple) from the Mollusc *Dicathais orbita* Gmelin. Tetrahedron Lett., 43 (1968).

1042. Fouquet, H., and H.-J. Bielig: Biological Precursors and Genesis of Tyrian Purple. Angew. Chem. Int. Ed. Engl. **10**, 816 (1971).

1043. Baker, J.T., and C.C. Duke: Isolation from the Hypobranchian Glands of Marine Molluscs of 6-Bromo-2,2-dimethylthioindolin-3-one and 6-Bromo-2-methylthioindoleninone as Alternative Precursors to Tyrian Purple. Tetrahedron Lett. 2481 (1973).

1044. Baker, J.T., and C.C. Duke: Chemistry of the Indoleninones, II: Isolation from the Hypobranchian Glands of Marine Molluscs of 6-Bromo-2,2-dimethylthioindolin-3-one and 6-Bromo-2-methylthioindoleninone as Alternative Precursors to Tyrian Purple. Aust. J. Chem. **26**, 2153 (1973).

1045. Baker, J.T., and C.C. Duke: Isolation of Choline and Choline Ester Salts of Tyrindoxyl Sulphate from the *Dicathais orbita* and *Mancinella keineri*. Tetrahedron Lett., 1233 (1976).

1046. Fujise, Y., K. Miwa, and S. Ito: Structure of Tyriverdin, the Immediate Precursor of Tyrian Purple. Chem. Lett., 631 (1980).

1047. Christophersen, C., F. Wätjen, O. Buchardt, and U. Anthoni: On the Formation of Indigotins. Tetrahedron Lett., 1747 (1977).

1048. Christophersen, C., F. Wätjen, O. Buchardt, and U. Anthoni: A Revised Structure of the Precursor of Tyrian Purple. Tetrahedron **34**, 2779 (1978).

1049. Higa, T., and P.J. Scheuer: Bisindoxyl-Derived Blue Marine Pigments. Heterocycles **4**, 227 (1976).

1050. Higa, T., and P.J. Scheuer: Synthesis and Properties of 6-Bromo-3-chloro- and of 3,5,7-Tribromoindole. Heterocycles **4**, 231 (1976).

1051. Higa, T., and P.J. Scheuer: 3-Chloroindole, Principal Odorous Constituent of the Hemichordate *Ptychodera flava laysanica*. Naturwiss. **62**, 395 (1975).

1052. Higa, T., T. Fujiyama, and P.J. Scheuer: Halogenated Phenol and Indole Constituents of Acorn Worms. Comp. Biochem. Physiol. **65B**, 525 (1980).

1053. HIGA, T., T. ICHIBA, and R.K. OKUDA: Marine Indoles of Novel Substitution Pattern from the Acorn Worm. Experientia **41**, 1487 (1985).

1054. BRENNAN, M.R., and K.L. ERICKSON: Polyhalogenated Indoles from the Marine Alga *Rhodophyllis membranacea* Harvey. Tetrahedron Lett. 1637 (1978).

1055. CARTER, G.T., K.L. RINEHART, L.H. LI, S.L. KUENTZEL, and J.L. CONNOR: Brominated Indoles from *Laurencia brongniartii*. Tetrahedron Lett., 4479 (1978).

1056. CAFIERI, F., E. FATTORUSSO, Y. MAHAJNAH, and A. MANGONI: 6-Bromo-5-hydroxy-3-indolecarboxyaldehyde from the Caribbean Sponge *Oceanapia bartschi*. Z. Naturforsch. **48B**, 1408 (1993).

1057. WRATTEN, S.J., M.S. WOLFE, R.J. ANDERSEN, and D.J. FAULKNER: Antibiotic Metabolites from a Marine Pseudomonad. Antimicrol. Agents Chemother. **11**, 411 (1977).

1058. GUELLA, G., I. MANCINI, H. ZIBROWIUS, and F. PIETRA: Aplysinopsin-Type Alkaloids from *Dendrophyllia* sp., a Scleratinian Coral of the Family Dendrophylliidae of the Philippines, Facile Photochemical (*Z*/*E*) Photoisomerization and Thermal Reversal. Helv. Chim. Acta **72**, 1444 (1989).

1059. GUELLA, G., I. MANCINI, D. DUHET, B. RICHER DE FORGES, and F. PIETRA: Ethyl 6-Bromo-3-indolecarboxylate and 3-Hydroxyacetal-6-bromoindole, Novel Bromoindoles from the Sponge *Pleroma menoui* of the Coral Sea. Z. Naturforsch. **44C**, 914 (1989).

1060. DUMDEI, E., and R.J. ANDERSEN: Igzamide, a Metabolite of the Marine Sponge *Plocamissma igzo*. J. Nat. Prod. **56**, 792 (1993).

1061. RASMUSSEN, T., J. JENSEN, U. ANTHONI, C. CHRISTOPHERSEN, and P.H. NIELSEN: Structure and Synthesis of Bromoindoles from the Marine Sponge *Pseudosuberites hyalinus*. J. Nat. Prod. **56**, 1553 (1993).

1062. SATO, A., and W. FENICAL: Gramine-Derived Bromo-Alkaloids from the Marine Bryozoan *Zoobotryon verticillatum*. Tetrahedron Lett. **24**, 481 (1983).

1063. ORTEGA, M.J., E. ZUBÍA, and J. SALVÁ: A New Brominated Indole-3-carbaldehyde from the Marine Bryozoan *Zoobotryon verticillatum*. J. Nat. Prod. **56**, 633 (1993).

1064. DELLAR, G., P. DJURA, and M.V. SARGENT: Structure and Synthesis of a New Bromoindole from a Marine Sponge. J. Chem. Soc., Perkin Trans. 1, 1679 (1981).

1065. KOBAYASHI, J., J. CHENG, S. YAMAMURA, T. SASAKI, and Y. OHIZUMI: Penaresin, a New Sarcoplasmic Reticulum Ca-Inducer from the Okinawan Marine Sponge *Penares* sp. Heterocycles **31**, 2205 (1990).

1066. GUERRIERO, A., M. D'AMBROSIO, F. PIETRA, C. DEBITUS, and O. RIBES: Pteridines, Sterols, and Indole Derivatives from the Lithistid Sponge *Corallistes undulatus* of the Coral Sea. J. Nat. Prod. **56**, 1962 (1993).

1067. TANAKA, J., T. HIGA, G. BERNARDINELLI, and C.W. JEFFORD: Itomanindoles A and B, Methylsulfinylindoles from *Laurencia brongniartii*. Tetrahedron Lett. **29**, 6091 (1988).

1068. TANAKA, J., T. HIGA, G. BERNARDINELLI, and C.W. JEFFORD: Sulfur-Containing Polybromoindoles from the Red Alga *Laurencia brongniartii*. Tetrahedron **45**, 7301 (1989).

1069. NORTON, R.S., and R.J. WELLS: A Series of Chiral Polybrominated Biindoles from the Marine Blue-Green Alga *Rivularia firma*. Application of ^{13}C NMR Spin-Lattice Relaxation Data and ^{13}C-^{1}H Coupling Constants to Structure Elucidation. J. Am. Chem. Soc. **104**, 3628 (1982).

1070. HODDER, A.R., and R.J. CAPON: A New Brominated Biindole from an Australian Cyanobacterium, *Rivularia firma*. J. Nat. Prod. **54**, 1661 (1991).

1071. VAN LEAR, G.E., G.O. MORTON, and W. FULMOR: New Antibacterial Bromoindole Metabolites from the Marine Sponge *Polyfibrospongia maynardii*. Tetrahedron Lett., 299 (1973)

1072. DJURA, P., D.B. STIERLE, B. SULLIVAN, D.J. FAULKNER, E. ARNOLD, and J. CLARDY: Some Metabolites of the Marine Sponges *Smenospongia aurea* and *Smenospongia echina* (= Polyfibrospongia). J. Org. Chem. **45**, 1435 (1980).

1073. TYMIAK, A.A., K.L. RINEHART, and G.J. BAKUS: Constituents of Morphologically Similar Sponges *Aplysina* and *Smenospongia* Species. Tetrahedron **41**, 1039 (1985).

1074. DEBITUS, C., D. LAURENT, and M. PAÏS: Alcaloïdes d'une Ascidie Neocaledonienne *Eudistoma fragum*. J. Nat. Prod. **51**, 799 (1988).

1075. FAHY, E., B.C.M. POTTS, D.J. FAULKNER, and K. SMITH: 6-Bromotryptamine Derivatives from the Gulf of California Tunicate *Didemnum candidum*. J. Nat. Prod. **54**, 564 (1991).

1076. RAVERTY, W.D., R.H. THOMSON, and T.J. KING: Metabolites from the Sponge *Pachymatisma johnstoni*; L-6-Bromohypaphorine, a New Amino Acid (and Its Crystal Structure). J. Chem. Soc., Perkin Trans. 1, 1204 (1977).

1077. KONDO, K., J. NISHI, M. ISHIBASHI, and J. KOBAYASHI: Two New Tryptophan-Derived Alkaloids from the Okinawan Marine Sponge *Aplysina* sp. J. Nat. Prod. **57**, 1008 (1994).

1078. LINDQUIST, N., and W. FENICAL: Polyandrocarpamides A–D, Novel Metabolites from the Marine Ascidian *Polyandrocarpa* sp. Tetrahedron Lett. **31**, 2521 (1990).

1079. BOBZIN, S.C., and D.J. FAULKNER: Aromatic Alkaloids from the Marine Sponge *Chelonaplysilla* sp. J. Org. Chem. **56**, 4403 (1991).

1080. ROLL, D.M., and C.M. IRELAND: Citorellamine, a New Bromoindole Derivative from *Polycitorella mariae*. Tetrahedron Lett. **26**, 4303 (1985).

1081. MORIARTY, R.M., D.M. ROLL, Y.-Y. KU, C. NELSON, and C.M. IRELAND: A Revised Structure for the Marine Bromoindole Derivative Citorellamine. Tetrahedron Lett. **28**, 749 (1987).

1082. TOMITA, M., Y. OKAMOTO, T. KIKUCHI, K. OSAKI, M. NISHIKAWA, K. KAMIYA, Y. SASAKI, K. MATOBA, and K. GOTO: Studies on the Alkaloids of Menispermaceous Plants, CCLIX: Alkaloids of *Menispermum dauricum* DC. (*Suppl.* 7). Structures of Acutumine and Acutumidine, Chlorine-Containing Alkaloids with a Novel Skeleton. Chem. Pharm. Bull. (Japan) **19**, 770 (1971).

1083. DOSKOTCH, R.W., and J.E. KNAPP: Alkaloids from *Menispermum canadense*. Lloydia **34**, 292 (1971).

1084. AJISAKA, M., K. KARIYONE, K. JOMON, H. YAZAWA, and K. ARIMA: Isolation of the Bromo Analogues of Pyrrolnitrin from *Pseudomonas pyrrolnitrica*. Agric. Biol. Chem. **33**, 294 (1969).

1085. DJURA, P., and D.J. FAULKNER: Metabolites of the Marine Sponge *Dercitus* sp. J. Org. Chem. **45**, 735 (1980).

1086. FATTORUSSO, E., V. LANZOTTI, S. MAGNO, and E. NOVELLINO: Tryptophan Derivatives from a Mediterranean Anthozoan, *Astroides calycularis*. J. Nat. Prod. **48**, 924 (1985).

1087. GUELLA, G., I. MANCINI, H. ZIBROWIUS, and F. PIETRA: Novel Aplysinopsin-Type Alkaloids from *Tribastraea* sp. *Leptopsammia pruvoti* Scleractinian Corals of the Family Dendrophylliidae of the Mediterranean and the Philippines. Configurational-Assignment Criteria, Stereospecific Synthesis, and Photoisomerization. Helv. Chim. Acta **71**, 773 (1988).

1088. SUN, H.H., and S. SAKEMI: A Brominated (Aminoimidazolinyl)indole from the Sponge *Discodermia polydiscus*. J. Org. Chem. **56**, 4307 (1991).

1089. BARTIK, K., J.-C. BRAEKMAN, D. DALOZE, C. STOLLER, J. HUYSECOM, G. VANDEVYVER, and R. OTTINGER: Topsentins, New Toxic Bis-Indole Alkaloids from the Marine Sponge *Topsentia genitrix*. Can. J. Chem. **65**, 2118 (1987).

1090. TSUJII, S., K.L. RINEHART, S.P. GUNASEKERA, Y. KASHMAN, S.S. CROSS, M.S. LUI, S.A. POMPONI, and M.C. DIAZ: Topsentin, Bromotopsentin, and Dihydrodeoxybromotopsentin: Antiviral and Antitumor Bis(indolyl)imidazoles from Caribbean Deep-Sea Sponges of the Family Halichondriidae. Structural and Synthetic Studies. J. Org. Chem. **53**, 5446 (1988).

1091. MORRIS, S.A., and R.J. ANDERSEN: Brominated Bis(indole) Alkaloids from the Marine Sponge *Hexadella* sp. Tetrahedron **46**, 715 (1990).

1092. KOHMOTO, S., Y. KASHMAN, O.J. MCCONNELL, K.L. RINEHART, A. WRIGHT, and F. KOEHN: Dragmacidin, a New Cytotoxic Bis(indole) Alkaloid from a Deep Water Marine Sponge, *Dragmacidon* sp. J. Org. Chem. **53**, 3116 (1988).

1093. SAKEMI, S., and H.H. SUN: Nortopsentins A, B, and C. Cytotoxic and Antifungal Imidazolediylbis[indoles] from the Sponge *Spongoscorites ruetzleri*. J. Org. Chem. **56**, 4304 (1991).

1094. WRIGHT, A.E., S.A. POMPONI, S.S. CROSS, and P. MCCARTHY: A New Bis(indole) Alkaloid from a Deep-Water Marine Sponge of the Genus *Spongosorites*. J. Org. Chem. **57**, 4772 (1992).

1095. SWERSEY, J.C., C.M. IRELAND, L.M. CORNELL, and R.W. PETERSON: Eusynstyelamide, a Highly Modified Dimer Peptide from the Ascidian *Eusynstyela misakiensis*. J. Nat. Prod. **57**, 842 (1994).

1096. YASUMOTO, T., and M. MURATA: Marine Toxins. Chem. Rev. **93**, 1897 (1993).

1097. KOSUGE, T., H. ZENDA, A. OCHIAI, N. MASAKI, M. NOGUCHI, S. KIMURA, and H. NARITA: Isolation and Structure Determination of a New Marine Toxin, Surugatoxin, from the Japanese Ivory Shell, *Babylonia japonica*. Tetrahedron Lett., 2545 (1972).

1098. KOSUGE, T., K. TSUJI, and K. HIRAI: Isolation and Structure Determination of a New Marine Toxin, Neosurugatoxin, from the Japanese Ivory Shell, *Babylonia japonica*, Tetrahedron Lett. **22**, 3417 (1981).

1099. KOSUGE, T., K. TSUJI, and K. HIRAI: Isolation of Neosurugatoxin from the Japanese Ivory shell, *Babylonia japonica*. Chem. Pharm. Bull. (Japan) **30**, 3255 (1982).

1100. KOSUGE, T., K. TSUJI, K. HIRAI, T. FUKUYAMA, H. NUKAYA, and H. ISHIDA: Isolation of a New Toxin, Prosurugatoxin, from the Toxic Japanese Ivory Shell, *Babylonia japonica*. Chem. Pharm. Bull. (Japan) **33**, 2890 (1985).

1101. KOSUGE, T., K. TSUJI, K. HIRAI, and T. FUKUYAMA: First Evidence of Toxin Production by Bacteria in a Marine Organism. Chem. Pharm. Bull. (Japan) **33**, 3059 (1985).

1102. CHRISTOPHERSEN, C.: Secondary Metabolites from Marine Bryozoans. Acta Chem. Scand. **B39**, 517 (1985).

1103. ANTHONI, U., P.H. NIELSEN, M. PEREIRA, and C. CHRISTOPHERSEN: Bryozoan Secondary Metabolites: A Chemotaxonomical Challenge. Comp. Biochem. Physiol. **96B**, 431 (1990).

1104. CARLÉ, J.S., and C. CHRISTOPHERSEN: Bromo-Substituted Physostigmine Alkaloids from a Marine Bryozoan *Flustra foliacea*. J. Am. Chem. Soc. **101**, 4012 (1979).

1105. CARLÉ, J.S., and C. CHRISTOPHERSEN: Marine Alkaloids, 2: Bromo Alkaloids from the Marine Bryozoan *Flustra foliacea*. Isolation and Structure Elucidation. J. Org. Chem. **45**, 1586 (1980).

1106. CARLÉ, J.S., and C. CHRISTOPHERSEN: Marine Alkaloids, 3: Bromo-Substituted Alkaloids from the Marine Bryozoan *Flustra foliacea*, Flustramine C and Flustraminol A and B. J. Org. Chem. **46**, 3440 (1981).

1107. LAYCOCK, M.V., J.L.C. WRIGHT, J.A. FINDLAY, and A.D. PATIL: New Physostigmine Related Bromoalkaloids from the Marine Bryozoan *Flustra foliacea*. Can. J. Chem. **64**, 1312 (1986).

1108. Holst, P.B., U. Anthoni, C. Christophersen, and P.H. Nielsen: Marine Alkaloids, 15: Two Alkaloids, Flustramine E and Debromoflustramine B, from the Marine Bryozoan *Flustra foliacea.* J. Nat. Prod. **57**, 997 (1994).

1109. Wulff, P., J.S. Carlé, and C. Christophersen: Marine Alkaloids, 5: Flustramide A and 6-Bromo-N_b-methyl-N_b-formyltryptamine from the Marine Bryozoan *Flustra foliacea.* Comp. Biochem. Physiol. **71B**, 523 (1982).

1110. Keil, P., E.G. Nielsen, U. Anthoni, and C. Christophersen: Marine Alkaloids, 11: Flustramide B and Flustrarine B from the Marine Bryozoan *Flustra foliacea.* Synthesis of Flustrarine B. Acta Chem. Scand. **B40**, 555 (1986).

1111. Wright, J.L.C.: A New Antibiotic from the Marine Bryozoan *Flustra foliacea.* J. Nat. Prod. **47**, 893 (1984).

1112. Wulff, P., J.S. Carlé, and C. Christophersen: Marine Alkaloids, Part 4: A Formamide, Flustrabromine, from the Marine Bryozoan *Flustra foliacea.* J. Chem. Soc., Perkin Trans. 1, 2895 (1981).

1113. Tsukamoto, S., H. Hirota, H. Kato, and N. Fusetani: Urochordamines A and B: Larval Settlement/Metamorphosis-Promoting, Pteridine-Containing Physostigmine Alkaloids from the Tunicate *Ciona savignyi.* Tetrahedron Lett. **34**, 4819 (1993).

1114. Blackman, A.J., T.W. Hambley, K. Picker, W.C. Taylor, and N. Thirasasana: Hinckdentine-A: A Novel Alkaloid from the Marine Bryozoan *Hincksinoflustra denticulata.* Tetrahedron Lett. **28**, 5561 (1987).

1115. Chevolot, L., A.-M. Chevolot, M. Gajhede, C. Larsen, U. Anthoni, and C. Christophersen: Chartelline A: A Pentahalogenated Alkaloid from the Marine Bryozoan *Chartella papyracea.* J. Am. Chem. Soc. **107**, 4542 (1985).

1116. Anthoni, U., L. Chevolot, C. Larsen, P.H. Nielsen, and C. Christophersen: Marine Alkaloids, 12: Chartellines, Halogenated β-Lactam Alkaloids from the Marine Bryozoan *Chartella papyracea.* J. Org. Chem. **52**, 4709 (1987).

1117. Anthoni, U., K. Bock, L. Chevolot, C. Larsen, P.H. Nielsen, and C. Christophersen: Chartellamide A and B, Halogenated β-Lactam Indole-Imidazole Alkaloids from the Marine Bryozoan *Chartella papyracea.* J. Org. Chem. **52**, 5638 (1987).

1118. Moore, R.E., C. Cheuk, and G.M.L. Patterson: Hapalindoles: New Alkaloids from the Blue-Green Alga *Hapalosiphon fontinalis.* J. Am. Chem. Soc. **106**, 6456 (1984).

1119. Moore, R.E., C. Cheuk, X.-Q.G. Yang, G.M.L. Patterson, R. Bonjouklian, T.A. Smitka, J.S. Mynderse, R.S. Foster, N.D. Jones, J.K. Swartzendruber, and J.B. Deeter: Hapalindoles, Antibacterial and Antimycotic Alkaloids from the Cyanophyte *Hapalosiphon fontinalis.* J. Org. Chem. **52**, 1036 (1987).

1120. Vaillancourt, V., and K.F. Albizati: Synthesis and Absolute Configuration of (+)-Hapalindole Q. J. Am. Chem. Soc. **115**, 3499 (1993).

1121. Bonjouklian, R., R.E. Moore, and G.M.L. Patterson: Acid-Catalyzed Reactions of Hapalindoles. J. Org. Chem. **53**, 5866 (1988).

1122. Moore, R.E., X.G. Yang, and G.M.L. Patterson: Fontonamide and Anhydrohapaloxindole A, Two New Alkaloids from the Blue-Green Alga *Hapalosiphon fontinalis.* J. Org. Chem. **52**, 3773 (1987).

1123. Bornemann, V., G.M.L. Patterson, and R.E. Moore: Isonitrile Biosynthesis in the Cyanophyte *Hapalosiphon fontinalis.* J. Am. Chem. Soc. **110**, 2339 (1988).

1124. Schwartz, R.E., C.F. Hirsch, J.P. Springer, D.J. Pettibone, and D.L. Zink: Unusual Cyclopropane-Containing Hapalindolinones from a Cultured Cyanobacterium. J. Org. Chem. **52**, 3704 (1987).

1125. Park, A., R.E. Moore, and G.M.L. Patterson: Fischerindole L, a New Isonitrile from the Terrestrial Blue-Green Alga *Fischerella muscicola.* Tetrahedron Lett. **33**, 3257 (1992).

1126. Smitka, T.A., R. Bonjouklian, L. Doolin, N.D. Jones, J.B. Deeter, W.Y. Yoshida, M.R. Prinsep, R.E. Moore, and G.M.L. Patterson: Ambiguine Isonitriles, Fungicidal Hapalindole-Type Alkaloids from Three Genera of Blue-Green Algae Belonging to the Stigonemataceae. J. Org. Chem. **57**, 857 (1992).

1127. Perry, N.B., J.W. Blunt, J.D. McCombs, and M.H.G. Munro: Discorhabdin C, a Highly Cytotoxic Pigment from a Sponge of the Genus *Latrunculia*. J Org. Chem. **51**, 5476 (1986).

1128. Perry, N.B., J.W. Blunt, and M.H.G. Munro: Cytotoxic Pigments from New Zealand Sponges of the Genus *Latrunculia*: Discorhabdins A, B and C. Tetrahedron **44**, 1727 (1988).

1129. Kobayashi, J., J. Cheng, M. Ishibashi, H. Nakamura, Y. Ohizumi, Y Hirata, T. Sasaki, H. Lu, and J. Clardy: Prianosin A, A Novel Antileukemic Alkaloid from the Okinawan Marine Sponge *Prianos melanos*. Tetrahedron Lett. **28**, 4939 (1987).

1130. Cheng, J., Y. Ohizumi, M.R. Wälchli, H. Nakamura, Y. Hirata, T. Sasaki, and J. Kobayashi: Prianosins B, C, and D, Novel Sulfur-Containing Alkaloids with Potent Antineoplastic Activity from the Okinawan Marine Sponge *Prianos melanos*. J. Org. Chem. **53**, 4621 (1988).

1131. Radisky, D.C., E.S. Radisky, L.R. Barrows, B.R. Copp, R.A. Kramer, and C.M. Ireland: Novel Cytotoxic Topoisomerase II Inhibiting Pyrroloiminoquinones from Fijian Sponge of the Genus *Zyzzya*. J. Am. Chem. Soc. **115**, 1632 (1993).

1132. Sakemi, S., H.H. Sun, C.W. Jefford, and G. Bernardinelli: Batzellines A, B, and C. Novel Pyrroloquinoline Alkaloids from the Sponge *Batzella* sp. Tetrahedron Lett. **30**, 2517 (1989).

1133. Sun, H.H., S. Sakemi, N. Burres, and P. McCarthy: Isobatzellines A, B, C, and D. Cytotoxic and Antifungal Pyrroloquinoline Alkaloids from the Marine Sponge *Batzella* sp. J. Org. Chem. **55**, 4964 (1990).

1134. De Jesus, A.E., P.S. Steyn, F.R. Van Heerden, R. Vleggaar, P.L. Wessels, and W.E. Hull: Structure and Biosynthesis of the Penitrems A–F, Six Novel Tremorgenic Mycotoxins from *Penicillium crustosum*. J. Chem. Soc., Chem. Commun., 289 (1981).

1135. De Jesus, A.E., P.S. Steyn, F.R. Van Heerden, R. Vleggaar, P.L. Wessels, and W.E. Hull: Tremorgenic Mycotoxins from *Pencillium crustosum*: Isolation of Penitrems A–F and the Structure Elucidation and Absolute Configuration of Penitrem A. J. Chem. Soc., Perkin Trans. 1, 1847 (1983).

1136. De Jesus, A.E., P.S. Steyn, F.R. Van Heerden, R. Vleggaar, P.L. Wessels, and W.E. Hull: Tremorgenic Mycotoxins from *Penicillium crustosum*: Structure Elucidation and Absolute Configuration of Penitrems B–F. J. Chem. Soc., Perkin Trans. 1, 1857 (1983).

1137. Penn, J., J.R. Biddle, P.G. Mantle, J.N. Bilton, and R.N. Sheppard: Pennigritrem, a Naturally-Occurring Penitrem A Analogue with a Novel Cyclization in the Diterpenoid Moiety. J. Chem. Soc., Perkin Trans. 1, 23 (1992).

1138. Musuku, A., M.I. Selala, T. de Bruyne, M. Claeys, P.J.C. Schepens, A. Tsatsakis, and M.I. Shtilman: Isolation and Structure Determination of a New Roquefortine-Related Mycotoxin from *Penicillium verrucosum* var. *cyclopium* Isolated from Cassava. J. Nat. Prod. **57**, 983 (1994).

1139. Cole, R.J., J.W. Kirksey, J. Clardy, N. Eickman, S.M. Weinreb, P. Singh, and D. Kim: Structures of Rugulovasine-A and -B and 8-Chlororugulovasine-A and -B. Tetrahedron Lett., 3849 (1976).

1140. Cole, R.J., J.W. Kirksey, H.G. Cutler, D.M. Wilson, and G. Morgan-Jones: Two Toxic Indole Alkaloids from *Penicillium islandicum*. Can. J. Microbiol. **22**, 741 (1976).

1141. REYNOLDS, V.L., J.P. MCGOVREN, and L.H. HURLEY: The Chemistry, Mechanism of Action and Biological Properties of CC-1065, a Potent Antitumor Antibiotic. J. Antibiot. **39**, 319 (1986).

1142. OHBA, K., H. WATABE, T. SASAKI, Y. TAKEUCHI, Y. KODAMA, T. NAKAZAWA, H. YAMAMOTO, T. SHOMURA, M. SEZAKI, and S. KONDO: Pyrindamycins A and B, New Antitumor Antibiotics. J. Antibiot. **41**, 1515 (1988).

1143. TAKAHASHI, I., K. TAKAHASHI, M. ICHIMURA, M. MORIMOTO, K. ASANO, I. KAWAMOTO, F. TOMITA, and H. NAKANO: Duocarmycin A, A New Antitumor Antibiotic from *Streptomyces*. J. Antibiot. **41**, 1915 (1988).

1144. YASUZAWA, T., T. IIDA, K. MUROI, M. ICHIMURA, K. TAKAHASHI, and H. SANO: Structures of Duocarmycins, Novel Antitumor Antibiotics Produced by *Streptomyces* sp. Chem. Pharm. Bull. (Japan) **36**, 3728 (1988).

1145. ISHII, S., M. NAGASAWA, Y. KARIYA, H. YAMAMOTO, S. INOUYE, and S. KONDO: Antitumor Activity of Pyrindamycins A and B. J. Antibiot. **42**, 1713 (1989).

1146. ASAI, A., S. NAGAMURA, and H. SAITO: A Novel Property of Duocarmycin and Its Analogues for Covalent Reaction with DNA. J. Am. Chem. Soc. **116**, 4171 (1994).

1147. OGAWA, T., M. ICHIMURA, S. KATSUMATA, M. MORIMOTO, and K. TAKAHASHI: New Antitumor Antibiotics, Duocarmycins B_1 and B_2. J. Antibiot. **42**, 1299 (1989).

1148. SYNGE, R.L.M., and E.P. WHITE: Sporidesmin: A Substance from *Sporidesmium bakeri* Causing Lesions Characteristic of Facial Eczema. Chem. Ind., 1546 (1959).

1149. FRIDRICHSONS, J., and A.McL. MATHIESON: The Structure of Sporidesmin: Causative Agent of Facial Eczema in Sheep. Tetrahedron Lett., 1265 (1962).

1150. FRIDRICHSONS, J., and A.McL. MATHIESON: The Structure of the Methylene Dibromide Adduct of Sporidesmin at − 150°C. Acta Cryst. **18**, 1043 (1965).

1151. BEECHAM, A.F., J. FRIDRICHSONS, and A.McL. MATHIESON: The Structure and Absolute Configuration of Gliotoxin and the Absolute Configuration of Sporidesmin. Tetrahedron Lett., 3131 (1966).

1152. HODGES, R., J.W. RONALDSON, A. TAYLOR, and E.P. WHITE: Sporidesmin and Sporidesmin-B. Chem. Ind. 42 (1963).

1153. RONALDSON, J.W., A. TAYLOR, E.P. WHITE, and R.J. ABRAHAM: Sporidesmins, Part I: Isolation and Characterization of Sporidesmin and Sporidesmin B. J. Chem. Soc., 3172 (1963).

1154. HODGES, R., and J.S. SHANNON: The Isolation and Structure of Sporidesmin C. Aust. J. Chem. **19**, 1059 (1966).

1155. JAMIESON, W.D., R. RAHMAN, and A. TAYLOR: Sporidesmins, Part VIII: Isolation and Structure of Sporidesmin-D and Sporidesmin-F. J. Chem. Soc. (C), 1564 (1969).

1156. RAHMAN, R., S. SAFE, and A. TAYLOR: Sporidesmins, Part IX: Isolation and Structure of Sporidesmin E. J. Chem. Soc. (C), 1665 (1969).

1157. FRANCIS, E., R. RAHMAN, S. SAFE, and A. TAYLOR: Sporidesmins, Part XII: Isolation and Structure of Sporidesmin G, a Naturally-Occurring 3,6-Epitetrathiopiperazine-2,5-dione. J. Chem. Soc., Perkin Trans. 1, 470 (1972).

1158. PRZYBYLSKA, M., E.M. GOPALAKRISHNA, A. TAYLOR, and S. SAFE: X-Ray Crystallographic Determination of the Stereochemistry of the Tetrathio-Bridge in Sporidesmin G. J. Chem. Soc., Chem. Commun., 554 (1973).

1159. RAHMAN, R., S. SAFE, and A. TAYLOR: Sporidesmins, Part 17: Isolation of Sporidesmin H and Sporidesmin J. J. Chem. Soc., Perkin Trans. 1, 1476 (1978).

1160. MARUMO, S., H. ABE, H. HATTORI, and K. MUNAKATA: Isolation of a Novel Auxin, Methyl 4-Chloroindoleacetate from Immature Seeds of *Pisum sativum*. Agric. Biol. Chem. **32**, 117 (1968).

1161. MARUMO, S., H. HATTORI, H. ABE, and K. MUNAKATA: Isolation of 4-Chloroindolyl-3-acetic Acid from Immature Seeds of *Pisum sativum*. Nature **219**, 959 (1968).

1162. ENGVILD, K.C., H. EGSGAARD, and E. LARSEN: Determination of 4-Chloroindole-3-acetic Acid Methyl Ester in *Lathyrus*, *Vicia* and *Pisum* by Gas Chromatography-Mass Spectrometry. Physiol. Plant. **48**, 499 (1980).

1163. HOFINGER, M., and M. BÖTTGER: Identification by GC-MS of 4-Chloroindolylacetic Acid and Its Methyl Ester in Immature *Vicia faba* Seeds. Phytochem. **18**, 653 (1979).

1164. PARK, K., T. YOKOTA, A. SAKURAI, and N. TAKAHASHI: Occurrence of Castasterone, Brassinolide and Methyl 4-Chloroindole-3-acetate in Immature *Vicia faba* Seeds. Agric. Biol. Chem. **51**, 3081 (1987).

1165. KATAYAMA, M., S.V. THIRUVIKRAMAN, and S. MARUMO: Identification of 4-Chloroindole-3-acetic Acid and Its Methyl Ester in Immature Seeds of *Vicia amurensis* (the Tribe *Vicieae*), and Their Absence from Three Species of *Phaseoleae*. Plant Cell Physiol. **28**, 383 (1987).

1166. ENGVILD, K.C., H. EGSGAARD, and E. LARSEN: Determination of 4-Chloroindoleacetic Acid Methyl Ester in *Vicieae* Species by Gas Chromatography-Mass Spectrometry. Physiol. Plant. **53**, 79 (1981).

1167. ERNSTSEN, A., and G. SANDBERG: Identification of 4-Chloroindole-3-acetic Acid and Indole-3-aldehyde in Seeds of *Pinus sylvestris*. Physiol. Plant. **68**, 511 (1986).

1168. PLESS, T., M. BÖTTGER, P. HEDDEN, and J. GRAEBE: Occurrence of 4-Cl-Indoleacetic Acid in Broad Beans and Correlation of Its Levels with Seed Development. Plant Physiol. **74**, 320 (1984).

1169. BÖTTGER, M., K.C. ENGVILD, and H. SOLL: Growth of *Avena* Coleoptiles and pH Drop of Protoplast Suspensions Induced by Chlorinated Indoleacetic Acids. Planta **140**, 89 (1978).

1170. THIRUVIKRAMAN, S.V., Y. SAKAGAMI, M. KATAYAMA, and S. MARUMO: *S*-4-Chlorotryptophan: Its Synthesis via Resolution, Determination of the Absolute Stereochemistry and Identification in the Crude Seed Protein of the Pea, *Pisum sativum*. Tetrahedron Lett. **29**, 2339 (1988).

1171. SAKAGAMI, Y., K. MANABE, T. AITANI, S.V. THIRUVIKRAMAN, and S. MARUMO: L-4-Chlorotryptophan from Immature Seeds of *Pisum sativum* and Reassignment of the Absolute Stereochemistry of *N*-Malonyl-4-Chlorotryptophan. Tetrahedron Lett. **34**, 1057 (1993).

1172. PIACEK-LLANES, B.G., and S.R. TANNENBAUM: Formation of an Activated *N*-Nitroso Compound in Nitrite-Treated Fava Beans (*Vicia faba*). Carcinogenesis **3**, 1379 (1982).

1173. YANG, D., S.R. TANNENBAUM, G. BÜCHI, and G.C.M. LEE: 4-Chloro-6-methoxyindole is the Precursor of a Potent Mutagen (4-Chloro-6-methoxy-2-hydroxy-1-nitroso-indolin-3-one Oxime) that Forms During Nitrosation of the Fava Bean (*Vicia faba*). Carcinogenesis **5**, 1219 (1984).

1174. BÜCHI, G., G.C.M. LEE, D. YANG, and S.R. TANNENBAUM: Direct Acting, Highly Mutagenic, α-Hydroxy *N*-Nitrosamines from 4-Chloroindoles. J. Am. Chem. Soc. **108**, 4115 (1986).

1175. BROWN, N.K., T.T. NGUYEN, K. TAGHIZADEH, J.S. WISHNOK, and S.R. TANNENBAUM: Mutagenic Decomposition Products of Nitrosated 4-Chloroindoles. Chem. Res. Toxicol. **5**, 797 (1992).

1176. REIMANN, H., and R. JARET: 5-Chloro-6-methoxy-1-methylisatin, A Metabolite of *Micromonospora carbonacea*. Chem. Ind., 2173 (1967).

1177. CARDELLINA, J.H., M.P. KIRKUP, R.E. MOORE, J.S. MYNDERSE, K. SEFF, and C.J. SIMMONS: Hyellazole and Chlorohyellazole, Two Novel Carbazoles from the Blue-Green Alga *Hyella caespitosa* Born. et Flah. Tetrahedron Lett., 4915 (1979).

1178. LUK, K.-C., L. STERN, M. WEIGELE, R.A. O'BRIEN, and N. SPIRT: Isolation and Identification of "Diazepam-Like" Compounds from Bovine Urine. J. Nat. Prod. **46**, 852 (1983).

1179. DEWAR, D., V. GLOVER, J. ELSWORTH, and M. SANDLER: Equol and Other Compounds from Bovine Urine as Monoamine Oxidase Inhibitors. J. Neural. Transm. **65**, 147 (1986).

1180. NETTLETON, D.E., T.W. DOYLE, B. KRISHNAN, G.K. MATSUMOTO, and J. CLARDY: Isolation and Structure of Rebeccamycin – A New Antitumor Antibiotic from *Nocardia aerocolonigenes*. Tetrahedron Lett. **26**, 4011 (1985).

1181. BUSH, J.A., B.H. LONG, J.J. CATINO, W.T. BRADNER, and K. TOMITA: Production and Biological Activity of Rebeccamycin, a Novel Antitumor Agent. J. Antibiot. **40**, 668 (1987).

1182. LAM, K.S., J. MATTEI, and S. FORENZA: Carbon Catabolite Regulation of Rebeccamycin Production in *Saccharothrix aerocolonigenes*. J. Indust. Microbiol. **4**, 105 (1989).

1183. ISIDOROV, V.A., I.G. ZENKEVICH, and B.V. IOFFE: Volatile Organic Compounds in Solfataric Gases. J. Atmos. Chem. **10**, 329 (1990).

1184. MATSON, J.A.: 4'-Deschlororebeccamycin Pharmaceutical Composition. Chem. Abstr. **103**, 159104 (1985).

1185. LAM, K.S., D.R. SCHROEDER, J.M. VEITCH, J.A. MATSON, and S. FORENZA: Isolation of a Bromo Analog of Rebeccamycin from *Saccharothrix aerocolonigenes*. J. Antibiot. **44**, 934 (1991).

1186. PEARCE, C.J., T.W. DOYLE, S. FORENZA, K.S. LAM, and D.R. SCHROEDER: The Biosynthetic Origins of Rebeccamycin. J. Nat. Prod. **51**, 937 (1988).

1187. MATSON, J.A., C. CLARIDGE, J.A. BUSH, J. TITUS, W.T. BRADNER, T.W. DOYLE, A.C. HORAN, and M. PATEL: AT2433-A1, AT2433-A2, AT2433-B1, and AT2433-B2. Novel Antitumor Antibiotic Compounds Produced by *Actinomadura melliaura*. Taxonomy, Fermentation, Isolation and Biological Properties. J. Antibiot. **42**, 1547 (1989).

1188. GOLIK, J., T.W. DOYLE, B. KRISHNAN, G. DUBAY, and J.A. MATSON: AT2433-A1, AT2433-A2, AT2433-B1, and AT2433-B2. Novel Antitumor Antibiotic Compounds Produced by *Actinomadura melliaura*, II: Structure Determination. J. Antibiot. **42**, 1784 (1989).

1189. BONJOUKLIAN, R., T.A. SMITKA, L.E. DOOLIN, R.M. MOLLOY, M. DEBONO, S.A. SHAFFER, R.E. MOORE, J.B. STEWART, and G.M.L. PATTERSON: Tjipanazoles. New Antifungal Agents from the Blue-Green Alga *Tolypothrix tjipanasensis*. Tetrahedron **47**, 7739 (1991).

1190. ASUMI, K., M. YOSHIMIZU, S. SUZUKI, Y. EZURA, and H. YOKOSAWA: Inhibitory Effect of Halocyamine, an Antimicrobial Substance from Ascidian Hemocytes, on the Growth of Fish Viruses and Marine Bacteria. Experientia **46**, 1066 (1990).

1191. DAVIDSON, B.S.: Ascidians: Producers of Amino Acid Derived Metabolites. Chem. Rev. **93**, 1771 (1993).

1192. RINEHART JR., K.L., J. KOBAYASHI, G.C. HARBOUR, J. GILMORE, M. MASCAL, T.G. HOLT, L.S. SHIELD, and F. LAFARGUE: Eudistomins A–Q, β-Carbolines from the Antiviral Caribbean Tunicate *Eudistoma olivaceum*. J. Am. Chem. Soc. **109**, 3378 (1987).

1193. KINZER, K.F., and J.H. CARDELLINA: Three New β-Carbolines from the Bermudian Tunicate *Eudistoma olivaceum*. Tetrahedron Lett. **28**, 925 (1987).

1194. BLUNT, J.W., R.J. LAKE, M.H.G. MUNRO, and T. TOYOKUNI: The Stereochemistry of Eudistomins C, K, E, F and L. Tetrahedron Lett. **28**, 1825 (1987).

1195. LAKE, R.J., J.D. MCCOMBS, J.W. BLUNT, M.H.G. MUNRO, and W.T. ROBINSON: Eudistomin K: Crystal Structure and Absolute Stereochemistry. Tetrahedron Lett. **29**, 4971 (1988).

1196. LAKE, R.J., M.M. BRENNAN, J.W. BLUNT, M.H.G. MUNRO, and L.K. PANNELL: Eudistomin K Sulfoxide – An Antiviral Sulfoxide from the New Zealand Ascidian *Ritterella sigillinoides*. Tetrahedron Lett. **29**, 2255 (1988).

1197. LAKE, R.J., J.W. BLUNT, and M.H.G. MUNRO: Eudistomins from the New Zealand Ascidian *Ritterella sigillinoides*. Aust. J. Chem. **42**, 1201 (1989).

1198. KOBAYASHI, J., M. ISHIBASHI, U. NAGAI, and Y. OHIZUMI: 9-Methyl-7-bromoeudistomin D, a Potent Inducer of Calcium Release from Sarcoplasmic Reticulum of Skeletal Muscle. Experientia **45**, 782 (1989).

1199. KOBAYASHI, J., M. TANIGUCHI, T. HINO, and Y. OHIZUMI: Eudistomin Derivatives, Novel Phosphodiesterase Inhibitors: Synthesis and Relative Activity. J. Pharm. Pharmacol. **40**, 62 (1988).

1200. KOBAYASHI, J., H. NAKAMURA, Y. OHIZUMI, and Y. HIRATA: Eudistomidin-A, a Novel Calmodulin Antagonist from the Okinawan Tunicate *Eudistoma glaucus*. Tetrahedron Lett. **27**, 1191 (1986).

1201. KOBAYASHI, J., J. CHENG, T. OHTA, S. NOZOE, Y. OHIZUMI, and T. SASAKI: Eudistomidins B, C, and D: Novel Antileukemic Alkaloids from the Okinawan Marine Tunicate *Eudistoma glaucus*. J. Org. Chem. **55**, 3666 (1990).

1202. MURATA, O., H. SHIGEMORI, M. ISHIBASHI, K. SUGAMA, K. HAYASHI, and J. KOBAYASHI: Eudistomidins E and F, New Alkaloids from the Okinawan Marine Tunicate *Eudistoma glaucus*. Tetrahedron Lett. **32**, 3539 (1991).

1203. ADESANYA, S.A., M. CHBANI, M. PAÏS, and C. DEBITUS: Brominated β-Carbolines from the Marine Tunicate *Eudistoma album*. J. Nat. Prod. **55**, 525 (1992).

1204. SHEN, G.Q., and B.J. BAKER: Biosynthetic Studies of Eudistomin H in the Tunicate *Eudistoma olivaceum*. Tetrahedron Lett. **35**, 4923 (1994).

1205. AIELLO, A., E. FATTORUSSO, S. MAGNO, and L. MAYOL: Brominated β-Carbolines from the Marine Hydroid *Aglaophenia pluma* Linnaeus. Tetrahedron **43**, 5929 (1987).

1206. CHBANI, M., M. PAÏS, J.-M. DELAUNEUX, and C. DEBITUS: Brominated Indole Alkaloids from the Marine Tunicate *Pseudodistoma arborescens*. J Nat. Prod. **56**, 99 (1993).

1207. LARSEN, L.K., R.E. MOORE, and G.M.L. PATTERSON: β-Carbolines from the Blue-Green Alga *Dichothrix baueriana*. J. Nat. Prod. **57**, 419 (1994).

1208. NAKAGAWA, A., Y. IWAI, H. HASHIMOTO, N. MIYAZAKI, R. OIWA, Y. TAKAHASHI, A. HIRANO, N. SHIBUKAWA, Y. KOJIMA, and S. OMURA: Virantmycin, a New Antiviral Antibiotic Produced by a Strain of *Streptomyces*. J. Antibiot. **34**, 1408 (1981).

1209. OMURA, S., and A. NAKAGAWA: Structure of Virantmycin, a Novel Antiviral Antibiotic. Tetrahedron Lett. **22**, 2199 (1981).

1210. MORIMOTO, Y., K. ODA, H. SHIRAHAMA, T. MATSUMOTO, and S. OMURA: Assignment of Absolute Configuration for Virantmycin and Synthesis of Its Antipode. Chem. Lett., 909 (1988).

1211. PEARCE, C.M., and J.K.M. SANDERS: Stereochemistry of (−)-Virantmycin. J. Chem. Soc., Perkin Trans. 1, 409 (1990).

1212. STAPLEY, E.O., D. HENDLIN, J.M. MATA, S. HERNANDEZ, A.K. MILLER, M. JACKSON, and H. WALLICK: Antibiotic MSD-819, I: Microbial Production and Biological Characteristics. Antimicrob. Agents Chemother., 249 (1968).

1213. MILLER, T.W., R.W. WALKER, N.R. TRENNER, B.A. ARISON, and F.J. WOLF: Antibiotic MSD-819, II: Isolation and Chemical Characterization as 6-Chloro-2-quinoxalinecarboxylic Acid 1,4-Dioxide. Antimicrob. Agents Chemother., 255 (1968).

1214. NIWA, H., Y. YOSHIDA, and K. YAMADA: A Brominated Quinazolinedione from the Marine Tunicate *Pyura sacciformis*. J. Nat. Prod. **51**, 343 (1988).

1215. LE-VAN, N., and S.J. WRATTEN: Compound 30.4, and Unusual Chlorinated 1,4-Benzoxazin-3-one Derivative from Corn (Zea Mays). Tetrahedron Lett. **25**, 145 (1984).

1216. PATEL, M., V. HEGDE, A.C. HORAN, V.P. GULLO, D. LOEBENBERG, J.A. MARQUEZ, G.H. MILLER, M.S. PUAR, and J.A. WAITZ: A Novel Phenazine Antifungal Antibiotic, 1,6-Dihydroxy-2-chlorophenazine. J. Antibiot. **37**, 943 (1984).

1217. GUREVICH, A.I., M.G. KARAPETYAN, M.N. KOLOSOV, V.V. ONOPRIENKO, and S.A. POPRAVKO: Structure of the Antibiotics Albofungin and Chloroalbofungin. Doklady Akad. Nauk (USSR) **207**, 1347 (1972); Chem. Abstr. **78**, 111232 (1972).

1218. REISCH, J., K. SZENDREI, ZS. RÓZSA, I. NOVÁK, and E. MINKER: Chlorine-Containing Acridone Alkaloids from *Ruta graveolens*. Phytochem. **11**, 2359 (1972).

1219. REISCH, J., ZS. RÓZSA, K. SZENDREI, I. NOVÁK, and E. MINKER: Studies in the Area of Natural Product Chemistry, Part 57: Furacridone, 1-Hydroxy-3-methoxy-*N*-methylacridone, and Isogravacridonechlorine from the Roots of *Ruta graveolens*. Phytochem. **16**, 151 (1977).

1220. AHOND, A., M.B. ZURITA, M. COLIN, C. FIZAMES, P. LABOUTE, F. LAVELLE, D. LAURENT, C. POUPAT, and J. PUSSET: Girolline, a New Antitumor Compound Extracted from the Sponge *Pseudaxinyssa cantharella* (n. sp. Axinellidae). C. R. Acad. Sci. Paris, Ser. 2, **307**, 145 (1988).

1221. ZURITA, M.B., A. AHOND, C. POUPAT, and P. POTIER: Premiere Syntheses Totale de la Girolline. Tetrahedron **45**, 6713 (1989).

1222. COMMERCON, A., and C. GUÉRÉMY: A Diastereoselective Synthesis of Girolline. Tetrahedron Lett. **32**, 1419 (1991).

1223. CHIARONI, A., C. RICHE, A. AHOND, C. POUPAT, M. PUSSET, and P. POTIER: Structure Cristalline et Configuration Absolue de la Girolline. C. R. Acad. Sci. Paris, Ser. 2, **312**, 49 (1991).

1224. SANGAMESWARAN, L., H.M. FALES, P. FRIEDRICH, and A.L. DE BLAS: Purification of a Benzodiazepine from Bovine Brain and Detection of Benzodiazepine-Like Immunoreactivity in Human Brain. Proc. Natl. Acad. Sci. **83**, 9236 (1986).

1225. WILDMANN, J., W. VETTER, U.B. RANALDER, K. SCHMIDT, R. MAURER, and H. MÖHLER: Occurrence of Pharmacologically Active Benzodiazepines in Trace Amounts in Wheat and Potato. Biochem. Pharmacol. **37**, 3549 (1988).

1226. MORTON, G.O., J.E. LANCASTER, G.E. VAN LEAR, W. FULMOR, and W.E. MEYER: The Structure of Nucleocidin, III (A New Structure). J. Am. Chem. Soc. **91**, 1535 (1969).

1227. JENKINS, I.D., J.P.H. VERHEYDEN, and J.G. MOFFATT: 4′-Substituted Nucleosides, 2: Synthesis of the Nucleoside Antibiotic Nucleocidin. J. Am. Chem. Soc. **98**, 3346 (1976).

1228. TAKAHASHI, E., and T. BEPPU: A New Nucleosidic Antibiotic AT-265. J. Antibiot. **35**, 939 (1982).

1229. ISONO, K., M. URAMOTO, H. KUSAKABE, N. MIYATA, T. KOYAMA, M. UBUKATA, S.K. SETHI, and J.A. MCCLOSKEY: Ascamycin and Dealanylascamycin, Nucleoside Antibiotics from *Streptomyces* sp. J. Antibiot. **37**, 670 (1984).

1230. KAZLAUSKAS, R., P.T. MURPHY, R.J. WELLS, J.A. BAIRD-LAMBERT, and D.D. JAMIESON: Halogenated Pyrrolo[2,3-*d*]pyrimidine Nucleosides from Marine Organisms. Aust. J. Chem. **36**, 165 (1983).

1231. SCHAUMBERG, J.P., G.C. HOKANSON, J.C. FRENCH, E. SMAL, and D.C. BAKER: 2′-Chloropentostatin, a New Inhibitor of Adenosine Deaminase. J. Org. Chem. **50**, 1651 (1985).

1232. OMURA, S., N. IMAMURA, H. KUGA, H. ISHIKAWA, Y. YAMAZAKI, K. OKANO, K. KIMURA, Y. TAKAHASHI, and H. TANAKA: Adechlorin, A New Adenosine Deaminase Inhibitor Containing Chlorine. J. Antibiot. **38**, 1008 (1985).

1233. TUNAC, J.B., and M. UNDERHILL: 2′-Chloropentostatin: Discovery, Fermentation and Biological Activity. J. Antibiot. **38**, 1344 (1985).

1234. WOODWARD, R.B., and V. SKARIC: A New Aspect of the Chemistry of Chlorins. J. Am. Chem. Soc. **83**, 4676 (1961).

1235. LIN, W., R.S. BURKHALTER, and R. TIMKOVICH: Isolation of Novel Meso-Substituted Uroporphyrins from Cultures of *Pseudomonas*. Heterocycles **36**, 2191 (1993).

1236. GITTERMAN, C.O., E.L. RICKES, D.E. WOLF, J. MADAS, S.B. ZIMMERMAN, T.H. STOUDT, and T.C. DEMNY: The Human Tumor-Egg Host System, IV: Discovery of a New Antitumor Agent, Compound 593A. J. Antibiot. **23**, 305 (1970).

1237. JENSEN, N.P., C.O. GITTERMAN, T.Y. CHEN, B.H. ARISON, and J.L. BECK: Antitumor Agent Identified. C & E News, April 17, p. 24 (1972).

1238. LOUSBERG, R.J.J.CH., and Y. TIRILLY: Structure of Furasterin, a Chlorinated Metabolite from the Fungus (Dounson) Burge et Isaac. Experientia **32**, 1394 (1976).

1239. MCGAHREN, W.J., and L.A. MITSCHER: Dihydroisocoumarins from a *Sporormia* Fungus. J. Org. Chem. **33**, 1577 (1968).

1240. TROFAST, J.: Chloromycorrhizinol A, a Furochroman from an Isolate of the Roots of *Monotropa hypopitys.* Phytochem. **17**, 1359 (1978).

1241. HABIB, A.M., D.K. HO, S. MASUDA, T. MCCLOUD, K.S. REDDY, M. ABOUSHOER, A. MCKENZIE, S.R. BYRN, C.-J. CHANG, and J.M. CASSADY: Structure and Stereochemistry of Psorospermin and Related Cytotoxic Dihydrofuranoxanthones from *Psorospermum febrifugum.* J. Org. Chem. **52**, 412 (1987).

1242. HIGA, T.: 2-(1-Chloro-2-hydroxyethyl)-4,4-dimethylcyclohexa-2,5-dienone: A Precursor of 4,5-Dimethylbenzo[*b*]furan from the Red Alga *Desmia hornemanni.* Tetrahedron Lett. **26**, 2335 (1985).

1243. KAZLAUSKAS, R., P.T. MURPHY, R.J. WELLS, and A.J. BLACKMAN: Macrocyclic Enol-Ethers Containing an Acetylenic Group from the Red Alga *Phacelocarpus labillardieri.* Aust. J. Chem. **35**, 113 (1982).

1244. BLACKMAN, A.J., J.B. BREMNER, A.M.C. PAANO, J.H. SKERRATT, and M.L. SWANN: A Further γ-Pyrone Derivative from the Marine Red Alga *Phacelocarpus labillardieri.* Aust. J. Chem. **43**, 1133 (1990).

1245. SHIN, J., V.J. PAUL, and W. FENICAL: New Macrocyclic *Alpha-* and *Gamma-*Pyrones from the Marine Red Alga *Phacelocarpus labillardieri.* Tetrahedron Lett. **27**, 5189 (1986).

1246. VAN DER MERWE, K.J., P.S. STEYN, L. FOURIE, D.B. SCOTT, and J.J. THERON: Ochratoxin A, a Toxic Metabolite Produced by *Aspergillus ochraceus* Wilh. Nature **205**, 1112 (1965).

1247. STEYN, P.S., and C.W. HOLZAPFEL: The Synthesis of Ochratoxins A and B. Metabolites of *Aspergillus ochraceus* Wilh. Tetrahedron **23**, 4449 (1967).

1248. VAN WALBEEK, W., P.M. SCOTT, J. HARWIG, and J.W. LAWRENCE: *Penicillium viridicatum* Westling: A New Source of Ochratoxin A. Can. J. Microbiol. **15**, 1281 (1965).

1249. CIEGLER, A., D.J. FENNELL, H.-J. MINTZLAFF, and L. LEISTNER: Ochratoxin Synthesis by *Penicillium* Species. Naturwiss. **59**, 365 (1972).

1250. HARWIG, J., and Y.-K. CHEN: Some Conditions Favoring Production of Ochratoxin A and Citrinin by *Penicillium viridicatum* in Wheat and Barley. Can. J. Plant Sci. **54**, 17 (1974).

1251. NORTHOLT, M.D., H.P. VAN EGMOND, and W.E. PAULSCH: Ochratoxin A Production by Some Fungal Species in Relation to Water Activity and Temperature. J. Food Prod. **42**, 485 (1979).

1252. LAI, M., G. SEMENIUK, and C.W. HESSELTINE: Conditions for Production of Ochratoxin A by *Aspergillus* Species in a Synthetic Medium. Appl. Microbiol. **19**, 542 (1970).

1253. VAN DER MERWE, K.J., P.S. STEYN, and L. FOURIE: Mycotoxins, Part II: The Constitution of Ochratoxins A, B, and C, Metabolites of *Aspergillus ochraceus* Wilh. J. Chem. Soc. (C), 7083 (1965).

1254. HUTCHINSON, R.D., and P.S. STEYN: The Isolation and Structure of 4-Hydroxyochratoxin A and 7-Carboxy-3,4-dihydro-8-hydroxy-3-methylisocoumarin from *Penicillium viridicatum*. Tetrahedron Lett., 4033 (1971).

1255. BRAZ FILHO, R., M.P.L. DE MORALES, and O.R. GOTTLIEB: Pterocarpans from *Swartzia laevicarpa*. Phytochem. **19**, 2003 (1980).

1256. BRAZ FILHO, R., C.A.S. MIRANDA, O.R. GOTTLIEB, and M.T. MAGALHAES: Chemistry of Brazilian Guttiferae, XXXVII: Chemical Constituents of *Tovomita brasiliensis*. Acta Amazonica **12**, 801 (1982); Chem. Abstr. **99**, 155230 (1983).

1257. ELLESTAD, G.A., F.M. LOVELL, N.A. PERKINSON, R.T. HARGREAVES, and W.J. MCGAHREN: New Zearalenone Related Macrolides and Isocoumarins from an Unidentified Fungus. J. Org. Chem. **43**, 2339 (1978).

1258. FUSETANI, N., T. SUGAWARA, S. MATSUNAGA, and H. HIROTA: Cytotoxic Metabolites of the Marine Sponge *Mycale adhaerens* Lambe. J. Org. Chem. **56**, 4971 (1991).

1259. SHARMA, P.N., A. SHOEB, R.S. KAPIL, and S.P. POPLI: Toddanol and Toddanone, Two Coumarins from *Toddalia asiatica*. Phytochem. **20**, 335 (1981).

1260. ITO, C., and H. FURUKAWA: Constituents of *Murraya exotica* L. Structure Elucidation of New Coumarins. Chem. Pharm. Bull. (Japan) **35**, 4277 (1987).

1261. ITO, C., H. FURUKAWA, H. ISHII, I. ISHIKAWA, and J. HAGINIWA: The Chemical Composition of *Murraya paniculata*. The Structure of Five New Coumarins and One New Alkaloid and the Stereochemistry of Murrangatin and Related Coumarins. J. Chem. Soc., Perkin Trans. 1, 2047 (1990).

1262. ABAUL, J., E. PHILOGENE, P. BOURGEOIS, C. POUPAT, A. AHOND, and P. POTIER: Contribution à la Connaissance des Rutacées Américaines: Études des Feuilles de *Triphasia trifolia*. J. Nat. Prod. **57**, 846 (1994).

1263. AVIAMENKO, L.G., and G.K. NIKONOV: Saxalin – A New Furocoumarin from the Roots of *Angelica saxatilis*. Khim. Prir. Soedin., 830 (1971).

1264. IVIE, G.W.: Linear Furocoumarins (Psoralens) from the Seed of Texas *Ammi majus* L. (Bishop's Weed). J. Agric. Food Chem. **26**, 1394 (1978).

1265. BEIER, R.C., G.W. IVIE, and E.H. OERTLI: Linear Furanocoumarins and Graveolone from the Common Herb Parsley. Phytochem. **36**, 869 (1994).

1266. GONZALEZ, A.G., J.T. BARROSO, J.R. LUIS, and F.R. LUIS: Umbelliferous Components, VII: New Coumarins from *Heracleum granatense*. An. Quim. **70**, 856 (1974).

1267. GONZALEZ, A.G., J.T. BARROSO, E.D. CHICO, J.R. LUIS, and F.R. LUIS: Components of the Umbelliferae, XI: New Coumarins from *Heracleum pyrenaicum* Lam. An. Quim. **72**, 584 (1976).

1268. SOOD, S., B.D. GUPTA, S.K. BANERJEE, and P.R. RAO: Coumarins from the Umbels of *Prangos pabularia*. J. Indian Chem. Soc. **55**, 850 (1978).

1269. ZHELEVA, A.B., M.M. MAHANDRU, and L. BUBEVA-IVANOVA: Four New Coumarins from the Roots of *Peucedanum arenarium*. Phytochem. **15**, 209 (1976).

1270. CHEXAL, K.K., CH. TAMM, K. HIROTSU, and J. CLARDY: Gilmaniellin and Dechlorogilmaniellin, Two Novel Dimeric Oxaphenalenones. Helv. Chim. Acta **62**, 1785 (1979).

1271. WULFF, P., J.S. CARLÉ, and C. CHRISTOPHERSEN: Marine Alkaloids, 6: The First Naturally Occurring Bromo-Substituted Quinoline from *Flustra foliacea*. Comp. Biochem. Physiol. **71B**, 525 (1982).

1272. ARSHAD, M., J.P. DEVLIN, W.D. OLLIS, and R.E. WHEELER: The Constitution of Sordidone and Its Relation to Thiophanic Acid. J. Chem. Soc., Chem. Commun., 154 (1968).

1273. FOX, C.H., and S. HUNECK: The Formation of Roccellic Acid, Eugenitol, Eugenetin, and Rupicolon by the Mycobiont *Lecanora rupicola*. Phytochem. **8**, 1301 (1969).

1274. DEVLIN, J.P., C.P. FALSHAW, W.D. OLLIS, and R.E. WHEELER: Phytochemical Examination of the Lichen, *Lecanora rupicola* (L.) Zahlbr. J. Chem. Soc. (C), 1318 (1971).

1275. RICHARDS, M., A.E. BIRD, and J.E. MUNDEN: Chlorflavonin, a New Antifungal Antibiotic. J. Antibiot. **22**, 388 (1969).

1276. BIRD, A.E., and A.C. MARSHALL: Structure of Chloroflavonin. J. Chem. Soc. (C), 2418 (1969).

1277. MUNDEN, J.E., D. BUTTERWORTH, G. HANSCOMB, and M.S. VERRALL: Production of Chloroflavonin, an Antifungal Metabolite of *Aspergillus candidus*. Appl. Microbiol. **19**, 718 (1970).

1278. MARCHELLI, R., and L.C. VINING: Biosynthesis of Flavonoid and Terphenyl Metabolites by the Fungus *Aspergillus candidus*. J. Chem. Soc., Chem. Commun., 555 (1973).

1279. MARCHELLI, R., and L.C. VINING: The Biosynthetic Origin of Chlorflavonin, a Flavonoid Antibiotic from *Aspergillus candidus*. Can. J. Biochem. **51**, 1624 (1973).

1280. KÖNIG, W.A., C. KRAUSS, and H. ZÄHNER: Stoffwechselprodukte von Mikroorganismen 6-Chlorgenistein und 6,3'-Dichlorgenistein. Helv. Chim. Acta **60**, 2071 (1977).

1281. ANYANWUTAKU, I.O., E. ZIRBES, and J.P.N. ROSAZZA: Isoflavonoids from Streptomycetes: Origins of Genistein, 8-Chlorogenistein, and 6,8-Dichlorogenistein. J. Nat. Prod. **55**, 1498 (1992).

1282. KOMIYAMA, K., S. FUNAYAMA, Y. ANRAKU, A. MITA, Y. TAKAHASHI, S. OMURA, and H. SHIMASAKI: Isolation of Isoflavonoids Possessing Antioxidant Activity from the Fermentation Broth of *Streptomyces* sp. J. Antibiot. **42**, 1344 (1989).

1283. SYRCHINA, A.I., G.G. ZAPESOCHNAYA, N.A. TYUKAYKINA, and M.G. VORONKOV: 6-Chloroapigenin from *Equisetum arvense* L. Chem. Nat. Comp. **16**, 356 (1980).

1284. KONDO, H., S. NAKAJIMA, N. YAMAMOTO, A. OKURA, F. SATOH, H. SUDA, M. OKANISHI, and N. TANAKA: BE-14348 Substances, New Specific Estrogen-Receptor Binding Inhibitors. Production, Isolation, Structure Determination and Biological Properties. J. Antibiot. **43**, 1533 (1990).

1285. SASSA, T., K. HORIGUCHI, and Y. SUZUKI: Chloromonilinic Acids A and B, Novel Catabolites of the Growth Self-Inhibitor Chloromonilicin Isolated from *Monilinia fructicola*. Agric. Biol. Chem. **53**, 1337 (1989).

1286. BOHLMANN, F., T. BURKHARDT, and C. ZDERO: Naturally Occurring Acetylenes. New York: Academic Press. 1973.

1287. BOHLMANN, F., and C. ZDERO: Thiophene and Its Derivatives, Pt. 1, Chap. III (S. Gronowitz, ed.). New York: Wiley. 1985.

1288. BOHLMANN, F., S. POSTULKA, and J. RUHNKE: Die Polyine der Gattung *Centaurea* L. Chem. Ber. **91**, 1642 (1958).

1289. ANDERSEN, A.B., J. LAM, and P. WRANG: Polyunsaturated Compounds of *Centaurea scabiosa*. Phytochem. **16**, 1829 (1977).

1290. BOHLMANN, F., S. KÖHN, and C. ARNDT: Die Polyine der Gattung *Carthamus* L. Chem. Ber. **99**, 3433 (1966).

1291. BOHLMANN, F., K.-M. RODE, and M. GRENZ: Über die Inhaltsstoffe von *Dicoma zeyheri.* Chem. Ber. **100**, 3201 (1967).
1292. BOHLMANN, F., and N.L. VAN: New Germacranolides from *Dicoma anomala.* Phytochem. **17**, 570 (1978).
1293. BOHLMANN, F., M. AHMED, M. GRENZ, R.M. KING, and H. ROBINSON: Bisabolene Derivatives and Other Constituents from *Coreopsis* Species. Phytochem. **22**, 2858 (1983).
1294. CHRISTENSEN, L.P., and J. LAM: Acetylenes and Related Compounds in Cynareae. Phytochem. **29**, 2753 (1990).
1295. CHRISTENSEN, L.P., and J. LAM: Acetylenes and Related Compounds in Heliantheae. Phytochem. **30**, 11 (1991).
1296. BOHLMANN, F., W. SUCROW, H. JASTROW, and H.-J. KOCH: Über weitere Polyine aus *Centaurea ruthenica* Lam. Chem. Ber. **94**, 3179 (1961).
1297. BOHLMANN, F., and C. ZDERO: Über weitere Inhaltsstoffe von *Centaurea ruthenica* Lam. Chem. Ber. **106**, 2140 (1973).
1298. BOHLMANN, F., W. SKUBALLA, C. ZDERO, T. KÜHLE, and P. STEIRL: Über eine neue Photoreaktion von Polyinenen und ein Cyclopropan-Derivat aus *Centaurea ruthenica* Lam. Liebigs Ann. Chem. **745**, 176 (1971).
1299. MCLACHLAN, D., T. ARNASON, and J. LAM: The Role of Oxygen in Photosensitizations with Polyacetylenes and Thiophene Derivatives. Photochem. Photobio. **39**, 177 (1984).
1300. KAGAN, J., T.P. WANG, I.A. KAGAN, R.W. TUVESON, G.-R. WANG, and J. LAM: Photosensitization by 2-Chloro-3,11-tridecadiene-5,7,9-triyn-1-ol: Damage to Erythrocyte Membranes, *Escherichia coli*, and DNA. Photochem. Photobio. **55**, 63 (1992).
1301. BOHLMANN, F., and U. HINZ: Biogenetische Beziehungen zwischen natürlich vorkommenden Polyinen. Chem. Ber. **97**, 520 (1964).
1302. BOHLMANN, F., and C. ARNDT: Über einen neuen Polyintyp aus *Anaphalis margaritacea* B. et H. Chem. Ber. **98**, 1416 (1965).
1303. BOHLMANN, F., C. ARNDT, and C. ZDERO: Über neue Enolätherpolyine aus *Anaphalis-* und *Gnaphalium*-Arten. Chem. Ber. **99**, 1648 (1966).
1304. BOHLMANN, F., and W. SKUBALLA: Synthese der Enolätherpolyine aus *Gnaphalium*-Arten. Chem. Ber. **104**, 1962 (1971).
1305. BOHLMANN, F., C. ARNDT, K.-M. KLEINE, and H. BORNOWSKI: Die Acetylenverbindungen der Gattung *Echinops* L. Chem. Ber. **98**, 155 (1965).
1306. BOHLMANN, F., K.-M. RODE, and C. ZDERO: Neue Polyine der Gattung *Centaurea* L. Chem. Ber. **99**, 3544 (1966).
1307. ABEGAZ, B.M., M. TADESSE, and R. MAJINDA: Distribution of Sesquiterpene Lactones and Polyacetylenic Thiophenes in *Echinops.* Biochem. Sys. Ecol. **19**, 323 (1991).
1308. LAM, J., L.P. CHRISTENSEN, and T. THOMASEN: Thiophene Derivatives from *Echinops* Species. Phytochem. **30**, 1157 (1991).
1309. BOHLMANN, F., M.A. METWALLY, and J. JAKUPOVIC: Eudesmanolides from *Pluchea dioscoridis.* Phytochem. **23**, 1975 (1984).
1310. BOHLMANN, F., and C. ZDERO: Über die Inhaltsstoffe aus *Eclipta erecta* L. Chem. Ber. **103**, 834 (1970).
1311. BOHLMANN, F., W.-R. ABRAHAM, R.M. KING, and H. ROBINSON: Thiophene Acetylenes and Flavanols from *Pterocaulon virgatum.* Phytochem. **20**, 825 (1981).
1312. BOHLMANN, F., N. BORTHAKUR, H. ROBINSON, and R.M. KING: Eudesmane Derivatives from *Epaltes brasiliensis.* Phytochem. **21**, 1795 (1982).

1313. BALZA, F., and G.H.N. TOWERS: Dithiacyclohexadiene Chlorohydrins and Related Sulphur Containing Polyynes from *Ambrosia chamissonis*. Phytochem. **29**, 2901 (1990).

1314. ATKINSON, R.E., R.F. CURTIS, and G.T. PHILLIPS: Naturally-Occurring Thiophens, Part II: 5-(4-Chloro-3-hydroxybut-1-ynyl)-2,2′-bithienyl from *Tagetes minuta* L. J. Chem. Soc. (C), 1101 (1966).

1315. BOHLMANN, F., R.N. BARUAH, and X. DOMINGUEZ: A Further Dithienyl Derivative from *Porophyllum scoparia*. Planta Med., 77 (1985).

1316. BOHLMANN, F., C. ZDERO, and W. GORDON: Über die Inhaltsstoffe von *Berkheya adlamii* Hook. Chem. Ber. **100**, 1193 (1967).

1317. BOHLMANN, F., N.L. VAN, T.V.C. PHAM, J. JACUPOVIC, A. SCHUSTER, V. ZABEL, and W.H. WATSON: β-Isocomen, ein neues Sesquiterpen aus *Berkheya*-Arten. Phytochem. **18**, 1831 (1979).

1318. BOHLMANN, F., and W.-R. ABRAHAM: Neue, chlorsubstituierte Thiophenacetylenverbindungen mit ungewöhnlicher Struktur aus *Helichrysum*-Arten. Phytochem. **18**, 839 (1979).

1319. FUJIMOTO, Y., and M. SATOH: A New Cytotoxic Chlorine-Containing Polyacetylene from the Callus of *Panax ginseng*. Chem. Pharm. Bull. (Japan) **36**, 4206 (1988).

1320. HIRAKURA, K., M. MORITA, K. NAKAJIMA, Y. IKEYA, and H. MITSUHASHI: Polyacetylenes from the Roots of *Panax ginseng*. Phytochem. **30**, 3327 (1991).

1321. CIMINO, G., and S. DE STEFANO: New Acetylenic Compounds from the Sponge *Reniera fulva*. Tetrahedron Lett., 1325 (1977).

1322. QUINN, R.J., and D.J. TUCKER: Further Acetylenic Acids from the Marine Sponge *Xestospongia testudinaria*. J. Nat. Prod. **54**, 290 (1991).

1323. BOURGUET-KONDRACKI, M.L., M.T. RAKOTOARISOA, M.T. MARTIN, and M. GUYOT: Bioactive Bromopolyacetylenes from the Marine Sponge *Xestospongia testudinaria*. Tetrahedron Lett. **33**, 225 (1992).

1324. PATIL, A.D., W.C. KOKKE, S. COCHRAN, T.A. FRANCIS, T. TOMSZEK, and J.W. WESTLEY: Brominated Polyacetylenic Acids from the Marine Sponge *Xestospongia muta*: Inhibitors of HIV Protease. J. Nat. Prod. **55**, 1170 (1992).

1325. FUSETANI, N., H. LI, K. TAMURA, and S. MATSUNAGA: Antifungal Brominated C_{18} Acetylenic Acids from the Marine Sponge, *Petrosia volcano* Hoshino. Tetrahedron **49**, 1203 (1993).

1326. WALKER, R.P., and D.J. FAULKNER: Chlorinated Acetylenes from the Nudibranch *Diaulula sandiegensis*. J. Org. Chem. **46**, 1475 (1981).

1327. LEE, M.D., T.S. DUNNE, M.M. SIEGEL, C.C. CHANG, G.O. MORTON, and D.B. BORDERS: Calichemicins, a Novel Family of Antitumor Antibiotics, 1: Chemistry and Partial Structure of Calichemicin γ_1^I. J. Am. Chem. Soc. **109**, 3464 (1987).

1328. LEE, M.D., T.S. DUNNE, C.C. CHANG, G.A. ELLESTAD, M.M. SIEGEL, G.O. MORTON, W.J. MCGAHREN, and D.B. BORDERS: Calichemicins, a Novel Family of Antitumor Antibiotics, 2: Chemistry and Structure of Calichemicin γ_1^I. J. Am. Chem. Soc. **109**, 3466 (1987).

1329. LEE, M.D., J.K. MANNING, D.R. WILLIAMS, N.A. KUCK, R.T. TESTA, and D.B. BORDERS: Calicheamicins, a Novel Family of Antitumor Antibiotics, 3: Isolation, Purification and Characterization of Calicheamicins β_1^{Br}, γ_1^{Br}, α_2^I, α_3^I, β_1^I, γ_1^I and δ_1^I. J. Antibiot. **42**, 1070 (1989).

1330. LEE, M.D., G.A. ELLESTAD, and D.B. BORDERS: Calicheamicins: Discovery, Structure, Chemistry, and Interaction with DNA. Accounts Chem. Res. **24**, 235 (1991).

1331. LAM, K.S., G.A. HESLER, D.R. GUSTAVSON, A.R. CROSSWELL, J.M. VEITCH, and S. FORENZA: Kedarcidin, a New Chromoprotein Antitumor Antibiotic, I: Taxonomy of Producing Organism, Fermentation, and Biological Activity. J. Antibiot. **44**, 472 (1991).

1332. HOFSTEAD, S.J., J.A. MATSON, A.R. MALACKO, and H. MARQUARDT: Kedarcidin, a New Chromoprotein Antitumor Antibiotic, II: Isolation, Purification and Physio-Chemical Properties. J. Antibiot. **45**, 1250 (1992).

1333. LEET, J.E., D.R. SCHROEDER, S.J. HOFSTEAD, J. GOLIK, K.L. COLSON, S. HUANG, S.E. KLOHR, T.W. DOYLE, and J.A. MATSON: Kedarcidin, a New Chromoprotein Antitumor Antibiotic: Structure Elucidation of Kedarcidin Chromophore. J. Am. Chem. Soc. **114**, 7946 (1992).

1334. LEET, J.E., D.R. SCHROEDER, D.R. LANGLEY, K.L. COLSON, S. HUANG, S.E. KLOHR, M.S. LEE, J. GOLIK, S.J. HOFSTEAD, T.W. DOYLE, and J.A. MATSON: Chemistry and Structure Elucidation of the Kedarcidin Chromophore. J. Am. Chem. Soc. **115**, 8432 (1993).

1335. ZEIN, N., A.M. CASAZZA, T.W. DOYLE, J.E. LEET, D.R. SCHROEDER, W. SOLOMON, and S.G. NADLER: Selective Proteolytic Activity of the Antitumor Agent Kedarcidin. Proc. Natl. Acad. Sci. USA **90**, 8009 (1993).

1336. HU, J., Y.-C. XUE, M.-Y. XIE, and R. ZHANG: A New Macromolecular Antitumor Antibiotic, C-1027, I: Discovery, Taxonomy of Producing Organism, Fermentation and Biological Activity. J. Antibiot. **41**, 1575 (1988).

1337. OTANI, T., Y. MINAMI, T. MARUNAKA, R. ZHANG, and M.-Y. XIE: A New Macromolecular Antitumor Antibiotic, C-1027, II: Isolation and Physico-Chemical Properties. J. Antibiot. **41**, 1580 (1988).

1338. ZHEN, Y., X. MING, B. YU, T. OTANI, H. SAITO, and Y. YAMADA: A New Macromolecular Antitumor Antibiotic, C-1027, III: Antitumor Activity. J. Antibiot. **42**, 1294 (1989).

1339. SUGIMOTO, Y., T. OTANI, S. OIE, K. WIERZBA, and Y. YAMADA: Mechanism of Action of a New Macromolecular Antitumor Antibiotic, C-1027. J. Antibiot. **43**, 417 (1990).

1340. MINAMI, Y., K. YOSHIDA, R. AZUMA, M. SAEKI, and T. OTANI: Structure of an Aromatization Product of C-1027 Chromophore. Tetrahedron Lett. **34**, 2633 (1993).

1341. YOSHIDA, K., Y. MINAMI, R. AZUMA, M. SAEKI, and T. OTANI: Structure and Cycloaromatization of a Novel Enediyne, C-1027 Chromophore. Tetrahedron Lett. **34**, 2637 (1993).

1342. IIDA, K., T. ISHII, M. HIRAMA, T. OTANI, Y. MINAMI, and K. YOSHIDA: Synthesis and Absolute Stereochemistry of the Aminosugar Moiety of Antibiotic C-1027 Chromophore. Tetrahedron Lett. **34**, 4079 (1993).

1343. KELLER-SCHIERLEIN, W., R. MUNTWYLER, W. PACHE, and H. ZÄHNER: Stoffwechselprodukte von Mikroorganismen Chlorothricin und Des-chlorothricin. Helv. Chim. Acta **52**, 127 (1969).

1344. BRUFANI, M., S. CERRINI, W. FEDELI, F. MAZZA, and R. MUNTWYLER: Kristallstrukturanalyse des Chlorothricolid-methylesters. Helv. Chim. Acta **55**, 2094 (1972).

1345. HOLZBACH, R., H. PAPE, D. HOOD, E.F. KREUTZER, C. CHANG, and H.G. FLOSS: Biosynthesis of the Macrolide Antibiotic Chlorothricin: Basic Building Blocks. Biochem. **17**, 556 (1978).

1346. MASCARETTI, O.A., C. CHANG, D. HOOK, H. OTSUKA, E.F. KREUTZER, and H.G. FLOSS: Biosynthesis of the Macrolide Antibiotic Chlorothricin. Biochem. **20**, 919 (1978).

1347. LEE, J.J., J.P. LEE, P.J. KELLER, C.E. COTTRELL, C. CHANG, H. ZÄHNER, and H.G. FLOSS: Further Studies on the Biosynthesis of Chlorothricin. J. Antibiot. **39**, 1123 (1986).

1348. KAWASHIMA, A., Y. NAKAMURA, Y. OHTA, T. AKAMA, M. YAMAGISHI, and

K. Hanada: New Cholesterol Biosynthesis Inhibitors MC-031 (*O*-Demethylchlorothricin), -032 (*O*-Demethylhydroxychlorothricin), -033 and -034. J. Antibiot. **45**, 207 (1992).

1349. Kupchan, S.M., Y. Komoda, A.R. Branfman, A.T. Sneden, W.A. Court, G.J. Thomas, H.P.J. Hintz, R.M. Smith, A. Karim, G.A. Howie, A.K. Verma, Y. Nagao, R.G. Dailey Jr., V.A. Zimmerly, and W.C. Sumner Jr.: The Maytansinoids. Isolation, Structure Elucidation and Chemical Interrelation of Novel Ansa Macrolides. J. Org. Chem. **42**, 2349 (1977).

1350. Wani, M.C., H.L. Taylor, and M.E. Wall: Plant Antitumour Agents: Colubrinol Acetate and Colubrinol, Antileukaemic Ansa Macrolides from *Colubrina texensis*. J. Chem. Soc., Chem. Commun., 390 (1973).

1351. Sneden, A.T., and G.L. Beemsterboer: Normaytansine, a New Antileukemic Ansa Macrolide from *Maytenus buchananii*. J. Nat. Prod. **43**, 637 (1980).

1352. Higashide, E., M. Asai, K. Ootsu, S. Tanida, Y. Kozai, T. Hasegawa, T. Kishi, Y. Sugino, and M. Yoneda: Ansamitocin, a Group of Novel Maytansinoid Antibiotics with Antitumor Properties from *Nocardia*. Nature **270**, 721 (1977).

1353. Asai, M., E. Mizuta, M. Izawa, K. Haibara, and T. Kishi: Isolation, Chemical Characterization and Structure of Ansamitocin, a New Antitumor Ansamycin Antibiotic. Tetrahedron **35**, 1079 (1979).

1354. Suwanborirux, K., C.-J. Chang, R.W. Spjut, and J.M. Cassady: Ansamitocin P-3, a Maytansinoid, from *Claopodium crispifolium* and *Anomodon attenuatus* or Associated Actinomycetes. Experientia **46**, 117 (1990).

1355. Powell, R.G., D. Weisleder, and C.R. Smith Jr.: Novel Maytansinoid Tumor Inhibitors from *Trewia nudiflora*: Trewiasine, Dehydrotrewiasine, and Demethyltrewiasine. J. Org. Chem. **46**, 4398 (1981).

1356. Powell, R.G., D. Weisleder, C.R. Smith Jr., J. Kozlowski, and W.K. Rohwedder: Treflorine, Trenudine, and *N*-Methyltrenudone: Novel Maytansinoid Tumor Inhibitors Containing Two Fused Macrocyclic Rings. J. Am. Chem. Soc. **104**, 4929 (1982).

1357. Sakai, K., T. Ichikawa, K. Yamada, M. Yamashita, M. Tanimoto, A. Hikita, Y. Ijuin, and K. Kondo: Antitumor Principles in Mosses: The First Isolation and Identification of Maytansinoids, Including a Novel 15-Methoxyansamitocin P-3. J. Nat. Prod. **51**, 845 (1988).

1358. Mascaretti, O., C. Chang, and H.G. Floss: Biosynthesis of the Antibiotic Chlorothricin: Assignment of the Carbon-13 Magnetic Resonance Spectrum. J. Nat. Prod. **42**, 455 (1979).

1359. Yamamoto, I., M. Nakagawa, Y. Hayakawa, K. Adachi, and E. Kobayashi: Hydroxychlorothricin, a New Antitumor Antibiotic. J. Antibiot. **40**, 1452 (1987).

1360. Mirrington, R.N., E. Ritchie, C.W. Shoppee, W.C. Taylor, and S. Sternhell: The Constitution of Radicicol. Tetrahedron Lett., 365 (1964).

1361. McCapra, F., A.I. Scott, P. Delmotte, J. Delmotte-Plaquée, and N.S. Bhacca: The Constitution of Monorden, an Antibiotic with Tranquilising Action. Tetrahedron Lett., 869 (1964).

1362. Ayer, W.A., S.P. Lee, A. Tsuneda, and Y. Hiratsuka: The Isolation, Identification, and Bioassay of the Antifungal Metabolites Produced by *Monocillium nordinii*. Can. J. Microbiol. **26**, 766 (1980).

1363. Cutler, H.G., R.F. Arrendale, J.P. Springer, P.D. Cole, R.G. Roberts, and R.T. Hanlin: Monorden from a Novel Source, *Neocosmospora tenuicristata*: Stereochemistry and Plant Growth Regulatory Properties. Agric. Biol. Chem. **51**, 3331 (1987).

1364. Ghisalberti, E.L., and C.Y. Rowland: 6-Chlorodehydrocurvularin, a New Metabolite from *Cochliobolus spicifer*. J. Nat. Prod. **56**, 2175 (1993).

1365. Boeynaems, J.M., D. Reagan, and W.C. Hubbard: Lactoperoxidase-Catalyzed Iodination of Arachidonic Acid: Formation of Macrolides. Lipids **16**, 246 (1981).

1366. Murakami, M., H. Matsuda, K. Makabe, and K. Yamaguchi: Oscillariolide, a Novel Macrolide from a Blue-Green Alga *Oscillatoria* sp. Tetrahedron Lett. **32**, 2391 (1991).

1367. Kato, Y., and P.J. Scheuer: Aplysiatoxin and Debromoaplysiatoxin, Constituents of the Marine Mollusk *Stylocheilus longicauda* (Quoy and Gaimard, 1824). J. Am. Chem. Soc. **96**, 2245 (1974).

1368. Kato, Y., and P.J. Scheuer: The Aplysiatoxins. Pure Appl. Chem. **41**, 1 (1975).

1369. Mynderse, J.S., R.E. Moore, M. Kashiwagi, and T.R. Norton: Antileukemic Activity in the Oscillatoriaceae: Isolation of Debromoaplysiatoxin from *Lyngbya*. Science **196**, 538 (1977).

1370. Mynderse, J.S., and R.E. Moore: Toxins from Blue-Green Algae: Structures of Oscillatoxin A and Three Related Bromine-Containing Toxins. J. Org. Chem. **43**, 2301 (1978).

1371. Moore, R.E., A.J. Blackman, C.E. Cheuk, J.S. Mynderse, G.K. Matsumoto, J. Clardy, R.W. Woodard, and J.C. Craig: Absolute Stereochemistries of the Aplysiatoxins and Oscillatoxin A. J. Org. Chem. **49**, 2484 (1984).

1372. Kobayashi, M., S. Aoki, H. Sakai, K. Kawazoe, N. Kihara, T. Sasaki, and I. Kitagawa: Altohyrtin A, a Potent Anti-Tumor Macrolide from the Okinawan Marine Sponge *Hyrtios altum*. Tetrahedron Lett. **34**, 2795 (1993).

1373. Kobayashi, M., S. Aoki, H. Sakai, N. Kihara, T. Sasaki, and I. Kitagawa: Altohyrtins B and C and 5-Desacetylaltohyrtin A, Potent Cytotoxic Macrolide Congeners of Altohyrtin A, from the Okinawan Marine Sponge *Hyrtios altum*. Chem. Pharm. Bull. (Japan) **41**, 989 (1993).

1374. Fusetani, N., K. Shinoda, and S. Matsunaga: Cinachyrolide A: A Potent Cytotoxic Macrolide Possessing Two Spiro Ketals from Marine Sponge *Cinachyra* sp. J. Am. Chem. Soc. **115**, 3977 (1993).

1375. Pettit, G.R., Z.A. Cichacz, F. Gao, C.L. Herald, M.R. Boyd, J.M. Schmidt, and J.N.A. Hooper: Isolation and Structure of Spongistatin 1. J. Org. Chem. **58**, 1302 (1993).

1376. Pettit, G.R., Z.A. Cichacz, F. Gao, C.L. Herald, and M.R. Boyd: Isolation and Structure of the Remarkable Human Cancer Cell Growth Inhibitors Spongistatins 2 and 3 from an Eastern Indian Ocean *Spongia* sp. J. Chem. Soc., Chem. Commun., 1166 (1993).

1377. Pettit, G.R., C.L. Herald, Z.A. Cichacz, F. Gao, J.M. Schmidt, M.R. Boyd, N.D. Christie, and F.E. Boettner: Isolation and Structure of the Powerful Human Cancer Cell Growth Inhibitors Spongistatins 4 and 5 from an African *Spirastrella spinispirulifera* (*Porifera*). J. Chem. Soc., Chem. Commun., 1805 (1993).

1378. Pettit, G.R., Z.A. Cichacz, C.L. Herald, F. Gao, M.R. Boyd, J.M. Schmidt, E. Hamel, and R. Bai: Antineoplastic Agents, 300: Isolation and Structure of the Rare Human Cancer Inhibitory Macrocyclic Lactones Spongistatins 8 and 9. J. Chem. Soc., Chem. Commun., 1605 (1994).

1379. Gremmen, J.: A New, Crystalline, Antibiotic Substance Produced by *Mollisia* Species (Discomycetes). Antonie van Leeuwenhoek J. Microbiol. Serol. **22**, 58 (1956).

1380. Van der Kerk, G.J.M., and J.C. Overeem: Mollisin, a Dichloronaphthoquinone Derivative Produced by the Fungus *Mollisia caesia*. Recueil **76**, 425 (1957).

1381. OVEREEM, J.C., and G.J.M. VAN DER KERK: Mollisin, a Naturally Occurring Chlorine-Containing Quinone, Part II: The Structure of Mollisin. Recueil **83**, 995 (1964).

1382. BENTLEY, R., and S. GATENBECK: Naphthoquinone Biosynthesis in Molds. The Mechanism for Formation of Mollisin. Biochem. **4**, 1150 (1965).

1383. BENDZ, G., and G. LINDBERG: Naphthoquinones and Anthocyanins from Two *Drosera* Species. Acta Chem. Scand. **22**, 2722 (1968).

1384. SIDHU, G.S., and A.V.B. SANKARAM: A New Biplumbagin and 3-Chloroplumbagin. Tetrahedron Lett., 2385 (1971).

1385. KREHER, B., A. NESZMÉLYI, and H. WAGNER: Naphthoquinones from *Dionaea muscipula*. Phytochem. **29**, 605 (1990).

1386. HIGA, M., K. HIMENO, S. YOGI, and K. HOKAMA: A New Brominated Naphthoquinone from *Diospyros maritima* Blume. Chem. Pharm. Bull. (Japan) **35**, 4366 (1987).

1387. BALERNA, M., W. KELLER-SCHIERLEIN, C. MARTIUS, H. WOLF, and H. ZÄHNER: Stoffwechselprodukte von Mikroorganismen, 72: Naphthomycin, ein Antimetabolit von Vitamin K. Arch. Mikrobiol. **65**, 303 (1969).

1388. WILLIAMS, T.H.: Naphthomycin, a Novel Ansa Macrocyclic Antimetabolite. Proton NMR Spectra and Structure Elucidation Using Lanthanide Shift Reagent. J. Antibiot. **28**, 85 (1975).

1389. BRUFANI, M., L. CELLAI, and W. KELLER-SCHIERLEIN: Degradation Studies of Naphthomycin. J. Antibiot. **32**, 167 (1979).

1390. KELLER-SCHIERLEIN, W., M. MEYER, L. CELLAI, S. CERRINI, D. LAMBA, A. SEGRE, W. FEDELI, and M. BRUFANI: Metabolites of Microorganisms, 229: Absolute Configuration of Naphthomycin A Determined by X-Ray Analysis and Chemical Degradation. J. Antibiot. **37**, 1357 (1984).

1391. MUKHOPADHYAY, T., C.M.M. FRANCO, G.C.S. REDDY, B.N. GANGULI, and H.W. FEHLHABER: A New Ansamycin Antibiotic, Naphthomycin H from a *Streptomyces* Species Y-83,40369. J. Antibiot. **38**, 948 (1985).

1392. KELLER-SCHIERLEIN, W., M. MEYER, A. ZEECK, M. DAMBERG, R. MACHINEK, H. ZÄHNER, and G. LAZAR: Isolation and Structural Elucidation of Naphthomycins B and C. J. Antibiot. **36**, 484 (1983).

1393. HENKEL, T., and A. ZEECK: Secondary Metabolites by Chemical Screening, 15: Structure and Absolute Configuration of Naphthomevalin, a New Dihydro-Naphthoquinone Antibiotic from *Streptomyces* sp. J. Antibiot. **44**, 665 (1991).

1394. SHOMURA, T., S. GOMI, M. ITO, J. YOSHIDA, E. TANAKA, S. AMANO, H. WATABE, S. OHUCHI, J. ITOH, M. SEZAKI, H. TAKEBE, and K. UOTANI: Studies on New Antibiotics SF 2415, I: Taxonomy, Fermentation, Isolation, Physico-Chemical Properties, and Biological Activities. J. Antibiot. **40**, 732 (1987).

1395. GOMI, S., S. OHUCHI, T. SASAKI, J. ITOH, and M. SEZAKI: Studies on New Antibiotics SF 2415, II: The Structural Elucidation. J. Antibiot. **40**, 740 (1987).

1396. SHIOMI, K., H. IINUMA, M. HAMADA, H. NAGANAWA, M. MANABE, C. MATSUKI, T. TAKEUCHI, and H. UMEZAWA: Novel Antibiotics Napyradiomycins. Production, Isolation, Physico-Chemical Properties and Biological Activity. J. Antibiot. **39**, 487 (1986).

1397. SHIOMI, K., H. NAKAMURA, H. IINUMA, H. NAGANAWA, K. ISSHIKI, T. TAKEUCHI, H. UMEZAWA, and Y. IITAKA: Structures of New Antibiotics Napyradiomycins. J. Antibiot. **39**, 494 (1986).

1398. SHIOMI, K., H. NAKAMURA, H. IINUMA, H. NAGANAWA, T. TAKEUCHI, H. UMEZAWA, and Y. IITAKA: New Antibiotic Napyradiomycins A2 and B4 and Stereochemistry of Napyradiomycins. J. Antibiot. **40**, 1213 (1987).

1399. HORI, Y., Y. ABE, N. SHIGEMATSU, T. GOTO, M. OKUHARA, and M. KOHSAKA: Napyradiomycins A and B1: Non-Steroidal Estrogen-Receptor Antagonists Produced by a *Streptomyces*. J. Antibiot. **46**, 1890 (1993).

1400. PATHIRANA, C., P.R. JENSEN, and W. FENICAL: Marinone and Debromomarinone: Antibiotic Sesquiterpenoid Naphthoquinones of a New Structure Class from a Marine Bacterium. Tetrahedron Lett. **33**, 7663 (1992).

1401. CAMERON, D.W., and M.D. SIDELL: 1,3,6,8,11,13-Hexachloro-4,10-dihydroxy-dinaphtho[2,1-*b*:1′,2′-*d*]furan-5,9-dione. A Polychloroquinone from Green Soils. Aust. J. Chem. **31**, 1323 (1978).

1402. CAMERON, D.W., C.L. RASTON, and A.H. WHITE: Crystal Structure of a Polychloroquinone Soil Pigment. Aust. J. Chem. **31**, 2441 (1978).

1403. BUTLER, J.H.A., D.T. DOWNING, and R.J. SWABY: Isolation of a Chlorinated Pigment from Green Soil. Aust. J. Chem. **17**, 817 (1964).

1404. BRINKMAN, L.C., F.R. LEY, and P.J. SEATON: Isolation and Synthesis of Benz[*a*]-anthraquinones Related to Antitumor Agent PD 116740. J. Nat. Prod. **56**, 374 (1993).

1405. DUGGAR, B.M.: Aureomycin, a Product of the Continuing Search for New Antibiotics. Ann. N.Y. Acad. Sci. **51**, 177 (1948).

1406. BROSCHARD, R.W., A.C. DORNBUSH, S. GORDON, B.L. HUTCHINGS, A.R. KOHLER, G. KRUPKA, S. KUSHNER, D.V. LEFEMINE, and C. PIDACKS: Aureomycin, a New Antibiotic. Science **109**, 199 (1949).

1407. STEPHENS, C.R., L.H. CONOVER, F.A. HOCHSTEIN, P.P. REGNA, F.J. PILGRIM, K.J. BRUNINGS, and R.B. WOODWARD: Terramycin, VIII: Structure of Aureomycin and Terramycin. J. Am. Chem. Soc. **74**, 4976 (1952).

1408. STEPHENS, C.R., L.H. CONOVER, R. PASTERNACK, F.A. HOCHSTEIN, W.T. MORELAND, P.P. REGNA, F.J. PILGRIM, K.J. BRUNINGS, and R.B. WOODWARD: The Structure of Aureomycin. J. Am. Chem. Soc. **76**, 3568 (1954).

1409. DOBRYNIN, V.N., A.I. GUREVICH, M.G. KARAPETYAN, M.N. KOLOSOV, and M.M. SHEMYAKIN: The Absolute Configuration of the Tetracyclines. Tetrahedron Lett., 901 (1962).

1410. DONOHUE, J.D., J. DUNITZ, K.N. TRUEBLOOD, and M.S. WEBSTER: The Crystal Structure of Aureomycin (Chlortetracycline) Hydrochloride. Configuration, Bond Distances and Conformation. J. Am. Chem. Soc. **85**, 851 (1963).

1411. MCCORMICK, J.R.D., N.O. SJOLANDER, U. HIRSCH, E.R. JENSEN, and A.P. DOERSCHUK: A New Family of Antibiotics: The Demethyltetracyclines. J. Am. Chem. Soc. **79**, 4561 (1957).

1412. MCCORMICK, J.R.D., P.A. MILLER, J.A. GROWICH, N.O. SJOLANDER, and A.P. DOERSCHUK: Two New Tetracycline-Related Compounds: 7-Chloro-5a(11a)-dehydrotetracycline and 5a-*epi*-Tetracycline. A New Route to Tetracycline. J. Am. Chem. Soc. **80**, 5572 (1958).

1413. MCCORMICK, J.R.D., E.R. JENSEN, S. JOHNSON, and N.O. SJOLANDER: Biosynthesis of the Tetracyclines, IX: 4-Aminodedimethylaminoanhydrodemethylchlortetracycline from a Mutant of *Streptomyces aureofaciens*. J. Am. Chem. Soc. **90**, 2201 (1968).

1414. MCCORMICK, J.R.D., and E.R. JENSEN: Biosynthesis of the Tetracyclines, XII: Anhydrodemethylchlortetracycline from a Mutant of *Streptomyces aureofaciens*. J. Am. Chem. Soc. **91**, 206 (1969).

1415. SENSI, P., G.A. DE FERRARI, G.G. GALLO, and G. ROLLAND: Bromotetracycline, an Antibiotic, I: Isolation and Characterization. Il Farmaco Ed. Sci. **10**, 337 (1955).

1416. PATEL, M., V.P. GULLO, V.R. HEDGE, A.C. HORAN, F. GENTILE, J.A. MARQUEZ, G.H. MILLER, M.S. PUAR, and J.A. WAITZ: A Novel Tetracycline from *Actinomadura brunnea*. Fermentation, Isolation and Structure Elucidation. J. Antibiot. **40**, 1408 (1987).

1417. SMITH, E.B., H.K. MUNAYYER, M.J. RYAN, B.A. MAYLES, V.R. HEGDE, and G.H. MILLER: Direct Selection of a Specifically Blocked Mutant of *Actinomadura brunnea*. Isolation of a Third 8-Methoxy Substituted Chlortetracycline. J. Antibiot. **40**, 1419 (1987).

1418. PATEL, M., V.P. GULLO, V.R. HEGDE, A.C. HORAN, J.A. MARQUEZ, R. VAUGHAN, M.S. PUAR, and G.H. MILLER: A New Tetracycline Antibiotic from a *Dactylosporangium* Species. Fermentation, Isolation and Structure Elucidation. J. Antibiot. **40**, 1414 (1987).

1419. WELLS, J.S., J. O'SULLIVAN, C. AKLONIS, H.A. AX, A.A. TYMIAK, D.R. KIRSCH, W.H. TREJO, and P. PRINCIPE: Dactylocyclines, Novel Tetracycline Derivatives Produced by a *Dactylosporangium* sp., I: Taxonomy, Production, Isolation, and Biological Activity. J. Antibiot. **45**, 1892 (1992).

1420. TYMIAK, A.A., H.A. AX, M.S. BOLGAR, A.D. KAHLE, M.A. PORUBCAN, and N.H. ANDERSEN: Dactylocyclines, Novel Tetracycline Derivatives Produced by a *Dactylosporangium* sp., II: Structure Determination. J. Antibiot. **45**, 1899 (1992).

1421. DEVASTHALE, P.V., L.A. MITSCHER, H. TELIKEPALLI, D.V. VELDE, J.-Y. ZOU, H.A. AX, and A.A. TYMIAK: Dactylocyclines, Novel Tetracycline Derivatives Produced by a *Dactylosporangium* sp., III: Absolute Stereochemistry of the Dactylocyclines. J. Antibiot. **45**, 1907 (1992).

1422. TYMIAK, A.A., C. AKLONIS, M.S. BOLGAR, A.D. KAHLE, D.R. KIRSCH, J. O'SULLIVAN, M.A. PORUBCAN, P. PRINCIPE, W.H. TREJO, H.A. AX, J.S. WELLS, N.H. ANDERSEN, P.V. DEVASTHALE, H. TELIKEPALLI, D. VAN DER VELDE, J.-Y. ZOU, and L.A. MITSCHER: Dactylocyclines: Novel Tetracycline Glycosides Active Against Tetracycline-Resistant Bacteria. J. Org. Chem. **58**, 535 (1993).

1423. MILES, D.H., N.V. MODY, J.P. MINYARD, and P.A. HEDIN: Constituents of Marsh Grass: Survey of the Essential Oils in *Juncus roemerianus*. Phytochem. **12**, 1399 (1973).

1424. PEREIRA, W.E., C.E. ROSTAD, and H.E. TAYLOR: Mount St. Helens, Washington, 1980 Volcanic Eruption. Geophys. Res. Lett. **7**, 953 (1980).

1425. SANDUJA, R., M.A. ALAM, and G.M. WELLINGTON: Secondary Metabolites of the Non-Symbiotic Coral *Tubastraea micrantha* (Ehrenberg): Isolation and Structure of Novel Anthraquinoid Derivatives. J. Chem. Res., 450 (1986).

1426. MILES, D.H., K. TUNSUWAN, V. CITTAWONG, P.A. HEDIN, U. KOKPOL, C.-Z. NI, and J. CLARDY: Agrochemical Activity and Isolation of *N*-(4'-Bromophenyl)-2,2-diphenylacetanilide from the Thai Plant *Arundo donax*. J. Nat. Prod. **56**, 1590 (1993).

1427. LI, M.K.W., and P.J. SCHEUER: Halogenated Blue Pigments of a Deep Sea Gorgonian. Tetrahedron Lett. **25**, 587 (1984).

1428. CHEN, J.L., R.E. MOORE, G.M.L. PATTERSON: Structures of the Nostocyclophanes A–D. J. Org. Chem. **56**, 4360 (1991).

1429. BOBZIN, S.C., and R.E. MOORE: Biosynthetic Origin of [7,7]Paracyclophanes from Cyanobacteria. Tetrahedron **49**, 7615 (1993).

1430. NILSSON, U.L., and C.E. ÖSTMAN: Chlorinated Polycyclic Aromatic Hydrocarbons: Method of Analysis and Their Occurrence in Urban Air. Environ. Sci. Technol. **27**, 1826 (1993).

1431. SNEDEN, A.T., W.C. SUMNER JR., and S.M. KUPCHAN: Normaytancyprine, a Minor Antileukemic Ansa Macrolide from *Putterlickia verrucosa*. J. Nat. Prod. **45**, 624 (1982).

1432. KEITH, L.H., and W.A. TELLIARD: Priority Pollutants, I: A Perspective View. Environ. Sci. Technol. **13**, 416 (1979).

1433. Ando, K., A. Kato, and S. Suzuki: Isolation of 2,4-Dichlorophenol from a Soil Fungus and Its Biological Significance. Biochem. Biophys. Res. Commun. **39**, 1104 (1970).

1434. Berger, R.S.: 2,6-Dichlorophenol, Sex Pheromone of the Lone Star Tick. Science **177**, 704 (1972).

1435. Berger, R.S.: Incorporation of Chloride from $Na^{36}Cl$ into 2,6-Dichlorophenol in Lone Star and Gulf Coast Ticks. Ann. Entomol. Soc. Am. **67**, 961 (1974).

1436. Kellum, D., and R.S. Berger: Relationship of the Occurrence and Function of 2,6-Dichlorophenol in Two Species of *Amblyomma* (Acari: Ixodidae). J. Med. Entomol. **13**, 701 (1977).

1437. Chow, Y.S., C.B. Wang, and L.C. Lin: Identification of a Sex Pheromone of the Female Brown Dog Tick, *Rhipicephalus sanguineus.* Ann. Entomol. Soc. Am. **68**, 485 (1975).

1438. Wood, W.F., M.G. Leahy, R. Galun, G.D. Prestwich, T. Meinwald, R.E. Purnell, and R.C. Payne: Phenols as Pheromones of Ixodid Ticks: A General Phenomenon. J. Chem. Ecol. **1**, 501 (1975).

1439. Sonenshine, D.E., R.M. Silverstein, E.L. Plummer, J.R. West, and T. McCullough: 2,6-Dichlorophenol, the Sex Pheromone of the Rocky Mountain Wood Tick, *Dermacentor andersoni* Stiles, and the American Dog Tick, *Dermacentor variabilis* (SAY). J. Chem. Ecol. **2**, 201 (1976).

1440. Chow, Y.S., F. Lu, C. Peng, and P. Cheng: Isolation of Lipids and Sex Pheromones from Hard Ticks. Bull. Inst. Zool. Acad. Sin. **11**, 1 (1972); Chem. Abstr. **79**, 113363 (1973).

1441. Khalil, G.M., S.A. Nada, and D.E. Sonenshine: Sex Pheromone Regulation of Mating Behavior in the Camel Tick *Hyalomma dromedarii* (Ixodoidea: Ixodidae). J. Parasitol. **67**, 70 (1981).

1442. Berger, R.S.: Occurrence of 2,6-Dichlorophenol in *Dermacentor albipictus* and *Haemaphysalis leporispalustris* (Acari: Ixodidae). J. Med. Entomol. **20**, 103 (1983).

1443. Eisner, T., L.B. Hendry, D.B. Peakall, and J. Meinwald: 2,5-Dichlorophenol (from Ingested Herbicide?) in Defensive Secretion of Grasshopper. Science **172**, 277 (1971).

1444. Kavanagh, F., A. Hervey, and W.J. Robbins: Antibiotic Substances from Basidiomycetes, IX: *Drosophila subatrata.* Proc. Natl. Acad. Sci. **38**, 555 (1952).

1445. Anchel, M.: Identification of Drosophilin A as *p*-Methoxytetrachlorophenol. J. Am. Chem. Soc. **74**, 2943 (1952).

1446. Singh, P., and S. Rangaswami: Occurrence of *O*-Methyldrosphilin A in *Fomes fastuosus* Lev. Tetrahedron Lett., 1229 (1966).

1447. Butruille, D., and X.A. Dominguez: Un Nouveau Produit Naturel: Dimethoxy-1,4-nitro-2-trichloro-3,5,6-benzène. Tetrahedron Lett., 211 (1972).

1448. Hsu, C.-S., M. Suziyi, and Y. Yamada: Chemical Constituents of Fungi, I: 1,4-Dimethoxy-2,3,5,6-tetrachlorobenzene (*O*-Methyldrosophilin A) from *Phellinus yucatensis.* Tokyo Yakka Daigaku Kenkyu Nempo, 635 (1970); Chem. Abstr. **75**, 115864 (1971).

1449. Grove, J.F.: Volatile Compounds from the Mycelium of the Mushroom *Agaricus bisporus.* Phytochem. **20**, 2021 (1981).

1450. Kamal, A., C.H. Jarboe, I.H. Qureshy, S.A. Husain, N. Murtaza, R. Noorani, and A.A. Qureshi: Biochemistry of Microorganism, VIII: Isolation and Characterization of *Penicillium martinsii* Biourge Metabolic Products. The Structure of Amudol. Pak. J. Sci. Ind. Res. **13**, 236 (1970).

1451. SÉQUIN-FREY, M., and CH. TAMM: Gentisinacetal und Chlorgentisinalkohol, zwei neue Metabolite einer *Phoma* Species. Helv. Chim. Acta **54**, 851 (1971).

1452. MCCORKINDALE, N.J., T.P. ROY, and S.A. HUTCHINSON: Isolation and Synthesis of 3-Chlorogentisyl Alcohol – A Metabolite of *Penicillium canadense*. Tetrahedron **28**, 1107 (1972).

1453. THALLER, V., and J.L. TURNER: Natural Acetylenes, Part XXXV: Polyacetylenic Acid and Benzenoid Metabolites from Cultures of the Fungus *Lepista diemii* Singer. J. Chem. Soc., Perkin Trans. 1, 2032 (1972).

1454. AHAD, A.M., Y. GOTO, F. KIUCHI, Y. TSUDA, K. KONDO, and T. SATO: Nematocidal Principles in "Oakmoss Absolute" and Nematocidal Activity of 2,4-Dihydroxybenzoates. Chem. Pharm. Bull. (Japan) **39**, 1043 (1991).

1455. OKA, H., K. ASAHI, H. MORISHIMA, M. SANADA, K. SHIRATORI, Y. IIMURA, T. SAKURAI, I. UZAWA, S. IWADARE, and N. TAKAHASHI: Differanisole A, a New Differentiation-Inducing Substance. J. Antibiot. **38**, 1100 (1985).

1456. BOHLMANN, F., W. KNAUF, and L.N. MISRA: Structure and Synthesis of Chlorophenol Derivatives from *Helichrysum* Species. Tetrahedron **40**, 4987 (1984).

1457. GRIMVALL, A., H. BORÉN, S. JONSSON, S. KARLSSON, and R. SÄVENHED: Organohalogens of Natural and Industrial Origin in Large Recipients of Bleach-Plant Effluents. Water Sci. Technol. **24**, 373 (1991).

1458. HODIN, F., H. BORÉN, A. GRIMVALL, and S. KARLSSON: Formation of Chlorophenols and Related Compounds in Natural and Technical Chlorination Processes. Water Sci. Technol. **24**, 403 (1991).

1459. NYSTRÖM, A., A. GRIMVALL, C. KRANTZ-RÜLCKER, R. SÄVENHED, and K. ÅKERSTRAND: Drinking Water Off-Flavour Caused by 2,4,6-Trichloroanisole. Water Sci. Technol. **25**, 241 (1992).

1460. SÄVENHED, R., G. ASPLUND, H. BORÉN, and A. GRIMVALL: Analysis of Naturally Produced Organohalogens in Surface Water. Organohalogen Compounds **2**, 419 (1990).

1461. PAASIVIRTA, J., J. KNUUTINEN, P. MAATELA, R. PAUKKU, J. SOIKKELI, and J. SÄRKKÄ: Oganic Chlorine Compounds in Lake Sediments and the Role of the Chlorobleaching Effluents. Chemosphere **17**, 137 (1988).

1462. REINECCIUS, G.: Off-Flavours in Foods. Critical Rev. in Food Sci. & Nutrition **29**, 381 (1991).

1463. SEVENANTS, M.R., and R.A. SANDERS: Anatomy of an Off-Flavor Investigation: The "Medicinal" Cake Mix. Analyt. Chem. **56**, 293A (1984).

1464. WHITFIELD, F.B., J.H. LAST, K.J. SHAW, and C.R. TINDALE: 2,6-Dibromophenol: The Cause of an Iodoform-Like Off-Flavour in Some Australian Crustacea. J. Sci. Food Agric. **46**, 29 (1988).

1465. WEBB, J.S.: Other Natural Bromo Compounds. Science **158**, 522 (1967).

1466. HIGA, T.: Phenolic Substances. In: Marine Natural Products, Vol. IV, Chap. 3 (P.J. Scheuer, ed.). New York: Academic Press. 1981.

1467. LEMAN, A.: Sur un Produit Phénolique Extrat des Algues. Bull. Chem. Soc. (France) **11**, 564 (1944).

1468. MASTAGLI, P., and J. AUGIER: The Structure of the Phenolic Compound Contained in *Polysiphonia fastigiata*. Compt. Rend. **229**, 775 (1949).

1469. AUGIER, J., and M.-H. HENRY.: Au Sujet du Bromé Dans les Rhodophycées. Bull. Soc. Bot. (France) **97**, 29 (1950); Chem. Abstr. **44**, 10064 (1950).

1470. AUGIER, J.: The Chemical Constitution of Some Red Algae (Rhodomelaceae). Rev. Gen. Botan. **60**, 257 (1953); Chem. Abstr. **50**, 1140 (1956).

1471. AUGIER, J., and P. MASTAGLI: Composition of Brominated Phenol Extracted from Red Alga *Halopitys incurvus*. Comp. Rend. **242**, 190 (1956).

1472. SAITO, T., and Y. ANDO: Bromine Compounds in Seaweed, I: A Bromophenolic Compound from the Red Alga, *Polysiphonia morrowii* Harv. Nippon Kagaku Zasshi **76**, 478 (1955); Chem. Abstr. **51**, 17810 (1957).

1473. HODGKIN, J.H., J.S. CRAIGIE, and A.G. MCINNES: The Occurrence of 2,3-Dibromobenzyl Alcohol 4,5-Disulfate, Dipotassium Salt, in *Polysiphonia lanosa*. Can. J. Chem. **44**, 74 (1966).

1474. CRAIGIE, J.S., and D.E. GRUENIG: Bromophenols from Red Algae. Science **157**, 1058 (1967).

1475. KATSUI, N., Y. SUZUKI, S. KITAMURA, and T. IRIE: 5,6-Dibromoprotocatechualdehyde and 2,3-Dibromo-4,5-dihydroxybenzyl Methyl Ether. Tetrahedron **23**, 1185 (1967).

1476. GLOMBITZA, K.-W., and H. STOFFELEN: 2,3-Dibrom-5-hydroxybenzyl-1',4-disulfat (Dikaliumsalz) aus Rhodomelaceen. Planta Med. **22**, 391 (1972).

1477. MCLACHLAN, J., and J.S. CRAIGIE: Antialgal Activity of Some Simple Phenols. J. Phycol. **2**, 133 (1966).

1478. CHANTRAINE, J.-M., G. COMBAUT, and J. TESTE: Phenols Bromés d'Une Algue Rouge, *Halopytis incurvus*: Acides Carboxyliques. Phytochem. **12**, 1793 (1973).

1479. STOFFELEN, H., K.-W. GLOMBITZA, U. MURAWSKI, J. BIELACZEK, and H. EGGE: Bromophenole aus *Polysiphonia lanosa* (L.) Tandy. Planta Med. **22**, 396 (1972).

1480. MATSUMOTO, T., and S. KAGAWA: 5,6-Dibromoprotocatechualdehyde and 2,3-Dibromo-4,5-dihydroxybenzyl Methyl Ether. Tetrahedron **23**, 1185 (1967).

1481. WEINSTEIN, B., T.L. ROLD, C.E. HARRELL JR., M.W. BURNS III, and J.R. WAALAND: Reexamination of the Bromophenols in the Red Alga *Rhodomela larix*. Phytochem. **14**, 2667 (1975).

1482. GLOMBITZA, K.-W., H. STOFFELEN, U. MURAWSKI, J. BIELACZEK, and H. EGGE: Antibiotics from Algae, 9: Bromophenols from Rhodomelaceae. Planta Med. **25**, 105 (1974).

1483. PEDERSÉN, M., P. SAENGER, and L. FRIES: Simple Brominated Phenols in Red Algae. Phytochem. **13**, 2273 (1974).

1484. PEDERSÉN, M., and L. FRIES: Bromophenols in *Fucus vesiculosus*. Z. Pflanzenphysiol. **74**, 272 (1975).

1485. KURATA, K., T. AMIYA, and N. NAKANO: 3,3'-Dibromo-4,4',5,5'-tetrahydroxybibenzyl, a New Bromophenol from the Red Alga, *Polysiphonia urceolata*. Chem. Lett., 821 (1976).

1486. CHEVOLOT–MAGUEUR, A.-M., A. CAVE, P. POTIER, J. TESTE, A. CHIARONI, and C. RICHE: Composés Bromés de *Rytiphlea tinctoria* (Rhodophyceae). Phytochem. **15**, 767 (1976).

1487. KURATA, K., and T. AMIYA: Two New Bromophenols from the Red Alga, *Rhodomela larix*. Chem. Lett., 1435 (1977).

1488. PEDERSÉN M.: Bromochlorophenols and a Brominated Diphenylmethane in Red Algae. Phytochem. **17**, 291 (1978).

1489. LUNDGREN, L., K. OLSSON, and O. THEANDER: Synthesis of Some Bromophenols Present in Red Algae. Acta Chem. Scand. **B33**, 105 (1979).

1490. KURATA, K., and T. AMIYA: A New Bromophenol from the Red Alga *Polysiphonia urceolata*. Bull. Chem. Soc. Japan **53**, 2020 (1980).

1491. SUZUKI, M., N. KOWATA, and E. KUROSAWA: Bromophenols from the Red Alga *Rhodomela larix*. Bull. Chem. Soc. Japan **53**, 2099 (1980).

1492. KURATA, K., and T. AMIYA: Disodium 2,3,6-Tribromo-5-hydroxybenzyl 1',4-Disulfate, a New Bromophenol from the Red Alga, *Symphyocladia latiuscula*. Chem. Lett., 279 (1980).

1493. DE NANTEUIL, G., and P. MASTAGLI: A Bromophenol in the Red Alga *Halopitys incurvus.* Phytochem. **20**, 1750 (1981).

1494. BLACKMAN, A.J., and D.J. MATTHEWS: Halogenated Phloroglucinols from *Rhabdonia verticillata.* Phytochem. **21**, 2141 (1982).

1495. FUJIMOTO, K., H. OHMURA, and T. KANEDA: Screening for Antioxygenic Compounds in Marine Algae and Bromophenols as Effective Principles in a Red Alga *Polysiphonia ulceolate.* Bull. Japan Soc. Sci. Fish. **51**, 1139 (1985).

1496. KUBO, I., M. OCHI, K. SHIBATA, F.J. HANKE, T. NAKATSU, K.-S. TAN, M. TANIGUCHI, T. KAMIKAWA, Y. YAMAGIWA, M. ARIZUKA, and W.F. WOOD: Effect of a Marine Algal Constituent on the Growth of Lettuce and Rice Seedlings. J. Nat. Prod. **53**, 50 (1990).

1497. SAENGER, P., M. PEDERSÉN, and K.S. ROWAN: Bromo-Compounds of the Red Alga *Lenormandia prolifera.* Phytochem. **15**, 1957 (1976).

1498. PEDERSÉN, M., P. SAENGER, K.S. ROWAN, and A.V. HOFSTEN: Bromine, Bromophenols and Floridorubin in the Red Alga *Lenormandia prolifera.* Physiol. Plant. **46**, 121 (1979).

1499. PEDERSÉN, M., and E.J. DA SILVA: Simple Brominated Phenols in the Bluegreen Alga *Calothrix brevissima.* Planta **115**, 83 (1973).

1500. PEDERSÉN, M., and N. FRIES: Bromophenols Identified in a Marine Ascomycete and Its Culture Medium. Z. Pflanzenphysiol. **82**, 363 (1977).

1501. GLOMBITZA, K.-W., and G. GERSTBERGER: Phlorotannins with Dibenzodioxin Structural Elements from the Brown Alga *Eisenia arborea.* Phytochem. **24**, 543 (1985).

1502. ASHWORTH, R.B., and M.J. CORMIER: Isolation of 2,6-Dibromophenol from the Marine Hemichordate *Balanoglossus biminiensis.* Science **155**, 1558 (1967).

1503. SHEIKH, Y.M., and C. DJERASSI: 2,6-Dibromophenol and 2,4,6-Tribromophenols – Antiseptic Secondary Metabolities of *Phoronopsis viridis.* Experientia **31**, 265 (1975).

1504. KING, G.M.: Inhibition of Microbial Activity in Marine Sediments by a Bromophenol from Hemichordate. Nature **323**, 257 (1986).

1505. KING, G.M.: Dehalogenation in Marine Sediments Containing Natural Sources of Halophenols. Appl. Environ. Microbiol. **54**, 3079 (1988).

1506. HIGA, T., and P.J. SCHEUER: Thelepin, a New Metabolite from the Marine Annelid *Thelepus setosus.* J. Am. Chem. Soc. **96**, 2246 (1974).

1507. HIGA, T., and P.J. SCHEUER: Constituents of the Marine Annelid *Thelepus setosus.* Tetrahedron **31**, 2379 (1975).

1508. WEBER, K., and W. ERNST: Occurrence of Brominated Phenols in the Marine Polychaete *Lanice conchilega.* Naturwiss. **65**, 262 (1978).

1509. NORTE, M., M.L. RODRIGUEZ, J.J. FERNÁNDEZ, L. EGUREN, and D.M. ESTRADA: Aplysinadiene and (*R,R*)-5-[3,5-Dibromo-4-[(2-oxo-5-oxazolidinyl)]methoxyphenyl]-2-oxazolidinone, Two Novel Metabolites from *Aplysina aerophoba.* Synthesis of Aplysinadiene. Tetrahedron **44**, 4973 (1988).

1510. HAMANN, M.T., P.J. SCHEUER, and M. KELLY-BORGES: Biogenetically Diverse, Bioactive Constituents of a Sponge, Order Verongida: Bromotyramines and Sesquiterpene-Shikimate Derived Metabolites. J. Org. Chem. **58**, 6565 (1993).

1511. BLACKMAN, A.J., and S.-L. FU: A β-Phenylethylamine-Derived Possible Biosynthetic Precursor to the Amathamides, Alkaloids from the Bryozoan *Amathia wilsoni.* J. Nat. Prod. **52**, 436 (1989).

1512. ARABSHAHI, L., and F.J. SCHMITZ: Brominated Tyrosine Metabolites from an Unidentified Sponge. J. Org. Chem. **52**, 3584 (1987).

1513. WATANABE, I., T. KASHIMOTO, and R. TATSUKAWA: Brominated Phenols and Anisoles in River and Marine Sediments in Japan. Bull. Environ. Contam. Toxicol. **35**, 272 (1985).

1514. ATLAS, E., K. SULLIVAN, and C.S. GIAM: Widespread Occurrence of Polyhalogenated Aromatic Ethers in the Marine Atmosphere. Atmos. Environ. **20**, 1217 (1986).

1515. SUN, H.H., V.J. PAUL, and W. FENICAL: Avrainvilleol, a Brominated Diphenylmethane Derivative with Feeding Deterrent Properties from the Tropical Green Alga *Avrainvillea longicaulis.* Phytochem. **22**, 743 (1983).

1516. COLON, M., P. GUEVARA, W.H. GERWICK, and D. BALLANTINE: 5′-Hydroxyisoavrainvilleol, a New Diphenylmethane Derivative from the Tropical Green Alga *Avrainvillea nigricans.* J. Nat. Prod. **50**, 368 (1987).

1517. CARTE, B.K., N. TROUPE, J.A. CHAN, J.W. WESTLEY, and D.J. FAULKNER: Rawsonol, an Inhibitor of HMG-CoA Reductase from the Tropical Green Alga *Avrainvillea rawsoni.* Phytochem. **28**, 2917 (1989).

1518. CHEN, J.L., W.H. GERWICK, R. SCHATZMAN, and M. LANEY: Isorawsonol and Related IMP Dehydrogenase Inhibitors from the Tropical Green Alga *Avrainvillea rawsonii.* J. Nat. Prod. **57**, 947 (1994).

1519. WIEMER, D.F., D.D. IDLER, and W. FENICAL: Vidalols A and B, New Anti-Inflammatory Bromophenols from the Caribbean Marine Red Alga *Vidalia obtusaloba.* Experientia **47**, 851 (1991).

1520. GREEN, D., Y. KASHMAN, and A. MIROZ: Colpol, a New Cytotoxic C_6-C_4-C_6 Metabolite from the Alga *Colpomenia sinuosa.* J. Nat. Prod. **56**, 1201 (1993).

1521. CURTIS, R.F., P.C. HARRIES, C.H. HASSALL, and J.D. LEVI: The Biosynthesis of Phenols, 5: The Relationships of Some Phenolic Metabolites of Mutants of *Aspergillus terreus* Thom, I.M.I. 16043. Biochem. J. **90**, 43 (1964).

1522. SHARMA, G.M., B. VIG, and P.R. BURKHOLDER: Antimicrobial Substances of Marine Sponges, IV. In: Food-Drugs from the Sea – Proceedings, pp. 307–310 (H.W. Youngken Jr., ed.). Washington, D.C.: Marine Technology Society. 1970.

1523. SHARMA, G.M., and B. VIG: Studies on the Antimicrobial Substances of Sponges, VI: Structures of Two Antibacterial Substances Isolated from the Marine Sponge *Dysidea herbacea.* Tetrahedron Lett., 1715 (1972).

1524. NORTON, R.S., and R.J. WELLS: Use of ^{13}C Spin-Lattice Relaxation Measurements to Determine the Structure of a Tetrabromo Diphenyl Ether from the Sponge *Dysidea herbacea.* Tetrahedron Lett. **21**, 3801 (1980).

1525. CARTÉ, B., and D.J. FAULKNER: Polybrominated Diphenyl Ethers from *Dysidea herbacea, Dysidea chlorea* and *Phyllospongia foliascens.* Tetrahedron **37**, 2335 (1981).

1526. NORTON, R.S., K.D. CROFT, and R.J. WELLS: Polybrominated Oxydiphenol Derivatives from the Sponge *Dysidea herbacea.* Tetrahedron **37**, 2341 (1981).

1527. UTKINA, N.K., M.V. KAZANTSEVA, and V.A. DENISENKO: Brominated Diphenyl Ethers from the Marine Sponge *Dysidea fragilis.* Khim. Prir. Soed. **23**, 603 (1987)

1528. SALVÁ, J., and D.J. FAULKNER: A New Brominated Diphenyl Ether from a Philippine *Dysidea* Species. J. Nat. Prod. **53**, 757 (1990).

1529. ELYAKOV, G.B., T. KUZNETSOVA, V.V. MIKHAILOV, I.I. MALTSEV, V.G. VOINOV, and S.A. FEDOREYEV: Brominated Diphenyl Ethers from a Marine Bacterium Associated with the Sponge *Dysidea* sp. Experientia **47**, 632 (1991).

1530. VOINOV, V.G., Y.N. EL'KIN, T.A. KUZNETSOVA, I.I. MAL'TSEV, V.V. MIKHAILOV, and V.A. SASUNKEVICH: Use of Mass Spectrometry for the Detection and Identification of Bromine-Containing Diphenyl Ethers. J. Chromatogr. **586**, 360 (1991).

1531. CAPON, R., E.L. GHISALBERTI, P.R. JEFFERIES, B.W. SKELTON, and A.H. WHITE: Structural Studies of Halogenated Diphenyl Ethers from a Marine Sponge. J. Chem. Soc., Perkin Trans. 1, 2464 (1981).

1532. DILLMAN, R.L.: Secondary Metabolites from the Bermudian Sponge *Tedania ignis*, Ph. D. Dissertation, Montana State University, Bozeman, MT 1990. We thank Dr. Cardellina and Professor Dillman for informing us of this work.

1533. KUNIYOSHI, M., K. YAMADA, and T. HIGA: A Biologically Active Diphenyl Ether from the Green Alga *Cladophora fascicularis*. Experientia **41**, 523 (1985).

1534. POCH, G.K., J.B. GLOER, and C.A. SHEARER: New Bioactive Metabolites from a Freshwater Isolate of the Fungus *Kirschsteiniothelia* sp. J. Nat. Prod. **55**, 1093 (1992).

1535. FALCH, B.S., G.M. KÖNIG, A.D. WRIGHT, O. STICHER, H. RÜEGGER, and G. BERNARDINELLI: Ambigol A and B: New Biologically Active Polychlorinated Aromatic Compounds from the Terrestrial Blue-Green Alga *Fischerella ambigua*. J. Org. Chem. **58**, 6570 (1993).

1536. GLOMBITZA, K.-W., M. KOCH, and G. ECKHARDT: Chlorierte Phlorethole aus *Laminaria ochroleuca*. Phytochem. **16**, 796 (1977).

1537. KOCH, M., and R.P. GREGSON: Brominated Phlorethols and Nonhalogenated Phlorotannins from the Brown Alga *Cystophora congesta*. Phytochem. **23**, 2633 (1984).

1538. GLOMBITZA, K.-W., and G. ZIEPRATH: Phlorotannins from the Brown Alga *Analipus japonicus*. Planta Med. **55**, 171 (1989).

1539. KURATA, K., and T. AMIYA: Bis(2,3,6-tribromo-4,5-dihydroxybenzyl) Ether from the Red Alga, *Symphyocladia latiuscula*. Phytochem. **19**, 141 (1980).

1540. DRECHSEL, E.: Beitrage zur Chemie einiger Seethiere. Z. Biol. **33**, 85 (1896).

1541. HENZE, M.: The Iodine Compound of the Naturally Occurring Iodine Proteins. The Constitution of Iodogorgoic Acid. Z. Physiol. Chem. **51**, 64 (1907).

1542. HENZE, M.: The History of Iodogorgoic Acid. Z. Physiol. Chem. **72**, 505 (1911).

1543. MÖRNER, C.TH.: Zur Kenntnis der organischen Gerüstsubstanz des Anthozoënskeletts. Z. Physiol. Chem. **88**, 138 (1913).

1544. CAMERON, A.J.: Contributions to the Biochemistry of Iodine, II: The Distribution of Iodine in Plant and Animal Tissues. J. Biol. Chem. **23**, 1 (1915).

1545. UVAROV, B.P.: Insect Nutrition and Metabolism. Trans. Ent. Soc. London **76**, 255 (1928).

1546. KENNEDY, G.R.: The Distribution and Nature of Iodine Compounds in Ascidians. Gen. Comp. Endocrinol. **7**, 500 (1966).

1547. HUNT, S.: Halogenated Tyrosine Derivatives in Invertebrate Scleroproteins: Isolation and Identification. In: Methods of Enzymology, Vol. 107, Posttranslational Modifications, Part B, pp. 413–438 (F. Wold and K. Moldave, eds.). New York: Academic Press. 1984.

1548. KENDALL, E.C.: The Isolation of Crystalline Form of the Compound Containing Iodin, Which Occurs in the Thyroid. J. Am. Med. Assoc., 2042 (1915).

1549. HARINGTON, C.R.: Chemistry of Thyroxine, I: Isolation of Thyroxine from the Thyroid Gland. Biochem. J. **20**, 293 (1926).

1550. ROCHE, J., and R. MICHEL: Nature, Biosynthesis and Metabolism of Thyroid Hormones. Physiol. Rev. **35**, 583 (1955).

1551. FOSTER, G.L.: The Isolation of 3,5-Diiodotyrosine from the Thyroid. J. Biol. Chem. **LXXXIII**, 345 (1929).

1552. HARINGTON, C.R., and S.S. RANDALL: The Isolation of *d*-3:5-Diiodotyrosine from the Thyroid Gland by the Action of Proteolytic Enzymes. Biochem. J. **25**, 1032 (1931).

1553. Roche, J., R. Michel, W. Wolf, and J. Nunez: Sur Deux Nouveaux Constituants Hormonaux du Corps Thyroide, La 3:3′-Diiodothyronine et la 3:3′:5′-Triiodothyronine. Biochim. Biophys. Acta **19**, 308 (1956).

1554. Ackermann, D., and E. Müller: Über das Vorkommen von Dibromtyrosin neben Dijodtyrosin im Spongin. Z. Physiol. Chem. **269**, 146 (1941).

1555. Low, E.M.: Halogenated Amino Acids of the Bath Sponge. J. Marine Res. **10**, 239 (1951).

1556. Roche, J.: Biochimie Comparée des Scléroprotéines Iodées des Anthozoaires et des Spongiaires. Experientia **8**, 45 (1952).

1557. Roche, J., G. Salvatore, and G. Rametta: Sur la Présence et la Biosynthése d'Hormones Thyroidiennes Chez un Tunicier, *Ciona intestinalis* L. Biochim. Biophys. Acta **63**, 154 (1962).

1558. Barrington, E.J.W., and A. Thorpe: Comparative Observations on Iodine Binding by *Saccoglossus horsti* Brambell and Goodhart, and by the Tunic of *Ciona intestinalis* (L). Gen. Comp. Endocrinol. **3**, 166 (1963).

1559. Barrington, E.J.W., and A. Thorpe: The Identification of Monoiodotyrosine, Diiodotyrosine and Thyroxine in Extracts of the Endostyle of the Ascidian, *Ciona intestinalis* L. Proc. Royal Soc., London **B163**, 136 (1965).

1560. Do Amaral, A.D., R. Morris, and E.J.W. Barrington: Thin-Layer Chromatography of Iodinated Compounds in the Ascidian *Dendrodoa grossularia* (van Beneden). Gen. Comp. Endocrinol. **19**, 370 (1972).

1561. Roche, J., R. Michel, and Y. Yagi: Sur la Presence de Monobromotyrosine dans les Gorgorines. Comp. Rend. **232**, 570 (1951).

1562. Roche, J., S. André, and G. Salvatore: Métabolisme de l'Iode et Formation de la Scléroprotéine (Gorgonine) du Squelette Corné Chez *Eunicella verrucosa* (Pallas). Comp. Biochem. Physiol. **1**, 286 (1960).

1563. Coulson, C.B.: Proteins of Marine Algae. Chem. Ind., 997 (1953).

1564. Wheeler, B.M.: Halogen Metabolism of *Drosophila gibberosa*. J. Exp. Zool. **115**, 83 (1950).

1565. Limpel, L.E., and J.E. Casida: Iodine Metabolism in Insects. J. Exp. Zool. **135**, 19 (1957).

1566. Hunt, S., and S.W. Breuer: Isolation of a New Naturally Occurring Halogenated Amine Acid: Monochloromonobromotyrosine. Biochim. Biophys. Acta **252**, 401 (1971).

1567. Welinder, B.S.: Halogenated Tyrosines from the Cuticle of *Limulus polyphemus* (L.). Biochim. Biophys. Acta **279**, 491 (1972).

1568. Welinder, B.S., P. Roepstorff, and S.O. Andersen: The Crustacean Cuticle, IV: Isolation and Identification of Cross-Links from *Cancer pagurus* Cuticle. Comp. Biochem. Phys. **53B**, 529 (1976).

1569. Andersen, S.O.: 3-Chlorotyrosine in Insect Cuticular Proteins. Acta Chem. Scand. **26**, 3097 (1972).

1570. Hunt, S., and S.W. Breuer: Chlorinated and Brominated Tyrosine Residues in Molluscan Scleroprotein. Biochem. Soc. Trans. **1**, 215 (1973).

1571. Albrizio, S., P. Ciminiello, E. Fattorusso, S. Magno, and M. Pansini: Chemistry of Verongida Sponges, I: Constituents of the Caribbean Sponge *Pseudoceratina crassa*. Tetrahedron **50**, 783 (1994).

1572. Sharma, G.M., and P.R. Burkholder: Antimicrobial Substance of Sponges, I: Isolation, Purification, and Properties of a New Bromine-Containing Antibacterial Substance. J. Antibiot. **A20**, 200 (1967).

1573. Sharma, G.M., and P.R. Burkholder: Studies on the Antimicrobial Substances of

Sponges, II: Structure and Synthesis of a Bromine-Containing Antibacterial Compound from a Marine Sponge. Tetrahedron Lett., 4147 (1967).

1574. D'AMBROSIO, M., A. GUERRIERO, and F. PIETRA: Novel, Racemic or Nearly-Racemic Antibacterial Bromo- and Chloroquinols and γ-Lactams of the Verongiaquinol and the Cavernicolin Type from the Marine Sponge *Aplysina* (= *Verongia*) *cavernicola.* Helv. Chim. Acta **67**, 1484 (1984).

1575. FATTORUSSO, E., L. MINALE, and G. SODANO: Aeroplysinin-I, a New Bromo-Compound from *Aplysina aerophoba.* J. Chem. Soc., Chem. Commun., 751 (1970).

1576. MAKARIEVA, T.N., V.A. STONIK, P. ALCOLADO, and Y.B. ELYAKOV: Comparative Study of the Halogenated Tyrosine Derivatives from Demospongiae (Porifera). Comp. Biochem. Physiol. **68B**, 481 (1980).

1577. CAPON, R.J., and J.K. MACLEOD: Two Epimeric Dibromo Nitriles from the Australian Sponge *Aplysina laevis.* Aust. J. Chem. **40**, 341 (1987).

1578. CRUZ, F., L. QUIJANO, F. GOMÉZ–GARIBAY, and T. RIOS: Brominated Metabolites from the Sponge *Aplysina* (*Verongia*) *thiona.* J. Nat. Prod. **53**, 543 (1990).

1579. ANDERSEN, R.J., and D.J. FAULKNER: A Novel Antibiotic from a Sponge of the Genus *Verongia.* Tetrahedron Lett., 1175 (1973).

1580. D'AMBROSIO, M., A. GUERRIERO, R. DE CLAUSER, G. DE STANCHINA, and F. PIETRA: Dichloroverongiaquinol, a New Marine Antibacterial Compound from *Aplysina cavernicola.* Isolation and Synthesis. Experientia **39**, 1091 (1983).

1581. GOO, Y.M.: Antimicrobial and Antineoplastic Tyrosine Metabolites from a Marine Sponge, *Aplysina fistularis.* Arch. Pharm. Res. **8**, 21 (1985).

1582. SHARMA, G.M., B. VIG, and P.R. BURKHOLDER: Studies on the Antimicrobial Substances of Sponges, IV: Structure of a Bromine-Containing Compound from a Marine Sponge. J. Org. Chem. **35**, 2823 (1970).

1583. IL'IN, S.G., M.V. RESHETNYAK, and A.N. SOBOLEV: Crystal and Molecular Structures of 4-Acetamido-2,6-dibromo-4-hydroxy-1,2-dimethoxycyclohexa-2,5-diene. Khim. Prir. Soed. **28**, 215 (1992).

1584. FATTORUSSO, E., L. MINALE, and G. SODANO: Aeroplysinin-1, An Antibacterial Bromo Compound from the Sponge *Verongia aerophoba.* J. Chem. Soc., Perkin Trans. 1, 16 (1972).

1585. MAZZARELLA, L., and R. PULITI: Crystal Structure and Absolute Configuration of Aeroplysinin-I. Gazz. Chim. Ital. **102**, 391 (1972).

1586. FULMOR, W., and G.E. VAN LEAR, G.O. MORTON, and R.D. MILLS: Isolation and Absolute Configuration of the Aeroplysinin I Enantiomorphic Pair from *Ianthella ardis.* Tetrahedron Lett., 4551 (1970).

1587. COSULICH, D.B., and F.M. LOVELL: An X-Ray Determination of an Antibacterial Compound from the Sponge, *Ianthella ardis.* J. Chem. Soc., Chem. Commun., 397 (1971).

1588. MINALE, L., G. SODANO, W.R. CHAN, and A.M. CHEN: Aeroplysinin-2, a Dibromolactone from Marine Sponges *Aplysina* (*Verongia*) *aerophoba* and *Ianthella* sp. J. Chem. Soc., Chem. Commun., 674 (1972).

1589. STEMPIEN JR., M.F., J.S. CHIB, R.F. NIGRELLI, and R.A. MIERZUA: In: Food-Drugs from the Sea, Proceedings 1972 (L.R. Worthen, ed.), pp. 105–110. Washington, D.C.: Marine Technology Society. 1973.

1590. GOO, Y.M., and K.L. RINEHART JR.: In: Drugs from the Sea; Myth or Reality (P.N. Kaul and C.J. Sindermann, eds.), p. 107. Norman, OK: University of Oklahoma. 1978.

1591. HOLST, P.B., U. ANTHONI, C. CHRISTOPHERSEN, and P.H. NIELSEN: Marine Alkaloids, 16: Reversible Conversion of Flustramine B *N*-1-Oxide to Flustrarine B. J. Nat. Prod. **57**, 1310 (1994).

1592. Quiñoá, E., and P. Crews: Phenolic Constituents of *Psammaplysilla*. Tetrahedron Lett. **28**, 3229 (1987).

1593. Chang, C.W.J., and A.J. Weinheimer: 2-Hydroxy-3,5-dibromo-4-methoxyphenylacetamide. A Dibromotyrosine Metabolite from *Psammoposilla purpurea*. Tetrahedron Lett., 4005 (1977).

1594. Kelecom, A., and G.J. Kannengiesser: Chemical Constituents of Verongia Sponges, II: Structure of Two Dibrominated Compounds Isolated from the Mediterranean Sponge *Verongia aerophoba*. An. Acad. Brasil Cienc. **51**, 639 (1979); Chem. Abstr. **93**, 4208 (1980).

1595. Krejcarek, G.E., R.H. White, L.P. Hager, W.O. McClure, R.D. Johnson, K.L. Rinehart Jr., J.A. McMillan, I.C. Paul, P.D. Shaw, and R.C. Brusca: A Rearranged Dibromotyrosine Metabolite from *Verongia aurea*. Tetrahedron Lett., 507 (1975).

1596. Norte, M., and J.J. Fernández: Isolation and Synthesis of Aplysinadiene, a New Rearranged Dibromotyrosine Derivative from *Aplysia aerophoba*. Tetrahedron Lett. **28**, 1693 (1987).

1597. D'Ambrosio, M., A. Guerriero, P. Traldi, and F. Pietra: Cavernicolin-1 and Cavernicolin-2, Two Epimeric Dibromolactams from the Mediterranean Sponge *Aplysina* (*Verongia*) *cavernicola*. Tetrahedron Lett. **23**, 4403 (1982).

1598. Guerriero, A., M. D'Ambrosio, P. Traldi, and F. Pietra: On the First Marine Natural Product Having Low Enantiomeric Purity. Naturwiss. **71**, 425 (1984).

1599. D'Ambrosio, M., C. Mealli, A. Guerriero, and F. Pietra: 7-Bromocavernicolenone, a New α-Bromoenone from the Mediterranean Sponge *Aplysina* (= *Verongia*) *cavernicola*. Implied, Unprecedented Involvement of a Halogenated Dopa in the Biogenesis of a Natural Product. Helv. Chim. Acta **68**, 1453 (1985).

1600. D'Ambrosio, M., A. Guerriero, and F. Pietra: *N*-Carbamoylpyrrolidine and 7-Chlorocavernicolenone, Two New Metabolites from the Mediterranean Sponge *Aplysina* (= *Verongia*) *cavernicola*. Comp. Biochem. Physiol. **B83**, 309 (1986).

1601. Cavazza, M., G. Guella, L. Nucci, F. Pergola, N. Bicchierini, and F. Pietra: Anodic Oxidation of 3,5-Dihalogenotyrosines as a Model Reaction for the Biogenesis of the Cavernicolins, Metabolites of the Verongid Sponge *Aplysina cavernicola*. J. Chem. Soc., Perkin Trans. 1, 3117 (1993).

1602. Tymiak, A.A., and K.L. Rinehart Jr.: Biosynthesis of Dibromotyrosine-Derived Antimicrobial Compounds by the Marine Sponge *Aplysina fistularis* (*Verongia aurea*). J. Am. Chem. Soc. **103**, 6763 (1981).

1603. Bicchierini, N., M. Cavazza, L. Nucci, F. Pergola, and F. Pietra: Halogenated *p*-Quinols of Marine Sponges. Synthesis via Anodic Oxidation of Phenols and NHI-Like Rearrangement. Tetrahedron Lett. **32**, 4039 (1991).

1604. Jiménez, C., and P. Crews: Novel Marine Sponge Derived Amino Acids, 13: Additional Psammaplin Derivatives from *Psammaplysilla purpurea*. Tetrahedron **47**, 2097 (1991).

1605. Sesin, D.F., and C.M. Ireland: Iodinated Phenethylamine Products from a Didemnid Tunicate. Tetrahedron Lett. **25**, 403 (1984).

1606. Borders, D.B., G.O. Morton, and E.R. Wetzel: Structure of a Novel Bromine Compound Isolated from a Sponge. Tetrahedron Lett., 2709 (1974).

1607. Litaudon, M., and M. Guyot: Ianthelline, un Nouveau Derivé de la Dibromo-3,5 Tyrosine, Isolé de l'Eponge *Ianthella ardis* (Bahamas). Tetrahedron Lett. **27**, 4455 (1986).

1608. Mierzwa, R., A. King, M.A. Conover, S. Tozzi, M.S. Puar, M. Patel, S.J. Coval, and S.A. Pomponi: Verongamine, a Novel Bromotyrosine-Derived Histamine H_3-Antagonist from the Marine Sponge *Verongula gigantea*. J. Nat. Prod. **57**, 175 (1994).

1609. RODRIGUEZ, A.D., R.K. AKEE, and P.J. SCHEUER: Two Bromotyrosine-Cysteine Derived Metabolites from a Sponge. Tetrahedron Lett. **28**, 4989 (1987).

1610. WU, H., H. NAKAMURA, J. KOBAYASHI, Y. OHIZUMI, and Y. HIRATA: Lipopurealins, Novel Bromotyrosine Derivatives with Long Chain Acyl Groups, from the Marine Sponge *Psammaplysilla purea.* Experientia **42**, 855 (1986).

1611. ISHIBASHI, M., M. TSUDA, Y. OHIZUMI, T. SASAKI, and J. KOBAYASHI: Purealidin A, a New Cytotoxic Bromotyrosine-Derived Alkaloid from the Okinawan Marine Sponge *Psammaplysilla purea.* Experientia **47**, 299 (1991).

1612. TSUDA, M., H. SHIGEMORI, M. ISHIBASHI, and J. KOBAYASHI: Purealidin D, a New Pyridine Alkaloid from the Okinawan Marine Sponge *Psammaplysilla purea.* Tetrahedron Lett. **33**, 2597 (1992).

1613. TSUDA, M., H. SHIGEMORI, M. ISHIBASHI, and J. KOBAYASHI: Purealidins E–G, New Bromotyrosine Alkaloids from the Okinawan Marine Sponge *Psammaplysilla purea.* J. Nat. Prod. **55**, 1325 (1992).

1614. XYNAS, R., and R.J. CAPON: Two New Bromotyrosine-Derived Metabolites from an Australian Marine Sponge, *Aplysina* sp. Aust. J. Chem. **42**, 1427 (1989).

1615. KOBAYASHI, J., M. TSUDA, K. AGEMI, H. SHIGEMORI, M. ISHIBASHI, T. SASAKI, and Y. MIKAMI: Purealidins B and C, New Bromotyrosine Alkaloids from the Okinawan Marine Sponge *Psammaplysilla purea.* Tetrahedron **47**, 6617 (1991).

1616. KERNAN, M.R., R.C. CAMBIE, and P.R. BERGQUIST: Chemistry of Sponges, VIII: Anomoian A, a Bromotyrosine Derivative from *Anomoianthella popeae.* J. Nat. Prod. **53**, 720 (1990).

1617. JAMES, D.M., H.B. KUNZE, and D.J. FAULKNER: Two New Brominated Tyrosine Derivatives from the Sponge *Druinella* (= *Psammaplysilla*) *purpurea.* J. Nat. Prod. **54**, 1137 (1991).

1618. GAVIN, J., G. NICOLLIER, and R. TABACCHI: Composants Volatils de la "Mousse de Chêne" (*Evernia prunastri* (L.) ACH.) 3[e] Communication. Helv. Chim. Acta **61**, 352 (1978).

1619. JUREK, J., W.Y. YOSHIDA, P.J. SCHEUER, and M. KELLY-BORGES: Three New Bromotyrosine-Derived Metabolites of the Sponge *Psammaplysilla purpurea.* J. Nat. Prod. **56**, 1609 (1993).

1620. YAGI, H., S. MATSUNAGA, and N. FUSETANI: Purpuramines A–I, New Bromotyrosine-Derived Metabolites from the Marine Sponge *Psammaplysilla purpurea.* Tetrahedron **49**, 3749 (1993).

1621. PAKRASHI, S.C., B. ACHARI, P.K. DUTTA, A.K. CHAKRABARTI, C. GIRI, and S. SAHA: Marine Products from Bay of Bengal: Constituents of the Sponge *Psammaplysilla purpurea.* Tetrahedron **50**, 12009 (1994).

1622., KASSÜHLKE, K.E., and D.J. FAULKNER: Two New Dibromotyrosine Derivatives from the Caribbean Sponge *Pseudoceratina crassa.* Tetrahedron **47**, 1809 (1991).

1623. FATTORUSSO, E., L. MINALE, G. SODANO, K. MOODY, and R.H. THOMSON: Aerothionin, a Tetrabromo-Compound from *Aplysina aerophoba* and *Verongia thiona.* J. Chem. Soc., Chem. Commun., 752 (1970).

1624. MOODY, K., R.H. THOMSON, E. FATTORUSSO, L. MINALE, and G. SODANO: Aerothionin and Homoaerothionin: Two Tetrabromo Spirocyclohexadienylisoxazoles from *Verongia* Sponges. J. Chem. Soc., Perkin Trans. 1, 18 (1972).

1625. MCMILLAN, J.A., I.C. PAUL, Y.M. GOO, K.L. RINEHART, W.C. KRUEGER, and L.M. PSCHIGODA: An X-Ray Study of Aerothionin from *Aplysina fistularis* (Pallas). Tetrahedron Lett. **22**, 39 (1981).

1626. GOPICHAND, Y., and F.J. SCHMITZ: Marine Natural Products: Fistularin-1, -2 and -3 from the Sponge *Aplysina fistularis* Forma *fulva.* Tetrahedron Lett., 3921 (1979).

1627. CIMINO, G., S. DE ROSA, S. DE STEFANO, R. SELF, and G. SODANO: The Bromo-Compounds of the True Sponge *Verongia aerophoba*. Tetrahedron Lett. **24**, 3029 (1983).

1628. GUNASEKERA, S.P., and S.S. CROSS: Fistularin 3 and 11-Ketofistularin 3. Feline Leukemia Virus Active Bromotyrosine Metabolites from the Marine Sponge *Aplysina archeri*. J. Nat. Prod. **55**, 509 (1992).

1629. KERNAN, M.R., R.C. CAMBIE, and P.R. BERGQUIST: Chemistry of Sponges, VII: 11, 19-Dideoxyfistularin 3 and 11-Hydroxyaerothionin, Bromotyrosine Derivatives from *Pseudoceratina durissima*, J. Nat. Prod. **53**, 615 (1990).

1630. CIMINIELLO, P., V. COSTANTINO, E. FATTORUSSO, S. MAGNO, A. MANGONI, and M. PANSINI: Chemistry of Verongida Sponges, II: Constituents of the Caribbean Sponge *Aplysina fistularis* Forma *fulva*. J. Nat. Prod. **57**, 705 (1994).

1631. ACOSTA, A.L., and A.D. RODRÍGUEZ: 11-Oxoaerothionin: A Cytotoxic Antitumor Bromotyrosine-Derived Alkaloid from the Caribbean Sponge *Aplysina lacunosa*. J. Nat. Prod. **55**, 1007 (1992).

1632. GUNASEKERA, M., and S.P. GUNASEKERA: Dihydroxyaerothionin and Aerophobin 1. Two Brominated Tyrosine Metabolites from the Deep Water-Marine Sponge *Verongula rigida*. J. Nat. Prod. **52**, 753 (1989).

1633. KÖNIG, G.M., and A.D. WRIGHT: Agelorins A and B, and 11-*epi*-Fistularin-3, Three New Antibacterial Fistularin-3 Derivatives from the Tropical Marine Sponge *Agelas oroides*. Heterocycles **36**, 1351 (1993).

1634. ROTEM, M., S. CARMELY, Y. KASHMAN, and Y. LOYA: Two New Antibiotics from the Red Sea Sponge *Psammaplysilla purpurea*. Tetrahedron **39**, 667 (1983).

1635. ROLL, D.M., C.W.J. CHANG, P.J. SCHEUER, G.A. GRAY, J.N. SHOOLERY, G.K. MATSUMOTO, G.D. VAN DUYNE, and J. CLARDY: Structure of the Psammaplysins. J. Am. Chem. Soc. **107**, 2916 (1985).

1636. COPP, B.R., C.M. IRELAND, and L.R. BARROWS: Psammaplysin C: A New Cytotoxic Dibromotyrosine-Derived Metabolite from the Marine Sponge *Druinella* (= *Psammaplysilla*) *purpurea*. J. Nat. Prod. **55**, 822 (1992).

1637. ICHIBA, T., P.J. SCHEUER, and M. KELLY-BORGES: Three Bromotyrosine Derivatives, One Terminating in an Unprecedented Diketocyclopentenylidene Enamine. J. Org. Chem. **58**, 4149 (1993).

1638. MORRIS, S.A., and R.J. ANDERSEN: Nitrogenous Metabolites from the Deep Water Sponge *Hexadella* sp. Can. J. Chem. **67**, 677 (1989).

1639. LONGEON, A., M. GUYOT, and J. VACELET: Araplysillins I and II: Biologically Active Dibromotyrosine Derivatives from the Sponge *Psammaplysilla arabica*. Experientia **46**, 548 (1990).

1640. RODRÍGUEZ, A.D., and I.C. PIÑA: The Structures of Aplysinamisines I, II, and III: New Bromotyrosine-Derived Alkaloids from the Caribbean Sponge *Aplysina cauliformis*. J. Nat. Prod. **56**, 907 (1993).

1641. MANCINI, I., G. GUELLA, P. LABOUTE, C. DEBITUS, and F. PIETRA: Hemifistularin 3: A Degraded Peptide or Biogenetic Precursor? Isolation from a Sponge of the Order Verongida from the Coral Sea or Generation from Base Treatment of 11-Oxofistularin 3. J. Chem. Soc., Perkin Trans. 1, 3121 (1993).

1642. NAKAMURA, H., H. WU, J. KOBAYASHI, Y. NAKAMURA, Y. OHIZUMI, and Y. HIRATA: Purealin, a Novel Enzyme Activator from the Okinawan Marine Sponge *Psammaplysilla purea*. Tetrahedron Lett. **26**, 4517 (1985).

1643. TAKITO, J., H. NAKAMURA, J. KOBAYASHI, Y. OHIMUZI, K. EBISAWA, and Y. NONOMURA: Purealin, a Novel Stabilizer of Smooth Muscle Myosin Filaments That Modulates ATPase Activity of Dephosphorylated Myosin. J. Biol. Chem. **261**, 13861 (1986).

1644. TAKITO, J., H. NAKAMURA, J. KOBAYASHI, and Y. OHIZUMI: Enhancement of the Actin-Activated ATPase Activity of Myosin from Canine Cardiac Ventricle by Purealin. Biochim. Biophys. Acta **912**, 404 (1987).

1645. TAKITO, J., Y. OHIZUMI, H. NAKAMURA, J. KOBAYASHI, K. EBISAWA, and Y. NONOMURA: The Mechanism of Inhibition of Light-Chain Phosphorylation by Purealin in Chicken Gizzard Myosin. Eur. J. Pharmacol. **142**, 189 (1987).

1646. NAKAMURA, Y., M. KOBAYASHI, H. NAKAMURA, H. WU, J. KOBAYASHI, and Y. OHIZUMI: Purealin, a Novel Activator of Skeletal Muscle Actomyosin ATPase and Myosin EDTA-ATPase That Enhanced the Superprecipitation of Actomyosin. Eur. J. Biochem. **167**, 1 (1987).

1647. KAZLAUSKAS, R., R.O. LIDGARD, P.T. MURPHY, and R.J. WELLS: Brominated Tyrosine-Derived Metabolites from the Sponge *Ianthella basta*. Tetrahedron Lett. **21**, 2277 (1980).

1648. KAZLAUSKAS, R., R.O. LIDGARD, P.T. MURPHY, R.J. WELLS, and J.F. BLOUNT: Brominated Tyrosine-Derived Metabolites from the Sponge *Ianthella basta*. Aust. J. Chem. **34**, 765 (1981).

1649. NISHIYAMA, S., and S. YAMAMURA: Total Synthesis of Bastadins. Tetrahedron Lett. **23**, 1281 (1982).

1650. MIAO, S., R.J. ANDERSEN, and T.M. ALLEN: Cytotoxic Metabolites from the Sponge *Ianthella basta* Collected in Papua, New Guinea. J. Nat. Prod. **53**, 1441 (1990).

1651. PORDESIMO, E.O., and F.J. SCHMITZ: New Bastadins from the Sponge *Ianthella basta*. J. Org. Chem. **55**, 4704 (1990).

1652. BUTLER, M.S., T.K. LIM, R.J. CAPON, and L.S. HAMMOND: The Bastadins Revisited: New Chemistry from the Australian Marine Sponge *Ianthella basta*. Aust. J. Chem. **44**, 287 (1991).

1653. ABRAHAMSSON, K., A. EKDAHL, J. COLLÉN, E. FAHLSTRÖM, N. SPORRONG, and M. PEDERSÉN: Marine Algae – A Source of Trichloroethylene and Perchloroethylene. Limnol. and Oceanog., submitted, August, 1994; private communication from Dr. Abrahamsson.

1654. CARNEY, J.R., P.J. SCHEUER, and M. KELLY-BORGES: A New Bastadin from the Sponge *Psammaplysilla purpurea*. J. Nat. Prod. **56**, 153 (1993).

1655. DEXTER, A.F., M.J. GARSON, and M.E. HEMLING: Isolation of a Novel Bastadin from the Temperate Marine Sponge *Ianthella* sp. J. Nat. Prod. **56**, 782 (1993).

1656. GULAVITA, N.K., A.E. WRIGHT, P.J. MCCARTHY, S.A. POMPONI, M. KELLY-BORGES, M. CHIN, and M.A. SILLS: Isolation and Structure Elucidation of 34-Sulfatobastadin 13, an Inhibitor of the Endothelin A Receptor, from a Marine Sponge of the Genus *Ianthella*. J. Nat. Prod. **56**, 1616 (1993).

1657. PARK, S.K., J. JUREK, J.R. CARNEY, and P.J. SCHEUER: Two More Bastadins, 16 and 17, from an Indonesian Sponge *Ianthella basta*. J. Nat. Prod. **57**, 407 (1994).

1658. ELIX, J.A., A.A. WHITTON, and M.V. SARGENT: Recent Progress in the Chemistry of Lichen Substances. Progr. Chem. Org. Nat. Prod. **45**, 103 (1984).

1659. ST. PFAU, A.: Zur Kenntnis der Flechtenbestandteile über Chlor-atranorin. Helv. Chim. Acta **17**, 1319 (1934).

1660. CULBERSON, C.F.: The Lichen Substances of the Genus *Evernia*. Phytochem. **2**, 335 (1963).

1661. TER HEIDE, R., N. PROVATOROFF, P.C. TRAAS, P.J. DE VALOIS, N. VAN DER PLASSE, H.J. WOBBEN, and R. TIMMER: Qualitative Analysis of the Odoriferous Fraction of Oakmoss (*Evernia prunastri* (L.) Ach.). J. Agric. Food Chem. **23**, 950 (1975).

1662. KOLLER, G., and K. PÖPL: Über einen chlorhaltigen Flechtenstoff, I. Monatsh. **64**, 106 (1934).

1663. KOLLER, G., and K. PÖPL: Über einen chlorhaltigen Flechtenstoff, II. Monatsh. **64**, 126 (1934).

1664. CULBERSON, C.F.: Chemical and Botanical Guide to Lichen Products. Chapel Hill, NC: University of North Carolina Press. 1969.

1665. ASAHINA, Y., and F. FUZIKAWA: Untersuchungen über Flechtenstoffe, L. Mitteil.: Über die Bestandteile von *Parmelia perlata* Ach. Chem. Ber. **68**, 634 (1935).

1666. HUNECK, S., and G. FOLLMAN: Lichen Components, XXIII: Chilean Lichens, 9: Components of *Lecanona dispersa*, *Parmelia perlata*, and *Parmelia pseudoreticulata*. Z. Naturforsch. **20B**, 1138 (1965).

1667. ASAHINA, Y., and F. FUZIKAWA: Untersuchungen über Flechtenstoffe, LXI. Mitteil.: Über Olivetorsäure (III). Chem. Ber. **68**, 2026 (1935).

1668. CULBERSON, C.F.: Some Constituents of the Lichen *Parmelia cryptochlorophaea*. J. Pharm. Sci. **54**, 1815 (1965).

1669. ELIX, J.A., and U. ENGKANINAN: 4,5-Di-*O*-methylhiascic Acid, a New Tridepside from the Lichens *Parmelia pseudofatiscens* and *Parmelia horrescens*. Aust. J. Chem. **29**, 2701 (1976).

1670. ELIX, J.A., V.K. JAYANTHI, and C.C. LEZNOFF: 2,4-Di-*O*-methylgyrophoric Acid and 2,4,5-Tri-*O*-methylhiascic Acid. New Tridepsides from *Parmelia damaziana*. Aust. J. Chem. **34**, 1757 (1981).

1671. CACCAMESE, S., R.M. TOSCANO, F. BELLESIA, and A. PINETTI: Methyl β-Orcinolcarboxylate and Depsides from *Parmelia furfuracea*. J. Nat. Prod. **48**, 157 (1985).

1672. NOLAN, T.J.: The Chemical Constituents of Lichens Found in Ireland. *Buellia canescens*. – Part I. Sci. Proc., Roy. Dublin Soc. **21**, 67 (1936).

1673. SPILLANE, P.A., J. KEANE, and T.J. NOLAN: The Chemical Constituents of Lichens Found in Ireland. *Buellia canescens*. – Part II. Sci. Proc., Roy. Dublin Soc. **21**, 333 (1936).

1674. CULBERSON, C.F.: Some Constituents of the Lichen *Ramalina siliquosa*. Phytochem. **4**, 951 (1965).

1675. LEE, P., and K. CHANTRAPROMMA: The Major Constituents of the Lichen *Parmelia tinctorum* Nyl. Warasan Songkhla Nakkharin **5**, 149 (1983); Chem. Abstr. **101**, 147817 (1984).

1676. HUNECK, S., and J. SANTESSON: Die Inhaltsstoffe von *Lecidea carpathica* (Koerb.) Szat. und die Struktur des Thuringions, eines neuen Xanthons. Z. Naturforsch. **24B**, 756 (1969).

1677. SEVILLA-SANTOS, P., and L.M. JOSON: Philippine Lichens, I: Chemical Constituents of *Physcia albicans* and *Physcia picta*. Philipp. J. Sci. **98**, 1 (1969); Chem. Abstr. **75**, 72756 (1971).

1678. ELIX, J.A.: 2′-*O*-Methylphysodic Acid and Hydroxyphysodic Acid: Two New Depsidones from the Lichen *Hypogymnia billardieri*. Aust. J. Chem. **28**, 849 (1975).

1679. HIRAYAMA, T., F. FUJIKAWA, I. YOSIOKA, and I. KITAGAWA: On the Constituents of the Lichen in the Genus *Menegazzia*. Menegazziaic Acid, a New Depsidone from *Menegazzia asahinae* (Yas. *ex* Zahlbr.) Sant. and *Menegazzia terebrata* (Hoffm.) Mass. Chem. Pharm. Bull. (Japan) **24**, 2340 (1976).

1680. ELIX, J.A., K.L. GAUL, and P.W. JAMES: α-Acetylhypoconstictic Acid, a New Depsidone from the Lichen *Menegazzia dispora*. Aust. J. Chem. **38**, 1735 (1985).

1681. HVEDING-BERGSETH, N., T. BRUUN, and H. KJOSEN: Isolation of 30-Nor-21α-hopan-22-one (Isoadiantone) from the Lichen *Platismatia glauca*. Phytochem. **22**, 1826 (1983).

1682. ELIX, J.A., and V.K. JAYANTHI: The Isolation and Synthesis of the Lichen Depside 4-*O*-Demethylmicrophyllinic Acid. Aust. J. Chem. **40**, 1851 (1987).

1683. DAVID, F., J.A. ELIX, and M.W.B. SAMSUDIN: Two New Aliphatic Acids from the Lichen *Parmotrema praesorediosum.* Aust. J. Chem. **43**, 1297 (1990).

1684. ELIX, J.A., G.A. JENKINS, and H.T. LUMBSCH: Chemical Variation in the Lichen *Lecanora epibryon* S. Ampl. (Lecanoraceae: Ascomycotina). Mycotaxon **35**, 169 (1989).

1685. BENDZ, G., J. SANTESSON, and C.A. WACHTMEISTER: Studies on the Chemistry of Lichens, 20: The Chemistry of the *Ramalina ceruchis* Group. Acta Chem. Scand. **19**, 1185 (1965).

1686. BENDZ, G., J. SANTESSON, and C.A. WACHTMEISTER: Chemistry of Lichens, XXI: Isolation and Synthesis of Methyl 3,5-Dichlorolecanorate, a New Depside from *Ramalina* spp. Acta Chem. Scand. **19**, 1188 (1965).

1687. HUNECK, S., and G. FOLLMANN: Lichen Constituents, XIV: Chilean Lichens, 5: Constituents of *Ramalina ceruchis* var. *tumidula.* Z. Naturforsch. **20B**, 611 (1965).

1688. HUNECK, S.: Die Struktur von Tumidulin, einem chlorhaltigen Depsid. Chem. Ber. **99**, 1106 (1966).

1689. HUNECK, S., G. HÖFLE, and C.F. CULBERSON: 3,5-Dichlor-2′-*O*-methylanziasäure, ein neues Depsid aus *Lecanora sulphurella.* Phytochem. **16**, 995 (1977).

1690. HUNECK, S., G. SUNDHOLM, and G. FOLLMANN: 3-Chlordivaricatsäure, ein neues Depsid aus *Thelomma*-Arten. Phytochem. **19**, 645 (1980).

1691. MAASS, W.S.G., and A.W. HANSON: Wrightiin, a New Chlorinated Depside from *Erioderma wrightii* Tuck (Ascolichenes). Z. Naturforsch. **41B**, 1589 (1986).

1692. ELIX, J.A., I. MAHADEVAN, J.H. WARDLAW, L. ARVIDSSON, and P.M. JORGENSEN: New Depsides from *Erioderma* Lichens. Aust. J. Chem. **40**, 1581 (1987).

1693. ELIX, J.A., D.O. CHESTER, K.L. GAUL, J.L. PARKER, and J.H. WARDLAW: The Identification and Synthesis of Further Lichen β-Orcinol *para*-Depsides. Aust. J. Chem. **42**, 1191 (1989).

1694. ELIX, J.A., A.L. WILKINS, and J.H. WARDLAW: Five New Fully Substituted Depsides from the Lichen *Pseudocyphellaria pickeringii.* Aust. J. Chem. **40**, 2023 (1987).

1695. ELIX, J.A., J.E. EVANS, and T.H. NASH III: New Depsides from *Dimelaena* Lichens. Aust. J. Chem. **41**, 1789 (1988).

1696. ZOPF, W.: Zur Kenntnis der Flechtenstoffe. Ann. **336**, 46 (1904).

1697. NOLAN, T.J.: A Lichen Substance Containing Chlorine. Chem. Ind., 512 (1934).

1698. NOLAN, T.J., J. ALGAR, E.P. MCCANN, W.A. MANAHAN, and N. NOLAN: The Chemical Constituents of Lichens Found in Ireland. *Buellia canescens,* Part 3: The Constitution of Diploicin. Sci. Proc., Roy. Dublin Soc. **24**, 319 (1948).

1699. HARDIMAN, J., J. KEANE, and T.J. NOLAN: Chemical Constituents of Lichens Found in Ireland, I: *Lecanora gangaleoides.* Sci. Proc., Roy. Dublin Soc. **21**, 141 (1935).

1700. DAVIDSON, V.E., J. KEANE, and T.J. NOLAN: The Chemical Constituents of Lichens Found in Ireland. *Lecanora gangaleoides,* Part 3: The Constituents of Gangaleoidin. Sci. Proc., Roy. Dublin Soc. **23**, 143 (1943).

1701. DEVLIN, J.P., M.M. MAHANDRU, W.D. OLLIS, and C. SMITH: Phytochemical Examination of the Lichen, *Lecanora gangaleoides* Nyl. J. Chem. Soc., Perkin Trans. 1, 1491 (1986).

1702. SARGENT, M.V., P. VOGEL, and J.A. ELIX: Structure of the Lichen Depsidone Gangaleoidin. J. Chem. Soc., Perkin Trans. 1, 1986 (1975).

1703. CHESTER, D.O., J.A. ELIX, and A.J. JONES: Lecideoidin and 3′-Dechlorolecideoidin, New Depsidones from an Australian *Lecidea* (Lichen). Aust. J. Chem. **32**, 1857 (1979).

1704. YOSIOKA, I.: On the Structure of Pannarin, a Component of the Lichen Genus *Pannaria.* J. Pharm. Soc. Japan **61**, 332 (1941).

1705. HUNECK, S.: On the Constituents of *Lecanora hercynica* Poelt and Ulrich, *Lecidea silacea* (Ach.) Ach. and *Acarospora* H. Magn. Z. Naturforsch. **21B**, 80 (1966).

1706. ELIX, J.A., U.A. JENIE, L. ARVIDSSON, P.M. JORGENSEN, and P.W. JAMES: New Depsidones from the Lichen Genus *Erioderma*. Aust. J. Chem. **39**, 719 (1986).

1707. JACKMAN, D.A., M.V. SARGENT, and J.A. ELIX: Structure of the Lichen Depsidone Pannarin. J. Chem. Soc., Perkin Trans. 1, 1979 (1975).

1708. HUNECK, S., and I.M. LAMB: 1′-Chloropannarin, a New Depsidone from *Argopsis friesiana*: Notes on the Structure of Pannarin and on the Chemistry of the Lichen Genus *Argopsis*. Phytochem. **14**, 1625 (1975).

1709. DOERING, W.E., R.J. DUBOS, D.S. NOYCE, and R. DREYFUS: Metabolic Products of *Aspergillus ustus*. J. Am. Chem. Soc. **68**, 725 (1946).

1710. DEAN, F.M., J.C. ROBERTS, and A. ROBERTSON: The Chemistry of Fungi, Part XXII: Nidulin and Nornidulin ("Ustin"): Chlorine-Containing Metabolic Products of *Aspergillus nidulans*. J. Chem. Soc., 1432 (1954).

1711. DEAN, F.M., A.D.T. ERNI, and A. ROBERTSON: The Chemistry of Fungi, Part XXVI: Dechloronornidulin. J. Chem. Soc., 3545 (1956).

1712. BEACH, W.F., and J.H. RICHARDS: The Structure and Biosynthesis of Nidulin. J. Org. Chem. **28**, 2746 (1963).

1713. KAMAL, A., Y. HAIDER, Y.A. KHAN, I.H. QURESHI, and A.A. QURESHI: Studies in the Biochemistry of Microorganisms, X: Isolation, Structure, and Stereochemistry of Yasimin and Other Metabolic Products of *Aspergillus unguis* Emile-Weil and Gaudin. Pak. J. Sci. Ind. Res. **13**, 244 (1970); Chem. Abstr. **74**, 95757 (1971).

1714. KAWAHARA, N., S. NAKAJIMA, Y. SATOH, M. YAMAZAKI, and K. KAWAI: Studies on Fungal Products, XVIII: Isolation and Structure of a New Fungal Depsidone Related to Nidulin and a New Phthalide from *Emericella unguis*. Chem. Pharm. Bull. (Japan) **36**, 1970 (1988).

1715. KAWAHARA, N., K. NOZAWA, S. NAKAJIMA, K. KAWAI, and M. YAMAZAKI: Isolation and Structures of Novel Fungal Depsidones, Emeguisins A, B, and C, from *Emericella unguis*. J. Chem. Soc., Perkin Trans. 1, 2611 (1988).

1716. DEAN, F.M., D.S. DEORHA, A.D.T. ERNI, D.W. HUGHES, and J.C. ROBERTS: The Structure of Nidulin, a Metabolite of *Aspergillus nidulans*. J. Chem. Soc., 4829 (1960).

1717. BEACH, W.F., and J.H. RICHARDS: The Nature of the Alkyl Groups in Nidulin. J. Org. Chem. **26**, 3011 (1961).

1718. BYCROFT, B.W., J.A. KNIGHT, and J.C. ROBERTS: Studies in Mycological Chemistry, Part XV: Synthesis of 2,5-Dihydroxy-3-methyl-6-*s*-butyl-1,4-benzoquinone and Its Bearing on the Structure of Nidulin. J. Chem. Soc., 5148 (1963).

1719. BYCROFT, B.W., and J.C. ROBERTS: The Structure of Nidulin. J. Org. Chem. **28**, 1429 (1963).

1720. SIERANKIEWICZ, J., and S. GATENBECK: The Biosynthesis of Nidulin and Trisdechloronornidulin. Acta Chem. Scand. **27**, 2710 (1973).

1721. NEELAKANTAN, S., T.R. SESHADRI, and S.S. SUBRAMANIAN: Constitution of Vicanicin from the Lichen *Teloschistes flavicans*. Tetrahedron Lett. No. 9, 1 (1959).

1722. NEELAKANTAN, S., T.R. SESHADRI, and S.S. SUBRAMANIAN: Chemical Investigation of Indian Lichens, XXVI: Constitution of Vicanicin from *Teloschistes flavicans*. Tetrahedron **18**, 597 (1962).

1723. YOSIOKA, I., K. HINO, M. FUJIO, and I. KITAGAWA: A New Trichlorodepsidone from a Lichen of the Genus *Caloplaca*. Chem. Pharm. Bull. (Japan) **19**, 1070 (1971).

1724. YOSIOKA, I., K. HINO, M. FUJIO, and I. KITAGAWA: The Structure of Caloploicin, a New Lichen Trichloro-depsidone. Chem. Pharm. Bull. (Japan) **21**, 1547 (1973).

1725. SARGENT, M.V., P. VOGEL, J.A. ELIX, and B.A. FERGUSON: Depsidone Synthesis, VII: Vicanicin and Norvicanicin. Aust. J. Chem. **29**, 2263 (1976).

1726. ELIX, J.A., L. LAJIDE, and D.J. GALLOWAY: Metabolites from the Lichen Genus *Psoroma*. Aust. J. Chem. **35**, 2325 (1982).

1727. QUILHOT, W., B. DIDYK, V. GAMBARO, and J.A. GARBARINO: Studies on Chilean Lichens, VI: Depsidones from *Erioderma chilense*. J. Nat. Prod. **46**, 942 (1983).
1728. HAMAT, A.L.B., L.B. DIN, M.W.B. SAMSUDIN, and J.A. ELIX: Two New Depsidones from the Lichen *Erioderma phaeorhizum* Vainio *sensu lato*. Aust. J. Chem. **46**, 153 (1993).
1729. ELIX, J.A., G.A. JENKINS, and D.A. VENABLES: New Chlorine-Containing Depsidones from Lichens. Aust. J. Chem. **43**, 197 (1990).
1730. SARGENT, M.V., and P. VOGEL: Depsidone Synthesis, IV: Caloploicin. Aust. J. Chem. **29**, 907 (1976).
1731. KAMAL, A., Y. HAIDER, A.A. QURESHI, and Y.A. KHAN: Studies in the Biochemistry of Microorganisms, XII: Isolation and Structures of Haiderin, Rubinin, Shirin, and Nasrin, Metabolic Products of *Aspergillus unguis* Emile-Weil and Gaudin. Pak. J. Sci. Ind. Res. **13**, 364 (1970); Chem. Abstr. **75**, 45797 (1971).
1732. BODO, B., and D. MOLHO: Structure de l'Argopsine, Nouvelle Chlorodepsidone du Lichen *Argopsis megalospora*. C. R. Acad. Sci. Paris, Ser. 2, **278**, 625 (1974).
1733. CONNOLLY, J.D., A.A. FREER, K. KALB, and S. HUNECK: Eriodermin, a Dichlorodepsidone from the Lichen *Erioderma physcioides* – Crystal Structure Analysis. Phytochem. **23**, 857 (1984).
1734. MAASS, W.S.G., A.G. MCINNES, D.G. SMITH, and A. TAYLOR: Lichen Substances, X: Physciosporin, a New Chlorinated Depsidone. Can. J. Chem. **55**, 2839 (1977).
1735. RENNER, B., A. HENSSEN, and E. GERSTNER: Methylvirensat und 5-Chlor-Methylvirensat aus Arten der Flechtengattung *Pseudocyphellaria*. Z. Naturforsch. **33C**, 826 (1978).
1736. GOH, E.M., and A.L. WILKINS: Structures of the Lichen Depsidones Granulatin and Chlorogranulatin. J. Chem. Soc., Perkin Trans. 1, 1656 (1979).
1737. SALA, T., M.V. SARGENT, and J.A. ELIX: Depsidone Synthesis, Part 15: New Metabolites of the Lichen *Buellia canescens* (Dicks). De Not: Novel Phthalide Catabolites of Depsidones. J. Chem. Soc., Perkin Trans. 1, 849 (1981).
1738. MAHANDRU, M.M., and A. TAJBAKHSH: Scensidin, a New Depsidone from the Lichen *Buellia canescens* (Dicks). De Not. J. Chem. Soc., Perkin Trans. 1, 413 (1983).
1739. ELIX, J.A., D.A. VENABLES, and L. BRAKO: New Chlorine-Containing Depsidones from the Lichen *Phyllopsora corallina* var. *ochroxantha*. Aust. J. Chem. **43**, 1953 (1990).
1740. MAHANDRU, M.M., and O.L. GILBERT: Chemical Studies in *Fulgensia*: Structures of Two New Chlorodepsidones. Bryologist **82**, 302 (1979).
1741. BÜCHI, G., and P.G. WILLIARD: The Total Synthesis of Mollicellin A. Heterocycles **11**, 437 (1978).
1742. HUNECK, S.: Lichen Metabolites, XXXII: Thiophanic Acid, a New Chlorine-Containing Xanthone from *Lecanora rupicola* (L.) Zahlbr. Tetrahedron Lett., 3547 (1966).
1743. ARSHAD, M., J.P. DEVLIN, and W.D. OLLIS: Synthesis of Sordidone and Thiophanic Acid, Two Chlorine-Containing Lichen Metabolites. J. Chem. Soc. (C), 1324 (1971).
1744. SANTESSON, J.: Chemical Studies on Lichens, 12: A New Lichen Xanthone from *Lecanora reuteri*. Acta Chem. Scand. **22**, 1698 (1968).
1745. SANTESSON, J.: Chemical Studies on Lichens, 16: The Xanthones of *Lecanora straminea* I. Arthothelin and Thiophanic Acid. Arkiv. Kemi. **30**, 449 (1969).
1746. SANTESSON, J.: Chemical Studies on Lichens, 28: The Pigments of Some Foliicolous Lichens. Acta Chem. Scand. **24**, 371 (1970).
1747. SUNDHOLM, E.G.: Total Syntheses of Lichen Xanthones. Tetrahedron **34**, 577 (1978).
1748. SUNDHOLM, E.G.: Syntheses and ^{13}C NMR Spectra of Some 5-Chloro-Substituted Lichen Xanthones. Acta Chem. Scand. **33B**, 475 (1979).

1749. SANTESSON, J.: Chemical Studies on Lichens, 20: The Xanthones of Some Crustaceous Lichens. Arkiv. Kemi. **31**, 57 (1969).

1750. SANTESSON, J.: Chemical Studies on Lichens, 17: The Xanthones of *Lecanora straminea* II. 2,7-Dichloronorlichexanthone. Arkiv. Kemi. **30**, 455 (1969).

1751. HUNECK, S., and G. HÖFLE: Struktur und ^{13}C-NMR-Spektroskopie von chlorhaltigen Flechtenxanthonen. Tetrahedron **34**, 2491 (1978).

1752. FITZPATRICK, L., T. SALA, and M.V. SARGENT: Further Total Syntheses of Chlorine-Containing Lichen Xanthones. J. Chem. Soc., Perkin Trans. 1, 85 (1980).

1753. SANTESSON, J.: Chemical Studies on Lichens, 10: Mass Spectrometry of Lichens. Arkiv. Kemi. **30**, 363 (1969).

1754. SANTESSON, J.: Chemical Studies on Lichens, 21: Two Novel Chlorinated Lichen Xanthones. Arkiv. Kemi. **31**, 121 (1969).

1755. SANTESSON, J.: Chemical Studies on Lichens, 18: The Xanthones of *Lecanora straminea* III. Nerlich Exanthone, 2-Chloronorlichexanthone, and 2,4-Dichloronorlichexanthone. Arkiv. Kemi. **30**, 461 (1969).

1756. SANTESSON, J., and C.A. WACHTMEISTER: Chemical Studies on Lichens, 15: 2,4-Dichloro-6-methoxy-1,3-dihydroxy-8-methylxanthone (Thiophaninic Acid) from *Pertusaria flavicans*. Arkiv. Kemi. **30**, 445 (1969).

1757. HUNECK, S., and G. FOLLMANN: Mitteilungen über Flechteninhaltsstoffe, LXXXIV: Zur Phytochemie und Chemotaxonomie der Lecanoraceengattung *Haematomma*. J. Hattori Bot. Lab., 319 (1972); Chem. Abstr. **79**, 50736 (1973).

1758. ELIX, J.A., H.W. MUSIDLAK, T. SALA, and M.V. SARGENT: Structure and Synthesis of Some Lichen Xanthones. Aust. J. Chem. **31**, 145 (1978).

1759. ELIX, J.A., H. JIANG, and V.J. PORTELLI: Structure and Synthesis of the Lichen Xanthone Isoarthothelin (2,5,7-Trichloronorlichexanthone). Aust. J. Chem. **43**, 1291 (1990).

1760. ELIX, J.A., and S.A. BENNETT: 6-*O*-Methylarthothelin and 1,3,6-Tri-*O*-methylarthothelin, Two New Xanthones from a *Dimelaena* Lichen. Aust. J. Chem. **43**, 1587 (1990).

1761. ELIX, J.A., H. JIANG, and J.H. WARDLAW: A New Synthesis of Xanthones. 2,4,7-Trichloronorlichexanthone and 4,5,7-Trichloronorlichexanthone, Two New Lichen Xanthones. Aust. J. Chem. **43**, 1745 (1990).

1762. ELIX, J.A., S.A. BENNETT, and H. JIANG: 2,5-Dichloro-6-*O*-methylnorlichexanthone and 4,5-Dichloro-6-*O*-methylnorlichexanthone, Two New Xanthones from an Australian *Dimelaena* Lichen. Aust. J. Chem. **44**, 1157 (1991).

1763. ELIX, J.A., H.-M. CHAPPELL, and H. JIANG: Four New Lichen Xanthones. The Bryologist **94**, 304 (1991).

1764. ELIX, J.A., and C.E. CROOK: The Joint Occurrence of Chloroxanthones in Lichens, and a Further Thirteen New Lichen Xanthones. The Bryologist **95**, 52 (1992).

1765. ELIX, J.A., K.L. GAUL, M. STERNS, M.W. BIN SAMSUDIN: The Structure of the Novel Lichen Xanthone Thiomelin and Its Cogenors. Aust. J. Chem. **40**, 1169 (1987).

1766. ELIX, J.A., K.L. GAUL, and H. JIANG: The Structure and Synthesis of Some Minor Xanthones from the Lichen *Rinodina thiomela*. Aust. J. Chem. **46**, 95 (1993).

1767. KACHI, H., H. HATTORI, and T. SASSA: A New Antifungal Substance, Bromomonilicin, and Its Precursor Produced by *Monilinia fructicola*. J. Antibiot. **39**, 164 (1986).

1768. SASSA, T., H. KACHI, M. NUKINA, and Y. SUZUKI: Chloromonilicin, a New Antifungal Metabolite Produced by *Monilinia fructicola*. J. Antibiot. **38**, 439 (1985).

1769. SASSA, T.: Overproduction of Moniliphenone, a Benzophenone Biosynthetic Intermediate of Chloromonilicin, by *Monilinia fructicola* and Its Anthraquinone Precursors. Agric. Biol. Chem. **55**, 95 (1991).

1770. KUPCHAN, S.M., D.R. STREELMAN, and A.T. SNEDEN: Psorospermin, a New Antileukemic Xanthone from *Psorospermum febrifugum*. J. Nat. Prod. **43**, 296 (1980).

1771. ABOU-SHOER, M., F.E. BOETTNER, C. CHANG, and J.M. CASSADY: Antitumour and Cytotoxic Xanthones of *Psorospermum febrifugum*. Phytochem. **27**, 2795 (1988).

1772. STEYN, P.S., and R. VLEGGAAR: Austocystins. Six Novel Dihydrofuro[3′, 2′:4,5]furo-[3,2-*b*]xanthenones from *Aspergillus ustus*. J. Chem. Soc., Perkin Trans. 1, 2250 (1974).

1773. STEYN, P.S., and R. VLEGGAAR: Dihydrofuro[3′, 2′:4,5]furo[3,2-*b*]xanthenones: The Structures of Austocystins G, H and I. J. S. Afr. Chem. Inst. **28**, 375 (1975); Chem. Abstr. **84**, 132390 (1976).

1774. DRAUTZ, H., W. KELLER-SCHIERLEIN, and H. ZÄHNER: Stoffwechselprodukte von Mikroorganismen, 149. Mitteilung: Lysolipin I, ein neues Antibioticum aus *Streptomyces violaceoniger*. Arch. Microbiol. **106**, 175 (1975).

1775. DOBLER, M., and W. KELLER-SCHIERLEIN: Metabolites of Microorganisms, 162nd Communication: The Crystal and Molecular Structure of Lysolipin I. Helv. Chim. Acta **60**, 178 (1977).

1776. BOCKHOLT, H., G. UDVARNOKI, J. ROHR, U. MOCEK, J.M. BEALE, and H.G. FLOSS: Biosynthetic Studies on the Xanthone Antibiotics Lysolipins X and I. J. Org. Chem. **59**, 2064 (1994).

1777. MILAT, M.-L., T. PRANGÉ, P.-H. DUCROT, J.-C. TABET, J. EINHORN, J.-P. BLEIN, and J.-Y. LALLEMAND: Structures of the Beticolins, the Yellow Toxins Produced by *Cercospora beticola*. J. Am. Chem. Soc. **114**, 1478 (1992).

1778. JALAL, M.A.F., M.B. HOSSAIN, D.J. ROBESON, and D. VAN DER HELM: *Cercospora beticola* Phytotoxins: Cebetins That Are Photoactive, Mg^{2+}-Binding, Chlorinated Anthraquinone–Xanthone Conjugates. J. Am. Chem. Soc. **114**, 5967 (1992).

1779. ARNONE, A., G. NASINI, L. MERTINI, E. RAGG, and G. ASSANTE: Secondary Mould Metabolites, Part 41: Structure and Biosynthesis of *Cercospora beticola* Toxin (CBT). J. Chem. Soc., Perkin Trans. 1, 145 (1993).

1780. MILAT, M.-L., J.-P. BLEIN, J. EINHORN, J.-C. TABET, P.-H. DUCROT, and J.-Y. LALLEMAND: The Yellow Toxins Produced by *Cercospora beticola*, Part II: Isolation and Structure of Beticolins 3 and 4. Tetrahedron Lett. **34**, 1483 (1993).

1781. DUCROT, P.-H., M.-L. MILAT, J.-P. BLEIN, and J.-Y. LALLEMAND: The Yellow Toxins Produced by *Cercospora beticola*. Revised Structures of Beticolin 1 and Beticolin 3. J. Chem. Soc., Chem. Commun., 2215 (1994).

1782. DUCROT, P.-H., J.-Y. LALLEMAND, M.-L. MILAT, and J.-P. BLEIN: The Yellow Toxins Produced by *Cercospora beticola*, Part VIII: Chemical Equilibrium Between Beticolins; Structures of Minor Compounds: Beticolin 6 and Beticolin 8. Tetrahedron Lett. **35**, 8797 (1994).

1783. HORIGUCHI, K., Y. SUZUKI, and T. SASSA: Biosynthetic Study of Chloromonilicin, a Growth Self-Inhibitor Having a Novel Lactone Ring, from *Monilinia fructicola*. Agric. Biol. Chem. **53**, 2141 (1989).

1784. RAISTRICK, H., and J. ZIFFER: Studies in the Biochemistry of Microorganisms, 84: The Colouring Matters of *Penicillium nalgiovensis* Laxa, Part 1: Nalgiovensin and Nalgiolaxin, Isolation, Derivatives and Partial Structures. Biochem. J. **49**, 563 (1951).

1785. BIRCH, A.J., and K.S.J. STAPLEFORD: The Structure of Nalgiolaxin. J. Chem. Soc. (C), 2570 (1967).

1786. BRUUN, T., D.P. HOLLIS, and R. RYHAGE: Constitution of Fragilin. Acta Chem. Scand. **19**, 839 (1965).

1787. BENDZ, G., G. BOHMAN, and J. SANTESSON: Chemical Studies on Lichens, 9: Chlorinated Anthraquinones from *Nephroma laevigatum*. Acta Chem. Scand. **21**, 2889 (1967).

1788. YAMAMOTO, Y., N. KIRIYAMA, and S. ARAHATA: Studies on the Metabolic Products of *Aspergillus fumigatus* (J-4). Chemical Structure of Metabolic Products. Chem. Pharm. Bull. (Japan) **16**, 304 (1968).

1789. YOSIOKA, I., H. YAMAUCHI, K. MORIMOTO, and I. KITAGAWA: Two New Chlorine Containing Anthraquinones from a Lichen, *Anaptychia obscurata* (NYL.) VAIN. Tetrahedron Lett., 1149 (1968).

1790. FOX, C.H., W.S.G. MAASS, and T.P. FORREST: Papulsoin, a Novel Chlorinated Anthraquinone from *Lasallia papulosa.* (Ach.) Llano. Tetrahedron Lett., 919 (1969).

1791. BOHMAN, G.: Chemical Studies on Lichens, 20: Anthraquinones from the Lichen *Lasallia papulosa* var. *rubiginosa* and the Fungus *Valsaria rubricosa.* Acta Chem. Scand. **23**, 2241 (1969).

1792. STEGLICH, W., W. LÖSEL, and V. AUSTEL: Anthrachinon-Pigmente aus *Dermocybe sanguinea* (Wulf. ex Fr.) Wünsche und *D. semisanguinea* (Fr.). Chem. Ber. **102**, 4104 (1969).

1793. BOHMAN, G.: Anthraquinones from the Genus *Caloplaca.* Phytochem. **8**, 1829 (1969).

1794. SANTESSON, J.: Anthraquinones in *Caloplaca.* Phytochem. **9**; 2149 (1970).

1795. BRIGGS, L.H., D.R. CASTAING, A.N. DENYER, E.F. ORGAIS, and C.W. SMALL: Chemistry of Fungi, Part VIII: Constituents of *Valsaria rubricosa* and the Identification of Papulosin with Valsarin. J. Chem. Soc., Perkin Trans. 1, 1464 (1972).

1796. NAKANO, H., T. KOMIYA, and S. SHIBATA: Anthraquinones of the Lichens of *Xanthoria* and *Caloplaca* and Their Cultivated Mycobionts. Phytochem. **11**, 3505 (1972).

1797. KELLER, G.: Pigmentationsuntersuchungen bei Europäischen Arten aus der Gattung *Dermocybe* (Fr.) Wünsche. Sydowia **35**, 110 (1982).

1798. KELLER, G., and J.F. AMMIRATI: Chemotaxonomic Significance of Anthraquinone Derivatives in North American Species of *Dermocybe*, Section *Sanguineae.* Mycotaxon **18**, 357 (1983).

1799. AYER, W.A., and L.S. TRIFONOV: Anthraquinones and a 10-Hydroxyanthrone from *Phialophora alba.* J. Nat. Prod. **57**, 317 (1994).

1800. YOSIOKA, I., H. YAMAUCHI, K. MORIMOTO, and I. KITAGAWA: Three New Chlorine-Containing Bisanthronyls from a Lichen, *Anaptychia obscurata.* Vain. Tetrahedron Lett., 3749 (1968).

1801. DE RICCARDIS, F., M. IORIZZI, L. MINALE, R. RICCIO, B.R. DE FORGES, and C. DEBITUS: The Gymnochromes: Novel Marine Brominated Phenanthroperylenequinone Pigments from the Stalked Crinoid *Gymnocrinus richeri.* J. Org. Chem. **56**, 6781 (1991).

1802. DAVIES, R.R.: Antifungal Chemotherapy, p. 149. New York: Wiley-Interscience. 1980.

1803. RAISTRICK, H., and G. SMITH: Studies in the Biochemistry of Micro-Organisms, LI: The Metabolic Products of *Aspergillus terreus* Thom., Part II: Two New Chlorine-Containing Mould Metabolic Products, Geodin and Erdin. Biochem. J. **30**, 1315 (1936).

1804. BARTON, D.H.R., and A.I. SCOTT: The Constitutions of Geodin and Erdin. J. Chem. Soc., 1767 (1958).

1805. OXFORD, A.E., H. RAISTRICK, and P. SIMONART: Studies in the Biochemistry of Micro-Organism Griseofulvin, $C_{17}H_{17}O_6Cl$, a Metabolic Product of *Penicillium griseo-fulvum* Dierckx. Biochem. J. **33**, 240 (1939).

1806. GROVE, J.F., and J.C. MCGOWAN: Identity of Griseofulvin and Curling-Factor. Nature **160**, 574 (1947).

1807. GROVE, J.F.: Griseofulvin and Some Analogues. Fortsch. Chem. Org. Naturstoffe **22**, 203 (1964).

1808. GROVE, J.F.: Griseofulvin. Quart. Rev. **17**, 1 (1963).

1809. KAMAL, A., S.A. HUSAIN, N. MURTAZA, R. NOORANI, I.H. QURESHI, and A.A. QURESHI: Studies in the Biochemistry of Microorganisms, IX: Structure of Amudane, Amudene and Amujane, Metabolic Products of *Penicillium martinsii* Biourge. Pak. J. Sci. Ind. Res. **13**, 240 (1970); Chem. Abstr. **74**, 75168 (1971).

1810. MACMILLAN, J.: Griseofulvin, Part XIV: Some Alcoholytic Reactions and the Absolute Configuration of Griseofulvin. J. Chem. Soc., 1823 (1959).

1811. BROWN, W.A.C., and G.A. SIM: Fungal Metabolites, Part I: The Stereochemistry of Griseofulvin: X-Ray Analysis of 5-Bromogriseofulvin. J. Chem. Soc., 1050 (1963).

1812. MCMASTER, W.J., A.I. SCOTT, and S. TRIPPETT: Metabolic Products of *Penicillium patulum*. J. Chem. Soc., 4628 (1960).

1813. HASSALL, C.H., and T.C. MCMORRIS: The Constitution of Geodoxin, a Metabolic Product of *Aspergillus terreus* Thom. J. Chem. Soc., 2831 (1959).

1814. NOMOTO, K., K. MIZUKAWA, Y. KATO, M. KUBO, T. KAMIKI, and Y. INAMORI: Chemical Structure of a New Metabolite Isolated from the Mycelium of *Aspergillus terreus* var. *aureus*. Chem. Pharm. Bull. (Japan) **32**, 4213 (1984).

1815. RHODES, A., B. BOOTHROYD, M.P. MCGONAGLE, and G.A. SOMERFIELD: Biosynthesis of Griseofulvin: The Methylated Benzophenone Intermediates. Biochem. J. **81**, 28 (1961).

1816. MORITA, H., S. NAGASHIMA, K. TAKEYA, and H. ITOKAWA: Conformational Analysis of an Antitumour Cyclic Pentapeptide, Astin B, from *Aster tataricus*. Tetrahedron **50**, 11613 (1994).

1817. INAMORI, Y., Y. KATO, M. KUBO, T. KAMIKI, T. TAKEMOTO, and K. NOMOTO: Studies on Metabolites Produced by *Aspergillus terreus* var. *aureus*, I: Chemical Structures and Antimicrobial Activities of Metabolites Isolated from Culture Broth. Chem. Pharm. Bull. (Japan) **31**, 4543 (1983).

1818. BIRCH, A.J., R.A. MASSY-WESTROPP, R.W. RICKARDS, and H. SMITH: Studies in Relation to Biosynthesis, Part XIII: Griseofulvin. J. Chem. Soc., 360 (1958).

1819. TAMURA, G., S. SUZUKI, A. TAKATSUKI, K. ANDO, and K. ARIMA: Ascochlorin, a New Antibiotic, Found by Paper-Disc Agar-Diffuson Method, I: Isolation, Biological and Chemical Properties of Ascochlorin (Studies on Antiviral and Antitumor Antibiotics, I). J. Antibiot. **21**, 539 (1968).

1820. NAWATA, Y., K. ANDO, G. TAMURA, K. ARIMA, and Y. IITAKA: The Molecular Structure of Ascochlorin. J. Antibiot. **22**, 511 (1969).

1821. NAWATA, Y., and Y. IITAKA: The Crystal Structure of Ascochlorin *p*-Bromobenzenesulfonate. Bull. Chem. Soc. Japan **44**, 2652 (1971).

1822. ELLESTAD, G.A., R.H. EVANS, and M.P. KUNSTMANN: Some New Terpenoid Metabolites from an Unidentified *Fusarium* Species. Tetrahedron **25**, 1323 (1969).

1823. MINATO, H., T. KATAYAMA, S. HAYAKAWA, and K. KATAGIRI: Identification of Ilicicolins with Ascochlorin and LL-Z 1272. J. Antibiot. **25**, 315 (1972).

1824. KATO, A., K. ANDO, G. TAMURA, and K. ARIMA: Cylindrochlorin, a New Antibiotic Produced by *Cylindrocladium*. J. Antibiot. **23**, 168 (1970).

1825. HAYAKAWA, S., H. MINATO, and K. KATAGIRI: The Ilicicolins, Antibiotics from *Cylindrocladium ilicicola*. J. Antibiot. **24**, 653 (1971).

1826. ALDRIDGE, D.C., A. BORROW, R.G. FOSTER, M.S. LARGE, H. SPENCER, and W.B. TURNER: Metabolites of *Nectria coccinea*. J. Chem. Soc., Perkin Trans. 1, 2136 (1972).

1827. SASAKI, H., T. HOSOKAWA, Y. NAWATA, and K. ANDO: Isolation and Structure of Ascochlorin and Its Analogs. Agric. Biol. Chem. **38**, 1463 (1974).

1828. KOSUGE, Y., A. SUZUKI, S. HIRATA, and S. TAMURA: Structure of Colletochlorin from *Colletotrichum nicotianae*. Agric. Biol. Chem. **37**, 455 (1973).

1829. KOSUGE, Y., A. SUZUKI, and S. TAMURA: Structures of Colletochlorin C, Colletorin A and Colletorin C from *Colletotrichum nicotianae*. Agric. Biol. Chem. **38**, 1265 (1974).

1830. KOSUGE, Y., A. SUZUKI, and S. TAMURA: Structure of Colletochlorin D from *Colletotrichum nicotianae*. Agric. Biol. Chem. **38**, 1553 (1974).

1831. SASAKI, H., T. HOSOKAWA, M. SAWADA, and K. ANDO: Isolation and Structure of Ascofuranone and Ascofuranol, Antibiotics with Hypolipidemic Activity. J. Antibiot. **26**, 676 (1973).

1832. CAGNOLI-BELLAVITA, N., P. CECCHERELLI, R. FRINGUELLI, and M. RIBALDI: Ascochlorin: A Terpenoid Metabolite from *Acremonium luzulae*. Phytochem. **14**, 807 (1975).

1833. SCHRAMM, G., W. STEGLICH, T. ANKE, and F. OBERWINKLER: Strobilurin A und B, Antifungische Stoffwechselprodukte aus *Strobilurus tenacellus*. Chem. Ber. **111**, 2779 (1978).

1834. MCCORKINDALE, N.J., S.A. HUTCHINSON, A.C. MCRITCHIE, and G.R. SOOD: Lamellicolic Anhydride, 4-*O*-Carbomethoxylamellicolic Anhydride and Monomethyl 3-Chlorolamellicolate, Metabolites of *Verticillium lamellicola*. Tetrahedron **39**, 2283 (1983).

1835. KIGOSHI, H., Y. IMAMURA, K. YOSHIKAWA, and K. YAMADA: Three New Cytotoxic Alkaloids, Aplaminone, Neoaplaminone and Neoaplaminone Sulfate from the Marine Mollusc *Aplysia kurodai*. Tetrahedron Lett. **31**, 4911 (1990).

1836. BERG, D.H., R.P. MASSING, M.M. HOEHN, L.D. BOECK, and R.L. HAMILL: A30641, a New Epidithiodiketopiperazine with Antifungal Activity. J. Antibiot. **29**, 394 (1976).

1837. SAKATA, K., H. MASAGO, A. SAKURAI, and N. TAKAHASHI: Isolation of Aspirochlorine (= Antibiotic A30641) Possessing a Novel Dithiodiketopiperazine Structure from *Aspergillus flavus*. Tetrahedron Lett. **23**, 2095 (1982).

1838. SAKATA, K., T. KUWATSUKA, A. SAKURAI, N. TAKAHASHI, and G. TAMURA: Isolation of Aspirochlorine (= Antibiotic A30641) as a True Antimicrobial Constituent of the Antibiotic Oryzachlorin, from *Aspergillus oryzae*. Agric. Biol. Chem. **47**, 2673 (1983).

1839. SAKATA, K., M. MARUYAMA, J. UZAWA, A. SAKURAI, H.S.M. LU, and J. CLARDY: Structural Revision of Aspirochlorine (= Antibiotic A30641), A Novel Epidithiopiperazine-2,5-dione Produced by *Aspergillus* sp. Tetrahedron Lett. **28**, 5607 (1987).

1840. MIKNIS, G.F., and R.M. WILLIAMS: Total Synthesis of ($\pm$)-Aspirochlorine. J. Am. Chem. Soc. **115**, 536 (1993).

1841. YANG, J., Y. CHEN, X. FENG, D. YU, and X. LIANG: Chemical Constituents of *Armillaria mella* Mycelium, I: Isolation and Characterization of Armillarin and Armillaridin. Planta Med. **50**, 288 (1984).

1842. ARNONE, A., R. CARDILLO, and G. NASINI: Structures of Melleolides B–D, Three Antibacterial Sesquiterpenoids from *Armillaria mellea*. Phytochem. **25**, 471 (1986).

1843. ARNONE, A., R. CARDILLO, G. NASINI, and S.V. MEILLE: Secondary Mould Metabolites, Part 19: Structure Elucidation and Absolute Configuration of Melledonals B and C, Novel Antibacterial Sesquiterpenoids from *Armillaria mellea*. X-Ray Molecular Structure of Melledonal C. J. Chem. Soc., Perkin Trans. 1, 503 (1988).

1844. YANG, J.S., Y.W. CHEN, X.Z. FENG, D.Q. YU, C.H. HE, Q.T. ZHENG, J. YANG, and X.T. LIANG: Chemical Constituents of *Armillaria mellea* Mycelium, Part 2: Isolation and Structure Elucidation of Armillaricin. Planta Med. **55**, 564 (1989).

1845. Itabashi, T., K. Nozawa, M. Miyaji, S. Udagawa, S. Nakajima, and K. Kawai: Falconensins A, B, C, and D, New Compounds Related to Azaphilone, from *Emericella falconensis.* Chem. Pharm. Bull. (Japan) **40**, 3142 (1992).
1846. Itabashi, T., K. Nozawa, S. Nakajima, and K. Kawai: A New Azaphilone, Falconensin H, from *Emericella falconensis.* Chem. Pharm. Bull. (Japan) **41**, 2040 (1993).
1847. Andrewes, A.G., S. Hertzberg, S. Liaaen-Jensen, and M.P. Starr: Xanthomonas Pigments, 2: The *Xanthomonas* "Carotenoids" – Non-Carotenoid Brominated Aryl-Polyene Esters. Acta Chem. Scand. **27**, 2383 (1973).
1848. Andrewes, A.G., C.L. Jenkins, M.P. Starr, J. Shepherd, and H. Hope: Structure of Xanthomonadin I, a Novel Dibrominated Aryl-Polyene Pigment Produced by the Bacterium *Xanthomonas juglandis.* Tetrahedron Lett., 4023 (1976).
1849. Starr, M.P., C.L. Jenkins, L.B. Bussey, and A.G. Andrewes: Chemotaxonomic Significance of the Xanthomonadins, Novel Brominated Aryl-Polyene Pigments Produced by Bacteria of the Genus *Xanthomonas.* Arch. Microbiol. **113**, 1 (1977).
1850. Achenbach, H., W. Kohl, and H. Reichenbach: 5-Chlorflexirubin, ein Nebenpigment aus *Flexibacter elegans.* Ann., 1 (1977).
1851. Achenbach, H., W. Kohl, S. Alexanian, and H. Reichenbach: Neue Pigmente vom Flexirubin-Typ aus *Cytophaga* Spec. Stamm *Samoa.* Chem. Ber. **112**, 196 (1979).
1852. Achenbach, H., W. Kohl, and H. Reichenbach: Die Konstitutionen der Pigmente vom Flexirubin-Typ aus *Cytophaga johnsonae* Cyjl. Chem. Ber. **112**, 1999 (1979).
1853. Achenbach, H., W. Kohl, A. Böttger-Vetter, and H. Reichenbach: Untersuchungen an Stoffwechselprodukten von Mikroorganismen, XXII: Untersuchung der Pigmente aus *Flavobacterium spec.* Stamm C1/2. Tetrahedron **37**, 559 (1981).
1854. Achenbach, H.: The Pigments of the Flexirubin-Type. A Novel Class of Natural Products. Progr. Chem. Org. Nat. Prod. **52**, 73 (1987).
1855. Mason, C.P., K.R. Edwards, R.E. Carlson, J. Pignatello, F.K. Gleason, and J.M. Wood: Isolation of Chlorine-Containing Antibiotic from the Freshwater Cyanobacterium *Scytonema hofmanni.* Science **215**, 400 (1982).
1856. Pignatello, J.J., J. Porwoll, R.E. Carlson, A. Xavier, F.K. Gleason, and J.M. Wood: Structure of the Antibiotic Cyanobacterin, a Chlorine-Containing γ-Lactone from the Freshwater Cyanobacterium *Scytonema hofmanni.* J. Org. Chem. **48**, 4035 (1983).
1857. Yang, X., Y. Shimizu, J.R. Steiner, and J. Clardy: Nostoclide I and II, Extracellular Metabolites from a Symbiotic Cyanobacterium, *Nostoc* sp., from the Lichen *Peltigera canina.* Tetrahedron Lett. **34**, 761 (1993).
1858. Marumoto, R., C. Kilpert, and W. Steglich: New Pulvinic Acid Derivatives from *Pulveroboletus* Species (Boletales). Z. Naturforsch. **41C**, 363 (1986).
1859. Liu, C., T.E. Hermann, B.L.T. Prosser, N.J. Palleroni, J.W. Westley, and P.A. Miller: X-14766A, a Halogen Containing Polyether Antibiotic Produced by *Streptomyces malachitofuscus* subsp. *downeyi* ATCC 31547. J. Antibiot. **34**, 133 (1981).
1860. Westley, J.W., R.H. Evans Jr., L.H. Sello, N. Troupe, C. Liu, J.F. Blount, R.G. Pitcher, T.H. Williams, and P.A. Miller: Isolation and Characterization of the First Halogen Containing Polyether Antibiotic X-14766A, a Product of *Streptomyces malachitofuscus* subsp. *downeyi.* J. Antibiot. **34**, 139 (1981).
1861. Cullen, W.P., W.D. Celmer, L.R. Chappel, L.H. Huang, H. Maeda, S. Nishiyama, R. Shibakawa, J. Tone, and P.C. Watts: CP-54,883 A Novel Chlorine-Containing Polyether Antibiotic Produced by a New Species of *Actinomadura*: Taxonomy of the Producing Culture, Fermentation, Physio-Chemical and Biological Properties of the Antibiotic. J. Antibiot. **40**, 1490 (1987).

1862. Bordner, J., P.C. Watts, and E.B. Whipple: Structure of the Natural Antibiotic Ionophore CP-54,883. J. Antibiot. **40**, 1496 (1987).

1863. Blackman, A.J., and R.D. Green: Further Amathamide Alkaloids from the Bryozoan *Amathia wilsoni*. Aust. J. Chem. **40**, 1655 (1987).

1864. Zhang, H.-P., H. Shigemori, M. Ishibashi, T. Kosaka, G.R. Pettit, Y. Kamano, and J. Kobayashi: Convolutamides A–F, Novel γ-Lactam Alkaloids from the Marine Bryozoan *Amathia convoluta*. Tetrahedron **50**, 10201 (1994).

1865. Combaut, G., J.-M. Chantraine, J. Teste, and K.-W. Glombitza: Phenols Bromés des Algues Rouges: Cyclotribromoveratrylene, Nouveau Derivé Obtenu au Cours de l'Extraction de *Halopytis pinastroïdes*. Phytochem. **17**, 1791 (1978).

1866. Combaut, G., J.-M. Chantraine, J. Teste, J. Soulier, and K.-W. Glombitza: Structure et Conformation du Dihydro-10,15-tribromo-1,6,11-hexamethoxy-2,3,7,8,12,13-5H-tribenzo[*a*, *d*, *g*]cyclononene. Tetrahedron Lett., 1699 (1978).

1867. McCormick, M.H., W.M. Stark, G.E. Pittenger, R.C. Pittenger, and J.M. McGuire: Vancomycin, a New Antibiotic, I: Chemical and Biological Properties. Antibiot. Annu. 1955–1956, 601 (1956).

1868. Geraci, J.E., and P.E. Hermans: Vancomycin. Mayo Clin. Proc. **58**, 88 (1983).

1869. Barna, J.C.J., and D.H. Williams: The Structure and Mode of Action of Glycopeptide Antibiotics of the Vancomycin Group. Ann. Rev. Microbiol. **38**, 339 (1984).

1870. Nagarajan, R.: Structure-Activity Relationships of Vancomycin-Type Glycopeptide Antibiotics. J. Antibiot. **46**, 1181 (1993).

1871. Walsh, C.T.: Vancomycin Resistance: Decoding the Molecular Logic. Science **261**, 308 (1993).

1872. Fan, C., P.C. Moews, C.T. Walsh, and J.R. Knox: Vancomycin Resistance: Structure of D-Alanine:D-Alanine Ligase at 2.3 Å Resolution. Science **266**, 439 (1994).

1873. Harris, C.M., R. Kannan, H. Kopecka, and T.M. Harris: The Role of the Chlorine Substituents in the Antibiotic Vancomycin: Preparation and Characterization of Mono- and Didechlorovancomycin. J. Am. Chem. Soc. **107**, 6652 (1985).

1874. Roberts, P.J., O. Kennard, K.A. Smith, and D.H. Williams: Concerning the Molecular Weight and Structure of the Antibiotic Vancomycin. J. Chem. Soc., Chem. Commun., 772 (1973).

1875. Smith, K.A., D.H. Williams, and G.A. Smith: Structural Studies on the Antibiotic Vancomycin; the Nature of the Aromatic Rings. J. Chem. Soc., Perkin Trans. 1, 2369 (1974).

1876. Smith, G.A., K.A. Smith, and D.H. Williams: Structural Studies on the Antibiotic Vancomycin: Evidence for the Presence of Modified Phenylglycine and β-Hydroxytyrosine Units. J. Chem. Soc., Perkin Trans. 1, 2108 (1975).

1877. Williams, D.H., and J.R. Kalman: Structural and Mode of Action Studies on the Antibiotic Vancomycin. Evidence from 270-MHz Proton Magnetic Resonance. J. Am. Chem. Soc. **99**, 2768 (1977).

1878. Sheldrick, G.M., P.G. Jones, O. Kennard, D.H. Williams, and G.A. Smith: Structure of Vancomycin and Its Complex with Acetyl-D-alanyl-D-alanine. Nature **271**, 223 (1978).

1879. Williamson, M.P., and D.H. Williams: Structure Revision of the Antibiotic Vancomycin. The Use of Nuclear Overhauser Effect Difference Spectroscopy. J. Am. Chem. Soc. **103**, 6580 (1981).

1880. Harris, C.M., H. Kopecka, and T.M. Harris: Vancomycin: Structure and Transformation to CDP-I. J. Am. Chem. Soc. **105**, 6915 (1983).

1881. WALTHO, J.P., D.H. WILLIAMS, D.J.M. STONE, and N.J. SKELTON: Intramolecular Determinants of Conformation and Mobility Within the Antibiotic Vancomycin. J. Am. Chem. Soc. **110**, 5638 (1988).

1882. GROVES, P., M.S. SEARLE, M.S. WESTWELL, and D.H. WILLIAMS: Expression of Electrostatic Binding Cooperativity in the Recognition of Cell-Wall Peptide Analogues by Vancomycin Group Antibiotics. J. Chem. Soc., Chem. Commun., 1519 (1994).

1883. WILLIAMS, D.H.: Structural Studies on Some Antibiotics of the Vancomycin Group, and on the Antibiotic-Receptor Complexes, by ^{1}H-NMR. Acct. Chem. Res. **17**, 364 (1984).

1884. WILLIAMS, D.H., M.S. SEARLE, J.P. MACKAY, U. GERHARD, and R.A. MAPLESTONE: Toward an Estimation of Binding Constants in Aqueous Solution: Studies of Associations of Vancomycin Group Antibiotics. Proc. Natl. Acad. Sci. USA **90**, 1172 (1993).

1885. MACKAY, J.P., U. GERHARD, D.A. BEAUREGARD, R.A. MAPLESTONE, and D.H. WILLIAMS: Dissection of the Contributions Toward Dimerization of Glycopeptide Antibiotics. J. Am. Chem. Soc. **116**, 4573 (1994).

1886. MACKAY, J.P., U. GERHARD, D.A. BEAUREGARD, M.S. WESTWELL, M.S. SEARLE, and D.H. WILLIAMS: Glycopeptide Antibiotic Activity and the Possible Role of Dimerization: A Model for Biological Signaling. J. Am. Chem. Soc. **116**, 4581 (1994).

1887. MUELLER, L., S.L. HEALD, J.C. HEMPEL, and P.W. JEFFS: Determination of the Conformation of Molecular Complexes of the Aridicin Aglycon with Ac_2-L-Lys-D-Ala-D-Ala and Ac-L-Ala-γ-D-Gln-L-Lys(Ac)-D-Ala-D-Ala: An Application of Nuclear Magnetic Resonance Spectroscopy and Distance Geometry in the Modeling of Peptides. J. Am. Chem. Soc. **111**, 496 (1989).

1888. NAGARAJAN, R., K.E. MERKEL, K.H. MICHEL, H.M. HIGGINS JR., M.M. HOEHN, A.H. HUNT, N.D. JONES, J.L. OCCOLOWITZ, A.A. SCHABEL, and J.K. SWARTZENDRUBER: M43 Antibiotics: Methylated Vancomycins and Unrearranged CDP-I Analogues. J. Am. Chem. Soc. **110**, 7896 (1988).

1889. HUNT, A.H., G.G. MARCONI, T.K. ELZEY, and M.M. HOEHN: A51568A: *N*-Demethylvancomycin. J. Antibiot. **37**, 917 (1984).

1890. BARDONE, M.R., M. PATERNOSTER, and C. CORONELLI: Teichomycins, New Antibiotics from *Actinoplanes teichomyceticus* Nov. Sp., II: Extraction and Chemical Characterization. J. Antibiot. **31**, 170 (1978).

1891. PARENTI, F., G. BERETTA, M. BERTI, and V. ARIOLI: Teichomycins, New Antibiotics from *Actinoplanes teichomyceticus* Nov. Sp., I: Description of the Producer Strain, Fermentation Studies and Biological Properties. J. Antibiot. **31**, 276 (1978).

1892. BORGHI, A., C. CORONELLI, L. FANIUOLO, G. ALLIEVI, R. PALLANZA, and G.G. GALLO: Teichomycins, New Antibiotics from *Actinoplanes teichomyceticus* Nov. Sp., IV: Separation and Characterization of the Components of Teichomycin (Teicoplanin). J. Antibiot. **37**, 615 (1984).

1893. CORONELLI, C., M.R. BARDONE, A. DE PAOLI, P. FERRARI, G. TUAN, and G.G. GALLO: Teicoplanin, Antibiotics from *Actinoplanes teichomyceticus* Nov. Sp., V: Aromatic Constituents. J. Antibiot. **37**, 621 (1984).

1894. HUNT, A.H., R.M. MOLLOY, J.L. OCCOLOWITZ, G.G. MARCONI, and M. DEBONO: Structure of the Major Glycopeptide of the Teicoplanin Complex. J. Am. Chem. Soc. **106**, 4891 (1984).

1895. BARNA, J.C.J., D.H. WILLIAMS, D.J.M. STONE, T.-W.C. LEUNG, and D.M. DODDRELL: Structure Elucidation of the Teicoplanin Antibiotics. J. Am. Chem. Soc. **106**, 4895 (1984).

1896. MALABARBA, A., P. FERRARI, G.G. GALLO, J. KETTENRING, and B. CAVALLERI: Teicoplanin, Antibiotics from *Actinoplanes teichomyceticus* Nov. Sp., VII: Preparation and NMR Characteristics of the Aglycone of Teicoplanin. J. Antibiot. **39**, 1430 (1986).

1897. BORGHI, A., P. ANTONINI, M. ZANOL, P. FERRARI, L.F. ZERILLI, and G.C. LANCINI: Isolation and Structure Determination of Two New Analogs of Teicoplanin, a Glycopeptide Antibiotic. J. Antibiot. **42**, 361 (1989).

1898. MCGÁHREN, W.J., J.H. MARTIN, G.O. MORTON, R.T. HARGREAVES, R.A. LEESE, F.M. LOVELL, and G.A. ELLESTAD: Avoparcin. J. Am. Chem. Soc. **101**, 2237 (1979).

1899. MCGAHREN, W.J., J.H. MARTIN, G.O. MORTON, R.T. HARGREAVES, R.A. LEESE, F.M. LOVELL, G.A. ELLESTAD, E. O'BRIEN, and J.S.E. HOLKER: Structure of Avoparcin Components. J. Am. Chem. Soc. **102**, 1671 (1980).

1900. ELLESTAD, G.A., R.A. LEESE, G.O. MORTON, F. BARBATSCHI, W.E. GORE, W.J. MCGAHREN, and I.M. ARMITAGE: Avoparcin and Epiavoparcin. J. Am. Chem. Soc. **103**, 6522 (1981).

1901. FESIK, S.W., I.M. ARMITAGE, G.A. ELLESTAD, and W.J. MCGAHREN: Nuclear Magnetic Resonance Studies on the Antibiotic Avoparcin. Conformational Properties in Relation to Antibacterial Activity. Mol. Pharmacol. **25**, 275 (1984).

1902. FESIK, S.W., I.M. ARMITAGE, G.A. ELLESTAD, and W.J. MCGAHREN: Nuclear Magnetic Resonance Studies on the Interaction of Avoparcin with Model Receptors of Bacterial Cell Walls. Mol. Pharmacol. **25**, 281 (1984).

1903. TAKEUCHI, M., R. ENOKITA, T. OKAZAKI, T. KAGASAKI, and M. INUKAI: Helvecardins A and B, Novel Glycopeptide Antibiotics, I: Taxonomy, Fermentation, Isolation and Physico-Chemical Properties. J. Antibiot. **44**, 263 (1991).

1904. TAKEUCHI, M., S. TAKAHASHI, M. INUKAI, T. NAKAMURA, and T. KINOSHITA: Helvecardins A and B, Novel Glycopeptide Antibiotics, II: Structural Elucidation. J. Antibiot. **44**, 271 (1991).

1905. OKAZAKI, T., R. ENOKITA, H. MIYAOKA, T. TAKATSU, and A. TORIKATA: Chloropolysporins A, B and C, Novel Glycopeptide Antibiotics from *Faenia interjecta* sp. Nov., I: Taxonomy of Producing Organism. J. Antibiot. **40**, 917 (1987).

1906. TAKATSU, T., M. NAKAJIMA, S. OYAJIMA, Y. ITOH, Y. SAKAIDA, S. TAKAHASHI, and T. HANEISHI: Chloropolysporins A, B and C, Novel Glycopeptide Antibiotics from *Faenia interjecta* sp. Nov., II: Fermentation, Isolation and Physio-Chemical Characterization. J Antibiot. **40**, 924 (1987).

1907. TAKATSU, T., S. TAKAHASHI, M. NAKAJIMA, T. HANEISHI, T. NAKAMURA, H. KUWANO, and T. KINOSHITA: Chloropolysporins A, B and C, Novel Glycopeptide Antibiotics from *Faenia interjecta* sp. Nov., III: Structure Elucidation of Chloropolysporins. J. Antibiot. **40**, 933 (1987).

1908. DEBONO, M., K.E. MERKEL, R.M. MOLLOY, M. BARNHART, E. PRESTI, A.H. HUNT, and R.L. HAMILL: Actaplanin, New Glycopeptide Antibiotics Produced by *Actinoplanes missouriensis*. The Isolation and Prelimary Chemical Characterization of Actaplanin. J. Antibiot. **37**, 85 (1984).

1909. HUNT, A.H., M. DEBONO, K.E. MERKEL, and M. BARNHART: Structure of the Pseudoaglycon of Actaplanin. J. Org. Chem. **49**, 635 (1984).

1910. HUNT, A.H., T.K. ELZEY, K.E. MERKEL, and M. DEBONO: Structures of the Actaplanins. J. Org. Chem. **49**, 641 (1984).

1911. HUBER, F.M., K.H. MICHEL, A.H. HUNT, J.W. MARTIN, and R.M. MOLLOY: Preparation and Characterization of Some Bromine Analogs of the Glycopeptide Antibiotic Actaplanin. J. Antibiot. **41**, 798 (1988).

1912. CHRISTENSEN, S.B., H.S. ALLAUDEEN, M.R. BURKE, S.A. CARR, S.K. CHUNG,

P. DEPHILLIPS, J.J. DINGERDISSEN, M. DI PAOLO, A.J. GIOVENELLA, S.L. HEALD, L.B. KILLMER, B.A. MICO, L. MUELLER, C.H. PAN, B.L. POEHLAND, J.B. RAKE, G.D. ROBERTS, M.C. SHEARER, R.D. SITRIN, L.J. NISBET, and P.W. JEFFS: Parvodicin, a Novel Glycopeptide from a New Species, *Actinomadura parvosata*: Discovery, Taxonomy, Activity and Structure Elucidation. J. Antibiot. **40**, 970 (1987).

1913. SHEARER, M.C., P. ACTOR, B.A. BOWIE, S.F. GRAPPEL, C.H. NASH, D.J. NEWMAN, Y.K. OH, C.H. PAN, and L.J. NISBET: Aridicins, Novel Glycopeptide Antibiotics, I: Taxonomy, Production and Biological Activity. J. Antibiot. **38**, 555 (1985).

1914. SITRIN, R.D., G.W. CHAN, J.J. DINGERDISSEN, W. HOLL, J.R.E. HOOVER, J.R. VALENTA, L. WEBB, and K.M. SNADER: Aridicins, Novel Glycopeptide Antibiotics, II: Isolation and Characterization. J. Antibiot. **38**, 561 (1985).

1915. SITRIN, R.D., G.W. CHAN, F. CHAPIN, A.J. GIOVENELLA, S.F. GRAPPEL, P.W. JEFFS, L. PHILLIPS, K.M. SNADER, and L.J. NISBET: Aridicins, Novel Glycopeptide Antibiotics, III: Preparation, Characterization, and Biological Activities of Aglycone Derivatives. J. Antibiot. **39**, 68 (1986).

1916. JEFFS, P.W., L. MUELLER, C. DE BROSSE, S.L. HEALD, and R. FISHER: Structure of Aridicin A. An Integrated Approach Employing 2D NMR, Energy Minimization, and Distance Constraints. J. Am. Chem. Soc. **108**, 3063 (1986).

1917. JEFFS, P.W., G. CHAN, R. SITRIN, N. HOLDER, G.D. ROBERTS, and C. DEBROSSE: The Structure of the Glycolipid Components of the Aridicin Antibiotic Complex. J. Org. Chem. **50**, 1726 (1985).

1918. SHEARER, M.C., A.J. GIOVENELLA, S.F. GRAPPEL, R.D. HEDDE, R.J. MEHTA, Y.K. OH, C.H. PAN, D.H. PITKIN, and L.J. NISBET: Kibdelins, Novel Glycopeptide Antibiotics, I: Discovery, Production, and Biological Properties. J. Antibiot. **39**, 1386 (1986).

1919. FOLENA-WASSERMAN, G., B.L. POEHLAND, E.W.-K. YEUNG, D. STAIGER, L.B. KILLMER, K. SNADER, J.J. DINGERDISSEN, and P.W. JEFFS: Kibdelins (AAD-609), Novel Glycopeptide Antibiotics, II: Isolation, Purification and Structure. J. Antibiot. **39**, 1395 (1986).

1920. WALTHO, J.P., D.H. WILLIAMS, E. SELVA, and P. FERRARI: Structure Elucidation of the Glycopeptide Antibiotic Complex A40926. J. Chem. Soc., Perkin Trans. 1, 2103 (1987).

1921. BOX, S.J., N.J. COATES, C.J. DAVIS, M.L. GILPIN, C.S.V. HOUGE-FRYDRYCH, and P.H. MILNER: MM55266 and MM55268, Glycopeptide Antibiotics Produced by a New Strain of *Amycolatopsis*. Isolation, Purification and Structure Determination. J. Antibiot. **44**, 807 (1991).

1922. TSUJI, N., M. KOBAYASHI, T. KAMIGAUCHI, Y. YOSHIMURA, and Y. TERUI: New Glycopeptide Antibiotics, I: The Structures of Orienticins. J. Antibiot. **41**, 819 (1988).

1923. TSUJI, N., T. KAMIGAUCHI, M. KOBAYASHI, and Y. TERUI: New Glycopeptide Antibiotics, II: The Isolation and Structures of Chloroorienticins. J. Antibiot. **41**, 1506 (1988).

1924. NAGARAJAN, R., D.M. BERRY, A.H. HUNT, J.L. OCCOLOWITZ, and A.A. SCHABEL: Conversion of Antibiotic A82846B to Orienticin A and Structural Relationships of Related Antibiotics. J. Org. Chem. **54**, 983 (1989).

1925. GAUSE, G.F., M.G. BRAZHNIKOVA, N.N. LOMAKINA, T.F. BERDNIKOVA, G.B. FEDOROVA, N.L. TOKAREVA, V.N. BORISOVA, and G.Y. BATTA: Eremomycin–New Glycopeptide Antibiotic: Chemical Properties and Structure. J. Antibiot. **42**, 1790 (1989).

1926. BATTA, G., F. SZTARICSKAI, K.E. KÖVÉR, C. RÜDEL, and T.F. BERDNIKOVA: An NMR Study of Eremomycin and Its Derivatives. J. Antibiot. **44**, 1208 (1991).

1927. RIVA, E., L. GASTALDO, M.G. BERETTA, P. FERRARI, L.F. ZERILLI, G. CASSANI, E. SELVA, B.P. GOLDSTEIN, M. BERTI, F. PARENTI, and M. DENARO: A42867, A Novel Glycopeptide Antibiotic. J. Antibiot. **42**, 497 (1989).

1928. BOX, S.J., A.L. ELSON, M.L. GILPIN, and D.J. WINSTANLEY: MM47761 and MM49721, Glycopeptide Antibiotics Produced by a New Strain of *Amycolatopsis orientalis*. J. Antibiot. **43**, 931 (1990).

1929. BOECK, L.D., F.P. MERTZ, and G.M. CLEM: A-41030, A Complex of Novel Glycopeptide Antibiotics Produced by a Strain of *Streptomyces virginiae*. Taxonomy and Fermentation Studies. J. Antibiot. **38**, 1 (1985).

1930. SKELTON, N.J., D.H. WILLIAMS, R.A. MONDAY, and J.C. RUDDOCK: Structure Elucidation of the Novel Glycopeptide Antibiotic UK-68,597. J. Org. Chem. **55**, 3718 (1990).

1931. TAKEUCHI, M., S. TAKAHASHI, R. ENOKITA, Y. SAKAIDA, H. HARUYAMA, T. NAKAMURA, T. KATAYAMA, and M. INUKAI: Galacardins A and B, New Glycopeptide Antibiotics. J. Antibiot. **45**, 297 (1992).

1932. NADKARNI, S.R., M.V. PATEL, S. CHATTERJEE, E.K.S. VIJAYAKUMAR, K.R. DESIKAN, J. BLUMBACH, B.N. GANGULI, and M. LIMBERT: Balhimycin, a New Glycopeptide Antibiotic Produced by *Amycolatopsis* sp. Y-86,21022. Taxonomy, Production, Isolation and Biological Activity. J. Antibiot. **47**, 334 (1994).

1933. CHATTERJEE, S., E.K.S. VIJAYAKUMAR, S.R. NADKARNI, M.V. PATEL, J. BLUMBACH, B.N. GANGULI, H.-W. FEHLHABER, H. KOGLER, and L. VERTESY: Balhimycin, a New Glycopeptide Antibiotic with an Unusual Hydrated 3-Amino-4-oxoaldopyranose Sugar Moiety. J. Org. Chem. **59**, 3480 (1994).

1934. SKELTON, N.J., D.H. WILLIAMS, M.J. RANCE, and J.C. RUDDOCK: Structure Elucidation of UK-72,051, a Novel Member of the Vancomycin Group of Antibiotics. J. Chem. Soc., Perkin Trans. 1, 77 (1990).

1935. ANG, S.-G., M.P. WILLIAMSON, and D.H. WILLIAMS: Structure Elucidation of a Glycopeptide Antibiotic, OA-7653. J. Chem. Soc., Perkin Trans. 1, 1949 (1988).

1936. JEFFS, P.W., B. YELLIN, L. MUELLER, and S.L. HEALD: Structure of the Antibiotic OA-7653. J. Org. Chem. **53**, 471 (1988).

1937. KANEKO, I., K. KAMOSHIDA, and S. TAKAHASHI: Complestatin, a Potent Anti-Complement Substance Produced by *Streptomyces lavendulae*, I: Fermentation, Isolation, and Biological Characterization. J. Antibiot. **42**, 236 (1989).

1938. NARUSE, N., O. TENMYO, S. KOBARU, M. HATORI, K. TOMITA, Y. HAMAGISHI, and T. OKI: New Antiviral Antibiotics, Kistamicins A and B, I: Taxonomy, Production, Isolation, Physico-Chemical Properties and Biological Activities. J. Antibiot. **46**, 1804 (1993).

1939. NARUSE, N., M. OKA, M. KONISHI, and T. OKI: New Antiviral Antibiotics, Kistamicins A and B, II: Structure Determination. J. Antibiot. **46**, 1812 (1993).

1940. WRIGHT, D.E.: The Orthosomycins, a New Family of Antibiotics: Tetrahedron **35**, 1207 (1979).

1941. GALMARINI, O.L., and V. DEULOFEU: Curamycin-I. Isolation and Characterization of Some Hydrolysis Products. Tetrahedron **15**, 76 (1961).

1942. NINET, L., F. BENAZET, Y. CHARPENTIE, M. DUBOST, J. FLORENT, J. LUNEL, D. MANCY, and J. PREUD'HOMME: Flambamycin, a New Antibiotic from *Streptomyces hygroscopicus* DS23 230. Experientia **30**, 1270 (1974).

1943. OLLIS, W.D., C. SMITH, I.O. SUTHERLAND, and D.E. WRIGHT: The Constitution of the Antibiotic Flambamycin. J. Chem. Soc., Chem. Commun., 350 (1976).

1944. OLLIS, W.D., C. SMITH, and D.E. WRIGHT: The Orthosomycin Family of Antibiotics – I. The Constitution of Flambamycin. Tetrahedron **35**, 105 (1979).

1945. BUZZETTI, F., F. EISENBERG, H.N. GRANT, W. KELLER-SCHIERLEIN, W. VOSER, and H. ZÄHNER: Avilamycin. Experientia **24**, 320 (1968).

1946. HEILMAN, W., E. KUPFER, W. KELLER-SCHIERLEIN, H. ZÄHNER, H. WOLF, and H.H. PETER: Stoffwechselprodukte von Mikroorganismen. Avilamycin C. Helv. Chim. Acta **62**, 1 (1979).

1947. KELLER-SCHIERLEIN, W., W. HEILMAN, W.D. OLLIS, and C. SMITH: Stoffwechselprodukte von Mikroorganismen. Die Avilamycine A und C: Chemischer Abbau und spektroskopische Untersuchungen. Helv. Chim. Acta **62**, 7 (1979).

1948. WAGMAN, G.H., G.M. LUEDEMANN, and M.J. WEINSTEIN: Fermentation and Isolation of Everninomicin. Antimicrob. Agents Chemother., 33 (1964).

1949. WEINSTEIN, M.J., G.M. LUEDEMANN, E.M. ODEN, and G.H. WAGMAN: Everninomicin, a New Antibiotic Complex from *Micromonospora carbonacea*. Antimicrob. Agents Chemother., 24 (1964).

1950. GANGULY, A.K., and A.K. SAKSENA: Structure of Everninomicin B. J. Antibiot. **28**, 707 (1975).

1951. GANGULY, A.K., and S. SZMULEWICZ: Structure of Everninomicin C. J. Antibiot. **28**, 710 (1975).

1952. GANGULY, A.K., O.Z. SARRE, D. GREEVES, and J. MORTON: Structure of Everninomicin D. J. Am. Chem. Soc. **97**, 1982 (1975).

1953. GANGULY, A.K., S. SZMULEWICZ, O.Z. SARRE, and V.M. GIRIJAVALLABHAN: Structure of Everninomicin-2. J. Chem. Soc., Chem. Commun., 609 (1976).

1954. GANGULY, A.T., O.Z. SARRE, A.T. MCPHAIL, and R.W. MILLER: X-Ray Crystal and Molecular Structure of Olgose, a Major Degradation Product of the Oligosaccharide Antibiotic Everninomicin D. J. Chem. Soc., Chem. Commun., 22 (1979).

1955. GOUGH, M.: Human Health Effects: What the Data Indicate: Sci. Total Environ. **104**, 129 (1991).

1956. EDULJEE, G.H.: Dioxins in the Environment. Chem. Brit. **24**, 1223 (1988).

1957. TRAVIS, C.C., H.A. HATTEMER-FREY, and E. SILBERGELD: Dioxin, Dioxin Everywhere. Environ. Sci. Technol. **23**, 1061 (1989).

1958. FIEDLER, H., D. HUTZINGER, and C.W. TIMMS: Dioxins: Sources of Environmental Load and Human Exposure. Toxicol. Environ. Chem. **29**, 157 (1990).

1959. TRAVIS, C.C., and H.A. HATTEMER-FREY: Human Exposure to Dioxin. Sci. Total Environ. **104**, 97 (1991).

1960. ZACK, J.A., and R.R. SUSKIND: The Mortality Experience of Workers Exposed to Tetrachlorodibenzodioxin in a Trichlorophenol Process Accident. J. Occup. Med. **22**, 11 (1980).

1961. TSCHIRLEY, F.H.: Dioxin. Sci. Amer. **254**, 29 (1986).

1962. FINGERHUT, M.A., W.E. HALPERIN, D.A. MARLOW, L.A. PIACITELLI, P.A. HONCHAR, M.H. SWEENEY, A.L. GREIFE, P.A. DILL, K. STEENLAND, and A.J. SURUDA: Cancer Mortality in Workers Exposed to 2,3,7,8-Tetrachlorodibenzo-*p*-dioxin. New England J. Med. **324**, 212 (1991).

1963. GOUGH, M.: Agent Orange: Exposure and Policy. Am. J. Pub. Health **81**, 289 (1991).

1964. WOLFE, W.H., J.E. MICHALEK, J.C. MINER, A. RAHE, J. SILVA, W.F. THOMAS, W.D. GRUBBS, M.B. LUSTIK, T.G. KARRISON, R.H. ROEGNER, and D.E. WILLIAMS: Health Status of Air Force Veterans Occupationally Exposed to Herbicides in Vietnam, I: Physical Health. J. Am. Med. Assoc. **264**, 1824 (1990).

1965. MICHALEK, J.E., W.H. WOLFE, and J.C. MINER: Health Status of Air Force Veterans Occupationally Exposed to Herbicides in Vietnam, II: Mortality. J. Am. Med. Assoc. **264**, 1832 (1990).

1966. KANG, H.K., K.K. WATANABE, J. BREEN, J. REMMERS, M.G. CONOMOS, J. STANLEY, and M. Flicker: Dioxins and Dibenzofurans in Adipose Tissue of US Vietnam Veterans and Controls. Am. J. Pub. Health **81**, 344 (1991).

1967. Bertazzi, P.A., C. Zocchetti, A.C. Pesatori, S. Guercilena, M. Sanarico, and L. Radice: Ten-Year Mortality Study of the Population Involved in the Seveso Incident in 1976. Amer. J. Epidemiol. **129**, 1187 (1989).

1968. Bertazzi, P.A.: Long-Term Effects of Chemical Disasters. Lessons and Results from Seveso. Sci. Total Environ. **106**, 5 (1991).

1969. Mocarelli, P., L.L. Needham, A. Marocchi, D.G. Patterson Jr., P. Brambilla, P.M. Gerthoux, L. Meazza, and V. Carreri: Serum Concentrations of 2,3,7,8-Tetrachlorodibenzo-*p*-dioxin and Test Results from Selected Residents of Seveso, Italy. J. Toxicol. Environ. Health **32**, 357 (1991).

1970. Gribble, G.W.: TCDD – A Deadly Molecule. Chemistry **47**, 15 (1974).

1971. Suskind, R.R.: Chloracne, "The Hallmark of Dioxin Intoxication." Scand J. Work. Environ. Health **11**, 165 (1985).

1972. Bumb, R.R., W.B. Crummett, S.S. Cutie, J.R. Gledhill, R.H. Hummel, R.O. Kagel, L.L. Lamparski, E.V. Luoma, D.L. Miller, T.J. Nestrick, L.A. Shadoff, R.H. Stehl, and J.S. Woods: Trace Chemistries of Fire: A Source of Chlorinated Dioxins. Science **210**, 385 (1980).

1973. Olie, K., M.v. d. Berg, and O. Hutzinger: Formation and Fate of PCDD and PCDF from Combustion Processes. Chemosphere **12**, 627 (1983).

1974. Chiu, C., R.S. Thomas, J. Lockwood, K. Li, R. Halman, and R.C.C. Lao: Polychlorinated Hydrocarbons from Power Plants, Wood Burning and Municipal Incinerators. Chemosphere **12**, 607 (1983).

1975. Tiernan, T.O., M.L. Taylor, J.H. Garrett, G.F. Van Ness, J.G. Solch, D.A. Deis, and D.J. Wagel: Chlorodibenzodioxins, Chlorodibenzofurans and Related Compounds in the Effluents from Combustion Processes. Chemosphere **12**, 595 (1983).

1976. Marklund, S., C. Rappe, M. Tysklind, and K.E. Egeback: Identification of Polychlorinated Dibenzofurans and Dioxins in Exhausts from Cars Run on Leaded Gasoline. Chemosphere **16**, 29 (1987).

1977. Harrad, S.J., A.R. Fernandes, C.S. Creaser, and E.A. Cox: Domestic Coal Combustion as a Source of PCDDs and PCDFs in the British Environment. Chemosphere **23**, 255 (1991).

1978. Ballschmiter, K., H. Buchert, R. Niemczyk, A. Munder, and M. Swerev: Automobile Exhausts Versus Municipal-Waste Incineration as Sources of the Polychlorodibenzodioxins (PCDD) and -furans (PCDF) Found in the Environment. Chemosphere **15**, 901 (1986).

1979. Junk, G.A., and J.J. Richard: Dioxins Not Detected in Effluents from Coal/Refuse Combustion. Chemosphere **10**, 1237 (1981).

1980. Eduljee, G.H., D.H.F. Atkins, and A.E. Eggleton: Observations and Assessment Relating to Incineration of Chlorinated Chemical Wastes. Chemosphere **15**, 1577 (1986).

1981. Smith, R.M., P. O'Keefe, K. Aldous, R. Briggs, D. Hilker, and S. Connor: Measurement of PCDFs and PCDDs in Air Samples and Lake Sediments at Several Locations in Upstate New York. Chemosphere **25**, 95 (1992).

1982. Nestrick, T.J., and L.L. Lamparski: Isomer-Specific Determination of Chlorinated Dioxins for Assessment of Formation and Potential Environment Emission from Wood Combustion. Anal. Chem. **54**, 2292 (1982).

1983. Clement, R.E., H.M. Tosine, and B. Ali: Levels of Polychlorinated Dibenzo-*p*-dioxins and Dibenzofurans in Wood Burning Stoves and Fireplaces. Chemosphere **14**, 815 (1985).

1984. SHEFFIELD, A.: Sources and Releases of PCDDs and PCDFs to the Canadian Environment. Chemosphere **14**, 811 (1985).

1985. ÖBERG, L.G., and K.G. PAUL: The Transformation of Chlorophenols by Lactoperoxidase. Biochim. Biophys. Acta **842**, 30 (1985).

1986. SVENSON, A., L.-O. KJELLER, and C. RAPPE: Enzyme-Mediated Formation of 2,3,7,8-Tetra-Substituted Chlorinated Dibenzodioxins and Dibenzofurans. Environ. Sci. Technol. **23**, 900 (1989).

1987. ÖBERG, L.G., B. GLAS, S.E. SWANSON, C. RAPPE, and K.G. PAUL: Peroxidase-Catalyzed Oxidation of Chlorophenols to Polychlorinated Dibenzo-*p*-dioxins and Dibenzofurans. Arch. Environ. Contam. Toxicol. **19**, 930 (1990).

1988. ÖBERG, L.G., and C. RAPPE: Biochemical Formation of PCDD/Fs from Chlorophenols. Chemosphere **25**, 49 (1992).

1989. MALONEY, S.W., J. MANEM, J. MALLEVIALLE, and F. FLESSINGER: Transformation of Trace Organic Compounds in Drinking Water by Enzymatic Oxidative Coupling. Environ. Sci. Technol. **20**, 249 (1986).

1990. ÖBERG, L.G., N. WÅGMAN, R. ANDERSSON, and C. RAPPE: *De Novo* Formation of PCDD/Fs in Compost and Sewage Sludge–A Status Report. Organohalogen Cmpds. **11**, 297 (1993).

1991. ÖBERG, L.G., R. ANDERSSON, and C. RAPPE: *De Novo* Formation of Hepta- and Octachlorodibenzo-*p*-dioxins from Pentachlorophenol in Municipal Sewage Sludge. Organohalogen Cmpds. **9**, 351 (1992).

1992. CHANCE, B., H. SIES, and A. BOVERIS: Hydroperoxide Metabolism in Mammalian Organs. Physiol. Rev. **59**, 527 (1979).

1993. SALA, T., and M.V. SARGENT: Photochemical Conversion of Grisadiendiones into Dibenzofurans. J. Chem. Soc., Chem. Commun., 1044 (1978).

1994. HASHIMOTO, S., T. WAKIMOTO, and R. TATSUKAWA: PCDDs in the Sediments Accumulated About 8120 Years Ago from Japanese Coastal Areas. Chemosphere **21**, 825 (1991).

1995. Humic Substances II. In: Search of Structure (M.H.B. Hayes, P. MacCarthy, R.I. Malcolm, and R.W. Swift, eds.). New York: Wiley-Interscience. 1989.

1996. WIGILIUS, B., B. ALLARD, H. BORÉN, and A. GRIMVALL: Determination of Adsorbable Organic Halogens (AOX) and Their Molecular Weight Distribution in Surface Water Samples. Chemosphere **17**, 1985 (1988).

1997. ASPLUND, G., A. GRIMVALL, and C. PETTERSSON: Naturally Produced Adsorbable Organic Halogens (AOX) in Humic Substances from Soil and Water. Sci. Total Environ. **81/82**, 239 (1989).

1998. ENELL, M., L. KAJ, and L. WENNBERG: Long-Distance Distribution of Halogenated Organic Compounds (AOX). In: River Basin Management–V, p. 29 (H. Laikari, ed.). Oxford: Pergamon Press PLC. 1989.

1999. GRIMVALL, A., and G. ASPLUND: Natural Halogenation of Organic Macromolecules. Proc. 3rd Int. Nordic Sym. on Humic Substances. HUMUS-uutiset, Finnish Humus News **3**, 41 (1991).

2000. ENELL, M., and L. WENNBERG: Distribution of Halogenated Organic Compounds (AOX)–Swedish Transport to Surrounding Sea Areas and Mass Balance Studies in Five Drainage Systems. Water Sci. Technol. **24**, 385 (1991).

2001. DAHLMAN, O., R. MÖRCK, P. LJUNGQUIST, A. REIMANN, C. JOHANSSON, H. BORÉN, and A. GRIMVALL: Chlorinated Structural Elements in High-Molecular-Weight Organic Matter from Unpolluted Waters and Bleached-Kraft Mill Effluents. Environ. Sci. Technol. **27**, 1616 (1993).

2002. Van Loon, W.M.G.M., J.J. Boon, B. de Groot, and A.-J. Bulterman: Natural AOX in the River Rhine: Modelling and Trace Analysis. International Conference on Naturally-Produced Organohalogens, Delft, The Netherlands. September, 1993.

2003. Hoekstra, E.J., and E.W.B. de Leer: High AOX-Levels in the River Rhine: 50 Percent of Natural Origin. International Conference on Naturally-Produced Organohalogens, Delft, The Netherlands. September, 1993.

2004. Manninen, P.K.G., and M. Lauren: Naturally Produced Organic Chlorine in the Finnish Aquatic Environment. International Conference on Naturally-Produced Organohalogens, Delft, The Netherlands. September, 1993.

2005. Francois, R.: Iodine in Marine Sedimentary Humic Substances. Sci. Total Environ. **62**, 341 (1987).

2006. Filip, Z., and J.J. Alberts: The Release of Humic Substances from *Spartina alterniflora* (Loisel.) into Sea Water as Influenced by Salt Marsh Indigenous Microorganisms. Sci. Total Environ. **73**, 143 (1988).

2007. Filip, Z., and J.J. Alberts: Humic Substances Isolated from *Spartina alterniflora* (Loisel.) Following Long-Term Decomposition in Sea Water. Sci. Total Environ. **83**, 273 (1989).

2008. Johansson, C., H. Borén, A. Grimvall, O. Dahlman, R. Mörck, A. Reimann, and R.L. Malcolm: Halogenated Structural Elements in Naturally Occurring High-Molecular-Weight Organic Matter. International Conference on Naturally-Produced Organohalogens, Delft, The Netherlands. September, 1993.

2009. Grøn, C., and B. Raben-Lange: Isolation and Characterization of a Haloorganic Soil Humic Acid. Sci. Total Environ. **113**, 281 (1992).

2010. Price, C.A., Sr. E.M. Lynch, B.A. Bowie, and D.J. Newman: Isolation and Identification of Chloramphenicol from the Moon Snail, *Lunatia heros*. J. Antibiot. **34**, 118 (1981).

2011. Rook, J.J.: Chlorination Reactions of Fulvic Acids in Natural Waters. Environ. Sci. Technol. **11**, 478 (1977).

2012. Boyce, S.D., and J.F. Hornig: Reaction Pathways of Trihalomethane Formation from the Halogenation of Dihydroxyaromatic Model Compounds for Humic Acid. Environ. Sci. Technol. **17**, 202 (1983).

2013. Hejzlar, J., H. Borén, and A. Grimvall: Structures of Aquatic Humic Substances Responsible for the Reaction with Chlorine. International Conference on Naturally-Produced Organohalogens, Delft, The Netherlands. September, 1993.

2014. Asplund, G., H. Borén, U. Carlsson, and A. Grimvall: Soil Peroxidase-Mediated Chlorination of Fulvic Acid. In: Humic Substances in the Aquatic and Terrestrial Environment. Proc. Int. Symp. Linköping, Sweden, 1989.

2015. Hoekstra, E.J., P. Lassen, J.G.E. van Leeuwen, E.W.B. de Leer, and L. Carlsen: The Identification and Quantitation of Low-Molecular Weight Chlorinated Organic Compounds in the Chloroperoxidase Mediated Reaction of Chloride and Humic Material. International Conference on Naturally-Produced Organohalogens, Delft, The Netherlands. September, 1993.

2016. Öberg, L., R. Andersson, N. Wågman, and C. Rappe: Formation of Polychlorinated Dibenzo-*p*-dioxins and Dibenzofurans from Chloroorganic Precursors in Activated Sewage Sludge and Garden Compost. International Conference on Naturally-Produced Organohalogens, Delft, The Netherlands. September, 1993.

2017. Bartha, R., and L. Bordeleau: Cell-Free Peroxidases in Soil. Soil Biol. Biochem. **1**, 139 (1969).

2018. Neidleman, S.L.: Microbial Halogenation. CRC Crit. Rev. Microbiol. **3**, 333 (1975).

2019. Morrison, M., and G.R. Schonbaum: Peroxidase-Catalyzed Halogenation. Ann. Rev. Biochem. **45**, 861 (1976).

2020. Neidleman, S.L., and J. Geigert: Biohalogenation: Principles, Basic Roles and Applications. New York: Halsted Press. 1986.

2021. Neidleman, S.L., and J. Geigert: Biological Halogenation: Roles in Nature, Potential in Industry. Endeavor, New Series **11**, 5 (1987).

2022. Wever, R., E. de Boer, H. Plat, and B.E. Krenn: Vanadium – An Element Involved in the Biosynthesis of Halogenation Compounds and Nitrogen Fixation. FEBS Lett. **216**, 1 (1987).

2023. Wever, R., M.G.M. Tromp, J.W.P.M. van Schijndel, E. Vollenbroek, R.L. Olsen, and E. Fogelqvist: Bromoperoxidases: Their Role in the Formation of HOBr and Bromoform by Seaweeds. In: Biogeochemistry of Global Change. Radiatively Active Trace Gases, pp. 811–824 (R.S. Oremland, ed.). New York: Chapman & Hall. 1991.

2024. Butler, A., and J.V. Walker: Marine Haloperoxidases. Chem. Rev. **93**, 1937 (1993).

2025. Scott, R.: Observations on the Iodo-Amino-Acids of Marine Algae Using Iodine-131. Nature **173**, 1098 (1954).

2026. Tong, W., and I.L. Chaikoff: Metabolism of I^{131} by the Marine Alga *Nereocystis luetkeana.* J. Biol. Chem. **215**, 473 (1955).

2027. Murphy, M.J., and C.Óh. Eocha: Peroxidase from the Green Alga *Enteromorpha linza.* Phytochem. **12**, 61 (1973).

2028. Murphy, M.J., and C.Óh. Eocha: Peroxidase Activity in the Brown Alga *Laminaria digitata.* Phytochem. **12**, 2645 (1973).

2029. Manley, S.L., and D.J. Chapman: Formation of 3-Bromo-4-hydroxybenzaldehyde from L-Tyrosine in Cell-Free Homogenates of *Odonthalia floccosa* (Rhodophyceae): A Proposed Biosynthetic Pathway for Brominated Phenols. FEBS Lett. **93**, 97 (1978).

2030. Manley, S.L., and D.J. Chapman: Metabolism of L-Tyrosine to 4-Hydroxybenzaldehyde and 3-Bromo-4-hydroxybenzaldehyde by Chloroplast-Containing Fractions of *Odonthalia floccosa* (Esp.) Falk. Plant Physiol. **64**, 1032 (1979).

2031. Manley, S.L., and D.J. Chapman: Metabolism of 4-Hydroxybenzaldehyde, 3-Bromo-4-hydroxybenzaldehyde and Bromide by Cell-Free Fractions of the Marine Red Alga *Odonthalia floccosa.* Phytochem. **19**, 1453 (1980).

2032. Vilter, H., K.-W. Glombitza, and A. Grawe: Peroxidases from Phaeophyceae, I: Extraction and Detection of the Peroxidases. Bot. Marina **26**, 331 (1983).

2033. Vilter, H.: Peroxidases from Phaeophyceae, IV: Fractionation and Location of Peroxidase Isoenzymes in *Ascophyllum nodosum* (L.) Le Jol. Bot. Marina **26**, 451 (1983).

2034. Vilter, H.: Peroxidases from Phaeophyceae: A Vanadium(V)-Dependent Peroxidase from *Ascophyllum nodosum.* Phytochem. **23**, 1387 (1983).

2035. Baden, D.B., and M.D. Corbett: Peroxidases Produced by the Marine Sponge *Iotrochota birotulata.* Comp. Biochem. Physiol. **64B**, 279 (1979).

2036. Serif, G.S., and S. Kirkwood: Enzyme Systems Concerned with the Synthesis of Monoiodotyrosine, II: Further Properties of the Soluble and Mitochondrial Systems. J. Biol. Chem. **233**, 109 (1958).

2037. Alexander, N.M.: Iodide Peroxidase in Rat Thyroid and Salivary Glands and Its Inhibition by Antithyroid Compounds. J. Biol. Chem. **234**, 1530 (1959).

2038. Ljunggren, J.-G.: The Catalytic Effect of Peroxidase on the Iodination of Tyrosine in the Presence of Hydrogen Peroxide. Biochim. Biophys. Acta **113**, 71 (1966).

2039. Coval, M.L., and A. Taurog: Purification and Iodinating Activity of Hog Thyroid Peroxidase. J. Biol. Chem. **242**, 5510 (1967).

2040. MAGNUSSON, R.P., A. TAUROG, and M.L. DORRIS: Mechanisms of Thyroid Peroxidase- and Lactoperoxidase-Catalyzed Reactions Involving Iodide. J. Biol. Chem. **259**, 13783 (1984).

2041. DUMONTET, C., and B. ROUSSET: Identification, Purification, and Characterization of a Non-Heme Lactoperoxidase in Bovine Milk. J. Biol. Chem. **258**, 14166 (1983).

2042. GEIGERT, J., S.L. NEIDLEMAN, D.J. DALIETOS, and S.K. DEWITT: Haloperoxidases: Enzymatic Synthesis of α, β-Halohydrins from Gaseous Alkenes. Appl. Environ. Microbiol. **45**, 366 (1983).

2043. SHAW, P.D., and L.P. HAGER: An Enzymatic Chlorination Reaction. J. Am. Chem. Soc. **81**, 1011 (1959).

2044. SHAW, P.D., and L.P. HAGER: Biological Chlorination, III: β-Ketoadipate Chlorinase: A Soluble Enzyme System. J. Biol. Chem. **234**, 2565 (1959).

2045. SHAW, P.D., and L.P. HAGER: Biological Chlorination, VI: Chloroperoxidase: A Component of the β-Ketoadipate Chlorinase System. J. Biol. Chem. **236**, 1626 (1961).

2046. MORRIS, D.R., and L.P. HAGER: Chloroperoxidase, I: Isolation and Properties of the Crystalline Glycoprotein. J. Biol. Chem. **241**, 1763 (1966).

2047. BECKWITH, J.R., and L.P. HAGER: Biological Chlorination, VIII: Late Intermediates in the Biosynthesis of Caldariomycin. J. Biol. Chem. **238**, 3091 (1963).

2048. HAGER, L.P., D.R. MORRIS, F.S. BROUN, and H. EBERWEIN: Chloroperoxidase, II: Utilization of Halogen Anions. J. Biol. Chem. **241**, 1769 (1966).

2049. NEIDLEMAN, S.L., P.A. DIASSI, B. JUNTA, R.M. PALMERE, and S.C. PAN: The Enzymatic Halogenation of Steroids. Tetrahedron Lett., 5337 (1966).

2050. BROUN, F.S., and L.P. HAGER: Chloroperoxidase, IV: Evidence for an Ionic Electrophilic Substitution Mechanism. J. Am. Chem. Soc. **89**, 719 (1967).

2051. GEIGERT, J., S.L. NEIDLEMAN, and D.J. DALIETOS: Novel Haloperoxidase Substrates. J. Biol. Chem. **258**, 2273 (1983).

2052. FRANSSEN, M.C.R., and H.C. VAN DER PLAS: A New Enzymatic Chlorination of Barbituric Acid and Its 1-Methyl and 1,3-Dimethyl Derivatives. Recueil **103**, 99 (1984).

2053. YAMADA, H., N. ITOH, and Y. IZUMI: Chloroperoxidase-Catalyzed Halogenation of *trans*-Cinnamic Acid and Its Derivatives. J. Biol. Chem. **260**, 11962 (1985).

2054. GRIFFIN, B.W., and R. HADDOX: Chlorination of NADH: Similarities of the HOCl-Supported and Chloroperoxidase-Catalyzed Reactions. Arch. Biochem. Biophys. **239**, 305 (1985).

2055. FRANSSEN, M.C.R., and H.C. VAN DER PLAS: The Chlorination of Barbituric Acid and Some of Its Derivatives by Chloroperoxidase. Bioorg. Chem. **15**, 59 (1987).

2056. WANNSTEDT, C., D. ROTELLA, and J.F. SIUDA: Chloroperoxidase Mediated Halogenation of Phenols. Bull. Environ. Contam. Toxicol. **44**, 282 (1990).

2057. WALTER, B., and K. BALLSCHMITER: Biohalogenation as a Source of Halogenated Anisoles in Air. Chemosphere **22**, 557 (1991).

2058. ALLAIN, E.J., L.P. HAGER, L. DENG, and E.N. JACOBSEN: Highly Enantioselective Epoxidation of Disubstituted Alkenes with Hydrogen Peroxide Catalyzed by Chloroperoxidase. J. Am. Chem. Soc. **115**, 4415 (1993).

2059. ITOH, N., Y. IZUMI, and H. YAMADA: Haloperoxidase-Catalyzed Halogenation of Nitrogen-Containing Aromatic Heterocycles Represented by Nucleic Bases. Biochemistry **26**, 282 (1987).

2060. TAUROG, A., and E.M. HOWELLS: Enzymatic Iodination of Tyrosine and Thyroglobulin with Chloroperoxidase. J. Biol. Chem. **241**, 1329 (1966).

2061. WALTER, B., and K. BALLSCHMITER: Formation of C_1/C_2-Bromo-/Chloro-Hydrocarbons by Haloperoxidase Reactions. Fresenius' J. Anal. Chem. **342**, 827 (1992).

2062. GRIFFIN, B.W.: Mechanism of Halide-Stimulated Activity of Chloroperoxidase. Evidence for Enzymatic Formation of Free Hypohalous Acid. Biochem. Biophys. Res. Commun. **116**, 873 (1979).

2063. LAMBEIR, A.-M., and H.B. DUNFORD: A Steady State Kinetic Analysis of the Reaction of Chloroperoxidase with Peracetic Acid, Chloride, and 2-Chlorodimedone. J. Biol. Chem. **258**, 13558 (1983).

2064. LIBBY, R.D., J.A. THOMAS, L.W. KAISER, and L.P. HAGER: Chloroperoxidase Halogenation Reactions. J. Biol. Chem. **257**, 5030 (1982).

2065. HAAS, C.N.: The Possibility for "Natural" Generation of Chlorinated Organic Compounds. Risk Anal. **14**, 143 (1994).

2066. WIESNER, W., K.-H. VAN PEE, and F. LINGENS: Detection of a New Chloroperoxidase in *Pseudomonas pyrrocinia*. FEBS Lett., **209**, 321 (1986).

2067. WIESNER, W., K.-H. VAN PÉE, and F. LINGENS: Purification and Characterization of a Novel Bacterial Non-Heme Chloroperoxidase from *Pseudomonas pyrrocinia*. J. Biol. Chem. **263**, 13725 (1988).

2068. WOLFFRAMM, C., F. LINGENS, R. MUTZEL, and K.-H. VAN PÉE: Chloroperoxidase-Encoding Gene from *Pseudomonas pyrrocinia*: Sequence, Expression in Heterologous Hosts, and Purification of the Enzyme. Gene **130**, 131 (1993).

2069. LIU, T.E., T. M'TIMKULU, J. GEIGERT, B. WOLF, S.L. NEIDLEMAN, D. SILVA, and J.C. HUNTER–CEVERA: Isolation and Characterization of a Novel Nonheme Chloroperoxidase. Biochem. Biophys. Res. Commun. **142**, 329 (1987).

2070. VAN SCHIJNDEL, J.W.P.M., E.G.M. VOLLENBROEK, and R. WEVER: The Chloroperoxidase from the Fungus *Curvularia inaequalis*; a Novel Vanadium Enzyme. Biochim. Biophys. Acta **1161**, 249 (1993).

2071. CHEN, Y.P., D.E. LINCOLN, S.A. WOODIN, and C.R. LOVELL: Purification and Properties of a Unique Flavin-Containing Chloroperoxidase from the Capitellid Polychaete *Notomastus lobatus*. J. Biol. Chem. **266**, 23909 (1991).

2072. ASPLUND, G., J.V. CHRISTIANSEN, and A. GRIMVALL: A Chloroperoxidase-Like Catalyst in Soil: Detection and Characterization of Some Properties. Soil Biol. Biochem. **25**, 41 (1993).

2073. HEWSON, W.D., and L.P. HAGER: Bromoperoxidase and Halogenated Lipids in Marine Algae. J. Phycol. **16**, 340 (1980).

2074. PEDERSÉN, M.: A Brominating and Hydroxylating Peroxidase from the Red Alga *Cystoclonium purpureum*. Physiol. Plant. **37**, 6 (1976).

2075. AHERN, T.J., G.G. ALLAN, and D.G. MEDCALF: New Bromoperoxidases of Marine Origin. Partial Purification and Characterization. Biochim. Biophys. Acta **616**, 329 (1980).

2076. ITOH, N., Y. IZUMI, and H. YAMADA: Purification of Bromoperoxidase from *Corallina pilulifera*. Biochem. Biophys. Res. Commun. **131**, 428 (1985).

2077. YAMADA, H., N. ITOH, S. MURAKAMI, and Y. IZUMI: New Bromoperoxidase from Coralline Algae That Brominates Phenol Compounds. Agric. Biol. Chem. **49**, 2961 (1985).

2078. ITOH, N., Y. IZUMI, and H. YAMADA: Characterization of Nonheme Type Bromoperoxidase in *Corallina pilulifera*. J. Biol. Chem. **261**, 5194 (1986).

2079. ITOH, N., Y. IZUMI, and H. YAMADA: Characterization of Nonheme Iron and Reaction Mechanism of Bromoperoxidase in *Corallina pilulifera*. J. Biol. Chem. **262**, 11982 (1987).

2080. ITOH, N., A.K.M.Q. HASAN, Y. IZUMI, and H. YAMADA: Substrate Specificity, Regiospecificity and Stereospecificity of Halogenation Reactions Catalyzed by Non-Heme-Type Bromoperoxidase of *Corallina pilulifera*. Eur. J. Biochem. **172**, 477 (1988).

2081. Krenn, B.E., Y. Izumi, H. Yamada, and R. Wever: A Comparison of Different (Vanadium) Bromoperoxidases; the Bromoperoxidase from *Corallina pilulifera* Is Also a Vanadium Enzyme. Biochim. Biophys. Acta **998**, 63 (1989).
2082. Yu, H., and J.W. Whittaker: Vanadate Activation of Bromoperoxidase from *Corallina officinalis*. Biochem. Biophys. Res. Commun. **160**, 87 (1989).
2083. Everett, R.R., J.R. Kanofsky, and A. Butler: Mechanistic Investigations of the Novel Non-Heme Vanadium Bromoperoxidases: Evidence for Singlet Oxygen Production. J. Biol. Chem. **265**, 4908 (1990).
2084. Shang, M., R.K. Okuda, and D. Worthen: Bromination of Phenols Using an Algal Bromoperoxidase. Phytochem. **37**, 307 (1994).
2085. Krenn, B.E., H. Plat, and R. Wever: The Bromoperoxidase from the Red Alga *Ceramium rubrum* Also Contains Vanadium as a Prosthetic Group. Biochim. Biophys. Acta **912**, 287 (1987).
2086. Wever, R., and K. Kustin: Vanadium: A Biologically Relevant Element. Adv. Inorg. Chem. **35**, 81 (1990).
2087. De la Rosa, R.I., M.J. Clague, and A. Butler: A Functional Mimic of Vanadium Bromoperoxidase. J. Am. Chem. Soc. **114**, 760 (1992).
2088. Tschirret-Guth, R.A., and A. Butler: Evidence for Organic Substrate Binding to Vanadium Bromoperoxidase. J. Am. Chem. Soc. **116**, 411 (1994).
2089. Baden, D.G., and M.D. Corbett: Bromoperoxidases from *Penicillus capitatus*, *Penicillus lamourouxii*, and *Rhipocephalus phoenix*. Biochem. J. **187**, 205 (1980).
2090. Manthey, J.A., and L.P. Hager: Purification and Properties of Bromoperoxidase from *Penicillus capitatus*. J. Biol. Chem. **256**, 11232 (1981).
2091. Nieder, M., and L. Hager: Conversion of α-Amino Acids and Peptides to Nitriles and Aldehydes by Bromoperoxidase. Arch. Biochem. Biophys. **240**, 121 (1985).
2092. Manthey, J.A., and L.P. Hager: Characterization of the Catalytic Properties of Bromoperoxidase. Biochem. **28**, 3052 (1989).
2093. Wever, R., H. Plat, and E. de Boer: Isolation Procedure and Some Properties of the Bromoperoxidase from the Seaweed *Ascophyllum nodosum*. Biochim. Biophys. Acta **830**, 181 (1985).
2094. De Boer, E., Y. van Kooyk, M.G.M. Tromp, H. Plat, and R. Wever: Bromoperoxidase from *Ascophyllum nodosum*: A Novel Class of Enzymes Containing Vanadium as a Prosthetic Group? Biochim. Biophys. Acta **869**, 48 (1986).
2095. De Boer, E., and R. Wever: The Reaction Mechanism of the Novel Vanadium-Bromoperoxidase. J. Biol. Chem. **263**, 12326 (1988).
2096. Krenn, B.E., M.G.M. Tromp, and R. Wever: The Brown Alga *Ascophyllum nodosum* Contains Two Different Vanadium Bromoperoxidases. J. Biol. Chem. **264**, 19287 (1989).
2097. Tromp, M.G.M., G. Olafsson, B.E. Krenn, and R. Wever: Some Structural Aspects of Vanadium Bromoperoxidase from *Ascophyllum nodosum*. Biochim. Biophys. Acta **1040**, 192 (1990).
2098. De Boer, E., M.G.M. Tromp, H. Plat, G.E. Krenn, and R. Wever: Vanadium(V) as an Essential Element for Haloperoxidase Activity in Marine Brown Algae: Purification and Characterization of a Vanadium(V)-Containing Bromoperoxidase from *Laminaria saccharina*. Biochim. Biophys. Acta **872**, 104 (1986).
2099. Jordan, P., and H. Vilter: Extraction of Proteins from Material Rich in Anionic Mucilages: Partition and Fractionation of Vanadate-Dependent Bromoperoxidases from the Brown Algae *Laminaria digitata* and *L. saccharina* in Aqueous Polymer Two-Phases Systems. Biochim. Biophys. Acta **1073**, 98 (1991).
2100. Jordan, P., B. Kloareg, and H. Vilter: Detection of Vanadate-Dependent

Bromoperoxidases in Protoplasts from the Brown Algae *Laminaria digitata* and *L. saccharina*. J. Plant Physiol. **137**, 520 (1991).

2101. SOEDJAK, H.S., and A. BUTLER: Characterization of Vanadium Bromoperoxidase from *Macrocystis* and *Fucus*: Reactivity of Vanadium Bromoperoxidase Toward Acyl and Alkyl Peroxides and Bromination of Amines. Biochem. **29**, 7974 (1990).

2102. BUTLER, A., H.S. SOEDJAK, M. POLNE-FULLER, A. GIBOR, C. BOYEN, and B. KLOAREG: Studies of Vanadium-Bromoperoxidase Using Surface and Cortical Protoplasts of *Macrocystis pyrifera* (Phaeophyta). J. Phycol. **26**, 589 (1990).

2103. PLAT, H., B.E. KRENN, and R. WEVER: The Bromoperoxidase from the Lichen *Xanthoria parientina* is a Novel Vanadium Enzyme. Biochem. J. **248**, 277 (1987).

2104. VAN PÉE, K.-H., and F. LINGENS: Detection of a Bromoperoxidase in *Streptomyces phaeochromogenes*. FEBS Lett. **173**, 5 (1984).

2105. VAN PÉE, K.-H., and F. LINGENS: Purification and Molecular and Catalytic Properties of Bromoperoxidase from *Streptomyces phaeochromogenes*. J. Gen. Microbiol. **131**, 1911 (1985).

2106. VAN PÉE, K.-H., G. SURY, and F. LINGENS: Purification and Properties of a Nonheme Bromoperoxidase from *Streptomyces aureofaciens*. Biol. Chem. Hoppe-Seyler **368**, 1225 (1987).

2107. KRENN, B.E., H. PLAT, and R. WEVER: Purification and Some Characteristics of a Non-Heme Bromoperoxidase from *Streptomyces aureofaciens*. Biochim. Biophys. Acta **952**, 255 (1988).

2108. SOBEK, H., T. HAAG, O. PFEIFER, D. SCHOMBURG, F. LINGENS, and K.-H. VAN PÉE: Crystallization and Preliminary X-Ray Data of Bromoperoxidase from *Streptomyces aureofaciens* ATCC 10762. J. Mol. Biol. **221**, 35 (1991).

2109. WENG, M., O. PFEIFER, S. KRAUSS, F. LINGENS, and K.-H. VAN PÉE: Purification, Characterization and Comparison of Two Non-Haem Bromoperoxidases from *Streptomyces aureofaciens* ATCC 10762. J. Gen. Microbiol. **137**, 2539 (1991).

2110. ZEINER, R., K.-H. VAN PÉE, and F. LINGENS: Purification and Partial Characterization of Multiple Bromoperoxidases from *Streptomyces griseus*. J. Gen. Microbiol. **134**, 3141 (1988).

2111. KNOCH, M., K.-H. VAN PÉE, L.C. VINING, and F. LINGENS: Purification, Properties and Immunological Detection of a Bromoperoxidase-Catalase from *Streptomyces venezuelae* and from a Chloramphenicol-Nonproducing Mutant. J. Gen. Microbiol. **135**, 2493 (1989).

2112. WIESNER, W., K.-H. VAN PÉE, and F. LINGENS: Purification and Properties of Bromoperoxidase from *Pseudomonas pyrrocinia*. Biol. Chem. Hoppe-Seyler **366**, 1085 (1985).

2113. VAN PÉE, K.-H., and F. LINGENS: Purification of Bromoperoxidase from *Pseudomonas aureofaciens*. J. Bacteriol. **161**, 1171 (1985).

2114. ITOH, N., N. MORINAGA, and A. NOMURA: A Variety of Catalases and Bromoperoxidases in Genus *Pseudomonas* and Their Characterization. Biochim. Biophys. Acta **1122**, 189 (1992).

2115. KLEBANOFF, S.J.: Myeloperoxidase-Halide-Hydrogen Peroxide Antibacterial System. J. Bacteriol. **95**, 2131 (1968).

2116. LEHRER, R.I., L.S. GOLDBERG, M.A. APPLE, and N.P. ROSENTHAL: Refractory Megaloblastic Anemia with Myeloperoxidase-Deficient Neutrophils. Ann. Int. Med. **76**, 447 (1972).

2117. MIGLER, R., L.R. DECHATELET, and D.A. BASS: Human Eosinophilic Peroxidase: Role in Bactericidal Activity. Blood **51**, 445 (1978).

2118. JONG, E.C., and S.J. KLEBANOFF: Eosinophil-Mediated Tumor Cell Cytotoxicity. Clinical Res. **27**, 387A (1979).

2119. JONG, E.C., W.R. HENDERSON, and S.J. KLEBANOFF: Bactericidal Activity of Eosinophil Peroxidase. J. Immunol. **124**, 1378 (1980).
2120. JONG, E.C., and S.J. KLEBANOFF: Eosinophil-Mediated Mammalian Tumor Cell Cytotoxicity: Role of the Peroxidase System. J. Immunol. **124**, 1949 (1980).
2121. WEVER, R., H. PLAT, and M.N. HAMERS: Human Eosinophil Peroxidase: A Novel Isolation Procedure, Spectral Properties and Chlorinating Activity. FEBS Lett. **123**, 327 (1981).
2122. LEHRER, R.I.: Antifungal Effects of Peroxidase Systems. J. Bacteriol. **99**, 361 (1969).
2123. BAKKENIST, A.R.J., R. WEVER, T. VULSMA, H. PLAT, and B.F. VAN GELDER: Isolation Procedure and Some Properties of Myeloperoxidase from Human Leucocytes. Biochim. Biophys. Acta **524**, 45 (1978).
2124. YAMADA, M., M. MORI, and T. SUGIMURA: Purification and Characterization of Small Molecular Weight Myeloperoxidase from Human Promyelocytic Leukemia HL-60 Cells. Biochem. **20**, 766 (1981).
2125. OLSEN, R.L., and C. LITTLE: Purification and Some Properties of Myeloperoxidase and Eosinophil Peroxidase from Human Blood. Biochem. J. **209**, 781 (1983).
2126. CARLSON, M.G.C., C.G.B. PETERSON, and P. VENGE: Human Eosinophil Peroxidase: Purification and Characterization. J. Immunol. **134**, 1875 (1985).
2127. BOLSCHER, B.G.J.M., H. PLAT, and R. WEVER: Some Properties of Human Eosinophil Peroxidase, a Comparison with Other Peroxidases. Biochim. Biophys. Acta **784**, 177 (1984).
2128. SMITH, Q.T., and C.H. YANG: Salivary Myeloperoxidase of Young Adult Humans. Proc. Soc. Exp. Biol. Med. **175**, 468 (1984).
2129. WEISS, S.J., S.T. TEST, C.M. ECKMANN, D. ROOS, and S. REGIANI: Brominating Oxidants Generated by Human Eosinophils. Science **234**, 200 (1986).
2130. SELVARAJ, R.J., J.M. ZGLICZYNSKI, B.B. PAUL, and A.J. SBARRA: Chlorination of Reduced Nicotinamide Adenine Dinucleotides by Myeloperoxidase: A Novel Bactericidal Mechanism. J. Reticuloendothelial Soc. **27**, 31 (1980).
2131. DEITS, T., M. FARRANCE, E.S. KAY, L. MEDILL, E.E. TURNER, P.J. WEIDMAN, and B.M. SHAPIRO: Purification and Properties of Ovoperoxidase, the Enzyme Responsible for Hardening the Fertilization Membrane of the Sea Urchin Egg. J. Biol. Chem. **259**, 13525 (1984).
2132. CASTRO, C.E.: Biodehalogenation. Environ. Health Perspect. **21**, 279 (1977).
2133. MÜLLER, R., and F. LINGENS: Microbial Degradation of Halogenated Hydrocarbons: A Biological Solution to Pollution Problems? Angew. Chem. Int. Ed. Engl. **25**, 779 (1986).
2134. BUMPUS, J.A., and S.D. AUST: Biodegradation of Environmental Pollutants by the White Rot Fungus *Phanerochaete chrysosporium*: Involvement of the Lignin Degrading System. Bioessays **6**, 166 (1987).
2135. REINECKE, W., and H.-J. KNACKMUSS: Microbial Degradation of Haloaromatics. Ann. Rev. Microbiol. **42**, 263 (1988).
2136. WACKETT, L.P.: Dehalogenation of Organohalide Pollutants by Bacterial Enzymes and Coenzymes. In: Applications of Enzyme Biotechnology (J.W. Kelly and T.O. Baldwin, eds.), pp. 191–200. New York: Plenum Press. 1991.
2137. HARDMAN, D.J.: Biotransformation of Halogenated Compounds. Crit. Rev. Biotech. **11**, 1 (1991).
2138. GOTTSCHALK, G., and H.-J. KNACKMUSS: Bacteria and the Biodegradation of Chemicals Achieved Naturally, by Combination, or by Construction. Angew. Chem. Int. Ed. Engl. **32**, 1398 (1993).
2139. WALKER, N., and G.H. WILTSHIRE: The Decomposition of 1-Chloro- and 1-Bromonaphthalene by Soil Bacteria. J. Gen. Microbiol. **12**, 478 (1955).

2140. GOLDMAN, P., G.W.A. MILNER, and M.T. PIGNATARO: Fluorine Containing Metabolites Formed from 2-Fluorobenzoic Acid by *Pseudomonas* Species. Arch. Biochem. Biophys. **118**, 178 (1967).

2141. GIBSON, D.T., J.R. KOCH, C.L. SCHULD, and R.E. KALLIO: Oxidative Degradation of Aromatic Hydrocarbons by Microorganisms, II: Metabolism of Halogenated Aromatic Hydrocarbons. Biochem. **7**, 3795 (1968).

2142. JOHNSTON, H.W., G.G. BRIGGS, and M. ALEXANDER: Metabolism of 3-Chlorobenzoic Acid by a Pseudomonad. Soil Biol. Biochem. **4**, 187 (1972).

2143. BALLSCHMITER, K., and CH. SCHOLZ: Mikrobieller Abbau von chlorierten Aromaten, VI: Bildung von Dichlorphenolen und Dichlorbenzkatechinen aus Dichlorbenzolen in mikromolarer Lösung durch *Pseudomonas* sp. Chemosphere **9**, 457 (1980).

2144. ROSSITER, J.T., S.R. WILLIAMS, A.E.G. CASS, and D.W. RIBBONS: Aromatic Biotransformations, 2: Production of Novel Chiral Fluorinated 3,5-Cyclohexadiene-*cis*-1,2-diol-1-carboxylates. Tetrahedron Lett. **28**, 5173 (1987).

2145. HUDLICKY, T., H. LUNA, J.D. PRICE, and F. RULIN: Microbial Oxidation of Chloroaromatics in the Enantiodivergent Synthesis of Pyrrolizidine Alkaloids: Trihydroxyheliotridanes. J. Org. Chem. **55**, 4683 (1990).

2146. BOYD, D.R., M.R.J. DORRITY, M.V. HAND, J.F. MALONE, N.D. SHARMA, H. DALTON, D.J. GRAY, and G.N. SHELDRAKE: Enantiomeric Excess and Absolute Configuration Determination of *cis*-Dihydrodiols from Bacterial Metabolism of Monocyclic Arenes. J. Am. Chem. Soc. **113**, 666 (1991).

2147. HUDLICKY, T., and B.P. MCKIBBEN: New Cycloaddition Chemistry of 1-Chloro-5,6-*cis*-isopropylidenedioxycyclohexa-1,3-diene Derived from the Oxidation of Halogenobenzenes by *Pseudomonas putida* 39D. J. Chem. Soc., Perkin Trans. 1, 485 (1994).

2148. HUDLICKY, T., M. MANDEL, J. ROUDEN, R.S. LEE, B. BACHMANN, T. DUDDING, K.J. YOST, and J.S. MEROLA: Microbial Oxidation of Aromatics in Enantiocontrolled Synthesis, Part 1: Expedient and General Asymmetric Synthesis of Inositols and Carbohydrates *via* an Unusual Oxidation of a Polarized Diene with Potassium Permanganate. J. Chem. Soc., Perkin Trans. 1, 1553 (1994).

2149. CAIN, R.B., E.K. TRANTER, and J.A. DARRAH: The Utilization of Some Halogenated Aromatic Acids by *Nocardia*. Biochem. J. **106**, 211 (1968).

2150. RIBBONS, D.W., A.E.G. CASS, J.T. ROSSITER, S.J.C. TAYLOR, M.P. WOODLAND, D.A. WIDDOWSON, S.R. WILLIAMS, P.B. BAKER, and R.E. MARTIN: Biotransformations of Fluoroaromatic Compounds. J. Fluorine Chem. **37**, 299 (1987).

2151. BOPP, L.H.: Degradation of Highly Chlorinated PCBs by *Pseudomonas* Strain LB400. J. Ind. Microbiol. **1**, 23 (1986).

2152. FURUKAWA, K., N. TOMIZUKA, and A. KAMIBAYASHI: Effect of Chlorine Substitution on the Bacterial Metabolism of Various Polychlorinated Biphenyls. Appl. Environ. Microbiol. **38**, 301 (1979).

2153. HANKIN, L., and B.L. SAWHNEY: Microbial Degradation of Polychlorinated Biphenyls in Soil. Soil Sci. **137**, 401 (1984).

2154. MAVOUNGOU, R., R. MASSÉ, and M. SYLVESTRE: Microbial Dehalogenation of 4,4'-Dichlorobiphenyl Under Anaerobic Conditions. Sci. Total Environ. **101**, 263 (1991).

2155. BROUN JR., J.F., D.L. BEDARD, M.J. BRENNAN, J.C. CARNAHAN, H. FENG, and R.E. WAGNER: Polychlorinated Biphenyl Dechlorination in Aquatic Sediments. Science **236**, 709 (1987).

2156. QUENSEN, J.F., J.M. TIEDJE, and S.A. BOYD: Reductive Dechlorination of Polychlorinated Biphenyls by Anaerobic Microorganisms from Sediments. Science **242**, 752 (1988).

2157. HARKNESS, M.R., J.B. MCDERMOTT, D.A. ABRAMOWICZ, J.J. SALVO, W.P. FLANAGAN, M.L. STEPHENS, F.J. MONDELLO, R.J. MAY, J.H. LOBOS, K.M. CARROLL, M.J. BRENNAN, A.A. BRACCO, K.M. FISH, G.L. WARNER, P.R. WILSON, D.K. DIETRICH, D.T. LIN, C.B. MORGAN, and W.L. GATELY: *In Situ* Stimulation of Aerobic PCB Biodegradation in Hudson River Sediments. Science **259**, 503 (1993).

2158. RHEE, Y., R.C. SOKOL, C.M. BETHONEY, and B. BUSH: Dechlorination of Polychlorinated Biphenyls by Hudson River Sediment Organisms: Specificity to the Chlorination Pattern of Congeners. Environ. Sci. Technol. **27**, 1190 (1993).

2159. EATON, D.C.: Mineralization of Polychlorinated Biphenyls by *Phanerochaete chrysosporium*: A Ligninolytic Fungus. Enzyme Microb. Technol. **7**, 194 (1985).

2160. BUMPUS, J.A., M. TIEN, D. WRIGHT, and S.D. AUST: Oxidation of Persistent Environmental Pollutants by a White Rot Fungus. Science **228**, 1434 (1985).

2161. ABRAMOWICZ, D.A.: Aerobic and Anaerobic Biodegradation of PCBs: A Review. Crit. Rev. Biotechnol. **10**, 241 (1990).

2162. LAMAR, R.T., M.J. LARSEN, and T.K. KIRK: Sensitivity to and Degradation of Pentachlorophenol by *Phanerochaete* spp. Appl. Environ. Microbiol. **56**, 3519 (1990).

2163. MILESKI, G.J., J.A. BUMPUS, M.A. JUREK, and S.D. AUST: Biodegradation of Pentachlorophenol by the White Rot Fungus *Phanerochaete chrysosporium*. Appl. Environ. Microbiol. **54**, 2885 (1988).

2164. HAMMELL, K.E., and P.J. TARDONE: The Oxidative 4-Dechlorination of Polychlorinated Phenols Is Catalyzed by Extracellular Fungal Lignin Peroxidases. Biochem. **27**, 6563 (1988).

2165. VALLI, K., and M.H. GOLD: Degradation of 2,4-Dichlorophenol by the Lignin-Degrading Fungus *Phanerochaete chrysosporium*. J. Bacteriol. **173**, 345 (1991).

2166. BUMPUS, J.A., and S.D. AUST: Biodegradation of DDT [1,1,1-Trichloro-2,2-bis-(4-chlorophenyl)ethane] by the White Rot Fungus *Phanerochaete chrysosporium*. Appl. Environ. Microbiol. **53**, 2001 (1987).

2167. GOLD, M.H., H. WARIISHI, and K. VALLI: Extracellular Peroxidases Involved in Lignin Degradation by the White Rot Basidiomycete *Phanerochaete chrysosporium*. In: Biocatalysis in Agricultural Biotechnology, ACS Symposium Series No. 389 (J.R. Whitaker and P.E. Sonnet, eds.). Copyright: American Chemical Society. 1989.

2168. CASTRO, C.E., and N.O. BELSER: Biodehalogenation. Reductive Dehalogenation of the Biocides Ethylene Dibromide, 1,2-Dibromo-3-chloropropane, and 2,3-Dibromobutane in Soil. Environ. Sci. Technol. **2**, 779 (1968).

2169. BOUWER, E.J., and P.L. MCCARTY: Transformations of 1- and 2-Carbon Halogenated Aliphatic Organic Compounds Under Metanogenic Conditions. Appl. Environ. Microbiol. **45**, 1286 (1983).

2170. BRUNNER, W., D. STAUB, and T. LEISINGER: Bacterial Degradation of Dichloromethane. Appl. Environ. Microbiol. **40**, 950 (1980).

2171. OMORI, T., and M. ALEXANDER: Bacterial and Spontaneous Dehalogenation of Organic Compounds. Appl. Environ. Microbiol. **35**, 512 (1978).

2172. KAWASAKI, H., K. MIYOSHI, and K. TONOMURA: Purification, Crystallization and Properties of Haloacetate Halidohydrolase from *Pseudomonas* Species. Agric. Biol. Chem. **45**, 543 (1981).

2173. WALKER, J.R.L., and B.C. LIEN: Metabolism of Fluoroacetate by a Soil *Pseudomonas* sp. and *Fusarium solani*. Soil Biol. Biochem. **13**, 231 (1981).

2174. MEYER, J.J.M., N. GROBBELAAR, and P.L. STEYN: Fluoroacetate-Metabolizing Pseudomonad Isolated from *Dichapetalum cymosum*. Appl. Environ. Microbiol. **56**, 2152 (1990).

2175. SJOBLAD, R.D , and J.-M. BOLLAG: Oxidative Coupling of Aromatic Pesticide Intermediates by a Fungal Phenol Oxidase. Appl. Environ. Microbiol. **33**, 906 (1977).
2176. BOLLAG, J.-M., R.D. SJOBLAD, and R.D. MINARD: Polymerization of Phenolic Intermediates of Pesticides by a Fungal Enzyme. Experientia **33**, 1564 (1977).
2177. BALLSCHMITER, K., C. UNGLERT, and P. HEINZMANN: Formation of Chlorophenols by Microbial Transformation of Chlorobenzenes. Angew. Chem. Int. Ed. Engl. **16**, 645 (1977).
2178. SMITH, A.E., and A.J. AUBIN: Transformation of [^{14}C]-2,4-Dichlorophenol in Saskatchewan Soils. J. Agric. Food Chem. **39**, 801 (1991).
2179. VANEK, Z., J. CUDLÍN, M. BLUMAUEROVÁ, and Z. HOSTÁLEK: How Many Genes Are Required for the Synthesis of Chlortetracycline. Folia Microbiol. **16**, 225 (1971).
2180. WILLIAMS, D.H., M.J. STONE, P.R. HAUCK, and S.K. RAHMAN: Why Are Secondary Metabolites (Natural Products) Biosynthesized? J. Nat. Prod. **52**, 1189 (1989).
2181. CHRISTOPHERSEN, C.: Evolution in Molecular Structure and Adaptive Variance in Metabolism. Comp. Biochem. Physiol. **98B**, 427 (1991).
2182. TONG, W., and I.L. CHAIKOFF: ^{131}I Utilization by an Aquarium Snail and the Cockroach. Biochim. Biophys. Acta **48**, 347 (1961).
2183. PAWLIK, J.R.: Marine Invertebrate Chemical Defenses. Chem. Rev. **93**, 1911 (1993).
2184. WALKER, R.P., J.E. THOMPSON, and D.J. FAULKNER: Exudation of Biologically-Active Metabolites in the Sponge *Aplysina fistularis*, II: Chemical Evidence. Mar. Biol. **88**, 27 (1985).
2185. BRAEKMAN, J.C., and D. DALOZE: Chemical Defence in Sponges. Pure Appl. Chem. **58**, 357 (1986).
2186. PAUL, V.J., and S.C. PENNINGS: Diet-Derived Chemical Defenses in the Sea Hare *Stylocheilus longicauda* (Quoy et Gaimard 1824). J. Exp. Mar. Biol. Ecol. **151**, 227 (1991).
2187. THOMPSON, J.E., R.P. WALKER, S.J. WRATTEN, and D.J. FAULKNER: A Chemical Defense Mechanism for the Nudibranch *Cadlina luteomarginata*. Tetrahedron **38**, 1865 (1982).
2188. MEBS, D.: Chemical Defense of a Dorid Nudibranch, *Glossodoris quadricolor*, from the Red Sea. J. Chem. Ecol. **11**, 713 (1985).
2189. PETTIT, G.R., R.H. ODE, C.L. HERALD, R.B. VON DREELE, and C. MICHEL: The Isolation and Structure of Dolatriol. J. Am. Chem. Soc. **98**, 4677 (1976).
2190. PAWLIK, J.R.: Chemical Ecology of the Settlement of Benthic Marine Invertebrates. Oceanogr. Mar. Biol. Ann. Rev. **30**, 273 (1992).
2191. BURKHOLDER, P.R., and G.M. SHARMA: Antimicrobial Agents from the Sea. Lloydia **32**, 466 (1969).
2192. SIMS, J.J., M.S. DONNELL, J.V. LEARY, and G.H. LACY: Antimicrobial Agents from Marine Algae. Antimicrob. Agents Chemother. **7**, 320 (1975).
2193. PAUL, V.J.: Feeding Deterrent Effects of Algal Natural Products. Bull. Mar. Sci. **41**, 514 (1987).
2194. FENICAL, W., and V.J. PAUL: Antimicrobial and Cytotoxic Terpenoids from Tropical Green Algae of the Family Udoteaceae. Hydrobiologica **116/117**, 135 (1984).
2195. HARPER, D.B., and J.T.G. HAMILTON: Biosynthesis of Chloromethane in *Phellinus pomaceus*. J. Gen. Microbiol. **134**, 2831 (1988).
2196. MCNALLY, K.J., J.T.G. HAMILTON, and D.B. HARPER: The Methylation of Benzoic and *n*-Butyric Acids by Chloromethane in *Phellinus pomaceus*. J. Gen. Microbiol. **136**, 1509 (1990).
2197. MCNALLY, K.J., and D.B. HARPER: Methylation of Phenol by Chloromethane in the Fungus *Phellinus pomaceus*. J. Gen. Microbiol. **137**, 1029 (1991).

2198. Coulter, C., J.T.G. Hamilton, and D.B. Harper: Evidence for the Existence of Independent Chloromethane- and *S*-Adenosylmethionine-Utilizing Systems for Methylation in *Phanerochaete chrysosporium*. Appl. Environ. Microbiol. **59**, 1461 (1993).

2199. Morris, H.R., G.W. Taylor, M.S. Masento, K.A. Jermyn, and R.R. Kay: Chemical Structure of the Morphogen Differentiation Inducing Factor from *Dictyostelium discoideum*. Nature **328**, 811 (1987).

2200. Morris, H.R., M.S. Masento, G.W. Taylor, K.A. Jermyn, and R.R. Kay: Structure Elucidation of Two Differentiation Inducing Factors (DIF-2 and DIF-3) from the Cellular Slime Mould *Dictyostelium discoideum*. Biochem. J. **249**, 903 (1988).

2201. Williams, G.M.: The Chlorine Controversy (Letter to the Editor). Science **262**, 15 (1993).

2202. Torii, S., K. Mitsumori, S. Inubushi, and I. Yanagisawa: The REM Sleep-Inducing Action of a Naturally-Occurring Organic Bromine Compound in the Encéphale Isolé Cat. Psychopharmacologia (Berlin) **29**, 65 (1973).

2203. Yanagisawa, I., M. Yamane, S. Torii, and S. Inubushi: Biosynthesis and Behavior of 1-Methylheptyl γ-Bromoacetoacetate in Animal Body. Bull. Jpn. Neurochem. Soc. **14**, 120 (1975).

2204. Yamane, M., K. Maeda, and I. Yanagisawa: Organic Bromine Compound in Animal Organs. Sleep Res. **9**, 60 (1980).

2205. Yamane, M., and I. Yanagisawa: Microdetermination of 1-Methylheptyl γ-Bromoacetoacetate in Human Blood. J. Biochem. **92**, 2009 (1982).

2206. Yamane, M., A. Abe, and I. Yanagisawa: Anticholinesterase Action of a Bromine Compound Isolated from Human Cerebrospinal Fluid. J. Neurochem. **42**, 1650 (1984).

2207. Williams, G.M., M.J. Iatropoulos, M.V. Djordjevic, and O.P. Kaltenberg: The Triphenylethylene Drug Tamoxifen Is a Strong Liver Carcinogen in the Rat. Carcinogenesis **14**, 315 (1993).

2208. Hard, G.C., M.J. Iatropoulos, K. Jordan, L. Radi, O.P. Kaltenberg, A.R. Imondi, and G.M. Williams: Major Difference in the Hepatocarcinogenicity and DNA Adduct Forming Ability Between Toremifene and Tamoxifen in Female Crl: CD(BR) Rats. Can. Res. **53**, 4534 (1993).

2209. Neudecker, T., K. Öhrlein, E. Eder, and D. Henschler: Effect of Methyl and Halogen Substitutions in the αC Position on the Mutagenicity of Cinnamaldehyde. Mutat. Res. **110**, 1 (1983).

2210. Collen, J., A. Ekdahl, K. Abrahamsson, and M. Pedersen: The Involvement of Hydrogen Peroxide in the Production of Volatile Halogenated Compounds by *Meristiella gelidium*. Phytochem. **36**, 1197 (1994).

2211. Scheuer, P.J.: Isocyanides and Cyanides as Natural Products. Acct. Chem. Res. **25**, 433 (1992).

2212. Fenical, W.: Molecular Aspects of Halogen-Based Biosynthesis of Marine Natural Products. Recent Adv. Phytochem. **13**, 219 (1979).

2213. Sassa, T.: Cotylenins, Leaf Growth Substances Produced by a Fungus, Part I: Isolation and Characterization of Cotylenins A and B. Agric. Biol. Chem. **35**, 1415 (1971).

2214. Aplin, T.E.H.: Poison Plants of Western Australia. The Toxic Species of the Genera Gastrolobium and Oxylobium. J. Agric. West. Aust. **8**, 42 (1967).

2215. Ignat'eva, N.S., V.V. Vandyshev, and M.G. Pimenov: Coumarins from the Roots of *Cachrys pubescens*. Khim. Prir. Soed. **8**, 388 (1972).

2216. Fock, A., J. Kettner, M. Böttger: The Occurrence of 4-Chlorotryptophane in *Vicia faba*. Phytochem. **31**, 2327 (1992).

2217. MIRRINGTON, R.M., E. RITCHIE, C.W. SHOPPLE, S. STERNHELL, and W.C. TAYLOR: Some Metabolites of *Nectria radicicola* Gerlach and Nilsson (syn. *Cylindrocarpon radicicola* Wr.) : The Structure of Radicicol (Monorden). Aust. J. Chem. **19**, 1265 (1966).

2218. DE JONG, E., J.A. FIELD, J.A.F.M. DINGS, J.B.P.A. WIJNBERG, and J.A.M. DE BONT: Denovo Biosynthesis of Chlorinated Aromatics by the White-Rot Fungus *Bjerkandera* sp. BOS55. FEBS Lett. **305**, 220 (1992).

2219. DELMOTTE, P., and J. DELMOTTE-PLAQUEE: A New Antifungal Substance of Fungal Origin. Nature **171**, 344 (1953).

2220. CIMINIELLO, P., E. FATTORUSSO, S. MAGNO, and M. PANSINI: Chemistry of Verongida Sponges, III: Constituents of a Caribbean *Verongula* sp. J. Nat. Prod. **57**, 1564 (1994).

2221. JASPARS, M., T. RALI, M. LANEY, R.C. SCHATZMAN, M.C. DIAZ, F.J. SCHMITZ, E.O. PORDESIMO, and P. CREWS: The Search for Inosine 5′-Phosphate Dehydrogenase (IMPDH) Inhibitors from Marine Sponges. Evaluation of the Bastadin Alkaloids. Tetrahedron **50**, 7367 (1994).

2222. HUNECK, S., C. DJERASSI, D. BECHER, M. BARBER, M. VON ARDENNE, K. STEINFELDER, and R. TÜMMLER: Massenspektrometrie von Depsiden, Depsidonen, Depsonen, Dibenzofuranen und Diphenylbutadienen mit positiven und negativen Ionen. Tetrahedron **24**, 2707 (1968).

2223. HUNECK, S., and J. SANTESSON: Über die Inhaltsstoffe von *Lecanora rupicola* (L.) zahlbr. und *Lecanora carpinea* (L.) ach. em. vain. und die Strukturaufklärung sowie Synthese von 8-Chlor-5.7-dihydroxy-2.6-dimethylchromon. Z. Naturforsch. **24B**, 750 (1969).

2224. KURUNG, J.M.: *Aspergillus ustus*. Science **102**, 11 (1945).

2225. POELT, J., and S. HUNECK: Association, Ecology, and Chemistry of *Lecanora vinetorum*. Österr. Bot. Z. **115**, 411 (1968); Chem. Abstr. **71**, 19536 (1969).

2226. JARMAN, W.M., M. SIMON, R.J. NORSTROM, S.A. BURNS, C.A. BACON, B.R.T. SIMONELT, and R.W. RISEBROUGH: Global Distribution of Tris(4-chlorophenyl)-methanol in High Trophic Level Birds and Mammals. Environ. Sci. Technol. **26**, 1770 (1992).

2227. KOMATSU, E.: Biochemistry of Geodin, I: Isolation of the Antibiotic Geodin from the Newly Isolated *Penicillium* sp. F 29. Nippon Nogei-Kagaku Kaishi **31**, 349 (1957); Chem. Abstr. **52**, 16473 (1958).

2228. KUNSTMANN, M.P., L.A. MITSCHER, J.N. PORTER, A.J. SHAY, and M.A. DARKEN: LL-AV290, a New Antibiotic, I: Fermentation, Isolation, and Characterization. Antimicrob. Agents Chemother., 242 (1968).

2229. ISIDOROV, V.A.: Organic Chemistry of the Earth's Atmosphere, p. 106. Berlin, Heidelberg: Springer. 1990.

2230. HUDLICKY, T., T. NUGENT, and W. GRIFFITH: Chemoenzymatic Synthesis of D-*erythro*-C_{18}- and L-*threo*-C_{18}-Sphingosines. J. Org. Chem. **59**, 7944 (1994).

2231. CARLSEN, L., and P. LASSEN: Enzymatically Mediated Formation of Chlorinated Humic Acids. Org. Geochem. **18**, 477 (1992).

2232. BOYCE, S.D., A.C. BAREFOOT, and J.F. HORNIG: 1,3-Dihydroxybenzene-2-^{13}C: Its Preparation and Reaction with Chlorine and Bromine in Aqueous Solution. J. Label. Cmpds. Radiopharm. **XX**, 243 (1983).

2233. SETO, H., T. FUJIOKA, K. FURIHATA, I. KANEKO, and S. TAKAHASHI: Structure of Complestatin, a Very Strong Inhibitor of Protease Activity of Complement in the Human Complement System. Tetrahedron Lett. **30**, 4987 (1989).

2234. Harper, D.B.: The Contribution of Natural Halogenation Processes to the Atmospheric Halomethane Burden. International Conference on Naturally-Produced Organohalogens, Delft, The Netherlands. September, 1993.

2235. Travis, C.C., and H.A. Hattemer-Frey: A Perspective on Dioxin Emissions from Municipal Solid Waste Incinerators. Risk Anal. **9**, 91 (1989).

2236. Zafiriou, O.C.: Reaction of Methyl Halides with Seawater and Marine Aerosols. J. Mar. Res. **33**, 75 (1975).

2237. Singh, H.B., L.J. Salas, H. Shigeishi, and E. Scribner: Atmospheric Halocarbons, Hydrocarbons, and Sulfur Hexafluoride: Global Distributions, Sources, and Sinks. Science **203**, 899 (1979).

2238. Crutzen, P.J., L.E. Heidt, J.P. Krasnec, W.H. Pollock, and W. Seiler: Biomass Burning as a Source of Atmospheric Gases CO, H_2, N_2O, NO, CH_3Cl and COS. Nature **282**, 253 (1979).

2239. Cicerone, R.J.: Halogens in the Atmosphere. Rev. Geophys. Space Sci. **19**, 123 (1981).

2240. Higa, T., and S. Sakemi: Environmental Studies on Natural Halogen Compounds, I: Estimation of Biomass of the Acorn Worm *Ptychodera flava* Eschscholtz (Hemichordata: Enteropneusta) and Excretion Rate of Metabolites at Kattore Bay, Kohama Island, Okinawa. J. Chem. Ecol. **9**, 495 (1983).

2241. Klotz, U.: Occurrence of Natural Benzodiazepines. Life Sci. **48**, 209 (1991).

2242. Scheuer, P.J.: Drugs from the Sea. Chem. Ind., 276 (1991).

2243. Faulkner, D.J.: Biomedical Uses for Natural Marine Chemicals. Oceanus **35**, 29 (1992).

2244. Marine Biotechnology, Vol. 1: Pharmaceutical and Bioactive Natural Products (D.H. Attaway, and O.R. Zaborsky, eds.). New York, London: Plenum. 1993.

2245. Hylin, J.W., R.E. Spenger, and F.A. Gunther: Potential Interferences in Certain Pesticide Residue Analyses from Organochlorine Compounds Occurring Naturally in Plants. Residue Rev. **26**, 127 (1969).

2246. Chameides, W.L., and D.D. Davis: Iodine: Its Possible Role in Tropospheric Photochemistry. J. Geophys. Res. **85**, 7383 (1980).

2247. Grimvall, A.: In: The Natural Chemistry of Chlorine in the Environment. Brussels: Euro Chlor. 1995.

2248. Asplund, G., and E.W.B. de Leer: In: The Natural Chemistry of Chlorine in the Environment. Brussels: Euro Chlor. 1995.

2249. Grøn, C.: In: The Natural Chemistry of Chlorine in the Environment. Brussels: Euro Chlor. 1995.

2250. Müller, G.: In: The Natural Chemistry of Chlorine in the Environment. Brussels: Euro Chlor. 1995.

2251. Cheng, X.-C., M. Varoglu, L. Abrell, P. Crews, E. Lobkovsky, and J. Clardy: Chloriolins A–C, Chlorinated Sesquiterpenes Produced by Fungal Cultures Separated from a *Jaspis* Marine Sponge. J. Org. Chem. **59**, 6344 (1994).

2252. Hautzel, R., H. Anke, and W.S. Sheldrick: Mycenon, a New Metabolite from a *Mycena* Species TA 87202 (Basidiomycetes) as an Inhibitor of Isocitrate Lyase. J. Antibiot. **43**, 1240 (1990).

2253. Paik, S., S. Carmeli, J. Cullingham, R.E. Moore, G.M.L. Patterson, and M.A. Tius: Mirabimide E, an Unusual *N*-Acylpyrrolinone from the Blue-Green Alga *Scytonema mirabile*: Structure Determination and Synthesis. J. Am. Chem. Soc. **116**, 8116 (1994).

2254. Baker, B.J., and P.J. Scheuer: The Punaglandins: 10-Chloroprostanoids from the Octocoral *Telesto riisei*. J. Nat. Prod. **57**, 1346 (1994).

2255. YOSHIDA, H., N. ARAI, M. SUGOH, J. IWABUCHI, K. SHIOMI, M. SHINOSE, Y. TANAKA, and S. OMURA: 4-Chlorothreonine, a Herbicidal Antimetabolite Produced by *Streptomyces* sp. OH-5093. J. Antibiot. **47**, 1165 (1994).

2256. MARTIN, D.G., D.J. DUCHAMP, and C.G. CHIDESTER: The Isolation, Structure, and Absolute Configuration of U42,126, a Novel Antitumor Antibiotic. Tetrahedron Lett., 2549 (1973).

2257. ISHIWATA, H., T. NEMOTO, M. OJIKA, and K. YAMADA: Isolation and Stereostructure of Doliculide, a Cytotoxic Cyclodepsipeptide from the Japanese Sea Hare *Dolabella auricularia.* J. Org. Chem. **59**, 4710 (1994).

2258. BEWLEY, C.A., and D.J. FAULKNER: Theonegramide, an Antifungal Glycopeptide from the Philippine Lithistid Sponge *Theonella swinhoei.* J. Org. Chem. **59**, 4849 (1994).

2259. CULVENOR, C.C.J., J.A. EDGAR, M.F. MACKAY, C.P. GORST-ALLMAN, W.F.O. MARASAS, P.S. STEYN, R. VLEGGAAR, and P.L. WESSELS: Structure Elucidation and Absolute Configuration of Phomopsin A, a Hexapeptide Mycotoxin Produced by *Phomopsis leptostromiformis.* Tetrahedron **45**, 2351 (1989).

2260. MORITA, H., S. NAGASHIMA, K. TAKEYA, and H. ITOKAWA: Cyclic Peptides from Higher Plants, Part 8: Three Novel Cyclic Pentapeptides, Astins F, G and H from *Aster tataricus.* Heterocycles **38**, 2247 (1994).

2261. CIABATTI, R., J.K. KETTENRING, G. WINTERS, G. TUAN, L. ZERILLI, and B. CAVALLERI: Ramoplanin (A-16686), a New Glycolipodepsipeptide Antibiotic, III: Structure Elucidation. J. Antibiot. **42**, 254 (1989).

2262. KETTENRING, J.K., R. CIABATTI, G. WINTERS, G. TAMBORINI, and B. CAVALLERI: Ramoplanin (A-16686), a New Glycolipodepsipeptide Antibiotic, IV: Complete Peptide Sequence Determination by Homonuclear 2D NMR Spectroscopy. J. Antibiot. **42**, 268 (1989).

2263. SALCHER, O., and F. LINGENS: Biosynthese von Pyrrolnitrin. Nachweis von 3-Chloroanthranilsäure und 7-Chlorindolessigsäure im Kulturmedium von *Pseudomonas aureofaciens.* Tetrahedron Lett., 3101 (1978).

2264. COPP, B.R., K.F. FULTON, N.B. PERRY, J.W. BLUNT, and M.H.G. MUNRO: Natural and Synthetic Derivatives of Discorhabdin C, a Cytotoxic Pigment from the New Zealand Sponge *Latrunculia* cf. bocagei. J. Org. Chem. **59**, 8233 (1994).

2265. GUNASEKERA, S.P., P.J. MCCARTHY, and M. KELLY-BORGES: Hamacanthins A and B, New Antifungal Bisindole Alkaloids from the Deep-Water Marine Sponge, *Hamacantha* sp. J. Nat. Prod. **57**, 1437 (1994).

2266. BIFULCO, G., I. BRUNO, L. MINALE, R. RICCIO, A. CALIGNANO, and C. DEBITUS: ($\pm$)-Gelliusines A and B, Two Diastereomeric Brominated Tris-Indole Alkaloids from a Deep Water New Caledonian Marine Sponge (*Gellius* or *Orina* sp.). J. Nat. Prod. **57**, 1294 (1994).

2267. MOORE, R.E., X.-Q.G. YANG, G.M.L. PATTERSON, R. BONJOUKLIAN, and T.A. SMITKA: Hapalonamides and Other Oxidized Hapalindoles from *Hapalosiphon fontinalis.* Phytochem. **28**, 1565 (1989).

2268. HARRISON, L., D.B. TEPLOW, M. RINALDI, and G. STROBEL: Pseudomycins, a Family of Novel Peptides from *Pseudomonas syringae* Possessing Broad-Spectrum Antifungal Activity. J. Gen. Microbiol. **137**, 2857 (1991).

2269. BALLIO, A., F. BOSSA, D. DI GIORGIO, P. FERRANTI, M. PACI, P. PUCCI, A. SCALONI, A. SEGRE, and G.A. STROBEL: Novel Bioactive Lipodepsipeptides from *Pseudomonas syringae*: The Pseudomycins. FEBS Lett. **355**, 96 (1994).

2270. STRATMANN, K., R.E. MOORE, R. BONJOUKLIAN, J.B. DEETER, G.M.L. PATTERSON, S. SHAFFER, C.D. SMITH, and T.A. SMITKA: Welwitindolinones, Unusual Alkaloids from the Blue-Green Algae *Hapalosiphon welwitschii* and *Westiella intricata*. Relationship to Fischerindoles and Hapalindoles. J. Am. Chem. Soc. **116**, 9935 (1994).

2271. SEARLE, P.A., and T.F. MOLINSKI: Five New Alkaloids from the Tropical Ascidian, *Lissoclinum* sp. Lissoclinotoxin A Is Chiral. J. Org. Chem. **59**, 6600 (1994).

2272. DOERNEMANN, D., and H. SENGER: Optical Properties and Structure of Chlorophyll RCI. Opt. Prop. Struct. Tetrapyrroles, Proc. Symp. 489 (1985); Chem. Abstr. **103**, 3664 (1985).

2273. DOLÀK, L.: The Structure of RP 18,631. J. Antibiot. **26**, 121 (1973).

2274. HIGA, T., and P.J. SCHEUER: Constituents of the Hemichordate *Ptychodera flava laysanica*. In: Marine Natural Products Chemistry (D.J. Faulkner and W.H. Fenical, eds.), pp. 35–43. New York: Plenum. 1977.

2275. CONSTANTINO, V., E. FATTORUSSO, A. MANGONI, and M. PANSINI: Three New Brominated and Iodinated Tyrosine Derivatives from *Iotrochota birotulata*, a Non-Verongida Sponge. J. Nat. Prod. **57**, 1552 (1994).

2276. MACK, M.M., T.F. MOLINSKI, E.D. BUCK, and I.N. PESSAH: Novel Modulators of Skeletal Muscle FKBP 12/Calcium Channel Complex from *Ianthella basta*. J. Biol. Chem. **269**, 23236 (1994).

2277. HUNT, A.H.: Structure of the Pseudoaglycon of A35512B. J. Am. Chem. Soc. **105**, 4463 (1983).

2278. MATSUZAKI, K., H. IKEDA, T. OGINO, A. MARSUMOTO, H.B. WOODRUFF, H. TANAKA, and S. OMURA: Chloropeptins I and II, Novel Inhibitors Against gp120-CD4 Binding from *Streptomyces* sp. J. Antibiot. **47**, 1173 (1994).

2279. YOUSSEF, D., and A.W. FRAHM: Constituents of the Egyptian *Centaurea scoparia*. Chlorinated Guaianolides of the Aerial Parts. Planta Med. **60**, 267 (1994).

2280. YOUSSEF, D., and A.W. FRAHM: Constituents of the Egyptian *Centaurea scoparia*, II: Guaianolides of the Aerial Parts. Planta Med. **60**, 573 (1994).

2281. SARG, T., S. EL-DAHMY, M. EL-DOMIATY, and A. ATEYA: Guaianolides and Other Constituents from *Centaurea Sinaica*. Acta Pharm. Hung. **58**, 129 (1988); Chem. Abstr. **109**, 115926t (1991).

2282. NOWAK, G., B. DROZDZ, M. HOLUB, M. BUDESINSKY, and D. SAMAN: Sesquiterpene Lactones, XXXI: New Guaianolides in *Centaurea bella* Trautv. and *Centaurea adjarica* Alb. Acta Soc. Bot. Pol. **55**, 227 (1986).

2283. NOWAK, G.: A Chemotaxonomic Study of Sesquiterpene Lactones from Subtribe Centaurinae of the Compositae. Phytochem. **31**, 2363 (1992).

2284. MASSIOT, G., A.-M. MORFAUX, L. LEMEN-OLIVIER, J. BOUGUANT, A. MADACI, A. MAHAMOUD, M. CHOPOVA, and P. ACLINOU: Guaianolides from the Leaves of *Centaurea incana*. Phytochem. **25**, 258 (1986).

2285. ÖKSÜZ, S., and E. PUTUN: Guaianolides from *Centaurea kotschyi*. Phytochem. **22**, 2615 (1983).

2286. NOWAK, G., B. DROZDZ, and M. HOLUB: Sesquiterpene Lactones, XXXII: Guaianolides in Species from the Genus *Chartolepis* Cass. Acta Soc. Bot. Pol. **55**, 233 (1986).

2287. NOWAK, G., B. DROZDZ, M. HOLUB, and A. LAGODZINSKA: Sesquiterpene Lactones, XXXIII: Guaianolides in the Subgenus *Psephellus* (Cass.) Schmalh., Genus *Centaurea L*. Acta Soc. Bot. Pol. **55**, 629 (1986).

2288. NOWAK, G., M. HOLUB, and M. BUDESINSKY: Sesquiterpene Lactones, XXXIV: Lactones in the Genus *Leuzea* DC. Acta Soc. Bot. Pol. **57**, 157 (1988).

(*Received February 7, 1995*)

Author Index

Page numbers printed in *italics* refer to References

Subject Index

SpringerChemistry

Fortschritte der Chemie organischer Naturstoffe

Progress in the Chemistry of Organic Natural Products

Founded by L. Zechmeister
Edited by W. Herz, G. W. Kirby, R. E. Moore, W. Steglich, and Ch. Tamm

Volume 67

1996. 28 figures and 1 coloured plate. VII, 176 pages.
Cloth DM 220,–, öS 1540,–
Subscription price: Cloth DM 198,–, öS 1386,–
ISBN 3-211-82695-5

Contents:
A. A. Leslie Gunatilaka: Triterpenoid Quinonemethides and Related Compounds (Celastroloids).
P. Walser-Volken and Ch. Tamm: The Spirostaphylotrichins and Related Microbial Metabolites.

Volume 66

1995. 6 figures. VII, 332 pages.
Cloth DM 290,–, öS 2030,–
Subscription price: Cloth DM 261,–, öS 1827,–
ISBN 3-211-82597-5

Contents:
T. Okuda, T. Yoshida, T. Hatano: Hydrolyzable Tannins and Related Polyphenols.
R. G. de Souza Berlinck: Some Aspects of Guanidine Secondary Metabolites.

P.O.Box 89, A-1201 Wien • New York, NY 10010, 175 Fifth Avenue
Heidelberger Platz 3, D-14197 Berlin • Tokyo 113, 3-13, Hongo 3-chome, Bunkyo-ku

SpringerChemistry

Fortschritte der Chemie organischer Naturstoffe
Progress in the Chemistry of Organic Natural Products

Founded by L. Zechmeister
Edited by W. Herz, G. W. Kirby, R. E. Moore, W. Steglich, and Ch. Tamm

Volume 65

1995. 2 figures. IX, 618 pages.
Cloth DM 440,–, öS 3080,–
Subscription price: Cloth DM 396,–, öS 2772,–
ISBN 3-211-82576-2

Contents:
Y. Asakawa: Chemical Constituents of the Bryophytes.

Volume 64

1995. 22 partly coloured figures. VII, 216 pages.
Cloth DM 250,–, öS 1750,–
Subscription price: Cloth DM 225,–, öS 1575,–
ISBN 3-211-82533-9

Contents:
A. G. González and J. Bermejo Barrera: Chemistry and Sources of Mono- and Bicyclic Sesquiterpenes from Ferula Species.
G. Prota: The Chemistry of Melanins and Melanogenesis.
H. J. M. Gijsen, J. B. P. A. Wijnberg, and Ae. de Groot: Structure, Occurrence, Biosynthesis, Biological Activity, Synthesis, and Chemistry of Aromadendrane Sesquiterpenoids.

P.O.Box 89, A-1201 Wien • New York, NY 10010, 175 Fifth Avenue
Heidelberger Platz 3, D-14197 Berlin • Tokyo 113, 3-13, Hongo 3-chome, Bunkyo-ku

Springer-Verlag and the Environment

WE AT SPRINGER-VERLAG FIRMLY BELIEVE THAT AN international science publisher has a special obligation to the environment, and our corporate policies consistently reflect this conviction.

WE ALSO EXPECT OUR BUSINESS PARTNERS – PRINTERS, paper mills, packaging manufacturers, etc. – to commit themselves to using environmentally friendly materials and production processes.

THE PAPER IN THIS BOOK IS MADE FROM NO-CHLORINE pulp and is acid free, in conformance with international standards for paper permanency.